MW01553595

NUMERICAL MATHEMATICS AND SCIENTIFIC COMPUTATION

Series Editors

G. H. GOLUB A. GREENBAUM
A. M. STUART E. SÜLI

NUMERICAL MATHEMATICS AND SCIENTIFIC COMPUTATION

Books in the series

Monographs marked with an asterisk (*) appeared in the series 'Monographs in Numerical Analysis' which has been folded into, and is continued by, the current series.

*P. Dierckx: *Curve and surface fittings with splines*

*J. H. Wilkinson: *The algebraic eigenvalue problem*

*I. Duff, A. Erisman, and J. Reid: *Direct methods for sparse matrices*

*M. J. Baines: *Moving finite elements*

*J. D. Pryce: *Numerical solution of Sturm–Liouville problems*

K. Burrage: *Parallel and sequential methods for ordinary differential equations*

Y. Censor and S. A. Zenios: *Parallel optimization: theory, algorithms, and applications*

M. Ainsworth, J. Levesley, W. Light, and M. Marletta: *Wavelets, multilevel methods, and elliptic PDEs*

W. Freeden, T. Gervens, and M. Schreiner: *Constructive approximation on the sphere: theory and applications to geomathematics*

Ch. Schwab: *p- and hp-finite element methods: theory and applications to solid and fluid mechanics*

J. W. Jerome: *Modelling and computation for applications in mathematics, science, and engineering*

Alfio Quarteroni and Alberto Valli: *Domain decomposition methods for partial differential equations*

G. E. Karniadakis and S. J. Sherwin: *Spectral/hp element methods for CFD*

I. Babuška and T. Strouboulis: *The finite element method and its reliability*

B. Mohammadi and O. Pironneau: *Applied shape optimization for fluids*

S. Succi: *The lattice Boltzmann equation for fluid dynamics and beyond*

P. Monk: *Finite element methods for Maxwell's equations*

A. Bellen and M. Zennaro: *Numerical methods for delay differential equations*

J. Modersitzki: *Numerical methods for image registration*

M. Feistauer, J. Felcman, and I. Straškraba: *Mathematical and computational methods for compressible flow*

W. Gautschi: *Orthogonal polynomials: computation and approximation*

M. K. Ng: *Iterative methods for Toeplitz systems*

Michael Metcalf, John Reid, and Malcolm Cohen: *Fortran 95/2003 explained*

George Em Karniadakis and Spencer Sherwin: *Spectral/hp element methods for CFD*, Second edition

Spectral/*hp* Element Methods for Computational Fluid Dynamics

SECOND EDITION

George Em Karniadakis

Division of Applied Mathematics
Brown University
Providence, RI 02912, USA

and

Spencer J. Sherwin

Department of Aeronautics
Imperial College London
South Kensington Campus
London, SW7 2AZ, UK

OXFORD
UNIVERSITY PRESS

OXFORD
UNIVERSITY PRESS

Great Clarendon Street, Oxford OX2 6DP

Oxford University Press is a department of the University of Oxford.
It furthers the University's objective of excellence in research, scholarship,
and education by publishing worldwide in

Oxford New York

Auckland Cape Town Dar es Salaam Hong Kong Karachi
Kuala Lumpur Madrid Melbourne Mexico City Nairobi
New Delhi Shanghai Taipei Toronto

With offices in

Argentina Austria Brazil Chile Czech Republic France Greece
Guatemala Hungary Italy Japan South Korea Poland Portugal
Singapore Switzerland Thailand Turkey Ukraine Vietnam

Oxford is a registered trade mark of Oxford University Press
in the UK and in certain other countries

Published in the United States
by Oxford University Press Inc., New York

© Oxford University Press, 2005

The moral rights of the authors have been asserted
Database right Oxford University Press (maker)

First published 2005

All rights reserved. No part of this publication may be reproduced,
stored in a retrieval system, or transmitted, in any form or by any means,
without the prior permission in writing of Oxford University Press,
or as expressly permitted by law, or under terms agreed with the appropriate
reprographics rights organization. Enquiries concerning reproduction
outside the scope of the above should be sent to the Rights Department,
Oxford University Press, at the address above

You must not circulate this book in any other binding or cover
and you must impose this same condition on any acquirer

British Library Cataloguing in Publication Data

Data available

Library of Congress Cataloging in Publication Data

Data available

ISBN 0 19 852869 8 9780198528692

1 3 5 7 9 10 8 6 4 2

Typeset by Julie M. Harris using LaTeX

Printed in Great Britain
on acid-free paper by
Biddles Ltd., King's Lynn, Norfolk

PREFACE TO SECOND EDITION

There has been significant progress in the development of multi-domain spectral methods both at the fundamental as well as at the application level in the last few years. We have, therefore, undertaken the 'non-trivial' task of updating the book in order to include these new developments. We also wanted to make these methods easier to comprehend and implement by the students, responding directly to the many requests and feedback we received after the publication of the first edition of our book. We are grateful to Oxford University Press, and in particular to the Mathematics and Statistics editor Dr Alison Jones, who gave us this opportunity.

The new developments are primarily in the discontinuous Galerkin methods, in non-tensorial nodal spectral element methods in simplex domains, and on stabilisation and filtering techniques. From the practical point of view, high-order solutions in complex geometries require high-order meshes and high-order post-processing, a subject that is often neglected in the everyday 'production' computing and simulation. Such subjects are now addressed in some detail in this new edition of the book. We have also seen the spectral/*hp* element method applied to less traditional fields, such as seismology, climate modelling, and magneto-hydrodynamics (MHD), and we have included some elements of modelling such applications in this revised version.

Finally, another objective in revising the book has been to provide more details on implementing various aspects of the method. To this end, we have put some emphasis on implementation and technical issues with exercises in the founding Chapters 2 to 5 to aid in implementing basic spectral element solvers, which can be used as building blocks for more complex application codes.

Overall, the book has been increased in new material by almost 50% in order to include all the aforementioned topics.

We would like to thank our many students and colleagues who have provided good critical feedback on the first version of the book. In particular, we would like to thank Drs H. Blackburn, J. Hesthaven, R. M. Kirby, J. Peiró, V. Theofilis, and Z. Yosibash who contributed to the new edition directly by providing plots and other information from their own work. We also want to thank our families who have supported us again during the course of this effort.

Boston, Massachusetts G. E. K.
London, England S. J. S.
Spring 2004

PREFACE TO FIRST EDITION

Our aim in writing this book is to introduce a wider audience to the use of spectral/*hp* element methods with particular emphasis on their application to unstructured meshes. These methods, as their name suggests, incorporate both multi-domain spectral methods (based on a development of original ideas by A. T. Patera) and also high-order finite element methods (based on original ideas of B. A. Szabó). In this book, we provide a unified description of both methods building on previously published works as well as on new material not previously published.

Although spectral methods have long been popular in direct and large eddy simulation of turbulent flows, their use in areas with complex-geometry computational domains has historically been much more limited. For example, in computational aerodynamics, which typically involve the use of unstructured meshes, the preferred methods have been low-order finite element and finite volume methods. More recently, however, the need to find accurate solutions to the viscous flow equations around complex aerodynamic configurations has led to the development of high-order discretisation procedures on unstructured meshes. High-order discretisation is also recognised as more efficient for solution of time-dependent oscillatory solutions over long time periods, for example, in the new field of computational electromagnetics in aerospace design.

Polynomial spectral methods were first introduced in Gottlieb and Orszag [199] and are covered extensively in Canuto *et al.* [86], Boyd [71], and Fornberg [164]. Szabó and Babuška [445] and Bernardi and Maday [45] deal with *hp* finite element and spectral element methods, respectively, and should be consulted for background reading. This book reviews most of the fundamental concepts but does assume the reader has some familiarity with the basic concepts of finite element discretisation and the Galerkin approximation technique.

The material contained in this book draws on the ideas and influence of many people, but particular thanks are due to A. T. Patera, S. A. Orszag, and D. Gottlieb, who have advised and collaborated on our research work over the past fifteen years. Much of this book is based on the doctoral thesis of the second author (SJS), as supervised by the first author (GEK). This book has also drawn heavily on doctoral theses of other students supervised by GEK, notably those of R. D. Henderson, I. G. Giannakouros, A. Beskok, C. H. Crawford, T. C. Warburton, I. Lomtev, and J. Trujillo, and also doctoral theses supervised by A. T. Patera, notably E. M. Ronquist, C. Mavriplis, and G. Anagnostou. We would like to thank J. Peiro for his help with the unstructured meshing, and are particularly grateful to our colleagues who provided original figures which we have included in this book. Thanks are also due to those who have given their time and effort to read the proof of this book.

GEK would like to acknowledge the sponsorship of his research program by the Office of Naval Research, the Air Force Office of Scientific Research, the Department of Energy, and the National Science Foundation.

Finally, we would like to thank our wives, Helen and Tracey, for their understanding and patience during the writing of this book. We would especially like to thank Tracey, who was the first person to read the book in its entirety, correcting our grammar and spelling, and also providing a notable contribution to the prose.

Boston, Massachusetts G. E. K.
London, England S. J. S.
Fall 1997

BOOK OUTLINE

In Chapter 1, we present reduced models of the compressible and incompressible Navier–Stokes equations which are used in the various discretisation concepts discussed in the following chapters. The convergence philosophy of spectral and finite element methods, the combination of which provides a dual path of convergence, is also introduced.

In Chapter 2, we present the fundamental concepts that are further developed in the remainder of the book, in the context of a one-dimensional formulation. In doing so, we illustrate the principles and underlying theory behind the construction of the spectral/*hp* element method. In this second revision we have included more details on implementing boundary conditions and used margin identifiers to highlight formulation details, using a '♣' symbol, and implementation details, using a '♦' symbol. We have also added new sections on nodal p-type expansions and on integration errors and polynomial aliasing. The series of exercises included at the end of the chapter solely focuses on developing a one-dimensional spectral element solver.

In Chapter 3, we consider the extension of the one-dimensional formulation to two and three dimensions by the development of expansion bases in standard regions such as triangles or rectangles in two dimensions, and tetrahedrons, prisms, pyramids, and hexahedrons in three dimensions. The construction of these bases uses a unified approach which permits the development of computationally efficient expansions. In this new revision we now formulate the modal basis as solutions to a generalised Sturm–Liouville problem. We also present optimal nodal points, the so-called Fekete points, as well as the electrostatic points on a simplex, and include related approximation results. The exercises at the end of the chapter target construction of multi-dimensional elemental matrices.

Compared to the first revision of the book, Chapter 4 has been restructured with extra emphasis on implementation aspects. In this chapter we complete the multi-dimensional formulation by explaining how the two- and three-dimensional expansions developed in Chapter 3 can be extended into a tessellation of multiple domains. These extensions are decomposed into three sections: local operations such as integration and differentiation; global operations such as the construction of global matrix systems; and pre- and post-processing issues such as boundary representation, high-order mesh generation, and particle tracking. The chapter introduces a matrix formulation to help illustrate the algebraic systems which need to be constructed when computationally implementing the spectral/*hp* method. Formulation of both Galerkin and collocation projections are considered in this manner. The exercises in this chapter include implementation of a two-dimensional spectral/*hp* element solver for a multi-element Galerkin projection.

In this revised version Chapter 5 now considers the diffusion equation; an implicit in time discretisation leads to the Helmholtz equation. This chapter discusses both the temporal discretisation and eigenspectra of second-order operators that dictate time-step restrictions. We also expand on appropriate preconditioning techniques for inversion of the stiffness matrix. The final part of this chapter discusses non-smooth solutions due to geometric singularities and this revised section now includes elements from recent advances in three-dimensional domains. The final exercises section focuses on building on the exercises of Chapters 3 and 4 to implement a two-dimensional standard Galerkin *hp* solution to the Helmholtz problem.

In Chapter 6, we focus on the scalar advection equation and develop a Galerkin discretisation using the techniques described in Chapter 4. We then include an extended presentation of the discontinuous Galerkin formulation for advection equations. Similar to Chapter 5, we also review eigenspectra of the advection operators in both two and three dimensions which are relevant for explicit time stepping. A further new addition is the discussion on two forms of a semi-Lagrangian method for advection (strong and auxiliary forms) that could potentially prove very effective in enhancing the speed and accuracy of spectral/*hp* element methods in advection-dominated problems. A further new section on stabilisation techniques is then introduced that discusses filters, spectral vanishing viscosity, and upwind collocation. This new material is important as it relates directly to the robustness of the method, especially in marginally resolved simulations.

In Chapter 7 (previously Chapter 5) we introduce and discuss the topic of non-conforming elements for second-order operators. This chapter has been extended to include a comprehensive presentation of the discontinuous Galerkin method with a comparison of different versions from theoretical, computational, and implementation standpoints.

In Chapter 8, we present different ways of formulating the incompressible Navier–Stokes equations based on primitive variables, that is, velocity and pressure, as well as velocity–vorticity algorithms. We consider both coupled, splitting, and least-squares formulations for primitive variables, and in this revision we now discuss both the Uzawa coupled algorithm and a new substructured solver. The discussion on primitive variables time-splitting has been rewritten to include recent theoretical advances in the pressure-correction and velocity-correction schemes as well as the rotational formulation of the pressure boundary condition. Whilst an important issue for primitive variable formulation is the efficient incorporation of the divergence-free constraint, an analogous problem for the velocity–vorticity formulation is the accurate imposition of the boundary conditions which is also discussed. The final section is devoted to nonlinear terms; it includes a discussion of spatial and temporal discretisation with focus on the semi-Lagrangian method for the incompressible Navier–Stokes equations.

In Chapter 9, we discuss numerical simulations of the incompressible Navier–Stokes equations. First, we present exact Navier–Stokes solutions that can be used as benchmarks to validate new codes and evaluate the accuracy of a particular discretisation. In the current revision this section has been restructured, with more benchmark solutions, to emphasise verification and validation of spectral/*hp* element solvers. A new section on three-dimensional stability (biglobal stability) is also included—spectral/*hp* elements are particularly effective in this field. The chapter continues by discussing some aspects of direct numerical simulation (DNS) and large-eddy simulation (LES). The issue of stabilisation at high Reynolds number is then presented using the concepts of dynamic subgrid modelling, over-integration, and spectral vanishing viscosity. A new parallel paradigm based on multi-level parallelism is introduced that can help realise adaptive refinement more easily; the final section includes a heuristic refinement method for Navier–Stokes equations.

Chapter 10 has been expanded to consider not only compressible Euler and Navier–Stokes equations but general hyperbolic conservation laws. This is an area in which high-order methods have had little success in the past. The principle issue is how to effectively use the high-order expansions of the spectral/*hp* method whilst honouring the inherent monotonicity and conservation properties of the analytic system. We consider different ways of dealing with these fundamental issues for both the Euler and the Navier–Stokes equations. A new section for the shallow water equations is also included and the section on the discontinuous Galerkin method has been rewritten. Finally, the last section discusses modelling of plasma flows, i.e., the so-called magneto-hydrodynamic (MHD) equations.

Finally, the appendices provide details on Jacobi and Askey polynomials as well as numerical integration and differentiation which are essential building blocks of the spectral/*hp* element techniques. A full description of commonly used expansion bases is provided, which now includes nodal points for non-tensorial electrostatic and Fekete point distribution in simplex domains. The final appendix also details Riemann solvers commonly used in the solution of the Euler equations.

CONTENTS

NOMENCLATURE

Sets

$\Re$	Real numbers
$\bigcup$	Set union
$\bigcap$	Set intersection
$\emptyset$	Empty set
$\in$	Is a member of; belongs to
$\notin$	Is not a member of; does not belong to
$\subset$	Is a subset of
$\not\subset$	Is not a subset of

Expansion basis notation

ϕ_{pq}, ϕ_{pqr}	Expansion basis
$\psi_p^a, \psi_{pq}^b, \psi_{pqr}^c$	Modified principal functions
$\widetilde{\psi}_p^a, \widetilde{\psi}_{pq}^b, \widetilde{\psi}_{pqr}^c$	Orthogonal principal functions
$h_p(\xi)$	One-dimensional Lagrange polynomial of order p
$L_i^{N_m}(\boldsymbol{\xi})$	Two-dimensional Lagrange polynomial through N_m nodes $\boldsymbol{\xi}_i$
P_i	Polynomial order in the ith direction
Q_i	Quadrature order in the ith direction
$x_1, x_2, x_3, \boldsymbol{x}$	Global Cartesian coordinates
$\xi_1, \xi_2, \xi_3, \boldsymbol{\xi}$	Local Cartesian coordinates
η_1, η_2, η_3	Local collapsed Cartesian coordinates
$\chi_i(\xi)$	Local Cartesian to global coordinate mapping

Various constants

N_{dof}	Number of global degrees of freedom
N_b	Number of global boundary degrees of freedom
N_m	Number of elemental degrees of freedom
N_{eof}	Total number of elemental degrees of freedom $N_{\text{eof}} \simeq N_{\text{el}} N_m$
N_Q	Total number of quadrature points $N_Q = Q_1 Q_2 Q_3$
N_{el}	Number of elements
λ	Helmholtz equation constant
e	Element number $1 \leqslant e \leqslant N_{\text{el}}$
i, j, k	General summation indices
p, q, r	General summation indices

Elemental arrays

$\boldsymbol{B}$	Basis matrix
$\boldsymbol{W}$	Diagonal weight/Jacobian matrix
$\boldsymbol{D}_\xi$	Elemental derivative matrix with respect to ξ
$\boldsymbol{\Lambda}(u)$	Diagonal matrix of $u(\xi_1, \xi_2)$ evaluated at quadrature points
$\boldsymbol{M}^e$	Elemental mass matrix

$\boldsymbol{L}^e$	Elemental Laplacian matrix
$\boldsymbol{H}^e$	Elemental Helmholtz matrix
$\boldsymbol{f}^e$	Force vector of the eth element
$\boldsymbol{u}^e$	Vector containing function evaluated at quadrature points
$\hat{\boldsymbol{u}}^e$	Vector of expansion coefficients

Global arrays

$\underline{\boldsymbol{W}^e}$	Block diagonal extension of matrix $\boldsymbol{W}^e$
$\underline{\boldsymbol{f}^e}$	Concatenation of elemental vector $\boldsymbol{f}^e$
$\boldsymbol{\mathcal{A}}^\top$	Matrix global assembly
$\boldsymbol{M}$	Mass matrix ($= \boldsymbol{\mathcal{A}}^\top \underline{\boldsymbol{M}^e} \boldsymbol{\mathcal{A}}$)
$\boldsymbol{L}$	Laplacian matrix ($= \boldsymbol{\mathcal{A}}^\top \underline{\boldsymbol{L}^e} \boldsymbol{\mathcal{A}}$)
$\boldsymbol{H}$	Helmholtz matrix ($= \boldsymbol{\mathcal{A}}^\top \underline{\boldsymbol{H}^e} \boldsymbol{\mathcal{A}}$)
$\hat{\boldsymbol{v}}_g$	List of all elemental coefficients ($= \underline{v^e}$)
$\hat{\boldsymbol{v}}_l$	Global list of coefficients

Solution domains

Ω	Solution domain
$\partial\Omega$	Boundary of Ω
Ω^e	Elemental region
$\partial\Omega^e$	Boundary of Ω^e
$\partial\Omega_{\mathcal{D}}$	Domain boundary with Dirichlet conditions
$\partial\Omega_{\mathcal{N}}$	Domain boundary with Neumann conditions
$\boldsymbol{n}$	Unit outward normal

Differential operators

∇^2	Laplacian
$\nabla \cdot$	Divergence
$\nabla \times$	Curl

Spaces

$\mathcal{X}$	Space of trial solutions
$\mathcal{X}^\delta$	Finite-dimensional space of trial solutions
$\mathcal{V}$	Space of test functions
$\mathcal{V}^\delta$	Finite-dimensional space of test functions
$\mathcal{P}_P(\Omega)$	Polynomial space of order P over Ω

Operators

$\mathcal{I}$	Interpolation operator
$\mathcal{I}^\delta$	Discrete interpolation operator
$\mathbb{P}$	Projection operator
$\mathbb{P}^\delta$	Discrete projection operator
$\mathbb{L}(u)$	Linear operator in u

Fluid variables

$\mathbf{v}$	Velocity $[u, v, w]^\top$
p	Pressure
$\boldsymbol{\omega}$	Vorticity
ρ	Density
μ, ν	Dynamic, kinematic viscosities

1

INTRODUCTION

1.1 The basic equations of fluid dynamics

Consider fluid flow in the non-deformable control volume Ω bounded by the control surface $\partial\Omega$ with $\boldsymbol{n}$ being the unit outward normal. The equations of motion can then be derived in an absolute reference frame by applying the principles of mechanics and thermodynamics [39]. They can be formulated in integral form for mass, momentum, and total energy, respectively, as

$$\frac{\mathrm{d}}{\mathrm{d}t}\int_{\Omega}\rho\,\mathrm{d}\Omega+\int_{\partial\Omega}\rho\mathbf{v}\cdot\boldsymbol{n}\,\mathrm{d}S=0\,, \tag{1.1.1a}$$

$$\frac{\mathrm{d}}{\mathrm{d}t}\int_{\Omega}\rho\mathbf{v}\,\mathrm{d}\Omega+\int_{\partial\Omega}[\rho\mathbf{v}(\mathbf{v}\cdot\boldsymbol{n})-\boldsymbol{n}\boldsymbol{\sigma}]\,\mathrm{d}S=\int_{\Omega}\boldsymbol{f}\,\mathrm{d}\Omega\,, \tag{1.1.1b}$$

$$\frac{\mathrm{d}}{\mathrm{d}t}\int_{\Omega}E\,\mathrm{d}\Omega+\int_{\partial\Omega}(E\mathbf{v}-\boldsymbol{\sigma}\mathbf{v}+\mathbf{q})\cdot\boldsymbol{n}\,\mathrm{d}S=\int_{\Omega}\boldsymbol{f}\cdot\mathbf{v}\,\mathrm{d}\Omega\,. \tag{1.1.1c}$$

Here $\mathbf{v}(\boldsymbol{x},t)=(u,v,w)$ is the velocity field, ρ is the density, and $E=\rho(e+1/2\mathbf{v}\cdot\mathbf{v})$ is the total specific energy where e represents the internal specific energy. Also, $\boldsymbol{\sigma}$ is the stress tensor, $\mathbf{q}$ is the heat flux vector, and $\boldsymbol{f}$ represents all external forces acting on this control volume. For Newtonian fluids, the stress tensor, which consists of the normal components (p for pressure) and the viscous stress tensor $\boldsymbol{\tau}$, is a *linear* function of the velocity gradient, that is,

$$\boldsymbol{\sigma}=-p\mathbf{I}+\boldsymbol{\tau}\,, \tag{1.1.2a}$$

$$\boldsymbol{\tau}=\mu[\nabla\mathbf{v}+(\nabla\mathbf{v})^{\top}]+\lambda(\nabla\cdot\mathbf{v})\mathbf{I}\,, \tag{1.1.2b}$$

where $\mathbf{I}$ is the unit tensor, and μ and λ are the first and second coefficients of viscosity, respectively. They are related by the Stokes hypothesis, that is, $2\mu+3\lambda=0$, which expresses local thermodynamic equilibrium. The heat flux vector is related to temperature gradients via the Fourier law of diffusion, that is,

$$\mathbf{q}=-k\nabla T\,, \tag{1.1.3}$$

where $k(T)$ is the thermal conductivity which may be a function of temperature T.

In the case of a deformable control volume, the velocity in the flux term should be recognised as in a frame of reference relative to the control surface and the appropriate time rate-of-change term be used. Considering, for example, the mass conservation equation, we have the forms

$$\frac{\mathrm{d}}{\mathrm{d}t}\int_{\Omega}\rho\,\mathrm{d}\Omega + \int_{\partial\Omega}\rho\mathbf{v}_r\cdot\boldsymbol{n}\,\mathrm{d}S = 0$$

or

$$\int_{\Omega}\frac{\partial\rho}{\partial t}\,\mathrm{d}\Omega + \int_{\partial\Omega}\rho\mathbf{v}_r\cdot\boldsymbol{n}\,\mathrm{d}S + \int_{\partial\Omega}\rho\mathbf{v}_{\mathrm{cs}}\cdot\boldsymbol{n}\,\mathrm{d}S = 0\,,$$

where $\mathbf{v}_{\mathrm{cs}}$ is the velocity of the control surface, $\mathbf{v}_r$ is the velocity of the fluid with respect to the control surface, and the total velocity of the fluid with respect to the chosen frame is $\mathbf{v} = \mathbf{v}_r + \mathbf{v}_{\mathrm{cs}}$. The above forms are equivalent but the first expression may be more useful in applications where the time history of the volume is of interest.

Equations (1.1.1a)–(1.1.1c) can be transformed into an equivalent set of partial differential equations by applying Gauss' theorem (assuming that sufficient conditions of differentiability exist), that is,

$$\frac{\partial\rho}{\partial t} + \nabla\cdot(\rho\mathbf{v}) = 0\,, \tag{1.1.4a}$$

$$\frac{\partial(\rho\mathbf{v})}{\partial t} + \nabla\cdot(\rho\mathbf{v}\mathbf{v} - \sigma) = \boldsymbol{f}\,, \tag{1.1.4b}$$

$$\frac{\partial E}{\partial t} + \nabla\cdot(E\mathbf{v} - \boldsymbol{\sigma}\mathbf{v} + \mathbf{q}) = \boldsymbol{f}\cdot\mathbf{v}\,. \tag{1.1.4c}$$

The momentum and energy equations can be rewritten in the following form by using the continuity equation (1.1.4a) and the constitutive equations (1.1.2a) and (1.1.2b)

$$\rho\frac{\mathrm{D}\mathbf{v}}{\mathrm{D}t} = -\nabla p + \nabla\cdot\boldsymbol{\tau} + \boldsymbol{f}\,, \tag{1.1.5a}$$

$$\rho\frac{\mathrm{D}e}{\mathrm{D}t} = -p\nabla\cdot\mathbf{v} - \nabla\cdot\mathbf{q} + \Phi\,, \tag{1.1.5b}$$

where $\Phi = \boldsymbol{\tau}\cdot\nabla\mathbf{v}$ is the dissipation function and $\mathrm{D}/\mathrm{D}t = \partial/\partial t + \mathbf{v}\cdot\nabla$ is the material derivative.

In addition to the governing conservation laws, an equation of state is required. For ideal gases, it has the simple form

$$p = \rho\mathcal{R}T\,, \tag{1.1.6}$$

where $\mathcal{R}$ is the ideal gas constant defined as the difference of the constant specific heats, that is, $\mathcal{R} = C_p - C_v$, where $C_v = (\partial e/\partial T)|_\rho$ and $C_p = \gamma C_v$, with γ being the adiabatic index. For ideal gases, the energy equation can be rewritten in terms of the temperature since $e = p/(\rho(\gamma-1)) = C_vT$, and so eqn (1.1.5b) becomes

$$\rho C_v\frac{\mathrm{D}T}{\mathrm{D}t} = -p\nabla\cdot\mathbf{v} + \nabla\cdot(k\nabla T) + \Phi\,. \tag{1.1.7}$$

The system of equations (1.1.4a), (1.1.5a), (1.1.6), and (1.1.7) is called the *compressible Navier–Stokes equations* and contains six unknown variables $(\rho, \mathbf{v}, p, T)$

with six scalar equations. This is an *incomplete parabolic* system as there are no second-order derivative terms in the continuity equation.

A hyperbolic system arises in the case of inviscid flow, that is, $\mu = 0$ (assuming that we also neglect heat losses by thermal diffusion, that is, $k = 0$). In that case we obtain the *Euler equations*, which in the absence of external forces or heat sources have the form

$$\frac{\partial \rho}{\partial t} + \nabla \cdot (\rho \mathbf{v}) = 0\,, \tag{1.1.8a}$$

$$\frac{\partial (\rho \mathbf{v})}{\partial t} + \nabla \cdot (\rho \mathbf{v}\mathbf{v}) = -\nabla p\,, \tag{1.1.8b}$$

$$\frac{\partial E}{\partial t} + \nabla \cdot [(E + p)\mathbf{v}] = 0\,. \tag{1.1.8c}$$

Appropriate boundary conditions will be discussed in Chapter 10. This system admits discontinuous solutions, and it can also describe the transition from a subsonic flow (where $|\mathbf{v}| < c$) to supersonic flow (where $|\mathbf{v}| > c$), where $c = (\gamma \mathcal{R} T)^{1/2}$ is the speed of sound. Typically, the transition is obtained through a shock wave, which represents a discontinuity in flow variables. In such a region the integral form of the equations should be used by analogy with eqns (1.1.1a)–(1.1.1c).

1.1.1 *Incompressible flow*

For an incompressible fluid, where $\mathrm{D}\rho/\mathrm{D}t = 0$, the mass conservation (or continuity) equation simplifies to

$$\nabla \cdot \mathbf{v} = 0\,. \tag{1.1.9a}$$

Typically, when we refer to an incompressible fluid we mean that $\rho =$ constant, but this is not necessary for a divergence-free flow; for example, in thermal convection the density varies with temperature variations. The corresponding momentum equation has the form

$$\rho \frac{\mathrm{D}\mathbf{v}}{\mathrm{D}t} = -\nabla p + \nabla \cdot \left[\mu[\nabla \mathbf{v} + (\nabla \mathbf{v})^{\top}]\right] + \boldsymbol{f}\,, \tag{1.1.9b}$$

where the viscosity $\mu(\boldsymbol{x}, t)$ may vary in space and time due to physics or a subgrid model in formulations of large-eddy simulations (see Chapter 9). The pressure $p(\boldsymbol{x}, t)$ is not a thermodynamic quantity but can be thought of as a constraint that projects the solution $\mathbf{v}(\boldsymbol{x}, t)$ onto a divergence-free space. In other words, the isentropic equation $p = C\rho^{\gamma}$ is no longer valid as it will make the incompressible Navier–Stokes system over-determined.

The acceleration terms can be written in various equivalent ways so that, in their discrete form, they conserve total linear momentum $\int_{\Omega} \mathbf{v}\, \mathrm{d}\Omega$ and total kinetic energy $\int_{\Omega} \mathbf{v} \cdot \mathbf{v}\, \mathrm{d}\Omega$ in the absence of viscosity and external forces. In particular, the following forms are often used:

- convective form: $\mathrm{D}\mathbf{v}/\mathrm{D}t = \partial \mathbf{v}/\partial t + (\mathbf{v} \cdot \nabla)\mathbf{v}$;

- conservative (flux) form: $\mathrm{D}\mathbf{v}/\mathrm{D}t = \partial\mathbf{v}/\partial t + \nabla \cdot (\mathbf{v}\mathbf{v})$;
- rotational form: $\mathrm{D}\mathbf{v}/\mathrm{D}t = \partial\mathbf{v}/\partial t - \mathbf{v} \times (\nabla \times \mathbf{v}) + 1/2\nabla(\mathbf{v} \cdot \mathbf{v})$;
- skew-symmetric form: $\mathrm{D}\mathbf{v}/\mathrm{D}t = \partial\mathbf{v}/\partial t + 1/2[(\mathbf{v} \cdot \nabla)\mathbf{v} + \nabla \cdot (\mathbf{v}\mathbf{v})]$.

In semi-discrete systems, that is, systems that are continuous in time but discrete in space, where inexact integration of nonlinear terms or pointwise discretisation (for example, collocation) is employed, only the rotational and skew-symmetric forms conserve both linear momentum and kinetic energy in the inviscid limit. The flux form conserves only linear momentum while the convective form conserves neither. Numerical experiments with turbulence simulations have shown that the skew-symmetric form is the most effective in minimising aliasing errors although it is computationally the most expensive. In three dimensions, it requires the calculation of eighteen derivatives versus six derivatives in the rotational form and nine derivatives in the convective form. A more detailed discussion on the discrete form of the advection terms is included in Chapter 9.

The incompressible Navier–Stokes equations (1.1.9a) and (1.1.9b) are written in terms of the primitive variables $(\mathbf{v}, p)$. An alternative form is to rewrite these equations in terms of the velocity $\mathbf{v}$ and vorticity $\boldsymbol{\omega} = \nabla \times \mathbf{v}$. This is a more general formulation than the standard vorticity streamfunction, which is limited to two dimensions. The following system is equivalent to eqns (1.1.9b) and (1.1.9a) assuming that ρ and μ are constant:

$$\rho\frac{\mathrm{D}\boldsymbol{\omega}}{\mathrm{D}t} = (\boldsymbol{\omega} \cdot \nabla)\mathbf{v} + \mu\nabla^2\boldsymbol{\omega} \quad \text{in } \Omega\,, \tag{1.1.10a}$$

$$\nabla^2\mathbf{v} = -\nabla \times \boldsymbol{\omega} \quad \text{in } \Omega\,, \tag{1.1.10b}$$

$$\nabla \cdot \mathbf{v} = 0 \quad \text{in } \Omega\,, \tag{1.1.10c}$$

$$\boldsymbol{\omega} = \nabla \times \mathbf{v} \quad \text{in } \Omega\,, \tag{1.1.10d}$$

where the elliptic equation for the velocity $\mathbf{v}$ is obtained using a vector identity and the divergence-free constraint. We also assume here that the domain Ω is simply connected. An equivalent system in terms of velocity and vorticity is studied in Chapter 8, where it is reformulated for easier implementation. The problem with the lack of direct boundary conditions for the vorticity also exists in the more-often-used vorticity-streamfunction formulation.

Finally, a note regarding non-dimensionalisation. Consider the free-stream flow U_0 past a body of characteristic size D in a medium of dynamic viscosity μ, as shown in Fig. 1.1. There are two characteristic time-scales in the problem, the first one representing the convective time-scale $t_c = D/U_0$, and the second one representing the diffusive time-scale $t_d = D^2/\nu$, where $\nu = \mu/\rho$ is the kinematic viscosity. If we non-dimensionalise all lengths with D, the velocity field with U_0, and the vorticity field with U_0/D, we obtain two different non-dimensional equations corresponding to the choice of the time non-dimensionalisation:

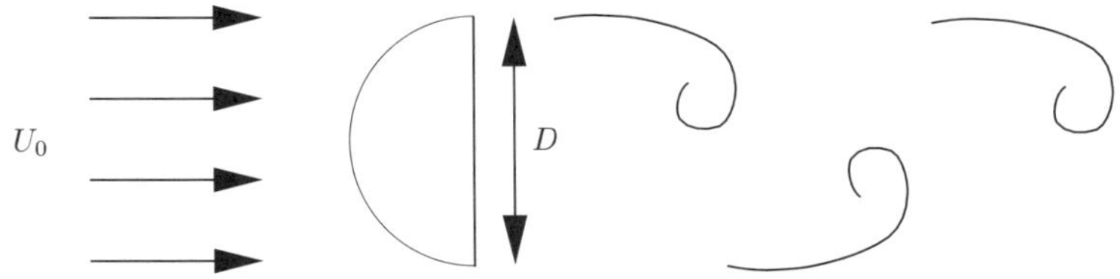

FIG. 1.1. Free-stream flow past a half-cylinder in a viscous fluid.

$$\frac{\partial \boldsymbol{\omega}}{\partial t_c^*} + \nabla \cdot (\mathbf{v}\boldsymbol{\omega}) = (\boldsymbol{\omega} \cdot \nabla)\mathbf{v} + \mathrm{Re}^{-1}\nabla^2 \boldsymbol{\omega}\,,$$

$$\frac{\partial \boldsymbol{\omega}}{\partial t_d^*} + \mathrm{Re}\nabla \cdot (\mathbf{v}\boldsymbol{\omega}) = \mathrm{Re}(\boldsymbol{\omega} \cdot \nabla)\mathbf{v} + \nabla^2 \boldsymbol{\omega}\,,$$

where t_c^* and t_d^* are the non-dimensionalised time variables with respect to t_c and t_d, respectively, and $\mathrm{Re} = U_0 D/\nu$ is the Reynolds number. Both forms are useful in simulations, the first in high Reynolds number simulations, and the second in low Reynolds number (creeping) flows.

When the nonlinear terms can be neglected we obtain the *Stokes equations*, which we can cast in the form

$$-\nu\nabla^2\mathbf{v} + \nabla p = \boldsymbol{f} \quad \text{in } \Omega\,, \tag{1.1.11a}$$

$$\nabla \cdot \mathbf{v} = 0 \quad \text{in } \Omega\,, \tag{1.1.11b}$$

along with appropriate boundary conditions for $\mathbf{v}$. This system is studied in detail in Chapter 8 as it provides the setting for the variational formulation of the Navier–Stokes equations.

The theory on incompressible Navier–Stokes equations is treated in several articles and books, including Constantin and Foias [107], Lions [306], and Temam [454]. The latter provides a theoretical framework for many of the implementation concepts that will be developed in this book in the context of variational formulation of the incompressible Navier–Stokes equations. It is fair to say, however, that the mathematical theory is fairly complete in two spatial dimensions and that several results regarding the existence of solutions, uniqueness, regularity, and continuous dependence on the data have been proved. However, there are several questions still unanswered regarding the Navier–Stokes equation in three dimensions, and little progress has been made in the theory since the work of Leray [293].

New concepts from modern dynamical systems theory introduced in the context of the incompressible Navier–Stokes system have helped in our understanding of the computational complexity of these equations. For example, the concept of determining modes [108] provided for the first time a rigorous theory for the classical Kolmogorov heuristic argument on the excited degrees of freedom in turbulence. Moreover, the analysis by Constantin *et al.* [108] showed that such an estimate is only a sufficient upper bound and applies to general unsteady flows for which boundedness of vorticity is assumed.

1.1.2 *Reduced models*

The mathematical nature of the Navier–Stokes equations varies depending on the flow that we model and the corresponding terms that dominate in the equations. For example, for an inviscid compressible flow, we obtain the Euler equations, which are of hyperbolic nature, whereas the incompressible Euler equations are of hybrid type corresponding to both real and imaginary eigenvalues. The unsteady incompressible Navier–Stokes equations are of mixed parabolic/hyperbolic nature, but the steady incompressible Navier–Stokes are of elliptic/parabolic type. To simplify the discretisation of Navier–Stokes and motivate the formulation adopted in this book we follow a hierarchical approach in reducing the Navier–Stokes equations to simpler equations, so that each introduces one new concept.

Take as an example the incompressible Navier–Stokes equations (1.1.9a) and (1.1.9b), a simpler model is the *unsteady Stokes* system. This retains all the complexity but not the nonlinear terms, that is,

$$\frac{\partial \mathbf{v}}{\partial t} = -\frac{\nabla p}{\rho} + \nu\nabla^2\mathbf{v} + \boldsymbol{f}\,,$$
$$\nabla \cdot \mathbf{v} = 0\,.$$

The Stokes system (eqns (1.1.11a) and (1.1.11b)) is recovered by dropping the time derivative. Alternatively, we can drop the divergence-free constraint and study the purely parabolic scalar equation for a variable u, that is,

$$\frac{\partial u}{\partial t} = \nu\nabla^2 u + f\,. \tag{1.1.12}$$

This equation is very helpful in studying the stability properties of the Navier–Stokes equations and analysing different time-stepping schemes. If we instead drop all terms on the right-hand side of eqn (1.1.9b), as well as the divergence-free constraint, we obtain a nonlinear advection equation. This equation also serves as a good model for studying time-stepping algorithms and issues associated with the stability and long-time integration of the Navier–Stokes equations. These topics are presented in Chapters 6 and 10.

Finally, by dropping the time derivative in the parabolic equation (1.1.12), we obtain the *Poisson* equation $-\nu\nabla^2 u = f$, which is useful in dictating the continuity requirements and corresponding functional spaces in the variational formulation context adopted in this book. A treatment of the one-dimensional problem in Chapter 2 and in multiple dimensions in Chapter 3 is based on the Poisson equation. In addition, the study of solution algorithms appropriate for the global system inversion required in the Navier–Stokes equations is motivated by the Poisson equation and is covered in Chapter 4.

1.2 Numerical discretisations

1.2.1 *The finite element method*

There are more than one hundred thousand references on the finite element method today, including textbooks, monographs, conference proceedings, and journals. The majority of these references are devoted to structural mechanics, but finite element methods have also proved very successful in fluid dynamics, although the initial developments did not target this field. The reason for this may be the difficulties with the nonlinear terms in the Navier–Stokes equations and the original difficulties with the application of finite element methods to non-symmetric operators.

The idea of building up a solution to a differential equation from a sequence of local approximations is an old one. While Courant [112] used a network of triangles to represent with piecewise linear interpolation an approximate solution to the Dirichlet problem, it was Argyris [14] who introduced the variational method of approximation. Patching the triangles or other subdomains (elements) together is an automatic procedure today known as 'global assembly', or 'direct stiffness assembly', and was introduced in analysing the structural behaviour of various components of an aircraft. The 1960s were the formative years of finite element methods, focusing primarily on linear plane elasticity problems. Most of the earlier finite element methods used a low-order polynomial approximation as expansion basis, with the exception of the work of Oden [341] who used Fourier series in an assembly of rectangular subdomains. This was perhaps the first attempt to develop spectral elements which provide high-order piecewise approximations. A comprehensive review of the significant developments in finite element methodology and its mathematical theory is given in [342]. There are also many textbooks that develop the fundamental ideas of the finite element discretisation, including the two volumes by Zienkiewicz and Taylor [511], the six-volume series by Carey and Oden [88], the standard textbook by Hughes [247], and the more theoretical book of Brenner and Scott [75].

Finite elements were introduced in fluid mechanics in the late 1970s and were used routinely in large-scale codes in flow simulations in the 1980s. A major contribution to these developments came from the theoretical work of Babuška [22] and Brezzi [76] on the so-called *inf–sup* condition that is very useful in studying constrained elliptic problems. The discretisation of the Stokes problem requires the satisfaction of such a condition to produce a stable finite element discretisation. The monograph by Girault and Raviart [191] presents an in-depth analysis of the Stokes problem on these issues.

Finite elements have been used very successfully in inviscid aerodynamic simulations [379] where the geometric complexity involved makes finite difference methods less efficient. The algorithms developed in computational aerodynamics allow for very efficient discretisation techniques and unstructured mesh generation strategies based on fast triangulation and tetrahedralisation algorithms. Since the accomplishment of the accurate solution of the Euler equations on unstructured meshes, however, interest has now shifted to the simulation of time-dependent Navier–Stokes equations requiring accurate resolution of boundary layers and minimum dispersion errors over a long-time integration interval.

1.2.2 *Spectral discretisation*

The formulation of modern spectral methods was first presented in the monograph of Gottlieb and Orszag [199]. Multi-dimensional discretisations were formulated as tensor products of one-dimensional constructs in separable domains, that is, orthogonal simply-connected domains. The textbook of Canuto *et al.* [86] focuses on fluid dynamics algorithms and includes both practical, as well as theoretical, aspects of global spectral methods.

Global spectral methods use a single representation of a function $u(x)$ throughout the domain via a truncated series expansion, for instance,

$$u(x) \approx u_N(x) = \sum_{n=0}^{N} \hat{u}_n \phi_n ,$$

where $\phi_n(x)$ are the basis functions. This series is then substituted into a differential (or integral) equation and upon the minimisation of the residual function the unknown coefficients $\hat{u}_n$ are computed. The basis functions may be the often-used Chebyshev polynomials $T_n(x)$, the Legendre polynomials $L_n(x)$, or another member of the family of the Jacobi polynomials $P_n^{\alpha,\beta}$ (see Appendix A).

Spectral methods can be broadly classified into two categories: the pseudo-spectral or collocation methods and the modal or Galerkin methods. The first category is associated with a grid, that is, a set of nodes, and that is why it is sometimes referred to as *nodal* methods. The unknown coefficients $\hat{u}$ are then obtained by requiring the residual function to be zero exactly at a set of nodes. The second category is associated with the method of weighted residuals where the residual function is weighted with a set of *test functions* and after integration is set to zero. The test functions are the same as the basis functions, but in the so-called Petro–Galerkin formulation they may be different. Spectral-tau methods are similar to Galerkin methods, but the boundary conditions are satisfied by a supplementary set of equations and not directly via the basis functions. Another difference is that in the collocation approach the coefficients represent the nodal value of the physical variable, unlike the Galerkin or the spectral-tau method.

The convergence of both Galerkin and pseudo-spectral method is exponential, similar to the Fourier spectral method. This property follows directly from the theory of singular Sturm–Liouville boundary value problems—the basis functions are such solutions. Unlike finite element and finite difference methods, the order of the convergence is not fixed and it is related to the maximum regularity of the solution. Exponential or *spectral convergence* for a very smooth solution, in practice, implies that as the number of collocation points or the number of modes is doubled, the error in the numerical solutions decreases by at least two orders of magnitude and not a fixed factor as in low-order methods. This fast convergence is easily lost if the solution has finite regularity or if the domain is irregular.

1.2.3 *Why high-order accuracy in CFD?*

High-order numerical methods, that is, spectral and implicit finite difference schemes, have been used almost exclusively in the direct numerical simulation of turbulent flows in the last two decades [265]. They provide fast convergence, small diffusion and dispersion errors, easier implementation of the *inf–sup* condition for the incompressible Navier–Stokes equations, better data volume-over-surface ratio for efficient parallel processing, and better input/output handling due to the smaller volume of data.

For many engineering applications where accuracy of the order of 10% is acceptable, quadratic convergence is usually sufficient for stationary problems. However, this may not be true in time-dependent flow simulations where long-time integration is required. Therefore, we must ask how *long-time integration* relates to the formal order of accuracy of a numerical scheme, and what is the corresponding computational cost? Consider the convection of a waveform at a constant speed. Let us now assume that there are $N^{(k)}$ grid points required per wavelength to reduce the error to a level ε, where k denotes the formal order of the scheme. In addition, let us assume that we integrate for M *time periods*. We can neglect temporal errors $\mathcal{O}(\Delta t)^J$ (where J is the order of the time integration) by assuming a sufficiently small time-step Δt. We wish to estimate the phase error in this simulation for second- $N^{(2)}$, fourth- $N^{(4)}$, and sixth- $N^{(6)}$ order finite difference schemes.

The following results can be obtained by following the analysis of Kreiss and Oliger [287]. Assuming an 'engineering accuracy' of $\varepsilon = 10\%$, we obtain

$$N^{(2)} \propto 20M^{1/2}, \quad N^{(4)} \propto 7M^{1/4}, \quad N^{(6)} \propto 5M^{1/6}.$$

We therefore see that the required resolution depends on the number of time periods M, and the lower the order of the scheme the stronger that dependence. To compare the corresponding computational cost $W^{(k)}$, we have to consider that higher-order schemes have wider stencils and thus a higher operation count. For this particular example, we find that the work required to achieve 10% accuracy is

$$W^{(2)} \propto 20M^{1/2}, \quad W^{(4)} \propto 14M^{1/4}, \quad W^{(6)} \propto 15M^{1/6},$$

where the superscripts on W refer to the order of the scheme. In Fig. 1.2 we compare the efficiency of these three different discretisations for the same phase error by plotting the computational work required to maintain an 'engineering' accuracy of $\varepsilon = 10\%$ versus the number of time periods for the integration. This comparison favours the fourth-order scheme for short times ($M \propto \mathcal{O}(1)$) over both the second-order and the sixth-order schemes. However, for long-time integration ($M \propto \mathcal{O}(100)$), even for this engineering accuracy of 10%, the sixth-order scheme is superior as the corresponding operation count $W^{(6)}$ is about 6 times lower than the operation count of the second-order scheme $W^{(2)}$, and half the work of the fourth-order scheme $W^{(4)}$.

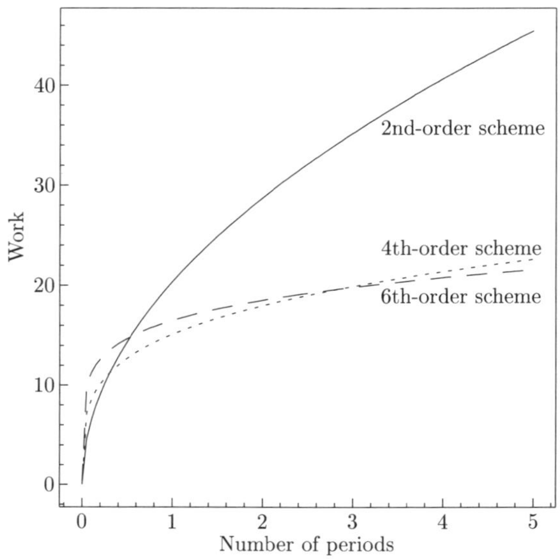

FIG. 1.2. Computational work (number of floating-point operations) required to integrate a linear advection equation for M periods while maintaining a cumulative phase error of $\varepsilon = 10\%$.

In a full Navier–Stokes simulation other considerations may be in place, including the time discretisation, which may change the 'break-even' point regarding the order of the scheme with the highest efficiency. The trend that we have established between resolution requirements and formal accuracy order is still valid for engineering accuracy despite, perhaps, our perception of the opposite! For an accuracy of 1% in the solution of this convection problem, the sixth-order scheme is superior even for short-time integration. For example, for one time period (for example, $M = 1$) the sixth-order scheme costs about 37% of the second-order scheme and 90% of the fourth-order scheme. In the limit of high accuracy and long-time integration, similar analysis suggests that spectral-based algorithms are computationally more efficient.

The reason why high accuracy is required in fluid dynamics simulations, even in stationary flow, can be demonstrated by considering the zero Reynolds number (Stokes) flow in a wedge with a driven lid (see Fig. 1.3). Using a similarity solution Moffatt [334] derived an asymptotic result for the strength and location of an 'infinite' number of eddies generated inside the wedge. The relations are dependent on the wedge angle; for example, for a wedge angle (28.1°) it is predicted that the strength of each eddy should asymptotically (that is, away from the forced top section) be about 406 times weaker than the previous eddy. This means that in order to resolve more than eight eddies, scales twenty orders of magnitude apart have to be captured. Moffatt's measure of the 'intensity' of

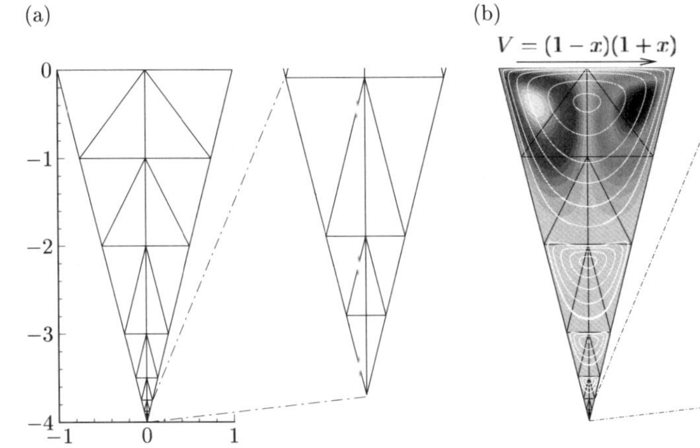

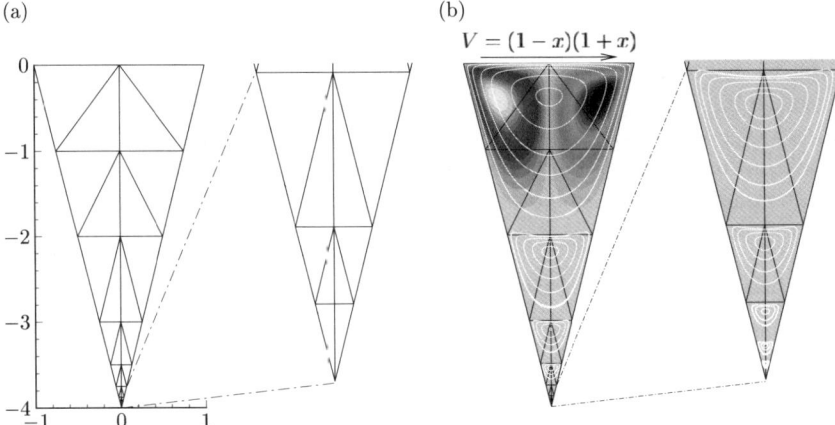

FIG. 1.3. (a) A wedge with an aspect ratio of 2 : 1 was discretised using thirty elements and an expansion order of $P = 17$. Stokes flow was then computed in this domain driven by a prescribed lid velocity. At steady state, nine eddies were observed, as indicated by the streamline plot in (b) (there are three eddies in the last two elements).

successive eddies was the ratio of the local maximum transverse velocity along the centre-line. Therefore, if we take a profile along the centre-line and plot the transverse (x_1 direction) velocity as a function of perpendicular distance from the top of wedge, then we obtain the distributions shown in Fig. 1.4 for the four resolutions using high-order expansions.

At the centre of an eddy we expect the transverse velocity to be zero, which

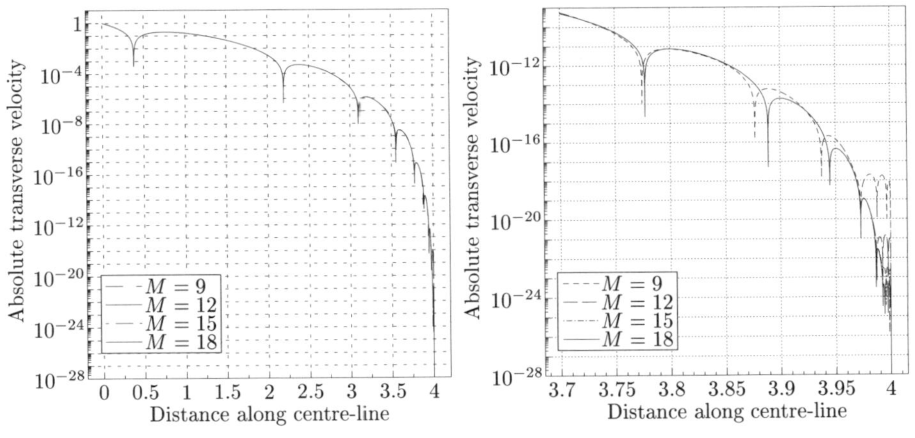

FIG. 1.4. The centre-line transverse velocity as a function of perpendicular height from the top of the wedge shown in Fig. 1.3. (Expansion order $P = M - 1$.)

is indicated by the spikes in Fig. 1.4. After each spike we note that there are local maxima which we use to determine the ratio of maximum velocities in order to evaluate Moffatt's eddy 'intensity'. We note from Fig. 1.4 that at all resolutions the first four eddies have been resolved to the accuracy of the plot. Refining the mesh at the lower corner (p-refinement, see next section) we are able to resolve up to nine eddies with only modest total resolution. We will revisit this example in Chapter 8.

1.2.4 *Structured versus unstructured discretisation*

The refinement procedure required in the wedge flow becomes more efficient if unstructured discretisation is employed. Features such as selective local refinement and efficient mesh adaptation are critically dependent on the flexibility of a discretisation in decomposing a computational domain into triangles and tetrahedra or other polymorphic elements in three dimensions or into *non-conforming* quadrilateral and hexahedral elements. This class of discretisations is what we characterise as unstructured. While structured discretisations have been the prevailing choice so far for static or quasi-static problems, it is inevitable that with the emphasis shifting towards time-dependent problems unstructured discretisation on non-fixed grids will be used almost exclusively in the future.

The theory required for non-conforming discretisations is presented in Chapter 7. Here we give an example for a two-dimensional unsteady flow past a half-cylinder (see Fig. 1.1) using a non-conforming and a conforming mesh in the near wake. The conforming mesh uses 276 elements, while the non-conforming mesh uses 176 elements. In Fig. 1.5 we plot vorticity contours for the two discretisations. The vorticity is obtained from $\nabla \times \mathbf{v}$, and thus the 'noisy' solution on the conforming mesh can be interpreted as a measure of the under-resolution that arises because of small-scale features unresolved by the mesh. The singular corner presents an additional difficulty in modelling this flow. The non-conforming discretisation of the domain isolates these features within a few elements close to the cylinder surface and resolves the fine scales present in the forming vortex. At the same time, it removes unnecessary elements away from the body where the solution is smooth, and maintains far-field boundaries at the same distance.

The second example is a triangulation of a complicated domain as shown in Fig. 1.6. This domain uses a variety of triangular elements of different aspect ratios and orientations in an almost random triangulation. Within this domain, we have solved an elliptic Helmholtz problem (see Chapter 5) of the form

$$\nabla^2 u(x_1, x_2) - u(x_1, x_2) = f(x_1, x_2)\,.$$

The exact solution considered was

$$u(x_1, x_2) = \sin\left[\frac{\pi}{4}(\sqrt{(x_1 - 15)^2 + (x_2 - 8)^2})\right].$$

Also shown in Fig. 1.6 is the H^1 error plotted with respect to the polynomial order of the expansion. Exponential convergence is observed, as indicated by the

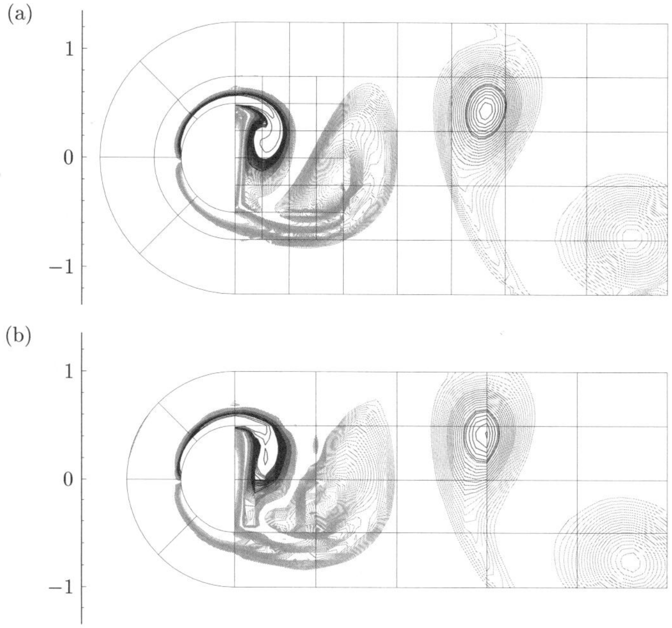

FIG. 1.5. Vortex shedding simulation using (a) non-conforming and (b) conforming discretisation. Shown are instantaneous vorticity contours in the near-wake.

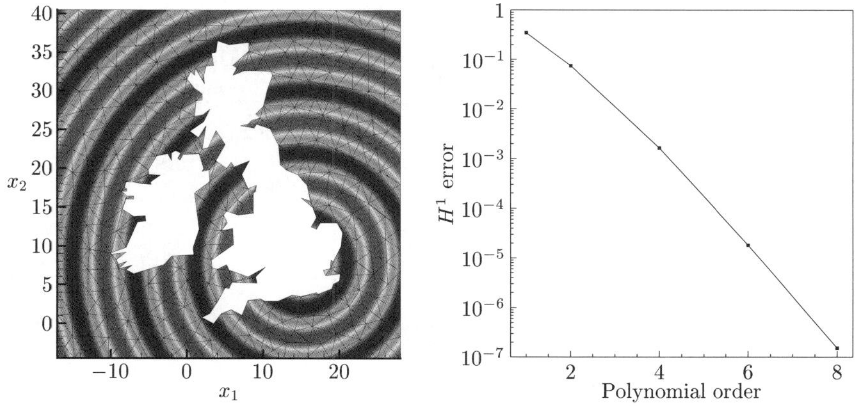

FIG. 1.6. Convergence to the Helmholtz problem with solution $u(x_1, x_2) = \sin(\pi(R(x_1, x_2) - R_0))$. Exponential convergence is obtained independently of the complexity of the discretisation. The exact solution is prescribed at all domain boundaries.

asymptotic linear behaviour of the curves on this linear–log plot. We note that the solution domain does not include the region immediately around $x_1 = 15$, $x_2 = 8$, since within this region it is not possible to bound all the derivatives of $u(x_1, x_2)$.

1.2.5 *What is hp convergence?*

The mathematical theory of finite elements in the 1970s has established rigorously the convergence of the h-version of the finite element. The error in the numerical solution decays algebraically by refining the mesh, that is, introducing more elements while keeping the (low) order of the interpolating polynomial fixed. An alternative approach is to keep the number of subdomains fixed and increase the order of the interpolating polynomials in order to reduce the error in the numerical solution. This is called p-type refinement and is typical of polynomial spectral methods [199]. For infinitely smooth solutions p-refinement usually leads to an exponential decay of the numerical error. Recognising the advantages of both types of convergence in mechanics problems, B. A. Szabó proposed and implemented a new method that he coined the *hp* version of the finite element. In this combined approach we simultaneously increase the number of subdomains (elements) and increase the interpolation order within the element either uniformly throughout the domain or selectively depending on the resolution requirements.

To give an example of an *hp* refinement we revisit the problems shown in Fig. 1.6. In the h-refinement strategy we refine the mesh as shown in Fig. 1.7(a)–(d), whereas in the p-refinement strategy we can fix the mesh and increase the

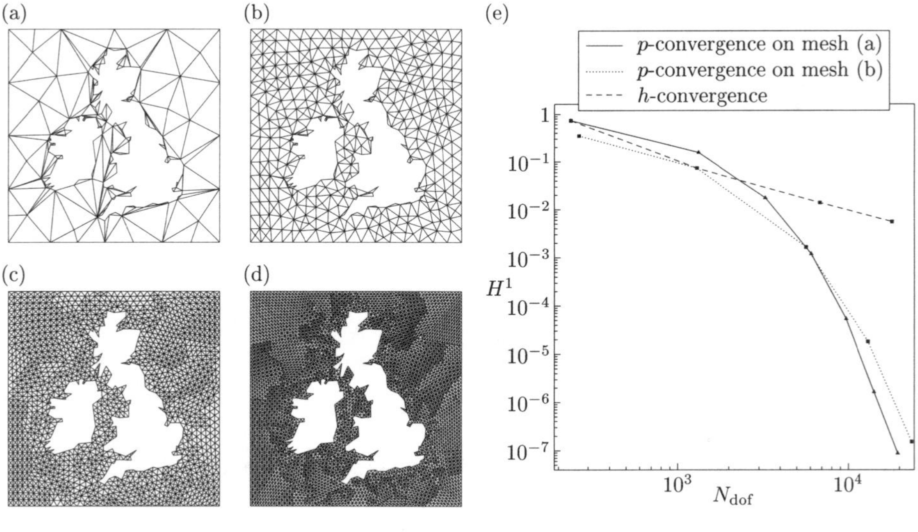

FIG. 1.7. Convergence history using h- and p-type refinement for the elliptic problem described in Fig. 1.6.

order of the polynomial expansion. The error in the H^1 norm as a function of the total degrees of freedom is shown in Fig. 1.7(e) for both the h-refinement with a fixed polynomial order of $P = 2$ and a p-refinement based on meshes (a) and (b). The h-refinement initially resolves the solution faster than the p-refinement on mesh (a); however, as the asymptotic exponential convergence is achieved the p-refinement takes over the h-refinement process. If we just consider the meshes shown in Fig. 1.7(a)–(d), the optimum convergence path as a function of degrees of freedom involves using both mesh (b) and mesh (a), thereby using both h- and p-refinement. In general, we would like to know the error as a function of computational cost, which is much harder to measure. However, for smooth solutions the concept of *hp* refinement still provides the optimal convergence strategy.

2

FUNDAMENTAL CONCEPTS IN ONE DIMENSION

In this chapter we illustrate the fundamental concepts behind the design and implementation of the spectral/*hp* element method for one-dimensional linear elliptic problems. The basic mechanics of this formulation will help to illustrate useful techniques for a variety of different types of mathematical problems, such as hyperbolic and parabolic equations, as well as different types of formulations such as the discontinuous Galerkin formulation discussed in Section 6.2.2. It will also provide the basis for understanding the multi-dimensional formulation which is discussed in Chapters 3 and 4.

The chapter starts by discussing the general framework of different formulations in the context of the method of weighted residuals in Section 2.1. This is followed by a more detailed description of the Galerkin method in Section 2.2. The efficiency of the spectral/*hp* element technique can be attributed to the elemental decomposition and implementation at this level. These ideas are introduced in Section 2.3 where we discuss the h-type elemental decomposition from a global expansion and then the p-type polynomial expansion within each elemental region. In Section 2.4 we then detail the principal elemental operations of numerical integration and differentiation. Finally, in Section 2.5 we outline some of the theory and results relating to error estimation of the technique, and in Section 2.6 we provide some exercises focused towards writing a one-dimensional spectral/*hp* element solver with examples.

♦[1] For the reader primarily interested in the implementation of the technique we have introduced margin identifiers to indicate key formulation ♣ and implementation ♦ details. The implementation details start with the descriptive formulation of the Galerkin problem in Section 2.2.1. This section also discusses the Galerkin implementation of Neumann boundary conditions through the weak form of the problem in Section 2.2.1.2 and the enforcement of Dirichlet boundary conditions through homogenisation of the solution in Section 2.2.1.3. Readers familiar with the finite element technique may already be aware of these techniques, although these sections also provide the authors' perspective on these issues. The next implementation-related topic is found in Section 2.3.1, which discusses the elemental (h-type) decomposition of the spectral/*hp* element approach. Section 2.3.1.4 also discusses how to transform from the elemental decomposition to the global problem, and Sections 2.3.3.3 and 2.3.4.2 present the most commonly used p-type polynomial expansion bases in one dimension. To

[1] Layout of chapter from an implementation point of view.

complete the implementation of the method requires knowledge of numerical integration and differentiation, which are discussed in Sections 2.4.1 and 2.4.2. Finally, in Section 2.6 we provide a structured series of exercises directed towards the implementation of the one-dimensional solver.

For a more detailed description, Section 2.2.3 discusses the Galerkin formulation in a more mathematical framework and Section 2.2.4 highlights the classical properties associated with the Galerkin formulation. After introducing the elemental decomposition in Section 2.3.1, we provide a more general discussion on the design and construction of p-type polynomial expansions in Section 2.3.2.1. To complement the discussion on numerical quadrature in Section 2.4.1 we also expand on this topic in Section 2.4.1.2 by discussing the effect of under-integration of nonlinear products of the polynomial solution, which is important when considering the nonlinear advection terms of the Navier–Stokes equations. Finally, in Section 2.5 we outline the basic formulation and error estimation results associated with one-dimensional spectral/hp element methods.

Although our principal focus will be on developing the Galerkin formulation for the spectral/hp element approach using higher-order polynomial expansions, the governing theory is taken from the traditional finite element technique which has been well documented in many texts, see, for example, [75, 247, 443, 511].

Historical setting of the finite element method

Structural engineers were responsible for the original implementation of the finite element method. It took approximately a decade before the method was recognised as a form of the Rayleigh–Ritz problem. The relation between these two techniques comes from considering the variational form of the problem [113]. For example, the quadratic functional

$$\mathcal{F}(u) = \int_0^1 \left[p(x)(u'(x))^2 + q(x)(u(x))^2 - 2f(x)u(x)\right] \mathrm{d}x \tag{2.0.1}$$

has a minimum with respect to a variation in $u(x)$ given by the Euler equation

$$-\frac{\mathrm{d}}{\mathrm{d}x}\left(p(x)\frac{\mathrm{d}u(x)}{\mathrm{d}x}\right) + q(x)u(x) = f(x)\,. \tag{2.0.2}$$

Therefore, instead of solving for the differential equation (2.0.2) to determine $u(x)$, an alternative but equivalent solution is to find the value of $u(x)$ which minimises the functional of eqn (2.0.1).

The Rayleigh–Ritz idea approximates the solution by a finite number of functions $u(x) = \sum_i^N q_i\Phi_i(x)$ to determine the unknown weights q_i, which minimise the functional of eqn (2.0.1). In the finite element method the solution is also approximated by a finite number of functions, which are typically local in nature as opposed to the global functions used in the Rayleigh–Ritz approach. However, the starting-point for a finite element method is the differential equation (2.0.2), which is formulated into an integral form (also known as the Galerkin formulation) so that the problem can be reduced to an algebraic system which can be solved numerically. The connection between the two methods was made when it was realised that the integral form of the finite element method was exactly the same as the functional form used in the Rayleigh–Ritz method for a linear problem. In structural mechanics it is also possible to form a functional directly from a statement of equilibrium without ever having to determine the Euler equation.

This relation between the finite element method and the Rayleigh–Ritz technique was very significant since it made the finite element technique mathematically respectable. It also ultimately proved to be somewhat misleading as it implied that a functional form was needed to formulate the problem. This is, in fact, not the case, as a more general formulation is possible using the method of weighted residuals which leads to the standard Galerkin formulation.

2.1 Method of weighted residuals

In approximating an exact solution numerically we are typically replacing an *infinite* expansion with a *finite* representation. Such approximation necessarily means that the differential equation cannot be satisfied everywhere in our region of interest and so we are only able to satisfy a finite number of *conditions.* It is the choice of the *conditions* which are to be satisfied that defines the type of numerical method or projection operator of the scheme. For example, the collocation method refers to a method where the differential equation is satisfied at a few distinct positions rather than at every point in the solution region.

The method of weighted residuals illustrates how the choice of different weight (or test) functions in an integral or weak form of the equation can be used to construct many of the common numerical methods.

To describe the method of weighted residuals we consider a *linear* differential equation in a domain Ω denoted by

$$\mathbb{L}(u) = 0\,, \tag{2.1.1}$$

subject to appropriate initial and boundary conditions. It is assumed that the solution $u(\boldsymbol{x}, t)$ can be accurately represented by the approximate solution of the form

$$u^{\delta}(\boldsymbol{x}, t) = u_0(\boldsymbol{x}, t) + \sum_{i=1}^{N_{\mathrm{dof}}} \hat{u}_i(t)\Phi_i(\boldsymbol{x})\,, \tag{2.1.2}$$

where $\Phi_i(\boldsymbol{x})$ are analytic functions called the *trial* (or *expansion*) *functions,* $\hat{u}_i(t)$ are the N_{dof} unknown coefficients, and $u_0(\boldsymbol{x}, t)$ is selected to satisfy the initial and boundary conditions. We note that, by definition, $\Phi_i(\boldsymbol{x})$ satisfies homogeneous boundary conditions (i.e., zero on Dirichlet boundaries) since the known function $u_0(\boldsymbol{x}, t)$ already satisfies the boundary conditions of the problem. Substitution of the approximation (2.1.2) into eqn (2.1.1) produces a nonzero residual, R, such that

$$\mathbb{L}(u^\delta) = R(u^\delta)\,. \tag{2.1.3}$$

The approximation has the form given by eqn (2.1.2) but we have no unique way of determining the coefficients $\hat{u}_i(t)$. To do so, we can place a restriction on the residual R which in turn will reduce eqn (2.1.3) to a system of ordinary differential equations in $\hat{u}_i(t)$. If the original eqn (2.1.1) is independent of time then the coefficients $\hat{u}_i$ can be determined directly from the solution of a system of algebraic equations.

To define the type of restriction to be placed on the residual R we must first introduce the Legendre inner product (f, g) over the domain Ω defined as

$$(f, g) = \int_\Omega f(\boldsymbol{x})g(\boldsymbol{x})\,\mathrm{d}\boldsymbol{x}\,. \tag{2.1.4}$$

The restriction placed on R is that the inner product of the residual with respect to a *weight* (or *test*) function is equal to zero, that is,

$$(v_j(\boldsymbol{x}), R) = 0\,, \quad j = 1, \ldots, N_{\mathrm{dof}}\,,$$

where the function $v_j(\boldsymbol{x})$ is the test or weight function. The weighted residual is then said to be zero and it is from this expression that the technique takes its name.

Upon convergence as $N_{\mathrm{dof}} \to \infty$, the residual $R(\boldsymbol{x})$ tends to zero since the approximate solution $u^\delta(\boldsymbol{x}, t)$ approaches the exact solution $u(\boldsymbol{x}, t)$. However, the nature of the scheme is determined by the choice of the expansion function $\Phi_i(\boldsymbol{x})$ and the test function v_j. A list of the most commonly used test functions and the computational method they produce is shown in Table 2.1 and will be briefly outlined in the following sections.

Collocation method

In the collocation method the test function is the Dirac delta function such that $v_j(\boldsymbol{x}) = \delta(\boldsymbol{x} - \boldsymbol{x}_j)$, where $\boldsymbol{x}_j$ denotes a set of given collocation points. At a collocation point the residual is set to zero ($R(\boldsymbol{x}_j) = 0$) and, accordingly, the differential equation is exactly satisfied at this point.

TABLE 2.1. Test functions $v_j(\boldsymbol{x})$ used in the method of weighted residuals and the method produced.

Test/weight function	Type of method
$v_j(\boldsymbol{x}) = \delta(\boldsymbol{x} - \boldsymbol{x}_j)$	Collocation
$v_j(\boldsymbol{x}) = \begin{cases} 1, & \text{inside } \Omega^j\,, \\ 0, & \text{outside } \Omega^j \end{cases}$	Finite volume (subdomain)
$v_j(\boldsymbol{x}) = \dfrac{\partial R}{\partial \hat{u}_j}$	Least-squares
$v_j(\boldsymbol{x}) = \Phi_j$	Galerkin
$v_j(\boldsymbol{x}) = \Psi_i\ (\neq \Phi_j)$	Petrov–Galerkin

Finite volume/subdomain methods

The finite volume or subdomain method is described by splitting the solution domain Ω into $N_{\rm dof}$ non-overlapping subdomains Ω^j and using a test function of the form

$$v_j = \begin{cases} 1\,, & \text{inside } \Omega^j\,, \\ 0\,, & \text{outside } \Omega^j\,, \end{cases}$$

where the union of Ω^j is equal to Ω (i.e., $\bigcup_{j=1}^{N_{\rm el}} \Omega_j = \Omega$). This method has been very popular in computational aerodynamics. It can be considered as a technique to recover a conservation statement from a partial differential equation.

Least-squares method

The least-squares method originates from the idea of least-squares estimation developed by Gauss. In this method the residual is set to $v_j = \partial R/\partial \hat{u}_j$. This choice determines the coefficients $\hat{u}_i$ which minimises (R, R). This formulation using a spectral/*hp* element discretisation has recently increased in popularity and is discussed further in Section 8.5.

Galerkin method

Finally, we consider the Galerkin method (also known as the Bubnov–Galerkin method). In this method the test functions are chosen to be the same as the trial or expansion functions such that $v_j = \Phi_j$. A broader class of the Galerkin method known as the Petrov–Galerkin method, or sometimes the generalised Galerkin method, uses test functions that may be similar, but not identical, to the trial functions ($v_j \neq \Phi_j$). The choice of Petrov–Galerkin test functions is typically based upon a perturbation of the trial functions, where the additional contribution is chosen to improve the numerical stability of the scheme or to impose an upwind condition [222,248]. For further details on the background of the methods of weighted residuals, please see Finlayson [160] and Fletcher [163].

This book is primarily concerned with Galerkin methods and for the majority of cases we shall be considering the standard Bubnov–Galerkin method. In Chapters 6 and 10 we will also discuss the *discontinuous* Galerkin methods in the context of solving hyperbolic conservation laws, and, in Chapter 7, in the context of second-order elliptic equations.

The method of weighted residuals illustrates how to construct different types of numerical techniques and defines the projection operator being employed in each method. It does not define the type of expansion function or approximation space, although the use of the terminology *spectral* or *finite element* does provide further insight. It is generally understood that spectral methods use a set of *global* expansion functions, that is, the expansion functions $\Phi_i(\boldsymbol{x})$ has a nonzero definition throughout the solution domain (i.e., like a sine or cosine function). The finite element or, alternatively, the finite volume technique uses a set of expansion functions $\Phi_i(\boldsymbol{x})$ which are only defined in a local 'finite' region. In the finite element expansion these regions are typically made up of non-overlapping tessellations of the total solution domain. Theoretically, both the spectral and finite-element-type expansions may be used with any of the numerical methodologies described above. The projection operators can also be mixed. This is commonly the case when considering nonlinear problems in spectral methods where the collocation projection is used to evaluate nonlinear products in the so-called *pseudo-spectral* method [199].

2.2 Galerkin formulation

Finite element methods typically use the Galerkin formulation introduced in the previous section. In this section, we describe how to formulate the Galerkin problem. We start in Section 2.2.1 by considering an informal formulation in order to solve the one-dimensional Poisson equation to introduce the basic concepts. The formulation is then illustrated by a worked example using linear finite elements in Section 2.2.2. A mathematical description of the formulation is presented in Section 2.2.3 and some important mathematical properties of the Galerkin formulation are discussed in Section 2.2.4.

2.2.1 *Descriptive formulation*

Our example problem is the Poisson equation ♣[1]

$$\mathbb{L}(u) \equiv \nabla^2 u + f = 0\,. \tag{2.2.1}$$

This equation arises in many areas of physics such as irrotational fluid flow and steady-state heat conduction, as well as in problems involving electrical and gravitational potentials. In one dimension, eqn (2.2.1) becomes

$$\mathbb{L}(u) \equiv \frac{\partial^2 u}{\partial x^2} + f = 0\,. \tag{2.2.2}$$

2.2.1.1 *Strong form and definition of boundary conditions*

For this problem to be well posed and thus have a unique solution we need to specify boundary conditions. If we consider the solution in a domain $\Omega = \{x \mid 0 \leqslant x \leqslant 1\}$, then we might consider the following boundary conditions:

$$u(0) = g_{\mathcal{D}}\,, \quad \frac{\partial u}{\partial x}(1) = g_{\mathcal{N}}\,,$$

where $g_{\mathcal{D}}$ and $g_{\mathcal{N}}$ are given constants.

The boundary condition $u(0) = g_{\mathcal{D}}$ specifies a condition on the solution and is referred to as a *Dirichlet* or *essential* boundary condition. The boundary condition $\partial u(1)/\partial x = g_{\mathcal{N}}$, however, specifies a condition on the derivative of the solution and is referred to as a *Neumann* or *natural* boundary condition.

[1] Galerkin problem statement and implementation of boundary conditions.

As we shall see, in the Galerkin formulation Dirichlet boundary conditions have to be specified explicitly whereas Neumann conditions are dealt with implicitly as part of the formulation. If the boundary conditions stated above are applied to eqn (2.2.2) it becomes a two-point boundary value problem and is said to be in the *strong* or *classical* form.

2.2.1.2 *Weak form and implementation of Neumann boundary conditions*

♣[2] To construct a weak approximation to eqn (2.2.2), we multiply this equation by a weight of test function $v(x)$, which by definition is *zero on all Dirichlet boundaries* $\partial\Omega_{\mathcal{D}}$, and integrate over the domain Ω to obtain the inner product of $\mathbb{L}(u)$ with respect to v:

$$(v, \mathbb{L}(u)) = \int_0^1 v\left(\frac{\partial^2 u}{\partial x^2} + f\right) \mathrm{d}x = 0\,. \tag{2.2.3}$$

We note that eqn (2.2.3) is equivalent to setting the weighted residual to zero. If u^δ is an approximation to u (recalling that $\mathbb{L}(u^\delta) = R(u^\delta)$) then eqn (2.2.3) is equivalent to the condition $(v, R) = 0$.

The next important step in the classical Galerkin spectral/*hp* element formulation is to integrate eqn (2.2.3) by parts to obtain

$$\int_0^1 \frac{\partial v}{\partial x}\frac{\partial u}{\partial x}\,\mathrm{d}x = \int_0^1 v f\,\mathrm{d}x + \left[v\frac{\partial u}{\partial x}\right]_0^1 . \tag{2.2.4}$$

In higher dimensions we would have used Gauss' divergence theorem to achieve an analogous result. As the test functions are defined to be zero on Dirichlet boundaries we know that $v(0) = 0$. Therefore, we can enforce the Neumann boundary condition $\partial u(1)/\partial x = g_{\mathcal{N}}$ by substitution into the last term of eqn (2.2.4), which simplifies to

$$\int_0^1 \frac{\partial v}{\partial x}\frac{\partial u}{\partial x}\,\mathrm{d}x = \int_0^1 v f\,\mathrm{d}x + v(1)g_{\mathcal{N}}\,. \tag{2.2.5}$$

In this last step we see how the Neumann boundary conditions are naturally included in the formulation through the action of the integration of parts. Note that for zero Neumann condition the last term vanishes; then, to impose the zero Neumann condition we do nothing! This operation not only reduces the order of the maximum derivative of the discrete problem but, as we shall see in Section 2.2.2, it also makes the resulting discrete matrix equation symmetric. The integral form of the problem given by eqns (2.2.4) and (2.2.5) is referred to as the *weak* form of the problem.

The Galerkin approximation of problem (2.2.2) is the solution to the weak form of the eqn (2.2.5) when the exact solution $u(x)$ is approximated by a finite

[2] Treatment of second-order differential operators and how to impose Neumann boundary conditions.

expansion denoted by $u^\delta(x)$. The function $v(x)$ in eqn (2.2.5) is also replaced by a finite expansion, denoted by $v^\delta(x)$, and so eqn (2.2.5) becomes

$$\int_0^1 \frac{\partial v^\delta}{\partial x}\frac{\partial u^\delta}{\partial x}\,\mathrm{d}x = \int_0^1 v^\delta f\,\mathrm{d}x + v^\delta(1)g_{\mathcal{N}}\,. \tag{2.2.6}$$

We recall that the set of functions used in the finite expansion of the solution u^δ are referred to as the *trial* functions, whereas the functions contained within v^δ are referred to as the *test* functions.

2.2.1.3 *Enforcing Dirichlet boundary conditions: lifting a known solution*

By definition, all Dirichlet boundary conditions are known. We can extend (or ♣[3] lift) these known functions on the boundary into the interior of the solution domain by any convenient function, which is contained within the solution space. The word 'lift' originates from the French 'relevement', as introduced by Lions [306]. The action of lifting a known solution is equivalent to decomposing the approximate solution u^δ into a known lifted function, $u^{\mathcal{D}}$, which satisfies the Dirichlet boundary conditions, and an unknown homogeneous function, $u^{\mathcal{H}}$, which is zero on the Dirichlet boundaries, i.e.,

$$u^\delta = u^{\mathcal{H}} + u^{\mathcal{D}}\,, \tag{2.2.7}$$

where

$$u^{\mathcal{H}}(\partial\Omega_{\mathcal{D}}) = 0\,, \quad u^{\mathcal{D}}(\partial\Omega_{\mathcal{D}}) = g_{\mathcal{D}}\,.$$

By substituting eqn (2.2.7) into our weak formulation of the problem given by eqn (2.2.6), we obtain

$$\int_0^1 \frac{\partial v^\delta}{\partial x}\left[\frac{\partial u^{\mathcal{D}}}{\partial x} + \frac{\partial u^{\mathcal{H}}}{\partial x}\right]\mathrm{d}x = \int_0^1 v^\delta f\,\mathrm{d}x + v^\delta(1)g_{\mathcal{N}}\,,$$

which can be rearranged to obtain

$$\int_0^1 \frac{\partial v^\delta}{\partial x}\frac{\partial u^{\mathcal{H}}}{\partial x}\,\mathrm{d}x = \int_0^1 v^\delta f\,\mathrm{d}x + v^\delta(1)g_{\mathcal{N}} - \int_0^1 \frac{\partial v^\delta}{\partial x}\frac{\partial u^{\mathcal{D}}}{\partial x}\,\mathrm{d}x\,. \tag{2.2.8}$$

Since $u^{\mathcal{D}}$ is a known function which satisfies the boundary conditions, all terms on the right-hand side of eqn (2.2.8) are known and this equation has been lifted in the sense that the unknown function u satisfies homogeneous (i.e., zero) Dirichlet boundary conditions.

As will be illustrated in Section 2.2.2, eqn (2.2.8) can be solved as a finite linear algebraic system as all the terms on the right-hand side are known, and the homogeneous solution $u^{\mathcal{H}}$ and the test function v^δ contain a finite number of functions. The Galerkin formulation has, therefore, reduced the differential

[3] Lifting a known solution to impose Dirichlet boundary conditions.

problem (2.2.2) to an algebraic matrix problem, which we can solve on a computer.

The process of lifting a known solution is an important part of the Galerkin spectral/*hp* formulation since we require that the same set of basis functions that are used to represent the test functions, v^δ, are also used in the representation of the solution u^δ. From Section 2.2.1.2 we recall that the expansion functions used for the test functions v^δ are defined to be zero on all Dirichlet boundaries and after the lifting step we can define the homogeneous solution vector $u^{\mathcal{H}}$ with the same expansion basis since this function also has zero boundary conditions on Dirichlet boundaries. This, therefore, permits us to use the same expansion space for $u^{\mathcal{H}}$ as we use for v^δ.

Although this decomposition may appear unnecessarily complicated, this step plays an important role in the Galerkin formulation. Without lifting a known solution out of our problem we will have more degrees of freedom in the test space than we have in the trial space. In implementation terms, this means that the algebraic system which results from evaluating the weak problem (2.2.8) will not be square.

An alternative approach to enforcing Dirichlet boundary conditions, commonly used in finite element methods, is to assemble a matrix system including all degrees of freedom in our approximation for both the test and trial functions. Dirichlet boundary conditions can then be enforced by zeroing rows which correspond to the known degrees of freedom, placing a unit term on the diagonal, and setting the right-hand side to the known value. The resulting matrix system will not be symmetric even if the original problem was symmetric. The matrix system which arises from the lifting approach is a submatrix of the full problem. The second right-hand-side term in eqn (2.2.8) can also be understood as another submatrix of the full problem multiplied by the known boundary conditions. In the lifted solution approach, the resulting matrix problem remains symmetric if the original problem was symmetric. The drawback of the lifting approach is that it requires a numbering system to reorder the matrix which is not required in the row-zeroing approach. We note, however, that the lifting approach also permits any known function satisfying the Dirichlet boundary conditions to be applied, which can be convenient when treating iterative solutions. Further, using the lifting approach, there is no need to assemble unnecessary components of the matrix problem, which can be quite costly in multiple dimensions. There is, however, a more complicated right-hand side to evaluate in eqn (2.2.8).

Finally, we note that to construct a matrix problem necessarily implies a linear problem. Quite often when treating nonlinear problems the nonlinear terms are treated explicitly in time or are linearised so that only linear terms are handled implicitly in time. As a result the above technique can still be applied.

2.2.1.4 *Mixed or Robin boundary conditions*

Another type of boundary condition is a *mixed* or *Robin* boundary condition. This commonly arises in convective heat transfer problems and is a linear combination of a Dirichlet and Neumann condition of the form ♣4

$$\alpha \frac{\partial u(1)}{\partial x} + \beta u(1) = g_{\mathcal{R}} \quad (\text{with } \alpha \neq 0),$$

where α, β, and $g_{\mathcal{R}}$ are known. To impose this condition we can substitute

$$\frac{\partial u(1)}{\partial x} = \frac{1}{\alpha}(g_{\mathcal{R}} - \beta u(1))$$

into eqn (2.2.6) and thus obtain

$$\int_0^1 \frac{\partial v^\delta}{\partial x}\frac{\partial u^\delta}{\partial x}\,dx + \frac{\beta}{\alpha} v^\delta(1) u^\delta(1) = \int_0^1 v^\delta f\,dx + \frac{v^\delta(1) g_{\mathcal{R}}}{\alpha}.$$

We have placed the term $\beta v^\delta(1)u^\delta(1)/\alpha$ on the left-hand side as the value of $u^\delta(1)$ has to be implicitly solved as part of the algebraic system. The practical difference between the implementation of the Robin and Neumann conditions is simply the modification of the matrix arising from the term $(\beta/\alpha)v^\delta(1)u^\delta(1)$.

2.2.2 *Two-domain linear finite element example*

Having defined the weak form of the problem and explained how to impose boundary conditions in Sections 2.2.1.2 to 2.2.1.4 we continue our overview of the Galerkin formulation with a worked example of a two-subdomain linear finite element solution. To illustrate the mechanics of the formulation we will use a globally-defined expansion basis. It should be noted, however, that in a practical implementation of a problem involving many elemental domains the normal practice is to use an elemental construction description of the basis, as discussed in Section 2.3. ♦2

Once again, we consider the one-dimensional Poisson equation in the interval $0 < x \leqslant 1$:

$$\mathbb{L}(u) \equiv \frac{\partial^2 u}{\partial x^2} + f = 0,$$

where $f(x)$ is a known function and the boundary conditions are

$$u(0) = g_{\mathcal{D}} = 1, \quad \frac{\partial u}{\partial x}(1) = g_{\mathcal{N}} = 1.$$

Following the formulation introduced in Sections 2.2.1.2 and 2.2.1.3, we construct our weak Galerkin approximation in two steps.

4 Imposing mixed or Robin boundary conditions.

2 Worked example of the Galerkin solution to an elliptic problem.

1. We start by considering the *weak form* by multiplying the problem by a discrete test space v^δ and integrating the second-order derivative by parts, which allows us to impose the Neumann boundary condition to arrive at

$$\int_0^1 \frac{\partial v^\delta}{\partial x}\frac{\partial u^\delta}{\partial x}\,\mathrm{d}x = \int_0^1 v^\delta f\,\mathrm{d}x + v^\delta(1)g_{\mathcal{N}}\,,$$

where u^δ is our discrete solution.

2. We now *lift* a known solution from the problem by decomposing u^δ into a known solution satisfying the Dirichlet boundary conditions $u^{\mathcal{D}}$ and a homogeneous solution $u^{\mathcal{H}}$ such that $u^\delta = u^{\mathcal{D}} + u^{\mathcal{H}}$, and so our weak solution becomes

$$\int_0^1 \frac{\partial v^\delta}{\partial x}\frac{\partial u}{\partial x}^{\mathcal{H}}\mathrm{d}x = \int_0^1 v^\delta f\,\mathrm{d}x + v^\delta(1)g_{\mathcal{N}} - \int_0^1 \frac{\partial v^\delta}{\partial x}\frac{\partial u}{\partial x}^{\mathcal{D}}\mathrm{d}x\,. \tag{2.2.9}$$

In our problem the solution is to be approximated by piecewise linear functions over two subdomains Ω^1 and Ω^2, as shown in Fig. 2.1. This type of approximation is known as an *h*-type approximation, where the h parameter represents the characteristic size of a subdomain (in one dimension, its length). Convergence to the exact solution is achieved by subdividing the solution domain Ω into smaller and smaller subdomains, so that $h \to 0$. We note, however, that

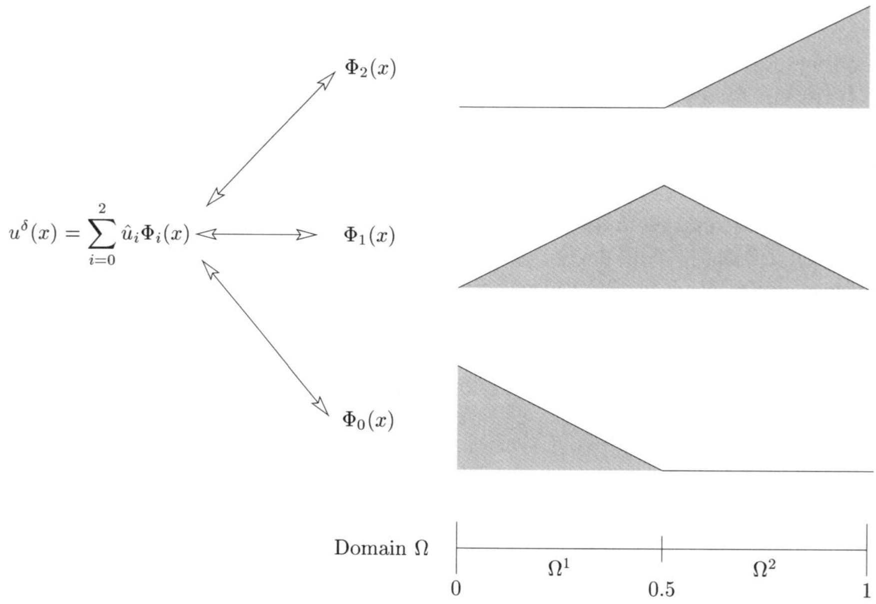

FIG. 2.1. Linear finite element approximation $u^\delta(x) = \sum_{i=0}^{2} \hat{u}_i\Phi_i(x)$, in a domain Ω, using two elemental subdomains, Ω^1 and Ω^2.

the elemental decomposition is still a vital component in the spectral/hp element decomposition. The main difference is that we use higher than linear order polynomials in each element. This is known as the p-type approximation.

For our linear two-subdomain case the approximate expansion has the form

$$u^\delta = \sum_{i=0}^{2} \hat{u}_i \Phi_i(x) ,$$

where $\Phi_i(x)$ is defined as

$$\Phi_0(x) = \begin{cases} 1 - 2x , & 0 \leqslant x \leqslant \frac{1}{2} , \\ 0 , & \frac{1}{2} \leqslant x \leqslant 1 , \end{cases} \qquad \Phi_1(x) = \begin{cases} 2x , & 0 \leqslant x \leqslant \frac{1}{2} , \\ 2(1-x) , & \frac{1}{2} \leqslant x \leqslant 1 , \end{cases}$$

$$\Phi_2(x) = \begin{cases} 0 , & 0 \leqslant x \leqslant \frac{1}{2} , \\ 2x - 1 , & \frac{1}{2} \leqslant x \leqslant 1 . \end{cases}$$

The only way to satisfy the Dirichlet boundary condition at $x = 0$ is to set $\hat{u}_0 = g_{\mathcal{D}}$ because $\phi_1(x)$ and $\phi_2(x)$ are zero at $x - 0$. Therefore, one choice for a lifted solution is to use the decomposition $u^\delta = u^{\mathcal{H}} + u^{\mathcal{D}}$, such that

$$\begin{aligned} u^{\mathcal{H}} &= \hat{u}_1 \Phi_1(x) + \hat{u}_2 \Phi_2(x) , \\ u^{\mathcal{D}} &= g_{\mathcal{D}} \Phi_0(x) , \end{aligned}$$

where $\hat{u}_1$ and $\hat{u}_2$ are still to be determined. Note that we could also have chosen $u^{\mathcal{D}}$ to be a known function of $\Phi_1(x)$ and $\Phi_2(x)$ if, for example, we had the solution to a previous problem. The expansion set used to define $u^{\mathcal{H}}$ contains the same functions used as homogeneous test functions. We can therefore define the test functions as

$$v^\delta(x) = \hat{v}_1 \Phi_1(x) + \hat{v}_2 \Phi_2(x) ,$$

where $\hat{v}_1$ and $\hat{v}_2$ are also unknown. As we shall see, these will never need to be determined. Finally, we need a representation of the function $f(x)$. This function is known explicitly and therefore it is theoretically possible to evaluate exactly any operations involving $f(x)$ with other known functions such as $\Phi_i(x)$. However, in practice, in order to treat an arbitrary function within an *efficient* computational implementation, the function is usually represented using the same expansion as applied to u^δ, i.e.,

$$f(x) = \sum_{i=0}^{2} \hat{f}_i \Phi_i(x) = \hat{f}_0 \Phi_0(x) + \hat{f}_1 \Phi_1(x) + \hat{f}_2 \Phi_2(x) .$$

Clearly, if $f(x)$ is a constant or a linear function then it will be exactly represented by this expression. For more complex functions the coefficients $\hat{f}_0$, $\hat{f}_1$, and $\hat{f}_2$ need to be determined and can be chosen to satisfy an interpolation approximation where $\hat{f}_0 = f(0)$, $\hat{f}_1 = f(0.5)$, and $\hat{f}_2 = f(1)$.

Evaluating the terms in eqn (2.2.9), we find

$$\int_0^1 \frac{\partial v^\delta}{\partial x}\frac{\partial u}{\partial x}^{\mathcal{H}} \mathrm{d}x = \int_0^{1/2} (2\hat{v}_1)(2\hat{u}_1)\,\mathrm{d}x + \int_{1/2}^1 (-2\hat{v}_1 + 2\hat{v}_2)(-2\hat{u}_1 + 2\hat{u}_2)\,\mathrm{d}x$$

$$= \begin{bmatrix}\hat{v}_1 & \hat{v}_2\end{bmatrix}\begin{bmatrix}4 & -2\\ -2 & 2\end{bmatrix}\begin{bmatrix}\hat{u}_1\\ \hat{u}_2\end{bmatrix}, \tag{2.2.10a}$$

$$\int_{\mathrm{C}}^1 v^\delta f\,\mathrm{d}x = \int_0^{1/2} (\hat{v}_1 2x)\left[\hat{f}_0(1-2x) + \hat{f}_1(2x)\right]\mathrm{d}x$$

$$+ \int_{1/2}^1 [\hat{v}_1 2(1-x) + \hat{v}_2(2x-1)]\left[\hat{f}_1 2(1-x) + \hat{f}_2(2x-1)\right]\mathrm{d}x$$

$$= \begin{bmatrix}\hat{v}_1 & \hat{v}_2\end{bmatrix}\begin{bmatrix}\frac{1}{12}\hat{f}_0 + \frac{1}{3}\hat{f}_1 + \frac{1}{12}\hat{f}_2\\ \frac{1}{12}\hat{f}_1 + \frac{1}{6}\hat{f}_2\end{bmatrix}, \tag{2.2.10b}$$

$$v^\delta(1)g_{\mathcal{N}} = [\hat{v}_1\Phi_1(1) + \hat{v}_2\Phi_2(1)]\,g_{\mathcal{N}} = \begin{bmatrix}\hat{v}_1 & \hat{v}_2\end{bmatrix}\begin{bmatrix}0\\1\end{bmatrix} g_{\mathcal{N}}, \tag{2.2.10c}$$

$$\int_0^1 \frac{\partial v^\delta}{\partial x}\frac{\partial u}{\partial x}^{\mathcal{D}} \mathrm{d}x = \int_0^{1/2} (2\hat{v}_1)(-2g_{\mathcal{D}})\,\mathrm{d}x = \begin{bmatrix}\hat{v}_1 & \hat{v}_2\end{bmatrix}\begin{bmatrix}-2g_{\mathcal{D}}\\ 0\end{bmatrix}. \tag{2.2.10d}$$

Therefore, eqn (2.2.9) can be written as the discrete matrix problem

$$\begin{bmatrix}\hat{v}_1 & \hat{v}_2\end{bmatrix}\left\{\begin{bmatrix}4 & -2\\ -2 & 2\end{bmatrix}\begin{bmatrix}\hat{u}_1\\ \hat{u}_2\end{bmatrix} - \begin{bmatrix}\frac{1}{12}\hat{f}_0 + \frac{1}{3}\hat{f}_1 + \frac{1}{12}\hat{f}_2\\ \frac{1}{12}\hat{f}_1 + \frac{1}{6}\hat{f}_2\end{bmatrix} - \begin{bmatrix}0\\ g_{\mathcal{N}}\end{bmatrix} + \begin{bmatrix}-2g_{\mathcal{D}}\\ 0\end{bmatrix}\right\} = 0\,.$$

For arbitrary choices of $\hat{v}_1$ and $\hat{v}_2$ we can solve this equation by evaluating the matrix equation in the curly brackets. Recalling that $g_{\mathcal{D}} = 1$ and $g_{\mathcal{N}} = 1$, the matrix equation becomes

$$\begin{bmatrix}4 & -2\\ -2 & 2\end{bmatrix}\begin{bmatrix}\hat{u}_1\\ \hat{u}_2\end{bmatrix} = \begin{bmatrix}2 + \frac{1}{12}\hat{f}_0 + \frac{1}{3}\hat{f}_1 + \frac{1}{12}\hat{f}_2\\ 1 + \frac{1}{12}\hat{f}_1 + \frac{1}{6}\hat{f}_2\end{bmatrix},$$

which has a solution

$$\begin{bmatrix}\hat{u}_1\\ \hat{u}_2\end{bmatrix} = \begin{bmatrix}\frac{3}{2} + \frac{1}{24}\hat{f}_0 + \frac{5}{24}\hat{f}_1 + \frac{1}{8}\hat{f}_2\\ 2 + \frac{1}{24}\hat{f}_0 + \frac{1}{4}\hat{f}_1 + \frac{5}{24}\hat{f}_2\end{bmatrix}.$$

The finite element approximation $u^\delta(x) = g_{\mathcal{D}}\Phi_0(x) + \hat{u}_1\Phi_1(x) + \hat{u}_2\Phi_2(x)$ is therefore

$$u^\delta = \begin{cases} 1 + x + \dfrac{x}{12}\hat{f}_0 + \dfrac{5x}{12}\hat{f}_1 + \dfrac{x}{4}\hat{f}_2\,, & 0 \leqslant x \leqslant \frac{1}{2}\,,\\[2ex] 1 + x + \dfrac{1}{24}\hat{f}_0 + \dfrac{2+x}{12}\hat{f}_1 + \dfrac{1+4x}{24}\hat{f}_2\,, & \frac{1}{2} \leqslant x \leqslant 1\,. \end{cases}$$

Having gone through the worked example we might now question what components are required to construct a general Galerkin approximation based on multiple elemental decompositions of the solution domain. Implicit in the implementation of this example was the assumption that we can differentiate and integrate the basis functions over the solution domain. In general, it is not practical analytically to integrate and differentiate, and so we adopt numerical rules to be discussed in Sections 2.4.1 and 2.4.2. However, to make this possible we must develop techniques to treat each element separately and thus permit us to automate the implementation. The construction of local elemental bases and the assembly of these into a global definition will be discussed in Section 2.3.

2.2.3 *Mathematical formulation*

In this section we shall construct the Galerkin approximation to a linear partial differential equation, similar to that discussed in Section 2.2.1, in a more mathematical framework. We consider the more general one-dimensional Helmholtz equation

$$\mathbb{L}(u) = \frac{\partial^2 u}{\partial x^2} - \lambda u + f = 0\,, \tag{2.2.11}$$

where λ is a real positive constant. The equation is presumed to be supplemented with appropriate boundary conditions such as

$$u(0) = g_{\mathcal{D}}\,, \quad \frac{\partial u}{\partial x}(l) = g_{\mathcal{N}}\,.$$

As indicated by the boundary conditions, we wish to determine the solution in the interval $0 < x < l$, which we shall denote by Ω.

Multiplying eqn (2.2.11) by an arbitrary test function $v(x)$, the properties of which are to be defined, and integrating over the domain Ω, we obtain

$$\int_0^l v\frac{\partial^2 u}{\partial x^2}\,\mathrm{d}x - \int_0^l \lambda v u\,\mathrm{d}x + \int_0^l v f\,\mathrm{d}x = 0\,.$$

Providing $u(x)$ and $v(x)$ are sufficiently smooth, we can integrate the first term by parts to arrive at

$$\int_0^l \frac{\partial v}{\partial x}\frac{\partial u}{\partial x}\,\mathrm{d}x + \int_0^l \lambda v u\,\mathrm{d}x = \int_0^l v f\,\mathrm{d}x + \left[v\frac{\partial u}{\partial x}\right]_0^l\,. \tag{2.2.12}$$

If we introduce the notation

$$a(v,u) = \int_0^l \left(\frac{\partial v}{\partial x}\frac{\partial u}{\partial x} + \lambda v u\right)\mathrm{d}x\,,$$

$$f(v) = \int_0^l v f\,\mathrm{d}x + \left[v\frac{\partial u}{\partial x}\right]_0^l\,,$$

then eqn (2.2.12) can be written as

$$a(v,u) = f(v)\,. \tag{2.2.13}$$

In structural mechanics, $a(u, u)$ is referred to as the *strain energy*, and the space of all functions which have a finite strain on Ω is called the *energy* space, which is denoted by $E(\Omega)$:

$$E(\Omega) = \{u \mid a(u, u) < \infty\}.$$

Associated with the energy space is the energy norm $||u||_E$ defined as

$$||u||_E = \sqrt{a(u, u)}. \tag{2.2.14}$$

Functions that belong to the energy space are called H^1 functions and satisfy the condition that the integral of the square of the function plus the square of its derivative are bounded.

We consider solutions to eqn (2.2.11) where the forcing function $f(x)$ is *well behaved* in the sense that $f(v)$ is finite. Therefore, we only consider candidate or *trial* solutions to eqn (2.2.12) which lie in the energy space and satisfy the Dirichlet boundary condition. This space is called the *trial space* and is denoted by $\mathcal{X}$. For our problem the trial space is defined by

$$\mathcal{X} = \{u \mid u \in H^1,\ u(0) = g_{\mathcal{D}}\}.$$

Similarly, we define the space of all test functions, denoted by $\mathcal{V}$, which are homogeneous on all Dirichlet boundaries, that is,

$$\mathcal{V} = \{v \mid v \in H^1,\ v(0) = 0\}.$$

The test space $\mathcal{V}$ is sometimes said to be in H_0^1, where the subscript 0 refers to the fact that it is in the homogeneous space. We can now define the generalised or weak formulation of eqn (2.2.11) as follows:

find $u \in \mathcal{X}$, such that

$$a(v, u) = f(v), \quad \forall\, v \in \mathcal{V}. \tag{2.2.15}$$

The weak problem is still an infinite-dimensional problem because the trial and test spaces, $\mathcal{X}$ and $\mathcal{V}$, contain an infinite number of functions. Therefore, we select subspaces $\mathcal{X}^\delta$ ($\mathcal{X}^\delta \subset \mathcal{X}$) and $\mathcal{V}^\delta$ ($\mathcal{V}^\delta \subset \mathcal{V}$) which contain a finite number of functions. In the spectral/*hp* element method we have two discretisation approaches, as denoted by h (element size) and p (polynomial order). We therefore interpret the use of δ in $\mathcal{X}^\delta$ and $\mathcal{V}^\delta$ to refer to these discretisation concepts, and so δ may be thought of as being a function of h (or similarly N_{el}) as well as p. The approximate form of the weak solution can then be stated as follows:

find $u^\delta \in \mathcal{X}^\delta$, such that

$$a(v^\delta, u^\delta) = f(v^\delta), \quad \forall\, v^\delta \in \mathcal{V}^\delta. \tag{2.2.16}$$

We note that in eqn (2.2.16) we have not imposed any Dirichlet boundary conditions. To impose Dirichlet boundary conditions we lift the solution by decomposing the function $u^\delta \in \mathcal{X}^\delta$ into a known component, $u^{\mathcal{D}}$, which lies in the trial space ($u^{\mathcal{D}} \in \mathcal{X}^\delta$) and satisfies the Dirichlet boundary condition, and an unknown component, $u^{\mathcal{H}}$, which lies in the test space ($u^{\mathcal{H}} \in \mathcal{V}^\delta$) and is homogeneous or zero on the Dirichlet boundary. In other words,

$$u^\delta = u^{\mathcal{H}} + u^{\mathcal{D}},$$

where

$$u^{\mathcal{H}}(0) = 0, \quad u^{\mathcal{D}}(0) = g_{\mathcal{D}}.$$

In the standard Galerkin approximation the same set of functions are used for both the test and trial functions. This is now possible since $u^{\mathcal{H}}$ and v^δ are both in $\mathcal{V}^\delta$. The Galerkin form of the problem can now be stated as follows:

find

$$u^\delta = u^{\mathcal{D}} + u^{\mathcal{H}}, \quad \text{where } u^{\mathcal{H}} \in \mathcal{V}^\delta,\ u^\delta \in \mathcal{X}^\delta,$$

such that

$$a(v^\delta, u^{\mathcal{H}}) = f^*(v^\delta), \quad \forall\, v^\delta \in \mathcal{V}^\delta, \tag{2.2.17}$$

where

$$f^*(v^\delta) = f(v^\delta) - a(v^\delta, u^{\mathcal{D}}).$$

For this linear equation another way of constructing the Galerkin solution is from a variational point of view. Equation (2.2.11) is the minimal solution to the functional

$$\mathcal{F}(v) = \int_0^l \left[\left(\frac{\partial v}{\partial x}\right)^2 + \lambda\,(v)^2 - 2vf \right] dx.$$

Therefore, if we minimise $\mathcal{F}(v)$ over the infinite-dimensional space $\mathcal{V}$ we will find the solution to eqn (2.2.11) which is the Euler equation of this functional. Replacing the variational problem by a finite-dimensional subspace $\mathcal{V}^\delta$ leads to the Ritz–Galerkin method (see Strang and Fix [443]).

2.2.4 *Mathematical properties of the Galerkin approximation*

In this section we introduce some significant properties of the Galerkin approximation. We consider the approximation u^δ to the solution u, where $u^\delta \in \mathcal{X}^\delta$ and satisfies

$$a(v^\delta, u^\delta) = f(v^\delta), \quad \forall\, v^\delta \in \mathcal{V}^\delta. \tag{2.2.18}$$

We mention that $a(v, u)$ is a *symmetric, bilinear form* which means

$$a(v, u) = a(u, v), \tag{2.2.19a}$$

$$a(c_1 v + c_2 w, u) = c_1 a(v, u) + c_2 a(w, u), \tag{2.2.19b}$$

where c_1 and c_2 are constants and u, v, and w are functions. Further, the operator $a(v, u)$ is said to be continuous (or bounded) if

$$|a(v, u)| \leqslant C_1 ||v||_1 ||u||_1, \tag{2.2.19c}$$

where $C_1 < \infty$ and the subscript denotes the norm in H^1. It is elliptic (or coercive) if

$$a(u,u) \geqslant C_2 ||u||_1^2 \,, \tag{2.2.19d}$$

where $C_2 > 0$.

We note that eqn (2.2.18) is equivalent to eqn (2.2.17) since $a(v^\delta, u^\delta) = a(v^\delta, u^{\mathcal{D}}) + a(v^\delta, u^{\mathcal{H}})$ using the bilinearity of $a(v,u)$ (eqn (2.2.19b)).

If $a(v^\delta, u^\delta)$ is a continuous, elliptic, bilinear form that is not necessarily symmetric and $f(v^\delta)$ is in the dual space of $\mathcal{V}^\delta$, then the Lax–Milgram theorem guarantees both *existence* and *uniqueness* of the solution of the Galerkin problem (2.2.18) (see Brenner and Scott [75]).

2.2.4.1 *Uniqueness*

To show that the solution u^δ is unique we assume that there are two distinct solutions u_1 and u_2 ($u_1, u_2 \in \mathcal{X}^\delta$) which satisfy

$$a(v^\delta, u_1) = f(v^\delta) \,, \quad \forall\, v^\delta \in \mathcal{V}^\delta \tag{2.2.20a}$$

and

$$a(v^\delta, u_2) = f(v^\delta) \,, \quad \forall\, v^\delta \in \mathcal{V}^\delta \,. \tag{2.2.20b}$$

Subtracting eqn (2.2.20a) from eqn (2.2.20b), we obtain

$$a(v^\delta, u_1) - a(v^\delta, u_2) = a(v^\delta, u_1 - u_2) = 0 \tag{2.2.20c}$$

using the bilinearity of $a(v,u)$. Now $u_1 - u_2 \in \mathcal{V}^\delta$ and therefore we can set $v^\delta = u_1 - u_2$, so eqn (2.2.20c) becomes

$$a(u_1 - u_2, u_1 - u_2) = 0 \,.$$

However, this implies that $||u_1 - u_2||_E = 0$. This is only possible if $u_1 = u_2$, which contradicts the assumption that they are distinct. We therefore conclude that there is only one unique solution. Strictly speaking, $||u_1 - u_2||_E = 0$ only implies that $u_1 = u_2$ if $\lambda \neq 0$. When $\lambda = 0$ the solution is only unique up to an arbitrary constant, that is, $u_1 - u_2 = C$. The constant, C, is necessarily zero if Dirichlet boundary conditions are specified, although the norm $||u_1 - u_2||_E$ cannot distinguish between functions that differ by an arbitrary constant when $\lambda = 0$.

2.2.4.2 *Orthogonality of the error to the test space in the energy norm*

The error between the exact and approximate solution, $\varepsilon = u - u^\delta$, is orthogonal to all functions in the finite-dimensional test space $\mathcal{V}^\delta$ in the energy norm, that is,

$$a(v^\delta, \varepsilon) = 0 \,, \quad \forall\, v^\delta \in \mathcal{V}^\delta \,. \tag{2.2.21a}$$

To prove this property we recall that the exact solution satisfies the weak equation (2.2.15); in other words,

$$a(v, u) = f(v), \quad \forall\, v \in \mathcal{V},$$

and the approximation satisfies eqn (2.2.18). The finite-dimensional test space $\mathcal{V}^\delta$ is a subspace of $\mathcal{V}$, and so the exact solution also satisfies

$$a(v^\delta, u) = f(v^\delta), \quad \forall\, v^\delta \in \mathcal{V}^\delta. \tag{2.2.21b}$$

Subtracting eqn (2.2.18) from eqn (2.2.21b) with $\varepsilon = u - u^\delta$ and using the bilinearity of $a(v, u)$ gives eqn (2.2.21a).

2.2.4.3 *Minimal property of error in the energy norm*

We can show that the finite element solution u^δ is the solution in $\mathcal{X}^\delta$ which minimises the energy norm of the error, that is,

$$||u - u^\delta||_E = \min_{w^\delta \in \mathcal{X}^\delta} ||u - w^\delta||_E. \tag{2.2.22a}$$

To demonstrate this result we let $\varepsilon = u - u^\delta$ and observe that for any $w^\delta \in \mathcal{X}^\delta$ we can write

$$||u - w^\delta||_E^2 = ||u - u^\delta + u^\delta - w^\delta||_E^2 = ||\varepsilon + v^\delta||_E^2,$$

where $v^\delta = u^\delta - w^\delta \in \mathcal{V}^\delta$. From the definition of the energy norm (2.2.14) and using the bilinearity of $a(v, u)$ (eqn (2.2.19b)), we obtain

$$||u - w^\delta||_E^2 = a(\varepsilon + v^\delta, \varepsilon + v^\delta) = a(\varepsilon, \varepsilon) + 2a(v^\delta, \varepsilon) + a(v^\delta, v^\delta). \tag{2.2.22b}$$

Now, since $v^\delta \in \mathcal{V}^\delta$, we know from eqn (2.2.21a) that $a(v^\delta, \varepsilon) = 0$. Therefore, if there were any choices of w^δ which gave a smaller error than $u - u^\delta$, in the energy norm, it would have to make the last term of eqn (2.2.22b) negative. However, if $v \neq 0$ then $a(v, v) > 0$, and so the minimising choice of w^δ is one that sets $v^\delta = 0$, thus implying that $w^\delta = u^\delta$ and proving eqn (2.2.22a).

2.2.4.4 *Equivalence of polynomial bases in the energy norm*

An almost trivial observation from the uniqueness of the Galerkin approximation is that any two linearly-independent expansions which span the same trial space $\mathcal{X}^\delta$ necessarily have the same approximate solution $u^\delta(x)$. So if we consider two solutions $u_1^\delta(x) = \sum_i^P \alpha_i \psi_i(x)$ and $u_2^\delta(x) = \sum_i^P \beta_i h_i(x)$, where both expansion functions are in a polynomial space of order P (i.e., $\psi_i(x), h_i(x) \in \mathcal{P}_P$), and if the solutions $u_1^\delta(x)$ and $u_2^\delta(x)$ are both determined as solutions to the Galerkin approximation (2.2.18) then we know that

$$u_1^\delta(x) = u_2^\delta(x) \quad \Rightarrow \quad \sum_{i=0}^{P} \alpha_i \psi_i(x) = \sum_{i=0}^{P} \beta_i h_i(x).$$

The important implication of this statement is that any error estimates are independent of the type of the polynomial expansion and only depend on the polynomial space. Nevertheless, different choices of polynomial expansion bases can have an important effect on the numerical conditioning of the algebraic systems resulting from the Galerkin approximation, as discussed in Section 2.3.2.1.

2.2.5 *Residual equation for the C^0 test and trial functions*

As we have seen in the example of Section 2.2.2, the finite element approximation u^δ to a second-order differential equation can be constructed from a class of functions which are C^0 continuous, that is, the approximation (but not the derivative of the approximation) is continuous everywhere in the domain Ω. Therefore, when the solution domain Ω is subdivided into finite elements, denoted by Ω^e, although the derivative is continuous within each element, at the boundary between elements the derivative may be discontinuous.

We note that for a C^0 approximation the substitution of u^δ into the weak equation (2.2.5) is not equivalent to setting the weighted residual (v^δ, R) equal to zero (i.e., $(v^\delta, R) \neq 0$, where $\mathbb{L}(u^\delta) = R$ from eqn (2.2.3)). To appreciate why, we can integrate the left-hand side of eqn (2.2.6) by parts and recover a form similar to eqn (2.2.3).

We observe that the integrand on the left-hand side of eqn (2.2.6) involves the derivatives of u^δ and v^δ, which are only piecewise continuous within each element when using a C^0 expansion. Therefore, to evaluate this integral we have to perform a series of integrals over each element Ω^e. If there are N_{el} elements we find

$$\begin{aligned}\int_0^1 \frac{\partial v^\delta}{\partial x}\frac{\partial u^\delta}{\partial x}\,\mathrm{d}x &= \sum_{e=1}^{N_{\mathrm{el}}}\int_{\Omega^e}\frac{\partial v^\delta}{\partial x}\frac{\partial u^\delta}{\partial x}\,\mathrm{d}x \\ &= -\sum_{e=1}^{N_{\mathrm{el}}}\int_{\Omega^e} v^\delta\frac{\partial^2 u^\delta}{\partial x^2}\,\mathrm{d}x + \sum_{e=1}^{N_{\mathrm{el}}}\left[v^\delta\frac{\partial u^\delta}{\partial x}\right]_{\Omega_L^e}^{\Omega_R^e} \\ &= -\int_0^1 v^\delta\frac{\partial^2 u^\delta}{\partial x^2}\,\mathrm{d}x + \sum_{e=1}^{N_{\mathrm{el}}}\left[v^\delta\frac{\partial u^\delta}{\partial x}\right]_{\Omega_L^e}^{\Omega_R^e}, \end{aligned} \qquad (2.2.23)$$

where Ω_R^e and Ω_L^e denote the x values of the left and right ends of the domain Ω^e, respectively. If the approximating function was C^1 continuous (that is, the function and its first derivative are continuous everywhere) then by definition

$$v^\delta\frac{\partial u^\delta}{\partial x}\bigg|_{\Omega_R^e} = v^\delta\frac{\partial u^\delta}{\partial x}\bigg|_{\Omega_L^{e+1}},$$

and we recover the standard 'integration by parts' result. However, since the finite element is globally only C^0 continuous, all the terms at the interior elemental boundaries remain. Now, by substituting eqn (2.2.23) into eqn (2.2.6) and rearranging, we obtain

$$\begin{aligned}&-\int_0^1 v^\delta\left(\frac{\partial^2 u^\delta}{\partial x^2} + f\right)\mathrm{d}x - v^\delta\frac{\partial u^\delta}{\partial x}\bigg|_{\Omega_L^1} \\ &+ \sum_{e=1}^{N_{\mathrm{el}}-1}\left[v^\delta\frac{\partial u^\delta}{\partial x}\bigg|_{\Omega_R^e} - v^\delta\frac{\partial u^\delta}{\partial x}\bigg|_{\Omega_L^{e+1}}\right] + \left[v^\delta\frac{\partial u^\delta}{\partial x}\bigg|_{\Omega_R^{N_{\mathrm{el}}}} - v^\delta(1)g_{\mathcal{N}}\right] = 0\,.\end{aligned} \qquad (2.2.24)$$

The first term is the standard weighted residual. The second term is zero since Ω_L^1 is a Dirichlet boundary and so $v^\delta(\Omega_L^1) = 0$. The third term represents the *jump* in the derivative of the approximation at the element boundaries in the interior of the domain, and the last term represents the difference between the exact and approximate Neumann boundary conditions. Upon convergence to the exact solution, the jump in the derivative must therefore become zero and the Neumann condition must also be exactly satisfied.

If we use a C^1 expansion which exactly satisfies the Neumann boundary conditions, then eqn (2.2.24) becomes the standard weighted residual, that is, $(v^\delta, R) = 0$.

2.3 One-dimensional expansion bases

Having defined the finite element framework in terms of the Galerkin formulation, we can now consider different types of one-dimensional expansion bases and provide some explanation of their construction.

An essential part of constructing different expansion bases will be to introduce a standard elemental region within which we will define the standard expansions. We will then discuss how to assemble the global expansion bases from these local definitions. This type of elemental construction also provides an efficient way to numerically implement the spectral/*hp* element technique once we have addressed how to numerically integrate and differentiate polynomial functions. This will be dealt with in Section 2.4; elemental expansion bases in multiple dimensions will be dealt with in Chapter 3.

At this stage we will only be concerned with polynomial expansions. Traditionally, the finite element method has always used polynomial expansions. This may be attributed to the historical use of Taylor series expansions which allow analytical functions to be expressed in terms of polynomials. Polynomial functions also have the added advantage of discrete integration rules which enable easy computer implementation.

In the h-type method, a fixed-order polynomial is used in every element and convergence is achieved by reducing the size of the elements. This is the so-called h-type extension, where h represents the characteristic size of an element, and was illustrated for two elements in Section 2.2.2. This type of extension aids in geometric flexibility, especially in high dimensions.

In the p-type method, a fixed mesh is used and convergence is achieved by increasing the order of the polynomial in every element. This is the so-called p-type extension, where p represents the expansion order in the elements. This type of extension aids rapid convergence for smooth problems. If the whole solution domain is treated as a single element then the p-type method becomes a spectral method.

The spectral/*hp* element method combines attributes from both the *h-type* and *p-type* extensions, permitting a combination of both approaches. We also note that, with the exception of a global spectral method, in most *p-type* methods there is an implied *h-type* decomposition to generate the initial mesh upon which the *p-type* extension is applied.

2.3.1 *Elemental decomposition: the h-type extension*

♣[5] One of the primary advantages of the finite element and finite volume methods is the ability to resolve complex geometries. This capability is inherently dependent on being able to decompose the solution domain into small subdomains or elements. In this section, we will demonstrate the technique of decomposing the expansion into elemental contributions, that is, the 'h-type extension' process. We note, however, that elemental decomposition is an integral part of both h- and p-type finite element methods as the p-type extension is based upon an initial mesh or h-type discretisation.

As discussed in the next three sections, the principal use of elemental representation is to enable the treatment of operations on a local elemental basis. This not only simplifies the implementation but also allows many operations to be performed more efficiently. For the one-dimensional case, the decomposition may seem unnecessarily involved; however, the same principles are applied to the decomposition in multiple dimensions. Therefore, the one-dimensional case is explained in detail as a building block for understanding decomposition in multiple dimensions. The decomposition is explained in terms of the linear finite element expansion; however, the same techniques can be applied with the higher-order p-type expansions discussed in Section 2.3.2.1.

2.3.1.1 *Partitioning of the solution domain*

When using an h-type method the solution domain is subdivided or partitioned into *non-overlapping* subdomains or elements within which a polynomial expansion is used.

Considering a solution domain Ω, we can partition it into a mesh containing N_{el} elements, denoted by Ω^e, such that the union of the non-overlapping elements equals the original domain, that is,

$$\Omega = \bigcup_{e=1}^{N_{\mathrm{el}}} \Omega^e , \quad \text{where} \bigcap_{e=1}^{N_{\mathrm{el}}} \Omega^e = \emptyset .$$

For the domain $\Omega = \{x \mid 0 < x < l\}$ a specific mesh can be denoted by the points

$$0 = x_0 < x_1 < \cdots < x_{N_{\mathrm{el}}-1} < x_{N_{\mathrm{el}}} = l .$$

Therefore, the eth element is defined as

$$\Omega^e = \{x \mid x_{e-1} < x < x_e\} .$$

[5] Decomposition of a solution domain into elemental regions: h-type extensions.

As an example we consider the case shown in Fig. 2.2 where the solution domain, $\Omega = \{x \mid 0 < x < l\}$, is subdivided into $N_{\rm el} = 3$ non-equal elements. The mesh is denoted by the $N_{\rm el} + 1$ points $x_0 = 0$, x_1, x_2, $x_3 = l$, and therefore the first element is defined as

$$\Omega^1 = \{x \mid x_0 < x < x_1\}.$$

2.3.1.2 *The standard element and the one-dimensional linear finite element expansion*

In Fig. 2.2 the global expansion modes for the linear finite element expansion over the $N_{\rm el} = 3$ elemental domains are also shown. As is typical in a linear finite element expansion, each mode has a unit value at the end of one of the elemental domains and decays linearly to zero across the neighbouring elements. Therefore, there are $N_{\rm dof} = 4$ degrees of freedom in this expansion which are $\Phi_0(x), \Phi_1(x), \Phi_2(x)$, and $\Phi_3(x)$. The global modes are nonzero on, at most, two elemental regions. It would therefore be very uneconomical to consider an expansion in terms of global modes, particularly when using a large number of elements.

We can see that on an elemental level each global mode only consists of two linearly-varying functions which are also shown on the right of Fig. 2.2. Therefore, if we introduce the one-dimensional standard element, $\Omega_{\rm st}$, such that

$$\Omega_{\rm st} = \{\xi \mid -1 \leqslant \xi \leqslant 1\},$$

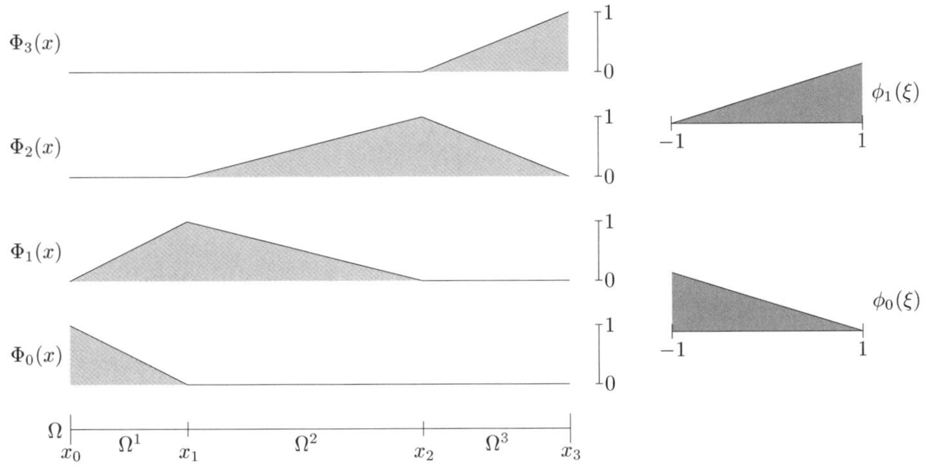

FIG. 2.2. Elemental decomposition of the solution domain Ω into three elements Ω^1, Ω^2, and Ω^3. Above this domain we show the global expansion modes $\Phi_0(x)$, $\Phi_1(x)$, $\Phi_2(x)$, and $\Phi_3(x)$ for a linear finite element expansion over the domain Ω. On the right are the local expansion bases $\phi_0(\xi)$ and $\phi_1(\xi)$ defined in the standard region $\Omega^{\rm st}$ which can be used to define the global expansion modes.

then we can define a similar linearly-varying function over $\Omega_{\rm st}$ in terms of the local coordinate ξ as

$$\phi_0(\xi) = \begin{cases} \dfrac{1-\xi}{2}, & \xi \in \Omega_{\rm st}\,, \\ 0, & \xi \notin \Omega_{\rm st}\,, \end{cases} \qquad \phi_1(\xi) = \begin{cases} \dfrac{1+\xi}{2}, & \xi \in \Omega_{\rm st}\,, \\ 0, & \xi \notin \Omega_{\rm st}\,. \end{cases}$$

The standard element $\Omega_{\rm st}$ can be mapped to any elemental domain Ω^e via the transformation $\chi^e(\xi)$, which expresses the global coordinate x in terms of the local coordinate ξ as

$$x = \chi^e(\xi) = \frac{1-\xi}{2} x_{e-1} + \frac{1+\xi}{2} x_e\,, \quad \xi \in \Omega_{\rm st}\,. \tag{2.3.1}$$

This mapping has an analytic inverse, $(\chi^e)^{-1}(x)$, of the form

$$\xi = (\chi^e)^{-1}(x) = 2\frac{x - x_{e-1}}{x_e - x_{e-1}} - 1\,, \quad x \in \Omega^e\,.$$

The global modes $\Phi_i(x)$ can now be represented in terms of the local elemental expansion modes $\phi_p(\xi)$ by mapping the standard element $\Omega_{\rm st}$ to each elemental domain Ω^e. For example, the first two global expansion modes $\Phi_0(x)$ and $\Phi_1(x)$ in Fig. 2.2 can be written as

$$\Phi_0(x) = \begin{cases} \dfrac{x - x_1}{x_0 - x_1}, & x \in \Omega^1\,, \\ 0, & x \notin \Omega^1\,, \end{cases} = \begin{cases} \phi_0(\xi) = \phi_0([\chi^1]^{-1}(x))\,, & x \in \Omega^1\,, \\ 0, & x \notin \Omega^1\,, \end{cases}$$

$$\Phi_1(x) = \begin{cases} \dfrac{x - x_0}{x_1 - x_0}, & x \in \Omega^1\,, \\ \dfrac{x - x_2}{x_1 - x_2}, & x \in \Omega^2\,, \\ 0, & \text{otherwise}\,, \end{cases} = \begin{cases} \phi_1(\xi) = \phi_1([\chi^1]^{-1}(x))\,, & x \in \Omega^1\,, \\ \phi_0(\xi) = \phi_0([\chi^2]^{-1}(x))\,, & x \in \Omega^2\,, \\ 0, & \text{otherwise}\,. \end{cases}$$

If a mapping for $\chi^e(\xi)$ other than the one given in eqn (2.3.1) has been used then the inverse mapping will not necessarily be analytic. This situation can arise in multiple dimensions where elements may be curved.

2.3.1.3 *Parametric mapping*

The transformation $\chi^e(\xi)$ given in eqn (2.3.1) maps the *local* coordinate ξ to the *global* coordinate x ($x \in \Omega_e$) and can be interpreted as expanding the global coordinate, x, in terms of a linear finite element expansion. It, therefore, could have been written as

$$x = \chi^e(\xi) = \phi_0(\xi)\, x_{e-1} + \phi_1(\xi)\, x_e\,, \quad \xi \in \Omega_{\rm st}\,.$$

This technique of expressing the global coordinate, x, in terms of the local expansion function is known as *parametric mapping*. Typically, we refer to the mapping as being *iso-parametric* if we use the same-order expansion to map the coordinates as we use to represent the dependent variables. If we use a higher- or lower-order mapping for the coordinates as compared to the dependent variable then the mapping is referred to as *super*- or *sub*-parametric, respectively. As we shall see in Section 4.1.3.2, parametric mappings provide a convenient way to express curved domains.

We note that the mapping in eqn (2.3.1) is linear and therefore so is its inverse. This means that the local expansion mode $\phi_p(\chi_e^{-1}(x))$ is a polynomial in x as well as in ξ, and therefore under the mapping (2.3.1) the global expansion modes are also polynomials in x. However, when a higher-order polynomial mapping is used, as is necessary for curved elements, the global expansion may not remain a polynomial in x although, by definition, it is always a polynomial in ξ.

2.3.1.4 *Global assembly/direct stiffness summation*

To relate the concepts of local and global expansion bases we need to introduce the concept of global assembly or direct stiffness summation, as it is sometimes known. In this section we shall describe the process for a one-dimensional linear basis, but the same idea can be used in higher-order expansions and multiple dimensions. Let us recall that the finite element approximation u^δ in terms of the global modes is written as ♦[3]

$$u^\delta(x) = \sum_{i=0}^{N_{\mathrm{dof}}-1} \hat{u}_i \Phi_i(x) .$$

We have seen in Section 2.3.1.2 that the global modes $\Phi_i(x)$ can be expressed in terms of the local expansion modes $\phi_p(\xi)$, and therefore we can express u^δ in terms of $\phi_p(\xi)$ as

$$u^\delta(x) = \sum_{i=0}^{N_{\mathrm{dof}}-1} \hat{u}_i \Phi_i(x) = \sum_{e=1}^{N_{\mathrm{el}}} \sum_{p=0}^{P} \hat{u}_p^e \phi_p^e(\xi) ,$$

where in this case P is the polynomial order of the expansion and $\phi_p^e(\xi) = \phi_p([\chi^e]^{-1}(x))$ (the superscript denotes the element in which the function is nonzero). As there are more of the local expansion coefficients, $\hat{u}_p^e$, than global expansion coefficients, $\hat{u}_i$, some further conditions are required to relate the local and global definitions of the solution $u^\delta(x)$.

For the linear finite element example shown in Fig. 2.3 where $P = 1$ and $N_{\mathrm{el}} = 3$, the constraint is that the global modes are continuous everywhere, which implies

$$\begin{aligned} \hat{u}_1^1 &= \hat{u}_0^2 , \\ \hat{u}_1^2 &= \hat{u}_0^3 . \end{aligned} \tag{2.3.2}$$

The relationship between the local and global expansion coefficients is therefore

3 Global assembly: assembling global bases and operations from local bases and operators.

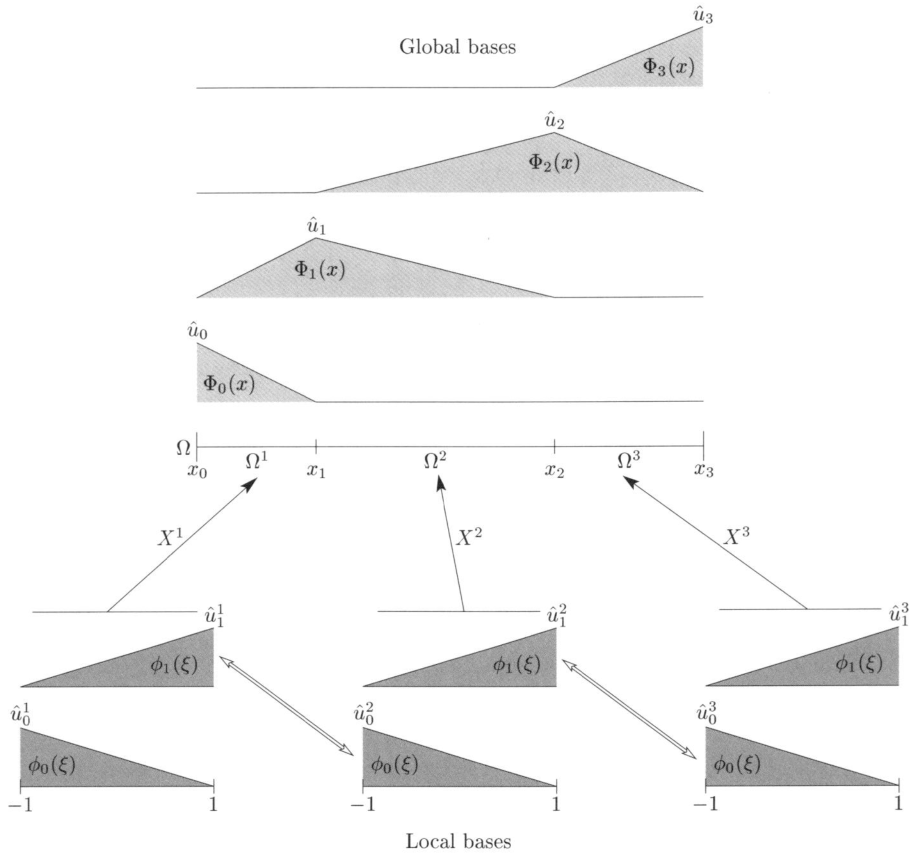

FIG. 2.3. Global and local expansion coefficients and bases in a three-element decomposition of the domain Ω.

$$
\begin{aligned}
\hat{u}_0^1 &= \hat{u}_0 \,, \\
\hat{u}_1^1 &= \hat{u}_0^2 = \hat{u}_1 \,, \\
\hat{u}_1^2 &= \hat{u}_0^3 = \hat{u}_2 \,, \\
\hat{u}_1^3 &= \hat{u}_3 \,.
\end{aligned}
$$

In this example it can be seen that the local representation of the function has six elemental degrees of freedom ($N_{\text{eof}} = N_{\text{el}} \cdot (P+1) = 6$) but only four global degrees of freedom ($N_{\text{dof}} = 4$). The two constraints shown in eqn (2.3.2) ensure that $u^\delta(x)$ is C^0 continuous, which is a sufficient condition to ensure that the expansion is in H^1 space and thereby can be an admissible function for the trial space $\mathcal{X}^\delta$ for a second-order elliptic problem.

To construct a more general description of the local to global mapping we let $\hat{\boldsymbol{u}}_g$ denote a vector of all global coefficients,

$$\hat{\boldsymbol{u}}_g = [\hat{u}_0, \ldots, \hat{u}_{N_{\text{dof}}-1}]^\top ,$$

and if $\hat{\boldsymbol{u}}^e$ is a vector of the local coefficients (that is, $\hat{\boldsymbol{u}}^e = [\hat{u}_0^e, \hat{u}_1^e]$) in element e, then the vector of all local coefficients, denoted by $\hat{\boldsymbol{u}}_l$, can be written as

$$\hat{\boldsymbol{u}}_l = \begin{bmatrix} \hat{\boldsymbol{u}}^1 \\ \hat{\boldsymbol{u}}^2 \\ \vdots \\ \hat{\boldsymbol{u}}^{N_{\text{el}}} \end{bmatrix} .$$

The relationship between the local degrees of freedom and the global degrees can be expressed in terms of an *assembly matrix* $\boldsymbol{\mathcal{A}}$ such that

$$\boxed{\hat{\boldsymbol{u}}_l = \boldsymbol{\mathcal{A}} \hat{\boldsymbol{u}}_g ,} \tag{2.3.3a}$$

where $\boldsymbol{\mathcal{A}}$ is a very sparse matrix whose entries are typically 1 (but may be -1 in multiple dimensions or even contain a submatrix for non-conforming elements). For our example in Fig. 2.3 the full form of eqn (2.3.3a) is

$$\hat{\boldsymbol{u}}_l = \begin{bmatrix} \hat{u}_0^1 \\ \hat{u}_1^1 \\ \hat{u}_0^2 \\ \hat{u}_1^2 \\ \hat{u}_0^3 \\ \hat{u}_1^3 \end{bmatrix} = \begin{bmatrix} 1 & & & \\ & 1 & & \\ & 1 & & \\ & & 1 & \\ & & 1 & \\ & & & 1 \end{bmatrix} \begin{bmatrix} \hat{u}_0 \\ \hat{u}_1 \\ \hat{u}_2 \\ \hat{u}_3 \end{bmatrix} .$$

We can also consider the *reverse* operation of constructing global operations from local operations. This is advantageous since we can perform operations locally within the elements and then assemble the global operation. In the Galerkin formulation this assembly process is typically associated with integral operations, which implies that we have to sum the local (elemental) contributions. For example, if we consider the integral of $u^\delta(x)$ with the global modes $\Phi_1(x)$ shown in Fig. 2.3, then we find

$$\int_\Omega \Phi_1(x) u^\delta(x)\, \mathrm{d}x = \int_{-1}^{1} \phi_1^1(\xi) u^\delta(\chi^1) \frac{\mathrm{d}\chi^1}{\mathrm{d}\xi}\, \mathrm{d}\xi + \int_{-1}^{1} \phi_0^2(\xi) u^\delta(\chi^2) \frac{\mathrm{d}\chi^2}{\mathrm{d}\xi}\, \mathrm{d}\xi .$$

From this we see that all integrals may also be computed in a standard region ($[-1, 1]$) which is convenient when performing numerical quadrature, as discussed in Section 2.4.1. The process of reassembling the global expansion from the local expansion on the elemental domains is called *global assembly* or *direct stiffness summation.*

The global assembly operation which constructs the global operations from the local operations is the *transpose* operation $\boldsymbol{\mathcal{A}}^\top$. To illustrate this operation we can consider the integration of the global base, $\Phi(x)$, with respect to the function $u^\delta(x)$. Following the same convention as the local and global degrees of freedom, we denote the integration with respect to the global basis as $\boldsymbol{I}_g$, i.e.,

$$\boldsymbol{I}_g[i] = \int_\Omega \Phi_i(x) u^\delta(x)\, \mathrm{d}x\,,$$

and the integration with respect to the local bases $\phi(x)$ in element e as $\boldsymbol{I}^e$, so that we can define a vector of local integral contributions, $\boldsymbol{I}_l$, such that

$$\boldsymbol{I}_l = \begin{bmatrix} \boldsymbol{I}^1 \\ \boldsymbol{I}^2 \\ \vdots \\ \boldsymbol{I}^{N_{\mathrm{el}}} \end{bmatrix}, \quad \text{where } \boldsymbol{I}^e = \begin{bmatrix} \displaystyle\int_{-1}^{1} \phi_0(\xi) u^\delta(\chi^e) \frac{\mathrm{d}\chi^e}{\mathrm{d}\xi}\, \mathrm{d}\xi \\ \vdots \\ \displaystyle\int_{-1}^{1} \phi_{P-1}(\xi) u^\delta(\chi^e) \frac{\mathrm{d}\chi^e}{\mathrm{d}\xi}\, \mathrm{d}\xi \end{bmatrix}.$$

The vector $\boldsymbol{I}_g$ can then be related to the local elemental vector $\boldsymbol{I}_l$ using the assembly matrix $\boldsymbol{\mathcal{A}}^\top$ by the operation

$$\boxed{\boldsymbol{I}_g = \boldsymbol{\mathcal{A}}^\top \boldsymbol{I}_l\,.} \qquad (2.3.3\mathrm{b})$$

This operation essentially performs a summation of the local modes into the global expansion.

We also note that

$$\hat{\boldsymbol{u}}_g \neq \boldsymbol{\mathcal{A}}^\top \boldsymbol{\mathcal{A}} \hat{\boldsymbol{u}}_g$$

as the operation $\boldsymbol{\mathcal{A}}$ *scatters* the global degrees of freedom to the local elements. However, $\boldsymbol{\mathcal{A}}^\top$ *assembles* the global contribution by summing together various terms of the local degrees of freedom.

Here $\boldsymbol{\mathcal{A}}$ and $\boldsymbol{\mathcal{A}}^\top$ represent key operations required in the construction of a Galerkin spectral/*hp* element method since they permit us to define a series of local operators which can then be assembled using these operators. It can be appreciated that only global modes which are split into elemental contributions will have multiple entries in the columns of the $\boldsymbol{\mathcal{A}}$ matrix. When using a higher-order p-type expansion, as discussed in Section 2.3.2.1, the extra interior modes are all global degrees of freedom and will not need to be assembled in this fashion.

♦4 In practice, we never construct the assembly matrix $\boldsymbol{\mathcal{A}}$ as it is very sparse and therefore numerically very inefficient to use as a matrix operator. An equivalent numerical operation is to use a *mapping array* for each element which contains the global location of every local degree of freedom. If we denote this array by

4 Construction of a mapping array for global to local scatter and local to global assembly.

'map[e][i]', where e denotes the element and i is the local mode index, then for the example in Fig. 2.3 the array would be defined as

$$\text{map}[1][i] = \begin{Bmatrix} 0 \\ 1 \end{Bmatrix}, \quad \text{map}[2][i] = \begin{Bmatrix} 1 \\ 2 \end{Bmatrix}, \quad \text{map}[3][i] = \begin{Bmatrix} 2 \\ 3 \end{Bmatrix}.$$

The scatter operation denoted by $\boldsymbol{\mathcal{A}}$ (see eqn (2.3.3a)) can then be evaluated as follows:

$$\left.\begin{array}{l} \text{do } e = 1, N_{\text{el}} \\ \quad \text{do } i = 0, N_m^e - 1 \\ \qquad \hat{\boldsymbol{u}}^e[i] = \hat{\boldsymbol{u}}_g[\text{map}[e][i]] \\ \quad \text{continue} \\ \text{continue} \end{array}\right\} \quad \Leftrightarrow \quad \hat{\boldsymbol{u}}_l = \boldsymbol{\mathcal{A}}\hat{\boldsymbol{u}}_g\,,$$

where $N_m^e = P^e + 1$. Alternatively, the global assembly operation may be written as follows:

$$\left.\begin{array}{l} \text{do } e = 1, N_{\text{el}} \\ \quad \text{do } i = 0, N_m^e - 1 \\ \qquad \hat{\boldsymbol{u}}_g[\text{map}[e][i]] = \hat{\boldsymbol{u}}_g[\text{map}[e][i]] + \hat{\boldsymbol{u}}^e[i] \\ \quad \text{continue} \\ \text{continue} \end{array}\right\} \quad \Leftrightarrow \quad \hat{\boldsymbol{u}}_g = \boldsymbol{\mathcal{A}}^\top \hat{\boldsymbol{u}}_l\,.$$

2.3.2 *Polynomial expansions: the p-type extension*

In multiple dimensions, complex domains make it difficult to identify global expansions analytically. The introduction of complex geometries can also generate different *scales* in the solution, which may have a very localised structure. Such considerations require the use of elemental decomposition, as discussed in Section 2.3.1. Therefore, if we decompose the solution domain into elemental regions which broadly capture either the geometry or the local scale of the problem then the application of the p-type extension can prove to be a numerically efficient approach to achieving a very accurate solution. In all that follows we will interpret the p-type extension as increasing the order of the polynomial expansion within an elemental region.

Before discussing the different types of p-type extension, we first define the *hp* element space in one dimension. Recalling the definition of the standard element, Ω_{st}, and the coordinate mapping $\chi^e(\xi)$ from Ω_{st} to an elemental region Ω^e, we start by denoting the space of all polynomials of degree P defined on the standard element Ω_{st} by $\mathcal{P}_P(\Omega_{\text{st}})$. The discrete *hp* expansion space $\mathcal{X}^\delta$ is the set of all functions $u^\delta(x)$ which exist in H^1 and that are polynomials in ξ within every element (e.g., $u^\delta(\chi^e(\xi)) \in \mathcal{P}_P(\Omega_{\text{st}})$), which is formally written as

$$\mathcal{X}^\delta = \{u^\delta \mid u^\delta \in H^1,\ u^\delta(\chi^e(\xi)) \in \mathcal{P}_{P^e}(\Omega_{\text{st}}),\ e = 1, \ldots, N_{\text{el}}\}\,. \tag{2.3.4}$$

This definition allows both the mapping $\chi^e(\xi)$ and the polynomial order P^e to vary within each element e thereby permitting both h-type refinement, which alters $\chi^e(\xi)$ and $N_{\rm el}$, and p-type refinement, which alters P^e.

In principle, all of the construction discussed in Section 2.3.1 applies equally well to an *hp* elemental decomposition. As we shall see in Section 2.3.2.2, the most standard polynomial decompositions have what is known as a boundary and interior decomposition, which permits us to directly use the construction adopted for linear elements in Section 2.3.1 for higher-order polynomial expansions. Examples of polynomial expansions with this type of decomposition will be discussed in Sections 2.3.3.3 and 2.3.4. However, before introducing these expansions in Section 2.3.2.1 we will first try to explain why certain forms of polynomial expansions are more favourable than others.

2.3.2.1 *Construction of a polynomial expansion*

In an *hp* elemental discretisation we can apply a polynomial expansion of any order within each elemental region. It is therefore appropriate to start our discussion of p-type methods by considering what makes an acceptable p-type expansion in a single domain.

The steps involved in designing an elemental p-type expansion, which we will also later adopt in constructing the unstructured basis in Section 3.2, are as follows.

- Determine a favourable expansion within a standard region.
- Modify the expansion so that it can easily be numerically implemented.

In the first step, a favourable expansion is typically an orthogonal or near-orthogonal set of functions within the standard regions. In the second step, the computational considerations of implementing this basis are taken into account and the basis is modified, if necessary, to facilitate this process. Typically, the basis is decomposed into contributions on the boundary and interior of the standard region since this simplifies the elemental decomposition process.

Modal and nodal expansions

Before discussing the benefits of different types of polynomial expansions, we first need to introduce the concepts of *modal* and *nodal* expansions. To illustrate the difference between a modal and a nodal polynomial expansion we introduce three expansion sets denoted by $\Phi_p^A(x)$, $\Phi_p^B(x)$, and $\Phi_p^C(x)$ $(0 \leqslant p \leqslant P)$, in the region $\Omega_{\rm st} = \{x \mid -1 \leqslant x \leqslant 1\}$. All of these expansions represent a complete set of polynomials up to order P and are mathematically defined as

$$\Phi_p^A(x) = x^p\,, \qquad p = 0, \dots, P\,,$$

$$\Phi_p^B(x) = \frac{\prod_{q=0,q\neq p}^{P}(x - x_q)}{\prod_{q=0,q\neq p}^{P}(x_p - x_q)}\,, \quad p = 0, \dots, P\,,$$

$$\Phi_p^C(x) = L_p(x)\,, \qquad p = 0, \dots, P\,.$$

The shape of these expansions can be seen in Fig. 2.4(a–c). The first expansion set simply increases the order of x in a monomial fashion and we shall refer to it as the *moment* expansion (each order contributing an extra moment to the expansion). This basis is referred to as a *modal* or a *hierarchical* expansion because the expansion set of order $P-1$ is contained within the expansion set of order P. There is a notion of hierarchy in the sense that higher-order expansion sets are built from the lower-order expansion sets. If we denote the trial space containing all the polynomials in $\Phi_p^A(x)$ up to order P by $\mathcal{X}_P^\delta$ then a hierarchical expansion is one where $\mathcal{X}_{P-1}^\delta \subset \mathcal{X}_P^\delta$; so if

$$\mathcal{X}_2^\delta = \{1, x, x^2\}$$

then

$$\mathcal{X}_3^\delta = \{1, x, x^2, x^3\} = \mathcal{X}_2^\delta \bigcup \{x^3\}.$$

The second polynomial $\Phi_p^B(x)$ is a Lagrange polynomial which is based on a series of $P+1$ nodal points x_q which are chosen beforehand and could be, for example, equispaced in the interval (for further details see Section 2.3.4). The

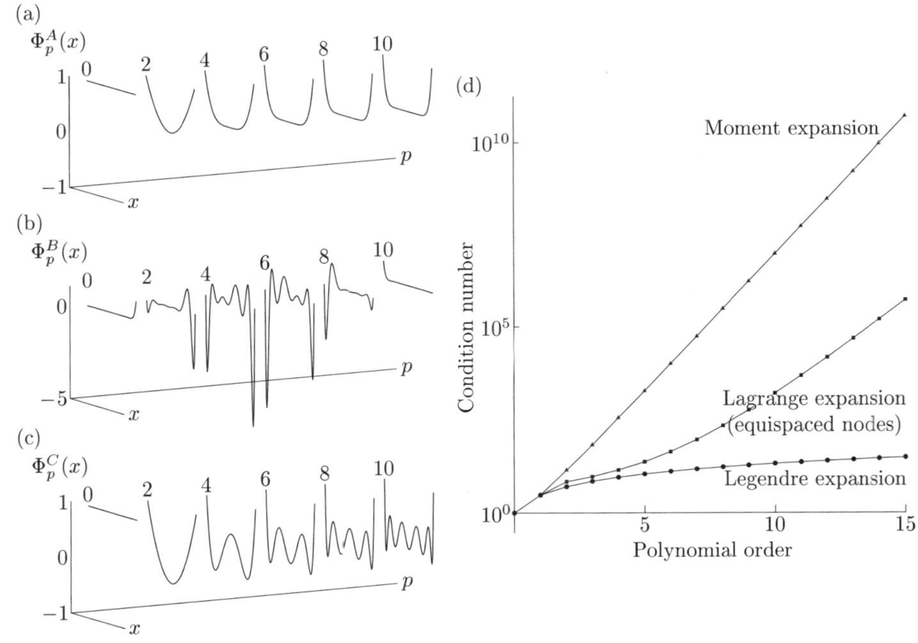

FIG. 2.4. Expansion modes (p even) in space for three expansion bases: (a) $\Phi_p^A(x)$(moment), (b) $\Phi_p^B(x)$ (Lagrange), and (c) $\Phi_p^C(x)$ (Legendre) of order $P = 10$ in the region $-1 \leqslant x \leqslant 1$. (d) Linear–log plot of the condition number of the mass matrix versus polynomial order for the bases $\Phi_p^A(x), \Phi_p^B(x)$, and $\Phi_p^C(x)$.

Lagrange polynomial is a non-hierarchical basis (that is, $\mathcal{X}_P^\delta \not\subset \mathcal{X}_{P+1}^\delta$) because it consists of $P+1$ polynomials of order P. This can be contrasted with the hierarchical expansion $\Phi_p^A(x)$ which consists of polynomials of increasing order. The Lagrange basis has the notable property that $\Phi_p^B(x_q) = \delta_{pq}$, where δ_{pq} represents the Kronecker delta. This property implies that

$$u^\delta(x_q) = \sum_{p=0}^{P} \hat{u}_p \Phi_p^B(x_q) = \sum_{p=o}^{P} \hat{u}_p \delta_{pq} = \hat{u}_q \,,$$

where we see that the expansion coefficient $\hat{u}_p$ can be defined in terms of the approximate solution at the point x_q.

The coefficients, therefore, have a physical interpretation in that they represent the approximate solution at the points x_q. The points x_q are referred to as *nodes* and the Lagrange expansion basis is referred to as a *nodal* expansion. Linear finite elements are an example of a nodal expansion where the nodal points are at the ends of the domain.

We draw a distinction between a *nodal* expansion and the *collocation* method (or collocation projection). In the collocation method, the equation being solved is *exactly* satisfied at the collocation points (see Section 2.1), whereas in a *nodal* expansion the expansion coefficients represent the *approximate* solution at a given set of nodes. However, a nodal expansion can be used in different types of methods such as the Galerkin or collocation method. It must be remembered that an approximate solution using a nodal expansion does not necessarily satisfy the equation exactly at the nodal points.

The final expansion, $\Phi_p^C(x)$, is also a hierarchical or modal expansion. However, in this case the expansion is the Legendre polynomial $L_p(x)$ (see Appendix A). By definition, this polynomial is orthogonal in the Legendre inner product

$$(L_p(x), L_q(x)) = \int_{-1}^{1} L_p(x) L_q(x)\, dx = \frac{2}{2p+1}\delta_{pq} \,.$$

As we shall see, orthogonality has important numerical implications for the Galerkin method.

As a final point, we should comment on a potential confusion over the use of the word *modes*. In general, we shall refer to all expansion sets, whether they are modal or nodal, as consisting of expansion modes (usually called *shape functions* in structural mechanics).

Choice of an expansion set

The choice of an expansion set is influenced by its numerical efficiency, conditioning, and the linear independence of the basis, as well as its approximation properties. To illustrate some of these factors we consider the three expansions $\Phi_p^A(x)$, $\Phi_p^B(x)$, and $\Phi_p^C(x)$ in a Galerkin projection.

The Galerkin or L^2 projection of a smooth function $f(x)$ in the domain $\Omega_{\rm st}$ onto the polynomial expansion $u^\delta(x)$ is the solution to the following problem:

find $u^\delta \in \mathcal{X}^\delta$ such that

$$(v^\delta, u^\delta) = (v^\delta, f), \quad \forall\, v^\delta \in \mathcal{V}, \tag{2.3.5}$$

where (u, v) is the Legendre inner product, see eqn (2.1.4).

In the absence of explicit boundary conditions, which need not be prescribed to obtain a solution for this problem, the trial and test space are both in the space of square integrable functions (that is, $\mathcal{X}^\delta = \mathcal{V}^\delta \subset L^2$). Letting $u^\delta(x) = v^\delta(x) = \sum_{p=0}^{P} \hat{u}_p \Phi_p(x)$, problem (2.3.5) is then equivalent to solving the matrix equation

$$\hat{\boldsymbol{v}}^\top [\boldsymbol{M}\hat{\boldsymbol{u}} = \boldsymbol{f}] \quad \Rightarrow \quad \boldsymbol{M}\hat{\boldsymbol{u}} = \boldsymbol{f},$$

where

$$\boldsymbol{M}_{pq} = (\Phi_p, \Phi_q), \quad \hat{\boldsymbol{u}} = [\hat{u}_0, \cdots, \hat{u}_P]^\top, \quad \boldsymbol{f}_p = (\Phi_p, f).$$

The matrix $\boldsymbol{M}$ is known as the *mass matrix*. It is a square non-singular matrix of order $P + 1$ and can be inverted to determine the solution

$$\hat{\boldsymbol{u}} = \boldsymbol{M}^{-1}\boldsymbol{f}.$$

The question of numerical efficiency for this problem is twofold. The first issue is the computational cost of constructing the matrix system, which may involve numerical integration (see Section 2.4.1). The second issue is the computational cost of inverting the matrix system to obtain the solution. It can be appreciated that the construction and inversion of the matrix $\boldsymbol{M}$ can be made far more efficient if there is some known structure to the matrix.

The moment expansion $\Phi_p^A(x)$ produces a mass matrix which has components $(0 \leqslant p, q \leqslant P)$ of the form

$$\boldsymbol{M}[p][q] = (\Phi_p^A, \Phi_q^A) = \int_{-1}^{1} x^p x^q \,\mathrm{d}x = \left[\frac{x^{p+q+1}}{p+q+1}\right]_{-1}^{1}$$
$$= \begin{cases} \dfrac{2}{p+q+1}, & p+q \text{ even}, \\ 0, & p+q \text{ odd}. \end{cases}$$

Therefore, when constructing $\boldsymbol{M}$ using this basis we need only calculate half of the components. However, the inverse will still be full and the cost of inverting the matrix is typically the dominant operation.

The second expansion $\Phi_p^B(x)$ is the Lagrange polynomial and so it is associated with a set of nodal points x_q. For the purpose of this example we shall define the nodes as being equispaced in the domain Ω_{st}, and so in the interval $x_q = 2q/P - 1$ $(0 \leqslant q \leqslant P)$ which is a common finite element nodal expansion. As we shall see in Section 2.3.4.2, a much better choice of points is at the Gaussian quadrature zeros. There is no explicit form for the mass matrix when using numerical integration and the matrix is full. Therefore, the construction of the mass matrix using $\Phi_p^B(x)$ is twice as expensive as $\Phi_p^A(x)$, although the matrix inversion is no more expensive.

The third expansion $\Phi_p^C(x)$ is the Legendre polynomial. By definition, this expansion produces a mass matrix which is diagonal because the components of the mass matrix are

$$\boldsymbol{M}[p][q] = (\Phi_p^C, \Phi_q^C) = \int_{-1}^{1} L_p(x) L_q(x)\, \mathrm{d}x = \frac{2}{2p+1}\delta_{pq}\,.$$

This matrix is very easy to construct and invert, and therefore might be considered to be numerically the most efficient of the three expansions. We note, however, that the basis cannot easily be extended to an elemental decomposition which is globally C^0 continuous since the continuity constraints destroy the orthogonality of the global matrix structure.

A further consideration is the conditioning of the matrix $\boldsymbol{M}$ which is related to the linear independence of the expansion. The condition number κ_2 is very important in the numerical inversion of matrix systems; a full discussion can be found in Isaacson and Keller [251]. The condition number κ_2 is defined as

$$\kappa_2 = ||\boldsymbol{M}||_2 \cdot ||\boldsymbol{M}^{-1}||_2\,,$$

where $||\boldsymbol{M}||_2$ denotes the matrix L^2 norm of $\boldsymbol{M}$.

When numerically inverting a matrix system there is an error associated with the inexact representation of the matrix due to round-off error. If a matrix system is ill-conditioned the round-off error in the matrix system can lead to large errors in the solution. Further, when using iterative techniques to invert the system the number of iterations required to perform the inversion typically depends on the conditioning of the matrix.

The condition number in the L^2 norm for the three types of expansion bases $\Phi_p^A(x), \Phi_p^B(x)$, and $\Phi_p^C(x)$ is shown in Fig. 2.4(d) as a function of polynomial order. We see that the condition number of the mass matrix for the moment expansion grows as $\kappa_2 \propto 10^P$. Initially, the conditioning of the equispaced Lagrange basis is relatively good; however, after about $P \approx 5$ the condition number also starts to grow as $\kappa_2 \propto 10^P$. In contrast, the Legendre basis is very well conditioned for all values of P. This is because the L^2 matrix norm for a real symmetric matrix is the ratio of the maximum to minimum eigenvalues, and so the condition number for the Legendre mass matrix is exactly $\kappa_2 = 2P + 1$.

The poor conditioning of the moment and Lagrange expansion reflects the fact that the basis is becoming numerically linearly dependent. This is particularly evident for $\Phi_p^A(x)$, as shown in Fig. 2.4(a), where we have plotted the even moment expansion modes as a function of x, for different polynomial orders p. We observe that the mode for $p = 8$ is practically indistinguishable from the mode when $p = 10$. Although each mode of $\Phi_p^B(x)$ is clearly distinguishable from the other in Fig. 2.4(b), the poor conditioning of this basis can be attributed to the high level of oscillations towards the end of the region, which can be seen in modes $p = 4$ and $p = 6$. As we shall see in Section 2.3.4.2, these oscillations are controlled by a better choice of nodal points, which makes it possible to obtain independently-shaped modes with well-behaved bounds, as shown by the modes of $\Phi_p^C(x)$ in Fig. 2.4(c).

Another set of orthogonal polynomials which have been extensively used in spectral methods are the Chebyshev polynomials (see Gottlieb and Orszag [199]). These polynomials have constant amplitude oscillations throughout the region, which leads to their optimal convergence property in the maximum norm.

2.3.2.2 *Boundary interior decomposition of polynomial bases*

From the discussion in Section 2.3.2.1, we might deduce that the 'best' choice for an expansion set is orthogonal polynomials such as the Legendre polynomials. This is true in so far as the hierarchy and orthogonality tend to lead to well-conditioned matrices. However, we also want to combine the expansion with the h-type elemental decomposition. The difficulty arises when we try to ensure a degree of continuity in the global expansion at elemental boundaries. For a second-order partial differential equation we have seen that it is sufficient to guarantee that the approximate solution u^δ is in H^1. Typically, in the finite element methods this is satisfied by imposing a C^0 continuity between elemental regions, that is, the global expansion modes are continuous everywhere in the solution domain although the derivatives may not be.

If we used the moment or the Legendre expansion basis in each elemental domain ($\phi_p(x) = \Phi_p^A(x)$ or $\phi_p(x) = \Phi_p^C(x)$) then the requirement that the approximation be in H^1 might be satisfied by prescribing an interface-matching condition of the form

$$\sum_{p=0}^{P} \hat{u}_p^e \phi_p^e(1) = \sum_{p=0}^{P} \hat{u}_p^{e+1} \phi_p^{e+1}(-1),$$

where the superscripts e and $e+1$ denote two adjacent domains.

Such a condition couples all of the degrees of freedom in one element with the modes in the adjacent element. Not only is this more difficult to implement than the standard finite element methods but it also destroys the orthogonality of the global matrix structure.

If the local expansions were constructed so that only a few expansion modes have magnitude at an elemental boundary then the matching condition can be imposed far more easily. For example, if we define an elemental expansion $\phi_p(\xi)$ in the region $-1 \leqslant \xi \leqslant 1$, where

$$\phi_p(-1) = \begin{cases} 1, & p = 0, \\ 0, & p \neq 0, \end{cases} \qquad \phi_p(1) = \begin{cases} 1, & p = P, \\ 0, & p \neq P, \end{cases}$$

then the C^0 continuity of the expansion is simply enforced by ensuring

$$\hat{u}_P^e \phi_P^e(1) = \hat{u}_0^{e+1} \phi_0^{e+1}(-1).$$

This type of decomposition is known as *boundary* and *interior* decomposition. Boundary modes have magnitude at one of the elemental boundaries and are zero at all other boundaries. Interior modes, sometimes known as *bubble* modes, only have magnitude in the interior of the element and are zero along all boundaries.

The equispaced Lagrange expansion $\Phi_p^B(x)$ already satisfies these conditions. If the nodal points did not include the end-points then the boundary/interior decomposition would not be possible. Therefore, for a nodal basis to be decomposed into boundary and interior modes it is sufficient to ensure that nodal points lie along the boundary of the element (that is, at the end-points in one dimension).

Note that the linear finite element introduced in Section 2.3.1 is an example of an expansion made up only of boundary modes. The p-type modal expansions are typically constructed with boundary and interior modes by adding second- and higher-order polynomials onto the linear finite element expansion. Since only polynomials of second order and higher are added, it is possible to ensure that they are zero at the elemental boundaries, thereby meeting the requirements for interior modes.

2.3.3 *Modal polynomial expansions*

In Section 2.3.2.1 we observed that it is advantageous to consider orthogonal polynomials when constructing p-type expansions. The most commonly used modal p-type elemental expansions are based upon the orthogonal set of polynomials called the Jacobi polynomials. Although the expansion is commonly expressed in terms of the integral of Legendre polynomials [343, 445], we note that these are, in fact, special cases of Jacobi polynomials. Accordingly, before defining the expansion bases in Section 2.3.3.3 we first introduce the orthogonal Jacobi polynomials.

2.3.3.1 *Jacobi polynomials*

Jacobi polynomials, denoted by $P_p^{\alpha,\beta}(x)$, represent the family of polynomial solutions to a singular Sturm–Liouville problem which, in the region $-1 < x < 1$, is written as

$$\frac{\mathrm{d}}{\mathrm{d}x}\left[(1-x)^{1+\alpha}(1+x)^{1+\beta}\frac{\mathrm{d}}{\mathrm{d}x}u_p(x)\right] = \lambda_p(1-x)^{\alpha}(1+x)^{\beta}u_p(x)\,, \tag{2.3.6}$$

where

$$u_p(x) = P_p^{\alpha,\beta}(x)$$

and

$$\lambda_p = -p(\alpha+\beta+p+1)\,.$$

An important property of these polynomials is their orthogonal relationship:

$$\int_{-1}^{1}(1-x)^{\alpha}(1+x)^{\beta}P_p^{\alpha,\beta}(x)P_q^{\alpha,\beta}(x)\,\mathrm{d}x = C\delta_{pq}\,, \tag{2.3.7}$$

where the value of C depends on α, β, and p and has the value

$$C = \frac{2^{\alpha+\beta+1}}{2p+\alpha+\beta+1} \frac{\Gamma(p+\alpha+1)\Gamma(p+\beta+1)}{p!\Gamma(p+\alpha+\beta+1)}.$$

This relation implies that $P_p^{\alpha,\beta}(x)$ is orthogonal to all polynomials of order less than p when integrated with respect to $(1-x)^\alpha(1+x)^\beta$. These polynomials can be constructed using a three-term recursion relationship, as shown in Appendix A.

A class of symmetric polynomials, known as the ultraspheric polynomials, correspond to the choice $\alpha = \beta$. Well-known ultraspheric polynomials are the Legendre polynomial ($\alpha = \beta = 0$) and the Chebyshev polynomial ($\alpha = \beta = -\frac{1}{2}$).

2.3.3.2 *Electrostatic interpretation of the zeros of Jacobi polynomials*

There is a connection between the minimum energy of electrostatic potentials in one dimension and the zeros of the Jacobi polynomial $P^{(\alpha,\beta)}$. This was first revealed by Stieltjes in 1885 [442] and also reported in the classical book on orthogonal polynomials by Szego [447].

In particular, Stieltjes considered the following problem.

In the interval $\xi \in [-1, 1]$ let two positive charges of strength $\tilde{\alpha} = (\alpha+1)/2$ and $\tilde{\beta} = (\beta+1)/2$ be located at $\xi = +1$ and $\xi = -1$, respectively. Assuming that we insert n charges between $\tilde{\alpha}$ and $\tilde{\beta}$ positioned at the locations $\xi_1, \xi_2, \ldots, \xi_n$, what is the distribution of these charges, in space, which minimises the energy $W_J(\xi_1, \ldots, \xi_n)$ defined as

$$W_J = -\sum_{i=1}^{n} \left[\frac{\alpha+1}{2} \log|\xi_i + 1| + \frac{\beta+1}{2} \log|\xi_i - 1| + \frac{1}{2} \sum_{j=1, j\neq i}^{n} \log|\xi_i - \xi_j| \right] ?$$

The solution to this minimisation problem determines the location of the charges corresponding to the condition of electrostatic equilibrium, and this position is uniquely determined. These minimising points are also the zeros of the Jacobi polynomial $P_n^{\alpha,\beta}$. As we shall see in Section 2.4.1.1, these are also the points used for Gauss quadrature.

The extension of the analogy between electrostatics and optimal position of interpolation points has been exploited by Hesthaven who applied these ideas to a triangle in [234] and to a tetrahedron in [238], as discussed in Chapter 3. A similar analogy can also be drawn for the Laguerre and Hermite polynomials; see Szego [447] for further details.

2.3.3.3 *Modal p-type basis*

In the standard interval $\Omega_{\rm st} = \{\xi \mid -1 \leqslant \xi \leqslant 1\}$ we denote by $\psi_p(\xi)$ the p-type modal expansion defined as

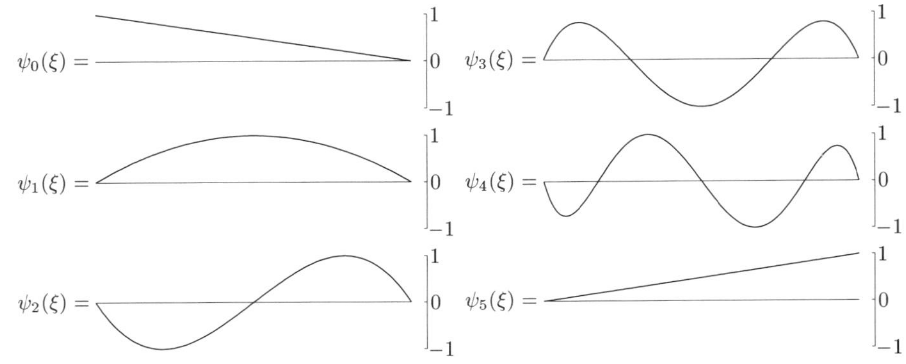

FIG. 2.5. Shape of modal expansion modes for a polynomial order of $P = 5$.

$$\phi_p(\xi) \mapsto \psi_p(\xi) = \begin{cases} \dfrac{1-\xi}{2}, & p = 0, \\ \left(\dfrac{1-\xi}{2}\right)\left(\dfrac{1+\xi}{2}\right) P_{p-1}^{1,1}(\xi), & 0 < p < P, \\ \dfrac{1+\xi}{2}, & p = P. \end{cases} \tag{2.3.8}$$

♣[6] Note that $\phi_p(\xi)$ is used to denote a general definition of a local polynomial basis, whereas $\psi(\xi)$ has the specific definition given above. The shape of all modes for $P = 5$, normalised to have a maximum value of one, are shown in Fig. 2.5. The lowest expansion modes, $\psi_0(x)$ and $\psi_P(x)$, are the same as the linear finite element expansion. These are boundary modes since they are the only modes which have magnitude at the ends of the interval. The remaining interior modes, by definition, are zero at the ends of the interval and increase in polynomial order, as is typical in a hierarchical expansion. Clearly, the only choice for the quadratic mode, $\psi_1(x)$, is the shape $[(1-\xi)/2]\,[(1+\xi)/2]$, and this is the usual hierarchical expansion for quadratic elements.

The shape of interior modes could be defined as any polynomial which satisfies the end conditions; however, using the Jacobi polynomial $P_{p-1}^{1,1}(\xi)$ maintains a high degree of orthogonality and generates a local mass matrix whose interior coupling produces a penta-diagonal system, as shown in Fig. 2.6(a). To appreciate why this is the case, we recall that the elemental mass matrix $\boldsymbol{M}^e$ has components given by $\boldsymbol{M}^e[p][q] = (\psi_p, \psi_q)$, where we have used the square brackets to denote the entry of the pth row and the qth column. For the basis defined by relation (2.3.8) we obtain

$$\boldsymbol{M}^e[p][q] = \int_{-1}^{1} \left(\frac{1-\xi}{2}\right)\left(\frac{1+\xi}{2}\right) P_{p-1}^{1,1}(\xi)\psi_q(\xi)\,\mathrm{d}\xi\,.$$

[6] Most commonly used modal polynomial expansion basis.

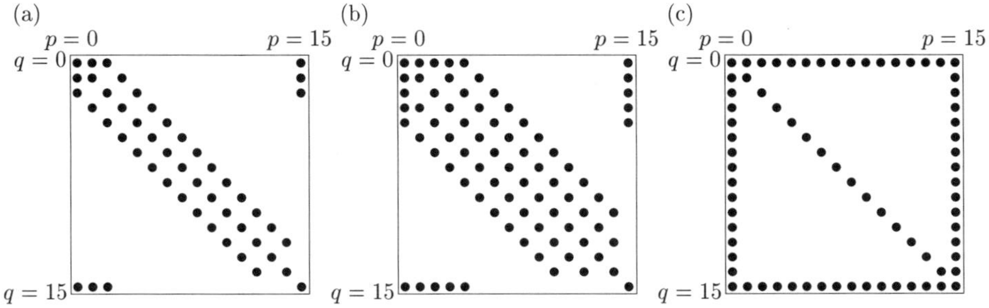

FIG. 2.6. Elemental mass matrix, $\boldsymbol{M}^e$, of the one-dimensional p-type expansion of order $P = 15$ in $\Omega_{\rm st} = \{-1 \leqslant x \leqslant 1\}$ using interior components of the form (a) $[(1-\xi)/2]\,[(1+\xi)/2]\,P^{1,1}_{p-1}(\xi)$, (b) $[(1-\xi)/2]\,[(1+\xi)/2]\,L_{p-1}(\xi)$, and (c) $[(1-\xi)/2]\,[(1+\xi)/2]\,P^{2,2}_{p-1}(\xi)$.

The function $[(1-\xi)/2]\,[(1+\xi)/2]$ is the weight function in the orthogonality relation given by eqn (2.3.7) when using the polynomial $P^{1,1}_p(\xi)$. Since $\psi_q(\xi)$ has interior components (i.e., $0 < q < P$) which are polynomials of order $q+1$, this integral will be zero when $p-1 > q+1$, which implies $p > q+2$. As the matrix is symmetric there are two upper and lower diagonals which are nonzero, making the matrix penta-diagonal.

As shown in Fig. 2.6(b), if we had chosen an interior expansion based on Legendre polynomials $L_p(\xi) = P^{0,0}_p(\xi)$ such as $[(1-\xi)/2]\,[(1+\xi)/2]\,L_{p-1}(\xi)$ then the interior components of the mass matrix are zero if $p > q+4$. This is a consequence of the fact that the function $[(1-\xi)/2]\,[(1+\xi)/2]$ is no longer the weight function in the orthogonality relation. We see from Fig. 2.6(c) that interior modes based on the form $[(1-\xi)/2]\,[(1+\xi)/2]\,P^{2,2}_{p-1}(\xi)$ produce a mass matrix with orthogonal interior components since the function $\psi_q(\xi)$ also contains a factor of the form $[(1-\xi)/2]\,[(1+\xi)/2]$ and the $P^{2,2}_p(\xi)$ polynomial is orthogonal in an inner product containing the square of this factor. Nevertheless, with these interior modes the coupling between the interior and the linear boundary modes is far stronger. Consideration of the Laplacian matrix illustrates why we do not use interior functions of this form.

The elemental Laplacian matrix is defined as a matrix with components $\boldsymbol{L}^e[p][q] = (\mathrm{d}\psi_p(\xi)/\mathrm{d}\xi, \mathrm{d}\psi_q(\xi)/\mathrm{d}\xi)$ and arises when solving a second-order elliptic problem. The elemental Laplacian matrix with interior modes of the form $\psi_p(\xi) = [(1-\xi)/2]\,[(1+\xi)/2]\,P^{1,1}_{p-1}(\xi)$ is therefore

$$\begin{aligned}\boldsymbol{L}^e[p][q] &= \int_{-1}^{1} \frac{\mathrm{d}}{\mathrm{d}\xi}\left[\left(\frac{1-\xi}{2}\right)\left(\frac{1+\xi}{2}\right)P^{1,1}_{p-1}(\xi)\right]\frac{\mathrm{d}}{\mathrm{d}\xi}\psi_q(\xi)\,\mathrm{d}\xi \\ &= -\int_{-1}^{1}\left(\frac{1-\xi}{2}\right)\left(\frac{1+\xi}{2}\right)P^{1,1}_{p-1}(\xi)\frac{\mathrm{d}^2}{\mathrm{d}\xi^2}\psi_q(\xi)\,\mathrm{d}\xi\,,\end{aligned}$$

where we are only considering the entries $0 < p, q < P$. The second equation is obtained by integrating by parts and applying the fact that the integrand is zero at the end-points. As illustrated in Fig. 2.7(a), if we use the orthogonality relationship (2.3.7) and recognise that $(\mathrm{d}^2/\mathrm{d}\xi^2)\psi_q(\xi)$ is a polynomial of order $q-1$ then we see that the components of the elemental Laplacian matrix are zero if $p-1 > q-1$ or, equivalently, $p > q$. The Laplacian matrix is also symmetric and so we deduce that it is diagonal except for these components.

Unfortunately, the linear modes (that is, $p = 0$ and $p = P$) are not orthogonal to each other. For interior components based on the Legendre function (i.e., $[(1-\xi)/2]\,[(1+\xi)/2]\,P^{0,0}_{p-1}(\xi)$), as illustrated in Fig. 2.7(b), we see that the elemental Laplacian has interior components which are penta-diagonal and so have zero contribution if $p > q+2$. Finally, if the interior components are of the form $[(1-\xi)/2]\,[(1+\xi)/2]\,P^{2,2}_{p-1}(\xi)$ then applying the same argument demonstrates that we will not have the correct weight function for the $P^{2,2}_p(\xi)$ polynomial, that is, $[(1-\xi)/2]^2\,[(1+\xi)/2]^2$, which means the interior matrix will be full, as shown in Fig. 2.7(c). Although this interior expansion is very attractive for the mass matrix system, when we couple it with the Laplacian this basis is not as attractive as either of the other two expansions.

As mentioned previously, the modal expansion is sometimes defined in terms of the integral of Legendre functions and to relate our definition to the form above we note that eqn (A.1.9) in Appendix A with $\alpha = \beta = 0$ becomes

$$2p\int_{-1}^{\xi} L_p(s)\,\mathrm{d}s = -(1-\xi)(1+\xi)P^{1,1}_{p-1}(\xi)\,.$$

2.3.4 *Nodal polynomial expansions*

As discussed in Section 2.3.2.1, polynomial nodal expansions are based upon the Lagrange polynomials which are associated with a set of *nodal* points. The nodal points must include the ends of the domain if the expansion is to be decomposed into boundary and interior modes. Apart from this restriction we

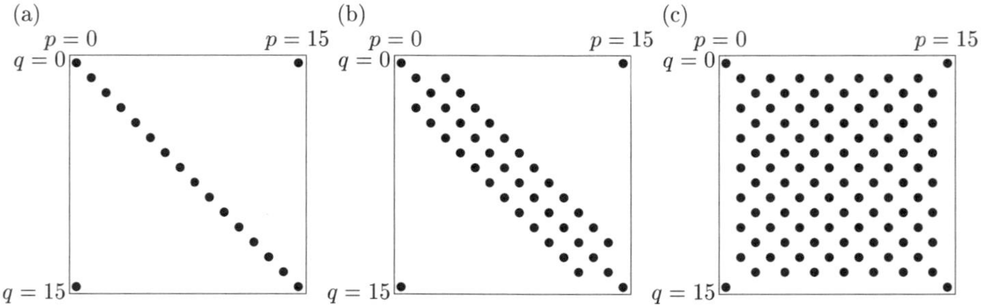

FIG. 2.7. Structure of the elemental Laplacian matrix, $\boldsymbol{L}^e$, for the one-dimensional p-type expansion of order $P = 15$ in $\Omega_{\rm st} = \{-1 \leqslant x \leqslant 1\}$ using interior components of the form (a) $[(1-\xi)/2]\,[(1+\xi)/2]\,P^{1,1}_{p-1}(\xi)$, (b) $[(1-\xi)/2]\,[(1+\xi)/2]\,L_{p-1}(\xi)$, and (c) $[(1-\xi)/2]\,[(1+\xi)/2]\,P^{2,2}_{p-1}(\xi)$.

are free to choose the location of the interior nodal points. The choice of these points, however, plays an important role in the stability of the approximation and the conditioning of the system. Using nodal points at the zeros of the Gauss–Legendre–Lobatto integration rule (see Section 2.4.1) produces a particularly efficient expansion which does not exhibit the oscillations seen when equispaced points are chosen, as shown in Fig. 2.4(b).

2.3.4.1 *Lagrange polynomials*

Given a set of $P+1$ nodal points, denoted by x_q $(0 \leqslant q \leqslant P)$, the Lagrange polynomial $h_p(x)$ is the unique polynomial of order P which has a unit value at x_p and is zero at x_q $(p \neq q)$. This definition can be written as

$$h_p(x_q) = \delta_{pq}\,, \tag{2.3.9}$$

where δ_{pq} is the Kronecker delta. The Lagrange polynomial can also be written in product form as

$$h_p(x) = \frac{\prod_{q=0,q\neq p}^{P}(x - x_q)}{\prod_{q=0,q\neq p}^{P}(x_p - x_q)}\,.$$

If we denote by $g(x)$ the polynomial of order $P+1$ with zeros at the $P+1$ nodal points x_q, then we can write $h_q(x)$ in the more compact form as

$$h_p(x) = \frac{g(x)}{g'(x_p)(x - x_p)}\,. \tag{2.3.10}$$

The Kronecker delta property shown in eqn (2.3.9) makes the Lagrange polynomial particularly useful as an interpolation basis. The Lagrange interpolant through the $P+1$ nodal points x_q is written as

$$\mathcal{I}u(x) = \sum_{p=0}^{P} \hat{u}_p h_p(x)\,.$$

The interpolation approximation requires that $\mathcal{I}u(x_q) = u(x_q)$ and, because of the property (2.3.9), this means that $\hat{u}_p = u(x_p)$. The interpolation approximation can therefore be written

$$\mathcal{I}u(x) = \sum_{p=0}^{P} u(x_p) h_p(x)\,.$$

If $u(x)$ is a polynomial of order P then the relationship is exact. This transformation from any polynomial to the Lagrange expansion is also useful when developing integration and differentiation solution schemes involving modal expansions (see Section 2.4.2).

2.3.4.2 *Nodal p-type bases: spectral elements*

A class of nodal p-type elements which have become known as 'spectral elements', due to Patera [363], use the Lagrange polynomial through the zeros of the Gauss–Lobatto polynomials. In the early version of the spectral element method the polynomials were Chebyshev type [263, 363], but in later versions Legendre polynomials were selected for more accurate numerical quadratures.

This type of integration is discussed in Section 2.4.1, where we see that using $P+1$ points gives nodal values at the roots of the polynomial $g(\xi) = (1-\xi)(1+\xi)L'_P(\xi)$. Substituting this into eqn (2.3.10) we obtain the nodal p-type expansion
♣[7] in the standard element $\Omega_{\rm st}$:

$$\phi_p(\xi) \mapsto h_p(\xi) = \begin{cases} 1, & \xi = \xi_p, \\ \dfrac{(\xi-1)(\xi+1)L'_P(\xi)}{P(P+1)L_P(\xi_p)(\xi_p-\xi)}, & \text{otherwise}, \end{cases} \qquad 0 \leqslant p \leqslant P, \tag{2.3.11}$$

where we have used eqn (A.1.2) (see Appendix A) to deduce $g'(\xi) = -P(P+1)L_P(\xi)$. The derivative of the Legendre polynomial $L'_P(\xi)$ can be related to the Jacobi polynomial $P^{1,1}_{P-1}(\xi)$ using eqn (A.1.8) (see Appendix A). Therefore, we could also consider the spectral element basis as a Lagrange polynomial with nodal points at the roots of the polynomial $g(\xi) = (1-\xi)(1+\xi)P^{1,1}_{P-1}(\xi)$. We recall from Section 2.3.3.2 that the zeros of the Jacobi polynomial can also be considered as the equilibrium location of electrostatic potential in the interval.

The shapes of the modes for an expansion with $P = 5$ are shown in Fig. 2.8. Unlike the modal p-type expansion shown in Fig. 2.5, all modes are polynomials of order P. The boundary modes are $h_0(\xi)$ and $h_5(\xi)$ whilst the interior modes, which are zero on the boundaries, are $h_1(\xi)$, $h_2(\xi)$, $h_3(\xi)$, and $h_4(\xi)$.

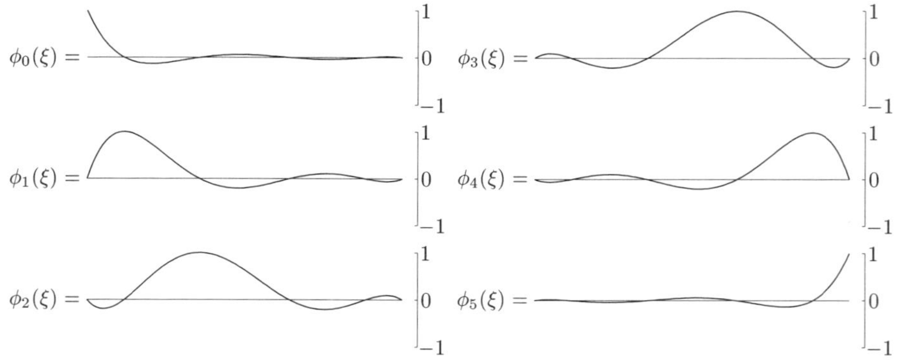

FIG. 2.8. Nodal expansion modes for a polynomial order of $P = 5$.

[7] Most commonly used nodal polynomial expansion basis.

As seen in Fig. 2.4(a–c), the equispaced Lagrange polynomial oscillates at the ends of the domains. The Lagrange polynomial through the Gauss–Lobatto–Legendre points does not exhibit this type of oscillation.

The elemental mass matrix using this expansion is full if we evaluate the inner product $\boldsymbol{M}^e[p][q] = (h_p, h_q)$ exactly. If, however, we use the Gauss–Legendre–Lobatto quadrature rule corresponding to the same choice of nodal points on which the expansion was defined, the mass matrix is diagonal due to the Kronecker delta property:

$$\boldsymbol{M}^e[p][q] = (h_p, h_q) \simeq \sum_{i=0}^{P} w_i h_p(\xi_i) h_q(\xi_i) = \sum_{i=0}^{P} w_i \delta_{pi} \delta_{qi} = w_p \delta_{pq} \, ,$$

where w_i are the weights for the Gauss–Legendre–Lobatto rule using $P{+}1$ points. If w_i corresponded to the quadrature rule using $P + 2$ points then the discrete integration would have been exact but the orthogonality would not be preserved.

This *discrete* orthogonality arises from the fact that the same nodal points are used to define the quadrature rule and the Lagrange polynomial. The quadrature rule using $Q = P + 1$ points is exact only for polynomials of order $2P - 1$ (see Section 2.4.1). The error incurred when using this reduced-order quadrature rule is, however, consistent with the approximation error of the expansion. That is to say, the error in approximating a smooth function using a polynomial expansion is of the same order as the error due to the inexact Gaussian integration rule. Further, Canuto and Quarteroni [87] have shown that the discrete norm is uniformly equivalent to the continuous norm.

The diagonal components of the elemental mass matrix using the reduced quadrature rule are equal to the row sum of the elemental mass matrix using exact integration. Summing the rows and using this as the entry of a diagonal matrix is common practice in finite elements and is known as *lumping* the mass matrix. In the standard finite element case, lumping the mass matrix is an approximation, but in the spectral element case this lumping has a direct equivalence. To illustrate this we consider the row sum

$$\sum_{q=0}^{P} \boldsymbol{M}^e[p][q] = \sum_{q=0}^{P} (h_p(\xi), h_q(\xi)) = \left(h_p(\xi), \sum_{q=0}^{P} h_q(\xi) \right) .$$

We note that the sum of the Lagrange basis over all modes is simply '1', that is,

$$\sum_{q=0}^{P} h_q(\xi) = 1 \, ,$$

and so the sum of the pth row becomes

$$\sum_{q=0}^{P} \boldsymbol{M}^e_{pq} = \left(h_p(\xi), \sum_{q=0}^{P} h_q(\xi) \right) = (h_p(\xi), 1) \, .$$

The last term $(h_p(\xi), 1)$ is, by definition, w_p, the weight corresponding to the pth point in the Gauss–Lobatto–Legendre quadrature rule using $P+1$ points. Since w_p is also the diagonal entry of the mass matrix using reduced integration, we see that mass lumping of the spectral element expansion is equivalent to constructing the mass matrix with a reduced integration rule.

As a final point, we note that the elemental Laplacian matrix using the spectral element expansion does not have any notable properties of this type and is full for all choices of quadrature order.

2.4 Elemental operations

We have seen in Section 2.3.1 that in the *hp* element decomposition the global expansion basis is decomposed into elemental subdomains that can then be mapped to standard regions. A polynomial basis is then defined in the standard region. To complete our Galerkin formulation we need to know how to integrate and differentiate the polynomial bases in the standard regions. In Section 2.4.1 we will discuss a particularly accurate form of numerical integration known as Gaussian quadrature. Gaussian quadrature defines a series of nodal or integration points upon which we need to know all values of the integrand. Therefore, when differentiating a function we typically require the value of the derivative at the quadrature points. A construction to determine the derivative at quadrature points is discussed in Section 2.4.2.

2.4.1 *Numerical integration*

In our Galerkin formulation we require a technique to evaluate, within each elemental domain, integrals of the form

$$\int_{-1}^{1} u(\xi)\,\mathrm{d}\xi\,, \tag{2.4.1}$$

where $u(\xi)$ may well be made up of products of polynomial bases, that is, $u(\xi) = \phi_p(\xi)\phi_q(\xi)$. Since the form of $u(\xi)$ is problem specific, we need an automated way to evaluate such integrals. This motivates the use of numerical integration or *quadrature*, as it is also known. The fundamental concept is the approximation of the integral by a finite summation of the form

$$\int_{-1}^{1} u(\xi)\,\mathrm{d}\xi \approx \sum_{i=0}^{Q-1} w_i u(\xi_i)\,,$$

where w_i are specified constants or *weights* and ξ_i represents an abscissa of Q distinct points in the interval $-1 \leqslant \xi_i \leqslant 1$. Although there are many different types of numerical integration, we shall restrict our attention to *Gaussian quadrature.*

2.4.1.1 *Gaussian quadrature*

Gaussian quadrature is a particularly accurate method for treating integrals where the integrand, $u(\xi)$, is smooth. In this technique the integrand is represented as a Lagrange polynomial (see Section 2.3.4.1) using the Q points ξ_i, which are to be specified, that is,

$$u(\xi) = \sum_{i=0}^{Q-1} u(\xi_i)h_i(\xi) + \varepsilon(u)\,, \tag{2.4.2}$$

where $\varepsilon(u)$ is the approximation error. If we substitute eqn (2.4.2) into eqn (2.4.1) then we obtain a representation of the integral as a summation:

$$\int_{-1}^{1} u(\xi)\,\mathrm{d}\xi = \sum_{i=0}^{Q-1} w_i u(\xi_i) + R(u)\,, \tag{2.4.3}$$

where

$$w_i = \int_{-1}^{1} h_i(\xi)\,\mathrm{d}\xi\,, \tag{2.4.4}$$

$$R(u) = \int_{-1}^{1} \varepsilon(u)\,\mathrm{d}\xi\,. \tag{2.4.5}$$

Or, equivalently,

given $w[i] = w_i$, $u[i] = u(\xi_i)$ for $0 \leqslant i \leqslant Q-1$, then $I = \int_{-1}^{1} u(\xi)\,\mathrm{d}\xi$ is numerically evaluated as follows: ♦5

```
set I = 0
   do i = 0, Q − 1
      I = I + w[i] u[i]
   continue
```

Equation (2.4.4) defines the weights w_i in terms of the integral of the Lagrange polynomial, but to perform this integration we need to know the location of the abscissae or zeros ξ_i. Since $u(\xi)$ is represented by a polynomial of order $Q-1$ we would expect the relation above to be exact if $u(\xi)$ is a polynomial of order $Q-1$ or less (that is, when $u(\xi) \in \mathcal{P}_{Q-1}([-1,1])$ then $R(u) = 0$). This would be true if, for example, we choose the points so that they are equispaced in the interval. There is, however, a better choice of zeros which permits the exact integration of polynomials of order higher than $Q-1$. This remarkable fact was first recognised by Gauss and is at the heart of Gaussian quadrature.

In this section we will present only the result of the Gauss quadrature for integrals of the type shown in eqn (2.4.3). This is known as Legendre integration.

5 Integration loop given definition of quadrature weights w_i and zeros ξ_i.

A more detailed description can be found in Appendix B. Also discussed in Appendix B are the Gauss-type quadrature rules for more general integrals of the form

$$\int_{-1}^{1} (1-\xi)^{\alpha}(1+\xi)^{\beta} u(\xi)\, \mathrm{d}\xi\,,$$

which are necessary to deal with the multi-dimensional bases.

There are three different types of Gauss quadrature known as Gauss, Gauss–Radau, and Gauss–Lobatto. The difference between the three types of quadrature lies in the choice of the zeros. Gauss quadrature uses zeros which have points that are interior to the interval, $-1 < \xi_i < 1$ for $i = 0, \ldots, Q-1$. In Gauss–Radau the zeros include one of the end-points of the interval, usually $\xi = -1$, and in Gauss–Lobatto the zeros include both end-points of the interval, that is, $\xi = \pm 1$.

Introducing $\xi_{i,P}^{\alpha,\beta}$ to denote the P zeros of the Pth-order Jacobi polynomial $P_P^{\alpha,\beta}$ (see Section 2.3.3.1) such that

$$P_P^{\alpha,\beta}(\xi_{i,P}^{\alpha,\beta}) = 0\,, \quad i = 0, 1, \ldots, P-1\,,$$

where

$$\xi_{0,P}^{\alpha,\beta} < \xi_{1,P}^{\alpha,\beta} < \cdots < \xi_{P-1,P}^{\alpha,\beta}\,,$$

we can define zeros and weights which approximate the integral

$$\int_{-1}^{1} u(\xi)\, \mathrm{d}\xi = \sum_{i=0}^{Q-1} w_i u(\xi_i) + R(u)$$

as follows.

1. *Gauss–Legendre*

$$\begin{aligned}
\xi_i &= \xi_{i,Q}^{0,0}\,, \quad i = 0, \ldots, Q-1\,,\\
w_i^{0,0} &= \frac{2}{1-(\xi_i)^2}\left[\frac{\mathrm{d}}{\mathrm{d}\xi}(L_Q(\xi))\bigg|_{\xi=\xi_i}\right]^{-2}, \quad i = 0, \ldots, Q-1\,,\\
R(u) &= 0 \quad \text{if } u(\xi) \in \mathcal{P}_{2Q-1}([-1,1])\,.
\end{aligned}$$

2. *Gauss–Radau–Legendre*

$$\begin{aligned}
\xi_i &= \begin{cases} -1\,, & i = 0\,,\\ \xi_{i-1,Q-1}^{0,1}\,, & i = 1, \ldots, Q-1\,,\end{cases}\\
w_i^{0,0} &= \frac{1-\xi_i}{Q^2[L_{Q-1}(\xi_i)]^2}\,, \quad i = 0, \ldots, Q-1\,,\\
R(u) &= 0 \quad \text{if } u(\xi) \in \mathcal{P}_{2Q-2}([-1,1])\,.
\end{aligned}$$

3. *Gauss–Lobatto–Legendre* ♦[6]

$$\xi_i = \begin{cases} -1, & i = 0, \\ \xi^{1,1}_{i-1,Q-2}, & i = 1, \dots, Q-2, \\ 1, & i = Q-1, \end{cases}$$

$$w^{0,0}_i = \frac{2}{Q(Q-1)[L_{Q-1}(\xi_i)]^2}, \quad i = 0, \dots, Q-1,$$

$$R(u) = 0 \quad \text{if } u(\xi) \in \mathcal{P}_{2Q-3}([-1,1]).$$

In all of the above quadrature formulae $L_Q(\xi)$ is the Legendre polynomial ($L_Q(\xi) = P^{0,0}_Q(\xi)$). The zeros of the Jacobi polynomial $\xi^{\alpha,\beta}_{i,m}$ do not have an analytic form and commonly the zeros and weights are tabulated. Tabulation of data can lead to copying errors and therefore a better way to evaluate the zeros is by the use of a numerical algorithm as illustrated in Appendix B.2. Having determined the zeros, the weights can be evaluated directly from the above formulae by generating the Legendre polynomial from the recursion relationship given in Appendix A. ♦[7]

2.4.1.2 *Integration errors: polynomial aliasing*

We have seen in Section 2.4.1.1 that when using Gauss–Lobatto–Legendre quadrature of order Q we can integrate a polynomial $u(\xi) \in \mathcal{P}_{2Q-3}$ exactly (to within machine precision computationally). From this rule, we can also deduce that for any given polynomial $u(\xi) \in \mathcal{P}_P$ the minimum number of quadrature points $Q_{\min}$, as a function of P, necessary for the quadrature to be exact (to within machine precision) is

$$Q_{\min} \geqslant \frac{P+3}{2}. \tag{2.4.6}$$

In Galerkin methods, we are often interested in numerically computing the inner product of two polynomials of the same (or lesser) degree, i.e., the inner product (ϕ_p, ϕ_q), where $\phi_p, \phi_q \in \mathcal{P}_P$. By rearranging eqn (2.4.6) we can therefore compute the number of quadrature points Q as a function of P which are required to integrate multiple powers of $u(\xi) \in \mathcal{P}_N$ exactly, as shown in Table 2.2.

TABLE 2.2. Number of quadrature points necessary for Gauss–Lobatto–Legendre quadrature to be exact in terms of the polynomial order, P, of the integrand.

Polynomial order P	Minimum quadrature order $Q_{\min}$
$[u(\xi)]^2 \in \mathcal{P}_{2P}$	$Q \geqslant P + 3/2$
$[u(\xi)]^3 \in \mathcal{P}_{3P}$	$Q \geqslant 3P/2 + 3/2$
$[u(\xi)]^4 \in \mathcal{P}_{4P}$	$Q \geqslant 2P + 3/2$

6 Most common choice of quadrature weights w_i and zeros ξ_i.

7 Approach to implement quadrature zeros and weights.

From Table 2.2 we observe that, for linear problems which only require the inner product of two polynomials of the same or lesser degree, only $Q_{\min} = P+2$ quadrature points are required to integrate the inner product exactly. In the case of quadratic nonlinearities, such as the convective contributions in the incompressible Navier–Stokes equations, $Q_{\min} = 3P/2+2$ points are needed. Finally, for cubic nonlinearities, such as those found in the compressible Navier–Stokes equations, $Q_{\min} = 2P+2$ points are needed for the numerical integrations to be exact.

In practice, most numerical solvers only use the number of quadrature points necessary to integrate the linear terms exactly. To understand the ramifications of this 'under-integration' of the nonlinear terms, we perform the following test similar to the one studied in [270].

Consider a function $u(\xi)$ in the standard region $\xi \in [-1, 1]$.

- Take as an initial condition the Legendre expansion or order $P = 10$ where all modal coefficients are one, i.e.,

$$u(\xi) = \sum_{p=0}^{P} L_p(\xi)\,.$$

  We also assume that $L_p(\xi)$ is normalised, so $(L_p(\xi), L_p(\xi)) = 1$. This is equivalent to a scaled summation of the type of modes shown in Fig. 2.4(c).
- Evaluate the function $u(\xi)$ on a set of Q zeros of the Gauss–Lobatto–Legendre quadrature, ξ_i, i.e.,

$$u(\xi_i) = \sum_{p=0}^{P} L_p(\xi_i)\,.$$

  This is equivalent to representing the solution on a Lagrange or spectral element basis.
- Evaluate the function $N(\xi) = u^2(\xi)$ by squaring, in a pointwise fashion, the values at the quadrature points:

$$N(\xi_i) = u^2(\xi_i) = u(\xi_i)\; u(\xi_i)\,,$$

  which is possible due to the Kronecker delta property of the Lagrange representation.
- Evaluate the inner product of N with basis $L_q(\xi)$, where $L_q(\xi) \in \mathcal{P}_{10}$, such that

$$(L_q, N)_Q = \sum_{i=0}^{Q-1} w_i L_q(\xi_i) N(\xi_i)\,,$$

  where w_i and ξ_i are the weights and zeros of the Q-order Gauss–Lobatto–Legendre quadrature rule, respectively.

The procedure above mimics the 'physical space' or pseudo-spectral evaluation of the inner product $(\phi_q(\xi), u^2(\xi))$ commonly used in spectral methods for evaluating nonlinear terms. This test was chosen because even in its simplicity it models the order of nonlinearity that occurs in the solution of the incompressible Navier–Stokes equations. The initial condition where all modes are set to *one* represents a case in which an element has under-resolved or marginally resolved the solution within the element.

In the test above, the only unspecified parameter is the number of quadrature points Q to be used. For exact integration of the inner product of $u(\xi)$ with $\phi_q(\xi)$, where both are polynomials of order of 10 or less, requires a minimum quadrature order of $Q_{\min} = 12$. However, exact integration of the inner product of $N(u) = u^2$ with $\phi_q(\xi)$ requires a quadrature of order $Q_{\min} = 17$. The ramifications of under-integration of this form are shown in Fig. 2.9(a) where we consider the case $\phi_q(\xi) = L_q(\xi)$. Since we have normalised the Legendre basis such that $(L_q(\xi), L_q(\xi)) = 1$, the inner product $\hat{N}_q = (L_q(\xi), N(\xi))_Q$ is analogous to the expansion coefficient of the expansion

$$N(\xi) = \sum_{p=0}^{20} \hat{N}_p L_p(\xi)\,.$$

From Fig. 2.9(a) we see the difference in the energy of the first ten modal coefficients when exact quadrature ($Q = 17$) is used compared to when a reduced quadrature ($Q = 12$) is used. From this figure it is evident that there is strong *aliasing* of the high-mode energy of the nonlinear term into the lower modes.

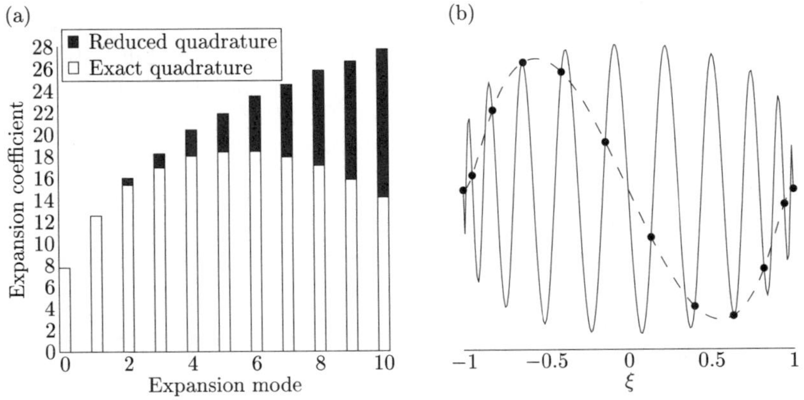

FIG. 2.9. (a) Comparison of the modal energy of a quadratic nonlinearity of a tenth-order polynomial expansion using reduced and exact numerical quadrature. (b) Aliasing error due to reduced integration: a twentieth-order Legendre polynomial (solid line) is evaluated at $Q = 12$ Gauss–Lobatto–Legendre quadrature points (solid circles) and the equivalent Lagrange polynomial through the quadrature points is shown by the dotted line.

To understand why this is the case we can return to the inner product

$$(\phi_q, N)_Q = \sum_{q=0}^{Q-1} w_i \phi_q(\xi_i) N(\xi_i), \tag{2.4.7}$$

where $\phi_q(\xi)$ is any polynomial basis such that $\phi_q \in \mathcal{P}_P$ and $N(\xi) \in \mathcal{P}_{2P}$. We can always expand $N(\xi)$ in terms of Legendre polynomials such that $N(\xi) = \sum_{p=0}^{2P} \hat{N}_p L_p(\xi)$. Substituting this form into eqn (2.4.7), we observe that

$$(\phi_q, N)_Q = \left(\phi_q, \sum_{p=0}^{P} \hat{N}_p L_p\right)_Q + \left(\phi_q, \sum_{p=P+1}^{2P} \hat{N}_p L_p\right)_Q . \tag{2.4.8}$$

If the quadrature order Q is sufficient for exact integration (i.e., $Q_{\min} > 3/2P + 3/2$) then the last term $(\phi_q, \sum_{p=P+1}^{2P} \hat{N} L_p)_Q$ will be identically zero since the Legendre polynomial $L_p(\xi)$ for $p > P$ is orthogonal to all polynomials of order P and less. However, if the numerical quadrature is only sufficient for integration of the linear terms (i.e., $Q_{\min} > P+3/2$) then the first part of the inner product, $(\phi_q, \sum_{p=0}^{P} \hat{N} L_p)_Q$, will be exactly integrated but the last term will not. It is the contribution of this last term which leads to the aliasing of modal energy seen in Fig. 2.9(a). To qualitatively appreciate why this is happening we can consider Fig. 2.9(b) which shows the highest energetic Legendre mode, $L_{20}(\xi)$, required to resolve the quadratic nonlinearity. When taking the inner product of this mode with $\phi_q(\xi)$ using quadrature of order $Q = 12$ we are only able to exactly integrate the Lagrange polynomial which passes through the quadrature zeros ξ_i as indicated by the filled points in Fig. 2.9(b). Therefore, we are only integrating the Lagrange polynomial represented by the dashed line in Fig. 2.9(b), and so the high-frequency energy can be considered as being *aliased* to this lower-energy mode.

Finally, we note there are two equivalent approaches to eliminating this error.

1. Exactly integrate the inner product (ϕ_q, N) using a quadrature order consistent with the order of the nonlinearity N. This is equivalent to setting $(\phi_q, \sum_{p=P+1}^{2P} \hat{N} L_p)_Q$ to zero since L_p is orthogonal to $\phi_q \in \mathcal{P}_P$ when $p > P$.
2. Filter the exact Legendre expansion of the nonlinearity N so that all modes above P are zero. This is equivalent to setting $(\phi_q, \sum_{p=P+1}^{2P} \hat{N}_p L_p)_Q$ to zero since by definition the filter $\hat{N}_p = 0$ for $p > P$.

Although the two approaches have a similar computational cost, the second, sharp cut-off, filtering approach is attractive from an implementation point of view since it allows us to maintain a quadrature order consistent with the treatment of a linear term.

2.4.2 *Differentiation*

Assuming a polynomial approximation of the form

$$u^\delta(x) = \sum_{p=0}^{P} \hat{u}_p \phi_p(\chi^{-1}) = \sum_{p=0}^{P} \hat{u}_p \phi_p(\xi),$$

where $\chi(\xi)$ is the mapping from the standard region $\xi \in \Omega^s$ to the region containing x in the interval $[a, b]$. we can differentiate $u(x)$ using the chain rule to obtain

$$\frac{\mathrm{d}u^\delta(\xi)}{\mathrm{d}x} = \frac{\mathrm{d}u^\delta(\xi)}{\mathrm{d}\xi}\frac{\mathrm{d}\xi}{\mathrm{d}x} = \sum_{p=0}^{P} \hat{u}_p \frac{\mathrm{d}\phi_p(\xi)}{\mathrm{d}\xi}\frac{\mathrm{d}\xi}{\mathrm{d}x}.$$

The differentiation of $u^\delta(x)$ is therefore dependent on evaluating $\mathrm{d}\phi_p(\xi)/\mathrm{d}\xi$ and $\mathrm{d}\xi/\mathrm{d}x$. In this section we shall consider the case where $\phi_p(\xi)$ is the Lagrange polynomial $h_p(\xi)$ and discuss how to evaluate $\mathrm{d}\phi_p(\xi)/\mathrm{d}\xi$. If $\chi(\xi)$ is an iso-parametric mapping this technique can also be used to evaluate $\mathrm{d}\chi/\mathrm{d}\xi = (\mathrm{d}\xi/\mathrm{d}x)^{-1}$. Differentiation of this form is often referred to as differentiation in physical space or *collocation differentiation*.

Any polynomial modal expansion can be represented in terms of Lagrange polynomials which can then be differentiated in a similar fashion. However, differentiation can also be performed in transformed space. This is particularly advantageous when $\mathrm{d}\phi_p(\xi)/\mathrm{d}\xi$ has an analytical form and may be expressed in terms of the function $\phi_p(\xi)$. A full discussion of differentiation in transformed space is beyond the scope of this book and we refer the reader to Gottlieb and Orszag [199] and Canuto *et al.* [86]. We would point out, however, that differentiation in physical and transformed space is not generally equivalent except when the function being differentiated is a polynomial which exists within the expansion space.

If we assume that $u^\delta(\xi)$ is a polynomial of order equal to or less than P (that is, $u^\delta(\xi) \in \mathcal{P}_P([-1,1])$), then it can be exactly expressed in terms of Lagrange polynomials $h_i(\xi)$ through a set of Q nodal points ξ_i $(0 \leqslant i \leqslant Q-1)$ as

$$u(\xi) = \sum_{i=0}^{Q-1} u(\xi_i) h_i(\xi), \quad h_i(\xi) = \frac{\Pi_{j=0, j\neq i}^{Q-1}(\xi - \xi_j)}{\Pi_{j=0, j\neq i}^{Q-1}(\xi_i - \xi_j)},$$

where $Q \geqslant P+1$. Therefore the derivative of $u(\xi)$ can be represented as

$$\frac{\mathrm{d}u(\xi)}{\mathrm{d}\xi} = \sum_{i=0}^{Q-1} u(\xi_i) \frac{\mathrm{d}}{\mathrm{d}\xi} h_i(\xi).$$

Typically, we only require the derivative at the nodal points ξ_i which is given by

$$\left.\frac{\mathrm{d}u(\xi)}{\mathrm{d}\xi}\right|_{\xi=\xi_i} = \sum_{j=0}^{Q-1} d_{ij} u(\xi_j),$$

where

$$d_{ij} = \left.\frac{\mathrm{d}h_j(\xi)}{\mathrm{d}\xi}\right|_{\xi=\xi_i}.$$

♦[8] Or, equivalently,

given $D[i][j] = d_{ij}$, $u[i] = u(\xi_i)$ for $0 \leqslant i, j \leqslant Q-1$, then $ud[i] = du/d\xi|_{\xi_i}$ is numerically evaluated as follows:

```
set ud[i] = 0
   do i = 0, Q-1
      do j = 0, Q-1
         ud[i] = ud[i] + D[i][j] u[j]
      continue
   continue
```

Following the derivation used by Funaro [168], an alternative representation of the Lagrange polynomial is

$$h_i(\xi) = \frac{g_Q(\xi)}{g'_Q(\xi_i)(\xi - \xi_i)}, \quad g_Q(\xi) = \prod_{j=0}^{Q-1}(\xi - \xi_j).$$

Taking the derivative of $h_i(\xi)$, we obtain

$$\frac{dh_i(\xi)}{d\xi} = \frac{g'_Q(\xi)(\xi - \xi_i) - g_Q(\xi)}{g'_Q(\xi_i)(\xi - \xi_i)^2}.$$

Finally, noting that because the numerator and denominator of this expression are zero as $\xi \to \xi_i$, and because $P_Q(\xi_i) = 0$ by definition,

$$\lim_{\xi \to \xi_i} \frac{dh_i(\xi)}{d\xi} = \lim_{\xi \to \xi_i} \frac{g''_Q(\xi)}{2g'_Q(\xi)} = \frac{g''_Q(\xi_i)}{2g'_Q(\xi_i)},$$

so we can write d_{ij} as

$$d_{ij} = \begin{cases} \dfrac{g'_Q(\xi_i)}{g'_Q(\xi_j)} \dfrac{1}{(\xi_i - \xi_j)}, & i \neq j, \\ \dfrac{g''_Q(\xi_i)}{2g'_Q(\xi_i)}, & i = j. \end{cases} \tag{2.4.9}$$

Equation (2.4.9) is the general representation of the derivative of the Lagrange polynomials evaluated at the nodal points ξ_i $(0 \leqslant i \leqslant Q-1)$. To proceed further we need to know specific information about the nodal points ξ_i which will allow us to deduce alternative forms of $g'_Q(\xi_i)$ and $g''_Q(\xi_i)$.

[8] Differentiation code given definition of differentiation matrix D_{ij} at the zeros ξ_i.

2.4.2.1 *Legendre formulae*

The most common differentiation matrices d_{ij} are those corresponding to the Gauss–Legendre quadrature points discussed in Section 2.4.1.1. In this section we illustrate the final form of the differential matrices that correspond to the use of Gauss–Legendre, Gauss–Radau–Legendre, and Gauss–Lobatto–Legendre quadrature points. The differentiation matrices for the general Gauss–Jacobi quadrature points are given in Appendix C. Denoting by $\xi_{i,P}^{\alpha,\beta}$ the P zeros of the Jacobi polynomial $P_P^{\alpha,\beta}(\xi)$ such that

$$P_P^{\alpha,\beta}(\xi_{i,P}^{\alpha,\beta}) = 0\,, \quad i = 0, 1, \ldots, P-1\,,$$

the derivative matrix d_{ij} used to evaluate $\mathrm{d}u(\xi)/\mathrm{d}\xi$ at ξ_i, that is,

$$\left.\frac{\mathrm{d}u(\xi)}{\mathrm{d}\xi}\right|_{\xi=\xi_i} = \sum_{j=0}^{Q-1} d_{ij} u(\xi_j)\,,$$

is defined as follows.

1. *Gauss–Legendre*

$$\xi_i = \xi_{i,Q}^{0,0}\,,$$

$$d_{ij} = \begin{cases} \dfrac{L'_Q(\xi_i)}{L'_Q(\xi_j)(\xi_i - \xi_j)}\,, & i \neq j, 0 \leqslant i, j \leqslant Q-1\,, \\ \dfrac{\xi_i}{1-\xi_i^2}\,, & i = j\,. \end{cases}$$

2. *Gauss–Radau–Legendre*

$$\xi_i = \begin{cases} -1\,, & i = 0\,, \\ \xi_{i-1,Q-1}^{0,1}\,, & i = 1, \ldots, Q-1\,, \end{cases}$$

$$d_{ij} = \begin{cases} \dfrac{-(Q-1)(Q+1)}{4}\,, & i = j = 0\,, \\ \dfrac{L_{Q-1}(\xi_i)}{L_{Q-1}(\xi_j)} \dfrac{1-\xi_j}{1-\xi_i} \dfrac{1}{\xi_i - \xi_j}\,, & i \neq j, 0 \leqslant i, j \leqslant Q-1\,, \\ \dfrac{1}{2(1-\xi_i)}\,, & 1 \leqslant i = j \leqslant Q-1\,. \end{cases}$$

♦[9] 3. *Gauss–Lobatto–Legendre*

$$\xi_i = \begin{cases} -1\,, & i = 0\,, \\ \xi^{1,1}_{i-1,Q-2}\,, & i = 1, \dots, Q-2\,, \\ 1\,, & i = Q-1\,, \end{cases}$$

$$d_{ij} = \begin{cases} \dfrac{-Q(Q-1)}{4}\,, & i = j = 0\,, \\ \dfrac{L_{Q-1}(\xi_i)}{L_{Q-1}(\xi_j)} \dfrac{1}{\xi_i - \xi_j}\,, & i \neq j, 0 \leqslant i,j \leqslant Q-1\,, \\ 0\,, & 1 \leqslant i = j \leqslant Q-2\,, \\ \dfrac{Q(Q-1)}{4}\,, & i = j = Q-1\,. \end{cases}$$

♦[10] In a similar way to the quadrature formulae, the construction of the differentiation matrices requires the quadrature zeros to be determined numerically as described in Appendix B.2. Having determined the zeros, the components of the differentiation matrix can be evaluated directly from the above formulae by generating the Legendre polynomial from the recursion relationship given in Appendix A.

2.5 Error estimates

In this section we review some of the basic error estimates for Galerkin spectral/*hp* element methods. In Section 2.5.1 we follow the classical analysis of Strang and Fix [443] for the h-convergence of linear finite element methods in the energy of the H^1 norm. Then in Section 2.5.2 we review the classical analysis of Gottlieb and Orszag [199] for the p-convergence of an orthogonal polynomial based on a singular Sturm–Liouville problem in a single standard element. Finally, in Section 2.5.3 we provide, without proofs, some more general error estimates for *hp* elements.

2.5.1 *h-convergence of linear finite elements*

Recalling the h-type extension to the global expansion basis discussed in Section 2.3.1, we consider the convergence of the linear ($P = 1$) finite element expansion on a regular mesh. In this case convergence is achieved using the h-type extension, that is, increasing the number of elements so that $h \to 0$.

Adopting the analysis used by Strang and Fix [443] and Szabó and Babuška [445], we consider the one-dimensional Helmholtz problem described in Section 2.2.3, which is stated in weak form as follows:

[9] Most common choice of differentiation matrix to go with equivalent quadrature rule.

[10] Approach to numerically implement the differentiation matrix.

find $u \in \mathcal{X}$ such that

$$\int_0^l \left[\frac{\partial v}{\partial x}\frac{\partial u}{\partial x} + \lambda v u\right] \mathrm{d}x = \int_0^l v f \, \mathrm{d}x\,, \quad \forall\, v \in \mathcal{V}\,. \tag{2.5.1}$$

We assume that $f(x)$ and λ are defined so that $u'' = \mathrm{d}^2u/\mathrm{d}x^2$ is bounded and continuous such that $|u''| \leqslant C$ in the interval $0 \leqslant x \leqslant l$. The energy norm is defined as $||u||_E = \sqrt{a(u,u)}$, where $a(v,u)$ is the left-hand side of eqn (2.5.1), i.e.,

$$(||u||_E)^2 = a(u,u) = \int_0^l \left[\frac{\partial u}{\partial x}\frac{\partial u}{\partial x} + \lambda u u\right] \mathrm{d}x\,.$$

To determine the error between the finite element solution u^δ and the exact solution u we consider the error between the linear interpolant $\mathcal{I}u$ and u. Since u^δ is the minimal solution to u for all functions in the trial space $\mathcal{X}^\delta$ (see eqn (2.2.22a)) the interpolation error will bound the error of the finite element solution.

Let us consider a uniform mesh as shown in Fig. 2.10, where the domain consists of N_{el} elements and each element is of equal length $h = l/N_{\mathrm{el}}$. The linear interpolant $\mathcal{I}u$ is a piecewise linear approximation to u such that

$$\mathcal{I}u(jh) = u(jh)\,, \quad j = 0, \ldots, N_{\mathrm{el}}\,.$$

Therefore, the interpolation error, $\bar{\varepsilon}(x) = u(x) - \mathcal{I}u(x)$, is zero at the end of each element Ω_e. We shall denote the interpolation error in each element by $\bar{\varepsilon}_e$:

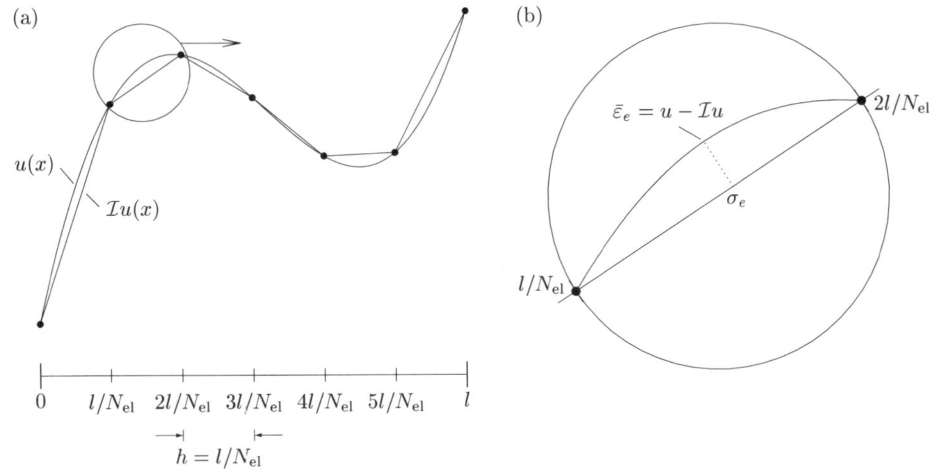

FIG. 2.10. (a) Linear interpolation, $\mathcal{I}u(x)$, of the function $u(x)$ in the region $0 \leqslant x \leqslant l$ using a uniform mesh with $N_{\mathrm{el}} = 6$ elemental domains. Each element is of size $h = l/N_{\mathrm{el}}$. (b) The linear interpolant is exact at the end-points of every element, and so there is a point σ_e where the error $\bar{\varepsilon} = u - \mathcal{I}u$ is a maximum.

$$\bar{\varepsilon}_e(x) = u(x) - \mathcal{I}u(x)\,, \quad (e-1)h \leqslant x \leqslant eh\,.$$

Since $\bar{\varepsilon}_e(x)$ vanishes at the ends of each element, there is a point in every element, σ_e, where $|\bar{\varepsilon}_e|$ is a maximum (see Fig. 2.10). At this point $\bar{\varepsilon}_e'(\sigma_e) = 0$ and so we can write

$$\bar{\varepsilon}_e'(x) = \int_{\sigma_e}^{x} \bar{\varepsilon}_e''(s)\,\mathrm{d}s = \int_{\sigma_e}^{x} u''(s)\,\mathrm{d}s\,, \quad (e-1)h \leqslant x \leqslant eh\,.$$

The last part of this relation uses the fact that the interpolant, $\mathcal{I}u(x)$, is linear and therefore $[\mathcal{I}u(x)]'' = 0$. As $|u''| \leqslant C$, we find

$$\max|\bar{\varepsilon}_e'(x)| \leqslant Ch\,, \quad (e-1)h \leqslant x \leqslant eh\,. \tag{2.5.2}$$

Assuming that σ_e lies nearer to eh than $(e-1)h$ so that $eh - \sigma_e \leqslant h/2$, we can, by applying the Taylor expansion to the error $\bar{\varepsilon}_e$ about the point σ_e, represent the error $\bar{\varepsilon}_e$ at eh as

$$\bar{\varepsilon}_e(eh) = \bar{\varepsilon}_e(\sigma_e) + (eh - \sigma_e)\bar{\varepsilon}_e'(\sigma_e) + \frac{(eh-\sigma_e)^2}{2}\bar{\varepsilon}_e''(s)\,, \tag{2.5.3}$$

where s is a point between σ_e and eh. (If σ_e had been closer to $(e-1)h$ than eh we could have written the Taylor expansion for the point $(e-1)h$ instead of that for eh and obtained the same result.)

Now, by definition, $\bar{\varepsilon}_e(eh) = 0$ and $\bar{\varepsilon}_e'(\sigma_e) = 0$. Accordingly, upon substitution into eqn (2.5.3) we deduce that

$$\max|\bar{\varepsilon}_e(\sigma_e)| \leqslant C\frac{h^2}{8}\,. \tag{2.5.4}$$

We recall that $\bar{\varepsilon}_e(\sigma_e)$ was defined as the point of maximum absolute error in the eth element, and so

$$\max|\bar{\varepsilon}_e(x)| \leqslant |\bar{\varepsilon}_e(\sigma_e)|\,.$$

Finally, the square of the error in the energy norm can be written as

$$\begin{aligned}(||\bar{\varepsilon}||_E)^2 = a(\bar{\varepsilon}, \bar{\varepsilon}) &= \int_0^l \left((\bar{\varepsilon}')^2 + \lambda\,(\bar{\varepsilon})^2\right)\mathrm{d}x \\ &= \sum_{e=1}^{N_{\mathrm{el}}} \int_{(e-1)h}^{eh} \left((\bar{\varepsilon}_e')^2 + \lambda\,(\bar{\varepsilon}_e)^2\right)\mathrm{d}x \\ &\leqslant hN_{\mathrm{el}}\left((Ch)^2 + \lambda\left(C\frac{h^2}{8}\right)^2\right).\end{aligned}$$

Noting that $l = h \cdot N_{\mathrm{el}}$, there is a constant K such that

$$(||\bar{\varepsilon}||_E)^2 \leqslant lKC^2h^2\,.$$

Due to the minimal property of the finite element solution, the error in the finite element approximation $\varepsilon = u - u^\delta$ is bounded by $\bar{\varepsilon}$, and so

$$||\varepsilon||_E \leqslant ||\bar{\varepsilon}||_E \leqslant K_1 Ch\,,$$

where C depends on f and λ but is independent of h.

2.5.2 L^2 *error of the p-type interpolation in a single element*

Following Gottlieb and Orszag [199] we now consider the approximation to a function $u(\xi)$ in the interval $-1 \leqslant \xi \leqslant 1$ by an expansion of Legendre polynomials, $L_p(\xi)$ of order P, that is,

$$u^\delta(\xi) = \sum_{p=0}^{P} \hat{u}_p L_p(\xi)\,.$$

Since the Legendre polynomials are orthogonal in this interval, a Galerkin projection to determine the expansion coefficients implies that

$$\hat{u}_p = \frac{(L_p(\xi), u(\xi))}{(L_p(\xi), L_p(\xi))} = (L_p(\xi), u(\xi))\,, \tag{2.5.5}$$

where we have assumed that the Legendre modes are orthonormalised so that $(L_p(\xi), L_p(\xi)) = 1$. Assuming the function $u(\xi)$ can be represented by an infinite expansion, the L^2 error between our finite expansion after P terms is

$$\begin{aligned} ||\varepsilon||_2 &= \left[\int_{-1}^{1} \left| u(\xi) - \sum_{p=0}^{P} \hat{u}_p L_p(\xi)\right|^2 \mathrm{d}\xi\right]^{1/2} \\ &= \left[\int_{-1}^{1} \left| \sum_{p=0}^{\infty} \hat{u}_p L_p(\xi) - \sum_{p=0}^{P} \hat{u}_p L_p(\xi)\right|^2 \mathrm{d}\xi\right]^{1/2} = \left[\sum_{p=P+1}^{\infty} \hat{u}_p^2\right]^{1/2}, \end{aligned} \tag{2.5.6}$$

where we have used the orthogonality of $L_p(\xi)$ or, equivalently, Parseval's identity to get the last relationship. The error ε in terms of the order of the expansion is therefore determined by the rate of decrease of $\hat{u}_p$ as $p \to \infty$.

Since $L_p(\xi) = P_p^{0,0}(\xi)$, we know from eqn (2.3.6) that

$$L_p(\xi) = -\frac{1}{\lambda_p}\frac{\mathrm{d}}{\mathrm{d}\xi}\left[(1-\xi^2)\frac{\mathrm{d}}{\mathrm{d}\xi}L_p(\xi)\right], \quad \lambda_p = p(p+1)\,. \tag{2.5.7}$$

Rearranging and substituting eqn (2.5.7) into eqn (2.5.5), we obtain

$$\hat{u}_p = -\frac{1}{\lambda_p}\int_{-1}^{1}\left[\frac{\mathrm{d}}{\mathrm{d}\xi} - 2\xi\frac{\mathrm{d}}{\mathrm{d}\xi}\right] L_p(\xi) u(\xi)\,\mathrm{d}\xi\,.$$

Now integrating by parts twice we obtain

$$\hat{u}_p = -\frac{1}{\lambda_p}\int_{-1}^{1} L_p(\xi)\left[\frac{\mathrm{d}}{\mathrm{d}\xi} - 2\xi\frac{\mathrm{d}}{\mathrm{d}\xi}\right] u(\xi)\,\mathrm{d}\xi\,. \tag{2.5.8}$$

Note that the integration by parts does not generate any boundary terms since the function $1 - \xi^2$ is identically zero at the ends of the region $\xi = \pm 1$. This is a property of solutions to singular Sturm–Liouville equations such as the Jacobi polynomials. We can therefore substitute eqn (2.5.7) into eqn (2.5.8) and again integrate by parts twice. Repeated application of this substitution and integration s times leads to

$$\hat{u}_p = (-1)^s \frac{1}{(\lambda_p)^s}\int_{-1}^{1} L_p(\xi)\left[\frac{\mathrm{d}}{\mathrm{d}\xi} - 2\xi\frac{\mathrm{d}}{\mathrm{d}\xi}\right]^s u(\xi)\,\mathrm{d}\xi\,.$$

Applying the Cauchy–Schwartz inequality, we observe that

$$\left[\int_{-1}^{1} L_p(\xi)\left[\frac{\mathrm{d}}{\mathrm{d}\xi} - 2\xi\frac{\mathrm{d}}{\mathrm{d}\xi}\right]^s u(\xi)\,\mathrm{d}\xi\right]^2 \leqslant \left[\int_{-1}^{1} \left[\frac{\mathrm{d}}{\mathrm{d}\xi} - 2\xi\frac{\mathrm{d}}{\mathrm{d}\xi}\right]^s u(\xi)\,\mathrm{d}\xi\right]^2,$$

since $\int_{-1}^{1} L_p^2(\xi)\,\mathrm{d}\xi = 1$. Therefore, utilising the last two equations in eqn (2.5.6), we can say that the L^2 error, ε, is bounded by

$$||\varepsilon||_2 \leqslant P^{-k}||u||_k\,, \tag{2.5.9}$$

where $k = 2s$ and $||u||_k$ denotes the H^k norm of the function $u(x)$. If $u(x)$ is sufficiently smooth then eqn (2.5.9) tells us that the error decreases faster than any power of P as $P \to \infty$. It can also be shown that the error in the H^1 norm is bounded as

$$||\varepsilon||_1 \leqslant P^{1-k}||u||_k\,. \tag{2.5.10}$$

For the derivation of this bound for global spectral expansions, please see [86].

2.5.3 *General error estimates for hp elements*

Once again we consider the one-dimensional Helmholtz problem with solution $u \in H^k(\Omega)$ described in Section 2.2.3. Assuming a discretisation on a uniform mesh of equispaced elements of size h, the general error estimate in the energy norm for the p- and h-type extension process can be written as (see Babuška and Suri [31])

$$||\varepsilon||_E \leqslant Ch^{\mu-1}P^{-(k-1)}||u||_k\,,$$

where $\varepsilon = u - u^\delta$, $\mu = \min(k, P+1)$, and C is independent of h, P, and u, but depends on k. We also assume that the forcing $f \in H^{k-2}(\Omega)$ so that it does not appear in the error estimate. Clearly, if the solution is smooth enough to have bounded derivatives for $k \geqslant P+1$ then this error estimate shows us that we can achieve exponential convergence as we increase the polynomial order P (p-type extension).

If the forcing f is chosen so that solutions that are not 'smooth' but are of the form

$$u(x) = x^{\alpha} - x\,, \quad \alpha > \frac{1}{2}\,,$$

and so $u(x) \in H^1(\Omega)$, Babuška and Suri [30,31] have shown that for the p-version the error in the energy norm is

$$||\varepsilon||_E \leqslant CP^{-2(\alpha-1/2)}\,,$$

where C depends on h and u. This convergence rate is dependent upon the singularity being centred between two elements. If this is not the case then the convergence rate is only of the form $P^{-(\alpha-1/2)}$, which is the same asymptotic convergence rate as would be expected in the h-type extension. This result was first proven by Babuška *et al.* [32].

A similar estimate is valid for the spectral element version. Assuming that the solution $u \in H_0^k(\Omega)$ and that the forcing $f \in H^{\rho}$, then the following error estimate is valid in the H^1 norm (equivalent to the energy norm):

$$||\varepsilon||_1 \leqslant C\left[P^{1-k}||u||_k + P^{-\rho}||f||_{\rho}\right].$$

Here we assume that the number of elements is constant and that homogeneous Dirichlet conditions are prescribed on $\partial\Omega$. This proof can be found in [319], and more details can be found in Bernardi and Maday [45], including different boundary conditions and variable coefficients in the elliptic problem of Section 2.2.3.

2.6 Implementation of a one-dimensional spectral/*hp* element solver

2.6.1 *Exercises*

To complement our introduction to spectral/*hp* element methods for elliptic problems we propose the following series of exercises and examples. The exercises are structured with a view towards developing a one-dimensional elliptic solver based on a standard Galerkin formulation. Although all of the concepts applied in this section have been discussed in the previous sections of this chapter, the exercises will demonstrate how each of the concepts can be used in practice. ♦11

Some useful codes are also available on the web page

`http://www.nektar.info/2nd_edition`

1. A good starting-point is to perform numerical integration on a polynomial function in $\xi \in [-1, 1]$. To do this you will require the quadrature weights w_i and zeros ξ_i which can either be generated as discussed in Appendix B.2 or the *Polylib* library which can be found on the web page given above. Having obtained the zeros and weights try the following tests using the Gauss–Lobatto–Legendre quadrature rule discussed in Section 2.4.1.1.

11 Implementation of one-dimensional spectral/hp element solver.

(a) Integrate $\int_{-1}^{1} \xi^6 \, d\xi = \sum_{i=0}^{Q-1} w_i [\xi_i]^6 = 2/7$ using $Q = 4$, 5, and 6.

(b) Integrate $\int_{-1}^{2} x^6 \, dx = 129/7$ using $Q = 4$, 5, and 6. Note that a mapping is required to map the interval $x \in [-1, 2]$ to $\xi \in [-1, 1]$.

(c) Calculate $\mathcal{I} = \int_0^{\pi/2} \sin x \, dx = 1$ for $2 \leqslant Q \leqslant 8$ and plot the error $\varepsilon = \mathcal{I} - 1$ versus Q on semi-log axes. In using Gaussian quadrature we are essentially approximating the smooth function $\sin x$ with polynomials of order $Q - 1$. The error, ε, will therefore be proportional to $\varepsilon \propto C^Q$. Therefore, plotting $\log \varepsilon$ versus Q should asymptotically give a straight line since by taking the log of the error relationship we have $\log \varepsilon \propto Q \log C$.

2. To introduce differentiation we have to construct the derivative matrix discussed in Section 2.4.2, which will again require the quadrature zeros used in Exercise 1. A routine is also available in the *Polylib* library on the web site. Having generated a routine to calculate the differentiation matrix, $D[i][j]$, try the following tests.

(a) Differentiate the function $u(\xi) = \xi^7$ at the quadrature points ξ_i to obtain $(du/d\xi)|_{\xi_i} = \sum_{j=0}^{Q-1} D[i][j](\xi_j)^7 = 7\xi_i^6$ using $Q = 7$, 8, and 9 quadrature points. Verify how many quadrature points are required to get an exact answer to numerical precision.

(b) Using the chain rule

$$\frac{du}{dx} = \frac{du}{d\xi}\frac{d\xi}{dx},$$

evaluate $(du/dx)|_{x_i} = 7x_i^6$ when $u(x) = x^7$ in the interval $x \in [2, 10]$, where $x_i = \chi(\xi_i)$ and χ is a linear mapping (see Section 2.3.1.2)

$$\chi(\xi) = 2\frac{1-\xi}{2} + 10\frac{1+\xi}{2}.$$

(c) Calculate $\mathcal{I} = -\int_0^{\pi/2} (d/dx) \cos x \, dx = 1$ for $2 \leqslant Q \leqslant 8$ and following Exercise 1(c) plot the error $\varepsilon = \mathcal{I} - 1$ versus Q on semi-log axes.

3. Having developed routines to numerically integrate and differentiate we can now start setting up a spectral/*hp* element problem beginning with a single element domain. Consider the projection problem $u^\delta(x) = f(x)$, where $f(x)$ is a known function, for example $f(x) = x^4$ or $f(x) = \sin x$. Projection problems are helpful at this stage since they do not require any boundary conditions to be imposed. Following Section 2.2, we state our Galerkin problem in the interval $\xi \in [-1, 1]$ as follows:

find $u^\delta \in \mathcal{X}^\delta$, such that

$$\int_{-1}^{1} v^\delta(\xi) u^\delta(\xi) \, d\xi = \int_{-1}^{1} v^\delta f(\xi) \, d\xi, \quad \forall \, v^\delta \in \mathcal{V}^\delta.$$

For a Galerkin expansion we define the expansion space $\mathcal{X}^\delta$ to be the same as the test space $\mathcal{V}^\delta$.

The definition of a discrete expansion $\phi_p(\xi)$, and accordingly a discrete solution $u^\delta(\xi) = \sum_{p=0}^{P} \hat{u}_p \phi_p(\xi)$, leads to the matrix problem

$$\boldsymbol{M}\hat{\boldsymbol{u}} = \hat{\boldsymbol{f}}\,,$$

where

$$\boldsymbol{M}[p][q] = \int_{-1}^{1} \phi_q(\xi)\phi_p(\xi)\,\mathrm{d}\xi\,, \quad \hat{\boldsymbol{u}}[p] = \hat{u}_p\,, \quad \hat{\boldsymbol{f}}[p] = \int_{-1}^{1} \phi_p(\xi) f(\xi)\,\mathrm{d}\xi\,.$$

In this exercise we will use the basis $\phi_p(\xi) = \psi_p(\xi)$ given by eqn (2.3.8) and shown in Fig. 2.5. This basis should be generated at the quadrature zeros ξ_i using the recursion relationship defined in Appendix A. Perform the following tasks.

(a) Construct the mass matrix $\boldsymbol{M}$, where

$$\boldsymbol{M}[p][q] = \int_{-1}^{1} \psi_p(\xi)\psi_q(\xi)\,\mathrm{d}\xi\,, \quad 0 \leqslant p, q \leqslant P$$

for $P = 8$ using Gauss–Lobatto–Legendre quadrature with $Q = 10$. Check that the matrix has the structure shown in Fig. 2.6(a).

(b) Try the last exercise using the spectral element basis $\phi_p(\xi) = h_p(\xi)$ given by eqn (2.3.11) and shown in Fig. 2.8. Confirm that when $Q = P + 1$ the matrix is diagonal.

(c) Construct the right-hand-side vector

$$\boldsymbol{f}[p] = \int_{-1}^{1} \phi_p(\xi) f(\xi)\,\mathrm{d}\xi\,, \quad 0 \leqslant p \leqslant P$$

using $f(\xi) = \xi^7$, $P = 8$, and $Q = 10$. Using a matrix inversion technique (for example, the Cholesky factorisation and solve routines *dpptrf* and *dpptrs* in the *LAPACK* library [12]) invert the symmetric mass matrix $\boldsymbol{M}$ and solve for $\hat{\boldsymbol{u}} = \boldsymbol{M}^{-1}\hat{\boldsymbol{f}}$. Verify your solution by evaluating $u^\delta(\xi) = \sum_{p=0}^{P} \hat{u}_p \phi_p(\xi)$ and checking it is equal to ξ^7.

(d) Impose Dirichlet boundary conditions by decomposing the discrete solution into an unknown homogeneous, $u^{\mathcal{H}}$, solution and a known function, $u^{\mathcal{D}}$, i.e., $u^\delta(x) = u^{\mathcal{D}} + u^{\mathcal{H}}$ (see Section 2.2.1.3), where

$$u^{\mathcal{D}} = u^\delta(-1)\phi_0(\xi) + u^\delta(1)\phi_P(\xi)\,, \quad u^{\mathcal{H}} = \sum_{p=1}^{P-1} \hat{u}_p \phi_p(\xi)\,.$$

Set up and solve for the matrix problem for the homogeneous solution. Note that the right-hand-side vector will now modified to be

$$\int_{-1}^{1} \phi_p(\xi) f(\xi)\,\mathrm{d}\xi - \int_{-1}^{1} \phi_p(\xi) u^{\mathcal{D}}(\xi)\,\mathrm{d}\xi\,, \quad 1 \leqslant p \leqslant P-1\,,$$

where $u^{\mathcal{D}}$ is a known function.

(e) Set up and solve the matrix problem in the interval $2 \leqslant x \leqslant 5$.

To extend the single element problem we now need to consider how to set up a multiple element spectral/*hp* element discretisation for the projection problem in Exercise 3. We will reuse the same elemental matrix, $\boldsymbol{M}^e$, although it will now relate to different elemental regions, and so will require a mapping from every element to the standard region, similar to Exercise 3(e). Note, however, that for a linear mapping $\chi^e(\xi)$ the matrices vary by only the constant factor of the Jacobian and therefore only a single matrix for every polynomial order P is required in practice.

To construct the global matrix we need to assemble the local matrices to a global structure, as discussed in Section 2.3.1.4. This process is illustrated in Fig. 2.11 where the assembly is shown for two assembly matrices, $\boldsymbol{\mathcal{A}}$. The top assembly matrix in Fig. 2.11 provides a global matrix with minimal bandwidth. The bottom assembly matrix illustrated in Fig. 2.11 provides a global matrix where the interior modes are separated from the boundary modes. This second form is particularly convenient when using the static condensation technique to solve the matrix, as will be discussed in Section 4.2.3. It is also advantageous when using higher polynomial orders since the local matrices of the interior modes are decoupled at an elemental level.

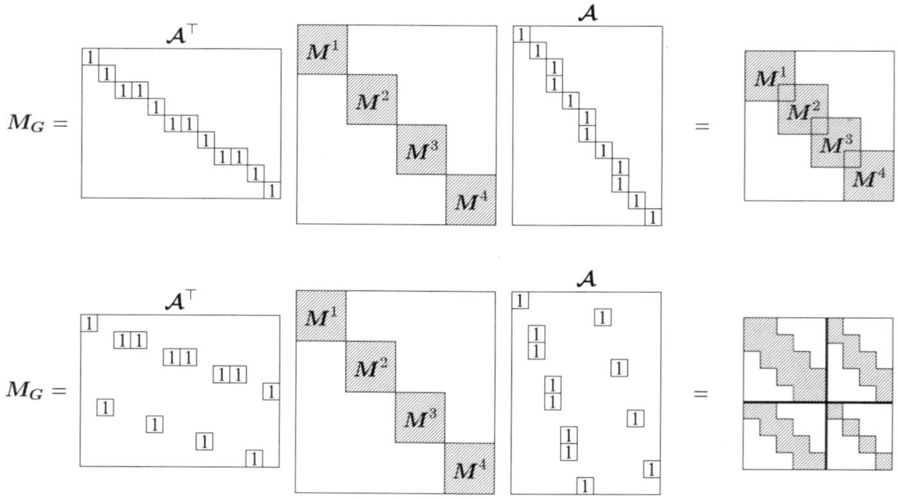

FIG. 2.11. Schematic illustration of the global matrix assembly $\boldsymbol{M}_G$ from the elemental matrices $\boldsymbol{M}^e$. Different choices of the assembly matrix $\boldsymbol{\mathcal{A}}$ lead to different filling of the final global matrix. The top assembly matrix leads to a global matrix with minimal bandwidth for this one-dimensional system, whereas the bottom assembly matrix separates the interior and elemental matrices which could then be inverted using static condensation.

Recall that rather than generating the assembly matrix $\boldsymbol{\mathcal{A}}$ it is numerically more efficient to use a mapping array. For example, in the case illustrated in the top of Fig. 2.11, where $N_{\rm el} = 4$ and $P = 2$, we can define the mapping array map[e][i] such that

$$\text{map}[1][p] = \begin{bmatrix} 0 \\ 1 \\ 2 \end{bmatrix}, \quad \text{map}[2][p] = \begin{bmatrix} 2 \\ 3 \\ 4 \end{bmatrix}, \quad \text{map}[3][p] = \begin{bmatrix} 4 \\ 5 \\ 6 \end{bmatrix}, \quad \text{map}[4][p] = \begin{bmatrix} 6 \\ 7 \\ 8 \end{bmatrix},$$

and then the global matrix can be assembled by setting all terms in $\boldsymbol{M_G}$ to zero and performing the following summation:

do $e = 1, N_{\rm el}$
  do $p = 0, P$
    do $q = 0, P$
      $\boldsymbol{M_G}[\text{map}[e][p]][\text{map}[e][q]] = \boldsymbol{M_G}[\text{map}[e][p]][\text{map}[e][q]] + \boldsymbol{M}^e[p][q]$
    continue
  continue
continue

4. Consider the following global problems.
   (a) Set up and solve the global matrix problem, without explicitly imposing boundary conditions, for the projection problem in Exercise 3 using $N_{\rm el} = 10$ elements in the interval $0 \leqslant x \leqslant 10$ with a local polynomial expansion of order $P = 8$ in each element. Take the forcing function to be $f(x) = \sin x$. Note that the right-hand-side vector $\boldsymbol{f}$ can also be constructed from the elemental contributions $\boldsymbol{f}^e$ and then assembled using $\boldsymbol{\mathcal{A}}^\top$, as shown in Section 2.3.1.4.
   (b) Modify your implementation to explicitly impose the Dirichlet boundary conditions $u(0) = 0$ and $u(10) = \sin 10$. Note that you can use the mapping array map[e][i] to reorder your global matrix so that the known degrees of freedom are listed at the end of the global vector.
5. Repeat Exercises 3 and 4 for the one-dimensional Helmholtz problem

$$\frac{\mathrm{d}^2 u}{\mathrm{d}x^2} - \lambda u(x) = f(x)$$

to generate the results discussed in Section 2.6.2. The mechanics of the construction are identical to the projection problem $u^\delta = f(x)$, except now we must also consider boundary conditions in our weak formulation and need to construct the elemental Laplacian matrix $\boldsymbol{L}^e$ defined as

$$\boldsymbol{L}^e[p][q] = \int_{\Omega^e} \frac{\mathrm{d}}{\mathrm{d}x}\phi_p(x) \, \frac{\mathrm{d}}{\mathrm{d}x}\phi_q(x) \, \mathrm{d}x \, .$$

To evaluate this matrix you will need to apply the chain rule to each derivative and then map it to the standard region $\Omega_{\rm st}$. Recall that the Helmholtz equation requires appropriate boundary conditions and that when a Neumann boundary condition is prescribed this modifies the right-hand-side vector, as discussed in Section 2.2.1.2.

2.6.2 *Convergence examples*

For the prototype problem

$$\nabla^2 u(x) - \lambda u(x) = f(x)\,, \quad \lambda \geqslant 0\,, \tag{2.6.1}$$

we consider two exact solutions, one of which is infinitely differentiable having the form $u(x) = \sin(\pi x)$ and one which is of finite regularity of the form $u(x) = x^\alpha$, where α is a non-integer value. For both cases we assume that $\lambda = 1$.

When the solution is of the form $u(x) = \sin(\pi x)$ we know that $f(x) = -(\pi^2 + \lambda)\sin(\pi x)$. The numerical solution $u^\delta(x)$ to this problem using both h- and p-type extensions in an interval $x = [-1, 1]$ with Dirichlet boundary conditions is shown in Fig. 2.12. The error is measured in the discrete energy (or equivalently H^1) norm and is normalised by the length of the domain l, that is,

$$||\varepsilon||_E = \frac{||u - u^\delta(x)||_E}{||1||^1} = \frac{||u - u^\delta(x)||_E}{\sqrt{l}}\,.$$

The h-type convergence tests were performed using a polynomial order $P = 1$ and the p-type convergence tests were performed using two elements.

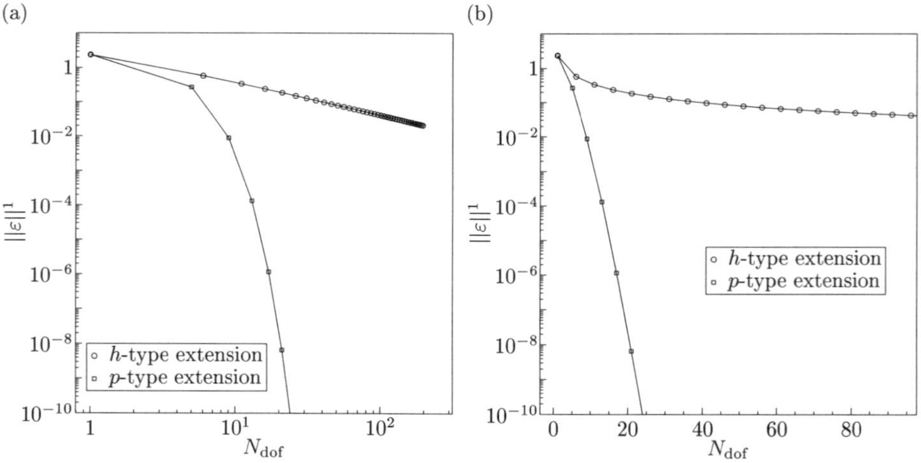

FIG. 2.12. Convergence in the discrete energy norm $||\varepsilon||_E$ as a function of degrees of freedom N_{dof}. Two tests were performed using the h-type extension with a fixed polynomial order $P = 1$ and the p-type extension with two elemental domains. (a) Error on a *log–log* axis, demonstrating the algebraic convergence of the h-type extension. (b) Error on a *semi-log* axis, demonstrating the exponential convergence of the p-type extension for smooth solutions.

In Fig. 2.12(a) we see the error plotted on a *log–log* axis as a function of the total degrees of freedom N_{dof}. As shown in Section 2.5.1, if the second derivative of the solution can be bounded by a constant ($|u''| < C$), then the error in the energy norm of the h-extension process is

$$||\varepsilon||_E \leqslant K_1 C h \, .$$

Since $h \approx 1/N_{\text{dof}}$, when P is fixed we expect the error to behave as a linear function of N_{dof} on the *log–log* plot. From the results given in Section 2.5.3, we see that the slope (or convergence rate) of the h-type extension process is related to the minimum of the polynomial order P plus one and the smoothness of the solution (more formally, the Sobolev norm $||u||_k$) is bounded. Since the solution $u(x) = \sin(\pi x)$ is an infinitely-differentiable function the polynomial order dictates the convergence rate. Also shown in Fig. 2.12(a) is the p-type extension process for a domain containing two elements. The smooth property of the solution means that an exponential rate of convergence is achieved in terms of the polynomial order. The number of elements is fixed to be $N_{\text{dof}} \approx P$, and so if we plot the error on a *semi-log* axis as shown in Fig. 2.12(b) then we observe the exponential decay in the error due to the p-type extension process.

When the solution is of the form $u(x) = x^\alpha$ it will only exist in the energy norm if $\alpha > 1/2$. This solution is not smooth in the sense that we can only bound the $u(x)$ in the Sobolev norm $||u||_k \equiv ||u||_{H^k(\Omega)}$, where k is a low number. As we mentioned in Section 2.5.3, it has been proven in [32] that the convergence rate of this type of solution is

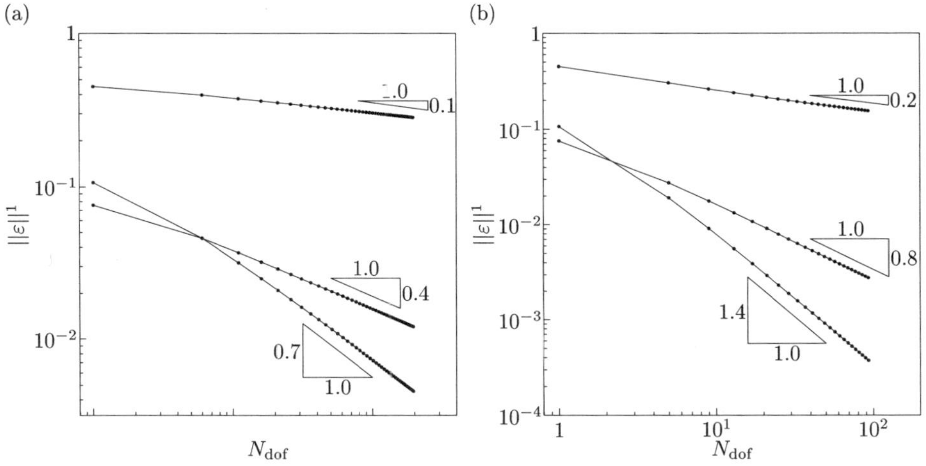

FIG. 2.13. Convergence in the energy norm for the solution $u(x) = x^\alpha$, $\alpha = 0.6$, 0.9, and 1.2 for (a) the h-type extension, and (b) the p-type extension. The asymptotic convergence rate can be seen to agree with the predictions given by eqns (2.6.2) and (2.6.3).

$$||\varepsilon||_E \leqslant CP^{-(\alpha-1/2)} \tag{2.6.2}$$

in the h-type extension, and

$$||\varepsilon||_E \leqslant CP^{-2(\alpha-1/2)} \tag{2.6.3}$$

in the p-type extension, provided that the singularity of the derivatives at $x = 0$ is placed between elemental regions. In Fig. 2.13, we see the error in the energy norm of the discrete solution to eqn (2.6.1) in the region $x \in [0, 1]$ when the exact solution is $u(x) = x^\alpha$ and $\alpha = 0.6$, 0.9, and 1.2. Figure 2.13(a) shows the convergence of the h-type extension process where the asymptotic convergence rate from eqn (2.6.2) is shown to be 0.1, 0.4, and 0.7 for the three values of α. Figure 2.13(b) shows the convergence of the p-type extension process where the asymptotic convergence rate from eqn (2.6.3) is shown to be 0.2, 0.8, and 1.4 for the three values of α. For all calculations the number of discrete quadrature points was set to be double the polynomial order.

3

MULTI-DIMENSIONAL EXPANSION BASES

In Chapter 2 we introduced the standard Galerkin technique for spectral/*hp* element discretisation in one dimension. Using this construction we illustrated how a global C^0 continuous expansion could be decomposed into either a modal or nodal expansion within an element domain in a standard region. In this chapter, we begin extending this concept to multiple dimensions in this chapter by discussing spectral/*hp* element multi-dimensional expansions in different standard regions. In Chapter 4 we extend the one-dimensional discussion from Chapter 2 on elemental operations such as integration and differentiation to multiple dimensions, and we also discuss the construction of a global expansion from the local expansions defined in this chapter.

Although the extension to multiple dimensions is analogous to the one-dimensional case, the aim of this chapter is to explain the reasons for the different choices of expansion bases as well as discussing their numerical implementation. All the expansions discussed in this chapter will be considered *within* a standard region Ω_{st}. In two or three dimensions, we now have a choice of different standard regions. In two dimensions we will consider standard regions that are either a quadrilateral or a triangle. In three dimensions we will consider standard regions that are either a hexahedron, prism, pyramid, or tetrahedron, which collectively we refer to as hybrid domains. Since we will only be referring to these standard regions, we shall not use the superscript e to denote the elemental domain within this chapter. The assembly of these expansions in multiple domains and the treatment of integration and differentiation is discussed in Chapter 4.

For tensorial bases we shall denote the polynomial bases by $\phi_{pq}(\xi_1, \xi_2)$ or $\phi_{pqr}(\xi_1, \xi_2, \xi_3)$ in two or three dimensions, respectively, where ξ_1, ξ_2, and ξ_3 are the standard Cartesian coordinates. This notation may equally well refer to a modal or nodal expansion within a triangular, quadrilateral, or any of the three-dimensional standard regions. Although there is a wide variety of expansion bases, particularly for the standard h-type finite element method, we shall be restricting our attention to those most commonly used in the spectral/*hp* element literature. The majority of the bases to be discussed can be expressed in terms of a product of one-dimensional functions or tensor product, for example,

$$\phi_{pq}(\xi_1, \xi_2) = \psi_p^a(\xi_1)\psi_q^a(\xi_2),$$

or

$$\phi_{pq}(\xi_1, \xi_2) = h_p(\xi_1)h_q(\xi_2).$$

Expansions which can be constructed with this form allow many numerical operations to be performed very efficiently using the sum factorisation or tensor product techniques discussed in Section 4.1.6.

Although most quadrilateral hexahedral expansions are typically constructed from a product of functions, this is not so common for expansions in triangular and tetrahedral domains. In Section 3.1 we give a comprehensive discussion of the tensorial extension for all hybrid regions and also introduce the underlying concepts which will be helpful when constructing a tensorial basis for the unstructured region, as covered in Section 3.1.1. Finally, in Section 2.3.4 we discuss non-tensorial, nodal bases in the triangular regions which are compatible with the standard nodal, tensor-based quadrilateral basis.

♦[12] From a purely implementation point of view, the following details may be of interest. The most commonly used spectral/hp element bases are those which can be expanded into a globally C^0 continuous expansion. The standard tensor product expansions for modal and nodal bases in quadrilateral and hexahedral domains are defined in Section 3.1. The decomposition of these expansions into an interior and boundary decomposition is discussed in Section 3.1.1.2. For the hybrid domains, that is triangular and quadrilateral regions in two dimensions and tetrahedral, pyramidic, prismatic, and hexahedral domains in three dimensions, a unified, C^0 continuous, generalised tensorial expansion is defined in Sections 3.2.3.1 and 3.2.3.2, and a full listing is also provided in Appendix D. This expansion makes use of the collapsed coordinate system discussed in Sections 3.2.1.1 and 3.2.1.2, and the assembly of these expansions is detailed in Section 3.2.3.3. Finally, two non-tensorial nodal sets of points in a triangular region, compatible with the nodal quadrilateral expansion, are introduced and defined in Sections 3.3.3 and 3.3.4 as well as Appendix D. For a discontinuous Galerkin formulation it is sometimes convenient to use orthogonal expansions in preference to the modified, globally C^0 continuous, expansion. Tensor-based orthogonal expansions in all two- and three-dimensional hybrid regions are defined in Section 3.2.2.1, which also uses the collapsed coordinate systems discussed in Sections 3.2.1.1 and 3.2.1.2.

Nomenclature

Although already specified in the general nomenclature section, we highlight again here in Table 3.1 some of the new notation adopted in this chapter. Similarly to Chapter 2, we will use ξ to denote the standard Cartesian directions with a subscript to identify different orthogonal directions depending on the dimension of the basis. Also, following the convention of Chapter 2, we will denote any general polynomial expansion as $\phi_{pq}(\xi_1, \xi_2)$ or $\phi_{pqr}(\xi_1, \xi_2, \xi_3)$ in two or three dimensions, respectively. The indices p, q, and r denote the different components of the tensorial expansion, which is introduced in Section 3.1. Two commonly

[12] Layout of the chapter from the implementation point of view.

TABLE 3.1. Notation for expansion bases.

$\boldsymbol{\xi} = (\xi_1, \xi_2, \xi_3)$	Local Cartesian coordinates
$\eta_1, \overline{\eta_1}, \eta_2, \eta_3$	Local collapsed coordinates
$\phi_{pq}(\xi_1, \xi_2)$	General expansion basis for any two-dimensional region
$\phi_{pqr}(\xi_1, \xi_2, \xi_3)$	General expansion basis for any three-dimensional region
$\psi^a_p(\eta_1), \psi^b_{pq}(\eta_2), \psi^c_{pqr}(\eta_3)$	Modified principal functions
$\widetilde{\psi}^a_p(\eta_1), \widetilde{\psi}^b_{pq}(\eta_2), \widetilde{\psi}^c_{pqr}(\eta_3)$	Orthogonal principal functions
$h_p(\xi)$	One-dimensional Lagrange polynomial of order p
$L^{N_m}_i(\boldsymbol{\xi})$	Two-dimensional Lagrange polynomial through N_m points $\boldsymbol{\xi}_i$
P_i	Polynomial order in the ith direction

used tensorial expansion bases, already introduced in Sections 2.3.3.3 and 2.3.4.2, are the modal modified basis, $\psi^a_p(\eta)$, and the nodal Lagrange basis, $h_p(\xi)$.

To define a generalised tensorial expansion in simplex domains, such as triangles and tetrahedrons, it will be necessary to introduce a new, non-orthogonal coordinate system. We will refer to this new coordinate system as *collapsed coordinates* and denote each ordinate as η_1, η_2, or η_3. The collapsed coordinate $\overline{\eta_1}$ will also be used in the pyramidic expansion bases. Its definition is analogous to η_2. Consistent with the introduction of the collapsed coordinates, we will introduce two generalised tensor bases for both a modified, C^0 continuous, basis $\psi^b_{pq}(\eta_2), \psi^c_{pqr}(\eta_3)$ and an orthogonal basis $\widetilde{\psi}^b_{pq}(\eta_2), \widetilde{\psi}^c_{pqr}(\eta_3)$.

In Section 2.3.4 we will discuss non-tensorial bases in triangular regions. Since there is no tensorial basis it is more convenient to use the basis notation with a single index, for example $\phi_i(\boldsymbol{\xi})$, where $\boldsymbol{\xi} = (\xi_1, \xi_2)$. We, however, understand the index i to sum over all of the two-dimensional modes. We shall also adopt the notation $L^{N_m}_i(\boldsymbol{\xi})$ to denote the multi-dimensional Lagrange polynomial through N_m nodal points $\boldsymbol{\xi}_i$.

Since for multi-dimensional bases the polynomial order can change in each Cartesian direction, we shall also adopt the notation P_i, where the subscript denotes the ith Cartesian direction.

3.1 Quadrilateral and hexahedral tensor product expansions

The extension to higher dimensions within quadrilateral or hexahedral regions is relatively straightforward, if rather more involved than the one-dimensional case discussed in Section 2.3.

We start by defining the two-dimensional standard region, $\mathcal{Q}^2$, as the bi-unit square,

$$\Omega_{\rm st} = \mathcal{Q}^2 = \{-1 \leqslant \xi_1, \xi_2 \leqslant 1\},$$

and the three-dimensional standard region, $\mathcal{Q}^3$ as the bi-unit cuboid,

$$\Omega_{\rm st} = \mathcal{Q}^3 = \{-1 \leqslant \xi_1, \xi_2, \xi_3 \leqslant 1\}.$$

In general, these regions will be referred to as $\Omega_{\rm st}$, which encompasses both $\mathcal{Q}^2$ and $\mathcal{Q}^3$. Since these regions are trivially defined by a standard Cartesian coordinate system, the most natural and straightforward way to construct the basis is by taking a product of the one-dimensional basis, which can be thought of as one-dimensional *tensors*. This type of extension may be applied equally well to either the modal- or nodal-type basis, and therefore we shall not distinguish between these two forms except where it is necessary or helpful.

3.1.1 *Standard tensor product extensions*

In Section 2.3.2.1 we introduced a variety of one-dimensional p-type expansion bases which we generally referred to as $\phi_p(\xi)$. Three of the most useful expansions are as follows.

♣[8] *Modal (C^0 continuous) basis*

$$\phi_p(\xi) = \begin{cases} \psi_0^a(\xi) = \dfrac{1-\xi}{2}, & p = 0, \\ \psi_p^a(\xi) = \left(\dfrac{1-\xi}{2}\right)\left(\dfrac{1+\xi}{2}\right) P_{p-1}^{1,1}(\xi), & 0 < p < P, \\ \psi_P^a(\xi) = \dfrac{1+\xi}{2}, & p = P. \end{cases} \tag{3.1.1}$$

♣[9] *Nodal (C^0 continuous) basis*

$$\phi_p(\xi) = h_p(\xi) = \frac{(\xi-1)(\xi+1)L'_P(\xi)}{P(P+1)L_P(\xi_p)(\xi-\xi_p)}, \quad 0 \leqslant p \leqslant P. \tag{3.1.2}$$

L^2 orthogonal basis

$$\phi_p(\xi) = \widetilde{\psi}_p^a(\xi) = L_p(\xi). \tag{3.1.3}$$

The use of definition (3.1.1) for $\phi_p(\xi)$ corresponds to the C^0 continuous hierarchical modal expansion and is the most commonly used hp-finite element expansion in quadrilateral domains. The original hierarchical expansions for p-type extensions bases were constructed by Peano [369]. This basis was then modified by using the integral of the Legendre polynomial to arrive at the widely adopted form given in eqn (3.1.1) (see Oden [343] and Szabó and Babuška [445]). Note that the integral of the Legendre polynomial is directly related to the $P_p^{1,1}(z)$ Jacobi polynomial, as seen from eqn (A.1.9) in Appendix A.

Definition (3.1.2) corresponds to the spectral element nodal basis originally developed using Chebyshev expansions by Patera [363]. Definition (3.1.3) is the Legendre polynomial which is orthogonal in the L^2 or Legendre inner product.

[8] Most commonly used modal polynomial expansion basis.

[9] Most commonly used nodal polynomial expansion basis.

In all these cases the expansion was denoted by a single subscript, p, and so may be considered as a one-dimensional 'tensor'. The two- and three-dimensional bases can be constructed by a simple product of the one-dimensional tensors in each of the Cartesian coordinate directions, that is,

$$\phi_{pq}(\xi_1, \xi_2) = \phi_p(\xi_1)\phi_q(\xi_2), \qquad 0 \leqslant p, q, \; p \leqslant P_1, \; q \leqslant P_2,$$
$$\phi_{pqr}(\xi_1, \xi_2, \xi_3) = \phi_p(\xi_1)\phi_q(\xi_2)\phi_r(\xi_3), \quad 0 \leqslant p, q, r, \; p \leqslant P_1, \; q \leqslant P_2, \; r \leqslant P_3.$$

We note that the polynomial order of the multi-dimensional expansions may differ in each coordinate direction, as denoted by the use of the bounds P_1, P_2, and P_3.

Figure 3.1 shows a diagrammatic representation of the tensor product extension to generate a two-dimensional expansion in the standard quadrilateral

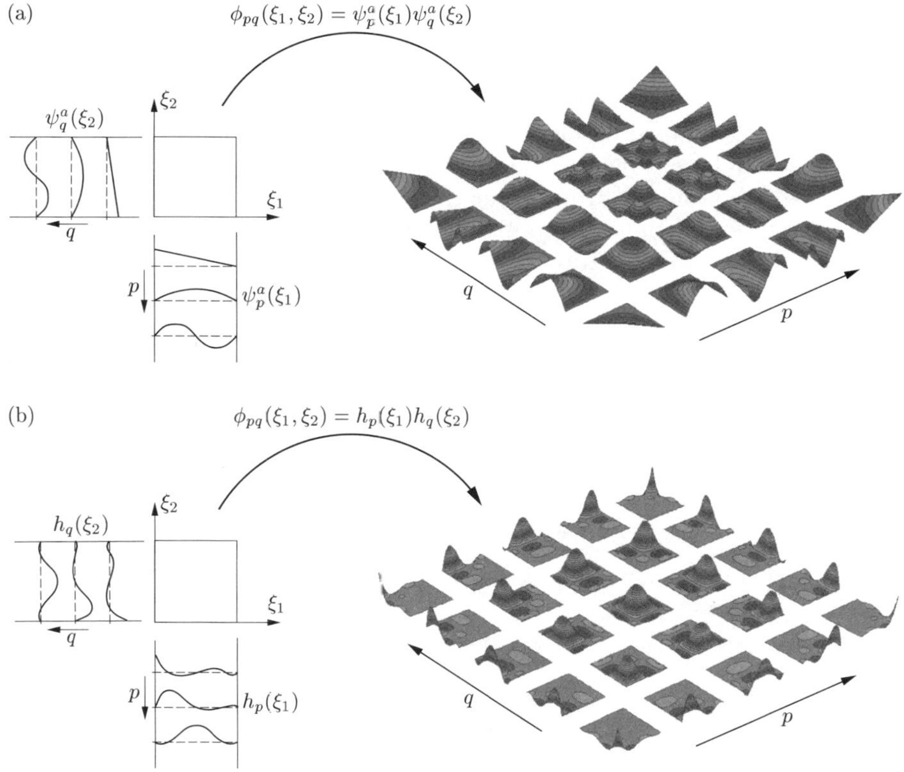

FIG. 3.1. Construction of a two-dimensional expansion basis from the product of two one-dimensional expansions of order $P = 4$. (a) Modal expansions using the one-dimensional expansion defined in eqn (3.1.1) (edge and face modes have been scaled by a factor of 4 and 16, respectively). (b) Nodal expansion using the one-dimensional Lagrange polynomial defined in eqn (3.1.2).

region using both the modal and nodal one-dimensional expansions. The modal basis shown in Fig. 3.1(a) was generated using the one-dimensional expansion $\phi_p(\xi) = \psi_p^a(\xi)$ (see eqn (3.1.1)) and the nodal basis expansion shown in Fig. 3.1(b) was generated using the one-dimensional expansion $\phi_p(\xi) = h_p(\xi)$ (see eqn (3.1.2)). The expansion modes shown in Fig. 3.1 represent a complete bi-linear expansion for fourth-order polynomials in both the ξ_1 and ξ_2 directions. Note that the modal expansion maintains a hierarchic form where the lower-order expansions are a subset of the higher-order expansions. In contrast, each component of the two-dimensional nodal expansion maintains the Kronecker delta form of the Lagrange polynomial where each mode has a unit value at a specified position within the region.

3.1.1.1 *Boundary/interior decomposition*

A significant property of the modal expansion based on definition (3.1.1) and the nodal expansion based on definition (3.1.2) is their inherent decomposition into boundary and interior modes. Boundary modes are defined as all the modes which have nonzero support on the boundary of the standard region; interior modes are all the modes which are zero on all boundaries. We recall from Section 2.3.2.2 that this type of decomposition is particularly convenient when a C^0 global expansion base is required since a global expansion can be generated from the local expansion simply by matching the shape of individual boundary modes.

An illustration of the decomposition is shown in Fig. 3.2, where we see all the modes shown in Fig. 3.1(a) decomposed into vertex, edge, and face modes. We define *vertex modes* as all modes which have a unit magnitude at one vertex and are zero at all other vertices; *edge modes* as all modes which have support along one edge and are zero at all other edges and vertices; and *face modes* as all

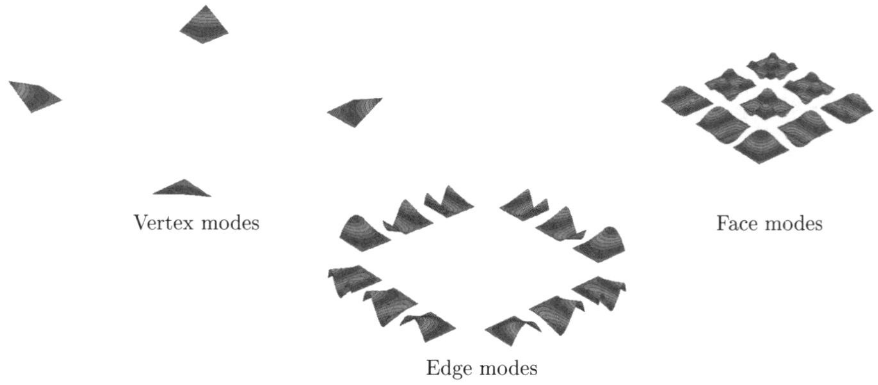

FIG. 3.2. Boundary/interior decomposition of the modal expansion shown in Fig. 3.1(a). The two-dimensional expansion is decomposed into boundary modes (vertex and edge modes) which have support along the boundary of the region, and interior modes which have zero support on all boundaries.

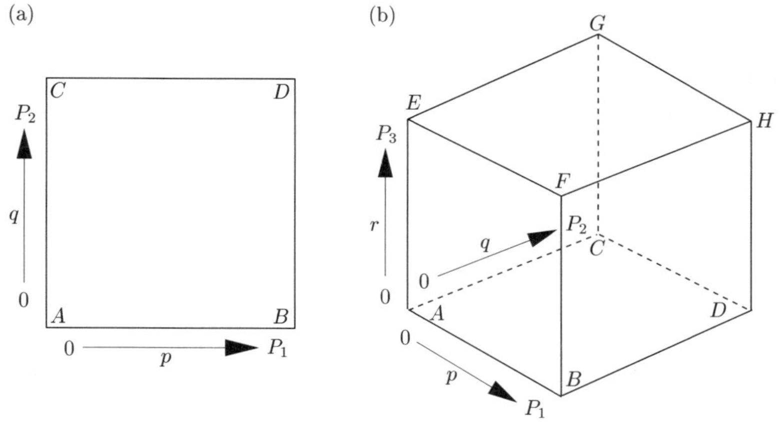

FIG. 3.3. Definition of the array indices for the (a) two-dimensional, and (b) three-dimensional tensor expansions.

modes which have magnitude along one face but are zero along all other faces, edges, and vertices. For a two-dimensional expansion the boundary modes are made up of the vertex and edge modes, whereas in three dimensions the boundary decomposition contains the vertex, edge, and face modes. The face modes of the three-dimensional expansions are analogous to the interior modes of the two-dimensional expansion. It should be appreciated that an identical decomposition is possible for the nodal basis shown in Fig. 3.1(b) under the definitions presented above. We further note that the vertex modes of the modal expansion are identical to the standard linear finite element basis for a quadrilateral subdomain.

3.1.1.2 *Construction procedure for the local expansion*

The definition of the two- and three-dimensional expansion bases ϕ_{pq} and ϕ_{pqr} as a tensor product of the one-dimensional functions defined in eqns (3.1.1) and (3.1.2) enables the boundary and interior modes to be interpreted intuitively, as discussed in the following sections.

Two-dimensional expansion

If we consider the two-dimensional basis, $\phi_{pq}(\xi_1, \xi_2)$, as a two-dimensional array spanning a similar region as the standard quadrilateral illustrated in Fig. 3.3(a), then the indices of the boundary modes correspond to their location within this array. For example, the vertex mode which has magnitude at the corner of the quadrilateral marked A corresponds to the indices $p = 0$, $q = 0$, and so $\phi_{0,0}(\xi_1, \xi_2) = \psi_0^a(\xi_1)\psi_0^a(\xi_2)$ is the appropriate mode for this vertex of a modal ♦13

13 Definition of boundary and interior modes for two-dimensional tensor product expansions.

expansion. The vertex mode at $p = 0, q = 0$ in the previously defined nodal expansion is $\phi_{0,0}(\xi_1, \xi_2) = h_0(\xi_1)h_0(\xi_2)$.

The four vertex modes labelled A, B, C, and D are therefore described as follows:

$$\begin{aligned}
\text{vertex } A: &\quad \phi_{0,0}(\xi_1, \xi_2) = \psi_0^a(\xi_1)\psi_0^a(\xi_2)\,,\\
\text{vertex } B: &\quad \phi_{P_1,0}(\xi_1, \xi_2) = \psi_{P_1}^a(\xi_1)\psi_0^a(\xi_2)\,,\\
\text{vertex } C: &\quad \phi_{0,P_2}(\xi_1, \xi_2) = \psi_0^a(\xi_1)\psi_{P_2}^a(\xi_2)\,,\\
\text{vertex } D: &\quad \phi_{P_1,P_2}(\xi_1, \xi_2) = \psi_{P_1}^a(\xi_1)\psi_{P_2}^a(\xi_2)\,.
\end{aligned}$$

The edge between A and C corresponds to the indices $p = 0$, $0 < q < P$, and so these edge modes of the expansion $\phi_{pq}(\xi_1, \xi_2)$ are defined as $\phi_{0,q}(\xi_1, \xi_2) = \psi_0^a(\xi_1)\psi_q^a(\xi_2)$ $(0 < q < P)$. The four edges of the quadrilateral expansions are therefore defined as follows:

$$\begin{aligned}
\text{edge } AB: &\quad \phi_{p,0}(\xi_1, \xi_2) = \psi_p^a(\xi_1)\psi_0^a(\xi_2)\,, &\quad 0 < p < P_1\,,\\
\text{edge } CD: &\quad \phi_{p,P_2}(\xi_1, \xi_2) = \psi_p^a(\xi_1)\psi_{P_2}^a(\xi_2)\,, &\quad 0 < p < P_1\,,\\
\text{edge } AC: &\quad \phi_{0,q}(\xi_1, \xi_2) = \psi_0^a(\xi_1)\psi_q^a(\xi_2)\,, &\quad 0 < q < P_2\,,\\
\text{edge } BD: &\quad \phi_{P_1,q}(\xi_1, \xi_2) = \psi_{P_1}^a(\xi_1)\psi_q^a(\xi_2)\,, &\quad 0 < q < P_2\,.
\end{aligned}$$

Finally, the interior modes to the quadrilateral expansion are analogously defined as

$$\text{interior modes:}\quad \phi_{pq}(\xi_1, \xi_2) = \psi_p^a(\xi_1)\psi_q^a(\xi_2)\,, \quad 0 < p, q,\ p < P_1,\ q < P_2\,.$$

An explicit listing of the nodal and modal quadrilateral basis is also given in Appendix D.

Three-dimensional expansion

A similar construction process to the two-dimensional expansion is possible for the three-dimensional basis; however, we now consider a three-dimensional array spanning the standard hexahedral region (also shown in Fig. 3.3). For this case the modes within the face $ABCD$ correspond to the indices $0 < p < P$, $0 < q < P$, $r = 0$, and so these face modes are defined as

$$\phi_{p,q,0}(\xi_1, \xi_2, \xi_3) = \psi_p^a(\xi_1)\psi_q^a(\xi_2)\psi_0^a(\xi_3)\,, \quad 0 < p < P_1,\ 0 < q < P_2\,.$$

To derive the same expansion for the nodal expansion we simply replace $\psi_p^a(\xi)$ with $h_p(\xi)$. For a full listing of the nodal and modal hexahedral bases see Appendix D.

3.1.2 *Polynomial space of tensor product expansions*

The expansions $\phi_{pq}(\xi_1, \xi_2)$ and $\phi_{pqr}(\xi_1, \xi_2, \xi_3)$ defined using the tensor product of any of the polynomials $\psi_p^a(\xi)$, $h_p(\xi)$, and $L_p(\xi)$ (see eqns (3.1.1), (3.1.2), and

(3.1.3)) all span the same multi-dimensional polynomial space. This polynomial space is mathematically defined in two dimensions as

$$\phi_{pq}(\xi_1, \xi_2) \subseteq \mathcal{P}_P(\mathcal{Q}^2) = \text{span}\{\xi_1^p \ \xi_2^q\}_{(p,q)\in\Upsilon},$$
$$\Upsilon = \{(p,q) \mid 0 \leqslant p \leqslant P_1, \ 0 \leqslant q \leqslant P_2\},$$

and in three dimensions as

$$\phi_{pqr}(\xi_1, \xi_2, \xi_3) \subseteq \mathcal{P}_P(\mathcal{Q}^3) = \text{span}\{\xi_1^p \ \xi_2^q \ \xi_3^r\}_{(p,q,r)\in\Upsilon},$$
$$\Upsilon = \{(p,q,r) \mid 0 \leqslant p \leqslant P_1, \ 0 \leqslant q \leqslant P_2, \ 0 \leqslant r \leqslant P_3\}.$$

In two dimensions, it is normal to consider the polynomial spaces in terms of Pascal's triangle. Figure 3.4 shows, for the expansion $\phi_{pq}(\xi_1, \xi_2)$, the space spanned when $P_1 = 4$ and $P_2 = 3$ as well as the hierarchical modes used in this expansion. The definition of the set Υ indicates the range over which the expansion modes $\phi_{pq}(\xi_1, \xi_2)$ and $\phi_{pqr}(\xi_1, \xi_2, \xi_3)$ must be assembled if they are to span the complete polynomial basis (that is, $0 \leqslant p \leqslant P_1$, $0 \leqslant q \leqslant P_2$, $0 \leqslant r \leqslant P_3$). Therefore, for an expansion to span the complete space up to $\xi_1^{P_1}\xi_2^{P_2}\xi_3^{P_3}$ requires $(P_1+1)(P_2+1)$ modes in two dimensions, and $(P_1+1)(P_2+1)(P_3+1)$ modes in three dimensions.

A broad range of modal expansions is possible due to the variety of combinations of edge, face, and interior modes. In the most general case we can describe a modal expansion which has a boundary/interior decomposition with an independent parameter to bound the modes along every edge, two independent parameters to bound the modes within every face, and another three parameters to bound the interior modes. This type of flexibility is desirable as it provides a relatively natural way to vary the expansion order from one elemental domain

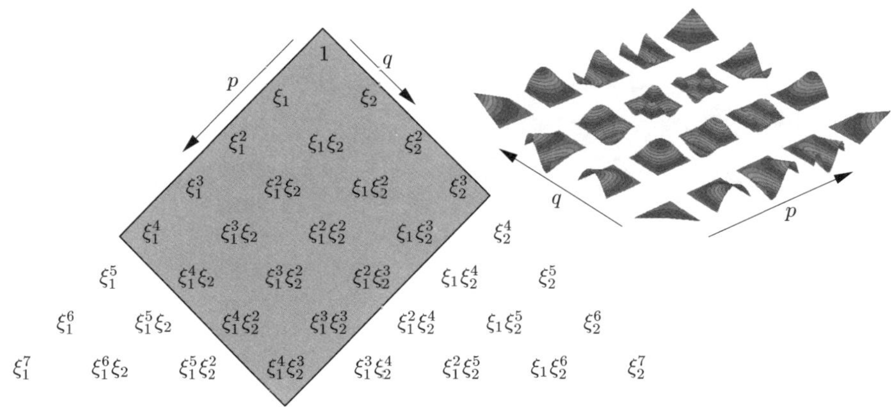

FIG. 3.4. Polynomial space in terms of the Pascal triangle of the full tensor product quadrilateral expansion with $P_1 = 4$ and $P_2 = 3$.

to another while maintaining C^0 continuity. The adaptivity of the polynomial space may also be introduced using the non-conforming techniques as discussed in Chapter 7.

3.1.2.1 *Serendipity expansion*

The hierarchical nature of the modal expansion gives rise to a greater flexibility in that it permits the use of a reduced number of expansion modes as compared with those in the full tensor space. One widely used modified expansion is the serendipity expansion which does not include the full tensor product of interior modes. In this expansion we only use the modes necessary to produce a horizontal level of the Pascal triangle, that is,

$$\mathcal{P}_P(\mathcal{Q}^2) = \text{span}\{\xi_1^p \; \xi_2^q\}_{(p,q)\in\Upsilon} \,,$$
$$\Upsilon = \{(p,q) \mid 0 \leqslant p,q \leqslant P, \; p+q \leqslant P\} \,.$$

This linear space is the natural space for a p-type expansion in a triangular region. The quadrilateral expansion cannot be reduced exactly to this space, although it can come very close. To achieve this we retain all the boundary modes and combine them with the interior modes up to the restriction $p + q \leqslant P - 2$. The Pascal triangle and modal shapes for this expansion when $P_1 = P_2 = P = 4$ are shown in Fig. 3.5. We observe that the reduced quadrilateral expansion is therefore almost identical to the triangular polynomial space, except for the two polynomials $\xi_1^4\xi_2$ and $\xi_1\xi_2^4$ which are introduced by the edge modes and cannot be removed because they are required for completeness of the expansion.

It is also possible to construct a serendipity nodal expansion which spans a similar space, but this requires the modification of the basis definition to permit different-order Lagrange polynomials in the interior of the expansion (see [75,247,

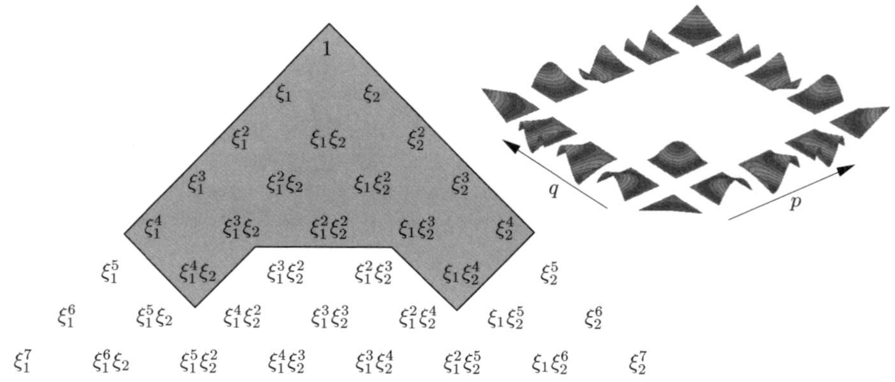

FIG. 3.5. Pascal triangle and modal shapes for a $P = 4$ serendipity expansion using all the boundary modes, with interior modes supported up to the limit $p + q \leqslant P - 2$ (recall that the interior mode is a polynomial of the form $\xi_1^2\xi_2^2$).

511] for further discussion). The construction of the Lagrange polynomial is such that a square number of modes are always needed to make a complete expansion, and so the polynomial space is slightly richer than the space illustrated in Fig. 3.5. However, a compatible nodal basis which does match this serendipity space may also be generated using a combination of the nodal and modal expansions [487]. This can destroy some of the inherent efficiency of the full nodal tensor product but permits greater expansion flexibility between domains.

3.2 Generalised tensor product modal expansions

To extend the tensor product expansion in quadrilateral and hexahedral domains discussed in Section 3.1 to simplex regions such as triangular and tetrahedral regions, we need to generalise the tensor product expansion concept. In this section we shall introduce modal expansions in subdomains typically associated with unstructured discretisation, which in two dimensions typically means the triangular region, and in three dimensions includes the tetrahedral region. A natural extension to the construction of a tetrahedral expansion will also lead to a unified basis which includes pentahedral regions such as prismatic and pyramidic shapes as well as the hexahedral modal expansion discussed previously. We shall refer to the bundle of mixed shapes as *hybrid* regions.

The use of triangular or tetrahedral spectral/*hp* element methods in computational fluid dynamics has been relatively limited when compared to quadrilateral and hexahedral spectral/*hp* element discretisations. An important consideration when using triangular or tetrahedral expansions for time-dependent computations, which typically arise in fluid dynamics, is the numerical efficiency of the algorithm in the context of *cost per time-step.* To be competitive, a triangular expansion must be as numerically efficient as the quadrilateral expansion. Since a great deal of the efficiency of the quadrilateral or hexahedral expansion (particularly at larger polynomial orders) arises from the tensor product construction, we would like to use a similar procedure to construct expansions within the triangular or tetrahedral domains.

An orthogonal, generalised tensor product, two-dimensional basis has been proposed by several authors, the first of which we believe to be Proriol in 1957 [386]. This basis has also been independently proposed by Karlin and McGregor [260] and Koornwinder [276] (who also constructed orthogonal polynomials for different kinds of domains) as well as more recently by Dubiner [143]. These expansions are also known to be solutions to a singular Sturm–Liouville problem [74, 285, 358, 483, 496]. Dubiner's paper also suggested a modified basis for C^0 continuous expansions and discussed the three-dimensional extension of the orthogonal expansion to a tetrahedral region. The derivation of a C^0 continuous expansion in a tetrahedral region based on Dubiner's orthogonal expansion was first presented by Sherwin [423] and Sherwin and Karniadakis [431]. Both the orthogonal and C^0 expansions were presented in a unified approach for hybrid elemental regions by Sherwin [424]. We shall be adopting the unified approach in this Section and we will discuss non-tensorial expansions for simplex regions in Section 2.3.4.

An interesting characteristic of these expansions is that the individual expansion modes are not rotationally symmetric in the standard regions. Rotational symmetry has historically been an important consideration when constructing unstructured polynomial expansion bases. The desire for rotational symmetry naturally motivates the use of rotationally-invariant barycentric coordinate systems (see Section 3.2.1.3). However, the use of the barycentric coordinate system can destroy much of the numerical efficiency associated with the standard tensor product expansion bases. One way to recover this efficiency is to design a coordinate system based on the mapping of a square to a triangle generating a *collapsed* coordinate system. The use of a collapsed coordinate system, as discussed in Section 3.2.1, regains some of this efficiency but inherently destroys the rotational symmetry of each mode of the expansion. Nevertheless, these expansions span an identical polynomial space as the traditional unstructured expansions using barycentric coordinates. Therefore, in the absence of any integration error, they are equivalent to any other polynomial expansion bases used in a Galerkin approximation. The lack of rotational symmetry does not affect the multi-domain construction of the triangular expansion, although for tetrahedral domains it does impose a restriction on orientation of the elemental regions which can be trivially satisfied, as will be explained in Section 4.2.1. Amongst other applications, these expansion bases have been applied to the incompressible and compressible Navier–Stokes equations (for example, see [309, 412, 423, 432, 433, 483]) as well as geophysical fluid dynamics problems [495].

Other modal expansions which are available in the literature are included in the book by Szabó and Babuška [445], which documents a modal triangular and tetrahedral expansion based on a barycentric coordinate system (see Section 3.2.1.3). These expansions are rotationally non-symmetric and have been applied to structural mechanics problems. Webb and Abouchacra [489] have also developed a hierarchical triangular expansion based on Jacobi polynomials, which is also rotationally non-symmetric and uses a barycentric coordinate system. Finally, Zumbusch [512] has proposed a symmetric hierarchical expansion for triangular and tetrahedral regions based on the barycentric coordinate system, but these bases depend on defining a new polynomial space; this basis has also been applied in the area of structural mechanics.

To understand the derivation of the generalised tensor product modal expansion we initially require the definition of a new *collapsed* coordinate system, as introduced in Section 3.2.1. Using the collapsed coordinate system we can then construct orthogonal polynomial expansions within both simplex regions and the standard quadrilateral and hexahedral regions, as discussed in Section 3.2.2. Finally, since the orthogonal expansions cannot easily be tessellated into C^0 expansions, we discuss in Section 3.2.3 a set of modified expansions which have an interior and boundary decomposition making them suitable for use in a global C^0 continuous expansion.

3.2.1 *Coordinate systems*

In the structured expansions discussed in Section 3.1 we generated a multi-dimensional expansion by forming a tensor product of one-dimensional expansions based on a Cartesian coordinate system. The one-dimensional expansion was defined between constant limits, and therefore an implicit assumption of the tensor extension was that the coordinates in the two-dimensional region were bounded between constant limits. As illustrated in Fig. 3.6, within the standard quadrilateral region, the Cartesian coordinates (ξ_1, ξ_2) are bounded by constant limits, that is,

$$\mathcal{Q}^2 = \{(\xi_1, \xi_2) \mid -1 \leqslant \xi_1, \xi_2 \leqslant 1\}.$$

However, as shown in Fig. 3.6(b), this is not the case in the standard triangular region as the bounds of the Cartesian coordinates (ξ_1, ξ_2) are dependent upon each other, that is,

$$\mathcal{T}^2 = \{(\xi_1, \xi_2) \mid -1 \leqslant \xi_1, \xi_2, \; \xi_1 + \xi_2 \leqslant 0\}.$$

Therefore, to develop a suitable tensorial-type basis within unstructured regions, such as the triangle, we need to develop a coordinate system where the local coordinates have independent bounds. The advantage of such a system is that we can then define one-dimensional functions upon which we can construct our multi-domain tensorial basis. It also defines an appropriate system upon which we can perform important numerical operations such as integration and differentiation, as discussed in Sections 2.4.1 and 2.4.2.

3.2.1.1 *Collapsed two-dimensional coordinate system*

A suitable coordinate system, which describes the triangular region between constant independent limits, is defined by the transformation

$$\begin{aligned} \eta_1 &= 2\frac{1+\xi_1}{1-\xi_2} - 1, \\ \eta_2 &= \xi_2, \end{aligned} \tag{3.2.1}$$

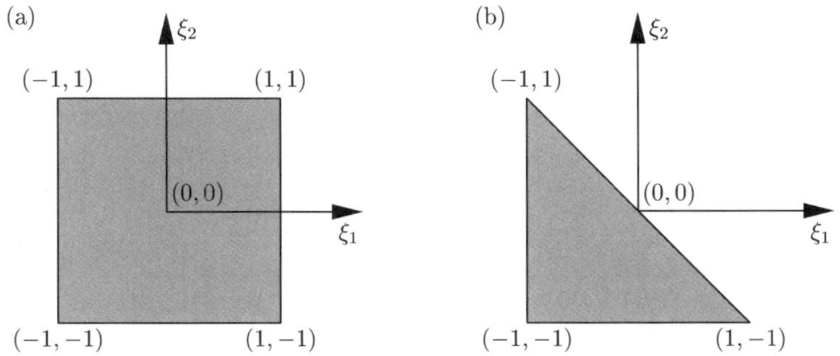

FIG. 3.6. Standard regions for the (a) quadrilateral, and (b) triangular expansion in terms of the Cartesian coordinates (ξ_1, ξ_2).

and has the inverse transformation

$$\begin{aligned} \xi_1 &= \frac{(1+\eta_1)(1-\eta_2)}{2} - 1\,, \\ \xi_2 &= \eta_2\,. \end{aligned} \tag{3.2.2}$$

These new local coordinates (η_1, η_2) define the standard triangular region by

$$\mathcal{T}^2 = \{(\eta_1, \eta_2) \mid -1 \leqslant \eta_1, \eta_2 \leqslant 1\}\,.$$

The definition of the triangular region in terms of the coordinate system (η_1, η_2) is identical to the definition of the standard quadrilateral region in terms of the Cartesian coordinates (ξ_1, ξ_2). This suggests that we can interpret the transformation (3.2.1) as a mapping from the triangular region to a rectangular one, as illustrated in Fig. 3.7. For this reason, we shall refer to the coordinate system (η_1, η_2) as the *collapsed coordinate system*. The transformation (3.2.1) maps the vertical lines in the rectangular domain (lines of constant η_1) onto lines radiating out of the top vertex ($\xi_1 = -1$, $\xi_2 = 1$) in the triangular domain. The triangular region is now described by a 'ray' coordinate, η_1, and the standard horizontal coordinate by $\xi_2 = \eta_2$. Another consequence of the transformation is that the 'ray' coordinate, η_1, is multi-valued at $\xi_1 = -1$, $\xi_2 = 1$. However, we can show that η_1 is bounded at this point by making a change of variables to (ϵ, θ), where $\xi_1 = -1 + \epsilon \sin\theta$ and $\xi_2 = 1 - \epsilon\cos\theta$. This change of variables simply expresses the Cartesian coordinates ξ_1 and ξ_2 in terms of a cylindrical system (ϵ, θ) centred on the singular point ($\xi_1 = -1$, $\xi_2 = 1$), where θ is defined in an anticlockwise sense from the vertical, as indicated in Fig. 3.7. Substituting these values into the definition of η_1 given by eqn (3.2.1), we can determine the limiting behaviour of the singularity as $\epsilon \to 0$, that is,

$$\eta_1\big|_{\xi_1=-1,\xi_2=1} = 2\frac{1-1+\epsilon\sin\theta}{1-1+\epsilon\cos\theta} - 1 = 2\tan\theta - 1\,.$$

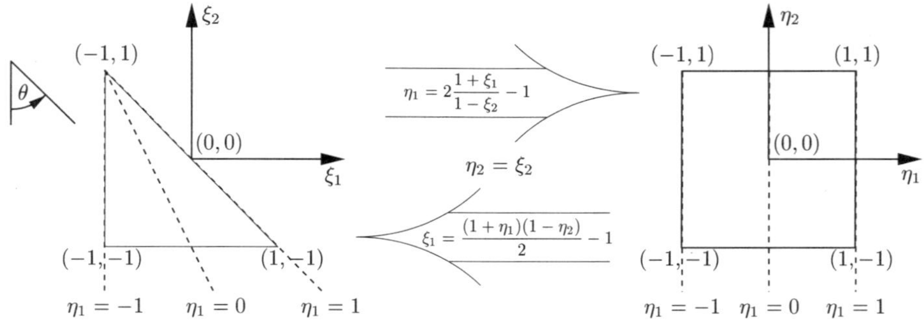

FIG. 3.7. Triangle-to-rectangle transformation.

Since $0 \leqslant \theta \leqslant \pi/4$ we know that $0 \leqslant \tan\theta \leqslant 1$, and so $-1 \leqslant \eta_1|_{\xi_1=-1,\xi_2=1} \leqslant 1$. Although the introduction of a singularity may seem unfavourable, such singularities occur naturally in cylindrical and spherical coordinate systems.

This type of coordinate system is sometimes referred to as *Duffy coordinates* [144] and is used in boundary element methods to handle the singular integrals.

3.2.1.2 *Collapsed three-dimensional coordinate systems*

The interpretation of a triangle-to-rectangle mapping of the two-dimensional local coordinate system, as illustrated in Fig. 3.7, is helpful in the construction of a new coordinate system for three-dimensional regions. If we consider the local coordinates (η_1, η_2) as independent axes (although they are not orthogonal) then the coordinate system spans a rectangular region. Therefore, if we start with a hexahedral region and apply the inverse transformation (3.2.2) then we can derive a new local coordinate system in the tetrahedral region $\mathcal{T}^3$ in three dimensions, where $\mathcal{T}^3$ is defined as

$$\mathcal{T}^3 = \{-1 \leqslant \xi_1, \xi_2, \xi_3,\ \xi_1 + \xi_2 + \xi_3 \leqslant -1\}.$$

To reduce the hexahedron to a tetrahedron requires repeated application of the transformation (3.2.2), as illustrated in Fig. 3.8. Initially, we consider a hexahedral domain defined in terms of the local coordinate system (η_1, η_2, η_3), where all three coordinates are bounded by constant limits, that is, $-1 \leqslant \eta_1, \eta_2, \eta_3 \leqslant 1$. Applying the rectangle-to-triangle transformation (3.2.2) in the (η_1, η_3) plane we obtain a new ordinate, $\overline{\eta_1}$, such that

$$\overline{\eta_1} = \frac{(1+\eta_1)(1-\eta_3)}{2} - 1,$$
$$\eta_3 = \eta_3 .$$

Treating the coordinates $(\overline{\eta_1}, \eta_2, \eta_3)$ as independent, the region which originally spanned a hexahedral domain is mapped to a triangular prism. If we now apply transformation (3.2.2) in the (η_2, η_3) plane, introducing the ordinates ξ_2 and ξ_3 defined as

$$\xi_2 = \frac{(1+\eta_2)(1-\eta_3)}{2} - 1,$$
$$\xi_3 = \eta_3 ,$$

we see that the coordinates $(-1 \leqslant \overline{\eta_1}, \xi_2, \xi_3 \leqslant 1)$ span a region of a square-based pyramid. The third and final transformation to reach the tetrahedral domain is slightly more subtle, as to reduce the pyramidic region to a tetrahedron we need to apply the mapping in every square cross-section parallel to the $(\overline{\eta_1}, \xi_2)$ plane. This means using the transformation (3.2.2) in the $(\overline{\eta_1}, \xi_2)$ plane to define the final ordinate, ξ_1, as

$$\xi_1 = \frac{(1+\overline{\eta_1})(1-\eta_2)}{2} - 1,$$
$$\xi_2 = \xi_2 .$$

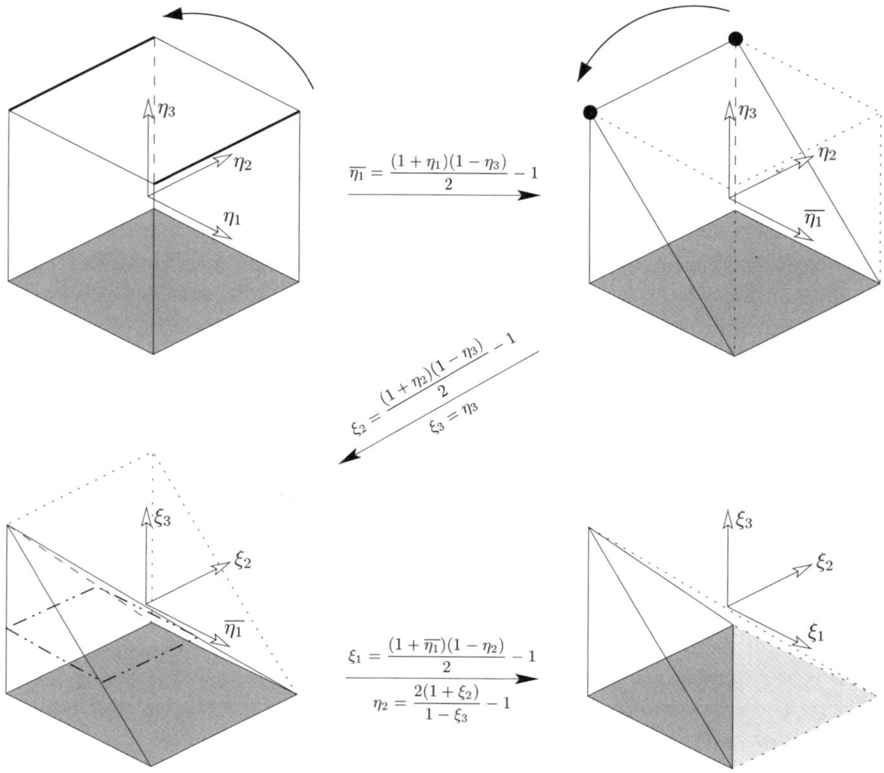

FIG. 3.8. Hexahedron-to-tetrahedron transformation by repeatedly applying the rectangle-to-triangle mapping (3.2.2).

If we choose to define the coordinate of the tetrahedron region (ξ_1, ξ_2, ξ_3) as the orthogonal Cartesian system, then by determining the hexahedral coordinates (η_1, η_2, η_3) in terms of the orthogonal Cartesian system, we obtain

$$\eta_1 = \frac{2(1+\xi_1)}{-\xi_2 - \xi_3} - 1, \quad \eta_2 = \frac{2(1+\xi_2)}{1-\xi_3} - 1, \quad \eta_3 = \xi_3, \tag{3.2.3}$$

which is a collapsed coordinate system for the tetrahedral domain, and is bounded by constant limits, so the region $\mathcal{T}^3$ can be defined as

$$\mathcal{T}^3 = \{-1 \leqslant \eta_1, \eta_2, \eta_3 \leqslant -1\}.$$

We also note that when $\xi_3 = -1$ this system reduces to the two-dimensional system defined in eqn (3.2.1).

By analogy with this technique, if we had chosen to define the coordinates in either the pyramidic or prismatic region as the orthogonal Cartesian system then evaluating the hexahedral coordinates in terms of these coordinates would

TABLE 3.2. The local collapsed coordinates which have constant bounds within the standard region may be expressed in terms of the Cartesian coordinates ξ_1, ξ_2, ξ_3. Each region may be defined in terms of the local coordinates as having a lower bound of $-1 \leqslant \xi_1, \xi_2, \xi_3$ and upper bound as indicated in the table. Each region and the planes of constant local coordinates are shown in Fig. 3.9.

Region	Upper bound	Local collapsed coordinate		
Hexahedron	$\xi_1, \xi_2, \xi_3 \leqslant 1$	ξ_1	ξ_2	ξ_3
Prism	$\xi_1 \leqslant 1, \xi_2 + \xi_3 \leqslant 0$	$\overline{\eta_1} = \dfrac{2(1+\xi_1)}{1-\xi_3} - 1$	ξ_2	ξ_3
Pyramid	$\xi_1 + \xi_3, \xi_2 + \xi_3 \leqslant 0$	$\overline{\eta_1} = \dfrac{2(1+\xi_1)}{1-\xi_3} - 1$	$\eta_2 = \dfrac{2(1+\xi_2)}{1-\xi_3} - 1$	$\eta_3 = \xi_3$
Tetrahedron	$\xi_1 + \xi_2 + \xi_3 \leqslant -1$	$\eta_1 = \dfrac{2(1+\xi_1)}{-\xi_2-\xi_3} - 1$	$\eta_2 = \dfrac{2(1+\xi_2)}{1-\xi_3} - 1$	$\eta_3 = \xi_3$

generate a collapsed coordinate system for these domains. Table 3.2 shows the local collapsed coordinate systems in all of the three-dimensional regions. A diagrammatic representation of the local collapsed coordinate system is shown in Fig. 3.9.

3.2.1.3 *Barycentric coordinate systems*

Barycentric coordinate systems, otherwise known as area/triangular or volume/tetrahedral coordinates, have historically been used in unstructured domains because of their rotational symmetry. Unlike the quadrilateral or hexahedral regions, in a simplex region such as the triangle and tetrahedron, maintaining symmetry requires an extra (dependent) coordinate. This makes the tensor process construction of expansions, as discussed in Sections 3.1 and 3.2, very difficult if not impossible. Barycentric coordinates will however be useful in defining the rotationally-symmetric non-tensorial expansions discussed in this section. We also define the relationship between the barycentric coordinates and volume coordinates and the collapsed coordinate systems discussed in Sections 3.2.1.1 and 3.2.1.2.

The area coordinate system is illustrated in Fig. 3.10(a) for the standard triangle. Any point in the triangle is described by three coordinates l_1, l_2, and l_3, which can be interpreted as the ratio of the areas A_1, A_2, and A_3 over the total area $A = A_1 + A_2 + A_3$, that is,

$$l_1 = \frac{A_1}{A}, \quad l_2 = \frac{A_2}{A}, \quad l_3 = \frac{A_3}{A}.$$

Therefore l_1, l_2, and l_3 have a unit value at the vertices marked 1, 2, and 3 in Fig. 3.10(a), respectively. By definition, these coordinates satisfy the relationship

$$l_1 + l_2 + l_3 = 1,$$

and they can be expressed in terms of Cartesian coordinates ξ_1, ξ_2 as follows:

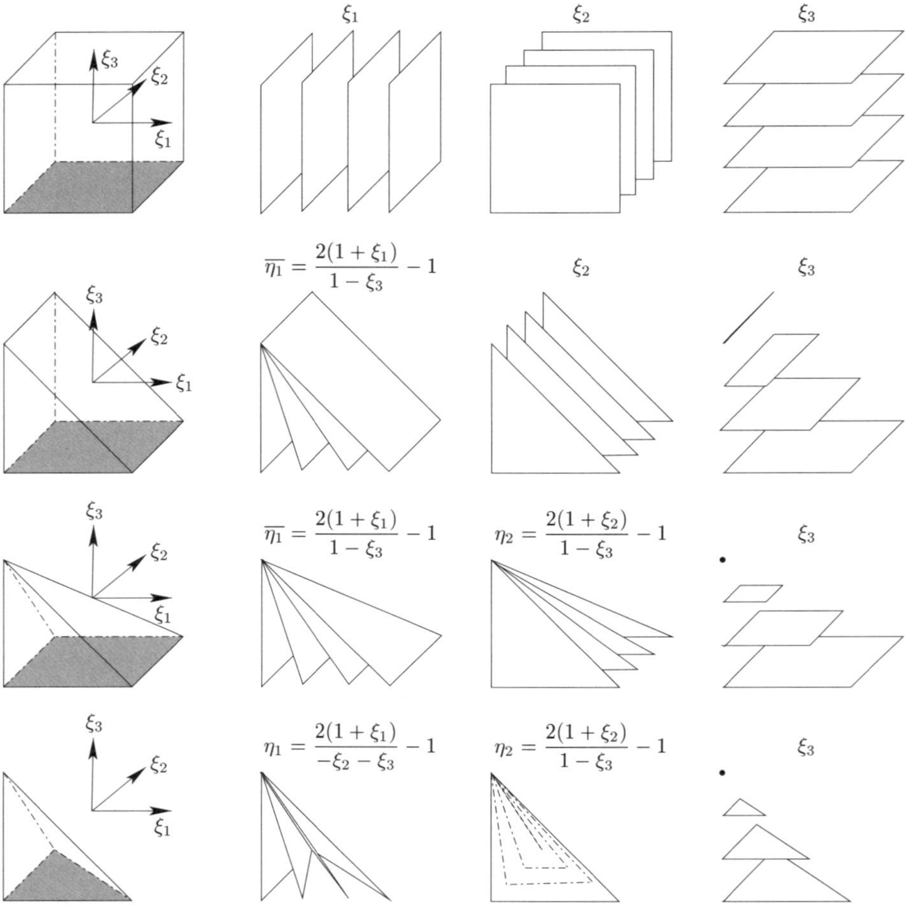

FIG. 3.9. Planes of constant value of the local collapsed Cartesian coordinate systems in the hexahedral, prismatic, pyramidic, and tetrahedral domains. In all but the hexahedral domain, the standard Cartesian coordinates ξ_1, ξ_2, ξ_3 describing the region have an upper bound which couples the coordinate system, as shown in Table 3.2. The local collapsed Cartesian coordinate system $\eta_1, \overline{\eta_1}, \eta_2, \eta_3$ represents a system of non-orthogonal coordinates which are bounded by a constant value within the region.

$$
\begin{aligned}
l_1 &= \tfrac{1}{2}(1-\xi_1) - \tfrac{1}{2}(1+\xi_2)\,,\\
l_2 &= \tfrac{1}{2}(1+\xi_1)\,,\\
l_3 &= \tfrac{1}{2}(1+\xi_2)\,.
\end{aligned}
$$

The two-dimensional collapsed coordinate system was defined in Sections 3.2.1.1 and 3.2.1.2 as

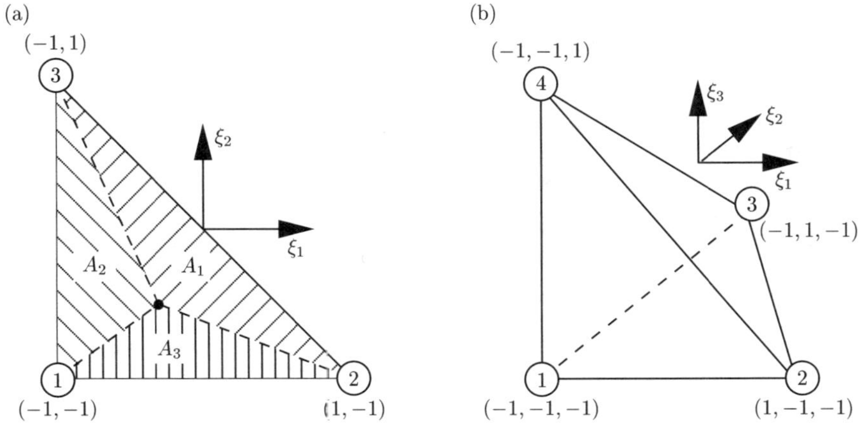

FIG. 3.10. (a) The area coordinate system in the standard triangular region. Each coordinate l_1, l_2, and l_3 can be interpreted as the ratio of areas A_1, A_2, and A_3 over the total area. (b) The standard tetrahedral region for the definition of volume coordinates.

$$\eta_1 = 2\frac{(1+\xi_1)}{1-\xi_2} - 1 \quad \text{and} \quad \eta_2 = \xi_2\,,$$

which can also be written in terms of the area coordinates as

$$\eta_1 = \frac{2l_2}{1-l_3} - 1 = \frac{l_2 - l_1}{1-l_3}, \qquad \eta_2 = 2l_3 - 1\,.$$

A similar construction follows for volume coordinates l_1, l_2, l_3, and l_4, which are defined as having a unit value at the vertices marked 1, 2, 3, and 4 in Fig. 3.10(b). In terms of the local Cartesian coordinates, the volume coordinate system is defined as

$$\begin{aligned} l_1 &= \frac{-(1+\xi_1+\xi_2+\xi_3)}{2}, & l_2 &= \frac{1+\xi_1}{2}, \\ l_3 &= \frac{1+\xi_2}{2}, & l_4 &= \frac{1+\xi_3}{2}\,. \end{aligned}$$

Finally, the three-dimensional collapsed coordinate system for the tetrahedron can be defined in terms of the volume coordinates as

$$\begin{aligned} \eta_1 &= \frac{2l_2}{1-l_3-l_4} = \frac{l_2-l_1}{1-l_3-l_4}, \\ \eta_2 &= \frac{2l_3}{1-l_4} - 1, \quad \eta_3 = 2l_4 - 1\,. \end{aligned}$$

3.2.2 *Orthogonal expansions*

We have previously seen in Section 2.3.2.1 that there are many considerations which motivate a *good* expansion basis. Typically, we are interested in developing a computationally efficient expansion which demonstrates attractive numerical properties such as matrix conditioning or, in the case of convection problems, appropriate explicit time-step restrictions (see Chapter 6). A reasonable starting-point in the development of a modal multi-dimensional expansion is to construct a set of polynomial expansions which are orthogonal in the Legendre inner product (or, indeed, any desired inner product) over each desired subdomain shape.

In the following section we will discuss a set of orthogonal polynomials in hybrid regions that have a tensor product form [143,260,276,386]. These orthogonal expansions have also been shown to be solutions to singular Sturm–Liouville problems. The first derivation of the Sturm–Liouville problem in a triangle was by Krall and Sheffer in 1967 [285]. Subsequently, this result has also been reported by Owens [358], Wingate and Taylor [496], Warburton [483], and Braess and Schwab [74]. The last three publications also deal with tetrahedral domains.

We saw in Section 3.1 how the structured expansions for a quadrilateral and hexahedral domain can be constructed using a product of two one-dimensional tensors. When we use the collapsed coordinate systems introduced in Section 3.2.1 we find that a similar extension process is possible for all the unstructured domains using a *warped* [143] or *generalised* product involving tensors of two and three dimensions. Unlike the structured two-dimensional tensor product form, where the expansion is constructed from the same one-dimensional basis, the L^2 orthogonal expansion has the form

$$\phi_{pq}(\xi_1, \xi_2) = L_p(\xi_1)L_q(\xi_2)\,,$$

where a more general product is used, combining a one-dimensional tensor $\widetilde{\psi}^a_p(z)$ with a two-dimensional tensor of the form $\widetilde{\psi}^b_{pq}(z)$, that is,

$$\phi_{pq}(\xi_1, \xi_2) = \widetilde{\psi}^a_p(\eta_1)\widetilde{\psi}^b_{pq}(\eta_2)\,.$$

Figure 3.11 illustrates the construction of the two-dimensional expansion modes using this more general form. To generate each mode the function $\widetilde{\psi}^a_p(\eta_1)$ is combined with $\widetilde{\psi}^b_{pq}(\eta_2)$. However, unlike the quadrilateral expansion, $\widetilde{\psi}^b_{pq}(\eta_2)$ now has a different form for every value of p of the principal function $\widetilde{\psi}^a_p(\eta_1)$.

This form still maintains the numerical efficiencies which can be achieved from the one-dimensional nature of the expansion using the sum-factorisation process discussed in Section 4.1.6. We shall refer to the functions $\widetilde{\psi}^a_p(z)$ and $\widetilde{\psi}^b_{pq}(z)$, as well as a third function $\widetilde{\psi}^c_{pqr}(z)$, as the *orthogonal principal functions* [424], where $\widetilde{\psi}^c_{pqr}(z)$ is required for the three-dimensional expansions. In Section 3.2.3 we shall introduce a modified version of the principal functions which are more suitable for multiple-domain expansions.

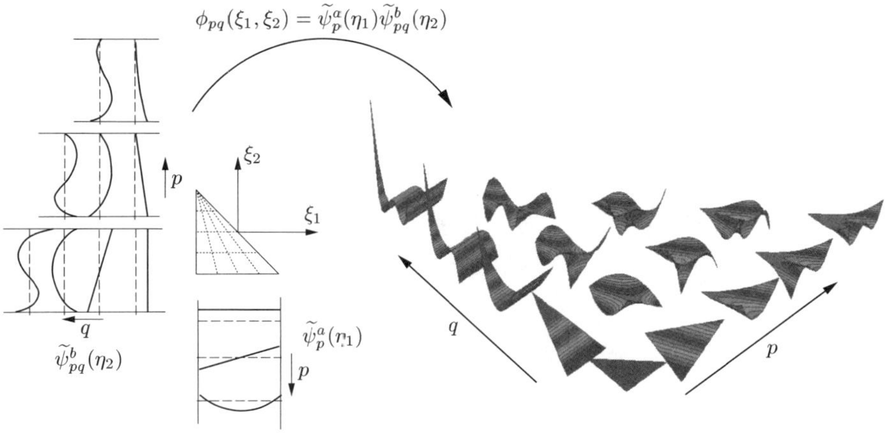

FIG. 3.11. Construction of two-dimensional expansion modes $\phi_{pq}(\xi_1, \xi_2)$ within a triangular region using the product of a one-dimensional tensor $\widetilde{\psi}^a_p(\eta_1(\xi_1, \xi_2))$ and a two-dimensional tensor $\widetilde{\psi}^b_{pq}(\eta_2(\xi_2))$.

3.2.2.1 *Orthogonal expansions in hybrid domains based on one-dimensional principal functions*

Recalling that the function $P_p^{\alpha,\beta}(z)$ denotes the pth-order Jacobi polynomial introduced in Section 2.3.3.1 (see also Appendix A), the principal functions, $\widetilde{\psi}^a_p(z)$, $\widetilde{\psi}^b_{pq}(z)$, and $\widetilde{\psi}^c_{pqr}(z)$, for orthogonal expansions in hybrid domains are

$$\widetilde{\psi}^a_p(z) = P_p^{0,0}(z), \quad \widetilde{\psi}^b_{pq}(z) = \left(\frac{1-z}{2}\right)^p P_q^{2p+1,0}(z),$$

$$\widetilde{\psi}^c_{pqr}(z) = \left(\frac{1-z}{2}\right)^{p+q} P_r^{2p+2q+2,0}(z).$$

The two-dimensional expansions in terms of the principal functions are defined as follows: ♣[10]

$$\text{quadrilateral expansion:} \quad \phi_{pq}(\xi_1, \xi_2) = \widetilde{\psi}^a_p(\xi_1)\widetilde{\psi}^a_q(\xi_2),$$

$$\text{triangular expansion:} \quad \phi_{pq}(\xi_1, \xi_2) = \widetilde{\psi}^a_p(\eta_1)\widetilde{\psi}^b_{pq}(\eta_2),$$

where

$$\eta_1 = \frac{2(1+\xi_1)}{1-\xi_2} - 1, \quad \eta_2 = \xi_2$$

are the two-dimensional collapsed coordinates illustrated in Fig. 3.7. The shape of all the triangular modes for a fourth-order polynomial expansion are shown in Fig. 3.11.

10 Definition of orthogonal modal expansions in two-dimensional standard hybrid regions.

The three-dimensional expansions are defined in terms of the principal func-
♣[11] tions as follows:

$$\begin{aligned}
&\text{hexahedral expansion:} && \phi_{pqr}(\xi_1,\xi_2,\xi_3) = \widetilde{\psi}^a_p(\xi_1)\widetilde{\psi}^a_q(\xi_2)\widetilde{\psi}^a_r(\xi_3)\,,\\
&\text{prismatic expansion:} && \phi_{pqr}(\xi_1,\xi_2,\xi_3) = \widetilde{\psi}^a_p(\overline{\eta_1})\widetilde{\psi}^a_q(\xi_2)\widetilde{\psi}^b_{pr}(\xi_3)\,,\\
&\text{pyramidic expansion:} && \phi_{pqr}(\xi_1,\xi_2,\xi_3) = \widetilde{\psi}^a_p(\overline{\eta_1})\widetilde{\psi}^a_q(\eta_2)\widetilde{\psi}^c_{pqr}(\eta_3)\,,\\
&\text{tetrahedral expansion:} && \phi_{pqr}(\xi_1,\xi_2,\xi_3) = \widetilde{\psi}^a_p(\eta_1)\widetilde{\psi}^b_{pq}(\eta_2)\widetilde{\psi}^c_{pqr}(\eta_3)\,,
\end{aligned}$$

where

$$\eta_1 = \frac{2(1+\xi_1)}{-\xi_2-\xi_3} - 1\,, \quad \overline{\eta_1} = \frac{2(1+\xi_1)}{1-\xi_3} - 1\,, \quad \eta_2 = \frac{2(1+\xi_2)}{1-\xi_3} - 1\,, \quad \eta_3 = \xi_3$$

are the three-dimensional collapsed coordinates illustrated in Fig. 3.9.

These expansions are all polynomials in terms of both their local collapsed coordinates and the Cartesian coordinates. The structured expansions in the quadrilateral and hexahedral domains are simply standard tensor products of Legendre polynomials in terms of Cartesian coordinates since $P_p^{0,0}(z) = L_p(z)$. The development of unstructured expansions using the local collapsed coordinate systems is linked to the use of the more general functions $\widetilde{\psi}^b_{pq}(z)$ and $\widetilde{\psi}^c_{pqr}(z)$. These functions both contain factors of the form $[(1-z)/2]^n$ which are necessary to keep the expansions as polynomials in terms of the Cartesian coordinates (ξ_1,ξ_2,ξ_3). For example, the coordinate η_1 in the triangular expansion necessitates the use of the function $\widetilde{\psi}^b_{pq}(\eta_2)$ (where $\eta_2 = \xi_2$), which introduces a factor of $[(1-\xi_2)/2]^p$. The product of this factor with $\widetilde{\psi}^a_p(\eta_1)$ is a polynomial function in ξ_1 and ξ_2. A similar argument supports the introduction of the local coordinate η_2 in the prismatic expansions. The local coordinate system in the pyramidic domains introduces a second collapsed coordinate $\overline{\eta_1}$ which requires the introduction of the principal function $\widetilde{\psi}^c_{pqr}(\eta_3)$. The expansion in the tetrahedral regions uses the additional collapsed coordinate $\eta_1 = 2(1+\xi_1)/(-\xi_2-\xi_3)-1$ (recall that this is the same as the two-dimensional definition when $\xi_3 = -1$). Noting that $-\xi_2-\xi_3$ can be expressed in terms of η_2 and η_3 as

$$-\xi_2 - \xi_3 = \frac{1}{2}(1-\eta_2)(1-\eta_3)\,,$$

we see that the polynomial $\widetilde{\psi}^a_p(\eta_1)$ becomes a polynomial in ξ_1, ξ_2, and ξ_3 if we multiply it by the factor $(1-\eta_2)^p(1-\eta_3)^p$. This factor is incorporated in the principal functions $\widetilde{\psi}^b_{pq}(\eta_2)$ and $\widetilde{\psi}^c_{pqr}(\eta_3)$.

The principal functions $\widetilde{\psi}^b_{pq}(\eta_2)$ and $\widetilde{\psi}^c_{pqr}(\eta_3)$ also contain a Jacobi polynomial of the form $P_i^{\alpha,0}(z)$. As can be seen from Fig. 3.12, for increasing values of α this polynomial has zeros which are declustered away from the point $z = +1$. As noted in [487], this declustering is important in maintaining the linear independence of the expansion, leading to well-conditioned numerical systems.

[11] Definition of orthogonal modal expansions in three-dimensional standard hybrid regions.

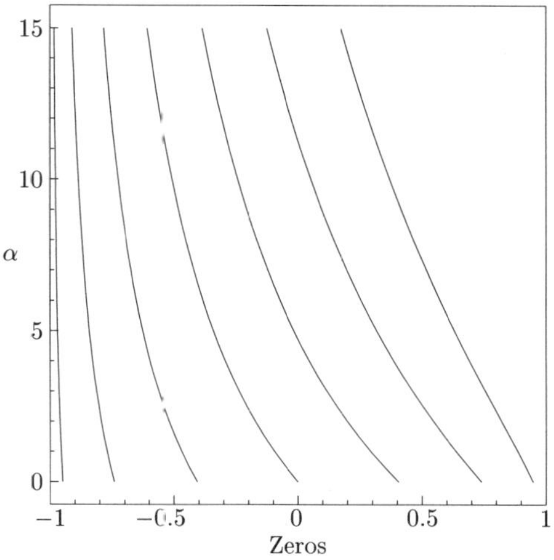

FIG. 3.12. Location of the zeros of the Jacobi polynomial $P_i^{\alpha,0}(z)$ $(i \leqslant 7)$ in the interval $[-1,1]$ for integer values of α, $0 \leqslant \alpha \leqslant 15$. For increasing α we see there is a declustering of the zeros away from the point $z = 1$.

3.2.2.2 *Demonstration of orthogonality*

To gain an insight into the use of the Jacobi polynomial $P^{\alpha,0}(z)$ in defining the principal functions $\widetilde{\psi}^b_{pq}(z)$ and $\widetilde{\psi}^c_{pqr}(z)$, we shall demonstrate the orthogonality of the triangular and tetrahedral expansions in the Legendre inner product, that is,

$$\int_{\Omega_{\mathrm{st}}} \phi_{pqr}(\xi_1,\xi_2,\xi_3)\phi_{ijk}(\xi_1,\xi_2,\xi_3)\,\mathrm{d}\xi_1\,\mathrm{d}\xi_2\,\mathrm{d}\xi_3\,, \tag{3.2.4}$$

where Ω_{st} denotes the appropriate standard region, i.e., triangle ($\mathcal{T}^2$) or tetrahedron ($\mathcal{T}^3$).

Orthogonality of the triangular expansion

Considering the triangular expansion in the standard region

$$\phi_{pq}(\xi_1,\xi_2) = \widetilde{\psi}^a_p(\eta_1)\widetilde{\psi}^b_{pq}(\eta_2)\,,$$

eqn (3.2.4) can be written in terms of the local coordinate system (η_1,η_2) as follows:

$$\int_{-1}^{1}\int_{-1}^{-\xi_2} \phi_{pq}\phi_{ij}\,\mathrm{d}\xi_1\,\mathrm{d}\xi_2 = \int_{-1}^{1}\int_{-1}^{1} \widetilde{\psi}^a_p(\eta_1)\widetilde{\psi}^b_{pq}(\eta_2)\widetilde{\psi}^a_i(\eta_1)\widetilde{\psi}^b_{ij}(\eta_2)\,J\,\mathrm{d}\eta_1\,\mathrm{d}\eta_2\,,$$

where

$$J = \frac{\partial(\xi_1,\xi_2)}{\partial(\eta_1,\eta_2)} = \frac{1-\eta_2}{2}\,.$$

Since the expansion is a product of polynomials in terms of the local coordinates η_1 and η_2, and because the Jacobian is only a function of η_2, the inner product can be expressed in terms of two one-dimensional integrals:

$$\int_{-1}^{1} \widetilde{\psi}_p^a \widetilde{\psi}_i^a \,\mathrm{d}\eta_1 \int_{-1}^{1} \widetilde{\psi}_{pq}^b \widetilde{\psi}_{ij}^b \frac{1-\eta_2}{2} \,\mathrm{d}\eta_2 \,. \tag{3.2.5}$$

Recalling the definitions of the principal functions,

$$\widetilde{\psi}_p^a(\eta_1) = P_p^{0,0}(\eta_1)\,, \quad \widetilde{\psi}_{pq}^b(\eta_2) = \left(\frac{1-\eta_2}{2}\right)^p P_q^{2p+1,0}(\eta_2)\,,$$

we see that the first integral in eqn (3.2.5) contains the standard Legendre polynomials ($P_p^{0,0}(z) = L_p(z)$) which are orthogonal in the interval $[-1,1]$. Accordingly, this integral is zero unless $p = i$ when it is equal to $2/(2p+1)$. The second integral can be written in full as

$$\int_{-1}^{1} \left(\frac{1-\eta_2}{2}\right)^p P_q^{2p+1,0}(\eta_2) \cdot \left(\frac{1-\eta_2}{2}\right)^i P_j^{2i+1,0}(\eta_2) \frac{1-\eta_2}{2} \,\mathrm{d}\eta_2 \,.$$

The Jacobi polynomial $P_r^{2p+1,0}(\eta_2)$ is orthogonal in this interval with respect to the weight function $((1-\eta_2)/2)^{2p+1}$ (see eqn (A.1.7) in Appendix A), and therefore if $p = i$ the integral is zero except when $q = j$. However, if $p \neq i$ the first integral in eqn (3.2.5) is necessarily zero, and so the expansion $\phi_{pq}(\xi_1, \xi_2)$ is orthogonal to $\phi_{ij}(\xi_1, \xi_2)$ in the standard region $\mathcal{T}^2$.

Orthogonality of the tetrahedral expansion

Orthogonality of the tetrahedral expansions follows from a construction similar to that previously shown for the triangular expansions. Expressing the basis ϕ_{pqr} in terms of its product form of the principal functions, the Legendre inner product (3.2.4) can be written as

$$\int_{-1}^{1} \int_{-1}^{-\xi_3} \int_{-1}^{-1-\xi_2-\xi_3} \phi_{pqr} \phi_{ijk} \,\mathrm{d}\xi_1 \,\mathrm{d}\xi_2 \,\mathrm{d}\xi_3$$
$$= \int_{-1}^{1} \int_{-1}^{1} \int_{-1}^{1} \widetilde{\psi}_p^a \widetilde{\psi}_{pq}^b \widetilde{\psi}_{pqr}^c \, \widetilde{\psi}_i^a \widetilde{\psi}_{ij}^b \widetilde{\psi}_{ijk}^c J \,\mathrm{d}\eta_1 \,\mathrm{d}\eta_2 \,\mathrm{d}\eta_3 \,,$$

where η_1, η_2, and η_3 are the three-dimensional collapsed coordinates and

$$J = \frac{\partial(\xi_1, \xi_2, \xi_3)}{\partial(\eta_1, \eta_2, \eta_3)} = \frac{1-\eta_2}{2} \left(\frac{1-\eta_3}{2}\right)^2 .$$

Since the basis is a product of the three principal functions and the Jacobian J can also be expressed as a product of two functions in terms of the local coordinates, the integral can be written as a product of the three one-dimensional integrals of the form

$$\int_{-1}^{1} \widetilde{\psi}_p^a \widetilde{\psi}_i^a \, d\xi_1 \mapsto \int_{-1}^{1} P_p^{0,0}(\eta_1) P_i^{0,0}(\eta_1) \, d\eta_1 \,, \tag{3.2.6a}$$

$$\int_{-1}^{1} \widetilde{\psi}_{pq}^b \widetilde{\psi}_{ij}^b \frac{1-\eta_2}{2} \, d\xi_2 \mapsto \int_{-1}^{1} P_p^{2p+1,0}(\eta_2) P_i^{2i+1,0}(\eta_2) \left(\frac{1-\eta_2}{2}\right)^{p+i+1} d\eta_2 \,, \tag{3.2.6b}$$

$$\int_{-1}^{1} \widetilde{\psi}_{pqr}^c \widetilde{\psi}_{ijk}^c \left(\frac{1-\eta_3}{2}\right)^2 d\xi_3 \mapsto$$
$$\int_{-1}^{1} P_r^{2p+2q+2,0}(\eta_3) P_k^{2i+2j+2,0}(\eta_3) \left(\frac{1-\eta_3}{2}\right)^{(p+q)+(i+j)+2} d\eta_3 \,. \tag{3.2.6c}$$

As discussed in the previous section, eqn (3.2.6a) is the inner product of the Legendre polynomial which is zero if $p \neq i$. The integral in eqn (3.2.6b) is zero if $q \neq j$ when $p = i$, since when $p = i$ this integral becomes the orthogonality relation for the Jacobi polynomials $P_p^{2p+1,0}(\eta_1)$, as given in Appendix A. Finally, integral (3.2.6c) is zero if $r \neq k$ when $p = i$ and $q = j$ because the integral becomes the orthogonality relation for the Jacobi polynomial $P_r^{2p+2q+2,0}(\eta_3)$.

We can appreciate that, unlike the structured hexahedral expansion where each one-dimensional integral gives an independent orthogonal relation, the tetrahedral expansion requires the orthogonality of the first integral to support the second, and the orthogonality of the first and second integral to support that of the third integral. This has implications for the ordering of the modes when dealing with the modified expansion in Section 3.2.3.

3.2.2.3 *Singular Sturm–Liouville equations of the orthogonal expansions*

We have previously noted that the Jacobi polynomials are the solution to a singular Sturm–Liouville problem. This is significant since, as discussed in Section 2.5.2, a numerical approximation based on expansion bases which satisfy singular Sturm–Liouville problems demonstrates spectral convergence if the approximated function is sufficiently smooth. The natural extension of this one-dimensional result to a quadrilateral region is the tensor product expansion

$$\phi_{pq}(\xi_1, \xi_2) = P_p^{0,0}(\xi_1) P_q^{0,0}(\xi_2)$$

which satisfies the singular Sturm–Liouville equation

$$\frac{\partial}{\partial \xi_1}\left[(1-\xi_1^2)\frac{\partial \phi_{pq}}{\partial \xi_1}\right] + \frac{\partial}{\partial \xi_2}\left[(1-\xi_2^2)\frac{\partial \phi_{pq}}{\partial \xi_2}\right] + \lambda_{pq}\phi_{pq} = 0 \,, \tag{3.2.7}$$

where $\lambda_{pq} = p(p+1) + q(q+1)$. This equation is invariant under all rotations and reflections in the square. A defining feature of this equation, as discussed in [496], is that, along each edge ($\xi_1 = \pm 1$, $\xi_2 = \pm 1$) where the derivative is tangential, there is a quadratic coefficient which goes to zero on all sides not tangential to the direction of differentiation. These features are important to make the differential operator self-adjoint.

The extension from a one-dimensional segment to a two-dimensional quadrilateral region is relatively intuitive. However, not quite as intuitive is the singular Sturm–Liouville for the triangular region, as investigated in [285, 358, 483, 496]. We start our discussion by noting that the orthogonal expansion in a triangle

$$\phi_{pq}(\xi_1, \xi_2) = P_p^{0,0}(\eta_1)\left(\frac{1-\eta_2}{2}\right)^p P_q^{2p+1,0}(\eta_2) \tag{3.2.8}$$

satisfies the singular Sturm–Liouville equation

$$\frac{2}{1-\eta_2}\left\{\frac{\partial}{\partial\eta_1}\left[(1-\eta_1^2)\frac{\partial\phi_{pq}}{\partial\eta_1}\right] + \frac{\partial}{\partial\eta_2}\left[(1-\eta_2^2)\frac{1-\eta_2}{2}\frac{\partial\phi_{pq}}{\partial\eta_2}\right]\right\} + \lambda_{pq}\phi_{pq} = 0\,, \tag{3.2.9}$$

where

$$\lambda_{pq} = (p+q)(p+q+2) \quad \text{and} \quad \eta_1 = \frac{2(1+\xi_1)}{1-\xi_2} - 1\,,\ \eta_2 = \xi_2\,.$$

We see from eqn (3.2.9) that the form of the triangular singular Sturm–Liouville equation, similar to eqn (3.2.7), contains quadratic or higher coefficients which are zero on the boundary of the region defined along $\eta_1 = \pm 1$, $\eta_2 = \pm 1$.

To demonstrate that the orthogonal basis (3.2.8) satisfies eqn (3.2.9) we follow the formulation used by Warburton [483] and initially consider the second differential term in eqn (3.2.9) which on substitution of expansion (3.2.8) can be written as

$$\begin{aligned}
&\frac{2}{1-\eta_2}\frac{\partial}{\partial\eta_2}\left[(1-\eta_2^2)\frac{1-\eta_2}{2}\frac{\partial\phi_{pq}}{\partial\eta_2}\right] \\
&= \frac{2P_p^{0,0}}{1-\eta_2}\frac{\partial}{\partial\eta_2}\left[-p(1+\eta_2)\left(\frac{1-\eta_2}{2}\right)^{p+1} P_q^{2p+1,0}\right. \\
&\qquad\qquad \left. + (1-\eta_2^2)\left(\frac{1-\eta_2}{2}\right)^{p+1}\frac{\partial P_q^{2p+1,0}}{\partial\eta_2}\right] \\
&= \left(\frac{1-\eta_2}{2}\right)^p P_p^{0,0}\left[(1-\eta_2^2)\frac{\partial^2 P_q^{2p+1,0}}{\partial\eta_2^2} + (-(2p+1)-(2p+3)\eta_2)\frac{\partial P_q^{2p+1,0}}{\partial\eta_2}\right. \\
&\qquad\qquad \left. - p(p+2)P_q^{2p+1,0} - \frac{2}{1-\eta_2}p(p+1)P_q^{2p+1,0}\right].
\end{aligned} \tag{3.2.10}$$

Now if we apply relationship (A.1.2) expressed in terms of the Jacobi polynomial $P_p^{0,0}(\eta)$, then we observe that the first differentiation term in eqn (3.2.9) can be written as

$$\frac{2}{1-b}\frac{\partial}{\partial \eta_1}\left[(1-\eta_1^2)\frac{\partial \phi_{pq}}{\partial \eta_1}\right] = \left(\frac{1-\eta_2}{2}\right)^{p-1} P_q^{2p+1,0}(\eta_2)\frac{\partial}{\partial \eta_1}\left[(1-\eta_1^2)\frac{\partial P_p^{0,0}}{\partial \eta_1}\right]$$
$$= -\frac{2}{1-\eta_2}p(p+1)\phi_{pq}\,. \qquad (3.2.11)$$

We now note that the last term in eqn (3.2.10) exactly cancels eqn (3.2.11). Finally, we apply eqn (A.1.1) in terms of the Jacobi polynomial $P_q^{2p+1,0}(\eta_2)$, to re-express the two differential terms in eqn (3.2.10) as

$$(1-\eta_2^2)\frac{\partial^2 P_q^{2p+1,0}}{\partial \eta_2^2}+(-(2p+1)-(2p+3)\eta_2)\frac{\partial P_q^{2p+1,0}}{\partial \eta_2} = -p(p+2p+2)P_q^{2p+1,0}a\,.$$

This term combined with $p(p+2)P_q^{2p+1,0}$ exactly balances the non-differential term in eqn (3.2.9), thereby demonstrating that the basis (3.2.8) is a solution to the Sturm–Liouville equation (3.2.9).

The eigenvalues of both the quadrilateral and triangular singular Sturm–Liouville equations do not uniquely correspond to a single eigenfunction as was the case for the one-dimensional equation. In the multi-dimensional problem a single eigenvalue $\lambda = \mathcal{C}$ corresponds to a family of eigenfunctions. As noted by Wingate and Taylor [496], the commonly used triangular polynomial space (see Section 3.2.2.4) corresponds to a union of these eigenspaces for all eigenvalues less than a constant. This can be appreciated from the eigenvalue definition since for $\lambda_{pq} = (p+q)(p+q+2)$ to be a constant requires $p+q$ to be constant. A fixed value of $p+q = P$ corresponds to all eigenfunctions of polynomial order P. Finally, we note that the singular Sturm–Liouville equation (3.2.9) can be written in terms of Cartesian coordinates as

$$\frac{\partial}{\partial \xi_1}\left((1+\xi_1)\left[(1-\xi_1)\frac{\partial \phi}{\partial \xi_1}-(1+\xi_2)\frac{\partial \phi}{\partial \xi_2}\right]\right)$$
$$+\frac{\partial}{\partial \xi_2}\left((1+\xi_2)\left[(1-\xi_2)\frac{\partial \phi}{\partial \xi_2}-(1+\xi_1)\frac{\partial \phi}{\partial \xi_1}\right]\right)+\lambda\phi = 0\,.$$

Similar singular Sturm–Liouville equations can also be derived for the hexahedral, prismatic, and tetrahedral expansions, as discussed in [483,496]. However, the pyramidic expansion is not encompassed in the same analysis.

3.2.2.4 *Polynomial space of bases and assembly of expansions*

The polynomial spaces, in Cartesian coordinates, for the two-dimensional expansions are

$$\mathcal{P} = \mathrm{Span}\{\xi_1^p\ \xi_2^q\ \}_{(pq)\in\Upsilon}\,, \qquad (3.2.12)$$

where Υ for each domain is as follows:

quadrilateral: $\Upsilon = \{(pq) \mid 0 \leqslant p,q,\ q \leqslant P_1,\ q \leqslant P_2\}$,
triangular: $\Upsilon = \{(pq) \mid 0 \leqslant p,q,\ q \leqslant P_1,\ p+q \leqslant P_2,\ P_1 \leqslant P_2\}$.

The polynomial spaces for the case when $P_1 = P_2 = 3$ for both the quadrilateral and triangular expansions are shown in Fig. 3.13.

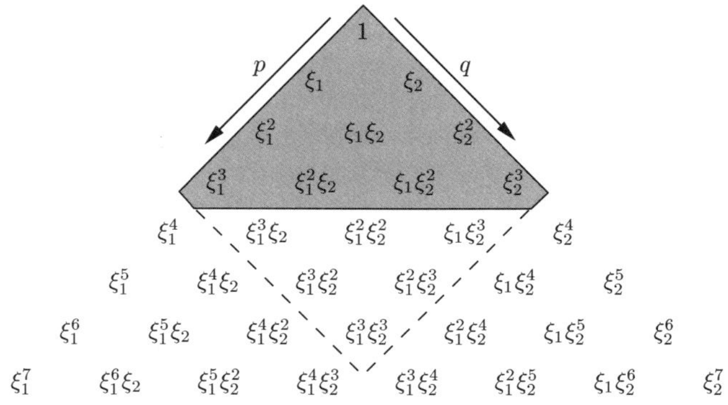

FIG. 3.13. Polynomial space in terms of Pascal's triangle for the triangular expansion (shaded region) and quadrilateral expansion (shaded region plus values within dotted line) when $P_1 = P_2 = 3$.

The polynomial spaces, in Cartesian coordinates, for the three-dimensional expansions are

$$\mathcal{P} = \text{Span}\{\xi_1^p\ \xi_2^q\ \xi_3^r\}_{(pqr)\in\Upsilon}\,, \tag{3.2.13}$$

where Υ for each domain is as follows:

hexahedron:

$$\Upsilon = \{(pqr) \mid 0 \leqslant p,q,r,\ p \leqslant P_1,\ q \leqslant P_2,\ r \leqslant P_3\}\,,$$

prism:

$$\Upsilon = \{(pqr) \mid 0 \leqslant p,q,r,\ p \leqslant P_1,\ q \leqslant P_2,\ p+r \leqslant P_3,\ P_1 \leqslant P_3\}\,,$$

pyramid:

$$\Upsilon = \{(pqr) \mid 0 \leqslant p,q,r,\ p \leqslant P_1,\ q \leqslant P_2,\ p+q+r \leqslant P_3,\ P_1,P_2 \leqslant P_3\}\,,$$

tetrahedron:

$$\Upsilon = \{(pqr) \mid 0 \leqslant p,q,r,\ p \leqslant P_1,\ p+q \leqslant P_2,\ p+q+r \leqslant P_3,\ P_1 \leqslant P_2 \leqslant P_3\}\,. \tag{3.2.14}$$

The range of p, q, and r indicate how the expansions should be assembled to generate an expansion with a complete polynomial space. As illustrated in Fig. 3.14, when $P_1 = P_2 = P_3$ the tetrahedral and pyramidic expansions span the same space and are in a subspace of the prismatic expansion, which is in turn a subspace of the hexahedral expansion.

3.2.3 *Modified C^0 expansions*

Although the orthogonality of the expansions in Section 3.2.2 is attractive, these expansions are not normally the most suitable expansions for a general spectral/*hp* element discretisation. For example, if we are using a standard Galerkin formulation then the global expansion is normally required to have

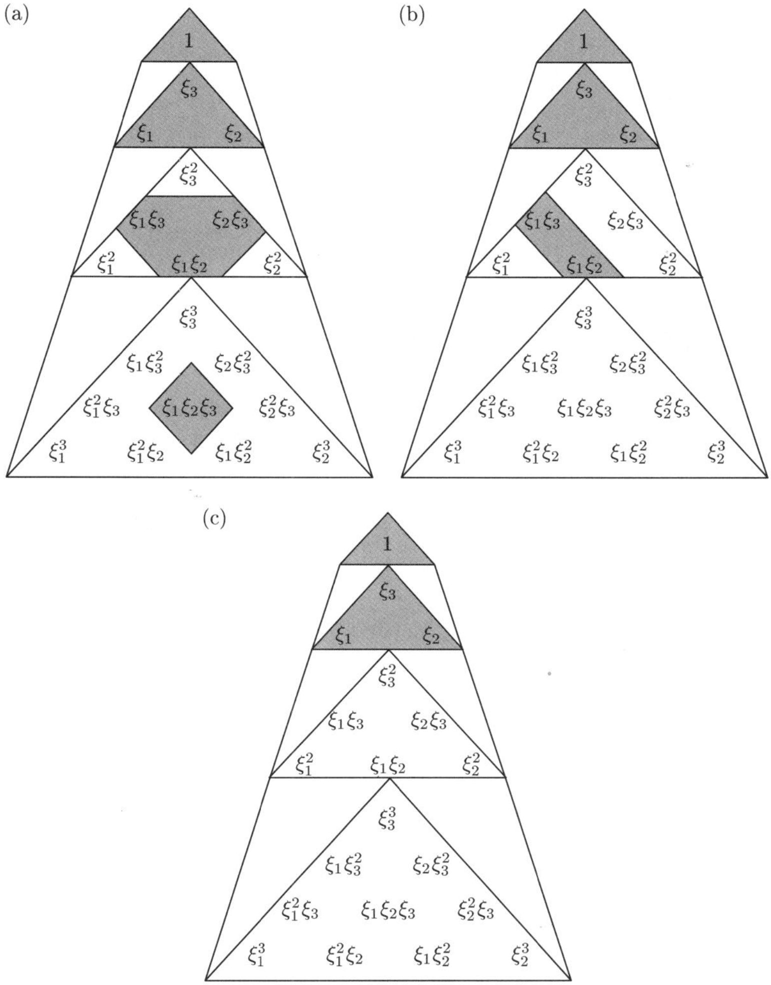

FIG. 3.14. Pascal's diagram demonstrating the polynomial space of the orthogonal expansion when $P_1 = P_2 = P_3 = 1$: (a) hexahedral expansion, (b) prismatic expansion, and (c) pyramidic and tetrahedral expansion.

C^0 continuity between elemental domains. The work by Dubiner [143] first discussed the modification of the triangular orthogonal expansions of Section 3.2.2 into a semi-orthogonal expansion suitable to generate C^0 continuous global expansions. The modified semi-orthogonal expansion was subsequently extended to three dimensions by Sherwin and Karniadakis [431] and Sherwin [424]. We note, however, that the orthogonal bases can still be useful when considering the discontinuous Galerkin formulation. Although it is theoretically possible to assemble the orthogonal expansion into multiple regions and enforce a degree of continuity between elemental regions, such an assembly will in practice destroy the orthogonality of the global expansion.

Similar to the discussion in Section 2.3.2.1, we can develop an expansion amenable to enforcing C^0 continuity globally by decomposing the orthogonal expansions into an interior and boundary contribution. We will require that the interior modes (or bubble functions) are zero on the boundary of the local

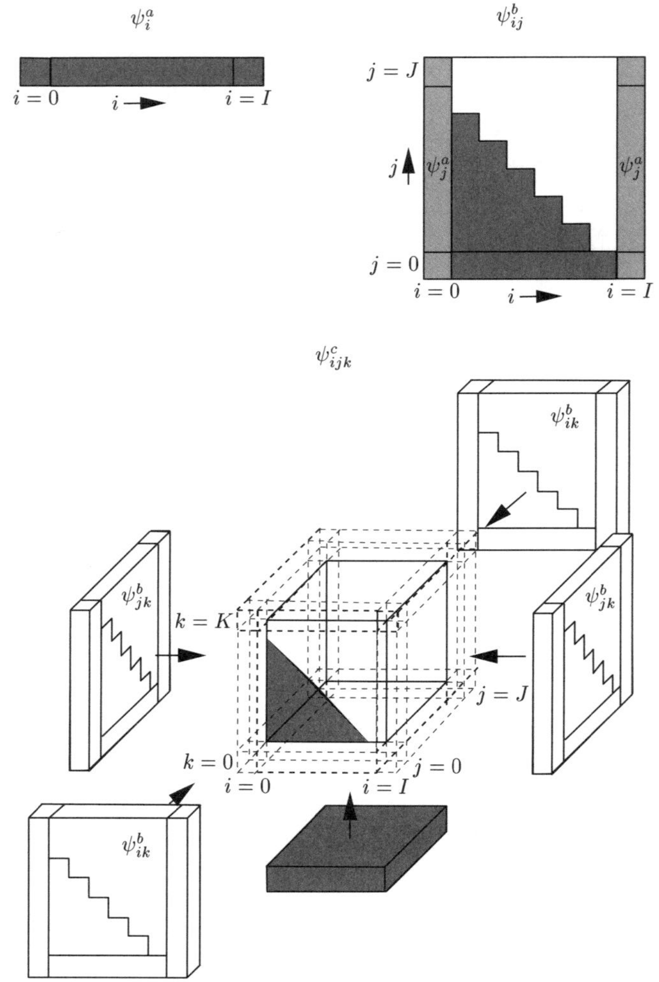

FIG. 3.15. Illustration of the structure of the arrays of the modified principal functions $\psi_i^a(z)$, $\psi_{ij}^b(z)$, and $\psi_{ijk}^c(z)$. These arrays are not globally close-packed, although any edge, face, or interior region of the array may be treated as such. The interior of the arrays $\psi_{ij}^b(z)$ and $\psi_{ijk}^c(z)$ have been shaded to indicate the minimum functions required for a complete triangular and tetrahedral expansion.

elemental domain. The completeness of the expansion is then ensured by adding boundary modes which consist of vertex, edge, and face contributions. The *vertex modes* have unit value at one vertex and decay to zero at all other vertices; *edge modes* have local support along one edge and are zero on all other edges and vertices; *face modes* have local support on one face and are zero on all other faces, edges, and vertices. Using this decomposition, C^0 continuity between elements can be enforced by matching similar-shaped boundary modes, see [433].

3.2.3.1 *Modified principal functions*

Analogous to orthogonal expansion, we define three principal functions denoted by $\psi_i^a(z)$, $\psi_{ij}^b(z)$, and $\psi_{ijk}^c(z)$ $(0 \leqslant i \leqslant I,\ 0 \leqslant j \leqslant J,\ 0 \leqslant k \leqslant K)$:

$$\psi_i^a(z) = \begin{cases} \dfrac{1-z}{2}, & i=0, \\ \dfrac{1-z}{2}\dfrac{1+z}{2}P_{i-1}^{1,1}(z), & 1 \leqslant i < I, \\ \dfrac{1+z}{2}, & i=I, \end{cases} \tag{3.2.15}$$

$$\psi_{ij}^b(z) = \begin{cases} \psi_j^a(z), & i=0,\ 0 \leqslant j \leqslant J, \\ \left(\dfrac{1-z}{2}\right)^{i+1}, & 1 \leqslant i < I,\ j=0, \\ \left(\dfrac{1-z}{2}\right)^{i+1}\dfrac{1+z}{2}P_{j-1}^{2i+1,1}(z), & 1 \leqslant i < I,\ 1 \leqslant j < J, \\ \psi_j^a(z), & i=I,\ 0 \leqslant j \leqslant J, \end{cases} \tag{3.2.16}$$

$$\psi_{ijk}^c(z) = \begin{cases} \psi_{jk}^b(z), & i=0,\ 0 \leqslant j \leqslant J,\ 0 \leqslant k \leqslant K, \\ \psi_{ik}^b(z), & 0 \leqslant i \leqslant I,\ j=0,\ 0 \leqslant k \leqslant K, \\ \left(\dfrac{1-z}{2}\right)^{i+j+1}, & 1 \leqslant i < I,\ 1 \leqslant j < J,\ k=0, \\ \left(\dfrac{1-z}{2}\right)^{i+j+1}\dfrac{1+z}{2}P_{k-1}^{2i+2j+1,1}(z), & \\ & 1 \leqslant i < I,\ 1 \leqslant j < J,\ 1 \leqslant k < K, \\ \psi_{ik}^b(z), & 0 \leqslant i \leqslant I,\ j=J,\ 0 \leqslant k \leqslant K, \\ \psi_{jk}^b(z), & i=I,\ 0 \leqslant j \leqslant J,\ 0 \leqslant k \leqslant K. \end{cases} \tag{3.2.17}$$

Figure 3.15 diagrammatically indicates the structure of the principal functions $\psi_i^a(z)$, $\psi_{ij}^b(z)$, and $\psi_{ijk}^c(z)$ as well as how the function $\psi_i^a(z)$ is incorporated into $\psi_{ij}^b(z)$, and similarly how $\psi_{ij}^b(z)$ is incorporated into $\psi_{ijk}^c(z)$. The function $\psi_i^a(z)$ has been decomposed into two linearly-varying components and a function which is zero at the end-points. This function is identical to the one-dimensional modal expansion which was used in the tensorial construction of the structured modal expansions. The linearly-varying components also generate vertex modes which are identical to the standard linear finite element expansion. The interior contributions of all the base functions (that is, $1 \leqslant i < I$, $1 \leqslant j < J$, $1 \leqslant k < K$) are similar to the orthogonal basis functions defined in Section 3.2.2. However, they are now pre-multiplied by a factor of the form $[(1-z)/2]\,[(1+z)/2]$ which ensures that these modes are zero on the boundaries of the domain. The values of α and β in the Jacobi polynomial $P_p^{\alpha,\beta}(z)$ have also been modified to maintain as much orthogonality as possible in the mass and Laplacian systems. As can be seen in Fig. 3.15, we have ordered the definitions of $\psi_i^a(z)$, $\psi_{ij}^b(z)$, and $\psi_{ijk}^c(z)$ in eqns (3.2.15)–(3.2.17) so that the modes can be interpreted according to the physical location of the modes. For example, the vertex modes correspond to the corner location of the arrays. We note, however, that an alternative technique would have been to order the arrays according to increasing polynomial order. Finally, we observe that there is a great deal of similarity between $\psi_{ij}^b(z)$ and $\psi_{ijk}^c(z)$. Not only does $\psi_{ijk}^c(z)$ contain $\psi_{ij}^b(z)$ along its boundary, but the interior contribution of $\psi_{ijk}^c(z)$ is related to the interior contribution of $\psi_{ij}^b(z)$ since

$$\psi_{ijk}^c(z) = \psi_{i+j,k}^b(z), \quad 1 \leqslant i, j, k,\ i < I,\ i+j < J,\ k < K .$$

3.2.3.2 *Definition of expansion bases*

In the same way as the orthogonal expansions, the two-dimensional expansions
♣[12] are defined in terms of the modified principal functions as follows:

$$\text{quadrilateral expansion:} \quad \phi_{pq}(\xi_1, \xi_2) = \psi_p^a(\xi_1)\psi_q^a(\xi_2)\,,$$
$$\text{triangular expansion:} \quad \phi_{pq}(\xi_1, \xi_2) = \psi_p^a(\eta_1)\psi_{pq}^b(\eta_2)\,,$$

where

$$\eta_1 = \frac{2(1+\xi_1)}{1-\xi_2} - 1\,, \quad \eta_2 = \xi_2$$

are the two-dimensional collapsed coordinates. In Fig. 3.16 we see all of the modified expansion modes for a fourth-order ($P = 4$) modified triangular expansion. From this figure it is immediately evident that the interior modes have zero support on the boundary of the element. This figure also illustrates that the shape of every boundary mode along a single edge is identical to one of the modes along the other two edges. This was not the case for the orthogonal expansion in Section 3.2.2, but is ensured in the modified expansion by the introduction of $\psi_i^a(z)$ into $\psi_{ij}^b(z)$. In the three-dimensional expansion an equivalent condition is ensured by the introduction of $\psi_{ij}^b(z)$ into $\psi_{ijk}^c(z)$.

[12] Definition of hierarchical modified C^0 expansions in two-dimensional standard regions.

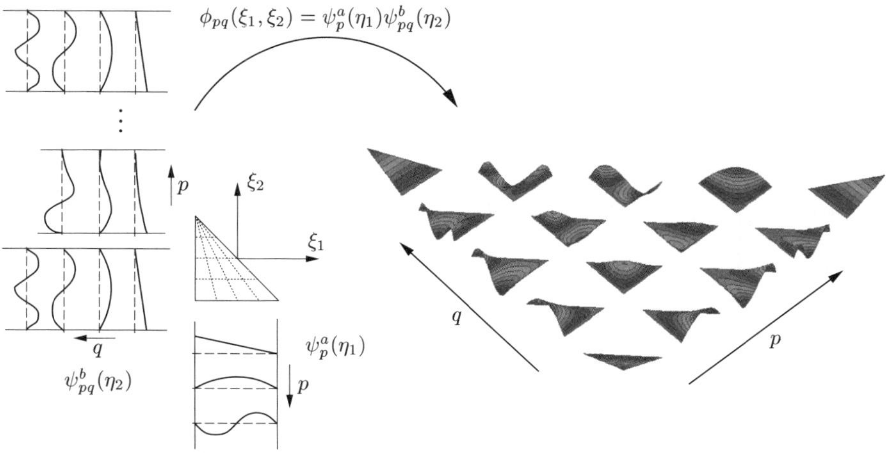

FIG. 3.16. Construction of a fourth-order ($P = 4$) triangular expansion using the product of two modified principal functions $\psi_p^a(\eta_1)$ and $\psi_{pq}^b(\eta_2)$. As compared with the orthogonal expansion shown in Fig. 3.11, the modes are now decomposed into interior and boundary contributions where the boundary modes have similar forms along each edge.

The three-dimensional expansions are defined in terms of the principal functions as follows: ♣[13]

$$
\begin{aligned}
&\text{hexahedral expansion:} && \phi_{pqr}(\xi_1,\xi_2,\xi_3) = \psi_p^a(\xi_1)\psi_q^a(\xi_2)\psi_r^a(\xi_3)\,,\\
&\text{prismatic expansion:} && \phi_{pqr}(\xi_1,\xi_2,\xi_3) = \psi_p^a(\overline{\eta_1})\psi_q^a(\xi_2)\psi_{pr}^b(\xi_3)\,,\\
&\text{pyramidic expansion:} && \phi_{pqr}(\xi_1,\xi_2,\xi_3) = \psi_p^a(\overline{\eta_1})\psi_q^a(\eta_2)\psi_{pqr}^c(\eta_3)\,,\\
&\text{tetrahedral expansion:} && \phi_{pqr}(\xi_1,\xi_2,\xi_3) = \psi_p^a(\eta_1)\psi_{pq}^b(\eta_2)\psi_{pqr}^c(\eta_3)\,,
\end{aligned}
$$

where

$$
\eta_1 = \frac{2(1+\xi_1)}{-\xi_2-\xi_3} - 1\,, \quad \overline{\eta_1} = \frac{2(1+\xi_1)}{1-\xi_3} - 1\,, \quad \eta_2 = \frac{2(1+\xi_2)}{1-\xi_3} - 1\,, \quad \eta_3 = \xi_3
$$

are the three-dimensional collapsed coordinates.

3.2.3.3 *Construction of a modified basis from principal functions*

Unlike the structured expansion in the quadrilateral and hexahedral domains, or even the orthogonal expansions introduced in Section 3.2.2, the modified principal functions for the unstructured regions are no longer in a close-packed form. That is to say, we cannot consecutively loop over the indices p, q, and r. The reason for this is that the introduction of the boundary/interior decomposition

[13] Definition of hierarchical modified C^0 expansions in three-dimensional standard hybrid regions.

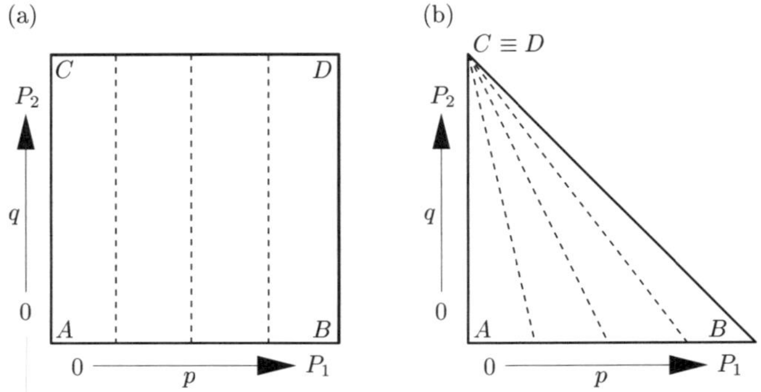

FIG. 3.17. The construction of the collapsed Cartesian coordinate system maps vertex D onto vertex C in (a). If we consider the quadrilateral region in (a) as describing a two-dimensional array in p and q then we can imagine an equivalent array within the triangular region, as shown in (b).

destroys the dense packing of the principal functions ψ^b_{pq} and ψ^c_{pqr}, although the indices corresponding to specific edge, face, or interior modes remain close-packed. Even though these arrays are not close-packed, their definition permits an intuitive construction of the expansion basis by considering each function to be part of an array within the local region.

Two-dimensional expansions

♦14 In Section 3.1, we demonstrated how the quadrilateral expansion may be constructed by considering the definition of the basis $\phi_{pq}(\xi_1, \xi_2)$ as a two-dimensional array within the standard quadrilateral region, with the indices $p = 0$, $q = 0$ corresponding to the lower left-hand corner, as indicated in Fig. 3.17(a). Using this diagrammatic form of the array we can construct the vertex and edge modes by determining the indices corresponding to the vertex or edge of interest. A similar approach is possible with the modified triangular expansion.

We recall that to construct the local coordinate system we used a collapsed Cartesian system where vertex D in Fig. 3.17(a) was collapsed onto vertex C, as shown in Fig. 3.17(b). Therefore, if we use the equivalent array system in the triangular region we can construct our triangular expansions. For example, the vertices marked A and B in Fig. 3.17(b) are defined as follows:

$$\text{vertex } A = \phi_{00}(\eta_1, \eta_2) \;\; = \psi^a_0(\eta_1)\psi^b_{00}(\eta_2),$$
$$\text{vertex } B = \phi_{P_1 0}(\eta_1, \eta_2) = \psi^a_{P_1}(\eta_1)\psi^b_{P_1 0}(\eta_2).$$

The vertex at the position marked CD in Fig. 3.17(b) was formed by collapsing the vertex D onto vertex C in Fig. 3.17(a). Therefore, this mode is generated by

[14] Details of how to construct a complete two-dimensional modified basis from principal functions.

adding the contribution from the indices corresponding to the vertices C and D, that is,

$$\text{vertex } CD = \phi_{0P_2}(\eta_1, \eta_2) + \phi_{P_1P_2}(\eta_1, \eta_2) = \psi_0^a(\eta_1)\psi_{0P_2}^b(\eta_2) + \psi_{P_1}^a(\eta_1)\psi_{P_1P_2}^b(\eta_2) .$$

From the definition of $\psi_{pq}^b(\eta_2)$ for the modified basis we see that $\psi_{0P_2}^b(\eta_2) = \psi_{P_1P_2}^b(\eta_2)$. This condition was necessary to ensure that all the boundary modes have a similar shape; however, we see that the definition of vertex CD can be simplified to

$$\text{vertex } CD = \left(\psi_0^a(\eta_1) + \psi_{P_1}^a(\eta_1)\right) \psi_{0P_2}^b(\eta_2) .$$

Finally, recalling the definition of $\psi_p^a(\eta_1)$, we find that

$$\psi_0^a(\eta_1) + \psi_{P_1}^a(\eta_1) = \frac{1-\eta_1}{2} + \frac{1+\eta_1}{2} = 1 ,$$

and therefore vertex CD is defined as

$$\text{vertex } CD = \psi_{CP_2}^t(\eta_2) = \frac{1+\eta_2}{2} \quad \left(= \frac{1+\xi_2}{2}\right) .$$

Although we could have gone straight to this answer, the construction using the analogy of the collapsed coordinate system to the rectangular system is helpful in assembling the three-dimensional basis.

For the triangular expansion the edge modes are defined as follows:

$$\begin{aligned}
\text{edge } AB\text{:} \quad & \phi_{p0}(\eta_1, \eta_2) = \psi_p^a(\eta_1)\psi_{p0}^b(\eta_2) , & 0 < p < P_1 , \\
\text{edge } AC\text{:} \quad & \phi_{0q}(\eta_1, \eta_2) = \psi_0^a(\eta_1)\psi_{0q}^b(\eta_2) , & 0 < q < P_2 , \\
\text{edge } BD\text{:} \quad & \phi_{P_1q}(\eta_1, \eta_2) = \psi_{P_1}^a(\eta_1)\psi_{P_1q}^b(\eta_2) , & 0 < q < P_2 .
\end{aligned}$$

In constructing the triangular region from the quadrilateral region as shown in Fig. 3.17, edge CD was eliminated. It does not, therefore, contribute to the triangular expansion.

Finally, the interior modes of the modified triangular expansion (which become the triangular-face modes in the three-dimensional expansions) are defined as

$$\text{interior:} \quad \phi_{pq}(\eta_1, \eta_2) = \psi_p^a(\eta_1)\psi_{pq}^b(\eta_2) , \quad 0 < p, q,\ p < P_1,\ p+q < P_2,\ P_1 \leqslant P_2 .$$

A full listing of the triangular basis in terms of the Jacobi polynomials can be found in Appendix D. There is a dependence of the interior modes in the p direction on the modes in the q direction, which ensures that each mode is a polynomial in terms of the Cartesian coordinates (ξ_1, ξ_2). This dependence requires that there should be as many modes in the q direction as there are in the p direction, and hence the restriction that $P_1 \leqslant P_2$. A complete polynomial expansion typically involves all the modes defined above. This expansion is optimal in the sense that it spans the widest possible polynomial space in (ξ_1, ξ_2) with the minimum number of modes. More interior or edge modes could be used, but if they are not increased in a consistent manner the polynomial space will not be increased. In Fig. 3.18 we see the structure of the mass matrix for a $P_1 = P_2 = 14$ polynomial order triangular expansion within the standard triangular region. The matrix is ordered so that the boundary modes are first followed by the interior system. It can be shown (see [432]) that if we order the interior system so that the q index runs fastest then the bandwidth of the interior system is $(P-2) + (P-3) + 1$.

As mentioned previously, the values of α and β in the Jacobi polynomials used in the principal functions $\psi_p^i(z)$ and $\psi_{ij}^b(z)$ were chosen to minimise the bandwidth in both the mass and Laplacian systems. However, as noted by [143, 495], the bandwidth of the interior system of the mass matrix can be made orthogonal by using $P_i^{2,2}(z)$ in the principal function $\psi_i^a(z)$ and $P_j^{2i+3,2}(z)$ in the principal function $P_j^{2i+3,2}(z)$. Nevertheless, as illustrated in Fig. 3.19, the coupling between the interior and boundary system is stronger.

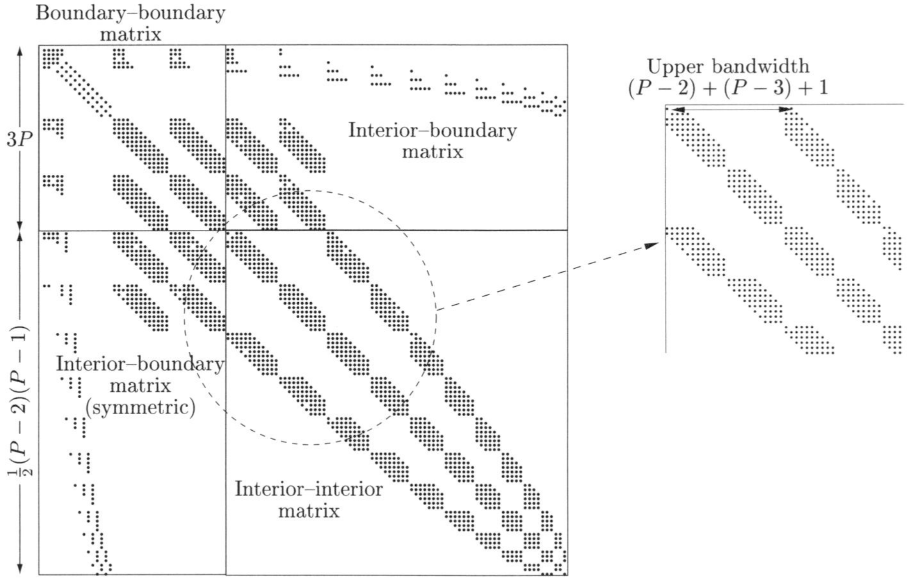

FIG. 3.18. The structure of the mass matrix for a triangular expansion $\phi_{pq} = \psi_p^a \psi_{pq}^b$ of order $P_1 = P_2 = 14$ within the standard region $\mathcal{T}^2$. The boundary modes have been ordered first, followed by the interior modes. If the q index is allowed to run faster then the interior matrix has a bandwidth of $(P-2) + (P-3) + 1$.

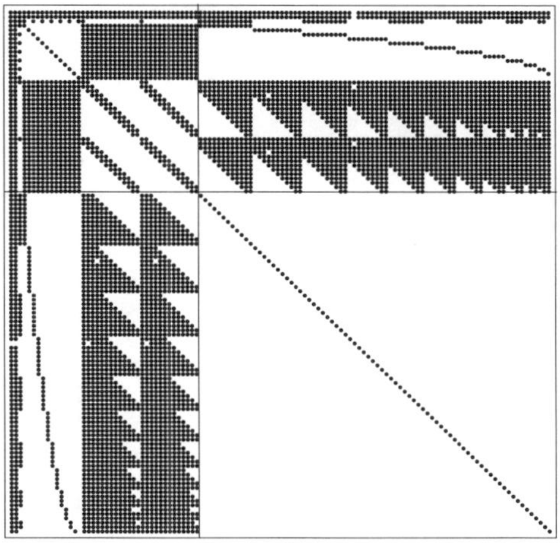

FIG. 3.19. The structure of the mass matrix for a triangular expansion of order $P_1 = P_2 = 14$ where the principal function $\psi_i^a(z)$ is defined using $P_i^{2,2}(z)$ and the principle function $\psi_{ij}^b(z)$ is defined using $P_j^{2i+3,2}(z)$. In this case the interior expansion is orthogonal; however, the boundary system has become more dense.

Three-dimensional expansions

As illustrated in Fig. 3.20, for the hexahedral domain the indices p, q, and r correspond directly to a three-dimensional array where all indices start from zero at the bottom left-hand corner. Therefore, the vertex mode labelled A is described by $\phi_{(000)} = \psi_0^a(\xi_1)\psi_0^a(\xi_2)\psi_0^a(\xi_3)$, similarly the vertex mode labelled H is described by $\phi_{(P_1,P_2,P_3)}$, and the edge modes between C and G correspond to $\phi_{0,P_2,r}$ $(1 < r < P_3)$. ♦[15]

When considering the prismatic domain we use the *equivalent* hexahedral indices. Accordingly, vertex A is now described by $\phi_{(000)} = \psi_0^a(\xi_1)\psi_0^a(\eta_2)\psi_{00}^b(\xi_3)$. In generating the new coordinate system, vertex G was mapped to vertex E and therefore the vertex mode, labelled EG in the prismatic domain, and is described by $\phi_{(0,0,P_3)} + \phi_{(0,P_2,P_3)}$ (that is, adding the two vertices from the hexahedral domain which form the new vertex in the prismatic domain). A similar addition process is necessary for the prismatic edge EG–FH which is constructed by adding the edge modes EF (that is, $\phi_{(p,0,P_3)}$) to the edge modes GH (that is, $\phi_{(p,P_2,P_3)}$). In degenerating from the hexahedral domain to the prismatic region, the edges EG and FH are removed and therefore do not contribute to the prismatic expansion.

This process can also be extended to construct the expansion for the pyramidic and tetrahedral domains. For both these cases the top vertex is constructed by summing the contributions of E, F, G, and H. In the tetrahedral domain edges

[15] Details of how to construct a complete three-dimensional modified basis from principal functions.

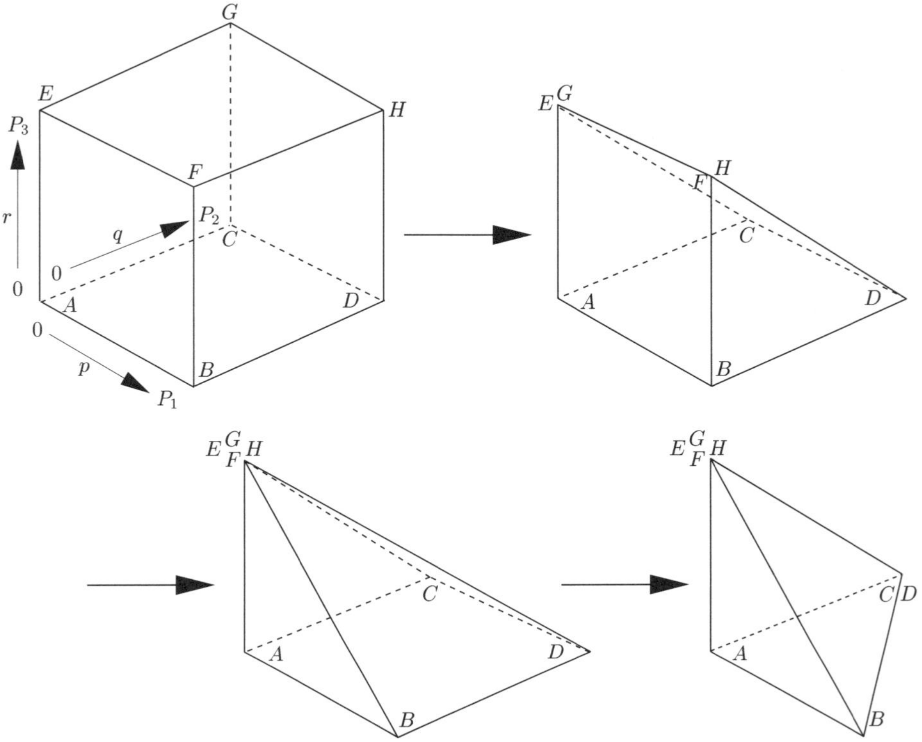

FIG. 3.20. Generation of the standard tetrahedral domains from repeated collapsing of a hexahedral region.

CG and DH are also added. Although the modified functions ψ^b_{ij} and ψ^c_{ijk} are not close-packed, every individual edge, face, and interior mode may be summed consecutively. A full listing of all the modes for the hexahedral, prismatic, pyramidic, and tetrahedral expansions is given in Appendix D.

The present construction has assumed that P_1, P_2, and P_3 have been fixed within the local expansion. In practice, every edge may have a different bound, every face can be described by two bounds, and the interior can be described by three bounds. However, to generate an expansion which spans as complete a polynomial space as possible, all edges which vary with the p index should be assembled from $0 < p < P_1$. Similarly, all edges which vary with q or r should be assembled from $0 < q < P_2$ or $0 < r < P_3$, respectively.

The face modes in all four regions are dependent on the index pairs (p, q), (p, r), and (q, r), where the third index is fixed for a given face. If we let (a, b) represent any one of these three index pairs and P_a and P_b represent the bounds upon which a face is dependent, then the quadrilateral face modes should be assembled over all the indices similar to the edge modes, that is, $0 < a < P_a$ and

$0 < b < P_b$. The triangular faces are dependent upon the function $\psi^b_{ij}(z)$ either directly or indirectly through $\psi^c_{ijk}(z)$. Therefore, the indices for a triangular face are similar to the two-dimensional expansion and should be assembled as $0 < a, b$, $a < P_a$, $a + b < P_b$, $P_a \leqslant P_b$.

Finally, the interior assembly depends upon which principal functions are used. In the hexahedral domain, the interior assembly is the same as that used for the edge and quadrilateral faces ($0 < p < P_1$, $0 < q < P_2$, $0 < r < P_3$). The prismatic domain contains the principal function $\psi^b_{qr}(\xi_3)$ and so the interior assembly is similar to the (q, r) face assembly and has the form $0 < p < P_1$, $0 < q, r$, $q < P_2$, $q + r < P_3$, $P_2 \leqslant P_3$. The interior modes for the pyramidic and tetrahedral expansions use the $\psi^c_{pqr}(\eta_3)$ principal function and should therefore be assembled up to the limits $0 < p, q, r$, $p < P_1$, $p + q < P_2$, $p + q + r < P_3$, $P_1 \leqslant P_2 \leqslant P_3$.

Providing all the modes used in the each edge, face, and interior assembly are consecutive, then the expansions are complete even though the polynomial space of the expansion may not contain as many monomials as the number of modes. With the exception of the pyramidic expansion, the above assembly procedure will give polynomial expansions which contain the same number of modes as the monomials in the polynomial space. The pyramidic function does not have this property since boundary/interior decomposition requires that there be a greater number of modes. The pyramidic expansion also differs in the fact that the individual modes are not polynomials in terms of the Cartesian coordinates (ξ_1, ξ_2, ξ_3). However, the linear combination of the modes does produce an expansion which is complete in terms of polynomials of the Cartesian coordinates.

Although the use of the modified expansion allows the hybrid domains to be tessellated into a global C^0 expansion, this process has reduced the orthogo-

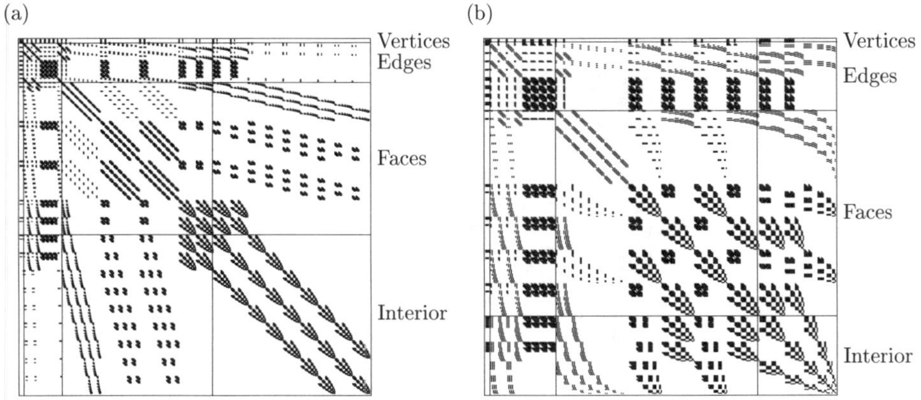

FIG. 3.21. Mass matrices for (a) the prismatic and (b) the pyramidic domains at a fixed polynomial order of $P = 10$. Both expansions have been plotted so that the vertex modes are first followed by the edge, face, and finally the interior modes.

nality of the modes as compared with the orthogonal versions given in Section 3.2.2. Nevertheless, the Jacobi polynomials used in the interior of the principal functions $\psi_i^a(z)$, $\psi_{ij}^b(z)$, and $\psi_{ijk}^c(z)$ have been chosen to maintain as much orthogonality in the expansions as possible. To realise this orthogonality the k index must run faster than the j index, which must run faster than the i index. Figure 3.21 shows all the nonzero entries of the mass matrix for the prismatic and pyramidic expansions within a single domain using an expansion of polynomial order $P = 10$. A high degree of sparsity is evident, particularly in the prismatic region, although it should be noted that for a fixed order of $P_1 = P_2 = P_3 = P = 10$ the prismatic expansion has 726 modes whereas the pyramidic expansion has only 386 modes. The structure of the mass matrix for the tetrahedral expansion is also shown in Fig. 3.22 for polynomial orders of $P = 4$, 9, and 19.

As a final point, we note that the use of the collapsed Cartesian coordinate system means that the coordinate system in the triangular faces, unlike the

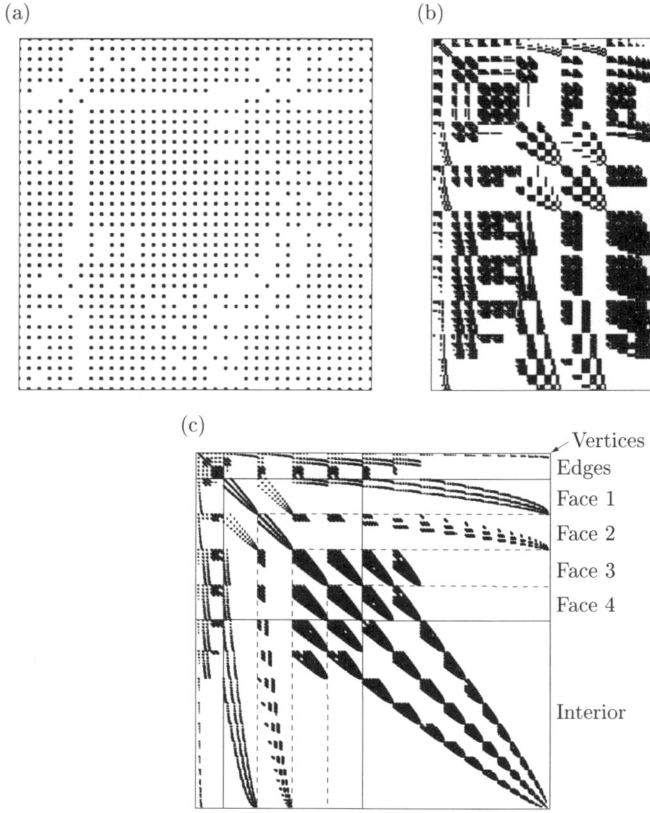

FIG. 3.22. The structure of the mass matrix of a tetrahedral expansion in $\mathcal{T}^3$ for (a) $P = 4$, (b) $P = 9$, and (c) $P = 19$.

quadrilateral faces, is not rotationally symmetric. This means that there is a restriction on how two triangular faces, in a multi-domain expansion, must be aligned. In Section 4.2.1.1 we show that this condition can easily be satisfied for all tetrahedral meshes, although some care must be taken when using a mixture of different elemental domains.

3.2.3.4 *Polynomial space of modified unstructured expansions*

The polynomial space of these expansions is the same as the orthogonal expansions (see eqn (3.2.13)), and with the exception of the pyramidic expansion the number of modes is equal to the size of the polynomial space. The pyramidic expansion requires extra modes to perform the interior/boundary decomposition as compared to the orthogonal expansion. Unfortunately, the introduction of these extra modes does not increase the polynomial space of the expansion.

A notable advantage of the boundary/interior decomposition in the modified expansions is the ability to use a different number of modes along every edge, within all faces as well as the interior. Similar to the structured expansions, this permits a high level of flexibility in the multi-domain extension where a different polynomial order can be used within every elemental region.

3.3 Non-tensorial nodal expansions in a simplex

The use of nodal spectral element methods in simplex regions such as triangles and tetrahedrons has also been very limited by comparison to nodal quadrilateral and hexahedral expansions. A nodal type basis was proposed by Mavriplis and Rosendale [326] in 1993, although this has been observed to lead to unstable schemes for $P > 3$ in C^0 continuous spectral elements. A variation of this was discussed in [487] but has not been used widely in any applications. A desirable feature of a nodal triangular basis is that it can be used in conjunction with the standard nodal spectral element. As discussed in Sections 2.3.4.2 and 3.1, the nodal spectral element basis commonly uses the tensor product of polynomials through the Gauss–Lobatto–Legendre quadrature points. In this section we shall review two non-tensorial bases within a triangular region, which have edges that match these Gauss–Lobatto–Legendre quadrature points. A recent, interesting technique for deriving a non-tensorial expansion in a triangular region based on a purely geometric point of view has been proposed by Blyth and Pozrikidis [61], but will not be discussed here.

Minimisation of electrostatic potential

Hesthaven [234] proposed an extension to the concept of using the minimisation of an electrostatic potential (see Section 2.3.3.2) to determine a nodal set of points in a triangle. Using the one-dimensional distribution of points to constrain the nodal distribution along the edges, the interior nodal points were then determined as the minimum of an electrostatic potential. By constraining the edge points in this manner, the nodes along all edges can be constrained to the Gauss–Lobatto–Legendre quadrature points. The nodal distribution, therefore, offers a compatible extension to the quadrilateral expansion where both quadrilateral and triangular regions can be assembled into a global C^0 expansion. This expansion has been used in the solution of conservation laws [237], incompressible Navier–Stokes equations [486], and computational electromagnetics [239]. The technique was extended to the tetrahedral region in Hesthaven and Teng [238].

Minimisation of the Lebesgue constant and the Fekete points

A reasonably good choice for the nodal points within the triangular or tetrahedral region are the points which minimise the Lebesgue constant. The Lebesgue constant is a measure of how close the polynomial approximation to a function is to the *best* polynomial approximation in the maximum norm. Based on this idea, an alternative to electrostatic minimisation is to search for a nodal set with a small Lebesgue constant by maximising the determinant of the Vandermonde matrix. These points are known as the Fekete points and this basis has been investigated by Bos [69], Chen and Babuška [93], and Taylor *et al.* [451]. The last authors applied this nodal basis to computational acoustics in [272]. In one dimension the Fekete points are also the Gauss–Lobatto–Legendre quadrature points which, as noted previously, are used in the standard quadrilateral spectral element basis. The Fekete distribution is, therefore, a good extension to the nodal quadrilateral expansion. We also note that a nodal distribution of points in a triangle and tetrahedron with an L^2-norm optimal Lebesgue constant were determined by Chen and Babuška [93,94], although these points do not have an edge distribution which can be identified with Gauss–Lobatto–Jacobi points.

The nodal basis for a triangular region cannot be defined in terms of a closed-form expression, as was the case for the tensorial expansions of Sections 3.1 and 3.2. Instead, we define the nodal basis as Lagrange polynomials, denoted by $L_i^{N_m}(\boldsymbol{\xi})$, through a set of Π_{N_m} nodal points in the triangular region, where $\boldsymbol{\xi}_i = (\xi_1^i, \xi_2^i)$ and

$$\Pi = \{\boldsymbol{\xi}_0, \boldsymbol{\xi}_1, \ldots, \boldsymbol{\xi}_{N_m}\}.$$

For the nodal bases to be complete in a linear space of order P, i.e.,

$$L_i^{N_m}(\xi) \subseteq \mathcal{P}_P = \operatorname{span} \mathcal{P}_P(\mathcal{T}^2) = \operatorname{span}\{\xi_1^p \; \xi_2^q\}_{(p,q)\in\Upsilon},$$
$$\Upsilon = \{(p,q) \mid 0 \leqslant p, q, \; p+q \leqslant P\},$$

it is then necessary that Π_{N_m} must contain $(P+1)(P+2)/2$ distinct nodal points $\boldsymbol{\xi}_i$.

Since there is not a closed-form expression for Lagrange polynomials, it is necessary to express the Lagrange basis in terms of another more easily defined polynomial, for example, the orthogonal expansion discussed in Section 3.2.2. Determining the Lagrange polynomial in terms of another polynomial expansion naturally leads us to the generalised Vandermonde matrix, which will be discussed in Section 3.3.2. However, before proceeding, we recall that a measure of the Lagrange basis through either the electrostatic or Fekete points is through the magnitude of the Lebesgue constant. Therefore, before introducing the nodal bases, we will first discuss the Lebesgue constant in the next section.

3.3.1 *The Lagrange polynomial and the Lebesgue constant*

Consider the problem of interpolating a function $f(\boldsymbol{\xi}_i) \equiv f(\xi_1^i, \xi_2^i)$ in the standard triangular region $\Omega_{\rm st} = \mathcal{T}^2 = \{-1 \leqslant \xi_1, \xi_2,\ \xi_1 + \xi_2 \leqslant 0\}$. Given a distinct set of points $\Pi = \{\boldsymbol{\xi}_0, \ldots, \boldsymbol{\xi}_{N_m}\}$, we assume a unique polynomial function $g(\boldsymbol{\xi})$ exists which satisfies

$$g(\boldsymbol{\xi}_i) = f(\boldsymbol{\xi}_i)\,, \quad \forall\ i,\ 0 \leqslant i < N_m\,.$$

This polynomial can be considered to be the interpolating polynomial such that

$$g(\boldsymbol{\xi}) = \mathcal{I}_{N_m} f(\xi)\,,$$

where $\mathcal{I}_{N_m}$ is the interpolation operator. Following [234, 451], the Lebesgue constant shows how well $\mathcal{I}_{N_m}$ approximates $f(\boldsymbol{\xi})$. We denote by $p^\star(\boldsymbol{\xi})$ the best approximating polynomial in the max norm, defined as

$$||f||_\infty = \max_{\boldsymbol{\xi} \in \Omega_{\rm st}} |f(\boldsymbol{\xi})|\,.$$

Since $p^\star(\boldsymbol{\xi})$ is in the same polynomial space as $\mathcal{I}_{N_m} f(\boldsymbol{\xi})$, we note that $p^\star = \mathcal{I}_{N_m} p^\star$. Therefore, we observe that

$$\begin{aligned}
||f - \mathcal{I}_{N_m} f||_\infty &= ||f - p^\star + \mathcal{I}_{N_m} p^\star - \mathcal{I}_{N_m} f||_\infty \\
&\leqslant ||f - p^\star||_\infty + ||\mathcal{I}_{N_m}||_\infty ||p^\star - f||_\infty \\
&\leqslant (1 + ||\mathcal{I}_{N_m}||_\infty) ||p^\star - f||_\infty\,,
\end{aligned}$$

where we understand that

$$||\mathcal{I}_{N_m}||_\infty = \max_{||f||_\infty = 1} ||\mathcal{I}_{N_m} f||_\infty\,.$$

The constant $\Lambda_{N_m} = ||\mathcal{I}_{N_m}||_\infty$ is known as the *Lebesgue constant.* We observe that the Lebesgue constant is a measure of how close the approximation, $\mathcal{I}_{N_m} f$, is to the best approximating polynomial $p^\star$ in the *max* norm. Choices such as equally-spaced points (in the square or triangle) are known to have Lebesgue constants that grow exponentially [451].

If we now represent our polynomial approximation in terms of the Lagrange interpolation or cardinal function at the nodal points, i.e.,

$$\mathcal{I}_{N_m} f(\boldsymbol{\xi}) = \sum_{i=0}^{N_m} f(\boldsymbol{\xi}_i) L_i^{N_m}(\boldsymbol{\xi})\,,$$

where

$$L_i^{N_m}(\boldsymbol{\xi}_j) = \delta_{ij}$$

and δ_{ij} is the Kronecker delta, then we observe that

$$\Lambda_{N_m} = ||\mathcal{I}_{N_m}||_\infty = \max_{||f||_\infty=1} ||\mathcal{I}_{N_m} f||_\infty = \max_{\boldsymbol{\xi}\in\Omega_{\rm st}} \sum_{0\leqslant i<N_m} |L_i^{N_m}(\boldsymbol{\xi})|. \tag{3.3.1}$$

Therefore, evaluating the Lagrange polynomials throughout the triangular region, $\boldsymbol{\xi} \in \Omega_{\rm st}$, allows us to get a graphical interpretation of the Lebesgue function, $\sum_{0\leqslant i<N_m} |L_i^{N_m}(\boldsymbol{\xi})|$. The maximum bound of this function over the region leads to the Lebesgue constant Λ_{N_m}. The Lebesgue function is illustrated in Fig. 3.23 where we show plots of $\sum_{0\leqslant i<N_m} |L_i^{N_m}(\boldsymbol{\xi})|$ for equispaced and Fekete (see Section 3.3.4) distributions of nodal points when $N_m = 15$ (or $P = 4$). In this case the equispaced and Fekete points have Lebesgue constants of $\Lambda_{N_m} = 3.47$ and $\Lambda_{N_m} = 2.72$, respectively.

3.3.2 *Generalised Vandermonde matrix*

Since there is not a closed-form expression for the Lagrange polynomial through an arbitrary set of points in the triangular region, it is necessary to express the Lagrange polynomial in terms of another polynomial which has a closed-form definition, for example, the orthogonal polynomial discussed in Section 3.2.2. The choice of the closed-form basis is important since ultimately a matrix inversion is involved and therefore the basis dictates the conditioning of the matrix and ultimately the computational stability. Following [486], we consider the interpolation of a polynomial function $p(\boldsymbol{\xi})$ through a set $\Pi_{N_m} = \{\boldsymbol{\xi}_0, \ldots, \boldsymbol{\xi}_{N_m-1}\}$ of $N_m = (P+1)(P+2)/2$ distinct points, where P is the polynomial order. The polynomial $p(\boldsymbol{\xi})$ can be exactly represented by any polynomial expansion, $\phi_i(\boldsymbol{\xi})$, which spans the same space, i.e.,

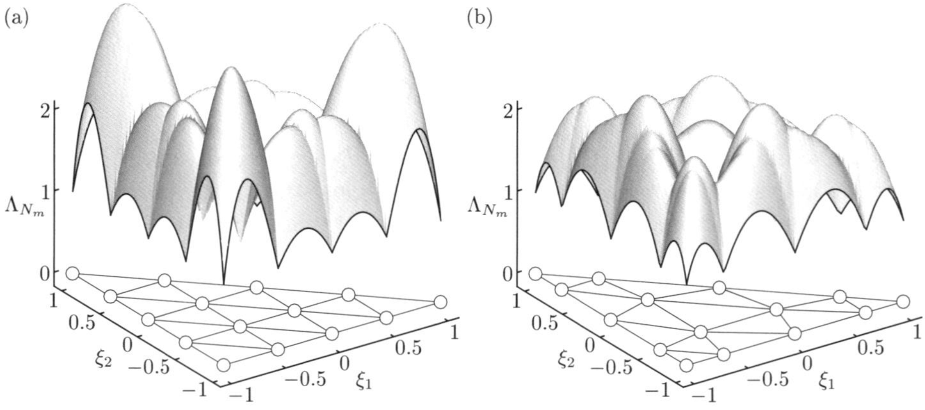

FIG. 3.23. Spatial distribution of the Lebesgue function, $\sum_{0\leqslant i<N_m} |L_i^{N_m}(\boldsymbol{\xi})|$, for (a) equispaced points $N_m = 15$ and (b) Fekete points $N_m = 15$. The maximum of the Lebesgue function gives the Lebesgue constant Λ_{N_m}. (Courtesy of T. C. Warburton.)

$$p(\boldsymbol{\xi}) = \sum_{0\leqslant i<N_m} \phi_i(\boldsymbol{\xi})\hat{f}_i\,,$$

where $\hat{f}_i$ represents the expansion coefficients associated with $\phi_i(\boldsymbol{\xi})$. We note that the use of the index i in the above expression denotes a summation over all modes in the expansion. In Sections 3.1 and 3.2 we adopted double indices (p, q) to represent a two-dimensional basis constructed from a tensor product of two one-dimensional expansions. We, therefore, understand that a summation over i represents a complete summation over the pair (p, q). Since $p(\boldsymbol{\xi})$ and $\phi_i(\boldsymbol{\xi})$ span the same polynomial space, any form of projection will recover the exact expansion coefficients, $\hat{f}_i$. We can, therefore, obtain the expansion coefficients by performing a collocation projection at the points $\boldsymbol{\xi}_i$ such that

$$\sum_{0\leqslant j<N_m} \phi_j(\boldsymbol{\xi}_i)\hat{f}_j = p(\boldsymbol{\xi}_i)\,, \quad \forall\ i,\ 0\leqslant i < N_m\,,$$

which can be written in matrix notation as

$$\boldsymbol{V}\hat{\boldsymbol{f}} = \boldsymbol{f}\,,$$

where $\boldsymbol{f}[i] = f(\boldsymbol{\xi}_i)$, $\boldsymbol{V}[i][j] = \phi_j(\boldsymbol{\xi}_i)$, and $\hat{\boldsymbol{f}}[i] = \hat{f}_i$. If $\phi_i(\boldsymbol{\xi})$ denotes a monomial basis $\phi_i(\boldsymbol{\xi}) = \phi_{i[p,q]}(\xi_1, \xi_2) = \xi_1^p\xi_2^q$ then the matrix $\boldsymbol{V}$ is the Vandermonde matrix. For a general basis $\phi_i(\boldsymbol{\xi})$ the matrix $\boldsymbol{V}$ is known as the generalised Vandermonde matrix. As recognised in [451, 486], the choice of the orthogonal expansion $(\phi_{i[p,q]}(\xi_1, \xi_2) = \psi_p^a(\xi_1)\psi_{pq}^b(\xi_2))$ discussed in Section 3.2.2 is particularly convenient due to the strong linear independence of the expansion. This property leads to a well-conditioned system which is then amenable to numerical inversion.

A further observation that is worth noting for our subsequent discussion of nodal bases is that if $\phi_p(\boldsymbol{\xi})$ is a known function, such as the orthogonal basis of Section 3.2.2, then $\boldsymbol{V}[i][j] = \phi_j(\boldsymbol{\xi}_i)$ and we find that

$$\boldsymbol{V}\begin{bmatrix} L_0(\boldsymbol{\xi}) \\ \vdots \\ L_{N_m-1}(\boldsymbol{\xi}) \end{bmatrix} = \begin{bmatrix} \phi_0(\boldsymbol{\xi}) \\ \vdots \\ \phi_{N_m-1}(\boldsymbol{\xi}) \end{bmatrix}$$

or

$$\begin{bmatrix} L_0(\boldsymbol{\xi}) \\ \vdots \\ L_{N_m-1}(\boldsymbol{\xi}) \end{bmatrix} = \boldsymbol{V}^{-1}\begin{bmatrix} \phi_0(\boldsymbol{\xi}) \\ \vdots \\ \phi_{N_m-1}(\boldsymbol{\xi}) \end{bmatrix}. \tag{3.3.2}$$

Therefore, given a set of points $\boldsymbol{\xi}_j$ $(0 \leqslant j < N_m)$ and polynomial functions $\phi_0(\boldsymbol{\xi}), \ldots, \phi_{N_m-1}(\boldsymbol{\xi})$ we can evaluate $L_0^{N_m}(\boldsymbol{\xi}), \ldots, L_{N_m-1}^{N_m}(\boldsymbol{\xi})$ using eqn (3.3.2). Further details on constructing the generalised Vandermonde matrix can also be found in Section 4.1.5.3.

3.3.3 *Electrostatic points*

As previously discussed in Section 2.3.3.2, Stieltjes [442] and Szego [447] showed the connection between the polynomial $(1-\xi)^\alpha(1+\xi)^\beta P_{P-1}^{\alpha,\beta}$ and the minimisation of the following problem. Assume that $P-1$ unit mass charges with unit charge are allowed to move freely inside the interval $[-1,1]$ between two fixed unit charges $\alpha \sim (\alpha+1)/2$ and $\beta \sim (\beta+1)/2$ held fixed at $\xi_1 = \pm 1$. The steady-state position of the charges that minimises the electrostatic energy

$$W = -\sum_{i=1}^{P-1} \left\{ \frac{\alpha+1}{2} \log|\xi_i+1| + \frac{\beta+1}{2} \log|\xi_i - 1| + \frac{1}{2} \sum_{j=1, j\neq i}^{P-1} \log|\xi_i - \xi_j| \right\}$$

is the distribution of the Gauss–Lobatto points. An analogous minimisation performed without the two fixed end-charges also leads to the zeros of the Jacobi polynomial $P_{P-1}^{\alpha,\beta}$ or, equivalently, the Gauss rather than the Gauss–Lobatto quadrature points. Since the Legendre polynomials $L_P(\xi) = P_P^{0,0}(\xi)$ and their derivatives $L_P'(\xi) = \frac{1}{2}(P-1)P_{P-1}^{1,1}(\xi)$ are widely used in both the modal and nodal expansions for quadrilateral and hexahedral domains, Hesthaven [234] adopted a similar approach to that in Stieltjes' problem to determine a set of nodal points in the simplex.

As the electrostatic points in one dimension can be constructed to comply with the Gauss–Lobatto–Jacobi quadrature points, a natural requirement for the distribution of nodes in the triangular region is that the boundary charges are located at these Gauss–Lobatto–Jacobi quadrature points along each edge. In this way it is possible to enforce that the nodal basis is aligned with the quadrilateral nodal basis. Hesthaven [234] assumed the potential from each edge 'e' contributed a potential at a point, ξ, in the triangular region of the form

$$\Psi_e(\xi) = \rho_e \int_0^1 \frac{1}{|\boldsymbol{\xi} - \boldsymbol{\xi}_e|}\, dt,$$

where $\boldsymbol{\xi}_e = \boldsymbol{v}_a + t(\boldsymbol{v}_b - \boldsymbol{v}_a)$ and $t \in [0,1]$ represents the coordinates along an edge between the vertices $\boldsymbol{v}_a$ and $\boldsymbol{v}_b$. He then assumed that the N_p unit charges were allowed to move, mutually interacting according to the potential

$$\Psi(\boldsymbol{\xi}_i, \boldsymbol{\xi}_j) = \frac{\rho_p^2}{|\boldsymbol{\xi}_i - \boldsymbol{\xi}_j|},$$

and posed the following minimisation problem, analogous to that of Stieltjes.

Problem *Let the line charge density be given by $\rho_e > 0$. Assume that N_p unit mass charges with unit charge, $\rho_p = 1$, are allowed to move freely inside the simplex. What is the steady-state position of the charges that minimises the electrostatic energy*

$$W(\boldsymbol{\xi}_1, \ldots, \boldsymbol{\xi}_{N_p}) = \sum_{i=1}^{N_p} \left(\sum_{i=1}^{3} \Psi_e(\boldsymbol{\xi}_i) + \frac{1}{2} \sum_{j=1, j\neq i}^{N_p} \Psi(\boldsymbol{\xi}_i, \boldsymbol{\xi}_j) \right) ?$$

In the minimisation the particles were constrained to have certain symmetries motivated by the symmetry of the domain and the likelihood is then that the optimal points will have strong symmetry. Therefore, the line charge ρ_e was set to equal values on all edges such that $\rho_1 = \rho_2 = \rho_3$. The value of ρ_e also depended on the choice of the Gauss–Lobatto–Jacobi quadrature nodes prescribed on the edges. Therefore, for a symmetrical distribution of edge nodes (i.e., $\alpha = \beta$) there was an additional parameter, α. If $\alpha = -1/2$ the edge points correspond to a Gauss–Lobatto–Chebyshev distribution, whereas $\alpha = 1$ corresponds to the Gauss–Lobatto–Legendre points commonly used in nodal spectral elements. Finally, for a P-order expansion there are $N_p = (P+1)(P+2)/2 - 3P$ points to be determined by the minimisation.

To solve the N_p-body minimisation problem, Hesthaven considered the steady-state solution of time-integrating Newton's second law stated as

$$\boldsymbol{\xi}_i'' = -\left(\sum_{i=1}^{3} \nabla \xi_e(\boldsymbol{\xi}_i) + \frac{1}{2} \sum_{j=1, j\neq i}^{N_p} \nabla \xi(\boldsymbol{\xi}_i, \boldsymbol{\xi}_j) \right) - \epsilon \boldsymbol{\xi}_i' .$$

The term $\boldsymbol{\xi}_i'$ corresponds to a friction term in order to make the problem slightly dissipative. In solving the problem, highly accurate time integration is required to reduce numerical dissipation; however, the value of ϵ does not alter the solution since we are only interested in the steady-state solution. The choice of initial conditions is important since finding the global minimum of the energy function ♣[14] is particularly complicated, especially for larger N_p. This was partly simplified by imposing a high degree of symmetry onto the nodal location; we refer the reader to [234] for more details.

The distribution of points corresponding to a choice of Gauss–Lobatto–Legendre quadrature distribution on the edges at polynomial orders of $P = 4$, 7, and 10 is shown in Fig. 3.24, and the values of the Lebesgue constants for these bases are shown in Table 3.3. The concept of the electrostatic distribution of points has been extended to a tetrahedral domain in the work of Hesthaven and Teng [238], and both the triangular and tetrahedral distributions of points are given in Appendix D.

3.3.4 *Fekete points*

Following the discussion in Section 3.3.1, the optimal distribution of nodal points in a simplex, from the point of view of interpolation, is one which minimises the Lebesgue constant. However, there does not appear to be a feasible method to compute these points. A tractable alternative is to use the Fekete points. Following Taylor *et al.* [451], Fekete points are a set of points $\Pi_{N_m} = \{\boldsymbol{\xi}_0, \ldots, \boldsymbol{\xi}_{N_m}\}$ which maximise (for a fixed basis) the determinant of the Vandermonde matrix, i.e.,

[14] Simplex nodal points that match the spectral element quadrilateral edge nodes. A full listing of the points is given in Appendix D.

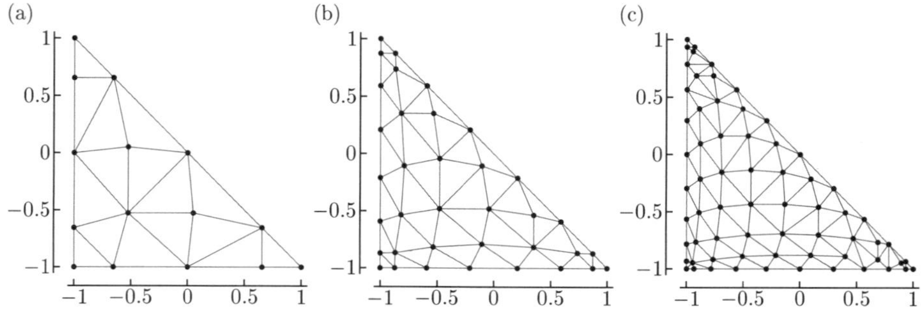

FIG. 3.24. Nodal points from electrostatic potential minimisation, with the boundary nodes constrained to be the Gauss–Lobatto–Legendre quadrature points at a polynomial order of (a) $P = 4$, (b) $P = 7$, and (c) $P = 10$. (Courtesy of J. Hesthaven.)

TABLE 3.3. Lebesgue constant Λ_{N_m} (where $N_m = (P+1)(P+2)/2$) for a set of points in the triangular region as a function of polynomial order P for the electrostatic points of Hesthaven [234], the Fekete points of Taylor *et al.* [451], and an equispaced distribution in the triangular region. Data was taken from [234,451] with the permission of the authors.

Polynomial order, P	Electrostatic points	Fekete points	Equispaced points
6	4.08	4.17	8.45
7	4.77	4.91	14.35
8	5.85	5.90	24.01
9	6.87	6.80	40.92
10	8.44	7.75	70.89
11	10.08	7.89	124.53
12	12.63	8.03	221.41
13	15.34	9.21	397.7
14	22.18	9.72	720.7
15	29.69	9.97	1315.9
16	41.73	12.10	2418.5

$$\max_{\boldsymbol{\xi}_i} |V(\boldsymbol{\xi}_0, \ldots, \boldsymbol{\xi}_{N_m})| \,. \tag{3.3.3}$$

These points are independent of the choice of basis, since a change of basis only multiplies the determinant by a constant independent of the points.

In the interval $[-1, 1]$, Fejér [158] showed that the Fekete points are the Gauss–Lobatto–Legendre quadrature points. This result was extended to the square and cube by Bos *et al.* [70], who showed that the Fekete points were the tensor product of Gauss–Lobatto–Legendre quadrature points. As discussed in Section 3.3.3, the one-dimensional Fekete points are also the minimum energy configuration of the point charges in the interval [442] when the choice of the electrostatic energy contains fixed charges at the end-points. However, in higher dimensions, Fekete points are not Gaussian-like quadrature points or the minimum-energy electrostatic points. In the triangular region Bos [69] conjectured that Fekete points contain the one-dimensional Gauss–Lobatto–Legendre point on the boundary, and this was numerically verified in [451]. This is significant since it means that the Fekete points will conform with the quadrilateral spectral elements.

Finally, to understand the connection between the Fekete points and the Lebesgue constant we recall eqn (3.3.2) where the Lagrange polynomial through a set of points $\Pi_{N_m} = \{\boldsymbol{\xi}_0, \ldots, \boldsymbol{\xi}_{N_m}\}$ in a triangle can be evaluated as

$$L_i^{N_m}(\boldsymbol{\xi}) = \frac{|\boldsymbol{V}(\boldsymbol{\xi}_0, \ldots, \boldsymbol{\xi}_{i-1}, \boldsymbol{\xi}, \boldsymbol{\xi}_{i+1}, \ldots, \boldsymbol{\xi}_{N_m-1})|}{|\boldsymbol{V}|}, \tag{3.3.4}$$

where $\boldsymbol{V}$ is the generalised Vandermonde matrix. By definition of the Fekete points (eqn (3.3.3)), the determinant in the denominator of eqn (3.3.4) is at its maximum value. Therefore, there is no value of $\boldsymbol{\xi}$ in the triangular region that can make the determinant in the numerator larger than the denominator. Only when $\boldsymbol{\xi} = \boldsymbol{\xi}_i$ will the numerator of eqn (3.3.4) equal the denominator, making the Lagrange polynomial equal to one. Therefore, Fekete points generate Lagrange polynomials which achieve their maximum in the triangular region at the associated Fekete point. This property also provides a bound on the Lebesgue constant since, from eqn (3.3.1), we observe that when $L_i^{N_m}(\boldsymbol{\xi})$ is evaluated at the Fekete points then

$$\Lambda_{N_m} = \max_{\boldsymbol{\xi} \in \Omega_{st}} \sum_{0 \leqslant i < N_m} |L_i^{N_m}(\boldsymbol{\xi})| \leqslant N_m .$$

In [451] it was numerically observed that the Lebesgue constant using the Fekete points behaved as $C\sqrt{N_m}$. In the one-dimensional case it is also known that the Lebesgue constant behaves as the logarithm of N_m.

3.3.4.1 *Evaluation of the Fekete points*

The earliest work on evaluating the Fekete points in a triangle was done by Bos [69], who derived the points up to polynomial order or $P = 3$ and an approximate solution up to $P = 7$. Chen and Babuška [93] improved and extended Bos's results up to $P = 13$. Subsequently, Taylor *et al.* [451] determined the points up to a polynomial order of $P = 19$, with improvement on the numerical points of Chen and Babuška [93] for $P > 10$.

To evaluate the Fekete points, Taylor *et al.* [451] used a steepest-ascent algorithm to determine the maximum determinant. This approach solved the ordinary differential system

$$\frac{\partial \boldsymbol{\xi}_i}{\partial t} = \frac{\partial |\boldsymbol{V}|}{\partial \boldsymbol{\xi}_i}, \quad \forall\, i,\ 0 \leqslant i < N_m , \tag{3.3.5}$$

where the points were evaluated by moving their location in the direction of the steepest ascent until an equilibrium was reached, subject to the constraint that the points could not leave the triangular region.

The Fekete point solution presented by Taylor *et al.* [451] has a very elegant construction and interpretation, and we therefore shall revisit it here. Recalling that we can use any definition of the Vandermonde matrix, $\boldsymbol{V}$, in eqn (3.3.5), we start by rewriting eqn (3.3.5) as

$$\frac{\partial \boldsymbol{\xi}_i}{\partial t} = \frac{\partial |\boldsymbol{V}|}{\partial \boldsymbol{\xi}_i} = \sum_{i,j} \frac{\partial |\boldsymbol{V}|}{\partial \boldsymbol{V}_{ij}} \frac{\partial \boldsymbol{V}_{ij}}{\partial \xi_i}, \tag{3.3.6}$$

where $\boldsymbol{V}_{ij}$ represents the (i,j) entry of $\boldsymbol{V}$. We then note that the partial derivative of the determinant of a matrix with respect to an entry $\boldsymbol{V}_{ij}$ is given by

$$\frac{\partial |\boldsymbol{V}|}{\partial \boldsymbol{V}_{ij}} = -1^{i+j} |\boldsymbol{A}_{ij}|,$$

where $\boldsymbol{A}_{ij}$ is the ij minor of $\boldsymbol{V}$. Now if we choose to define the Vandermonde matrix $\boldsymbol{V}$ with respect to a Lagrange basis then the matrix $\boldsymbol{V}$ is the identity matrix. Therefore, the derivative of $|\boldsymbol{V}|$ with respect to any matrix element is only nonzero for diagonal elements. In this case the ith diagonal element is also the ith Lagrange function, $L_i^{N_m}(\boldsymbol{\xi})$, evaluated at the points ξ_i and the determinant of the minor $|\boldsymbol{A}_{ii}| = 1$. Therefore, eqn (3.3.6) becomes

$$\frac{\partial \boldsymbol{\xi}_i}{\partial t} = \frac{\partial L_i^{N_m}}{\partial \boldsymbol{\xi}_i}.$$

The above algorithm has a very simple geometric interpretation which is illustrative of what the Fekete points are trying to achieve in the triangular space. Since we would like the Lagrange functions to approximate a delta function at ξ_i, the maximum of the function should be achieved at the nodal point ξ_i. The steepest-ascent algorithm simply moves each point towards the maximum of its ♣[15] associated Lagrange function. The iterative nature of the algorithm comes into play because the cardinal functions change with every change in the nodal points, and therefore we have to recompute the basis function at every iteration.

To complete the algorithm we also require a technique to evaluate $\partial L_i^{N_m} / \partial \boldsymbol{\xi}_i$. If we denote by $\boldsymbol{V}_\phi$ the Vandermonde matrix between a basis $\phi_i(\boldsymbol{\xi})$ and the Lagrange basis $L_i(\boldsymbol{\xi})$ then differentiating eqn (3.3.2) and evaluating the ith entry we obtain

$$\frac{\partial L_i^{N_m}}{\partial \boldsymbol{\xi}_i}(\boldsymbol{\xi}_i) = \sum_{j=0}^{N_m-1} \boldsymbol{V}_\phi{}^{-1}[i][j] \; \phi_i(\boldsymbol{\xi}_i).$$

Although any basis with a closed-form definition is mathematically suitable, numerically it is advantageous to use a basis which is well conditioned. The

[15] Simplex nodal points that match the spectral element quadrilateral edge nodes. A full listing of the points is given in Appendix D.

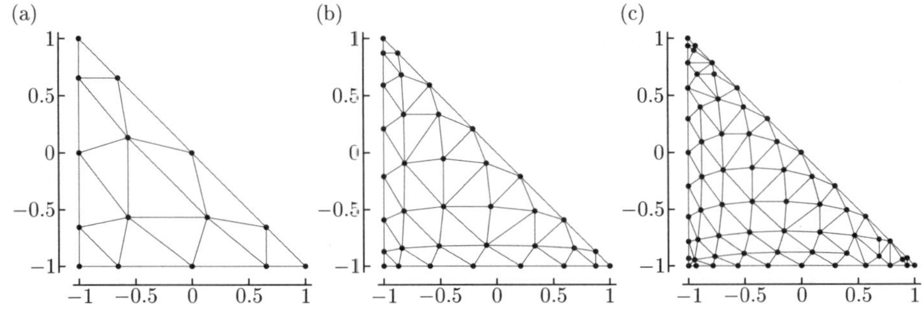

FIG. 3.25. Nodal Fekete points at a polynomial order of (a) $P = 4$, (b) $P = 7$, and (c) $P = 10$. (Courtesy of T. C. Warburton.)

orthogonal basis discussed in Section 3.2.2 is therefore a good choice and was adopted in [451, 486].

In a similar manner to the electrostatic problem of Section 3.3.3, the choice of initial conditions is important. In [451] it was found that the best initial distribution was one which generated a density of points which approximates the extremal measure for a triangle. We refer the reader to [451] for more details.

In Fig. 3.25 we show the distribution of Fekete points in a right-handed triangle, the nodal value of which is provided in Appendix D. Currently, no points are available for the tetrahedral region. Also shown in Table 3.3 are the associated Lebesgue constants of the Lagrange polynomial through the Fekete points. We note that for $P \leqslant 9$ the electrostatic points have a lower Lebesgue constant, but for $P > 10$ the Fekete points are better with a significant improvement for $P > 13$ [451].

3.4 Other useful tensor product extensions

In Sections 3.1 and 3.2 we focused on tensor product expansions within standard hybrid regions using either modal or nodal expansions. However, the great power of using a tensor product extension is that we can mix different expansion types according to the problems we are interested in discretising.

3.4.1 *Nodal elements in a prismatic region*

In Section 2.3.4 we discussed a triangular nodal expansion which is compatible with the standard quadrilateral nodal spectral element discretisation discussed in Section 3.1. As we have discussed previously, the quadrilateral expansion $\phi^{2D}_{pq}(\xi_1, \xi_2) = h_p(\xi_1)h_q(\xi_2)$ can be extended to the hexahedral domain by making a tensor product of ϕ^{2D}_{pq} with the Lagrange polynomial in the ξ_3 direction, i.e., $\phi_{pqr} = \phi^{2D}_{pq}(\xi_1, \xi_2)h_r(\xi_3)$. We can perform an analogous extension to the triangular nodal expansions in a similar way to the modal prismatic expansions discussed in Section 3.2.3. If we denote the triangular nodal expansion as $L^{N_m}_i(\boldsymbol{\xi}) = L^{N_m}_i(\xi_1, \xi_2)$, then a prismatic nodal basis which is compatible in terms of nodal location with the hexahedral nodal expansions is $\phi_{pqr}(\xi_1, \xi_2, \xi_3) = L_i^{N_m}(\xi_1, \xi_2) h_r(\xi_3)$.

3.4.2 *Expansions in homogeneous domains*

For a wide range of applications such as flow between parallel plates or past a circular cylinder, the problem of interest contains at least one homogeneous direction. If the homogeneous direction is in the ξ_3 coordinate then we can construct a three-dimensional basis $\phi_{pqr}(\xi_1, \xi_2, \xi_3)$ in terms of any two-dimensional expansions discussed previously, denoted by $\phi^{2D}_{pq}(\xi_1, \xi_2)$, multiplied by a complete expansion in ξ_3, which we shall denote by $\varphi_r(\xi_3)$, that is,

$$\phi_{pqr}(\xi_1, \xi_2, \xi_3) = \phi^{2D}_{pq}(\xi_1, \xi_2) \varphi_r(\xi_3) .$$

The above expansion is clearly a tensor product of the two-dimensional basis with the expansion $\varphi_r(\xi_3)$. The absence of any range of scales in the ξ_3 direction encourages us to use a purely spectral, or p-type, expansion which spans the complete homogeneous direction rather than a multi-domain, or h-type, extension.

However, the most convenient choice of $\varphi_r(\xi_3)$ is dependent upon the boundary conditions at the ends of the homogeneous direction. If Dirichlet or Neumann boundary conditions are required at the ends of the homogeneous direction then a variety of polynomial expansions might be used, including the Legendre polynomial $\widetilde{\psi}^a_r(\xi_3) = L_r(\xi_3)$ or the Chebyshev polynomials $T_r(\xi_3) = P_r^{-1/2,-1/2}(\xi_3)$. Each of these expansions have their own desirable properties which are appropriate for a given application (see [199]).

If the domain is periodic in the homogeneous direction then by far the most widely used expansion for $\varphi_r(\xi_3)$ is the Fourier expansion, that is,

$$\varphi_r(\xi_3) = \mathrm{e}^{\mathrm{i}r\beta\xi_3} ,$$

where

$$\beta = \frac{2\pi}{L_{\xi_3}}$$

and L_{ξ_3} is the periodic length. A great attraction of this expansion is the use of the fast Fourier transform to go between Fourier and physical space. Furthermore, when considering linear differential operators we note that

$$\nabla \phi_{pqr}(\xi_1, \xi_2, \xi_3) = [\tilde{\nabla}_r] \phi_{pqr}(\xi_1, \xi_2, \xi_3) = \begin{bmatrix} \dfrac{\partial}{\partial \xi_1} \\ \dfrac{\partial}{\partial \xi_2} \\ \mathrm{i}r\beta \end{bmatrix} \phi_{pqr}(\xi_1, \xi_2, \xi_3) ,$$

$$\nabla^2 \phi_{pqr}(\xi_1, \xi_2, \xi_3) = [\tilde{\nabla^2}_r] \phi_{pqr}(\xi_1, \xi_2, \xi_3) = \left(\frac{\partial^2}{\partial \xi_1^2} + \frac{\partial^2}{\partial \xi_2^2} - r^2 \beta^2 \right) \phi_{pqr}(\xi_1, \xi_2, \xi_3) .$$

The introduction of the operators $\tilde{\nabla}$ and $\tilde{\nabla}^2$ means that we can reduce a three-dimensional linear differential problem to a series of r two-dimensional problems over the Fourier planes. See Chapter 9 for details of the application of this expansion to the incompressible Navier–Stokes equations.

3.4.3 *Cylindrical domains*

Similar to the Cartesian homogeneous domains, any problem with a rotational symmetry can often be conveniently expressed in cylindrical coordinates. Using cylindrical coordinates allows a Fourier expansion to be imposed in the azimuthal direction which has the favourable properties mentioned in Section 3.4.2. However, cylindrical coordinates also introduce a radial geometric singularity at the cylinder axis. In the solution of the Navier–Stokes system this geometric singularity can be conveniently manipulated by multiplying the equations through by a factor of the radius [351]. Introducing an extra factor of r into the Galerkin framework can lead to a reduction in the convergence rate of the spectral element expansion. For example, Tomboulides *et al.* [467] and Gerritsma and Phillips [181] have developed a nodal spectral element expansion based on the zeros of the $P_p^{0,1}$ Jacobi polynomials in the radial direction. The two-dimensional expansion is then constructed by a tensor product of the standard Lagrange polynomial through the zeros of the $P_p^{1,1}$ Jacobi polynomial in the axial direction, see [181] for further details. It is, however, also possible to use a standard one-dimensional modal or nodal expansion in the radial direction provided that the expansion has a boundary–interior decomposition. As discussed in Blackburn and Sherwin [58], this construction still maintains the exponential convergence of the spectral/*hp* element method.

3.5 Exercises: construction of multi-dimensional elemental mass matrices

Building upon the implementation of a one-dimensional spectral/*hp* element solver presented in Section 2.6, we begin an analogous multi-dimensional formulation as part of the exercises given in this section. Although there are more concepts we need to appreciate for a full multi-dimensional formulation and implementation, as a starting-point we can consider how to numerically construct the multi-dimensional modal and nodal bases. To provide an application for using the multi-dimensional bases we consider the construction of an elemental mass matrix $\boldsymbol{M}^e$ (see Sections 2.3.2.1 and 4.1.5.3) for the different expansion bases. The mass matrix requires the evaluation of the inner product of the modes in an expansion basis with themselves, i.e.,

$$(\phi_{pq}, \phi_{rs})_{\Omega^e} = \int_{\Omega^e} \phi_{pq}(\xi_1, \xi_2)\phi_{rs}(\xi_1, \xi_2)\, d\xi_1\, d\xi_2 .$$

In evaluating the integrals we require some basic integration and numbering concepts which will be introduced in more detail in Chapter 4. In Section 4.4 we will provide exercises to extend the elemental matrices into a global matrix and also discuss how to impose boundary conditions. As will also be discussed in the last part of Section 4.1.5.1, the computation of a mass matrix for a non-tensorial nodal basis can be implemented by considering the Lagrange basis in terms of an orthogonal tensorial expansion. Therefore, in the final exercise in this section we will discuss the construction of the generalised Vandermonde matrix which expresses a Lagrange polynomial in terms of another polynomial basis.

Some useful codes are also available on the web page

`http://www.nektar.info/2nd_edition`

♦[16] 1. We start by constructing the elemental mass matrix $\boldsymbol{M}^e$ for a modal quadrilateral expansion $\phi_{pq}(\xi_1, \xi_2) = \psi_p^a(\xi_1)\psi_q^a(\xi_2)$ of order $P_1 = P_2 = P$, as defined in Section 3.1.1. The first part of the implementation is to generate an array containing each component of the tensorial basis, $\psi_p^a(\xi_1)$ and $\psi_q^a(\xi_2)$, at a set of discrete points ξ_{1i}, ξ_{2j}. Typically, we only require the basis at the discrete quadrature points. Since we have fixed the polynomial order in both expansion directions to P, we can also choose to fix the quadrature order in both directions to $Q_1 = Q_2 = Q$. Depending on the type of Gaussian quadrature adopted, we can set Q such that the integration is exact. For example, when using Gauss–Lobatto–Legendre integration the choice $Q = P+2$ will mean each component of the elemental mass matrix is exactly integrated (see Section 2.4.1). We now define two arrays base1[p][i] and base2[q][j] for $0 \leqslant p, q \leqslant P$, $0 \leqslant i, j < Q$ such that

$$\begin{aligned}\text{base1}[p][i] &= \psi_p^a(\xi_{1i}),\\ \text{base2}[q][j] &= \psi_q^a(\xi_{2j}),\end{aligned}$$

where we recall that $\psi_p^a(\xi)$ is dependent upon the Jacobi polynomials defined in Appendix A. A code in C and C++ is used to evaluate the quadrature points and the Jacobi polynomials can also be found on the web page above in the *Polylib* library. If we are using the same quadrature type, for example Gauss–Lobatto–Legendre (see Section 2.4.1.1), in both the ξ_1 and ξ_2 directions the arrays base1[p][i] and base2[q][j] will be identical and only one array need be defined. Although multi-dimensional integration is discussed in Section 4.1.1, we have already seen how it can be applied in one dimension in Section 2.4.1. The tensor-based expansion in a quadrilateral region is the product of two one-dimensional integrals, i.e.,

$$\begin{aligned}&\boldsymbol{M}^e[n(p,q)][m(r,s)]\\ &\quad= \int_{-1}^{1}\int_{-1}^{1} \phi_{pq}(\xi_1,\xi_2)\phi_{rs}(\xi_1,\xi_2)\,\mathrm{d}\xi_1\,\mathrm{d}\xi_2\\ &\quad= \int_{-1}^{1} \psi_p(\xi_1)\psi_r(\xi_2)\,\mathrm{d}\xi_1 \times \int_{-1}^{1} \psi_q(\xi_2)\psi_s(\xi_2)\,\mathrm{d}\xi_2\end{aligned}$$

[16] Implementation of two-dimensional quadrilateral elemental mass matrices.

$$= \left[\sum_{i=0}^{i=Q} w[i]\,\text{base1}[p][i]\,\text{base1}[r][i]\right] \times \left[\sum_{j=0}^{j=Q} w[j]\,\text{base2}[q][j]\,\text{base2}[s][j]\right],$$

where $w[i] = w_i^{0,0}$ and $w[j] = w_j^{0,0}$ are the quadrature weights. In the above operation $n(p,q)$ and $m(r,s)$ denote a mapping from the pair of one-dimensional indices (p,q) and (r,s) in a single unique numbering which represents the location of each two-dimensional mode in the array $\boldsymbol{M}^e$. There are clearly many different choices of numbering arrays and one of the most straightforward numbering scheme can be constructed as follows:

$$n(p,q) = p \times (P+1) + q\,, \quad m(r,s) = r \times (P+1) + s\,.$$

An alternative numbering scheme would be to place the boundary modes before the interior modes.

(a) To validate your implementation of the two-dimensional Gaussian integration rule, integrate the function $f(x,y) = \xi_1^4 \xi_2^5$ in the standard quadrilateral domain using $Q = 7$ points. The function can be numerically evaluated as

$$\int_{-1}^{1}\int_{-1}^{1} \xi_1^4\xi_2^5\, d\xi_1\, d\xi_2 = \left[\sum_{i=1}^{i=Q} (\xi_{1i})^4\, w[i]\right] \times \left[\sum_{j=1}^{j=Q} (\xi_{2j})^5\, w[j]\right].$$

(b) Construct the two-dimensional mass matrix $\boldsymbol{M}^e$ of rank $N_m = (P+1)^2$ for $P = 7$, $Q = 9$ and plot the structure of the matrix to obtain a similar plot to Fig. 3.18. Note that in this figure the matrix numbering $n(p,q)$ and $m(r,s)$ has been ordered so that the boundary modes appear first followed by the interior modes.

(c) Generate the mass matrix $\boldsymbol{M}^e$ for the orthogonal basis $\phi_{pq}(\xi_1,\xi_2) = \widetilde{\psi}_p^a(\xi_1)\widetilde{\psi}_q^a(\xi_2)$ defined in Section 3.2.2.1 for $P = 7$, $Q = 9$. Note that this is a good debugging exercise since we know it should lead to a diagonal matrix.

(d) Finally, construct the mass matrix for the nodal basis $\phi_{pq}(\xi_1,\xi_2) = h_p(\xi_1)h_q(\xi_2)$ defined in Section 3.1.1. In this case the matrix is only 'discretely' orthogonal when $Q = P+1$. To observe this feature assemble the matrix for $P = 7$, $Q = 9$ and $P = 7$, $Q = 8$.

2. We now consider the construction of the elemental mass matrix for the tensorial-based triangular expansion $\phi_{pq}(\xi_1,\xi_2) = \psi_p^a(\eta_1)\psi_{pq}^b(\eta_2)$ defined in Section 3.2.3.1. In general, this follows a very similar construction to the quadrilateral case, but we now need to use the collapsed coordinate system defined in ♦[17]

[17] Implementation of two-dimensional triangular elemental mass matrices.

Section 3.2.1.1. As will be discussed in more detail in Section 4.1.1.2, the integration of the basis ϕ_{pq} over the triangular region $\Omega_{\rm st} = \mathcal{T}^2 = \{(\xi_1, \xi_2)|-1 \leqslant \xi_1, \xi_2,\ \xi_1 + \xi_2 \leqslant 0\}$ can be expressed as

$$
\begin{aligned}
&\boldsymbol{M}^e[n(p,q)][m(r,s)] \\
&\quad = \int_{-1}^{1} \int_{-1}^{-\xi_1} \phi_{pq}(\xi_1,\xi_2)\phi_{rs}(\xi_1,\xi_2)\, \mathrm{d}\xi_1\, \mathrm{d}\xi_2 \\
&\quad = \int_{-1}^{1} \int_{-1}^{1} \psi_p^a(\eta_1)\psi_{pq}^b(\eta_2)\psi_r^a(\eta_1)\psi_{rs}^b(\eta_2) \left(\frac{1-\eta_2}{2}\right) \mathrm{d}\eta_1\, \mathrm{d}\eta_2 \\
&\quad = \int_{-1}^{1} \psi_p^a(\eta_1)\psi_r^a(\eta_1)\, \mathrm{d}\eta_1 \int_{-1}^{1} \psi_{pq}^b(\eta_2)\psi_{rs}^b(\eta_2) \left(\frac{1-\eta_2}{2}\right) \mathrm{d}\eta_2\,,
\end{aligned}
$$

where

$$
\eta_1 = \frac{2(1+\xi_1)}{1-\xi_2} - 1, \quad \eta_2 = \xi_2\,.
$$

Once again, we observe that a generalised tensor product basis can be evaluated as two one-dimensional type integrals. We can also use different types of Gauss–Jacobi quadrature which automatically absorb the factor of $(1-\xi_2)/2$ into the quadrature weights (see Section 4.1.1.2). The zeros and weight which contain this factor are denoted by $\xi_i^{1,0}$ and $w_i^{1,0}$ and can also be generated using the *Polylib* library available on the web page. To continue the construction of the mass matrix we define two arrays, base1$[p][i]$ and base2$[p][q][j]$, for $0 \leqslant p, q \leqslant P$, $0 \leqslant i, j < Q$ such that

$$
\begin{aligned}
\text{base1}]p][i] &= \psi_p^a(\xi_{1i}^{0,0})\,, \\
\text{base2}[p]]q][j] &= \psi_{pq}^b(\xi_{2j}^{1,0})\,,
\end{aligned}
$$

where $\psi_p^a(\xi)$ and $\psi_{pq}^b(\xi)$ depend upon Jacobi polynomials defined in Appendix A and can be numerically determined using the *Polylib* library. As discussed in Section 3.2.3.1, the array base2$[p][q][i]$ does not have to contain all entries since only some components are required to evaluate the full bases $\phi_{pq}(\xi_1,\xi_2) = \psi_p^a(\eta_1)\psi_{pq}^b(\eta_2)$. Having constructed base1$[p][i]$ and base2$[p][q][j]$, we can discretely construct the mass matrix as

$$
\begin{aligned}
&\boldsymbol{M}^e[n(p,q)][m(r,s)] \\
&\quad = \int_{-1}^{1} \psi_p^a(\eta_1)\psi_r^a(\eta_1)\, \mathrm{d}\eta_1 \int_{-1}^{1} \psi_{pq}^b(\eta_2)\psi_{rs}^b(\eta_2) \left(\frac{1-\eta_2}{2}\right) \mathrm{d}\eta_2 \\
&\quad = \sum_{i=0}^{i=Q} w_1[i]\, \text{base1}[p][i]\, \text{base1}[r][i] \sum_{j=0}^{j=Q} w_2[j]\, \text{base2}[p][q][i]\, \text{base2}[r][s][i]\,,
\end{aligned}
$$

where $w_1[i] = w_i^{0,0}$ and $w_2[j] = w_j^{1,0}/2$. Unlike the quadrilateral expansion, for this case the index arrays $n(p,q)$ and $m(r,s)$ cannot be defined in close-packed form. This point is discussed in Section 4.1.5.1 and highlighted in Fig. 4.6. As also discussed in Section 3.2.3.3, special attention must be taken when constructing the top vertex which is decomposed into two contributions from the base[p][q][j] array.

(a) Construct the two-dimensional mass matrix $\boldsymbol{M}^e$ of rank $N_m = (P+1)(P+2)/2$ for $P = 14$, $Q = 16$ and plot the structure of the matrix to recover Fig. 3.18. Recall that in this figure the matrix numbering $n(p,q)$ and $m(r,s)$ has been ordered so that the boundary modes appear first followed by the interior modes. The indices q and s must run faster than the indices p and r to observe the semi-orthogonal structure of the matrix.

(b) Construct the two-dimensional mass matrix $\boldsymbol{M}^e$ for the orthogonal triangular expansion $\phi_{pq}(\xi_1,\xi_2) = \widetilde{\psi}_p^a(\eta_1)\widetilde{\psi}_{pq}^b(\eta_2)$ and verify that your matrix is diagonal.

3. In this final exercise we will consider how to construct the Lagrange polynomial $L_i^{N_m}(\boldsymbol{\xi})$ through a set of nodal points $\boldsymbol{\xi}_i = [\xi_{1i},\xi_{2i}]^\top$ using the orthogonal triangular tensorial basis $\phi_{pq}(\boldsymbol{\xi}) = \widetilde{\psi}_p^a(\xi_1)\widetilde{\psi}_{pq}^b(\xi_2)$. From Section 3.3.2, we recall that the Lagrange polynomial can be evaluated using the generalised Vandermonde matrix $\boldsymbol{V}$ as

$$\begin{bmatrix} L_0(\boldsymbol{\xi}) \\ \vdots \\ L_{N_m-1}(\boldsymbol{\xi}) \end{bmatrix} = \boldsymbol{V}^{-1} \begin{bmatrix} \phi_0(\boldsymbol{\xi}) \\ \vdots \\ \phi_{N_m-1}(\boldsymbol{\xi}) \end{bmatrix}. \tag{3.5.1}$$

Whilst we can use any polynomial expansion which spans the same polynomial space as $L_i^{N_m}$, the choice of the orthogonal expansion leads to a well-conditioned generalised Vandermonde matrix which is important for numerical inversion. For a discrete set of nodal points $\boldsymbol{\xi}_i$, $0 \leqslant i \leqslant N_m$, through which we define the Lagrange polynomial, we can construct the generalised Vandermonde matrix

$$\boldsymbol{V}[m(p,q)][i] = \widetilde{\psi}_p^a(\eta_{1i})\widetilde{\psi}_{pq}^b(\eta_{2i}),$$

where

$$\eta_{1i} = \frac{2(1+\xi_{i1})}{1-\xi_{2i}}, \qquad \eta_{2i} = \xi_{2i},$$

and $m(p,q)$ denotes a mapping between the index pair (p,q) and the unique index m, for example,

$$m(pq) = q + \frac{p(2P+1-p)}{2}.$$

After inverting $\boldsymbol{V}$, for example using *LAPACK* [12], we can then evaluate the Lagrange polynomials $L_i^{N_m}(\boldsymbol{\xi})$ at any desired location, $\boldsymbol{\xi}$, using eqn (3.5.1).

(a) Determine the $P = 4$ order Lagrange polynomial expansion using $N_m = (P+1)(P+2)/2 = 15$ modes through a set of equispaced points

$$\xi_{1i} = \frac{2i}{P+1} - 1, \quad \xi_{2j} = \frac{2j}{P+1} - 1, \quad 0 \leqslant i, j, \; i+j \leqslant P+1.$$

To plot the function evaluate the Lagrange polynomials, $L_i^{N_m}$, using eqn (3.5.1) and evaluate the Lagrange functions at $P = 8$ ($N_m = 45$) equispaced points.

(b) Determine the $P = 4$ order Lagrange polynomial expansion using $N_m = 15$ modes through the electrostatic points defined in Appendix D and evaluate the Lagrange functions at $P = 8$ ($N_m = 45$) equispaced points.

(c) Evaluate the Lebesgue function $\sum_{0 \leqslant i < N_m} |L_i^{N_m}(\boldsymbol{\xi})|$ for the Lagrange polynomial defined at equispaced and electrostatic nodal points for $P = 4$ and compare your function with Fig. 3.23.

4

MULTI-DIMENSIONAL FORMULATION

In this chapter we consider the implementation and formulation details used in constructing multi-dimension spectral/*hp* element approximations. This chapter builds directly upon the expansion bases within *standard* regions defined in Chapter 3. Although more involved, the construction of a multiple dimensions is very similar to the one-dimensional approach discussed in Chapter 2. By analogy with the one-dimensional formulation, the operations of integration and differentiation of a known function can be performed at an elemental level and may therefore be considered to be *local* or elemental operations. We therefore start our multi-dimensional discussion in Section 4.1 by considering local operations. To extend these techniques to a C^0 multi-dimensional basis, as typically adopted in a Galerkin construction, we then require *global* operations such as matrix numbering, connectivity, and assembly. These topics are introduced in Section 4.2. In Section 4.3 we conclude our formulation by discussing more specialised topics relating to the pre- and post-processing aspects of a general multi-dimensional solver such as boundary condition representation, curvilinear mesh generation, and consistent particle tracking in high-order elements. Finally, in Section 4.4, we suggest a series of programming exercises to help construct a multi-dimensional Galerkin approximation.

Although integration and differentiation are important local operations, the elemental mapping which allows us to generalise the local operations in a standard region to elements of general shape is equally important. The structure of Section 4.1 therefore initially introduces integration and differentiation in the standard regions in Sections 4.1.1 and 4.1.2. Subsequently, in Section 4.1.1, we introduce the concept of an elemental mapping which then allows us to discuss how to perform integration and differentiation in general elemental regions. Having defined these concepts, we are then able to discuss elemental transformation in Section 4.1.5 for representing general functions over an elemental region using either collocation or Galerkin-type projections. Within this section we also introduce a matrix notation which is helpful in illustrating the necessary operations to numerically perform many of the local operations. The matrix construction is convenient to clarify many of the numerical operations; however, when using tensorial-based operations, it is computationally more efficient to use the sum-factorisation technique, as detailed in Section 4.1.6. Since all the techniques discussed in Section 4.1 only apply on a single elemental region, they may equally well be used on the modified or orthogonal bases discussed in Chapter 3. Many of the local operations are also relevant to the non-tensorial expansion bases for simplexes, as defined in Chapter 3. When using the non-tensorial basis we cannot use the sum-factorisation technique, and so matrix operators can be applied. Finally, we note that most of the local operations are equally valid for the standard Galerkin formulation introduced in Chapter 2 or the discontinuous Galerkin formulation which is introduced in Chapters 6 and 7.

For the standard Galerkin formulation we have seen in Chapter 2 that normally a global C^0 continuous expansion is required. This was also the motivation behind the derivation of modified expansion bases defined in Chapter 3. To construct a multiple-domain C^0 expansion from the elemental contributions in a computationally efficient manner requires the introduction of suitable global operations. We, therefore, start our discussion on global operations in Section 4.2.1 by introducing the concept of global assembly. Using the previously defined vector notation we illustrate, from a matrix operation point of view, how the elemental degrees of freedom are assembled (or combined) to make a global C^0 system. Implicit in this discussion is how local degrees of freedom have to be numbered and orientated. The global assembly operation can be used for an explicit and implicit implementation of the problem of interest. However, implicit implementation typically leads to global matrix systems, the construction of which is discussed in Section 4.2.2 using the elemental matrix notation introduced in Section 4.1.5. Finally, having constructed a global matrix system in Section 4.2.3, we discuss the *static condensation* technique which makes use of natural boundary/interior decomposition of the spectral/*hp* element expansion to decouple the interior modes from the boundary degrees of freedom. We then complete our discussion of global operations in Section 4.2.4 by presenting how to set up a global numbering scheme and then using it to impose Dirichlet boundary conditions, with examples in two dimensions.

4.1 Local elemental operations

As a motivation for the topics to be introduced in this section, we recall that to solve the Galerkin formulation of the Laplace equation we need to evaluate within every elemental region of our mesh the inner products of the form

$$(\nabla\phi_i \cdot \nabla\phi_j) = \int_{\Omega^e} \nabla\phi_i \cdot \nabla\phi_j \, \mathrm{d}\boldsymbol{x} = \int_{\Omega_{\mathrm{st}}} \nabla\phi_i \cdot \nabla\phi_j J \, \mathrm{d}\boldsymbol{\xi}\,,$$

where Ω^e denotes the element region, Ω_{st} denotes the standard elemental region such that $\xi \in \Omega_{\mathrm{st}}$, and J is the Jacobian of the mapping between these two regions. From the structure of this inner product, we note that there are three important concepts which we need to understand how to implement. Firstly, we need to know how to integrate within Ω_{st}, then we need to know how to differentiate, initially in the standard region Ω_{st} and then in the elemental region Ω^e. To perform the differentiation (and integration) within the elemental region we need to define a mapping between these regions which is the third concept we will discuss in this section.

In Section 2.4.1 we discussed an accurate form of one-dimensional numerical integration known as Gaussian integration. This is generally preferred in spectral/*hp* element methods as polynomial integrands of order $2P$ can be exactly integrated using a summation over $\mathcal{O}(P)$ points. In Section 4.1.1 we will review how the one-dimensional Gaussian integration can be applied to the standard regions, $\Omega_{\rm st}$, for multi-dimensional expansions.

Similarly, in Section 2.4.2 we illustrated the differentiation of polynomial functions in physical space using the derivatives of a Lagrange polynomial through the Gaussian quadrature zeros. In Section 4.1.2, we extend this concept to the standard multi-dimensional regions, $\Omega_{\rm st}$.

Having defined integration and differentiation in the standard regions in Section 4.1.3, we extend these ideas to an arbitrary region by introducing appropriate elemental mappings. In Section 4.1.5, we illustrate how to represent a multi-dimensional function within a general elemental region, thereby defining the idea of a forward and backward transformation. To this end, we introduce a matrix notation. Finally, in Section 4.1.6 we discuss the use of a numerically efficient technique, known as sum factorisation, to evaluate the salient numerical operations required in a Galerkin *hp*/spectral element formulation based on a tensor product expansion.

4.1.1 *Integration within the standard region* $\Omega_{\rm st}$

The one-dimensional integral of a smooth function may be approximated using Gaussian quadrature as a summation of the form (see also Section 2.4.1.1 and Appendix B)

$$\int_{-1}^{1} u(\xi)\,\mathrm{d}\xi = \sum_{i=0}^{Q-1} w_i u(\xi_i) + R(u),$$

where ξ_i are the Q discrete quadrature points or *zeros* at which the function $u(\xi)$ is evaluated and w_i is the set of coefficients of *weights*. The term $R(u)$ denotes the approximation error which, providing that a sufficient number of quadrature points are used, will be zero if the integrand is a polynomial. For example, if $u(\xi)$ represents the local expansion basis $u(\xi) = \phi_p(\xi)$ then, providing that there are sufficient quadrature points, there will be no approximation error. We recall that the classical Gauss–Legendre quadrature does not include any zeros at the ends of the interval. If both end-points are included then the integration is referred to as *Gauss–Lobatto*, and if only one end-point is included then the integration is referred to as *Gauss–Radau*. A more general form of quadrature involving a weight function in the integrand is referred to as Gauss–Jacobi (see Appendix B).

4.1.1.1 *Quadrilateral and hexahedral regions*

A trivial extension of the one-dimensional Gaussian rule is to the two-dimensional standard quadrilateral region, and similarly to the three-dimensional hexahedral ♦18

[18] Numerical integration in $\Omega_{\rm st}$: quadrilateral and hexahedral regions.

region. Integration over $\mathcal{Q}^2 = \{-1 \leqslant \xi_1, \xi_2 \leqslant 1\}$ is mathematically defined as two one-dimensional integrals of the form

$$\int_{\mathcal{Q}^2} u(\xi_1, \xi_2)\, \mathrm{d}\xi_1\, \mathrm{d}\xi_2 = \int_{-1}^{1} \left\{ \int_{-1}^{1} u(\xi_1, \xi_2)|_{\xi_2}\, \mathrm{d}\xi_1 \right\} \mathrm{d}\xi_2 .$$

So if we replace the right-hand-side integrals with our one-dimensional Gaussian integration rules we obtain

$$\int_{\mathcal{Q}^2} u(\xi_1, \xi_2)\, \mathrm{d}\xi_1\, \mathrm{d}\xi_2 \simeq \sum_{i=0}^{Q_1-1} w_i \left\{ \sum_{j=0}^{Q_2-1} w_j\, u(\xi_{1i}, \xi_{2j}) \right\},$$

where Q_1 and Q_2 are the number of quadrature points in the ξ_1 and ξ_2 directions, respectively. This expression will be exact if $u(\xi_1, \xi_2)$ is a polynomial and Q_1 and Q_2 are chosen appropriately. To numerically evaluate this expression the summation over i must be performed Q_1 times at every ξ_{2i} point, that is,

$$\int_{\mathcal{Q}^2} u(\xi_1, \xi_2)\, \mathrm{d}\xi_1\, \mathrm{d}\xi_2 \simeq \sum_{i=0}^{Q_1-1} w_i\, f(\xi_{1i}),$$

$$f(\xi_{1i}) = \sum_{j=0}^{Q_2-1} w_j\, u(\xi_{1i}, \xi_{2j}).$$

The corresponding three-dimensional numerical integral for $Q^3 = \{-1 \leqslant \xi_1, \xi_2, \xi_3 \leqslant 1\}$ is

$$\int_{\mathcal{Q}^3} u(\xi_1, \xi_2, \xi_3)\, \mathrm{d}\xi_1\, \mathrm{d}\xi_2\, \mathrm{d}\xi_3 \simeq \sum_{i=0}^{Q_1-1} w_i \left\{ \sum_{j=0}^{Q_2-1} w_j \left\{ \sum_{i=0}^{Q_3-1} w_k u(\xi_{1i}, \xi_{2j}, \xi_{3k}) \right\} \right\},$$

which is evaluated in a similar fashion to the two-dimensional case, except that the innermost summation must be evaluated $Q_1 \cdot Q_2$ times. Although any type of Gauss–Legendre quadrature may be used, the preferred distributions which include the end-points as boundary conditions may then be more easily imposed.

♦[19] 4.1.1.2 *Simplex and hybrid regions*

Triangular region

Unlike the structured regions, the standard triangular regions $\mathcal{T}^2 = \{-1 \leqslant \xi_1 \leqslant \xi_1, \xi_2,\ \xi_1 + \xi_2 \leqslant 0\}$ (see Fig. 3.6) expressed in Cartesian coordinates ξ_1, ξ_2 are not very conveniently represented in terms of one-dimensional Gaussian integration as the upper boundary of the region is described in terms of both coordinates. Since the one-dimensional rule is expressed in terms of an interval with constant

[19] Numerical integration in Ω_{st}: simplex and hybrid regions.

bounds (that is, $[-1,1]$), we need to perform a coordinate transformation before we can apply this technique.

The transformation of a triangular region into a region bounded by constants is equivalent to mapping the triangular region into a quadrilateral, as described in Section 3.2.1. The collapsed Cartesian system is therefore suitable for Gauss integration in the unstructured regions. For modal expansions, integration in terms of the collapsed Cartesian system remains accurate as the generalised tensorial bases are polynomials in both the Cartesian and the collapsed Cartesian systems. Although alternative integration schemes using the barycentric coordinate system (see Section 3.2.1) have been developed by various researchers including Dunavant [146] and Yu [503], this type of integration does not take advantage of the tensorial construction of the unstructured basis, as shown in Section 4.1.6. The order of these schemes also tends to be restricted by the numerical process of evaluating the quadrature weights.

The two-dimensional collapsed Cartesian system (see Section 3.2.1) is defined by the coordinate transformation

$$\eta_1 = 2\frac{1+\xi_1}{1-\xi_2} - 1, \quad \eta_2 = \xi_2 .$$

If we express our integral over the region $\mathcal{T}^2$, the collapsed Cartesian system (η_1, η_2), then we obtain

$$\begin{aligned} \int_{\mathcal{T}^2} u(\xi_1,\xi_2)\, d\xi_1\, d\xi_2 &= \int_{-1}^{1}\int_{-1}^{-\xi_2} u(\xi_1,\xi_2)\, d\xi_1\, d\xi_2 \\ &= \int_{-1}^{1}\int_{-1}^{1} u(\eta_1,\eta_2) \left| \frac{\partial(\xi_1,\xi_2)}{\partial(\eta_1,\eta_2)} \right| d\eta_1\, d\eta_2 , \end{aligned} \tag{4.1.1}$$

where $\partial(\xi_1,\xi_2)/\partial(\eta_1,\eta_2)$ is the Jacobian of the Cartesian to local coordinate transformation and can be expressed in terms of η_1 and η_2 by

$$\frac{\partial(\xi_1,\xi_2)}{\partial(\eta_1,\eta_2)} = \frac{1-\eta_2}{2} .$$

The last term in eqn (4.1.1) can be approximated using one-dimensional Gaussian quadrature rules to arrive at

$$\int_{-1}^{1}\int_{-1}^{1} u(\eta_1,\eta_2)\frac{1-\eta_2}{2}\, d\eta_1\, d\eta_2 = \sum_{i=0}^{Q_1-1} w_i \left\{ \sum_{j=0}^{Q_2-1} w_j\, u(\eta_{1i},\eta_{2j})\frac{1-\eta_{2j}}{2} \right\}, \tag{4.1.2}$$

where η_{1i} and η_{2j} are the quadrature points in the η_1 and η_2 directions, respectively. The weights w_i and w_j used in eqn (4.1.2) correspond to the standard Gauss–Legendre rule which may or may not include the end-points. However, a more general quadrature rule, which we shall refer to as Gauss–Jacobi quadrature, includes the factor $(1-\xi)^\alpha(1+\xi)^\beta$ in the integrand, that is,

$$\int_{-1}^{1}(1-\xi)^\alpha(1+\xi)^\beta u(\xi)\,\mathrm{d}\xi = \sum_{i=0}^{Q-1} w^{\alpha,\beta}u(\xi_i^{\alpha,\beta}),$$

where $w^{\alpha,\beta}$ and $\xi_i^{\alpha,\beta}$ are the weights and zeros which correspond to the choice of the exponents α and β (see Ghizzetti and Ossicini [182]). If $\alpha=\beta=0$ then we recover the standard Gauss–Legendre quadrature rules. The Gauss–Jacobi quadrature rules can be derived for a Lobatto and Radau distribution of zeros as well as the classical Gaussian distribution (see Appendix B).

The Gauss–Jacobi rules are convenient in evaluating the integral (4.1.2) since we are able to include the Jacobian term $\partial(\xi_1,\xi_2)/\partial(\eta_1,\eta_2)=(1-\eta_2)/2$ directly in the quadrature weights by setting $\alpha=1$, $\beta=0$. Accordingly, the integration scheme over $\mathcal{T}^2$ becomes

$$\int_{-1}^{1}\int_{-1}^{1} u(\eta_1,\eta_2)\frac{1-\eta_2}{2}\,\mathrm{d}\eta_1\,\mathrm{d}\eta_2 = \sum_{i=0}^{Q_1-1} w_i^{0,0}\left\{\sum_{j=0}^{Q_2-1}\hat{w}_j^{1,0}u(\eta_{1i},\eta_{2j})\right\},$$

where

$$\hat{w}_j^{1,0} = \frac{w_j^{1,0}}{2}.$$

The Gauss–Jacobi rule therefore uses fewer quadrature points than the standard Gauss–Legendre quadrature rule to achieve an equivalent accuracy.

When choosing a distribution of points on which to integrate, the Lobatto-type quadrature is preferred since it includes the end-points of the interval $[-1,1]$, which is helpful when setting boundary conditions. However, when integrating over a triangular region we note that the use of the Radau distribution in the η_2 direction (which includes the point at $\eta_2=-1$) is advantageous as it avoids the need for explicit calculation of any information at the degenerate vertex ($\eta_1=-1$, $\eta_2=1$). Although this vertex does not cause any problems when integrating over $\mathcal{T}^2$, it does present added complications when differentiating in $\mathcal{T}^2$ (see Section 4.1.2). The distribution of quadrature points in $\mathcal{T}^2$ for $Q_1=Q_2=7$ using a Gauss–Lobatto–Legendre scheme in the η_1 direction and a Gauss–Radau–Jacobi scheme in the η_2 direction is shown in Fig. 4.1.

Tetrahedral region

To integrate over $\mathcal{T}^3=\{-1\leqslant\xi_1,\xi_2,\xi_3,\ \xi_1+\xi_2+\xi_3\leqslant-1\}$ we use the collapsed Cartesian coordinate system for the tetrahedron defined as

$$\eta_1=\frac{2(1+\xi_1)}{-\xi_2-\xi_3}-1,\quad \eta_2=\frac{2(1+\xi_2)}{1-\xi_3}-1,\quad \eta_3=\xi_3.$$

Using this system the integration becomes

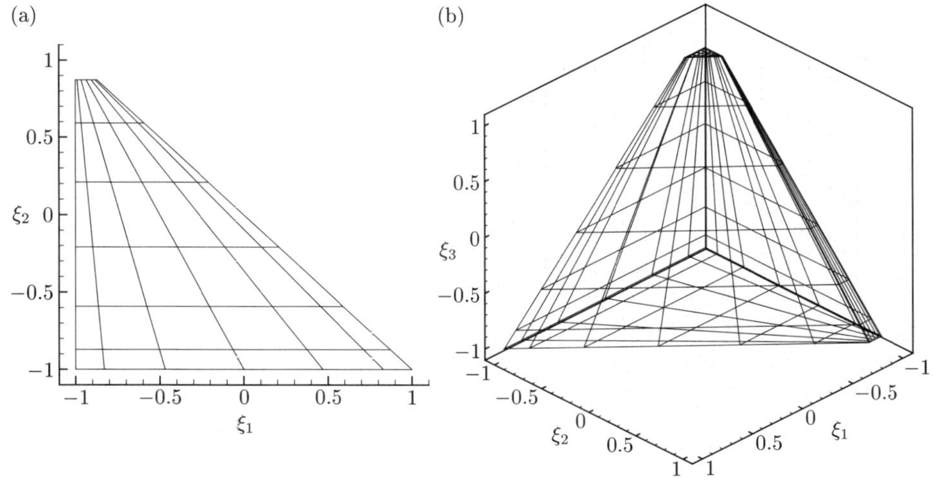

FIG. 4.1. Quadrature points in the standard (a) triangular $\mathcal{T}^2$, and (b) tetrahedral $\mathcal{T}^3$ space, with $Q_1 = Q_2 = Q_3 = 7$. In the 'η_1' direction a Gauss–Lobatto–Legendre distribution has been used, and in the 'η_2' and 'η_3' directions a Gauss–Radau–Jacobi distribution was used.

$$\int_{\mathcal{T}^3} u(\xi_1, \xi_2, \xi_3)\, d\xi_1\, d\xi_2\, d\xi_3 = \int_{-1}^{1}\int_{-1}^{1}\int_{-1}^{1} u(\eta_1, \eta_2, \eta_3) J\, d\eta_1\, d\eta_2\, d\eta_3\,,$$

where the Jacobian, J, is

$$J = \frac{\partial(\xi_1, \xi_2, \xi_3)}{\partial(\eta_1, \eta_2, \eta_3)} = \frac{1-\eta_2}{2}\left(\frac{1-\eta_3}{2}\right)^2.$$

As in the triangular integration, we can include the Jacobian in the quadrature weights by using the Gauss–Jacobi integration rules with $\alpha = 0$, $\beta = 0$ in the 'η_1' direction (i.e., Gauss–Legendre quadrature), $\alpha = 1$, $\beta = 0$ in the 'η_2' direction, and $\alpha = 2$, $\beta = 0$ in the 'η_3' direction. The integration rule over $\mathcal{T}^3$ then becomes

$$\int_{-1}^{1}\int_{-1}^{1}\int_{-1}^{1} u(\eta_1, \eta_2, \eta_3)\frac{1-\eta_2}{2}\left(\frac{1-\eta_3}{2}\right)^2 d\eta_1\, d\eta_2\, d\eta_3$$
$$= \sum_{i=0}^{Q_1-1}\sum_{j=0}^{Q_2-1}\sum_{k=0}^{Q_3-1} u(\eta_{1i}^{0,0}, \eta_{2j}^{1,0}, \eta_{3k}^{2,0}) w_i^{0,0} \hat{w}_j^{1,0} \hat{w}_k^{2,0}\,,$$

where

$$\hat{w}_j^{1,0} = \frac{w_j^{1,0}}{2}\,, \quad \hat{w}_k^{2,0} = \frac{w_k^{2,0}}{4}\,,$$

and Q_1, Q_2, and Q_3 are the number of quadrature points in the η_1, η_2, and η_3 directions, respectively.

Once again, we can choose any type of point distribution in the quadrature rule (that is, Gauss, Gauss–Radau, or Gauss–Lobatto). There are no explicit restrictions as to what type we use. The Gauss–Lobatto can be convenient as it has zeros at the ends of the integration domain, thus allowing greater ease in imposing the boundary conditions. Its use does mean, however, that we have multiple quadrature points at the vertices ($\xi_1 = -1, \xi_2 = -1, \xi_3 = 1$) and ($\xi_1 = -1, \xi_2 = 1, \xi_3 = -1$) as well as along the edge between these vertices. As in the two-dimensional case, this is undesirable because of the redundancy of quadrature points and the difficulty in evaluating the derivatives at these points. The use of Gauss–Radau quadrature, which includes a zero at -1 in the 'η_2' and 'η_3' directions, circumvents this problem. The use of Gauss–Lobatto–Legendre quadrature in the 'η_1' direction means that, as shown in Fig. 4.1, there are quadrature points along all of the boundaries of $\mathcal{T}^3$, except at the vertices ($\xi_1 = -1, \xi_2 = -1, \xi_3 = 1$) and ($\xi_1 = -1, \xi_2 = 1, \xi_3 = -1$) and the edge between them.

Prismatic and pyramidic regions

To complete the integration section we summarise the quadrature rules in the prismatic and pyramidic regions, as shown in Fig. 3.9. For both these regions it is preferable to use Lobatto-type quadrature in the first two ordinates and Radau-type in the third ordinate. This generates as many quadrature points as possible on the boundary of the region while avoiding any complications with differentiation in the region.

For the *prismatic region* we use the local coordinate system

$$\overline{\eta}_1 = \frac{2(1+\xi_1)}{1-\xi_3} - 1\,, \quad \xi_2\,, \quad \xi_3\,.$$

Integration within this region can be approximated by

$$\int_{-1}^{1}\int_{-1}^{1}\int_{-1}^{1} u(\overline{\eta}_1, \xi_2, \xi_3)\frac{1-\xi_3}{2}\,\mathrm{d}\overline{\eta}_1\,\mathrm{d}\xi_2\,\mathrm{d}\xi_3 = \sum_{i=0}^{Q_1-1}\sum_{j=0}^{Q_2-1}\sum_{k=0}^{Q_3-1} w_i^{0,0} w_j^{0,0} \hat{w}_k^{1,0} u(\overline{\eta}_{1i}^{0,0}, \xi_{2j}^{0,0}, \xi_{3k}^{1,0})\,,$$

where

$$\hat{w}_k^{1,0} = \frac{w^{1,0}}{2}\,.$$

For the *pyramidic region* the local coordinate system is

$$\overline{\eta}_1 = \frac{2(1+\xi_1)}{1-\xi_3} - 1\,, \quad \eta_2 = \frac{2(1+\xi_2)}{1-\xi_3} - 1\,, \quad \eta_3 = \xi_3\,,$$

and integration within this region can be approximated by

$$\int_{-1}^{1}\int_{-1}^{1}\int_{-1}^{1} u(\overline{\eta}_1,\eta_2,\eta_3)\left(\frac{1-\eta_3}{2}\right)^2 \mathrm{d}\overline{\eta}_1\,\mathrm{d}\eta_2\,\mathrm{d}\eta_3$$
$$= \sum_{i=0}^{Q_1-1}\sum_{j=0}^{Q_2-1}\sum_{k=0}^{Q_3-1} w_i^{0,0} w_j^{0,0} \hat{w}_k^{2,0} u(\overline{\eta}_{1i}^{0,0}, \eta_{2j}^{0,0}, \eta_{3k}^{2,0}),$$

where

$$\hat{w}_k^{2,0} = \frac{w^{2,0}}{4}.$$

4.1.2 *Differentiation in the standard region* Ω_{st}

In formulating a differential problem we typically require the derivative in terms of the general Cartesian coordinates such as $\partial u/\partial x$. We note, however, that the use of a mapping from an elemental region $\boldsymbol{x} \in \Omega_e$ to the standard region $\boldsymbol{\xi} \in \Omega_{\mathrm{st}}$ allows us to evaluate the derivative using the chain rule of the form

$$\frac{\partial u}{\partial x_1} = \frac{\partial u}{\partial \xi_1}\frac{\partial \xi_1}{\partial x_1} + \frac{\partial u}{\partial \xi_2}\frac{\partial \xi_2}{\partial x_1} + \frac{\partial u}{\partial \xi_3}\frac{\partial \xi_3}{\partial x_1}.$$

In Section 4.1.5 we will introduce the mapping which will allow us to determine the geometric factors $\partial\xi_1/\partial x_1$, $\partial\xi_2/\partial x_1, \ldots$ to evaluate the full derivatives in terms of Cartesian coordinates.

In this section we first discuss how to differentiate a polynomial function in the local region Ω_{st} to evaluate derivatives of the form $\partial u/\partial \xi_1$, $\partial u/\partial \xi_2, \ldots$. As for the one-dimensional case (see Section 2.4.2) we shall restrict ourselves to differentiation in physical space, which means that the polynomial function is represented by Lagrange polynomials through a set of discrete points which are typically the quadrature points.

There is no need to distinguish between the modal or nodal expansions discussed in Chapter 3 since a modal expansion can always be represented as a Lagrange polynomial of equivalent order, that is,

$$u^\delta(\xi) = \sum_{p=0}^{P} \hat{u}\phi_p(\xi) = \sum_{p=0}^{P} u_p h_p(\xi),$$

where $h_p(\xi)$ is the Lagrange polynomial which passes through a set of $P+1$ points. Due to the collocation property of the Lagrangian representation (that is, $h_i(\xi_j) = \delta_{ij}$), the coefficients u_p are the values of the approximation at the nodal point, for example in one dimension

$$u_p = u^\delta(\xi_p).$$

This collocation property also implies that the derivative of $u^\delta(\xi)$ at the nodal points, ξ_p, can also be described in terms of an expansion in Lagrange polynomials, that is,

$$\frac{\partial u^\delta}{\partial \xi}(\xi) = \sum_{p=0}^{P} u_p \frac{\partial h_p}{\partial \xi}(\xi) = \sum_{p=0}^{P} u'_p h_p(\xi)\,,$$

where

$$u'_p = \sum_{q=0}^{P} u_q \left.\frac{\partial h_q}{\partial \xi}(\xi)\right|_{\xi_p}.$$

This is very significant when calculating nonlinear terms such as the advection operator in the Navier–Stokes equation. For example, to determine the value of the nonlinear product

$$u^\delta(\xi)\frac{\partial u^\delta}{\partial \xi}(\xi)$$

at a point ξ_i, we have

$$\begin{aligned} u^\delta(\xi_i)\frac{\partial u^\delta}{\partial \xi}(\xi_i) &= \left(\sum_{p=0}^{P} u_p h_p(\xi_i)\right)\left(\sum_{q=0}^{P} u_q \frac{\partial h_q}{\partial \xi}(\xi_i)\right) \\ &= \left(\sum_{p=0}^{P} u_p h_p(\xi_i)\right)\left(\sum_{q=0}^{P} u'_q h_q(\xi_i)\right). \end{aligned}$$

Since $h_p(\xi_i) = \delta_{pi}$ and $h_q(\xi_i) = \delta_{qi}$, then we have

$$u^\delta(\xi_i)\frac{\partial u^\delta}{\partial \xi}(\xi_i) = u_i u'_i\,.$$

Finally, we can represent our nonlinear product in terms of an expansion of Lagrange polynomials as

$$u^\delta(\xi)\frac{\partial u^\delta}{\partial \xi}(\xi) \simeq \sum_{p=0}^{P} u_p u'_p h_p(\xi)\,.$$

We note, however, that if $u^\delta(\xi)$ is a polynomial of order P then the nonlinear product $u^\delta(\xi)\,\partial u^\delta/\partial\xi(\xi)$ is a polynomial of order $2P-1$, and so it cannot be exactly represented by the Lagrange polynomial expansion of order P. At the nodal points the coefficient $u_p u'_p$ will be identical to the value of $u^\delta(\xi_p)\,\partial u^\delta/\partial\xi(\xi_p)$. Nevertheless, projecting the nonlinear terms to a lower polynomial order in this fashion can lead to aliasing errors, as discussed in Section 2.4.1.2.

Although this example is in one dimension, the same properties apply in multiple dimensions provided that the expansion can be represented by a tensor product of Lagrange polynomials. Using the collapsed Cartesian coordinates systems described in Section 3.2.1, it is possible to represent any polynomial expansion as a tensor product of one-dimensional Lagrange polynomials.

4.1.2.1 *Two-dimensional differentiation in the standard regions, $\Omega_{\rm st}$* ♦[20]
Quadrilateral region
To differentiate an expansion within the standard quadrilateral region Q^2 of the form

$$u^\delta(\xi_1,\xi_2) = \sum_{p=0}^{P_1}\sum_{q=0}^{P_2} \hat{u}_{pq}\phi_{pq}(\xi_1,\xi_2)\,,$$

we first represent the function in terms of Lagrange polynomials, so

$$u^\delta(\xi_1,\xi_2) = \sum_{p=0}^{Q_1-1}\sum_{q=0}^{Q_2-1} u_{pq}\, h_p(\xi_1)h_q(\xi_2)\,,$$

where

$$u_{pq} = u^\delta(\xi_{1p},\xi_{2q})\,,\quad Q_1 > P_1\,,\quad Q_2 > P_2\,,$$

and ξ_p and ξ_q are typically the zeros of an appropriate Gaussian quadrature. The operation of evaluating u_{pq} from $\hat{u}_{pq}$ is a backwards transformation which we discuss further in Section 4.1.5. The partial derivative with respect to ξ_1 is therefore

$$\frac{\partial u^\delta}{\partial \xi_1}(\xi_1,\xi_2) = \sum_{p=0}^{P_1}\sum_{q=0}^{P_2} u_{pq}\frac{\mathrm{d}h_p(\xi_1)}{\mathrm{d}\xi_1}h_q(\xi_2)\,. \tag{4.1.3}$$

A procedure for evaluating $\mathrm{d}h_p(\xi)/\mathrm{d}\xi$ at the Gaussian quadrature points is illustrated in Section 2.4.2 and Appendix C. From eqn (4.1.3) we can see that to evaluate the partial derivative at an arbitrary point in (ξ_1,ξ_2) we need to perform an $\mathcal{O}(P^2)$ summation over p and q. If we evaluate the derivative at a nodal point (ξ_{1i},ξ_{2j}) of the Lagrange polynomial then the operation count is only $\mathcal{O}(P)$ since $h_q(\xi_{2j}) = \delta_{qj}$, that is,

$$\frac{\partial u^\delta}{\partial \xi_1}(\xi_{1i},\xi_{2j}) = \sum_{p=0}^{P_1}\sum_{q=0}^{P_2}\left\{ u_{pq}\left.\frac{\mathrm{d}h_p(\xi_1)}{\mathrm{d}\xi_1}\right|_{\xi_{1i}}\delta_{qj}\right\} = \sum_{p=0}^{P_1} u_{pj}\left.\frac{\mathrm{d}h_p(\xi_1)}{\mathrm{d}\xi_1}\right|_{\xi_{1i}}.$$

For a Galerkin formulation we normally only require the derivatives at the nodal points of the Gaussian quadrature since we typically have to evaluate inner products of the form $(\nabla\phi,\nabla\phi)$. The total cost of evaluating the derivative at $\mathcal{O}(P^2)$ quadrature points will therefore be $\mathcal{O}(P^3)$. The partial derivative with respect to ξ_2 can be evaluated in a similar fashion, to arrive at

$$\frac{\partial u^\delta}{\partial \xi_2}(\xi_{1i},\xi_{2j}) = \sum_{q=0}^{P_2} u_{iq}\left.\frac{\mathrm{d}h_q(\xi_2)}{\mathrm{d}\xi_2}\right|_{\xi_{2j}}.$$

[20] Numerical differentiation in $\Omega_{\rm st}$: quadrilateral and triangular regions.

Triangular region

For the triangular region, $\mathcal{T}^2$, we can also represent any polynomial expansion in terms of the Lagrange polynomial using the collapsed coordinates η_1 and η_2:

$$u^\delta(\xi_1,\xi_2) = \sum_{p,q}^{P_1,P_2} \hat{u}_{pq}\,\phi_{pq}(\eta_1,\eta_2) = \sum_{p=0}^{P_1}\sum_{q=0}^{P_2} u_{pq}h_p(\eta_1)h_q(\eta_2)\,,$$

where

$$u_{pq} = u^\delta(\eta_{1p},\eta_{2q})\,,\quad \eta_1 = \frac{2(1+\xi_1)}{1-\xi_2} - 1\,,\quad \eta_2 = \xi_2\,,$$

and η_{1p} and η_{2q} refer to the nodal points of the Lagrange polynomial. The summation over the indices p and q for the modified triangular expansion is dependent upon P_1 and P_2, but does not have a *close-packed* form and so it cannot be summed consecutively. However, if $Q_1 > P_1$ and $Q_2 > P_2$ then the polynomial space of the basis $\phi_{pq}(\eta_1,\eta_2)$ is a subset of the space spanned by the Lagrange polynomials $h_p(\eta_1)h_q(\eta_2)$. The partial derivative with respect to the Cartesian system ξ_1 and ξ_2 may be determined by applying the chain rule:

$$\begin{pmatrix} \dfrac{\partial}{\partial\xi_1} \\ \dfrac{\partial}{\partial\xi_2} \end{pmatrix} = \begin{pmatrix} \dfrac{2}{1-\eta_2}\dfrac{\partial}{\partial\eta_1} \\ 2\dfrac{1+\eta_1}{1-\eta_2}\dfrac{\partial}{\partial\eta_1} + \dfrac{\partial}{\partial\eta_2} \end{pmatrix}. \tag{4.1.4}$$

Similarly, to differentiate in the quadrilateral region, the value of the partial derivative with respect to η_1 and η_2 at the nodal points is given by

$$\frac{\partial u^\delta}{\partial\eta_1}(\eta_{1i},\eta_{2j}) = \sum_{p=0}^{P_1} u_{pj}\left.\frac{\mathrm{d}h_p(\eta_1)}{\mathrm{d}\eta_1}\right|_{\eta_{1i}}, \tag{4.1.5a}$$

$$\frac{\partial u^\delta}{\partial\eta_2}(\eta_{1i},\eta_{2j}) = \sum_{q=0}^{P_2} u_{iq}\left.\frac{\mathrm{d}h_q(\eta_2)}{\mathrm{d}\eta_2}\right|_{\eta_{2j}}. \tag{4.1.5b}$$

Finally, substituting eqns (4.1.5a) and (4.1.5b) into eqn (4.1.4) we obtain the partial derivatives of the function u with respect to the Cartesian coordinates (ξ_1,ξ_2) (see Section 2.4.2 and Appendix C for the formula to determine $\mathrm{d}h_q(\eta)/\mathrm{d}\eta$).

Before considering the three-dimensional cases, we note that determining the derivative at $\eta_2 = 1$ presents a problem as the coefficients of the $\partial/\partial\eta_1$ term in eqn (4.1.4) have a numerator which tends to zero as $\eta_2 \to 0$. Although the derivative is well defined at this point, from a computational viewpoint special treatment is necessary. This issue may, however, be circumvented by using a Radau-type Gaussian quadrature in the η_2 direction. An analytic form of the derivative at $(\xi_1 = -1, \xi_2 = 1)$ may be constructed from the modal representation of the expansion, for example, the partial derivative with respect to ξ_1 may be written

$$\frac{\partial u}{\partial \xi_1}(\eta_1, \eta_2) = \sum_{p,q}^{P_1,P_2} \hat{u}_{pq} \frac{2}{1-\eta_2} \frac{\partial \phi_{pq}}{\partial \eta_1}(\eta_1, \eta_2).$$

This summation still appears to be singular at $(-1, 1)$, but we note that all of the modes ϕ_{pq}, except one, contain the factor $1-\eta_2$ which cancels the $1/(1-\eta_2)$ term. The only mode which does not contain this factor corresponds to the top-vertex mode which is independent of η_1, and therefore is necessarily zero when we take the partial derivative with respect to η_1. Furthermore, this summation need not be evaluated over all $\mathcal{O}(P^2/2)$ modes as many modes contain a factor of $(1-\eta_2)^k$, where $k > 1$, and therefore, even when differentiated, will be zero at $\eta_2 = 1$.

4.1.2.2 *Three-dimensional differentiation in the standard regions, $\Omega_{\rm st}$* ♦[21]

The three-dimensional derivatives are analogous to the two-dimensional case, save that the function is now described in terms of a tensor product of three one-dimensional Lagrange polynomials, that is,

$$u^\delta(\xi_1, \xi_2, \xi_3) = \sum_p^{P_1} \sum_q^{P_2} \sum_r^{P_3} u_{pqr} h_p(\xi_1) h_q(\xi_2) h_r(\xi_3),$$

and so the partial derivative with respect to ξ_1 is

$$\frac{\partial u^\delta}{\partial \xi_1}(\xi_1, \xi_2, \xi_3) = \sum_p^{P_1} \sum_q^{P_2} \sum_r^{P_3} u_{pqr} \frac{\partial h_p(\xi_1)}{\partial \xi_1} h_q(\xi_2) h_r(\xi_3),$$

which when evaluated at the nodal points of the Lagrange polynomials becomes

$$\frac{\partial u^\delta}{\partial \xi_1}(\xi_{1i}, \xi_{2j}, \xi_{3k}) = \sum_p^{P_1} u_{pjk} \left.\frac{\partial h_p(\xi_1)}{\partial \xi_1}\right|_{\xi_{1i}}. \tag{4.1.6a}$$

Similarly, the other two partial derivatives at nodal points are given by

$$\frac{\partial u^\delta}{\partial \xi_2}(\xi_{1i}, \xi_{2j}, \xi_{3k}) = \sum_q^{P_2} u_{iqk} \left.\frac{\partial h_q(\xi_2)}{\partial \xi_2}\right|_{\xi_{2j}}, \tag{4.1.6b}$$

$$\frac{\partial u^\delta}{\partial \xi_3}(\xi_{1i}, \xi_{2j}, \xi_{3k}) = \sum_r^{P_3} u_{ijr} \left.\frac{\partial h_r(\xi_3)}{\partial \xi_3}\right|_{\xi_{3k}}. \tag{4.1.6c}$$

Each of these expressions takes an $\mathcal{O}(P)$ operation to evaluate a single derivative, which implies that the cost of evaluating a derivative at $\mathcal{O}(P^3)$ quadrature points within the region is $\mathcal{O}(P^4)$. The formula for evaluating $\mathrm{d}h_p(\xi)/\mathrm{d}\xi$ can be found in Section 2.4.2 and Appendix C.

[21] Numerical differentiation in $\Omega_{\rm st}$: three-dimensional regions.

Equations (4.1.6a)–(4.1.6c) clearly define the partial derivatives for the *hexahedral region.* If, however, we replace the Cartesian coordinates by any of the local collapsed Cartesian coordinates then we can interpret these equations as representing the partial derivatives with respect to the local coordinates. Just as in the two-dimensional case for the triangular region, we can use the chain rule to obtain the local partial derivatives with respect to the Cartesian coordinates. For hybrid three-dimensional domains the partial derivatives are evaluated using the following expressions.

Tetrahedral region

$$\begin{pmatrix} \dfrac{\partial}{\partial \xi_1} \\ \dfrac{\partial}{\partial \xi_2} \\ \dfrac{\partial}{\partial \xi_3} \end{pmatrix} = \begin{pmatrix} \dfrac{4}{(1-\eta_2)(1-\eta_3)} \dfrac{\partial}{\partial \eta_1} \\ \dfrac{2(1+\eta_1)}{(1-\eta_2)(1-\eta_3)} \dfrac{\partial}{\partial \eta_1} + \dfrac{2}{1-\eta_3} \dfrac{\partial}{\partial \eta_2} \\ \dfrac{2(1+\eta_1)}{(1-\eta_2)(1-\eta_3)} \dfrac{\partial}{\partial \eta_1} + \dfrac{1+\eta_2}{1-\eta_3} \dfrac{\partial}{\partial \eta_2} + \dfrac{\partial}{\partial \eta_3} \end{pmatrix},$$

where

$$\eta_1 = 2\frac{1+\xi_1}{-\xi_2-\xi_3} - 1, \quad \eta_2 = 2\frac{1+\xi_2}{1-\xi_3} - 1, \quad \eta_3 = \xi_3 .$$

Pyramidic region

$$\begin{pmatrix} \dfrac{\partial}{\partial \xi_1} \\ \dfrac{\partial}{\partial \xi_2} \\ \dfrac{\partial}{\partial \xi_3} \end{pmatrix} = \begin{pmatrix} \dfrac{2}{1-\eta_3} \dfrac{\partial}{\partial \overline{\eta_1}} \\ \dfrac{2}{1-\eta_3} \dfrac{\partial}{\partial \eta_2} \\ \dfrac{1+\overline{\eta_1}}{1-\eta_3} \dfrac{\partial}{\partial \overline{\eta_1}} + \dfrac{1+\eta_2}{1-\eta_3} \dfrac{\partial}{\partial \eta_2} + \dfrac{\partial}{\partial \eta_3} \end{pmatrix},$$

where

$$\overline{\eta_1} = 2\frac{1+\xi_1}{1-\xi_3} - 1, \quad \eta_2 = 2\frac{1+\xi_2}{1-\xi_3} - 1, \quad \eta_3 = \xi_3 .$$

Prismatic region

$$\begin{pmatrix} \dfrac{\partial}{\partial \xi_1} \\ \dfrac{\partial}{\partial \xi_2} \\ \dfrac{\partial}{\partial \xi_3} \end{pmatrix} = \begin{pmatrix} \dfrac{2}{1-\eta_3} \dfrac{\partial}{\partial \overline{\eta}_1} \\ \dfrac{\partial}{\partial \xi_2} \\ \dfrac{1+\overline{\eta}_1}{1-\eta_3} \dfrac{\partial}{\partial \overline{\eta}_1} + \dfrac{\partial}{\partial \eta_3} \end{pmatrix},$$

where

$$\overline{\eta}_1 = 2\frac{1+\xi_1}{1-\xi_3} - 1, \quad \eta_3 = \xi_3 .$$

For the tetrahedral, prismatic, and pyramidic regions there is a potential problem when $\eta_2 = 1$ or $\eta_3 = 1$ due to the factors $1/[(1-\eta_2)(1-\eta_3)]$ and $1/(1-\eta_3)$. As before, evaluation of the derivative at this point can be avoided by the use of Gauss–Radau quadrature in both the 'η_2' and 'η_3' directions for the tetrahedral region, and in the 'η_3' direction for the prismatic and pyramidic regions. The derivatives are well defined in these regions, although evaluation at $\eta_2 = 1$ or $\eta_3 = 1$ does require analytic differentiation of the expansion modes or interpolation from the derivatives evaluated at the Gauss–Radau distribution of points.

4.1.3 *Operations within general-shaped elements*

We have seen how to integrate and differentiate within the standard region Ω_{st}; however, in practice we need to perform these operations in the elemental regions, Ω^e, which may be of a generalised shape and orientation, as illustrated in Fig. 4.2. To consider these cases we need to define a one-to-one mapping between the Cartesian coordinates (x_1, x_2) and the local Cartesian coordinates (ξ_1, ξ_2), which are denoted by

$$x_1 = \chi_1^e(\xi_1, \xi_2), \quad x_2 = \chi_2^e(\xi_1, \xi_2)$$

in two dimensions, and similarly

$$x_1 = \chi_1^e(\xi_1, \xi_2, \xi_3), \quad x_2 = \chi_2^e(\xi_1, \xi_2, \xi_3), \quad x_3 = \chi_3^e(\xi_1, \xi_2, \xi_3)$$

in three dimensions.

In Section 4.1.3.1 we start by discussing how to define a mapping χ_i^e, from the elemental region to the standard region for straight-sided elements, which simply requires information about the vertices of an element. Elements can also be curvilinear, although in this case some information about how the edges (or faces in three dimensions) are curved is also required. When this is known, we can define a more complex elemental mapping, as discussed in Section 4.1.3.2. Regardless of the complexity of the mapping, we still need to know how to integrate and differentiate in a general domain. This is discussed in Sections 4.1.3.3 and 4.1.3.4. Finally, in Section 4.1.4 we discuss how to evaluate the Jacobian of the mapping between a curved boundary and a standard region, as typically required in a surface integral term.

4.1.3.1 *Elemental mappings for general straight-sided elements*

For elemental shapes with straight sides a simple mapping may be constructed using the linear vertex modes of a modified hierarchical/modal expansion. For example, to map a triangular region (as in Fig. 4.2(a)) assuming that the global coordinates of the triangle $\{(x_1^A, x_2^A), (x_1^B, x_2^B), (x_1^C, x_2^C)\}$ are known (with C being the collapsed vertex), we can use

$$\begin{aligned} x_i &= \chi_1^e(\eta_1, \eta_2) \\ &= x_i^A \frac{1-\eta_1}{2}\frac{1-\eta_2}{2} + x_i^B \frac{1+\eta_1}{2}\frac{1-\eta_2}{2} + x_i^C \frac{1+\eta_2}{2}, \quad i = 1, 2. \end{aligned} \tag{4.1.7}$$

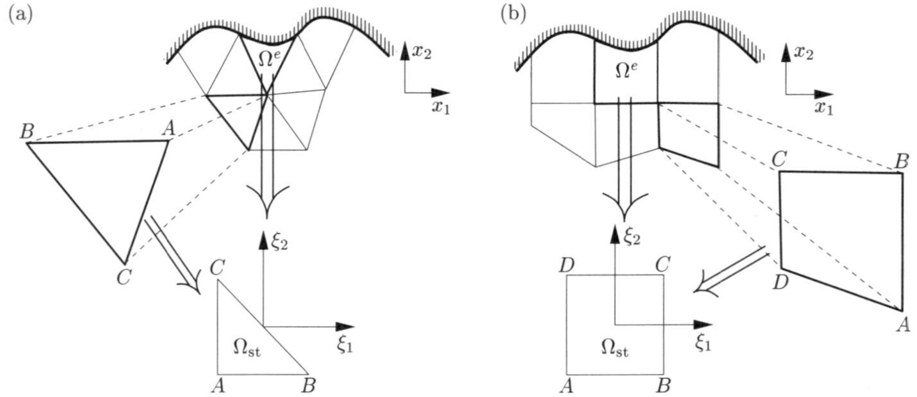

FIG. 4.2. To construct a C^0 expansion from multiple elements of specified shapes (for example, triangles or rectangles), each elemental region Ω^e is mapped to a standard region Ω_{st} in which all local operations are evaluated.

Equation (4.1.7) is expressed in terms of collapsed Cartesian coordinates, but can easily be expressed in terms of the Cartesian coordinates by recalling that $\eta_1 = 2(1+\xi_1)/(1-\xi_2) - 1$ and $\eta_2 = \xi_2$, which on substitution into eqn (4.1.7) gives

$$x_i = \chi(\xi_1, \xi_2) = x_i^A \frac{-\xi_2 - \xi_1}{2} + x_i^B \frac{1+\xi_1}{2} + x_i^C \frac{1+\xi_2}{2}, \quad i = 1, 2. \tag{4.1.8}$$

A similar approach leads to the bilinear mapping for an arbitrary-shaped straight-sided quadrilateral where only the vertices need to be prescribed. For the straight-sided quadrilateral with vertices labelled as shown in Fig. 4.2(b), the mapping is

$$x_i = \chi_1(\xi_1, \xi_2) = x_i^A \frac{1-\xi_1}{2}\frac{1-\xi_2}{2} + x_i^B \frac{1+\xi_1}{2}\frac{1-\xi_2}{2} + x_i^D \frac{1-\xi_1}{2}\frac{1+\xi_2}{2} + x_i^C \frac{1+\xi_1}{2}\frac{1+\xi_2}{2}, \quad i = 1, 2. \tag{4.1.9}$$

When developing a mapping it is important to ensure that the Jacobian of the mapping to the standard region is nonzero and of the same sign. To ensure this condition is satisfied when using the mappings given above, we require all elemental regions to have internal corners with angles that are less than 180° and so are convex. It is not, in fact, possible to generate a straight-sided triangular or tetrahedral region which does not satisfy this condition, but, as shown in Fig. 4.3, it is certainly possible within a quadrilateral or other three-dimensional region.

4.1.3.2 *Elemental mappings for general curvilinear elements*

For a general straight-sided elemental domain we have seen in Section 4.1.3.1 that a one-to-one linear mapping can be constructed onto the standard region

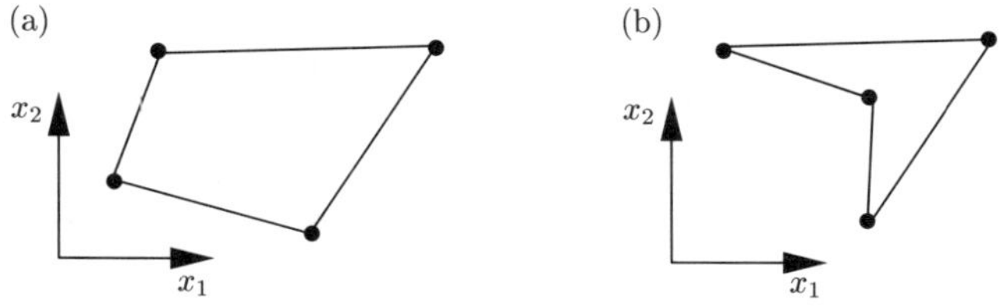

FIG. 4.3. (a) Illustration of a permissible quadrilateral region, with all interior angles less than 180°. (b) Illustration of a non-permissible quadrilateral region, which has a re-entrant corner and so has interior angles which are greater than 180°.

using the vertex modes of the hierarchical modal expansion. For example, for a quadrilateral domain of the form shown in Fig. 4.2(b) the mapping can be defined by eqn (4.1.9).

We note that this simply involves the vertex modes of the modified hierarchical expansion basis within a quadrilateral domain (see Section 3.1.1). We could, therefore, have written the expansion as

$$x_i = \chi_i(\xi_1, \xi_2) = \sum_{p=0}^{p=P_1} \sum_{q=0}^{q=P_2} \hat{x}^i_{pq} \phi_{pq}(\xi_1, \xi_2), \qquad (4.1.10)$$

where $\phi_{pq} = \psi^a_p(\xi_1)\psi^a_q(\xi_2)$ and $\hat{x}^i_{pq} = 0$, except for the vertex modes which have a value of

$$\hat{x}^i_{00} = x^A_i, \quad \hat{x}^i_{P_1 0} = x^B_i, \quad \hat{x}^i_{P_1 P_2} = x^C_i, \quad \hat{x}^i_{0 P_2} = x^D_i.$$

The construction of a mapping based upon the expansion modes in this form can be extended to include curved-sided regions using an *iso-parametric* representation. In this technique the geometry is represented with an expansion of the same form and polynomial order as the unknown variables.

To describe a straight-sided region we only need to know the values of the vertex locations. To describe a curved region, however, requires more information. As illustrated in Fig. 4.4, we typically expect to have a definition of a mapping of the shape of each edge in terms of the local coordinates, which we denote by $f^A_i(\xi_1)$, $f^B_i(\xi_2)$, $f^C_i(\xi_1)$, and $f^D_i(\xi_2)$. The process of defining the mapping functions can be considered as part of the mesh generation process, the discussion of which is in Section 4.3.3.

Knowing the definition of the edges (or faces in three dimensions), a mapping for a curvilinear domains can be determined using the iso-parametric form of eqn (4.1.10) to include more nonzero expansion coefficients than simply the vertex contributions. In two dimensions we wish to use the coefficient along each edge of the element, and in three dimensions we can use the face coefficients as well. Along each edge we therefore need to approximate the shape mapping $f_i(\xi)$ if

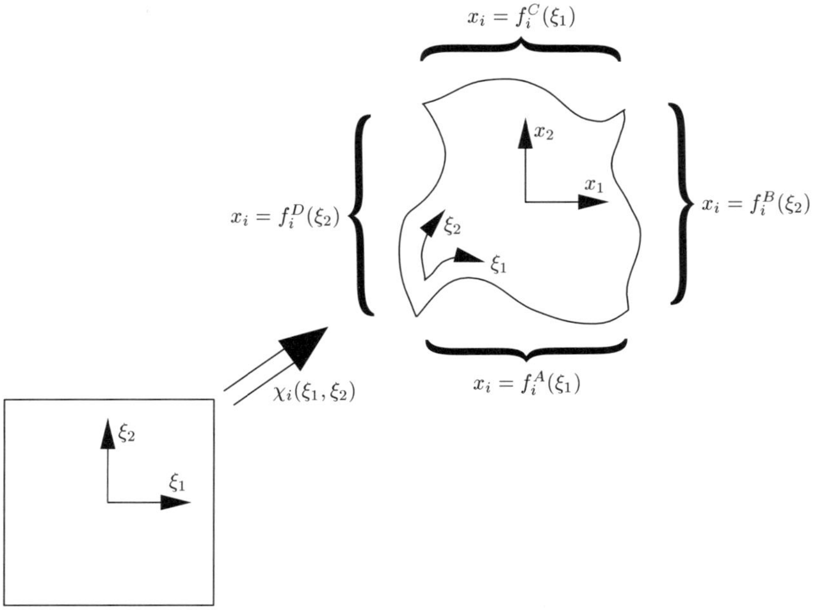

FIG. 4.4. A general curved element can be described in terms of a series of parametric functions $f^A(\xi_1)$, $f^B(\xi_2)$, $f^C(\xi_1)$, and $f^D(\xi_2)$. Representing these functions as a discrete expansion, we can construct an iso-parametric mapping $\chi_i(\xi_1, \xi_2)$ relating the standard region (ξ_1, ξ_2) to the deformed region (x_1, x_2).

it is not already represented by a polynomial of appropriate order. We might therefore consider the following approximations for $f_i^A(\xi_1)$:

$$f^A(\xi_1) \simeq \sum_{p=0} f^A(\xi_i, i) h_p(\xi_1) \tag{4.1.11a}$$

$$\simeq \sum_{p=0} \hat{x}_{p0}^i \psi_p(\xi_1)\,. \tag{4.1.11b}$$

One important feature of the approximation, and consequently the mapping χ_i, is that the vertices of each element coincide so that elements remain continuous. One way to ensure this is to use a collocation projection where the collocation points include the end-points $\xi_1 \pm 1$. The Lagrange representation of eqn (4.1.11a) is therefore a consistent way of approximating $f_i^A(\xi)$. Using the Gauss–Lobatto–Legendre quadrature points for our collocation projection is attractive due to their small Lebesgue constant associated with this interpolation, as discussed in Section 3.3.1. Similarly, within a triangular face of a three-dimensional element the Fekete or electrostatic points are a favourable choice for collocation points. By making collocation projections at a series of nodal points, we have then represented the function f^A as a polynomial which can be equivalently expressed in terms of a hierarchical expansion, $\psi_p(\xi)$, to obtain the coefficients $\hat{x}^i_{p0}$ in eqn (4.1.11b). This final transformation can be performed either by a collocation or Galerkin projection if the polynomials span the same space. If more collocation points are used then a modified Galerkin projection, as discussed in Section 4.3.1, can be applied. Having determined the coordinate expansion coefficients, $\hat{x}^i_{p0}$, we can then evaluate eqn (4.1.10) to determine the iso-parametric mapping from the standard region to the curvilinear region.

We note that the form of the boundary–interior decomposition of the modal quadrilateral and hexahedral expansion is discretely equivalent to using a linear blending function, as originally proposed by Gordon and Hall [196]. For the quadrilateral region shown in Fig. 4.4 the linear blending function is given by

$$\begin{aligned}\chi_i(\xi_1,\xi_2) &= f^A(\xi_1)\frac{1-\xi_2}{2} + f^C(\xi_1)\frac{1+\xi_2}{2}\\ &+ f^D(\xi_2)\frac{1-\xi_1}{2} + f^B(\xi_2)\frac{1+\xi_1}{2}\\ &- \frac{1-\xi_1}{2}\frac{1-\xi_2}{2}f^A(-1) - \frac{1+\xi_1}{2}\frac{1-\xi_2}{2}f^A(1)\\ &- \frac{1-\xi_1}{2}\frac{1+\xi_2}{2}f^C(-1) - \frac{1+\xi_1}{2}\frac{1+\xi_2}{2}f^C(1),\end{aligned} \tag{4.1.12}$$

where the vertex points are continuous (for example, $f^A(-1) = f^D(-1)$). If we replace the analytic curves $f^A(\xi_1)$, $f^B(\xi_2)$, $f^C(\xi_1)$, and $f^D(\xi_2)$ in expression (4.1.12) with an approximate expansion of the type given in eqn (4.1.11a) and rearrange, then we can obtain an expansion of the form given by eqn (4.1.10). The blending function (4.1.12) with approximations of the form (4.1.11a) to the mapped edges has traditionally been applied in spectral element methods. However, we note the aforementioned similarity with the modal expansion approach.

As a final point, we note that for curved triangular or tetrahedral elements, the linear blending function expressed in terms of the local collapsed coordinates η_1, η_2, and η_3 should not be used as this can generate a non-smooth Jacobian (at the singular vertices) and cause a loss in exponential convergence of smooth functions in curved domains. The use of the iso-parametric representation of the coordinates (eqn (4.1.10)) generates a mapping with a sufficiently smooth Jacobian.

4.1.3.3 *Integration within a general-shaped elemental region*

We denote an arbitrary triangular or quadrilateral region by Ω^e, which is a function of the global Cartesian coordinate system (x_1, x_2) in two dimensions. To integrate over Ω^e we transform this region into the standard region $\Omega_{\rm st}$ defined in terms of (ξ_1, ξ_2) and we have ♣[16]

$$\int_{\Omega^e} u(x_1,x_2)\,dx_1\,dx_2 = \int_{\Omega_{\rm st}} u(\xi_1,\xi_2)|J_{\rm 2D}|\,d\xi_1\,d\xi_2\,,$$

[16] Integration within a general region: definition of the two-dimensional and three-dimensional Jacobians.

where J_{2D} is the two-dimensional Jacobian due to the transformation, defined as

$$J_{2D} = \begin{vmatrix} \frac{\partial x_1}{\partial \xi_1} & \frac{\partial x_1}{\partial \xi_2} \\ \frac{\partial x_2}{\partial \xi_1} & \frac{\partial x_2}{\partial \xi_2} \end{vmatrix} = \frac{\partial x_1}{\partial \xi_1}\frac{\partial x_2}{\partial \xi_2} - \frac{\partial x_1}{\partial \xi_2}\frac{\partial x_2}{\partial \xi_1}. \qquad (4.1.13)$$

As we have assumed that we know the form of the mapping (i.e., $x_1 = \chi_1(\xi_1, \xi_2)$ and $x_2 = \chi_2(\xi_1, \xi_2)$), we can evaluate all the partial derivatives required to determine the Jacobian, as discussed in Section 4.1.2. If the elemental region is straight-sided then we have seen that a mapping from $(x_1, x_2) \to (\xi_1, \xi_2)$ is given by eqns (4.1.8) and (4.1.9). The simple form of these mappings means that the partial derivatives, and therefore the Jacobian, are constant for quadrilateral regions with similar shape and orientation to the standard region, as well as for all *triangular* regions. For deformed regions the Jacobian may be evaluated and stored at the quadrature points. This essentially represents the Jacobian as a polynomial function and can therefore increase the polynomial order of the integrand.

We note that integration over the triangular region now involves two transformations, that is, $(x_1, x_2) \to (\xi_1, \xi_2)$ and $(\xi_1, \xi_2) \to (\eta_1, \eta_2)$. However, the second transformation $(\xi_1, \xi_2) \to (\eta_1, \eta_2)$ may be absorbed entirely into the quadrature weights, as discussed in Section 4.1.1. This is preferable since the polynomial order, which may be exactly integrated, is higher. Having evaluated the Jacobian at the quadrature points, we multiply the integrand by the Jacobian and evaluate the integral in the same manner as that discussed in Section 4.1.1.

Integration over a three-dimensional region Ω^e is performed in an analogous fashion, where the three-dimensional Jacobian now has the form

$$\begin{aligned} J_{3D} &= \begin{vmatrix} \frac{\partial x_1}{\partial \xi_1} & \frac{\partial x_1}{\partial \xi_2} & \frac{\partial x_1}{\partial \xi_3} \\ \frac{\partial x_2}{\partial \xi_1} & \frac{\partial x_2}{\partial \xi_2} & \frac{\partial x_2}{\partial \xi_3} \\ \frac{\partial x_3}{\partial \xi_1} & \frac{\partial x_3}{\partial \xi_2} & \frac{\partial x_3}{\partial \xi_3} \end{vmatrix} \\ &= \frac{\partial x_1}{\partial \xi_1}\left(\frac{\partial x_2}{\partial \xi_2}\frac{\partial x_3}{\partial \xi_3} - \frac{\partial x_2}{\partial \xi_3}\frac{\partial x_3}{\partial \xi_2}\right) \\ &\quad - \frac{\partial x_1}{\partial \xi_2}\left(\frac{\partial x_2}{\partial \xi_1}\frac{\partial x_3}{\partial \xi_3} - \frac{\partial x_2}{\partial \xi_3}\frac{\partial x_3}{\partial \xi_1}\right) + \frac{\partial x_1}{\partial \xi_3}\left(\frac{\partial x_2}{\partial \xi_1}\frac{\partial x_3}{\partial \xi_2} - \frac{\partial x_2}{\partial \xi_2}\frac{\partial x_3}{\partial \xi_1}\right). \end{aligned} \qquad (4.1.14)$$

4.1.3.4 *Differentiation within a general-shaped elemental region*

To differentiate a function within the arbitrary region Ω^e, as illustrated in Fig. 4.2, we once again apply the chain rule which, for the two-dimensional case, gives ♣[17]

$$\nabla = \begin{bmatrix} \dfrac{\partial}{\partial x_1} \\ \dfrac{\partial}{\partial x_2} \end{bmatrix} = \begin{bmatrix} \dfrac{\partial \xi_1}{\partial x_1}\dfrac{\partial}{\partial \xi_1} + \dfrac{\partial \xi_2}{\partial x_1}\dfrac{\partial}{\partial \xi_2} \\ \dfrac{\partial \xi_1}{\partial x_2}\dfrac{\partial}{\partial \xi_1} + \dfrac{\partial \xi_2}{\partial x_2}\dfrac{\partial}{\partial \xi_2} \end{bmatrix}. \tag{4.1.15}$$

In Section 4.1.1 we illustrated differentiation with respect to ξ_1 and ξ_2, but we now also need to evaluate partial derivatives of the form $\partial\xi_1/\partial x_1$. For the linear mapping case given by eqns (4.1.8) and (4.1.9) it is possible to obtain an analytic formula, but in general we need a technique to handle a curvilinear elemental region. To do this, we express the partial derivatives such as $\partial\xi_1/\partial x_1$ in terms of partial derivatives with respect to ξ_1 and ξ_2, which we already know how to evaluate. For a general function dependent on two variables, $u(\xi_1, \xi_2)$, we know from the chain rule that the total change in $u(\xi_1, \xi_2)$ is

$$\mathrm{d}u(\xi_1, \xi_2) = \frac{\partial u}{\partial \xi_1}\,\mathrm{d}\xi_1 + \frac{\partial u}{\partial \xi_2}\,\mathrm{d}\xi_2 \,. \tag{4.1.16}$$

If we replace $u(\xi_1, \xi_2)$ by $x_1 = \chi_1(\xi_1, \xi_2)$ and $x_2 = \chi_2(\xi_1, \xi_2)$ then we obtain the matrix system

$$\begin{bmatrix} \mathrm{d}x_1 \\ \mathrm{d}x_2 \end{bmatrix} = \begin{bmatrix} \dfrac{\partial x_1}{\partial \xi_1} & \dfrac{\partial x_1}{\partial \xi_2} \\ \dfrac{\partial x_2}{\partial \xi_1} & \dfrac{\partial x_2}{\partial \xi_2} \end{bmatrix} \begin{bmatrix} \mathrm{d}\xi_1 \\ \mathrm{d}\xi_2 \end{bmatrix},$$

which can be inverted to obtain

$$\begin{bmatrix} \mathrm{d}\xi_1 \\ \mathrm{d}\xi_2 \end{bmatrix} = \frac{1}{J} \begin{bmatrix} \dfrac{\partial x_2}{\partial \xi_2} & -\dfrac{\partial x_1}{\partial \xi_2} \\ -\dfrac{\partial x_2}{\partial \xi_1} & \dfrac{\partial x_1}{\partial \xi_1} \end{bmatrix} \begin{bmatrix} \mathrm{d}x_1 \\ \mathrm{d}x_2 \end{bmatrix}, \tag{4.1.17}$$

where J is the Jacobian defined by eqn (4.1.13). However, as the mapping is assumed to be one-to-one and have an inverse, we assume that $\xi_1 = (\chi_1)^{-1}(x_1, x_2)$ and $\xi_2 = (\chi_2)^{-1}(x_1, x_2)$, and so we can apply the chain rule directly to ξ_1 and ξ_2 to obtain

$$\begin{bmatrix} \mathrm{d}\xi_1 \\ \mathrm{d}\xi_2 \end{bmatrix} = \frac{1}{J} \begin{bmatrix} \dfrac{\partial \xi_1}{\partial x_1} & -\dfrac{\partial \xi_1}{\partial x_2} \\ -\dfrac{\partial \xi_2}{\partial x_1} & \dfrac{\partial \xi_2}{\partial x_2} \end{bmatrix} \begin{bmatrix} \mathrm{d}x_1 \\ \mathrm{d}x_2 \end{bmatrix}. \tag{4.1.18}$$

[17] Differentiation within a general region in terms of local Cartesian coordinates: geometric factors.

Finally, equating eqns (4.1.17) and (4.1.18), we see that

$$\frac{\partial \xi_1}{\partial x_1} = \frac{1}{J}\frac{\partial x_2}{\partial \xi_2}, \quad \frac{\partial \xi_1}{\partial x_2} = -\frac{1}{J}\frac{\partial x_1}{\partial \xi_2}, \quad \frac{\partial \xi_2}{\partial x_1} = -\frac{1}{J}\frac{\partial x_2}{\partial \xi_1}, \quad \frac{\partial \xi_2}{\partial x_2} = \frac{1}{J}\frac{\partial x_1}{\partial \xi_1}.$$

We can now evaluate the two-dimensional gradient operator in eqn (4.1.15) as all the partial derivatives can be expressed in terms of differentials with respect to ξ_1 and ξ_2, which may be evaluated using the Lagrange polynomial representation explained in Section 4.1.2.

For the three-dimensional gradient operator we assume that the coordinates x_1, x_2, and x_3 are also known in terms of mappings dependent upon ξ_1, ξ_2, and ξ_3. So applying the chain rule, we obtain

$$\nabla = \begin{bmatrix} \frac{\partial}{\partial x_1} \\ \frac{\partial}{\partial x_2} \\ \frac{\partial}{\partial x_3} \end{bmatrix} = \begin{bmatrix} \frac{\partial \xi_1}{\partial x_1}\frac{\partial}{\partial \xi_1} + \frac{\partial \xi_2}{\partial x_1}\frac{\partial}{\partial \xi_2} + \frac{\partial \xi_3}{\partial x_1}\frac{\partial}{\partial \xi_3} \\ \frac{\partial \xi_1}{\partial x_2}\frac{\partial}{\partial \xi_1} + \frac{\partial \xi_2}{\partial x_2}\frac{\partial}{\partial \xi_2} + \frac{\partial \xi_3}{\partial x_2}\frac{\partial}{\partial \xi_3} \\ \frac{\partial \xi_1}{\partial x_3}\frac{\partial}{\partial \xi_1} + \frac{\partial \xi_2}{\partial x_3}\frac{\partial}{\partial \xi_2} + \frac{\partial \xi_3}{\partial x_3}\frac{\partial}{\partial \xi_3} \end{bmatrix}. \tag{4.1.19}$$

Following a similar analysis to that used for the two-dimensional case, we express the partial derivatives with respect to x_1, x_2, and x_3 in terms of derivatives with respect to ξ_1, ξ_2, and ξ_3 using the relations

$$\frac{\partial \xi_1}{\partial x_1} = \frac{1}{J_{3D}}\left(\frac{\partial x_2}{\partial \xi_2}\frac{\partial x_3}{\partial \xi_3} - \frac{\partial x_2}{\partial \xi_3}\frac{\partial x_3}{\partial \xi_2}\right), \quad \frac{\partial \xi_1}{\partial x_2} = -\frac{1}{J_{3D}}\left(\frac{\partial x_1}{\partial \xi_2}\frac{\partial x_3}{\partial \xi_3} - \frac{\partial x_1}{\partial \xi_3}\frac{\partial x_3}{\partial \xi_2}\right),$$
$$\frac{\partial \xi_1}{\partial x_3} = \frac{1}{J_{3D}}\left(\frac{\partial x_1}{\partial \xi_2}\frac{\partial x_2}{\partial \xi_3} - \frac{\partial x_1}{\partial \xi_3}\frac{\partial x_2}{\partial \xi_2}\right),$$
$$\frac{\partial \xi_2}{\partial x_1} = -\frac{1}{J_{3D}}\left(\frac{\partial x_2}{\partial \xi_1}\frac{\partial x_3}{\partial \xi_3} - \frac{\partial x_2}{\partial \xi_3}\frac{\partial x_3}{\partial \xi_1}\right), \quad \frac{\partial \xi_2}{\partial x_2} = \frac{1}{J_{3D}}\left(\frac{\partial x_1}{\partial \xi_1}\frac{\partial x_3}{\partial \xi_3} - \frac{\partial x_1}{\partial \xi_3}\frac{\partial x_3}{\partial \xi_1}\right),$$
$$\frac{\partial \xi_2}{\partial x_3} = -\frac{1}{J_{3D}}\left(\frac{\partial x_1}{\partial \xi_1}\frac{\partial x_2}{\partial \xi_3} - \frac{\partial x_1}{\partial \xi_3}\frac{\partial x_2}{\partial \xi_1}\right),$$
$$\frac{\partial \xi_3}{\partial x_1} = \frac{1}{J_{3D}}\left(\frac{\partial x_2}{\partial \xi_1}\frac{\partial x_3}{\partial \xi_2} - \frac{\partial x_2}{\partial \xi_3}\frac{\partial x_3}{\partial \xi_1}\right), \quad \frac{\partial \xi_3}{\partial x_2} = -\frac{1}{J_{3D}}\left(\frac{\partial x_1}{\partial \xi_1}\frac{\partial x_3}{\partial \xi_2} - \frac{\partial x_1}{\partial \xi_2}\frac{\partial x_3}{\partial \xi_1}\right),$$
$$\frac{\partial \xi_3}{\partial x_3} = \frac{1}{J_{3D}}\left(\frac{\partial x_1}{\partial \xi_1}\frac{\partial x_2}{\partial \xi_2} - \frac{\partial x_1}{\partial \xi_2}\frac{\partial x_2}{\partial \xi_1}\right),$$

where J_{3D} is as defined in eqn (4.1.14).

4.1.4 *Discrete evaluation of the surface Jacobian*

As a final local operation which combines the concepts of elemental mapping, integration, and differentiation, we consider the evaluation of the surface Jacobian. Surface integrals of the form

$$\langle v, g_{\mathcal{N}} \rangle = \int_{\partial\Omega_{\mathcal{N}}} v g_{\mathcal{N}} \, \mathrm{d}S = \sum_{e=1}^{N_{\mathrm{el}}} \int_{\partial\Omega_{\mathcal{N}} \bigcap \partial\Omega^e} v^e g_{\mathcal{N}} \, \mathrm{d}S^e \qquad (4.1.20)$$

typically arise in the Galerkin discretisation of the Laplacian operators due to the application of the divergence theorem. Since these types of integrals normally appear as right-hand-side contributions in our matrix problems, they can be evaluated as a series of integrals over different elemental boundaries. Therefore, in a similar way to integration in the elemental region, our strategy for evaluating eqn (4.1.20) is to transform each elemental contribution to a standard interval and perform the integration using Gaussian quadrature. This transformation to a standard region necessarily introduces a Jacobian which we shall refer to as the *surface Jacobian*.

4.1.4.1 *Surface Jacobian of a two-dimensional region*

In two dimensions the surface integral is simply a line integral of the form

$$\int_a^b f(x_1, x_2) \, \mathrm{d}s \,, \qquad (4.1.21)$$

where $\mathrm{d}s = \sqrt{(\mathrm{d}x_1)^2 + (\mathrm{d}x_2)^2}$ is the differential length. The region is naturally broken into elemental regions $\partial\Omega^e \bigcap s$ within which we want to evaluate each segment of the integral (4.1.21). We know that the global coordinates x_1 and x_2 within each element are related to the local coordinates ξ_1 and ξ_2 in terms of the mapping, that is,

$$x_1 = \chi_1(\xi_1, \xi_2) \,, \quad x_2 = \chi_2(\xi_1, \xi_2) \,.$$

We can therefore relate the differential change in x_1 and x_2 in terms of a differential change in ξ_1 and ξ_2 using the chain rule:

$$\mathrm{d}x_1 = \frac{\partial x_1}{\partial \xi_1} \mathrm{d}\xi_1 + \frac{\partial x_1}{\partial \xi_2} \mathrm{d}\xi_2 \,,$$
$$\mathrm{d}x_2 = \frac{\partial x_2}{\partial \xi_1} \mathrm{d}\xi_1 + \frac{\partial x_2}{\partial \xi_2} \mathrm{d}\xi_2 \,,$$

where the partial derivatives may be evaluated as shown in Section 4.1.2. Along the boundary of any element the edge is completely parameterised by either ξ_1 or ξ_2 as the other local coordinate is a constant having a value of 1 or −1. For example, in Fig. 4.22 the edge touching the domain boundary in element 'e' is parameterised by ξ_1, since $\xi_2 = -1$. For this case we can relate the differential length $\mathrm{d}s$ in terms of a differential change $\mathrm{d}\xi_1$ by

$$\mathrm{d}s = \sqrt{(\mathrm{d}x_1)^2 + (\mathrm{d}x_2)^2} = \sqrt{\left(\left. \frac{\partial x_1}{\partial \xi_1} \right|_{\xi_2=-1} \right)^2 (\mathrm{d}\xi_1)^2 + \left(\left. \frac{\partial x_2}{\partial \xi_1} \right|_{\xi_2=-1} \right)^2 (\mathrm{d}\xi_1)^2} \,,$$

where the partial derivatives $\partial x_1/\partial \xi_1$ and $\partial x_2/\partial \xi_1$ are evaluated at $\xi_2 = -1$. The contribution of element 'e' to the integral in eqn (4.1.21) can now be written as

$$\int_{\partial\Omega^e \bigcap s} f(x_1, x_2)\,\mathrm{d}s = \int_{-1}^{1} f(\xi_1, \xi_2)\sqrt{\left(\frac{\partial x_1}{\partial \xi_1}\right)^2 + \left(\frac{\partial x_2}{\partial \xi_1}\right)^2}\,\mathrm{d}\xi_1\,,$$

which can be evaluated using standard Gaussian quadrature. The whole surface is then generated as a summation of elemental contributions, although in general the integrand contains the test function which only has nonzero support in, at most, two elements.

4.1.4.2 *Surface Jacobian of a three-dimensional region*

In three dimensions we want to evaluate a two-dimensional surface integral of the form

$$\int_{\partial\Omega} f(x_1, x_2, x_3)\,\mathrm{d}S\,. \qquad (4.1.22)$$

Just as a one-dimensional surface can be completely parameterised in terms of a single parameter, a two-dimensional surface can be described by a doubly-infinite set of parametric curves.

As indicated in Fig. 4.5, we have already parameterised the surface in terms of two of the local Cartesian coordinates (ξ_1, ξ_2, ξ_3) depending on which face we are considering. If we consider the face $\xi_3 = -1$, as shown in Fig. 4.5, then we

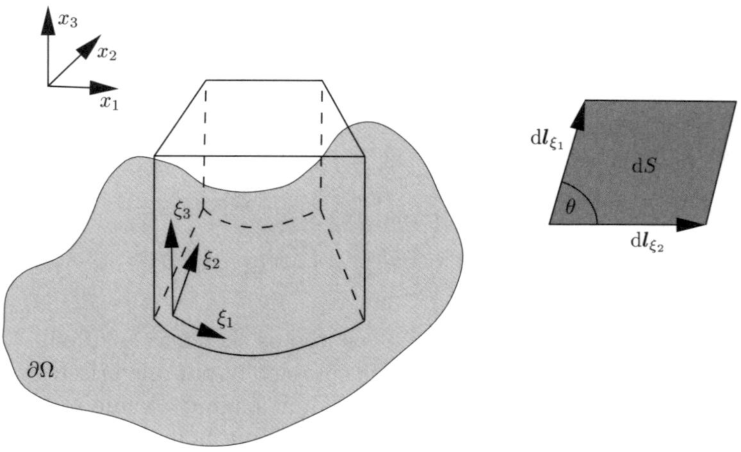

FIG. 4.5. The surface of the solution domain $\partial\Omega$ is represented as a tessellation of rectangular (or triangular) partitions within which we have a double parameterisation (that is, ξ_1 and ξ_2). We can use this parameterisation to relate the differential surface area $\mathrm{d}S$ to the differential change in ξ_1 and ξ_2 via the surface tangent vectors $\mathrm{d}\boldsymbol{l}_{\xi_1}$ and $\mathrm{d}\boldsymbol{l}_{\xi_2}$.

know that our surface can be completely described by the two parameters ξ_1 and ξ_2, that is,

$$x_1 = \chi_1(\xi_1, \xi_2, -1), \quad x_2 = \chi_2(\xi_1, \xi_2, -1), \quad x_3 = \chi_3(\xi_1, \xi_2, -1).$$

To relate the differential surface area $\mathrm{d}S$ in terms of the parametric coordinates ξ_1 and ξ_2, we note that the change in surface position $\mathrm{d}\boldsymbol{x} = [\mathrm{d}x_1, \mathrm{d}x_2, \mathrm{d}x_3]^\top$ due to a change in the parametric coordinates is given by

$$\mathrm{d}\boldsymbol{x} = \frac{\partial \boldsymbol{x}}{\partial \xi_1}\,\mathrm{d}\xi_1 + \frac{\partial \boldsymbol{x}}{\partial \xi_2}\,\mathrm{d}\xi_2\,,$$

where $\partial \boldsymbol{x}/\partial \xi_1$ and $\partial \boldsymbol{x}/\partial \xi_2$ are the surface tangent vectors along lines of constant ξ_2 and ξ_1, respectively. Therefore, along a line of constant ξ_2 the change in differential length $\mathrm{d}\boldsymbol{l}_{\xi_2}$ due to a differential change in ξ_1 is

$$\mathrm{d}\boldsymbol{l}_{\xi_2} = \left.\frac{\partial \boldsymbol{x}}{\partial \xi_1}\right|_{\xi_2} \mathrm{d}\xi_1 = \begin{bmatrix} \dfrac{\partial x_1}{\partial \xi_1} \\ \dfrac{\partial x_2}{\partial \xi_1} \\ \dfrac{\partial x_3}{\partial \xi_1} \end{bmatrix} \mathrm{d}\xi_1\,.$$

Similarly, along a line of constant ξ_1 the change in differential length $\mathrm{d}\boldsymbol{l}_{\xi_1}$ for a differential change in ξ_2 is

$$\mathrm{d}\boldsymbol{l}_{\xi_1} = \left.\frac{\partial \boldsymbol{x}}{\partial \xi_2}\right|_{\xi_1} \mathrm{d}\xi_2 = \begin{bmatrix} \dfrac{\partial x_1}{\partial \xi_2} \\ \dfrac{\partial x_2}{\partial \xi_2} \\ \dfrac{\partial x_3}{\partial \xi_2} \end{bmatrix} \mathrm{d}\xi_2\,.$$

We note that $\mathrm{d}\boldsymbol{l}_{\xi_2}$ and $\mathrm{d}\boldsymbol{l}_{\xi_1}$ only involve derivatives of the coordinates with respect to the local coordinates (ξ_1, ξ_2), and so they can be evaluated as explained in Section 4.1.2. In general $\mathrm{d}\boldsymbol{l}_{\xi_2}$ and $\mathrm{d}\boldsymbol{l}_{\xi_1}$ are not orthogonal and, as shown in Fig. 4.5, the differential surface area $\mathrm{d}S$ is given by a parallelogram

$$\mathrm{d}S = |\mathrm{d}\boldsymbol{l}_{\xi_2}||\mathrm{d}\boldsymbol{l}_{\xi_1}|\sin\theta\,,$$

where θ is the angle between the vectors $\mathrm{d}\boldsymbol{l}_{\xi_2}$ and $\mathrm{d}\boldsymbol{l}_{\xi_1}$. The magnitudes $|\mathrm{d}\boldsymbol{l}_{\xi_2}|$ and $|\mathrm{d}\boldsymbol{l}_{\xi_1}|$ are also known as the *Lamé parameters*.

This expression may also be written as a cross-product in the form

$$\mathrm{d}S = |\mathrm{d}\boldsymbol{l}_{\xi_2} \times \mathrm{d}\boldsymbol{l}_{\xi_1}| = \left|\frac{\partial \boldsymbol{x}}{\partial \xi_1} \times \frac{\partial \boldsymbol{x}}{\partial \xi_2}\right| \mathrm{d}\xi_1\,\mathrm{d}\xi_2\,.$$

Finally, we are in a position to evaluate the surface integral given in eqn (4.1.22) over the elemental face as

$$\int_{\partial\Omega^e \bigcap \partial\Omega} f(x_1, x_2, x_3)\, \mathrm{d}S = \int_{-1}^{1}\int_{-1}^{1} f(\xi_1, \xi_2, -1) \left| \frac{\partial \boldsymbol{x}}{\partial \xi_1} \times \frac{\partial \boldsymbol{x}}{\partial \xi_2} \right| \mathrm{d}\xi_1\, \mathrm{d}\xi_2 \,,$$

which can be evaluated using Gaussian quadrature.

Clearly, for different elemental faces we need to change the parametric coordinates. The shape of the surface region is not dependent on the derivation and therefore the same argument can be applied to a triangular surface, although the integral is usually evaluated using the local collapsed Cartesian coordinates η_1 and η_2. This type of construction more commonly arises in structural mechanics for thin-shell theory and further discussion may be found in Akin [6].

4.1.5 *Elemental projections and transformations*

As discussed in Chapter 3, there are a variety of choices of expansion bases, $\phi_m(\boldsymbol{\xi})$, which correspond to different-shaped regions as well as different constructions such as modal and nodal or tensorial and non-tensorial expansions. In general, we assume the notation

$$u^\delta(\boldsymbol{x}) = \sum_{m=0}^{N_m-1} \hat{u}_m\, \phi_m(\boldsymbol{\xi}) = \sum_{m(pqr)=0}^{N_m-1} \hat{u}_{pqr}\phi_{pqr}(\boldsymbol{\xi}) \,, \quad \boldsymbol{x} \in \Omega^e \,, \tag{4.1.23}$$

where $\boldsymbol{x} = [x_1, x_2, x_3]$, $\boldsymbol{\xi} = [\xi_1, \xi_2, \xi_3]$, and $m(pqr)$ represents the connection between the tensorial indices p, q, and r and a global index m which will be discussed in more detail below. We also recall that the local, $\boldsymbol{\xi}$, and global, $\boldsymbol{x}$, coordinates are related through the mappings

$$\xi_1 = (\chi_1^e)^{-1}(x_1, x_2, x_3)\,, \quad \xi_2 = (\chi_2^e)^{-1}(x_1, x_2, x_3)\,, \quad \xi_3 = (\chi_3^e)^{-1}(x_1, x_2, x_3)\,.$$

By analogy with the Fourier transform, we will refer to $u^\delta(\boldsymbol{x})$ as representing a *physical* space variable and $\hat{u}_m$ and $\hat{u}_{pqr}$ as being in *transformed* space.

In this section we discuss how to obtain the transformed coefficients $\hat{u}_{pqr}$ from the physical representation, $u^\delta(\boldsymbol{x})$, using collocation and Galerkin projections, and thereby define the *forward transformation*. Although we shall later discuss it in more detail, we note that eqn (4.1.23) represents the *backward transformation* from the coefficient to the physical values.

4.1.5.1 *Elemental matrix and vector notation*

The formulation in multiple dimensions and using different types of expansion bases necessitates the introduction of a large number of superscript and subscript notations. To help clarify the operations to be discussed in this section as well as those in Section 4.2 we, therefore, introduce a matrix and vector notation to represent operations such as integration and differentiation. This notation is helpful in illustrating the construction of discrete operations such as the transformation from physical to transformed space. It is also useful in providing an insight into the fundamental operations when implementing a numerical scheme. We note, however, that explicit construction of these matrices is not always necessary or sometimes even desirable in practice. Many of the matrices are diagonal or very sparse, and therefore explicit construction is unnecessarily costly from the computational point of view. Where the expansion bases are based upon a tensor or generalised tensor product construction we can apply the sum-factorisation technique discussed in Section 4.1.6. When using higher-order polynomial expansions (i.e., $P > 6$) this technique requires far less memory storage and a far lower operation count than the equivalent operations using the full matrices. Nevertheless, as we shall highlight in our discussion, when using a non-tensorial basis it may be necessary to explicitly construct and store some matrices.

Finally, we note that all definitions in this section refer to a single element. In Section 4.2 we will extend the notation to include multiple elements. To distinguish between the two uses we will introduce the superscript 'e'.

Vectors: $\boldsymbol{u}$ *and* $\hat{\boldsymbol{u}}$

Typically, we require the value of a function at a nodal set of points $\boldsymbol{\xi}_m$, $0 \leqslant m < N$, and so we define the vector $\boldsymbol{u}$ to denote the evaluation of $u(\boldsymbol{\xi})$ at these points, i.e.,

$$\boldsymbol{u}[m] = u(\boldsymbol{\xi}_m)\,,$$

where we have dropped the δ superscript for notational convenience.

When using a tensor-based expansion we typically require the function to be evaluated at the quadrature points. In this case we can take our set of nodes to be the set of one-dimensional quadrature points; for example, in the hexahedral region we can say $\boldsymbol{\xi}_{m(i,j,k)} = [\xi_{1i}, \xi_{2j}, \xi_{3k}]$, where i, j, and k are the indices over the one-dimensional quadrature nodes. If we have Q_1, Q_2, and Q_3 as the quadrature points in each direction then the vector $\boldsymbol{u}$ becomes

$$\boldsymbol{u}[m(ijk)] = u(\xi_{1i}, \xi_{2j}, \xi_{3k})\,,$$

where

$$m(ijk) = i + j \cdot Q_1 + k \cdot Q_1 \cdot Q_2\,.$$

In the above, array $m(ijk)$ is an integer value which runs consecutively from 0 to $N_Q = (Q_1 \cdot Q_2 \cdot Q_3)$ as the indices i, j, and k run through the ranges $0 \leqslant i \leqslant Q_1 - 1$, $0 \leqslant j \leqslant Q_2 - 1$, $0 \leqslant k \leqslant Q_3 - 1$. By convention, we let the ξ_1 coordinate run fastest, followed by the ξ_2, and finally the ξ_3 coordinate. Written explicitly, the vector $\boldsymbol{u}$ is

$$\boldsymbol{u} = [u(\xi_{10}, \xi_{20}, \xi_{30}), \ldots, u(\xi_{1Q_1}, \xi_{20}, \xi_{30}), u(\xi_{10}, \xi_{21}, \xi_{30}), \ldots,$$
$$u(\xi_{1Q_1}, \xi_{2Q_2}, \xi_{3Q_3})]^\top\,.$$

The two-dimensional case is analogously defined with $k = 0$. In the case of a non-tensorial basis the index m simply lists the set of discrete nodal points $\boldsymbol{\xi}_m$.

To represent the expansion coefficient $\hat{u}_m$ in vector form we adopt a similar notation but with a circumflex (that is, $\hat{\boldsymbol{u}}$), such that

$$\hat{\boldsymbol{u}}[m] = \hat{u}_m \,.$$

The number of the expansion coefficients is clearly related to the number of expansion modes, which has previously been denoted as N_m. For a non-tensorial nodal expansion, m would also correspond to the expansion coefficient of the Lagrange polynomial with unit value at $\boldsymbol{\xi}_m$.

For tensor-based expansions the storage convention for the vector $\hat{\boldsymbol{u}}$ depends on the local indices of the one-dimensional functions, usually denoted as p, q, and r. Therefore, for the orthogonal tensorial expansions we can define the 'packing process' which relates $m(p,q,r)$ to p, q, and r. For example, an orthogonal expansion $\hat{\boldsymbol{u}}$ in the hexahedral region of polynomial order P_1, P_2, P_3 is defined as

$$\hat{\boldsymbol{u}}[m(pqr)] = \hat{u}_{pqr} \,,$$

where

$$m(pqr) = r + q(P_2+1) + p(P_2+1)(P_3+1)$$

or, equivalently,

$$\hat{\boldsymbol{u}} = \left[\hat{u}_{000}, \hat{u}_{001}, \ldots, \hat{u}_{00P_3}, \hat{u}_{010}, \ldots, \hat{u}_{01P_3}, \ldots, \hat{u}_{020}, \ldots, \hat{u}_{P_1P_2P_3}\right]^\top .$$

We note that the skipping factors are $P+1$ as it requires $P+1$ modes to represent a polynomial of order P. By convention, we let the index r run fastest, followed by q, and finally p. Although this convention would appear to be the reverse of the nodal point definition, the ordering is necessary to take advantage of the partial orthogonality of the generalised tensor product expansions, as illustrated in Fig. 3.21.

The orthogonal expansion within a triangular region, $\phi_{pq} = \phi^a_p(\eta_1)\phi^b_{pq}(\eta_2)$, of order P_1, P_2 defined in Section 3.2.2 would have an index system defined by

$$m(pq) = q + \frac{p(2P_2+1-p)}{2} \,,$$

where the index range of p and q is

$$0 \leqslant p, q, \;\; p \leqslant P_1, \;\; p+q \leqslant P_2, \;\; P_1 \leqslant P_2 \,.$$

For the orthogonal, generalised tensor product expansions, it is possible to define $m(pqr)$ analytically since the indices are *close-packed.* However, for the modified generalised tensor product expansion bases defined in Section 3.2.3 there is not a *close-packed* form for $m(pqr)$ due to the way we have defined the expansion modes. This is also true for the serendipity and variable expansion in quadrilateral and hexahedral regions. In general, we interpret the form of $m(pqr)$ as reordering the N_m modes of an elemental expansion into a consecutive order. A typical storage scheme might place the indices of the vertex modes first, followed by the edge, face, and finally the interior modes, where we shall still allow r to run fastest, followed by q, and then p within each group of modes. An example of this ordering for quadrilateral and triangular expansions of order $P_1, P_2 = 4$ is shown in Fig. 4.6.

Weight and basis matrices: $\boldsymbol{W}$ *and* $\boldsymbol{B}$

A matrix of dimension $(m+1) \times (n+1)$ is understood to be a rectangular array of the form

$$\boldsymbol{A} = \begin{bmatrix} a_{00} & a_{01} & \cdots & a_{0n} \\ a_{10} & a_{11} & \cdots & a_{0n} \\ \vdots & \vdots & & \vdots \\ a_{m0} & a_{m1} & \cdots & a_{mn} \end{bmatrix}.$$

To complement the vectors $\boldsymbol{u}$ and $\hat{\boldsymbol{u}}$ we introduce two matrices $\boldsymbol{W}$ and $\boldsymbol{B}$. The weight matrix, $\boldsymbol{W}$, is a diagonal matrix containing the Gaussian quadrature weights multiplied by the Jacobian at the quadrature points, and is defined to be consistent with $\boldsymbol{u}$ evaluated at the set of quadrature points. This matrix is not defined for an arbitrary set of nodal points. If J_{ijk} represents the discrete value of the Jacobian $J(\xi_{1i}, \xi_{2j}, \xi_{3k})$ then $\boldsymbol{W}$ in three dimensions is defined as

$$\boldsymbol{W}[m(ijk)][n(rst)] = J_{ijk}\, w_i\, w_j\, w_k\, \delta_{mn}\,,$$

where δ_{mn} is the Dirac delta function, and $m(ijk)$ and $n(rst)$ are defined in a consistent fashion to $\boldsymbol{u}$ so

$$m(ijk) = n(ijk) = i + j \cdot Q_1 + k \cdot Q_1 \cdot Q_2\,.$$

The quadrature weights are defined according to the elemental region so that for a hexahedral region $w_i = w_i^{0,0}$, $w_j = w_j^{0,0}$, and $w_k = w_k^{0,0}$, whereas for a

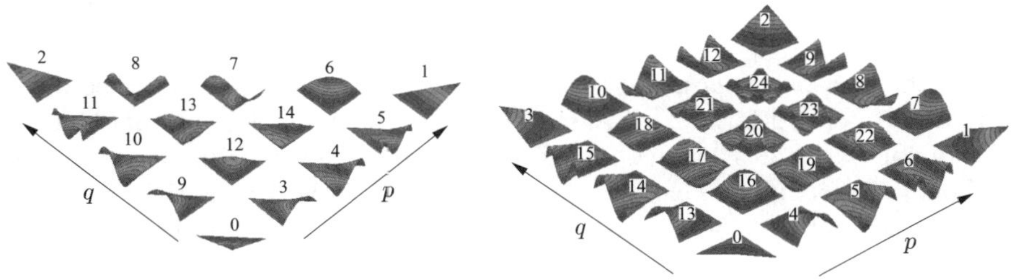

FIG. 4.6. Ordering of the expansion coefficients for the modified expansion in a triangular and quadrilateral region with $P_1 = P_2 = 4$. In this ordering we label the vertices first, followed by the edges, and finally the interior. For the interior modes the index q is allowed to run fastest.

tetrahedral region $w_i = w_i^{0,0}$, $w_j = \hat{w}_j^{1,0}$, and $w_k = \hat{w}_k^{2,0}$. The explicit form of the weight matrix in a tetrahedral region is therefore

$$\boldsymbol{W} = \begin{bmatrix} w_0^{0,0}\hat{w}_0^{1,0}\hat{w}_0^{2,0}J_{ijk} & 0 & 0 & 0 & 0 \\ 0 & \ddots & 0 & 0 & 0 \\ 0 & 0 & w_{Q_1}^{0,0}\hat{w}_0^{1,0}\hat{w}_0^{2,0}J_{ijk} & 0 & 0 \\ 0 & 0 & 0 & \ddots & 0 \\ 0 & 0 & 0 & 0 & w_{Q_1-1}^{0,0}\hat{w}_{Q_2-1}^{1,0}\hat{w}_{Q_3-1}^{2,0}J_{ijk} \end{bmatrix}.$$

The basis matrix, $\boldsymbol{B}$, is a type of Vandermonde matrix as introduced in Section 3.3.2. The matrix $\boldsymbol{B}$ is therefore defined as having columns which are fixed expansion modes $\phi_n(\boldsymbol{\xi})$ evaluated at all the nodal points $\boldsymbol{\xi}_m$, that is,

$$\boldsymbol{B}[m][n] = \phi_n(\boldsymbol{\xi}_m).$$

Using the previously defined indexing for the tensorial expansions, we can define the basis matrix for a tensor expansion, $\phi_{m(pqr)} = \phi_{pqr}$, evaluated at the quadrature points $\boldsymbol{\xi}_{n(ijk)} = [\xi_{1i}, \xi_{2j}, \xi_{3k}]$ as

$$\boldsymbol{B}[m(ijk)][n(pqr)] = \phi_{pqr}(\xi_{1i}, \xi_{2j}, \xi_{3k}).$$

The matrix is formulated so that its columns (looping over $m(ijk)$) are ordered in a consistent fashion to $\boldsymbol{u}$ and its rows (looping over $n(pqr)$) are ordered in a consistent fashion to $\hat{\boldsymbol{u}}$, that is,

$$\boldsymbol{B} = \begin{bmatrix} \phi_{000}(\xi_{10}, \xi_{20}, \xi_{30}) & \cdots & \phi_{00P_3}(\xi_{10}, \xi_{20}, \xi_{30}) & \cdots & \phi_{pqr}(\xi_{10}, \xi_{20}, \xi_{30}) \\ \vdots & \vdots & \vdots & \vdots & \vdots \\ \phi_{000}(\xi_{1Q_1}, \xi_{20}, \xi_{30}) & \cdots & \phi_{00P_3}(\xi_{1Q_1}, \xi_{20}, \xi_{30}) & \cdots & \phi_{pqr}(\xi_{1Q_1}, \xi_{20}, \xi_{30}) \\ \vdots & \vdots & \vdots & \vdots & \vdots \\ \phi_{000}(\xi_{1Q_1}, \xi_{2Q_2}, \xi_{3Q_3}) & \cdots & \phi_{00P_3}(\xi_{1Q_1}, \xi_{2Q_2}, \xi_{3Q_3}) & \cdots & \phi_{pqr}(\xi_{1Q_1}, \xi_{2Q_2}, \xi_{3Q_3}) \end{bmatrix}.$$

For the *non-tensorial nodal basis* in simplex regions discussed in Section 3.3 there is not a closed-form expression to evaluate the multi-dimensional Lagrange polynomial $L_n(\boldsymbol{\xi})$ at arbitrary points, $\boldsymbol{\xi}_m$. It is therefore not immediately obvious how to construct the basis matrix $\boldsymbol{B}$ in this case. Following the formulation of Warburton *et al.* [486], we recall from Section 3.3.2 that the triangular (or tetrahedral) Lagrange polynomial can be expressed in terms of another basis $\phi_{pq}(\boldsymbol{\xi})$ (such as the orthogonal tensorial expansion in a simplex) using the transformation

$$\begin{bmatrix} L_0(\boldsymbol{\xi}) \\ \vdots \\ L_{N_m-1}(\boldsymbol{\xi}) \end{bmatrix} p = \boldsymbol{B}_{\mathcal{N}}^{-\top} \begin{bmatrix} \phi_0(\boldsymbol{\xi}) \\ \vdots \\ \phi_{N_m-1}(\boldsymbol{\xi}) \end{bmatrix}, \tag{4.1.24}$$

where

$$\boldsymbol{B}_{\mathcal{N}}[n'][n(pq)] = \phi_{pq}(\boldsymbol{\xi}_{n'})$$

and $\xi_{n'}$ $(0 \leqslant n' < N_m)$ are the set of nodes which define the Lagrange polynomial $L_n(\xi)$. Note that we have used $\boldsymbol{B}_{\mathcal{N}}$, whereas we previously used $\boldsymbol{V}$ in eqn (3.3.2) of Section 3.3.2. We can now evaluate the Lagrange polynomial basis at an arbitrary set of points by evaluating $\phi_{pq}(\xi)$ at the desired points. Therefore, the matrix $\boldsymbol{B}[m][n] = L_n(\boldsymbol{\xi}_m)$ of the non-tensorial triangular Lagrange basis evaluated at the quadrature points in the triangle, $\boldsymbol{\xi}_{m(ij)} = [\eta_{1i}, \eta_{2j}]$, can be determined by

$$\boldsymbol{B}^{\top} = \boldsymbol{B}_{\mathcal{N}}^{-\top} \boldsymbol{B}_{\mathcal{T}}^{\top}$$

or, equivalently,

$$\boldsymbol{B} = \boldsymbol{B}_{\mathcal{T}} \boldsymbol{B}_{\mathcal{N}}^{-1}\,, \tag{4.1.25}$$

where

$$\boldsymbol{B}_{\mathcal{T}}[m(ij)][n(pq)] = \phi_{pq}(\boldsymbol{\xi}_{m(ij)})\,.$$

Using this construction in the following section we will be able to integrate and differentiate exactly the non-tensorial expansions in general-shaped elemental domains.

Differentiation matrix: $\boldsymbol{D}_{\xi_i}$

The final matrix we require to complete our set of discrete operators is a matrix representing the action of differentiation. Recall that partial differentiation with respect to the local coordinates ξ_1 defined in physical space can be written as

$$\frac{\partial u}{\partial \xi_1}(\xi_{1i}, \xi_{2j}, \xi_{3k}) = \sum_{r=0}^{Q_1} \sum_{s=0}^{Q_2} \sum_{t=0}^{Q_3} \left.\frac{\mathrm{d}h_r(\xi_1)}{\mathrm{d}\xi_1}\right|_{\xi_{1i}} h_s(\xi_{2j}) h_t(\xi_{3k})\; u_{rst}\,,$$

where $h_r(\xi)$ is the one-dimensional Lagrange polynomial through the Q_1 quadrature points and $u_{rst} = u(\xi_{1r}, \xi_{2s}, \xi_{3t})$. A similar definition follows for the other partial derivatives (see Section 4.1.2). We can, therefore, define a derivative matrix $\boldsymbol{D}_{\xi_1}$ which acts on $\boldsymbol{u}$ evaluated at the quadrature nodes such that $\boldsymbol{D}_{\xi_1}\boldsymbol{u}$ is the derivative at the quadrature points $\boldsymbol{u}_{\xi_1} = \partial \boldsymbol{u}/\partial \xi_1$, that is,

$$\frac{\partial \boldsymbol{u}}{\partial \xi_1} = \boldsymbol{D}_{\xi_1} \boldsymbol{u}\,,$$

where

$$\boldsymbol{D}_{\xi_1}[m(ijk)][m'(rst)] = \left.\frac{\mathrm{d}h_r(\xi_1)}{\mathrm{d}\xi_1}\right|_{\xi_{1i}} h_s(\xi_{2j}) h_t(\xi_{3k}) \tag{4.1.26}$$

and

$$m(ijk) = m'(ijk) = i + j \cdot Q_1 + k \cdot Q_1 \cdot Q_2\,.$$

The differentiation matrix can clearly be used to evaluate the derivative of the expansion bases, ϕ_n, at the quadrature points, $\xi_{m(i,j,k)}$, if applied to the basis matrix $\boldsymbol{B}$. Therefore, to evaluate the derivative with respect to ξ_1 of the basis $\phi_{n(pqr)}$ at the quadrature nodes $\boldsymbol{\xi}_{m(ijk)} = [\xi_{1i}, \xi_{2j}, \xi_{3k}]$ we can write

$$\frac{\partial \phi_n}{\partial \xi_1}(\boldsymbol{\xi}_m) = (\boldsymbol{D}_{\xi_1}\boldsymbol{B})\,[m][n]\,.$$

For the non-tensorial nodal basis expansion $\phi_n = L_n$ the derivative is evaluated in an identical fashion, with the basis matrix $\boldsymbol{B}$ evaluated using eqn (4.1.25).

The matrix $\boldsymbol{D}_{\xi_1}$ is very sparse, as can be appreciated from the fact that $h_s(\xi_{2j}) = \delta_{sj}$ and $h_t(\xi_{3k}) = \delta_{tk}$. This is particularly evident in the case of $\boldsymbol{D}_{\xi_1}$ because of the way we have chosen to order the quadrature points. Since the ξ_1 coordinate index runs fastest we find that $\boldsymbol{D}_{\xi_1}$ is a block diagonal matrix made from the one-dimensional derivative matrix, that is,

$$\boldsymbol{D}_{\xi_1} = \begin{bmatrix} \boldsymbol{D}^{1d} & 0 & 0 & 0 \\ 0 & \boldsymbol{D}^{1d} & 0 & 0 \\ & & & \\ 0 & 0 & \ddots & 0 \\ 0 & 0 & 0 & \boldsymbol{D}^{1d} \end{bmatrix},$$

where

$$\boldsymbol{D}^{1d}[i][r] = \left.\frac{\mathrm{d}h_r(\xi_1)}{\mathrm{d}\xi_1}\right|_{\xi_{1i}}, \qquad 0 \leqslant i, r \leqslant Q_1\,.$$

This illustrates the potential inefficiency of forming the matrix $\boldsymbol{D}_{\xi_1}$, since the matrix vector product $\boldsymbol{D}_{\xi_1 u}$ is an $\mathcal{O}\left((Q_1 \cdot Q_2 \cdot Q_3)^2\right)$ operation. However, if we perform the equivalent operation using the one-dimensional matrix $\boldsymbol{D}^{1d}$ along all lines of constant ξ_2 and ξ_3 then it is an $\mathcal{O}\left((Q_1)^2 \cdot Q_2 \cdot Q_3\right)$ operation. This reduction in cost is a direct consequence of the tensor construction of the basis and can also be considered as a sum-factorisation operation, as discussed in Section 4.1.6.

Diagonal coefficient matrices: $\boldsymbol{\Lambda}(c)$

We have seen previously that the derivative with respect to the global coordinates x_1, x_2, and x_3 can be obtained via the chain rule from the derivative with respect to the local coordinates (see eqn (4.1.19)). The equivalent matrix operation requires us to pre-multiply the derivative matrices by a diagonal matrix containing factors such as $\partial\xi_1/\partial x_1$, $\partial\xi_1/\partial x_2$, $\partial\xi_1/\partial x_3, \ldots$ evaluated at the quadrature points. To denote these diagonal coefficient matrices we introduce the notation $\boldsymbol{\Lambda}\left(f(\xi_1, \xi_2, \xi_3)\right)$ to be a diagonal matrix whose diagonal components are the evaluation of the function $f(\xi_1, \xi_2, \xi_3)$ at the quadrature points, that is,

$$\boldsymbol{\Lambda}(f(\xi_1, \xi_2, \xi_3))[m(ijk)][n(rst)] = f(\xi_{1i}, \xi_{2j}, \xi_{3k})\delta_{mn}\,, \tag{4.1.27}$$

where

$$m(ijk) = n(ijk) = i + j \cdot Q_1 + k \cdot Q_1 \cdot Q_2\,.$$

Therefore, since we know that

$$\frac{\partial u}{\partial x_1} = \frac{\partial \xi_1}{\partial x_1}\frac{\partial u}{\partial \xi_1} + \frac{\partial \xi_2}{\partial x_1}\frac{\partial u}{\partial \xi_2} + \frac{\partial \xi_3}{\partial x_1}\frac{\partial u}{\partial \xi_3}$$

we can evaluate $\boldsymbol{u}_{x_1} = \partial \boldsymbol{u}/\partial x_1$ at the quadrature nodes using the notation

$$\boldsymbol{u}_{x_1} = \left[\boldsymbol{\Lambda}\left(\frac{\partial \xi_1}{\partial x_1}\right)\boldsymbol{D}_{\xi_1} + \boldsymbol{\Lambda}\left(\frac{\partial \xi_2}{\partial x_1}\right)\boldsymbol{D}_{\xi_2} + \boldsymbol{\Lambda}\left(\frac{\partial \xi_3}{\partial x_1}\right)\boldsymbol{D}_{\xi_3}\right]\boldsymbol{u}\,.$$

The introduction of the diagonal matrix notation also allows us to represent the derivatives with respect to the local collapsed coordinate systems. For example, within a triangular region we can express the derivative with respect to the local Cartesian coordinates ξ_1, ξ_2 in terms of the local collapsed coordinates η_1, η_2 as

$$\boldsymbol{D}_{\xi_1} = \boldsymbol{\Lambda}\left(\frac{2}{1-\eta_1}\right)\boldsymbol{D}_{\eta_2}\,,$$
$$\boldsymbol{D}_{\xi_2} = \boldsymbol{\Lambda}\left(\frac{1+\eta_1}{1-\eta_2}\right)\boldsymbol{D}_{\eta_1} + \boldsymbol{D}_{\eta_2}\,.$$

4.1.5.2 *Backward transformation*

The backward transformation from coefficient space $\hat{u}_{pqr}$ to physical space $u(\xi_1, \xi_2, \xi_3)$ is defined in eqn (4.1.23) and simply involves the summation of the coefficients multiplied by the modes. We shall primarily be concerned with the discrete backward transform where the function $u(\xi_1, \xi_2, \xi_3)$ is evaluated at the quadrature points:

$$u^\delta(\xi_{1i}, \xi_{2j}, \xi_{3k}) = \sum_{m(pqr)} \hat{u}_{pqr}\,\phi_{pqr}(\xi_{1i}, \xi_{2j}, \xi_{3k})\,.$$

This operation can be represented in terms of the vectors $\boldsymbol{u}$ and $\hat{\boldsymbol{u}}$ and the matrix $\boldsymbol{B}$ as ♣[18]

$$\boldsymbol{u} = \boldsymbol{B}\hat{\boldsymbol{u}}\,. \tag{4.1.28}$$

The two-dimensional case is given simply by $k = 0$ and the summation is performed over p and q only. Although we have represented the solution in terms of the local coordinates (ξ_1, ξ_2, ξ_3), we note that this is equivalent to evaluating the functions at the global coordinates (x_1, x_2, x_3) under the mapping $x_i = \chi_i^e(\xi_1, \xi_2, \xi_3)$. In general, the discrete backward transformation may be evaluated at any set of discrete points depending on the definition of the matrix $\boldsymbol{B}$. For example, a more evenly distributed set of points may be required graphically to display the solution. For most computational needs, however, we will normally evaluate the basis at the quadrature points.

[18] Matrix formulation of the elemental backward transformation.

Nodal expansions

We note that when using a nodal expansion basis there are special cases of the backward transform matrix where the matrix becomes the identity matrix. If the matrix $\boldsymbol{B}$ is generated using a Lagrange polynomial through a set of nodal points and the basis is evaluated at the same nodal points, then $\boldsymbol{B} = \boldsymbol{I}$. This arises when using either the non-tensorial simplex basis or the quadrilateral and hexahedral nodal expansions. When $\boldsymbol{B} = \boldsymbol{I}$ we observe that

$$\boldsymbol{u} = \boldsymbol{B}\hat{\boldsymbol{u}} = \boldsymbol{I}\hat{\boldsymbol{u}} = \hat{\boldsymbol{u}}\,.$$

As discussed previously, this demonstrates that, for the classical spectral element method and Lagrange expansion in simplexes, the expansion coefficients are simply the values of the solution at the nodal points. We note, however, that when the basis is evaluated at points other than the nodal points the matrix $\boldsymbol{B}$ is full. Such a situation would arise even in the tensorial quadrilateral/hexahedral nodal basis if the quadrature order is not exactly equivalent to the polynomial order plus one (i.e., $Q = P + 1$).

4.1.5.3 *Elemental forward transformation*

In this section we discuss the formulation of the forward transformation using the previously introduced matrix and vector notation. The action of the forward transformation is that, given either a continuous $u(\boldsymbol{\xi})$ or discrete $u^\delta(\boldsymbol{\xi})$ function, we determine the expansion coefficients $\hat{\boldsymbol{u}}$. There are two commonly applied approaches using either a *collocation* or *Galerkin* projection, both of which can be formulated from the method of weighted residuals statement which we will first review. A similar construction was outlined in one dimension in Section 2.3.2.1.

If we consider a two-dimensional function $u(\xi_1, \xi_2)$ which does *not* lie within the polynomial space of the expansion basis then there will be an approximation error between our approximation $u^\delta = \sum_{p,q} \hat{u}_{pq}\phi_{pq}$ and the function $u(\boldsymbol{\xi})$ which we denote by $R(u)$, that is,

$$u^\delta(\xi_1, \xi_2) - u(\xi_1, \xi_2) = R(u) \tag{4.1.29}$$

or, equivalently,

$$\sum_{pq} \hat{u}_{pq}\phi_{pq}(\xi_1, \xi_2) - u(\xi_1, \xi_2) = R(u)\,.$$

Following the method of weighted residuals, we represent the inner product of both sides of this equation by a presently undefined function $v(\xi_1, \xi_2)$:

$$\left(v, \sum_{p,q} \hat{u}_{pq}\phi_{pq}\right) - (v, u) = (v, R(u))$$

and set the right-hand-side term to zero, $(v, R(u)) = 0$, to obtain

$$\left(v, \sum_{p,q} \hat{u}_{pq}\phi_{pq}\right) = (v, u)\,. \tag{4.1.30}$$

The choice of $v(\boldsymbol{\xi})$ will define the type of projection. This is illustrated in the next two sections for the case of a collocation and Galerkin projection.

Collocation projection—interpolation

We note that the matrix $\boldsymbol{B}$ introduced in the previous section is directly analogous to the generalised Vandermonde matrix $\boldsymbol{V}$ introduced in Section 3.3.2. In particular, if we consider the case where the set of distinct points $\boldsymbol{\xi}_i$ is of the same dimension as the expansion basis then the matrix $\boldsymbol{B}$ is identical to our previous definition of $\boldsymbol{V}$ in Section 3.3.2. In this case $\boldsymbol{B}$ is square and invertible, and the inversion of this matrix is equivalent to a collocation projection. To see how this fits into the statement of the method of weighted residuals we write our approximation as

$$u^{\delta}(\boldsymbol{\xi}) = \sum_{n=1}^{N_m-1} \hat{u}_n\phi_n(\boldsymbol{\xi})\,,$$

then the method of weighted residuals eqn (4.1.30) implies that

$$\int_{\Omega} v_m u^{\delta}(\boldsymbol{\xi})\,\mathrm{d}\boldsymbol{\xi} = \int_{\Omega} v_m \sum_{n=1}^{N_m-1} \hat{u}_n\phi_n(\boldsymbol{\xi})\,\mathrm{d}\boldsymbol{\xi}\,, \quad m = 0,\ldots,N_m-1\,. \tag{4.1.31}$$

In the collocation method we set $v_m = \delta(\boldsymbol{\xi}_m)$, where $\delta(\boldsymbol{\xi}_m)$ is the Dirac delta function at the N_m discrete nodal points $\boldsymbol{\xi}_m$, and implies that $R(u(\boldsymbol{\xi}_m)) = 0$. The action of the Dirac delta functions on the integrals means that eqn (4.1.31) can be evaluated as

$$u^{\delta}(\boldsymbol{\xi}_m) = \sum_{n=1}^{N_m-1} \hat{u}_n\phi_n(\boldsymbol{\xi}_m)\,, \quad n = 0,\ldots,N_m-1\,,$$

which can be written in matrix form to obtain

$$\boldsymbol{u} = \boldsymbol{B}_{\mathcal{N}}\hat{\boldsymbol{u}} \quad \text{or} \quad \hat{\boldsymbol{u}} = \boldsymbol{B}_{\mathcal{N}}^{-1}\boldsymbol{u}\,, \tag{4.1.32}$$

where $\boldsymbol{B}_{\mathcal{N}}[m][n] = \phi_n(\boldsymbol{\xi}_m)$.

In general, the above formulation can be applied to any function $u(\boldsymbol{\xi})$ and not just the approximation $u^{\delta}(\boldsymbol{\xi})$, and thus we can use it as a technique to interpolate a function within the region $\boldsymbol{\xi} \in \Omega_{\mathrm{st}}$.

Discrete Galerkin projection

The discrete forward transform which determines the modal coefficients $\hat{u}_{pq}$ from a prescribed function $u(\boldsymbol{\xi})$ evaluated at a set of nodal or quadrature points can be expressed in terms of the matrix formulation previously introduced as

$$\hat{\boldsymbol{u}} = \left(\boldsymbol{B}^{\top}\boldsymbol{W}\boldsymbol{B}\right)^{-1}\boldsymbol{B}^{\top}\boldsymbol{W}\boldsymbol{u}\,, \tag{4.1.33}$$

where we note that the inverted matrix $\boldsymbol{BWB}$ is symmetric, in contrast to the collocation matrix $\boldsymbol{B}_{\mathcal{N}}$. We refer to this as a *discrete* forward transform since, if

$u(\boldsymbol{\xi})$ is not a polynomial function of similar space as the projecting polynomial, then a collocation projection is being performed when evaluating the function at the quadrature points to obtain $\boldsymbol{u}$. We note, however, that the collocation projection may be onto a richer polynomial space than the Galerkin projection, and so the error associated with the Galerkin projection typically dominates.

In the Galerkin projection we choose the weight function in the method of weighted residuals (4.1.30) to be the same as the expansion basis, so that $v(\xi_1, \xi_2) = \phi_{rs}(\xi_1 \xi_2)$. Equation (4.1.30) can therefore be written as

$$\left(\phi_{rs}, \sum_{p,q} \hat{u}_{pq} \phi_{pq}\right) = (\phi_{rs}, u)\,.$$

Noting that the coefficients $\hat{u}_{pq}$ are independent of ξ_1 and ξ_2, we can rewrite this equation as

$$\sum_{p,q} (\phi_{rs}, \phi_{pq}) \hat{u}_{pq} = (\phi_{rs}, u)\,, \tag{4.1.34}$$

which is a scalar equation. If we test this equation versus all N_m modes ϕ_{rs} we then have N_m scalar equations to solve for the N_m unknown degrees of freedom $\hat{u}_{pq}$.

Equation (4.1.34) is the functional representation of a linear system which can be solved to determine $\hat{u}_{pq}$, where the term (ϕ_{rs}, ϕ_{pq}) represents the components of the two-dimensional elemental mass matrix $\boldsymbol{M}$, which has a rank equal to N_m. An identical procedure to the one outlined above using ϕ_{pqr} would have led to the three-dimensional system.

Although eqn (4.1.34) represents the forward transformation, it is not immediately clear how to construct the matrix system. First, we need to make an approximation to represent the inner product discretely by using Gaussian quadrature. The integral in the inner product on the left-hand side may be evaluated exactly using Gaussian quadrature, provided that a sufficient number of quadrature points are used. The right-hand-side inner product in eqn (4.1.34) involves the arbitrary function $u(\xi_1, \xi_2, \xi_3)$, which may not be a polynomial. Nevertheless, providing that the function u is sufficiently smooth, the error in evaluating the integral will be consistent with the approximation error. Similarly, by evaluating the continuous function at the collocation points we are performing a collocation projection onto the quadrature points. The discrete form of eqn (4.1.34) can be written as

$$\sum_{pq} (\phi_{rs}, \phi_{pq})_\delta \, \hat{u}_{pq} = (\phi_{rs}, u)_\delta\,, \quad \forall\ r, s\,. \tag{4.1.35}$$

This equation represents a system of N_m scalar equations for every ϕ_{rs}.

We can now illustrate the use of the matrix and vector notation by initially considering the inner product of a function $v(x_1, x_2, x_3)$ with a function $u(x_1, x_2, x_3)$ and is defined as

$$(v, u)_\delta = \int v(\xi_1, \xi_2, \xi_3) u(\xi_1, \xi_2, \xi_3) |J| \, d\xi_1 \, d\xi_2 \, d\xi_3 .$$

Representing the integral using Gaussian quadrature, we have a discrete approximation such that

$$(v, u)_\delta = \sum_{i=0}^{Q_1-1} \sum_{j=0}^{Q_2-1} \sum_{k=0}^{Q_3} w_i w_j w_k \, v(\xi_{1i}, \xi_{2j}, \xi_{3k}) \, u(\xi_{1i}, \xi_{2j}, \xi_{3k}) \, |J_{ijk}| , \qquad (4.1.36)$$

where

$$(v, u) = (v, u)_\delta + \varepsilon$$

and ε is the error due to the numerical integration or collocation projection, as defined in Appendix B. If the functions u and v are sufficiently smooth in the sense that the first Q derivatives are bounded, then ε will be of the same order as the approximation error, which is important if this error is not to dominate [87].

The operation in eqn (4.1.36) can be evaluated using the vectors $\boldsymbol{v}$ and $\boldsymbol{u}$ and the matrix $\boldsymbol{W}$ as

$$(v, u)_\delta = \boldsymbol{v}^\top \boldsymbol{W} \boldsymbol{u} . \qquad (4.1.37)$$

Now, to assemble eqn (4.1.35) into a matrix system we note that when $v(\boldsymbol{\xi}) = \phi_{rs}(\boldsymbol{\xi})$ in eqn (4.1.37) we have the right-hand side of eqn (4.1.35) for a single expansion mode. To evaluate the complete right-hand side of the matrix system we need to evaluate this inner product over all N_m expansion modes to produce a vector of length N_m. The columns of the matrix $\boldsymbol{B}$ represent the expansion modes ϕ_{rs} at the quadrature points, and so to evaluate the inner product with respect to all modes we replace $\boldsymbol{v}$ in eqn (4.1.37) with $\boldsymbol{B}$, that is,

$$\boldsymbol{B}^\top \boldsymbol{W} \boldsymbol{u}[m(rs)] = (\phi_{rs}, u)_\delta ,$$

where $m(rs)$ represents the consecutive ordering of the expansion modes whose indices run over r and s. The next step is to express the left-hand side of eqn (4.1.35) in a similar fashion. The left-hand side is the inner product of every expansion mode with respect to every other expansion mode, which leads to the elemental mass matrix $\boldsymbol{M}$:

$$\boldsymbol{M}[m(rs)][n(pq)] = \boldsymbol{B}^\top \boldsymbol{W} \boldsymbol{B}[m(rs)][n(pq)] = (\phi_{rs}, \phi_{pq})_\delta ,$$

where $n(pq)$ is analogous to $m(rs)$ and represents the consecutive ordering of expansion modes. Finally, the summation in eqn (4.1.35) is identical to multiplying the mass matrix $\boldsymbol{M}$ by the coefficient vector $\hat{\boldsymbol{u}}$ to obtain the matrix representation of the system (4.1.35):

$$(\boldsymbol{B}^\top \boldsymbol{W} \boldsymbol{B}) \hat{\boldsymbol{u}} = \boldsymbol{B}^\top \boldsymbol{W} \boldsymbol{u} .$$

The solution to this system determines the vector of expansion coefficients $\hat{\boldsymbol{u}}$

♣[19] from the values of a function at the quadrature points denoted by $\boldsymbol{u}$, that is,

$$\hat{\boldsymbol{u}} = \left(\boldsymbol{B}^\top \boldsymbol{W} \boldsymbol{B}\right)^{-1} \boldsymbol{B}^\top \boldsymbol{W} \boldsymbol{u} = (\boldsymbol{M})^{-1} \boldsymbol{B}^\top \boldsymbol{W} \boldsymbol{u},$$

which is the *discrete forward transform*. We note that if $\boldsymbol{u}$ spans the same space as the polynomial basis used to evaluate $\boldsymbol{B}$ and integration is exact then the above projection would give an identical answer to eqn (4.1.32) up to numerical precision.

Galerkin projection with nodal expansions

Just as the nodal expansion basis was a special case for the backward transformation, it is also true for the forward transformation. We recall that for the spectral element method the nodal expansion is defined by the Lagrange polynomials through the quadrature points, implying that

$$\boldsymbol{B} = \boldsymbol{B}^\top = \boldsymbol{I}.$$

Therefore, the discrete forward transform becomes

$$\begin{aligned}\hat{\boldsymbol{u}} &= \left(\boldsymbol{B}^\top \boldsymbol{W} \boldsymbol{B}\right)^{-1} \boldsymbol{B}^\top \boldsymbol{W} \boldsymbol{u} \\ &= (\boldsymbol{I} \boldsymbol{W} \boldsymbol{I})^{-1} \boldsymbol{W} \boldsymbol{I} \boldsymbol{u} \\ &= \boldsymbol{W}^{-1} \boldsymbol{W} \boldsymbol{u} = \boldsymbol{u}.\end{aligned}$$

Analogous to the interpretation of the forward transformation, we observe that the modal coefficients, $\hat{\boldsymbol{u}}$, are simply the values of the solution at the nodal points, $\boldsymbol{u}$.

Positive-definiteness of the elemental mass matrix

Working back from the matrix formulation to the functional form of the integral operator, it is possible to show that the elemental mass matrix

$$\boldsymbol{M} = \boldsymbol{B}^\top \boldsymbol{W} \boldsymbol{B}$$

is positive definite. A sufficient condition for the matrix $\boldsymbol{M}$ to be positive definite is that

$$\hat{\boldsymbol{u}}^\top \boldsymbol{M} \hat{\boldsymbol{u}} > 0, \quad \forall \text{ nonzero vectors } \hat{\boldsymbol{u}}.$$

If we replace $\boldsymbol{M}$ by its full matrix components then we find

$$\hat{\boldsymbol{u}}^\top \boldsymbol{M} \hat{\boldsymbol{u}} = \hat{\boldsymbol{u}}^\top \left(\boldsymbol{B}^\top \boldsymbol{W} \boldsymbol{B}\right) \hat{\boldsymbol{u}}.$$

Now, from the definition of the backward transformation (4.1.28), we see that

$$\hat{\boldsymbol{u}}^\top \left(\boldsymbol{B}^\top \boldsymbol{W} \boldsymbol{B}\right) \hat{\boldsymbol{u}} = (\boldsymbol{B} \hat{\boldsymbol{u}})^\top \boldsymbol{W} (\boldsymbol{B} \hat{\boldsymbol{u}}) = \boldsymbol{u}^\top \boldsymbol{W} \boldsymbol{u}.$$

[19] Matrix formulation of the elemental Galerkin forward transformation.

A comparison with eqn (4.1.37) shows that the last expression is simply the inner product of $u^\delta(\xi_1, \xi_2)$ with itself, that is,

$$\boldsymbol{u}^\top \boldsymbol{W} \boldsymbol{u} = (u^\delta, u^\delta)_\delta \, .$$

Provided that the quadrature order is sufficiently high, the integration will be exact since $u^\delta(\xi_1, \xi_2)$ is a polynomial and, therefore,

$$(u^\delta, u^\delta)_\delta = \int (u^\delta)^2 \, \mathrm{d}\xi_1 \, \mathrm{d}\xi_2 \geqslant 0 \, ,$$

which is positive for any nonzero value of $u^\delta(\xi_1, \xi_2)$, thereby proving that $\boldsymbol{M}$ is positive definite.

Discrete Galerkin projection to physical space

We have previously considered the projection of the continuous function $u(\boldsymbol{\xi})$ evaluated at the quadrature points $\boldsymbol{u}$ onto the polynomial space to obtain a set of expansion coefficients $\hat{\boldsymbol{u}}$, for example

$$\hat{\boldsymbol{u}} = (\boldsymbol{M})^{-1} \boldsymbol{B}^\top \boldsymbol{W} \boldsymbol{u} \, .$$

However, if we now want to re-evaluate the projected function at the same quadrature points then we can perform a backwards transform or, equivalently, multiply by the basis matrix $\boldsymbol{B}$ such that

$$\boldsymbol{u}^\delta = \boldsymbol{P}^\delta \boldsymbol{u} = \boldsymbol{B} (\boldsymbol{M})^{-1} \boldsymbol{B}^\top \boldsymbol{W} \boldsymbol{u} \, .$$

This entire process, denoted by $\boldsymbol{P}^\delta$, can be considered to be a discrete Galerkin projection to physical space and has the property that $\boldsymbol{P}^\delta \boldsymbol{P}^\delta \boldsymbol{u} = \boldsymbol{P}^\delta \boldsymbol{u}$, which is easily demonstrated since

$$\begin{aligned}
\boldsymbol{P}^\delta \boldsymbol{P}^\delta \boldsymbol{u} &= \boldsymbol{B}^\top (\boldsymbol{M})^{-1} \boldsymbol{B}^\top \boldsymbol{W} \boldsymbol{B} (\boldsymbol{M})^{-1} \boldsymbol{B}^\top \boldsymbol{W} \boldsymbol{u} \\
&= \boldsymbol{B}^\top (\boldsymbol{M})^{-1} \boldsymbol{M} (\boldsymbol{M})^{-1} \boldsymbol{B}^\top \boldsymbol{W} \boldsymbol{u} \\
&= \boldsymbol{B}^\top (\boldsymbol{M})^{-1} \boldsymbol{B}^\top \boldsymbol{W} \boldsymbol{u} \\
&= \boldsymbol{P}^\delta \boldsymbol{u} \, .
\end{aligned}$$

4.1.5.4 *Differential operators: weak Laplacian*

To complete this section we illustrate how to construct a matrix system from a differential problem and thereby construct the weak elemental Laplacian matrix.

We wish to consider the Galerkin approximation of the two-dimensional Poisson equation within an elemental region:

$$\nabla^2 u(\boldsymbol{x}) = f(\boldsymbol{x}) \, , \quad \boldsymbol{x} \in \Omega^e \, .$$

The one-dimensional formulation of this equation has been dealt with in Section 2.2.1 and a complete multi-dimensional formulation can be found in Chapter 5.

To recap the Galerkin approximation of this equation, we take the inner product with respect to a continuous function $v(\boldsymbol{x})$ to obtain

$$(v, \nabla^2 u) = (v, f)\,.$$

Applying the divergence theorem to the left-hand side, we obtain

$$(\nabla v, \nabla u) = \int_{\partial\Omega} v \nabla u \cdot \boldsymbol{n}\, \mathrm{d}S - (v, f)\,,$$

where $\partial\Omega$ is the boundary of the problem and $\boldsymbol{n}$ is the unit normal along the boundary. The term $(\nabla v, \nabla u)$ is the weak Laplacian and written in full in two dimensions has the form

$$(\nabla v, \nabla u) = \left(\frac{\partial v}{\partial x_1}, \frac{\partial u}{\partial x_1}\right) + \left(\frac{\partial v}{\partial x_2}, \frac{\partial u}{\partial x_2}\right).$$

In a Galerkin formulation the same functions are used to approximate $v(\boldsymbol{x})$ and $u(\boldsymbol{x})$. Approximating all integrals with Gaussian quadrature, the matrix form of the elemental weak Laplacian operator $\boldsymbol{L}^e$ becomes

$$\begin{aligned}\boldsymbol{L}^e = {} & \left[\left(\boldsymbol{\Lambda}\left(\frac{\partial \xi_1}{\partial x_1}\right)\boldsymbol{D}_{\xi_1} + \boldsymbol{\Lambda}\left(\frac{\partial \xi_2}{\partial x_1}\right)\boldsymbol{D}_{\xi_2}\right)\boldsymbol{B}\right]^\top \boldsymbol{W}\left[\boldsymbol{\Lambda}\left(\frac{\partial \xi_1}{\partial x_1}\right)\boldsymbol{D}_{\xi_1}\right.\\ & \qquad\qquad\qquad\qquad \left. + \boldsymbol{\Lambda}\left(\frac{\partial \xi_2}{\partial x_1}\right)\boldsymbol{D}_{\xi_2}\right]\boldsymbol{B}\\ & + \left[\left(\boldsymbol{\Lambda}\left(\frac{\partial \xi_1}{\partial x_2}\right)\boldsymbol{D}_{\xi_1} + \boldsymbol{\Lambda}\left(\frac{\partial \xi_2}{\partial x_2}\right)\boldsymbol{D}_{\xi_2}\right)\boldsymbol{B}\right]^\top \boldsymbol{W}\left[\boldsymbol{\Lambda}\left(\frac{\partial \xi_1}{\partial x_2}\right)\boldsymbol{D}_{\xi_1}\right.\\ & \qquad\qquad\qquad\qquad \left. + \boldsymbol{\Lambda}\left(\frac{\partial \xi_2}{\partial x_2}\right)\boldsymbol{D}_{\xi_2}\right]\boldsymbol{B}\,,\end{aligned}$$

which can be rearranged into the form

$$\begin{aligned}\boldsymbol{L}^e = {} & \boldsymbol{B}^\top\left[\boldsymbol{D}_{\xi_1}^\top\boldsymbol{\Lambda}\left(\frac{\partial \xi_1}{\partial x_1}\right) + \boldsymbol{D}_{\xi_2}^\top\boldsymbol{\Lambda}\left(\frac{\partial \xi_2}{\partial x_1}\right)\right]\boldsymbol{W}\left[\boldsymbol{\Lambda}\left(\frac{\partial \xi_1}{\partial x_1}\right)\boldsymbol{D}_{\xi_1}\right.\\ & \qquad\qquad\qquad\qquad \left. + \boldsymbol{\Lambda}\left(\frac{\partial \xi_2}{\partial x_1}\right)\boldsymbol{D}_{\xi_2}\right]\boldsymbol{B}\\ & + \boldsymbol{B}^\top\left[\boldsymbol{D}_{\xi_1}^\top\boldsymbol{\Lambda}\left(\frac{\partial \xi_1}{\partial x_2}\right) + \boldsymbol{D}_{\xi_2}^\top\boldsymbol{\Lambda}\left(\frac{\partial \xi_2}{\partial x_2}\right)\right]\boldsymbol{W}\left[\boldsymbol{\Lambda}\left(\frac{\partial \xi_1}{\partial x_2}\right)\boldsymbol{D}_{\xi_1}\right.\\ & \qquad\qquad\qquad\qquad \left. + \boldsymbol{\Lambda}\left(\frac{\partial \xi_2}{\partial x_2}\right)\boldsymbol{D}_{\xi_2}\right]\boldsymbol{B}\,,\end{aligned}$$

which also demonstrates the symmetry of this matrix system.

4.1.6 *Sum-factorisation/tensor product operations*

The sum-factorisation or tensor product technique was first recognised by Orszag [352] and is considered to be the key to the efficiency of spectral methods. It is based on the fact that the expansion is a tensor product of one-dimensional functions, which means that many important numerical operations may be numerically evaluated with a notable reduction in operation count as compared to a non-tensorial expansion.

To demonstrate this technique we consider the evaluation of a summation over r and s of an array f_{rs}, with the functions h_{ir} and h_{js} for all indices i and j, that is,

$$U_{ij} = \sum_r^P \sum_s^P f_{rs} h_{ir} h_{js} \,, \quad \forall\, i, j \,. \tag{4.1.38}$$

If $f_{rs} = f(\xi_{1r}, \xi_{2s})$ and $h_{ir} = h_r(\xi_{1i})$ and $h_{js} = h_s(\xi_{2j})$ then this summation would represent the interpolation, using Lagrange polynomials, from a set of points (ξ_{1r}, ξ_{2s}) to a set of points (ξ_{1i}, ξ_{2j}). If all indices i, j, k, and l are assumed to be of $\mathcal{O}(P)$ then the evaluation of this whole operation reduces to an $\mathcal{O}(P^2)$ summation over k and l for each one of the $\mathcal{O}(P^2)$ indices i and j, and so the total operation count would be $\mathcal{O}(P^4)$. However, noting that we can factor the h_{ir} term out of the second summation:

$$U_{ij} = \sum_r^P h_{ir} \left(\sum_s^P f_{rs} h_{js} \right) , \quad \forall\, i, j \,,$$

we can then evaluate the summation over 's' and replace the terms in brackets by

$$\bar{f}_{jr} = \sum_s^P f_{rs} h_{js} \,.$$

To construct $\bar{f}_{jr}$ is an $\mathcal{O}(P^3)$ operation since we are evaluating an $\mathcal{O}(P)$ summation over 's' for all the $\mathcal{O}(P^2)$ indices j and r. The original summation (4.1.38) can now be written as

$$U_{ij} = \sum_r^P h_{ir} \bar{f}_{jr} \,, \quad \forall\, i, j \,,$$

which is also an $\mathcal{O}(P^3)$ operation as we are evaluating an $\mathcal{O}(P)$ summation over the index r for all $\mathcal{O}(P^2)$ points i and j. We therefore see that this factorisation has reduced the cost from an $\mathcal{O}(P^4)$ operation to an $\mathcal{O}(P^3)$ operation, which is the typical reduction for a two-dimensional summation. In three dimensions it is possible to reduce an $\mathcal{O}(P^6)$ operation to an $\mathcal{O}(P^4)$ operation.

To illustrate this technique we consider the sum-factorisation applied to the backward transformation and inner product when using both tensor product and generalised tensor product expansions. Another important operation is differentiation. However, we note that in the example above if we let $h_{ir} = \mathrm{d}h_r/\mathrm{d}\xi_1(\xi_{1i})$

then the summation would have represented the numerical differentiation of the function $f(\xi_1, \xi_2)$ with respect to ξ_1.

4.1.6.1 *Backward transformation example*

Recall that the two-dimensional backward transformation evaluated at the quadrature points for a general basis $\phi_{pq}(\xi_1, \xi_2)$ is

$$u(\xi_{1i}, \xi_{2j}) = \sum_p \sum_q \hat{u}_{pq} \phi_{pq}(\xi_{1i}, \xi_{2j}), \quad \forall\, i, j, \tag{4.1.39}$$

which can be described by the matrix operation

$$\boldsymbol{u} = \boldsymbol{B}\hat{\boldsymbol{u}}, \tag{4.1.40}$$

where $\boldsymbol{u}$ is a vector of length N_Q and denotes the evaluation of $u(\xi_1, \xi_2)$ at the quadrature points, $\hat{\boldsymbol{u}}$ is a vector of length N_m which contains all the elemental expansion coefficients, and $\boldsymbol{B}$ is a matrix of dimension $N_Q \cdot N_m$ whose columns are constructed from the expansion modes evaluated at the quadrature points. We recall that to evaluate the summation (4.1.39) at all the quadrature points (ξ_{1i}, ξ_{2j}), or alternatively perform the matrix–vector multiplication, would be an $\mathcal{O}(P^4)$ operation. This is because each quadrature point involves a summation over $\mathcal{O}(P^2)$ modes and there are typically $\mathcal{O}(P^2)$ quadrature points. In three dimensions the equivalent operation would be $\mathcal{O}(P^6)$.

Standard tensorial expansion

For the quadrilateral region the tensorial expansion basis can be written as $\phi_{pq}(\xi_1, \xi_2) = \psi_p^a(\xi_1)\psi_q^a(\xi_2)$. Putting this definition into eqn (4.1.39) and factoring out the term $\psi_p^a(\xi_{1i})$, we obtain

$$u(\xi_{1i}, \xi_{2j}) = \sum_{p=0}^{P_1} \psi_p^a(\xi_{1i}) \left\{ \sum_{q=0}^{P_2} \hat{u}_{pq} \psi_q^a(\xi_{2j}) \right\}. \tag{4.1.41}$$

We note that to evaluate $u(\xi_1, \xi_2)$ at an arbitrary point is still an $\mathcal{O}(P^2)$ operation. However, if we wish to evaluate the summation at all the $\mathcal{O}(P^2)$ quadrature points ξ_{1i} and ξ_{2j} then we can use two steps:

$$f_p(\xi_{2j}) = \sum_{q=0}^{P_2} \hat{u}_{pq} \psi_q^a(\xi_{2j}), \tag{4.1.42a}$$

$$u(\xi_{1i}, \xi_{2j}) = \sum_{p=0}^{P_1} \psi_p^a(\xi_{1i}) f_p(\xi_{2j}). \tag{4.1.42b}$$

Inserting (4.1.42a) into (4.1.42b) recovers eqn (4.1.41). In step (4.1.42a) the array $f_p(\xi_{2j})$ is evaluated by summing the modal coefficients, multiplied by the second part of the tensor expansion $\psi_q^a(\xi_{2j})$ over q at every ξ_{2j} point and for every p index. This operation is equivalent to performing a one-dimensional backward transform and requires extra storage for an array of size $\mathcal{O}(P^2)$. The summation for every entry in this array is an $\mathcal{O}(P)$ operation, and therefore the total cost to generate $f_p(\xi_{2j})$ is an $\mathcal{O}(P^3)$ operation. However, the second step (4.1.42b) is now independent of the summation in q and so it is also an $\mathcal{O}(P^3)$ operation involving an $\mathcal{O}(P)$ summation at $\mathcal{O}(P^2)$ points ξ_{1i} and ξ_{2j}. In summary, we have replaced the $\mathcal{O}(P^4)$ operation (4.1.41) with two $\mathcal{O}(P^3)$ operations (4.1.42a) and (4.1.42b) and an extra array of size $\mathcal{O}(P^2)$.

In three dimensions the backward transformation for a hexahedral expansion is

$$u(\xi_{1i}, \xi_{2j}, \xi_{3k}) = \sum_{p=0}^{P_1} \psi_p^a(\xi_{1i}) \left\{ \sum_{q=0}^{P_2} \psi_q^a(\xi_{2j}) \left\{ \sum_{r=0}^{P_3} \hat{u}_{pqr} \psi_r^a(\xi_{3k}) \right\} \right\}. \tag{4.1.43}$$

This summation can be evaluated at all the $\mathcal{O}(P^3)$ quadrature points in an $\mathcal{O}(P^4)$ operation using a three-step process of the form

$$f_{pq}(\xi_{3k}) = \sum_{r=0}^{P_3} \hat{u}_{pqr} \psi_r^a(\xi_{3k}), \tag{4.1.44a}$$

$$\bar{f}_p(\xi_{2j}, \xi_{3k}) = \sum_{q=0}^{P_2} \psi_q^a(\xi_{2j}) f_{pq}(\xi_{3k}), \tag{4.1.44b}$$

$$u(\xi_{1i}, \xi_{2j}, \xi_{3k}) = \sum_{p=0}^{P_1} \psi_p^a(\xi_{1i}) \bar{f}_p(\xi_{2j}, \xi_{3k}). \tag{4.1.44c}$$

In this three-dimensional case we have replaced the $\mathcal{O}(P^6)$ operation (4.1.43) with three $\mathcal{O}(P^4)$ operations to evaluate the steps (4.1.44a)–(4.1.44c). This also requires memory for two $\mathcal{O}(P^3)$ arrays to store $f_{pq}(\xi_{3k})$ and $\bar{f}_p(\xi_{2j})$.

Generalised tensorial expansion

For the hybrid regions, where the expansion bases are of the form $\phi_{pqr} = \psi_p^a \psi_{pq}^b \psi_{pqr}^c$, the sum-factorisation technique may still be applied in a very similar fashion to the standard tensor product regions. However, whereas for the quadrilateral and hexahedral expansions it did not matter which part of the tensor product we factored out, for the generalised hybrid expansions there is only one choice of factorisation which maintains the efficiency. To illustrate this point, consider the backward transformation of the triangular expansion $\phi_{pq}(\xi_1, \xi_2) = \psi_p^a(\eta_1)\psi_{pq}^b(\eta_2)$, where $\eta_1 = 2(1+\xi_1)/(1-\xi_2)$ and $\eta_2 = \xi_2$:

$$u(\eta_{1i}, \eta_{2j}) = \sum_p \sum_q \hat{u}_{pq}\, \psi_p^a(\eta_{1i}) \psi_{pq}^b(\eta_{2j}).$$

We are unable to factor out the term $\psi_{pq}^b(\eta_{2j})$ because it is dependent upon both indices p and q, and so we can only factor the $\psi_p^a(\eta_{1i})$ term to arrive at

$$u(\eta_{1i}, \eta_{2j}) = \sum_p \psi_p^a(\eta_{1i}) \left\{ \sum_q \hat{u}_{pq} \, \psi_{pq}^b(\eta_{2j}) \right\}, \tag{4.1.45}$$

which can be evaluated in two steps as

$$f_p(\xi_{2j}) = \sum_q \hat{u}_{pq} \, \psi_{pq}^b(\xi_{2j}), \tag{4.1.46a}$$

$$u(\xi_{1i}, \xi_{2j}) = \sum_p \psi_p^a(\xi_{1i}) f_p(\xi_{2j}). \tag{4.1.46b}$$

Since the expansion coefficients $\hat{u}_{pq}$ depend on both indices p and q, there is no extra expense in evaluating a tensor product of this form as compared with the structured case. However, we have deliberately omitted any limits on the summation indices over p and q. For the orthogonal triangular expansion the indices are close-packed and the range for p and q is $0 \leqslant p, q,\, p \leqslant P_1,\, p+q \leqslant P_2$; however, for the modified triangular expansion the indices, and therefore $\hat{u}_{pq}$, are not close-packed. Nevertheless, it is possible to produce a sparse array of coefficients $\hat{u}_{pq}$ which will allow us to perform the summation (4.1.46a) over the range $0 \leqslant p \leqslant P_1$, $0 \leqslant q \leqslant P_2$ or, more efficiently, over every nonzero entry of $\hat{u}_{pq}$.

To illustrate how to generate the non-sparse form of $\hat{u}_{pq}$ we consider the example shown in Fig. 4.7. This figure shows all $(P+1)(P+2)/2$ expansion modes for the modified triangular expansion when $P = P_1 = P_2 = 4$. We recall that each mode is constructed from the product of two one-dimensional functions $\psi_p^a(\eta_1)\psi_{pq}^b(\eta_2)$ (see Section 3.2.3).

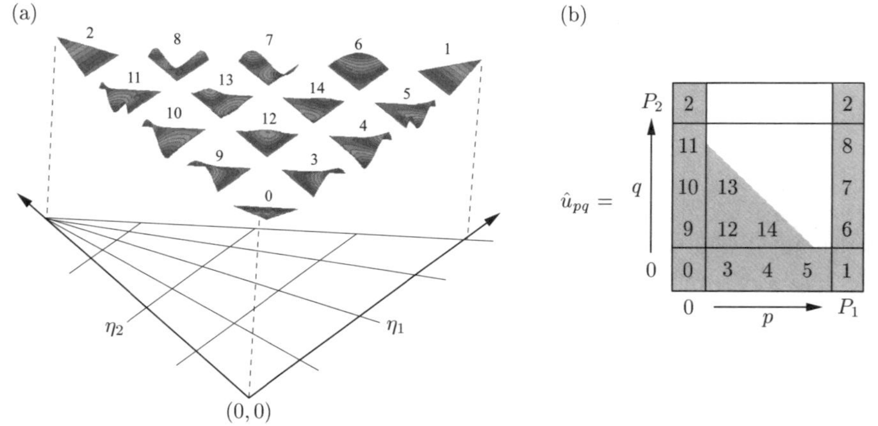

FIG. 4.7. Illustration of the mapping procedure for the physical expansion coefficients corresponding to the modes shown in (a) to the two-dimensional array $\hat{u}_{pq}$ shown in (b).

In Fig. 4.7 we also see a numbering system for every physical mode which can be arbitrarily defined. In this case we have adopted the convention that the vertex modes are labelled first, followed by edge modes, and then the interior modes. The location of the modal coefficient within the array corresponds to the p and q indices used to generate the principal function mode $\psi^b_{pq}(\xi_2)$, as shown in Fig. 4.7(b). Therefore the construction of this mapping has a physical interpretation since if we consider the bottom-left corner of the array as being at $\eta_1 = 0$, $\eta_2 = 0$ then this is also the corner where the vertex mode has a unit value. Similarly, the edges along which the modes have a nonzero value relate to their location within the two-dimensional array. Finally, the interior modes are related to the interior of the array where q runs fastest. ♦[22]

The final point to note is that the degenerate vertex mode, labelled vertex '2' in Fig. 4.7(a), must be entered twice within the array $\hat{u}_{pq}$ as it is generated by combining the shape of two expansion modes $\phi_{0,P_2}(\eta_1,\eta_2) + \phi_{P_1,P_2}(\eta_1,\eta_2)$. This condition arises from the fact that

$$\phi_{0,P_2}(\xi_1,\xi_2) = \frac{1+\xi_2}{2}$$

is being represented in terms of the principal functions $\psi^a_p(\eta_1)$ and $\psi^b_{pq}(\eta_2)$ as

$$\phi_{0,P_2}(\xi_1,\xi_2) = \left(\frac{1-\eta_1}{2} + \frac{1+\eta_1}{2}\right)\frac{1+\eta_2}{2} = \left[\psi^a_0(\eta_1) + \psi^a_{P_1}(\eta_1)\right]\psi^b_{0P_2}(\eta_2)\,.$$

The *prismatic* expansion is analogous to the triangular expansion plus a tensor product of the function $\psi^a_r(\xi_3)$, and so has a similar mapping to this example. However, for the *pyramidic* expansion the top vertex is constructed from four components of ϕ_{pqr}. Accordingly, the coefficient of this vertex will need to be mapped to the four locations. Similarly, for the tetrahedral expansion the top vertex relates to four locations, whereas the base degenerate vertex and the degenerate edge both have double entries in the unpacked array.

4.1.6.2 *Inner product*

The inner product with respect to all two-dimensional expansion modes requires the evaluation of the summation

$$(\phi_{pq}, u)_\delta = \sum_i \sum_j \phi_{pq}(\xi_{1i},\xi_{2j})\, w_i w_j\, J(\xi_{1i},\xi_{2j})\, \hat{u}_{pq}(\xi_{1i},\xi_{2j})\,, \quad \forall\,(p,q)\,, \tag{4.1.47}$$

where w_i and w_j are the weights in the ξ_1 and ξ_2 directions, respectively, and $J(\xi_{1i},\xi_{2j})$ is the Jacobian. The backward transform and the inner product (4.1.47) are closely related and may be considered as the transpose of each other. To appreciate this we can consider operation (4.1.47) in matrix notation:

[22] Ordering of the modified basis expansion coefficient for application of sum-factorisation technique.

$$\boldsymbol{B}^\top [\boldsymbol{W}\boldsymbol{u}] , \tag{4.1.48}$$

where $\boldsymbol{u}$ is the vector of function values at the quadrature points, $\boldsymbol{B}$ is the basis matrix, and $\boldsymbol{W}$ is a diagonal matrix containing the quadrature weights multiplied by the appropriate Jacobian. Since $\boldsymbol{W}$ is a diagonal matrix, the evaluation of the product $\boldsymbol{W}\boldsymbol{u}$ involves a multiplication at every quadrature point. We can, therefore, consider the bracketed term as a new vector $\boldsymbol{f}$ ($\boldsymbol{f} = (\boldsymbol{W}\boldsymbol{u})$), and so the principal operation is the matrix multiplication $\boldsymbol{B}^\top \boldsymbol{f}$ which is the transpose operation to the backward transformation (4.1.40). As with the backward transform, to evaluate (4.1.47) or the matrix multiplication $\boldsymbol{B}^\top \boldsymbol{f}$ would require an $\mathcal{O}(P^4)$ operation in two dimensions and an $\mathcal{O}(P^6)$ operation in three dimensions.

For standard tensorial expansions the application of the sum-factorisation technique is analogous to the previous description of the backwards transformation. The generalised tensor product is also similar. The ordering of the factorisation, however, needs to be reversed and so we will consider this case in more detail.

Generalised tensorial expansion

When considering a generalised tensorial expansion of the form $\phi_{pqr} = \psi_p^a \psi_{pq}^b \psi_{pqr}^c$ the sum-factorisation process may still be applied, although we are again restricted as to which product may be factored. Considering the case of a triangular expansion where $\phi_{pq}(\xi_1, \xi_2) = \psi_p^a(\eta_1)\psi_{pq}^b(\eta_2)$, $\eta_1 = 2(1+\xi_1)/(1-\xi_2)$, and $\eta_2 = \xi_2$, the inner product becomes

$$(\phi_{pq}, u)_\delta = \sum_{i=0}^{Q_1-1} \sum_{j=0}^{Q_2-1} \psi_p^a(\eta_{1i})\psi_{pq}^b(\eta_{2j}) w_i w_j J(\eta_{1i}, \eta_{2j}) u(\eta_{1i}, \eta_{2j}) ,$$

where $w_j = w_j^{1,0}/2$, which accounts for the transformation of the coordinates from (ξ_1, ξ_2) to (η_1, η_2). If we factor out the term $\psi_p^a(\eta_{1i})$ then the innermost summation will still involve a sum over i for every η_{2j} points as well as all the modes over p and q. This would involve an $\mathcal{O}(P^4)$ operation and can be as expensive as the unfactored case. However, if we factor out the $\psi_{pq}^b(\eta_{2j})$ term then the summation becomes

$$(\phi_{pq}, u)_\delta = \sum_{j=0}^{Q_2-1} \psi_{pq}^b(\eta_{2j}) \left\{ \sum_{i=0}^{Q_1-1} \psi_p^a(\eta_{1i}) w_i w_j J(\eta_{1i}, \eta_{2j}) u(\eta_{1i}, \eta_{2j}) \right\} ,$$

which can be evaluated in two steps:

$$f_p(\xi_{2j}) = \sum_{i=0}^{Q_1-1} \psi_p^a(\xi_{1i}) u(\xi_{1i}, \xi_{2j}) w_i w_j J(\xi_{1i}, \xi_{2j}) , \tag{4.1.49a}$$

$$(\phi_{pq}, u)_\delta = \sum_{j=0}^{Q_2-1} \psi_{pq}^b(\xi_{2j}) f_p(\xi_{2j}) , \tag{4.1.49b}$$

where both the steps (4.1.49a) and (4.1.49b) are $\mathcal{O}(P)$ operations requiring $\mathcal{O}(P^2)$ extra memory for the array $f_p(\xi_{2j})$. If we were using an orthogonal expansion then the range for p and q would be $0 \leqslant p, q,\ p \leqslant P_1,\ p + q \leqslant P_2$. For the modified C^0 continuous expansion, however, we need to sum over p and q according to the local sparsity. The sparseness of the $\hat{u}_{pq}$ when $P = P_1 = P_2 = 4$ is shown in Fig. 4.7(b), where we interpret $\hat{u}_{pq}$ as the inner product $\hat{u}_{pq} = (\phi_{pq}, u)_\delta$. Similar to the backward transformation, when using an unstructured expansion special attention must be paid to the degenerate vertices.

4.2 Global operations

The operations described in Section 4.1 were all local in the sense that they only involved a single element and no information was coupled from any other element. In general, however, we are interested in solving second-order partial differential equations which require that some form of continuity is maintained between elemental regions. Our primary focus in this section will be the classical Galerkin method where continuity is typically imposed by making the approximation globally C^0 continuous. Alternative continuity requirements are imposed in techniques such as the mortar method or discontinuous Galerkin methods, but we leave discussion of these techniques to Sections 7.4 and 7.5. We note, however, that all the local operations described in the previous section are equally applicable to all spectral/*hp* element formulations.

To construct a globally C^0 continuous expansion from elemental or local contributions we need to introduce a local to global assembly process, often referred to as *direct stiffness summation* or *global assembly*. This process was introduced in one dimension in Section 2.3.1.4. This type of assembly is particularly important in setting up global matrices such as the mass and Laplacian systems. In Section 2.3.1 we also saw how, in one dimension, the global linear finite element can be decomposed into elemental contributions of two similarly-shaped linearly-varying modes.

In what follows, we will adopt a similar formulation for the multi-dimensional expansions to that defined in Chapter 3. We recall that an important part of the construction of the elemental bases for Galerkin formulations was the decomposition of the basis into modes which contribute to the expansion on the boundary of an element (*boundary modes*) and the remaining modes which were zero on all boundaries (*interior modes*) (see Section 3.1.1.1). As shown in Fig. 4.8, this boundary/interior decomposition allows us to construct a C^0 expansion by matching boundary modes of a similar shape. This figure illustrates all the modes used in a quadrilateral modal expansion of order $P_1 = P_2 = 4$, and we can appreciate that to construct the global expansion we can simply match the vertex and edge modes of similar shape.

In a practical implementation, it is advantageous to perform most operations in a local environment within each element and then assemble the local contributions to form the global system. However, to enable us to do this we need a

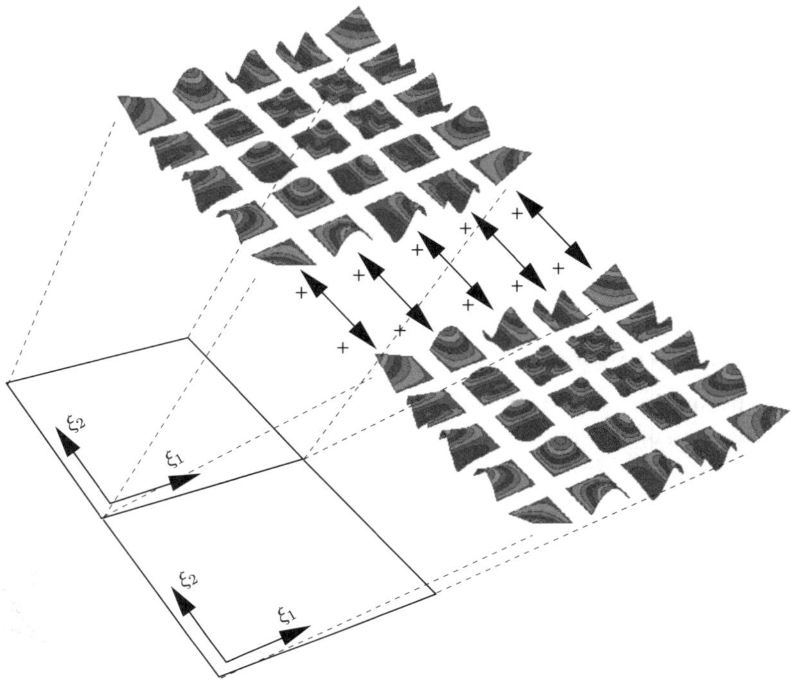

FIG. 4.8. Illustration of the construction of a C^0 global expansion from two local modal expansions of order $P_1 = P_2 = 4$. The C^0 continuity condition can be simply ensured by matching the vertex and boundary modes of similar shape.

mapping which relates to the global system from the local system. The definition of this mapping is central to the global assembly process and is discussed in Section 4.2.1. Having constructed an assembly procedure, we then illustrate the construction of a global matrix system in Section 4.2.2 using an analogous matrix and vector notation which is introduced in Section 4.1.5.1. Finally, in Section 4.2.3, we introduce a matrix manipulation technique known as static condensation which takes advantage of the global matrix structure that arises for many spectral/*hp* element expansions and allows more efficient inversion of the global matrix.

4.2.1 *Global assembly and connectivity*

Before describing the global assembly procedure, we need to define the local expansion modes $\phi_{pq}(\xi_1, \xi_2)$ within our global solution domain Ω. If Ω is divided into N_{el} contiguous elemental regions denoted by Ω^e then the expansion modes $\phi^e_{pq}(\xi_1, \xi_2)$ are defined as

$$\phi^e_{pq}(\xi_1, \xi_2) = \begin{cases} \phi_{pq}(\xi_1, \xi_2), & (x_1, x_2) \in \Omega^e, \\ 0, & \text{otherwise}, \end{cases}$$

where

$$\xi_1 = [\chi^e_1]^{-1}(x_1, x_2), \quad \xi_2 = [\chi^e_2]^{-1}(x_1, x_2),$$

and χ^e_i represents a bijective mapping from (ξ_1, ξ_2) onto $(x_1, x_2) \in \Omega^e$ (as introduced in Section 4.1.3). We see that the local expansions within Ω^e are extended to the global domain Ω by having zero support everywhere except in the region Ω^e. Clearly, the elemental boundary modes cannot be C^0 continuous in the global region Ω. The interior modes, however, which are, by definition, zero on the boundary of the elements $\partial\Omega^e$, are C^0 continuous in the global region. This implies that the interior elemental degrees of freedom are also global degrees of freedom.

From a practical point of view, it is preferable to treat all operations locally within the standard elemental region where it is easier to define all the salient operations like integration and differentiation. This is possible if we also construct a mapping procedure which assembles our global system from the local systems defined on each element. The process is as follows.

1. Formulate a Galerkin elemental problem with respect to a set of global modes which constitute our trial space $\mathcal{X}$.
2. Split each global mode into local contributions over every element where all operations are performed.
3. Reassemble the global system.

When we split the global expansion modes, as shown in Fig. 4.9(a), into their local elemental contributions the expansion coefficient $\hat{u}$ is transmitted to both of the elemental regions. However, when we need to integrate this global expansion mode with respect to some function $u(x_1, x_2)$ as shown in Fig. 4.9(b), this may be performed locally with respect to the elemental modes and then summed together to obtain the integral of $u(x_1, x_2)$ with respect to the global mode.

We recall that the Galerkin method is constructed from the weak problem which is an integral form. We do not need to explicitly assemble the global expansion modes as we can treat the integration locally and sum the elemental contributions. Nevertheless, in order to describe the solution within the elemental region we will first need to perform the one-to-many mapping which takes the global system to the local elemental system. The assembly process is referred to as *direct stiffness summation* or *global assembly*. The word summation is somewhat misleading as only adjacent boundary modes of similar shape need to be added together, and so we shall refer to the process as *global assembly*.

We define the *local degrees of freedom* as all the elemental expansion coefficients over all elements. We have previously introduced the vector $\hat{\boldsymbol{u}}$ to represent a consecutive list of all expansion modes within an elemental region. If we now

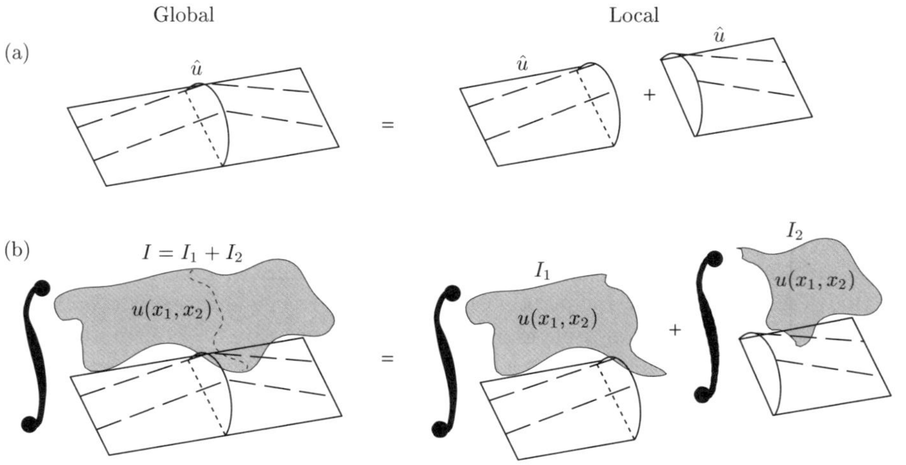

FIG. 4.9. Illustration of local to global assembly. If we have a global expansion as represented in (a) it can be decomposed into two elemental contributions multiplied by the same global coefficient $\hat{u}$. To integrate a function $u(x_1, x_2)$ with respect to the global mode, as illustrated in (b), the integration in the global region is the sum of the integration in the local regions.

use a superscript e to denote the elemental vector of expansion coefficients $\hat{\boldsymbol{u}}^e$ then the vector of all the local degrees of freedom, denoted by $\hat{\boldsymbol{u}}_l$, is

$$\hat{\boldsymbol{u}}_l = \underline{\hat{\boldsymbol{u}}}^e = \begin{bmatrix} \hat{\boldsymbol{u}}^1 \\ \hat{\boldsymbol{u}}^2 \\ \vdots \\ \hat{\boldsymbol{u}}^{N_{\text{el}}} \end{bmatrix}, \tag{4.2.1}$$

which is of dimension N_{eof}. We also introduce the notation that an underlined vector implies the extension over all elemental regions. In Section 4.2.2 we will see that an underlined matrix denotes a block diagonal extension of the matrix. To complement $\hat{\boldsymbol{u}}_l$ we define $\hat{\boldsymbol{u}}_g$ to denote the global degrees of freedom, which is a vector of dimension N_{dof}. The many-to-one mapping from global to local degrees of freedom can be represented by the matrix operation $\boldsymbol{\mathcal{A}}$, that is,

$$\hat{\boldsymbol{u}}_l = \boldsymbol{\mathcal{A}}\hat{\boldsymbol{u}}_g\,. \tag{4.2.2}$$

The matrix $\boldsymbol{\mathcal{A}}$ is a very sparse rectangular matrix of dimension $N_{\text{eof}} \times N_{\text{dof}}$ whose values may typically be either 1 or -1 depending on the shape of connecting modes. For a nodal expansion all entries are positive. Typically, only one entry will appear on any given row of the matrix. However, for different types of continuity conditions, such as the constrained approximation where two geometrically non-conforming elements meet (see Section 7.3), multiple entries may appear on rows and columns of the assembly matrix. To illustrate the form of the assembly matrix $\boldsymbol{\mathcal{A}}$ we consider the case shown in Fig. 4.10 where we have a domain containing two triangular elements. In this example we are considering an expansion order of $P_1 = P_2 = 2$ which only contains boundary modes. Therefore, the number of modes in each element is $N_m = (P_1 + 1)(P_2 + 2)/2 = 6$ and the total number of local degrees of freedom is $N_{\text{eof}} = 2N_m = 12$. In Fig. 4.10(a) we see the local numbering of the $N_m = 6$ elemental modes. This is dependent upon the orientation of the local coordinate system within the triangle as indicated by the arrow system. We have numbered the local degrees of freedom according to the convention where vertices are labelled first, followed by edges, then faces (in three dimensions), and finally the interior modes.

To enforce C^0 continuity between the two triangles we must match the boundary modes $(1, 4, 2)$ in triangle 1 with the boundary modes $(1, 4, 2)$ in triangle 2. This is achieved by assigning a global numbering scheme of $N_{\text{dof}} = 9$ global degrees of freedom, as shown in Fig. 4.10(b). Similar to the local numbering scheme, the global numbering convention applied is to number all global vertices first, followed by all global edges, faces (in three dimensions), and finally the interior modes where interior elemental blocks are numbered consecutively. This type of global numbering scheme, particularly when interior modes are numbered consecutively, is also convenient for the static condensation technique described in Section 4.2.3.

The assembly matrix $\boldsymbol{\mathcal{A}}$ which relates the local degrees of freedom $\hat{\boldsymbol{u}}_l$ to the global degrees of freedom $\hat{\boldsymbol{u}}_g$ is shown in Fig. 4.11. In this figure we see that every row of the matrix $\boldsymbol{\mathcal{A}}$ contains only one entry, signifying the fact that each local degree of freedom is related to one global degree of freedom. Every column of the matrix $\boldsymbol{\mathcal{A}}$ contains at least one entry, although for geometrically non-conforming elements or mortar constructions there may be more than one entry. In the case where every row contains only one entry the summation of the absolute value

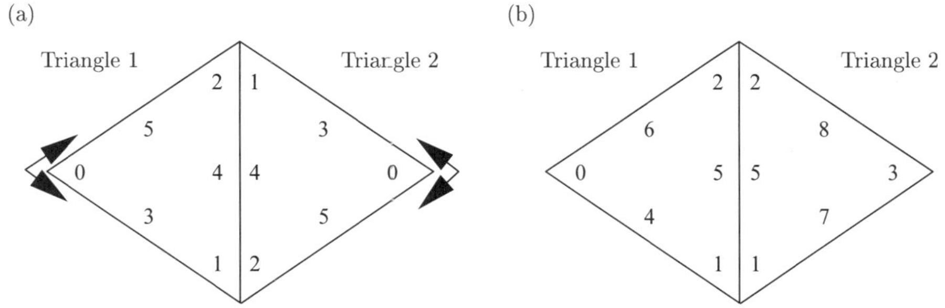

FIG. 4.10. Illustration of (a) local and (b) global numbering schemes for a region containing two triangular elements. The numbering corresponds to a triangular modal expansion where $P_1 = P_2 = 2$, which only includes boundary modes. The orientation of the local coordinate system for each triangle is indicated by the arrow system.

$$
\hat{\boldsymbol{u}}_l = \begin{bmatrix} \hat{\boldsymbol{u}}^1[0] \\ \hat{\boldsymbol{u}}^1[1] \\ \hat{\boldsymbol{u}}^1[2] \\ \hat{\boldsymbol{u}}^1[3] \\ \hat{\boldsymbol{u}}^1[4] \\ \hat{\boldsymbol{u}}^1[5] \\ \cdots \\ \hat{\boldsymbol{u}}^2[0] \\ \hat{\boldsymbol{u}}^2[1] \\ \hat{\boldsymbol{u}}^2[2] \\ \hat{\boldsymbol{u}}^2[3] \\ \hat{\boldsymbol{u}}^2[4] \\ \hat{\boldsymbol{u}}^2[5] \end{bmatrix} = \begin{bmatrix} 1 & & & & & & & & \\ & 1 & & & & & & & \\ & & 1 & & & & & & \\ & & & & 1 & & & & \\ & & & & & 1 & & & \\ & & & & & & 1 & & \\ \cdots & \cdots & \cdots & \cdots & \cdots & \cdots & \cdots & \cdots & \cdots \\ & & & 1 & & & & & \\ & & 1 & & & & & & \\ & 1 & & & & & & & \\ & & & & & & & & 1 \\ & & & & 1 & & & & \\ & & & & & & & 1 & \end{bmatrix} \begin{bmatrix} \hat{\boldsymbol{u}}_g[0] \\ \hat{\boldsymbol{u}}_g[1] \\ \hat{\boldsymbol{u}}_g[2] \\ \hat{\boldsymbol{u}}_g[3] \\ \hat{\boldsymbol{u}}_g[4] \\ \hat{\boldsymbol{u}}_g[5] \\ \hat{\boldsymbol{u}}_g[6] \\ \hat{\boldsymbol{u}}_g[7] \\ \hat{\boldsymbol{u}}_g[8] \end{bmatrix}
$$

FIG. 4.11. Relation between the local $\hat{\boldsymbol{u}}_l$ and global $\hat{\boldsymbol{u}}_g$ degrees of freedom using the assembly matrix $\boldsymbol{\mathcal{A}}$.

of the columns tells us how many local modes contribute to construct a global degree of freedom, which is known as the *multiplicity* of the mode.

We are now in a position to define the assembly process from local to global degrees of freedom. The action of the assembly process can be mathematically expressed as the transpose of $\boldsymbol{\mathcal{A}}$, but is heuristically captured by the integral operation, similar to the case shown in Fig. 4.9. If we consider the inner product function $u(x_1, x_2)$ with respect to the global basis $\Phi_n(x_1, x_2)$, i.e.,

$$\hat{\boldsymbol{I}}_g[n] = \int_\Omega u(x_1, x_2)\Phi_n(x_1, x_2)\, dx_1\, dx_2\,, \quad 0 \leqslant n < N_{\text{dof}}\,,$$

this series of integrals can be expressed as elemental contributions, such that

$$\hat{\boldsymbol{I}}_g[n] = \int_\Omega u(x_1, x_2)\Phi_n(x_1, x_2)\, dx_1\, dx_2 = \int_{\Omega^e} u(x_1, x_2)\phi_m(x_1, x_2)\, dx_1\, dx_2\,, \tag{4.2.3}$$

where $n(m, e)$ represents a unique global indexing of each elemental modal contribution m over each element e. This will be defined in terms of a mapping array map[e][i] shortly. The evaluation of the integrals (4.2.3) over all global modes $0 \leqslant n \leqslant N_{\text{dof}}$ can be represented in matrix form as

$$\hat{\boldsymbol{I}}_g = \boldsymbol{\mathcal{A}}^\top \hat{\boldsymbol{I}}_l = \boldsymbol{\mathcal{A}}^\top \underline{\hat{\boldsymbol{I}}}^e\,.$$

In the above equation $\hat{\boldsymbol{I}}_l$ is analogous to the definition (4.2.1), where

$$\hat{\boldsymbol{I}}^e[m] = \int_{\Omega^e} u(x_1, x_2)\phi_m(x_1, x_2)\, dx_1\, dx_2\,,$$

and m denotes the summation over all elemental modes which may involve a tensor product basis $\phi_m(p, q) = \phi_{pq}$ (see Section 4.1.5.1).

We note that the matrix operations $\boldsymbol{\mathcal{A}}$ and $\boldsymbol{\mathcal{A}}^\top$ are not the inverse of each other, and therefore

$$\hat{\boldsymbol{u}}_g \neq \boldsymbol{\mathcal{A}}^\top \boldsymbol{\mathcal{A}} \hat{\boldsymbol{u}}_g \,.$$

The operation of $\boldsymbol{\mathcal{A}}$ is a scatter from a global to a local system, whereas the operation of $\boldsymbol{\mathcal{A}}^\top$ is a global assembly or summation procedure. The inverse of the $\boldsymbol{\mathcal{A}}$ matrix would normally be considered as a standard 'gather'-type procedure. The operations of $\boldsymbol{\mathcal{A}}$ and $\boldsymbol{\mathcal{A}}^\top$ are the key constructs to form a global system when using the Galerkin technique.

As an aside, we note that all the boundary modes touching the solution domain boundary have been treated as global degrees of freedom. As we shall see in Section 4.3.1, boundaries with Neumann conditions are typically treated in this fashion. However, boundaries associated with Dirichlet conditions are not part of the Galerkin test space and therefore some reordering is required to remove them from the global degrees of freedom.

In a numerical implementation it is not practical, or even desirable, to construct $\boldsymbol{\mathcal{A}}$ explicitly due to the size and sparsity of the matrix. The operation may be numerically implemented by setting up a mapping array which we will denote by $n(e, i) = \text{map}[e][i]$. This array is of dimension $N_{\text{el}} \times \max_e(N_m{}^e)$, where $N_m{}^e$ is the number of expansion modes in an elemental expansion. Typically, $N_m{}^e$ will be fixed over all elements, although, in general, the value may change between elements. The array map[e][i] contains the global value of the ith expansion coefficient within the eth element. The example shown in Fig. 4.10 would therefore have an array map[e][i] of the form

$$\text{map}[1][i] = \begin{cases} 0 \\ 1 \\ 2 \\ 4 \\ 5 \\ 6 \end{cases}, \quad \text{map}[2][i] = \begin{cases} 3 \\ 2 \\ 1 \\ 8 \\ 5 \\ 7 \end{cases}.$$

The total number of entries of the map[e][i] is the same as the number of nonzero entries in $\boldsymbol{\mathcal{A}}$. The scatter operation $\boldsymbol{\mathcal{A}}$ from $\hat{\boldsymbol{u}}_g$ to $\hat{\boldsymbol{u}}_l$ can now be evaluated by ♦[23]

$$\left.\begin{array}{l} \text{do } e = 1, N_{\text{el}} \\ \quad \text{do } i = 0, N_m{}^e - 1 \\ \qquad \hat{\boldsymbol{u}}^e[i] = \text{sign}[e][i] \cdot \hat{\boldsymbol{u}}_g[\text{map}[e][i]] \\ \quad \text{continue} \\ \text{continue} \end{array}\right\} \quad \Leftrightarrow \quad \hat{\boldsymbol{u}}_l = \boldsymbol{\mathcal{A}} \hat{\boldsymbol{u}}_g \,, \qquad (4.2.4)$$

where sign[e][i] is an array of similar dimensions to map[e][i] containing 1 or -1 entries depending on the modal connectivity between two elements, as discussed

[23] Numerical implementation of the global to local gather operation denoted by $\boldsymbol{\mathcal{A}}$.

in Section 4.2.1.1. For a nodal expansion sign$[e][i]$ would only contain positive entries and so they may be removed from the loop. The global assembly operation
♦[24] can be evaluated as

$$\left.\begin{array}{l} \text{do } e = 1, N_{\text{el}} \\ \quad \text{do } i = 0, {N_m}^e - 1 \\ \qquad \hat{\boldsymbol{I}}_g[\text{map}[e][i]] = \hat{\boldsymbol{I}}_g[\text{map}[e][i]] \\ \qquad\qquad\qquad\qquad\quad + \text{sign}[e][i] \cdot \hat{\boldsymbol{I}}^e[i] \\ \quad \text{continue} \\ \text{continue} \end{array}\right\} \Leftrightarrow \quad \hat{\boldsymbol{I}}_g = \boldsymbol{\mathcal{A}}^\top \hat{\boldsymbol{I}}_l . \tag{4.2.5}$$

If the inner summation did not contain the $\hat{\boldsymbol{v}}_g[\text{map}[e][i]]$ term on the right-hand side it would be the standard 'gather' operation.

4.2.1.1 *Local to global boundary mapping: global boundary assembly*

We have seen that the global assembly procedure primarily involves boundary mode connectivity as the interior modes may be independently numbered as global degrees of freedom. We shall also see in Section 4.2.3 that the assembly procedure need only involve the boundary modes as the interior modes may be removed from the full matrix problem using static condensation. In this case we only require a boundary mapping bmap$[e][i]$ rather than the full numbering system map$[e][i]$.

We assume that the local degrees of freedom are ordered so that the boundary modes are listed first. If there are $n_b[e]$ boundary modes in the eth element then
♦[25] the local to global assembly process is numerically evaluated as

$$\left.\begin{array}{l} \text{do } e = 1, N_{\text{el}} \\ \quad \text{do } i = 0, n_b[e] - 1 \\ \qquad \hat{\boldsymbol{I}}_g[\text{bmap}[e][i]] = \hat{\boldsymbol{I}}_g[\text{bmap}[e][i]] \\ \qquad\qquad\qquad\qquad\quad + \text{sign}[e][i] \cdot \hat{\boldsymbol{I}}^e[i] \\ \quad \text{continue} \\ \text{continue} \end{array}\right\} \Leftrightarrow \quad \hat{\boldsymbol{I}}_g = \boldsymbol{\mathcal{A}}_b^\top \underline{\hat{\boldsymbol{I}}_b^e} , \tag{4.2.6}$$

which is identical to map$[e][i]$ in eqn (4.2.5) save that we are now only using the first $n_b[e]$ elements of the mapping. Here $\hat{\boldsymbol{I}}^e$ and $\hat{\boldsymbol{I}}_g$ have their usual meaning referring to the local and global degrees of freedom, even though only the entries corresponding to the boundary modes are being used. Similar to 'bmap$[e][i]$', the array sign$[e][i]$ need only be of dimension N_{el} by the maximum size of $n_b[e]$. We see from eqn (4.2.6) that the equivalent matrix operation is denoted by $\boldsymbol{\mathcal{A}}_b^\top$,

[24] Numerical implementation of the local to global assembly operation denoted by $\boldsymbol{\mathcal{A}}^\top$.

[25] Numerical implementation of the local to global boundary assembly operation denoted by $\boldsymbol{\mathcal{A}}_b^\top$.

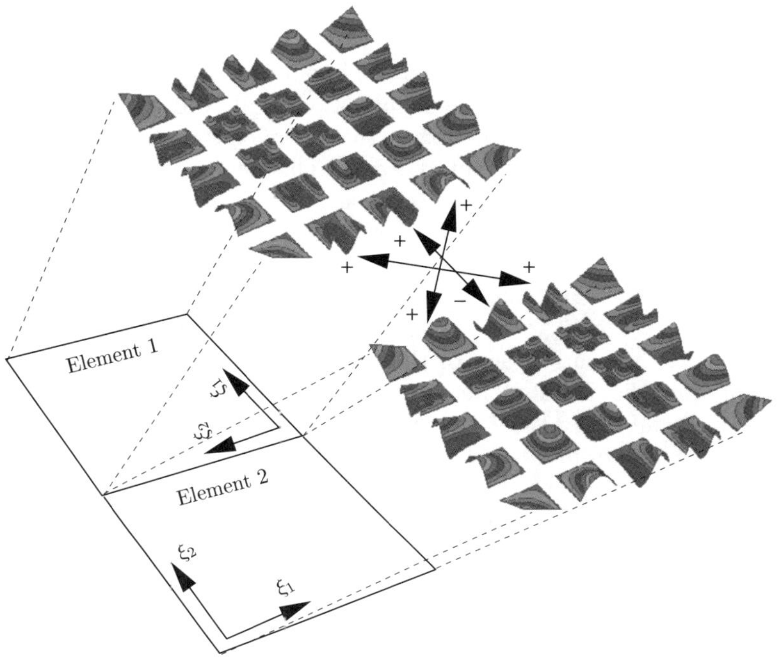

FIG. 4.12. Illustration of the construction of a C^0 global expansion from two local *modal* expansions of order $P_1 = P_2 = 4$. To ensure C^0 continuity the boundary modes of similar shape need to be matched. Depending on the orientation of the local coordinate systems, the modes of odd order may also need to be negated.

which is a submatrix of $\boldsymbol{\mathcal{A}}^\top$ and operates on the vector of all local boundary degrees of freedom denoted by $\underline{\hat{\boldsymbol{v}}_b^e}$. If we know how to construct bmap[e][i], then it is a straightforward extension to generate map[e][i] by adding a unique block of global degrees of freedom equal in length to the number of interior modes within the element.

Modal edge connectivity ♦26

In Fig. 4.12 we see all the modes for an order $P_1 = P_2 = 4$ expansion in two quadrilateral elements. Note that each mode is to be interpreted as spanning the entire element. When considering an expansion with more than one edge mode we need to consider the local orientation of the element. As shown in Fig. 4.12, depending on the orientations of the local coordinate systems within the element the sign of odd-ordered modes may need to be reversed. This is in contrast to Fig. 4.8 where the cubic edge mode had a similar shape on either side of the intersecting edge. The reason for the sign negation is that the elemental modal

26 Modal basis sign determination for inter-element edge global assembly.

shapes are defined with respect to the local coordinate system (ξ_1, ξ_2). If the local systems are orientated so that the two neighbouring coordinates are in opposite directions then the sign of one odd-shaped mode will need to be reversed.

It also appears from Fig. 4.12 that the *order* of the edge modes needs to be reversed. This is, however, not the case. The hierarchical boundary modes only have a physical interpretation insofar as they are associated with a physical vertex or edge within the region. Therefore, we number the edge modes according to their polynomial order (that is, the lowest polynomial order mode has the lowest edge number). For the example shown in Fig. 4.12, the numbering of the local modes is shown in Fig. 4.13 where we have placed all numbering for a given edge at the centre-point of the edge. If we follow a similar convention when numbering the global modes, as shown in Fig. 4.13(b), then the modes of similar polynomial order (which we need to match) will have the same global number and so we are just left with the issue of sign reversal. In this example, the mode of cubic order needs to have its sign reversed in one element as the local coordinate system has an opposite direction along the intersecting edge. By convention, we assume that the element with the lowest number has precedence and therefore mode 14 in triangle 1 will be negated. Therefore, when assembling the array sign$[e][i]$ we require that sign$[1][11] = -1$.

In general, we need an automatic procedure to identify which edges need to have odd modes negated. Such a procedure may be constructed by considering the sign of the inner product between two vectors representing the global coordinate direction of an edge. Since we always know the vertices which define an element, we let $\boldsymbol{\Delta x}^e_{\rm edg}$ denote a vector parallel to an edge in an element 'e' oriented according to the local coordinate direction of ξ_1 or ξ_2. For example, along the bottom edge where $\xi_2 = -1$ we have

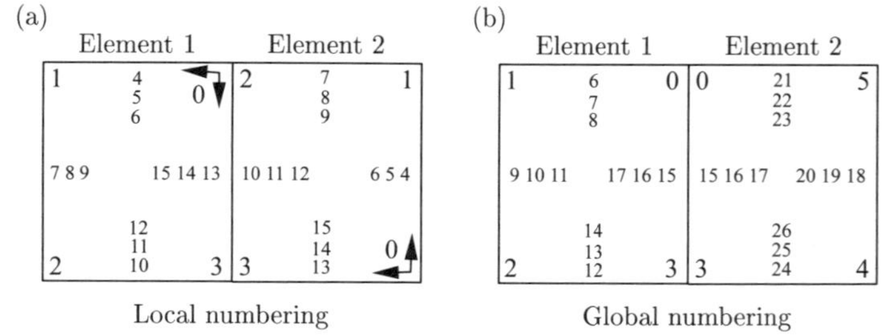

FIG. 4.13. Numbering system for a hierarchical quadrilateral expansion of order $P_1 = P_2 = 4$, where the arrows indicate the local coordinate system. The individual edge modes have no physical location associated with the expansion and so are listed at the centre-point of the edge. The number nearest the edge corresponds to the edge mode of the lowest polynomial order. Following a similar ordering for the global numbering means that modes of similar order are automatically matched.

$$\boldsymbol{\Delta x}^e_{\text{edg}} = \begin{bmatrix} \chi_1(1,-1) - \chi_1(-1,-1) \\ \chi_2(1,-1) - \chi_2(-1,-1) \end{bmatrix}, \quad x_1 = \chi_1(\xi_1,\xi_2), \quad x_2 = \chi_2(\xi_1,\xi_2),$$

or along the edge where $\xi_1 = -1$ we have

$$\boldsymbol{\Delta x}^e_{\text{edg}} = \begin{bmatrix} \chi_1(-1,1) - \chi_1(-1,-1) \\ \chi_2(-1,1) - \chi_2(-1,-1) \end{bmatrix}, \quad x_1 = \chi_1(\xi_1,\xi_2), \quad x_2 = \chi_2(\xi_1,\xi_2).$$

To determine whether the odd edge modes need to have their sign reversed on an edge between element e and element k we apply the following test:

if

$$\boldsymbol{\Delta x}^e_{\text{edg}} \cdot \boldsymbol{\Delta x}^k_{\text{edg}} < 0 \tag{4.2.7}$$

and

$$k > e$$

then negate odd modes.

The extra criterion $k > e$ simply ensures that only one of the two edge modes is negated and implies that we have some information about the edge connectivity. This test only identifies whether the local coordinate systems along a specific edge are in the same or opposite directions. An identical procedure may be applied to edges of a triangular region by use of η_1, η_2 instead of ξ_1, ξ_2. For edges in a three-dimensional mesh an analogous test may be set up where $\boldsymbol{\Delta x}^e_{\text{edg}}$ is now a three-dimensional vector. In this case, the number of elemental domains along an edge will typically be greater than two and therefore the test needs to be performed relative to the edge from the element with the lowest number.

Nodal edge connectivity ♦27

When using a nodal expansion the modes may be identified with a physical location of the nodal points where the modes have a unit value. In the nodal expansion, we are not concerned with matching edge modes of similar order (as in the hierarchical expansion case) but with matching modes with the same nodal location, as illustrated in Fig. 4.14. The physical interpretation of an expansion mode being associated with a nodal position makes it easier to generate a numbering system, by numbering the location of the nodal points along an edge as shown in Fig. 4.15.

In Fig. 4.13 we have chosen to locally number the elemental degrees of freedom using an anticlockwise convention where the vertex modes are listed first. Using an anticlockwise convention ensures that one side of the elemental matching is always reversed with respect to the global numbering. For example, the modes in element 1 have locally increasing numbers $(13, 14, 15)$ corresponding to globally increasing numbers $(15, 16, 17)$, whereas the modes in element 2 have locally increasing numbers $(10, 11, 12)$ corresponding to globally decreasing numbers $(17, 16, 15)$. If we had ordered the local edge modes according to the direction

27 Nodal basis numbering for inter-element edge global assembly.

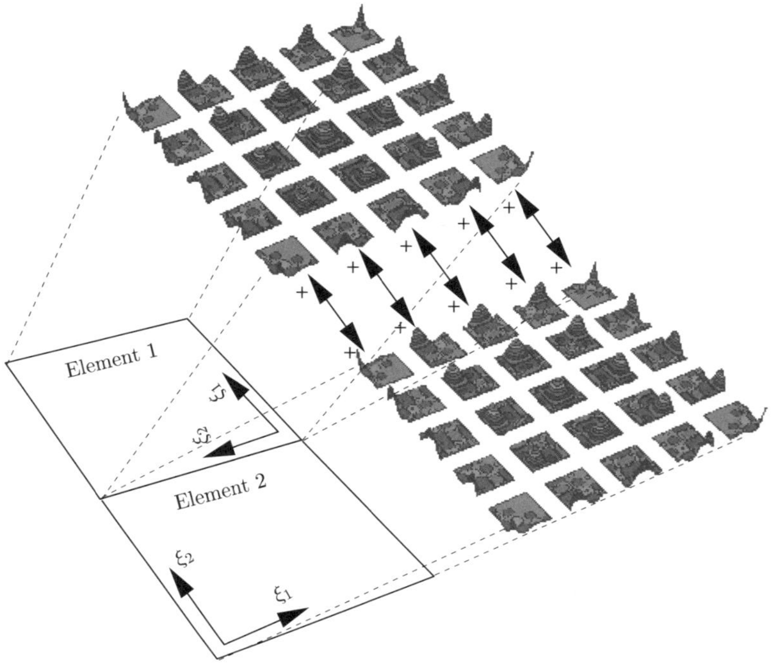

FIG. 4.14. Illustration of the construction of a C^0 global expansion from two local *nodal* expansions of order $P_1 = P_2 = 4$. To ensure C^0 continuity the boundary modes with similar nodal locations need to be matched.

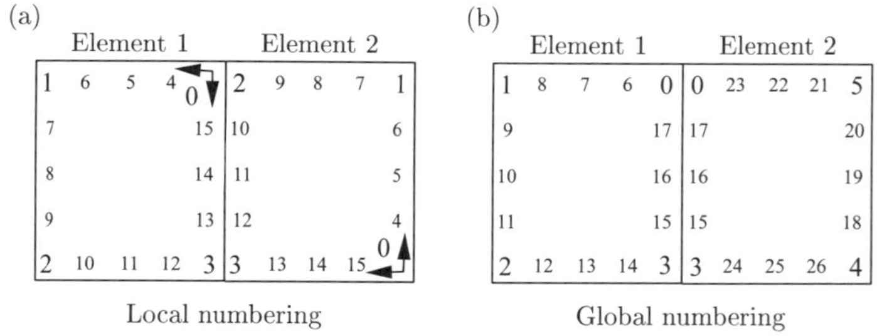

FIG. 4.15. Numbering system for a nodal quadrilateral expansion of polynomial order $P_1 = P_2 = 4$. The arrows indicate the orientation of the local coordinate system (ξ_1, ξ_2). In a nodal expansion the edge mode can be physically identified with the nodal points of the Lagrange polynomial. Numbering each nodal location therefore provides a global numbering scheme which will ensure C^0 continuity.

of the local coordinate system, then we would have a similar situation to the hierarchical expansions where we would have to determine if we needed to reverse the ordering depending on the direction of the local edge coordinate. For the nodal expansion in two dimensions, the use of an anticlockwise local numbering scheme implies that the ordering is always reversed between two elements, and therefore no extra test is required.

Modal and nodal quadrilateral face orientation and connectivity ♦[28]

One of the complexities of matching three-dimensional shapes with similar-shaped faces as compared to the two-dimensional edge matching is the number of different orientations that can be imposed. This issue is highlighted in Fig. 4.16 where we show some of the different local face alignments between the quadrilateral faces of two hexahedral elements. In this figure the $(1, 2, 3)$ axes systems denote the local Cartesian system of each hexahedral element. Similarly, the (a, b) axis system denotes the local face coordinates, where the a direction is aligned to the lowest local Cartesian coordinate within the local face. From this figure we see that, as in the two-dimensional case, local coordinates can be aligned in opposite directions. However, unlike the two-dimensional case, we also have a possible situation where local coordinates can be transposed (i.e., the a direction in one element is aligned to the b direction in another).

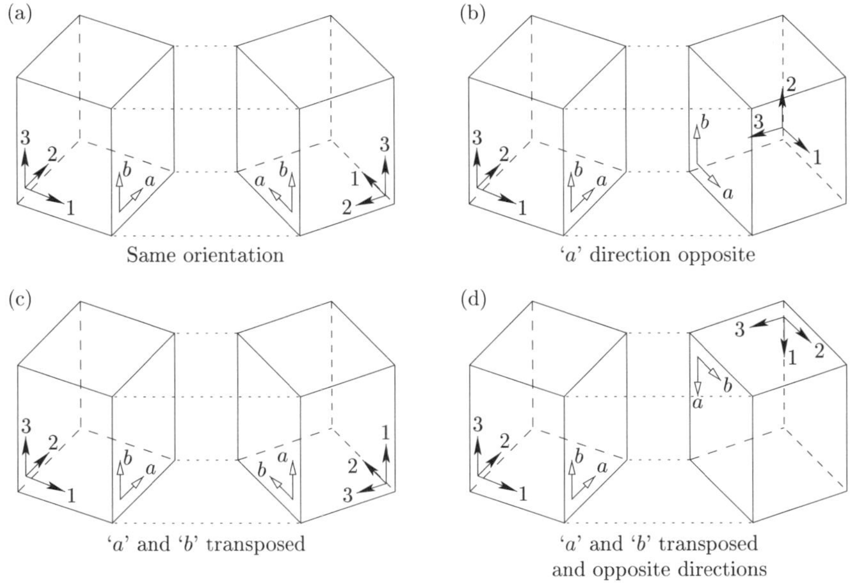

FIG. 4.16. Example of some of the different local coordinate alignments when two hexahedral elements are matched. The local Cartesian coordinates are denoted by the $(1, 2, 3)$ axis and the local face coordinates are denoted by the (a, b) axis.

[28] Orientation requirement and connectivity issues for global assembly between adjacent quadrilateral faces.

From the previous two subsections we have observed that when the local coordinate direction is reversed then, in a modal expansion, we expect to use the same number scheme in bmap[e][i], although there will be sign changes in the array sign[e][i]. However, for a nodal expansion the reversing of the local coordinate direction necessitates a change in the local array numbering, i.e., bmap[e][i] but does not influence sign[e][i]. Therefore, considering Fig. 4.16(a) we expect the same bmap[e][i] and sign[e][i] in both faces. In Fig. 4.16(b) the modal numbering is not altered between the two faces, although there is a sign change due to the local 'a' coordinate reversing direction between the two faces. For this situation there would have been a numbering alteration if we were using a nodal case. Finally, in Fig. 4.16(c,d) we require a transposition of the numbering scheme for both modal and nodal expansions due to the transposing of the local coordinate system.

In assembling bmap[e][i] and sign[e][i] between two modal quadrilateral faces we therefore need to know how a face is orientated and then whether each local axis is aligned in the same or opposite directions. Determining how the local face coordinates are orientated with respect to each other permits us to locally number the face degrees of freedom and therefore construct the component of bmap[e][i]. Determining whether the local axes of the two faces are aligned in the same or opposite directions informs us whether or not the odd order components of the expansion basis in each coordinate direction need to be negated in one of the adjacent faces and so permitting the construction of sign[e][i].

There are many different techniques which could be applied to determine the orientation and sign of the local face coordinates. As a potential example we can consider the following construction. If we are considering a face defined by $\xi_3 = -1$ in element e then linear two-edge vectors can be defined as

$$\boldsymbol{\Delta x}_a^e = \begin{bmatrix} \chi_1(1,-1,-1) - \chi_1(-1,-1,-1) \\ \chi_2(1,-1,-1) - \chi_2(-1,-1,-1) \\ \chi_3(1,-1,-1) - \chi_3(-1,-1,-1) \end{bmatrix},$$

$$\boldsymbol{\Delta x}_b^e = \begin{bmatrix} \chi_1(-1,1,-1) - \chi_1(-1,-1,-1) \\ \chi_2(-1,1,-1) - \chi_2(-1,-1,-1) \\ \chi_3(-1,1,-1) - \chi_3(-1,-1,-1) \end{bmatrix},$$

where

$$x_1 = \chi_1(\xi_1, \xi_2, \xi_3), \quad x_2 = \chi_2(\xi_1, \xi_2, \xi_3), \quad x_3 = \chi_3(\xi_1, \xi_2, \xi_3).$$

We then analogously define the two-edge vectors $\boldsymbol{\Delta x}_a^k$ and $\boldsymbol{\Delta x}_b^k$ in the adjacent face of element k under the constraint that they have the same common vertex, $(\chi_1^e(-1,-1,-1), \chi_2^e(-1,-1,-1), \chi_3^e(-1,-1,-1))$. Whether the $(a,b)^e$ coordinate system in element e has a transpose orientation to the $(a,b)^k$ coordinate system of element k can be determined by an inner product test of the following form:

if

$$\Delta x^e_a \cdot \Delta x^k_a = |\Delta x^e_a||\Delta x^k_a| \qquad (4.2.8)$$

then

$$\Delta x^e_a \parallel \Delta x^k_a \quad \Rightarrow \quad a^e \parallel a^k,\ b^e \parallel b^k$$

else

$$\Delta x^e_a \parallel \Delta x^k_b \quad \Rightarrow \quad a^e \parallel b^k,\ b^e \parallel a^k\ .$$

In a modal expansion if Δx^e_a is parallel to Δx^k_a then the modes in each face can be numbered in an identical manner. However, if Δx^e_a is parallel to Δx^k_b then we have to set up a transposed numbering system in one of the two adjacent faces. For example, if $\phi^e_{pq}(\xi_1, \xi_2, -1)$ represents the $(P-1)(P-1)$ modes within a face of element e then the matching modes in a face of element k must be

$$\phi^k_{pq0}(\xi_1, \xi_2, -1) = \pm\phi^e_{qp0}(\xi_1, \xi_2, -1)\,, \quad 0 < p, q < P\,, \qquad (4.2.9)$$

where we have assumed that the connecting face in element k is defined by $\xi_3 = -1$. Therefore, if the mapping bmap[e][i] for ϕ^e_{pq0} is initially chosen then the mapping in bmap[k][i] must be ordered so that condition (4.2.9) is satisfied.

Finally, we need to determine the possible sign change between modes to set up the array sign[e][i] for the global assembly process. Providing the local edge vectors $\Delta x^e_a, \Delta x^k_a, \ldots$ are defined so that a positive vector is aligned in the positive direction of the local axes, then the sign of the matching modes can be determined by applying test (4.2.7). Clearly, the local edge vectors used in this test depend on whether the coordinates are transposed or not. Testing first the a and then the b direction of a reference face indicates whether the odd numbered p and q index modes require negating within the reference face for the two face modes to match identically.

For a *nodal face* we could adopt a similar procedure where we use the orientation test (4.2.8) to determine how the nodal points in a face are orientated relative to the adjacent face. However, a simpler, although more expensive, approach is to number the nodal locations in one face and then determine the global numbering of the adjacent face by matching every nodal (x_1, x_2, x_3) position in one face with the other. As the numbering procedure may be considered as an overhead cost at the pre-processing stage there is generally no concern over the cost of this approach.

Modal triangular face orientation and connectivity ♦[29]

Similar to the quadrilateral face, in matching two triangular faces we also have to consider the orientation of a face. To generate a C^0 global expansion we want to match face modes of similar shape. For modal expansions the use of the collapsed Cartesian system means that triangular faces have a local coordinate system which is not rotationally symmetric. This point is illustrated in Fig. 4.17

[29] Orientation requirement and connectivity issues for global assembly between adjacent triangular faces.

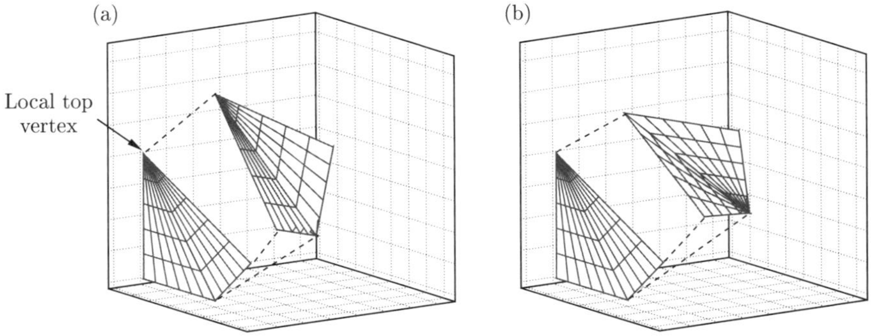

FIG. 4.17. To ensure that the modal expansion can be assembled to form a C^0 expansion, the collapsed coordinate system in the triangular faces must be aligned as shown in (a). Connection of two faces where the local coordinate system is oriented, as shown in (b), is not permitted.

where we see the two tetrahedral regions marked with their surface coordinate lines. To be able to match modes within a triangular face we require that these coordinate lines are orientated as shown in Fig. 4.17(a), but not as shown in Fig. 4.17(b). This would appear to be rather constraining. We are, however, free to specify how the local coordinate systems within an element are orientated. For an arbitrary conforming tetrahedral mesh, it is possible to orient all the local coordinate system so that the coordinate lines are consistent. Referring to the two vertices where the coordinate system degenerates as the *local base vertex* ($\xi_1 = -1, \xi_2 = 1, \xi_3 = -1$) and the *local top vertex* ($\xi_1 = -1, \xi_2 = -1, \xi_3 = 1$), the following is an orientation algorithm suggested by Warburton [483].

Assuming that every global vertex has a unique number then for every element we have four vertices with unique global numbers.

(i) Place the *local top vertex* at the global vertex with the lowest global number.

(ii) Place the *local base vertex* at the global vertex with the second lowest global number.

(iii) Orient the last two vertices to be consistent with the local rotation of the element (typically anticlockwise).

This algorithm is local to each element and can be implemented at a pre-processing stage. Although it is possible to guarantee this connectivity for tetrahedral meshes, it is not possible for a general mesh using tetrahedrons, prisms, and pyramids. Nevertheless, enough permutations of connectivity still exist to provide a wide range of flexibility even when using all the three-dimensional hybrid elements [483].

The orientation criteria simplify the numbering and sign evaluation process in a triangular face as all faces have a similar orientation. Therefore, for a hierarchical/modal expansion the global numbering on one face may be directly applied to the adjacent face without any index swapping. There is still, however, the possibility for modes which vary in the base direction (i.e., η_1 direction on face 1 or 2, and η_2 direction on face 3 or 4) to require their sign to be reversed. This may be determined by using the edge test (4.2.7).

The non-tensorial nodal expansions are rotationally symmetric and therefore do not have to comply with the above orientation criteria. However, application of these criteria in the case of a nodal tetrahedral expansion can still reduce the potential orientations of adjacent faces and thus simplify the construction of a numbering scheme. In this case the edge test (4.2.7) would indicate whether the number should be reversed in one direction between two adjacent faces. Nevertheless, similar to the quadrilateral faces, with a nodal expansion a comparison of the coordinates of nodal points can again be used to determine an appropriate face ordering for bmap[e][i].

4.2.2 *Global matrix system*

♣[20]

Now that we have a way of assembling the global system from a local system, we can apply this technique to generate global matrices from the elemental matrices. To illustrate this process we can consider a global forward transformation which is similar to the elemental transformation described in Section 4.1.5, except that we now want to project a function into a C^0 continuous global expansion space. To recall the notation introduced in Section 4.1.5, where the superscript e now refers to the element number, we have the following.

- $\hat{\boldsymbol{u}}^e$ Vector of length N_m containing the expansion coefficients corresponding to the order of the basis matrix $\boldsymbol{B}$.
- $\boldsymbol{u}^e$ Vector of length N_Q containing the function $u(\xi_1, \xi_2)$ evaluated at the quadrature points.
- $\boldsymbol{B}^e$ $N_Q \times N_m$ basis matrix whose columns contain the basis $\phi_{pq}(\xi_1, \xi_2)$ evaluated at the quadrature points.
- $\boldsymbol{W}^e$ $N_Q \times N_Q$ diagonal weight matrix containing the quadrature weights multiplied by the Jacobian at the quadrature points.

For a nodal non-tensorial basis the vector $\hat{\boldsymbol{u}}^e$ can also be interpreted as the solution at the nodal points of the Lagrange expansion. In Section 4.2.1 we also introduced the following notation for global systems.

- $\hat{\boldsymbol{u}}_g$ Vector of length N_{dof} containing the global expansion coefficients.
- $\hat{\boldsymbol{u}}_l, \underline{\hat{\boldsymbol{u}}}^e$ Vector of length N_{eof} which is the concatenation of the local expansion coefficients $\hat{\boldsymbol{u}}^e$ over all N_{el} elements.
- $\boldsymbol{u}_l, \underline{\boldsymbol{u}}^e$ $N_{\text{el}} \cdot N_Q$ vector which is the concatenation of the local vectors $\boldsymbol{u}^e$ over all N_{el} elements.

[20] Construction of global matrix systems from elemental contributions.

$\boldsymbol{\mathcal{A}}$ $N_{\text{eof}} \times N_{\text{dof}}$ permutation matrix which constructs the local vector $\hat{\boldsymbol{u}}_l$ from the global vector $\hat{\boldsymbol{u}}_g$.
($\boldsymbol{\mathcal{A}}^\top$ represents the global assembly process.)

In the above $\boldsymbol{u}_l$ has been introduced as an analogous vector to $\hat{\boldsymbol{u}}_l$ and contains the function evaluation at the quadrature points over all elements. We also note that

$$N_{\text{eof}} = \sum_{e=1}^{N_{\text{el}}} N_m{}^e ,$$

where $N_m{}^e$ is the number of expansion modes in the element e. Typically, this is constant, in which case $N_{\text{eof}} = N_{\text{el}} \cdot N_m$.

We recall that to determine $\hat{\boldsymbol{u}}^e$ given a vector $\boldsymbol{u}^e$ we performed a discrete elemental forward transformation which involved the solution of the matrix system

$$\boldsymbol{M}^e \hat{\boldsymbol{u}}^e = (\boldsymbol{B}^e)^\top \boldsymbol{W}^e \boldsymbol{B}^e \hat{\boldsymbol{u}}^e = (\boldsymbol{B}^e)^\top \boldsymbol{W}^e \boldsymbol{u}^e , \tag{4.2.10}$$

where $\boldsymbol{M}^e$ is the elemental mass matrix. We now want to set up an analogous system to determine the solution for the global vector $\hat{\boldsymbol{u}}_g$.

We can represent the elemental forward transformation over all elements in terms of a global matrix process by assembling diagonal matrices of the form

$$\underline{\boldsymbol{M}^e} = \begin{bmatrix} \boldsymbol{M}^1 & 0 & 0 & 0 \\ 0 & \boldsymbol{M}^2 & 0 & 0 \\ & & & \\ 0 & 0 & \ddots & 0 \\ 0 & 0 & 0 & \boldsymbol{M}^{N_{\text{el}}} \end{bmatrix},$$

$$\underline{(\boldsymbol{B}^e)^\top \boldsymbol{W}^e} = \begin{bmatrix} (\boldsymbol{B}^1)^\top \boldsymbol{W}^1 & 0 & 0 & 0 \\ 0 & (\boldsymbol{B}^2)^\top \boldsymbol{W}^2 & 0 & 0 \\ & & & \\ 0 & 0 & \ddots & 0 \\ 0 & 0 & 0 & (\boldsymbol{B}^{N_{\text{el}}})^\top \boldsymbol{W}^{N_{\text{el}}} \end{bmatrix}.$$

In constructing the above matrix systems we have adopted the notation that an underlined matrix $\underline{\boldsymbol{M}^e}$ denotes the local matrices $\boldsymbol{M}^1, \boldsymbol{M}^2, \ldots$ as block diagonal entries to a larger matrix system. This is equivalent to the concatenation of $\boldsymbol{u}^e$ and $\hat{\boldsymbol{u}}^e$ into $\boldsymbol{u}_l = \underline{\boldsymbol{u}^e}$ and $\hat{\boldsymbol{u}}_l = \underline{\hat{\boldsymbol{u}}^e}$, respectively. The matrix system

$$\underline{\boldsymbol{M}^e}\ \hat{\boldsymbol{u}}_l = \underline{(\boldsymbol{B}^e)^\top W^e}\ \boldsymbol{u}_l \tag{4.2.11}$$

represents the local forward transformation over all N_{el} elements and is an equivalent statement to eqn (4.2.10). This system is invertible since the local systems are decoupled and invertible, but there is no guarantee of continuity between elements. Nevertheless, we know from Section 4.2.1 that

$$\hat{\boldsymbol{u}}_l = \boldsymbol{\mathcal{A}} \hat{\boldsymbol{u}}_g , \tag{4.2.12}$$

which determines the local degrees of freedom from the global degrees of freedom. Substituting eqn (4.2.12) into eqn (4.2.11), we have

$$\underline{\boldsymbol{M}^e}\,\boldsymbol{\mathcal{A}}\hat{\boldsymbol{u}}_g = \underline{(\boldsymbol{B}^e)^\top \boldsymbol{W}^e}\,\boldsymbol{u}_l. \tag{4.2.13}$$

The effect of post-multiplying the matrix $\underline{\boldsymbol{M}^e}$ by $\boldsymbol{\mathcal{A}}$ is to globally assemble the rows of this matrix from their local contributions. The matrix $\underline{\boldsymbol{M}^e}\boldsymbol{\mathcal{A}}$ is not square as the columns of this system have not been globally assembled. This may be achieved by pre-multiplying the entire eqn (4.2.13) by $\boldsymbol{\mathcal{A}}^\top$ to obtain the global square system

$$\left[\boldsymbol{\mathcal{A}}^\top \underline{\boldsymbol{M}^e}\,\boldsymbol{\mathcal{A}}\right]\hat{\boldsymbol{u}}_g = \boldsymbol{\mathcal{A}}^\top \underline{(\boldsymbol{B}^e)^\top \boldsymbol{W}^e}\,\boldsymbol{u}_l\,, \tag{4.2.14}$$

which may be solved to determine the global forward transformation. The matrix in square brackets is the global mass matrix $\boldsymbol{M}$, that is,

$$\boldsymbol{M} = \boldsymbol{\mathcal{A}}^\top \underline{\boldsymbol{M}^e}\,\boldsymbol{\mathcal{A}}.$$

Although we have set up a global system for the forward transformation using the local matrix $\boldsymbol{M}^e$, an identical procedure follows for any local matrix system such as the weak Laplacian system $\boldsymbol{L}^e$ defined in Section 4.1.3. Therefore, the global weak Laplacian matrix $\boldsymbol{L}$ is

$$\boldsymbol{L} = \boldsymbol{\mathcal{A}}^\top \underline{\boldsymbol{L}^e}\,\boldsymbol{\mathcal{A}}.$$

Finally, we note that global matrix systems such as (4.2.14) can only be directly assembled for relatively small problems (i.e., a low number of elements of low polynomial order). The matrix formulation, however, provides insight into understanding the important steps in constructing the global system. In practice, for a large problem we recall that the explicit inner product operation $\underline{(\boldsymbol{B}^e)^\top \boldsymbol{W}^e}\,\boldsymbol{u}_l$ can be evaluated on an elemental level. If we use a tensorial basis then the sum-factorisation technique discussed in Section 4.1.6 can also be applied. This produces a vector, equal in length to the local degrees of freedom, which may be assembled into the global form using a mapping array, as shown by the operation (4.2.5). Therefore, we see that the left-hand side of eqn (4.2.14) can be efficiently evaluated using local elemental operations.

The idea of locally assembling the elemental matrices and then using the global assembly operator can equally well be applied to generate the global matrix system $\boldsymbol{M}$ from the local matrices $\boldsymbol{M}^e$. This system is of dimension $N_{\rm dof} \times N_{\rm dof}$ but is typically very sparse. Nevertheless, it is usually too large to invert, or even factor directly. In the next section we see how we can reduce this global matrix system into smaller components based on the natural decomposition of the spectral/*hp* element method.

4.2.3 *Static condensation/substructuring*

Although the following technique may be applied to a general non-symmetric matrix system, we shall restrict our attention to symmetric matrix systems which typically arise in the Galerkin discretisation of symmetric operators.

We therefore assume that we have a system of the form

$$\boldsymbol{M}\boldsymbol{x} = \boldsymbol{\mathcal{A}}^\top \underline{\boldsymbol{M}}^e \boldsymbol{\mathcal{A}}\boldsymbol{x} = \boldsymbol{f}, \tag{4.2.15}$$

where $\boldsymbol{x}$ is a vector of global unknowns, typically the global vector of expansion coefficients $\hat{\boldsymbol{u}}_g$, and $\underline{\boldsymbol{M}}^e$ is a block diagonal matrix which may have been formed from either the local mass or Laplacian matrices, or indeed a combination of the two. The matrix $\boldsymbol{M}$ is typically very sparse, although it may have a full bandwidth. It is therefore very inefficient, and potentially impossible, to store the full matrix so that it may be directly inverted. A far more efficient approach is to use the structure of the spectral/*hp* element discretisation, which is the motivation behind *static condensation* or *substructuring*.

Each of the elemental matrices $\boldsymbol{M}^e$ can be split into components containing boundary and interior contributions, that is,

$$\boldsymbol{M}^e = \begin{bmatrix} \boldsymbol{M}_b^e & \boldsymbol{M}_c^e \\ (\boldsymbol{M}_c^e)^\top & \boldsymbol{M}_i^e \end{bmatrix},$$

where $\boldsymbol{M}_b^e$ represents the components of $\boldsymbol{M}^e$ resulting from boundary–boundary mode interactions, $\boldsymbol{M}_c^e$ represents the components of $\boldsymbol{M}^e$ resulting from coupling between the boundary–interior modes, and $\boldsymbol{M}_i^e$ represents the components of $\boldsymbol{M}^e$ resulting from interior–interior mode interactions.

As mentioned previously, if we know the value of $\boldsymbol{x}$ then we can perform the matrix–vector operation $\boldsymbol{\mathcal{A}}^\top \underline{\boldsymbol{M}}^e \boldsymbol{\mathcal{A}}\boldsymbol{x}$ far more efficiently by considering separately the local operations represented by the $\boldsymbol{\mathcal{A}}$ and $\underline{\boldsymbol{M}}^e$ matrices. This is one way of solving the system iteratively. However, if we need to directly invert the matrix $\boldsymbol{\mathcal{A}}^\top \underline{\boldsymbol{M}}^e \boldsymbol{\mathcal{A}}$ then we cannot perform each operation independently.

We recall that the assembly process, denoted by $\boldsymbol{\mathcal{A}}^\top$, may be viewed as a mapping and partial summation process where we are free to specify the order in which the global system is chosen. Previously, we have stated that the global system is ordered so that the global boundary degrees of freedom (that is, those constructed from the local boundary modes) are listed first, followed by the global interior degrees of freedom (that is, those constructed from the interior modes of the elemental construction). In addition, the global interior degrees of freedom were numbered consecutively. This ordering is important to make maximum use of the structure of the discretisation in the static condensation process. If we adopt this ordering, the global system has the form shown in Fig. 4.18.

In this figure the matrix $\boldsymbol{M}_b$ corresponds to the global assembly of the elemental boundary–boundary mode interaction from $\boldsymbol{M}_b^e$, and similarly $\boldsymbol{M}_c, \boldsymbol{M}_i$ correspond to the global assembly of the elemental boundary–interior coupling and interior–interior systems $\boldsymbol{M}_c^e, \boldsymbol{M}_i^e$. A notable feature of the global system is that the global boundary–boundary, $\boldsymbol{M}_b$, matrix is sparse and may be reordered to reduce the bandwidth using, for example, a reverse Cuthill–McKee algorithm [119], or refactored in a multi-level Schur complement solver as discussed in Section 4.2.3.1. The global boundary–interior coupling matrix, $\boldsymbol{M}_c$, is very sparse,

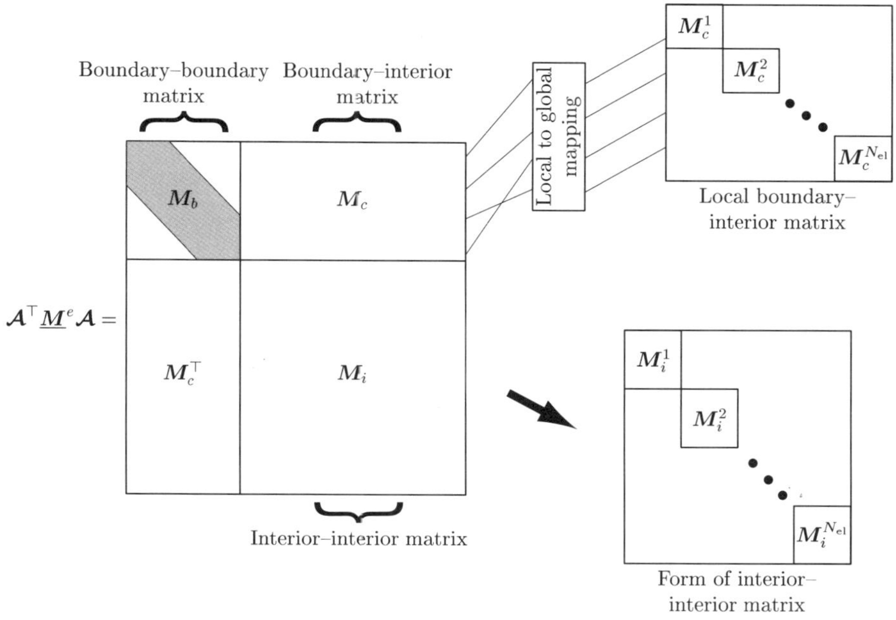

FIG. 4.18. Depending on the global numbering scheme the global matrix system has a very distinctive structure, as shown here. The boundary–boundary matrix is usually banded; the boundary–interior matrix can be constructed from the sparse local boundary–interior matrices; and the interior–interior matrix consists of smaller uncoupled matrices.

but as it only operates on known vectors we only need to store the local matrices $\boldsymbol{M}_c^e$. Finally, the natural form of $\boldsymbol{M}_i$ is a block diagonal matrix which is very inexpensive to evaluate since each block may be inverted individually. It is the structure of $\boldsymbol{M}_i$ which makes the static condensation technique so effective. This arises from the fact that the interior modes are non-overlapping and are therefore orthogonal at an elemental level.

Now, if we distinguish between the boundary and interior components of $\boldsymbol{x}$ and $\boldsymbol{f}$ using $\boldsymbol{x}_b, \boldsymbol{x}_i$ and $\boldsymbol{f}_b, \boldsymbol{f}_i$, respectively, that is,

$$\boldsymbol{x} = \begin{bmatrix} \boldsymbol{x}_b \\ \boldsymbol{x}_i \end{bmatrix}, \quad \boldsymbol{f} = \begin{bmatrix} \boldsymbol{f}_b \\ \boldsymbol{f}_i \end{bmatrix},$$

then eqn (4.2.15) can be written in its constituent parts as

$$\begin{bmatrix} \boldsymbol{M}_b & \boldsymbol{M}_c \\ \boldsymbol{M}_c^\top & \boldsymbol{M}_i \end{bmatrix} \begin{bmatrix} \boldsymbol{x}_b \\ \boldsymbol{x}_i \end{bmatrix} = \begin{bmatrix} \boldsymbol{f}_b \\ \boldsymbol{f}_i \end{bmatrix}. \tag{4.2.16}$$

To solve this system we perform a block elimination by pre-multiplying this system by the matrix

$$\begin{bmatrix} \boldsymbol{I} & -\boldsymbol{M}_c\boldsymbol{M}_i^{-1} \\ 0 & \boldsymbol{I} \end{bmatrix},$$

to arrive at

$$\begin{bmatrix} \boldsymbol{M}_b - \boldsymbol{M}_c\boldsymbol{M}_i^{-1}\boldsymbol{M}_c^\top & 0 \\ \boldsymbol{M}_c^\top & \boldsymbol{M}_i \end{bmatrix} \begin{bmatrix} \boldsymbol{x}_b \\ \boldsymbol{x}_i \end{bmatrix} = \begin{bmatrix} \boldsymbol{f}_b - \boldsymbol{M}_c\boldsymbol{M}_i^{-1}\boldsymbol{f}_i \\ \boldsymbol{f}_i \end{bmatrix}. \tag{4.2.17}$$

The equation for the boundary unknowns is therefore

$$\left(\boldsymbol{M}_b - \boldsymbol{M}_c\boldsymbol{M}_i^{-1}\boldsymbol{M}_c^\top\right)\boldsymbol{x}_b = \boldsymbol{f}_b - \boldsymbol{M}_c\boldsymbol{M}_i^{-1}\boldsymbol{f}_i\,.$$

Once $\boldsymbol{x}_b$ is known we can determine $\boldsymbol{x}_i$ from the second row of eqn (4.2.17) since

$$\boldsymbol{x}_i = \boldsymbol{M}_i^{-1}\boldsymbol{f}_i - \boldsymbol{M}_i^{-1}\boldsymbol{M}_c^\top\boldsymbol{x}_b\,. \tag{4.2.18}$$

The solution of eqn (4.2.15) has been split into three operations. The first is to evaluate and invert $\boldsymbol{M}_b - \boldsymbol{M}_c\boldsymbol{M}_i^{-1}\boldsymbol{M}_c^\top$, which is also known as the *Schur complement.* The second is to evaluate $\boldsymbol{M}_i^{-1}$, and the final operation is the evaluation of $\boldsymbol{M}_c\boldsymbol{M}_i^{-1} = \left[\boldsymbol{M}_i^{-1}\boldsymbol{M}_c^\top\right]^\top$.

The second and third operations can both be performed at a local elemental level. Since $\boldsymbol{M}_i$ is made up of block diagonals of the local matrices $\boldsymbol{M}_i^e$ (i.e., $\boldsymbol{M}_i = \underline{\boldsymbol{M}_i^e}$) the inverse of $\boldsymbol{M}_i$ is

$$\boldsymbol{M}_i^{-1} = \left[\underline{\boldsymbol{M}_i^e}\right]^{-1} = \underline{[\boldsymbol{M}_i^e]^{-1}}\,,$$

which is still a block diagonal matrix that can be evaluated locally within every element. The products $\boldsymbol{M}_c\boldsymbol{M}_i^{-1}\boldsymbol{f}_i$ and $\boldsymbol{M}_i^{-1}\boldsymbol{M}_c^\top\boldsymbol{x}_b$ can also treated as local operations because they only involve the matrix–vector products of a known vector (that is, $\boldsymbol{f}_i$ and $\boldsymbol{x}_b$). To illustrate this operation in its matrix form we define the matrix $\boldsymbol{\mathcal{A}}_b$, which is the boundary version of $\boldsymbol{\mathcal{A}}$ and is equivalent to the bmap[e][i] operation discussed in Section 4.2.1. The operation $\boldsymbol{\mathcal{A}}_b$ therefore scatters the global boundary degrees of freedom to the local boundary degrees of freedom, that is,

$$\begin{bmatrix} \boldsymbol{x}_b^1 \\ \boldsymbol{x}_b^2 \\ \vdots \\ \boldsymbol{x}_b^{N_{\rm el}} \end{bmatrix} = \boldsymbol{\mathcal{A}}_b\boldsymbol{x}_b\,,$$

where $\boldsymbol{x}_b^e$ contains the components of $\boldsymbol{x}_b$ in element e. Similarly, the operation $\boldsymbol{\mathcal{A}}_b^\top$ assembles the global boundary degrees of freedom from the local boundary degrees of freedom. The boundary–interior matrix $\boldsymbol{M}_c$ can now be written as

$$\boldsymbol{M}_c = \boldsymbol{\mathcal{A}}_b^\top\underline{\boldsymbol{M}_c^e}\,,$$

and so the products $\boldsymbol{M}_c\boldsymbol{M}_i^{-1}\boldsymbol{f}_i$ and $\boldsymbol{M}_i^{-1}\boldsymbol{M}_c^\top\boldsymbol{x}_b$ become

$$\boldsymbol{M}_c\boldsymbol{M}_i^{-1}\boldsymbol{f}_i = \boldsymbol{\mathcal{A}}_b^\top \underline{\boldsymbol{M}_c^e}\ \underline{[\boldsymbol{M}_i^e]^{-1}}\boldsymbol{f}_i\,,$$
$$\boldsymbol{M}_i^{-1}\boldsymbol{M}_c^\top\boldsymbol{x}_b = \underline{[\boldsymbol{M}_i^e]^{-1}}\ \underline{(\boldsymbol{M}_c^e)^\top}\boldsymbol{\mathcal{A}}_b\boldsymbol{x}_b\,.$$

As both $\underline{[\boldsymbol{M}_i^e]^{-1}}$ and $\underline{\boldsymbol{M}_c^e}$ are essentially local matrices, these products can be evaluated on an elemental level with respect to a vector which is either globally scattered, as represented by $\boldsymbol{\mathcal{A}}_b$, or globally assembled, as represented by $\boldsymbol{\mathcal{A}}_b^\top$. The boundary scatter and assembly operation also illustrates how to construct the boundary–boundary system since

$$\boldsymbol{M}_b = \boldsymbol{\mathcal{A}}_b^\top\underline{\boldsymbol{M}_b}\boldsymbol{\mathcal{A}}_b\,.$$

Therefore, the Schur complement system may be written as

$$\begin{aligned}\boldsymbol{M}_b - \boldsymbol{M}_c\boldsymbol{M}_i^{-1}\boldsymbol{M}_c^\top &= \boldsymbol{\mathcal{A}}^\top\underline{\boldsymbol{M}_b^e}\boldsymbol{\mathcal{A}}_b - \boldsymbol{\mathcal{A}}_b^\top\underline{\boldsymbol{M}_c^e}\ \underline{[\boldsymbol{M}_i^e]^{-1}}\ \underline{(\boldsymbol{M}_c^e)^\top}\boldsymbol{\mathcal{A}}_b \\ &= \boldsymbol{\mathcal{A}}_b^\top\left[\underline{\boldsymbol{M}_b^e - \boldsymbol{M}_c^e[\boldsymbol{M}_i^e]^{-1}(\boldsymbol{M}_c^e)^\top}\right]\boldsymbol{\mathcal{A}}_b\,,\end{aligned}$$

which shows that the global Schur complement system may be generated from the local elemental Schur complement system $\boldsymbol{M}_i^e - \boldsymbol{M}_c^e[\boldsymbol{M}_i^e]^{-1}(\boldsymbol{M}_c^e)^\top$. We now appreciate that global assembly is only necessary for the boundary system when using static condensation. Once the boundary solution is known, the solution for the interior elemental modes given by eqn (4.2.18) can be performed at an elemental level. In this formulation, we have assumed that all the global boundary conditions are unknown values since the Schur complement matrix is of size N_b. However, in general we have some known Dirichlet boundary conditions which can be dealt with by numbering the global system so that they are ordered after the unknown degrees of freedom. We then use the appropriate submatrix of $\boldsymbol{M}_i - \boldsymbol{M}_c[\boldsymbol{M}_i]^{-1}(\boldsymbol{M}_c)^\top$, as shown in Section 4.2.4.

Since the majority of the storage requirement in this technique is used in the global Schur complement system, an alternative approach is to solve this system iteratively where only storage for the local Schur complements $\boldsymbol{M}_b^e - \boldsymbol{M}_c^e[\boldsymbol{M}_i^e]^{-1}(\boldsymbol{M}_c^e)^\top$ is required. The Schur complement system is also better conditioned than the complete system, which also makes this approach more attractive (see Chapter 5).

As a final point, we note that storage savings can also be made when evaluating the Laplacian and mass matrix systems by noting that the matrices $(\boldsymbol{M}_i^e)^{-1}$ and $\boldsymbol{M}_c^e(\boldsymbol{M}_i^e)^{-1}$ are similar for straight-sided elements of the same size and orientation. However, to exploit this feature, some degree of structure in the mesh is required.

4.2.3.1 *Multi-level static condensation*

The motivation behind using static condensation was the natural decoupling of the interior degrees of freedom within each element, leading to a global system which contained a block diagonal submatrix. This decoupling can be mathematically attributed to the fact that the interior degrees of freedom in one element are orthogonal to the interior degrees of freedom of another simply because these modes are non-overlapping. To take advantage of this block diagonal submatrix we have to construct the Schur complement system $\boldsymbol{M}_b^e - \boldsymbol{M}_c^e[\boldsymbol{M}_i^e]^{-1}(\boldsymbol{M}_c^e)^\top$ for the boundary degrees of freedom which may be evaluated locally.

The effect of constructing each of the local Schur complement matrices $\boldsymbol{M}_b^e - \boldsymbol{M}_c^e[\boldsymbol{M}_i^e]^{-1}(\boldsymbol{M}_c^e)^\top$ is to orthogonalise the boundary modes from the interior modes. However, the inverse matrix $[\boldsymbol{M}_i^e]^{-1}$ is typically full, which means that the boundary modes become tightly coupled. It is this coupling which dictates the bandwidth of the globally assembled Schur complement system. To compute the bandwidth we simply need to find the maximum difference between the global numbering of the boundary modes within every local element. Even though the boundary modes are coupled to all other boundary modes within the element and the boundary modes of neighbouring elements, they are not coupled with the boundary modes within non-neighbouring elements. Therefore, an appropriate numbering of the boundary system will lead to a Schur complement matrix which also contains a submatrix that is block diagonal, and so the static condensation technique can be reapplied. This technique has been more commonly used in the structural mechanics field and is also known as substructuring [439].

To illustrate this ordering we consider the triangular mesh shown in Fig. 4.19(a) using $N_{\rm el} = 32$ elements. The construction of the global Schur complement $\boldsymbol{M}_b^e - \boldsymbol{M}_c^e[\boldsymbol{M}_i^e]^{-1}(\boldsymbol{M}_c^e)^\top$ requires us to globally number all of the boundary degrees of freedom, as indicated by the open circles (○) and squares (□).

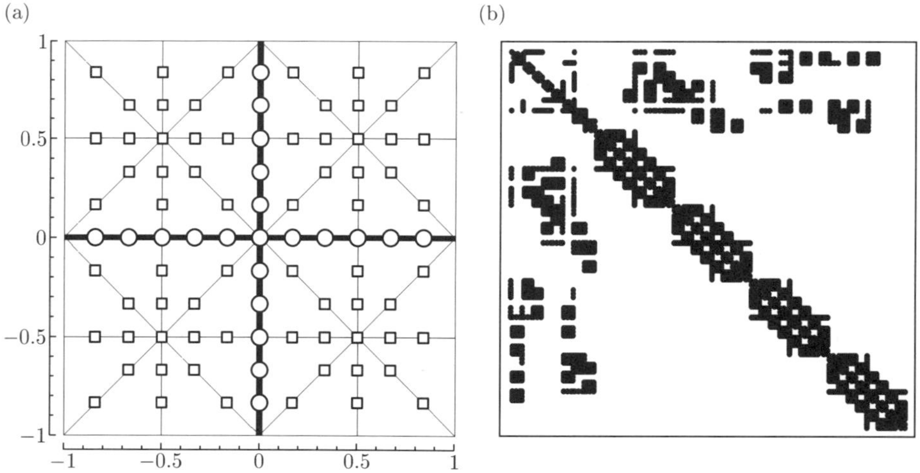

FIG. 4.19. The boundary degrees of freedom, on the mesh shown in (a), are ordered so that the boundary modes indicated by open squares (□) are first, followed by the boundary modes indicated by open circles (○) within each quadrant. Using this ordering the resulting Schur complement matrix has a block diagonal submatrix, as shown in (b).

We have not included the boundary of the domain as we assume that we are applying Dirichlet boundaries, and so these values are not part of our Galerkin matrix system. If we order the numbering of the elemental boundary degrees of freedom so that the vertex and edge modes indicated by the open circles (○) are first followed by the vertex and edge modes within each quadrant, indicated by the open squares (□), then the resulting Schur complement system of the mass matrix for a polynomial expansion of $p = 6$ is shown in Fig. 4.19(b). The block diagonal structure of the matrix is due to the fact that, even after constructing the elemental Schur complement systems, the boundary degrees of freedom in each quadrant do not overlap and so are orthogonal.

We can now construct another Schur complement system to solve for the (○) degrees of freedom and decoupling each quadrant of (□) degrees of freedom. This technique can be repeated provided that there is more than one region of non-overlapping data. The approach is clearly independent of the elemental shape and may equally well be applied to quadrilateral regions or any hybrid shape in three dimensions.

4.2.4 *Global boundary system numbering and ordering to enforce Dirichlet boundary conditions*

♦[30]

We have previously discussed in Section 4.2.1 the issues relating a local elemental numbering scheme to a global numbering, taking account of inter-element connectivity. The aim of Section 4.2.1 was to assemble the local expansions into a global C^0 continuous expansion where a global numbering scheme map$[e][i]$ or bmap$[e][i]$ was assumed to be known. In this section we outline the generation of this numbering scheme and then demonstrate how the numbering scheme can be reordered to enforce Dirichlet boundary conditions.

Heuristically, a global numbering scheme for the boundary can be generated in the following manner. We assume, as a starting-point, the output of a finite element or volume mesh generator where the global Cartesian coordinates of each vertex are known. We start our numbering scheme by assigning a unique number to every unique vertex defined through its global coordinates. In the spectral/*hp* element method we also require a global numbering of all boundary degrees of freedom, including all edge modes and the face modes in three dimensions. Using the global vertex coordinates, every unique edge can be identified using the global coordinates of the vertices at the ends of each edge. The global numbering of the degrees of freedom along each global edge can then be defined. An example of this type of global numbering is shown in Fig. 4.20(a) where we note that the global vertices are numbered first, followed by the global edges, according to the element numbering (given in Fig. 4.20(c)). A similar strategy can also be followed to number global degrees of freedom associated with every face. If a full numbering scheme is required then the interior degrees of freedom can

[30] Generation of a global numbering scheme and ordering of the boundary numbering to strongly enforce Dirichlet conditions in a matrix system.

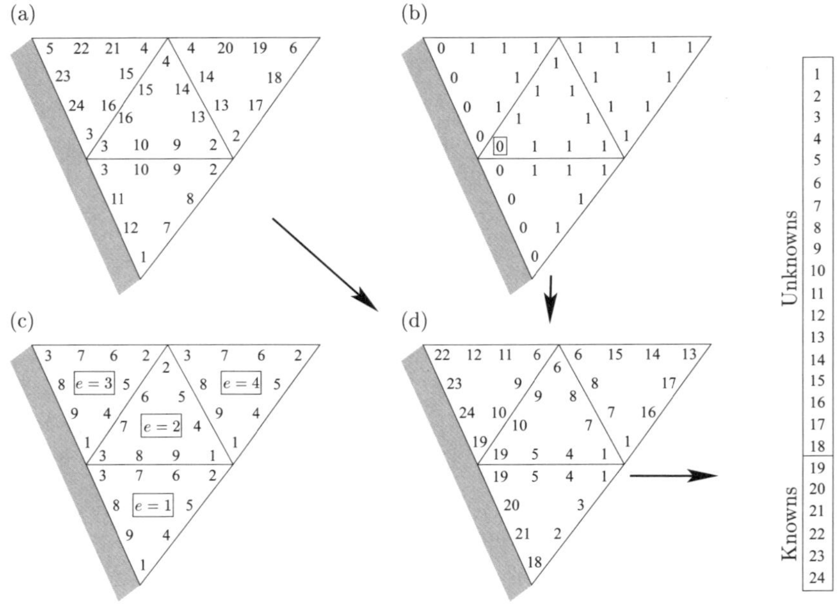

FIG. 4.20. Given a global numbering scheme as shown in (a), we can reorder the system so that known degrees of freedom on a Dirichlet boundary are listed after the unknown degrees of freedom as shown in (d). A reordering strategy using the elemental information, as given in (c), is to generate a global mask array as shown in (b).

finally be independently numbered by looping over all elements. This type of global numbering is unlikely to lead to an ordering which will give a minimal bandwidth for global matrix problems. However, once one global numbering has been obtained, it is then possible to use standard algorithms and packages to reorder the scheme to minimise the difference in numbers between coupled degrees of freedom in order to reduce bandwidth or maximise matrix infill [119, 266].

It may also be necessary to enforce Dirichlet boundary conditions. In this case it is advantageous to have a global numbering scheme which orders the unknown boundary degrees of freedom first, followed by the known degrees of freedom which lie on Dirichlet boundaries.

A procedure to perform the boundary reordering adopted by Henderson [224] is illustrated in Fig. 4.20. In this example we consider a mesh of four triangular elements and start by assuming that we have obtained a global numbering scheme as shown in Fig. 4.20(a) for a $P = 3$ polynomial expansion within each element. From the numbering in Fig. 4.20(a), we can define elemental arrays bmap[e][i] which relate all the element boundary degrees of freedom to the global numbering scheme. This type of array was used in Section 4.2.1 to define the global assembly operation. In the example of Fig. 4.20, if we assume that the local vertex boundary degrees of freedom are ordered first, followed by the edge degrees of freedom in each element, as shown in Fig. 4.20(c) then the array bmap[e][i] can be defined as ($1 \leqslant i \leqslant 9$)

$$\begin{aligned}
\text{bmap}[1][i] &\Rightarrow [1,2,3,7,8,9,10,11,12]\,,\\
\text{bmap}[2][i] &\Rightarrow [2,4,3,13,14,15,16,10,9]\,,\\
\text{bmap}[3][i] &\Rightarrow [3,4,5,16,15,21,22,23,24]\,,\\
\text{bmap}[4][i] &\Rightarrow [2,6,4,17,18,19,20,14,13]\,.
\end{aligned}$$

If we assume that the boundary is of a Dirichlet type then we would like to place all the global numbers along this boundary at the end of a new global numbering system. Although this would appear to be straightforward in the example, devising an automatic implementation using elemental information is a bit more involved and a suggested implementation is provided below.

From the point of view of implementation, a convenient way of specifying ♦[31] boundary conditions, in two dimensions, is to identify which local *edges* of the elements touch the given domain Dirichlet boundary. It is then possible to identify the degrees of freedom along these edges which have Dirichlet values and associate with them an elemental mask array 'mask[e][i]', as shown in Fig. 4.20(b). This mask array is set so that all Dirichlet degrees of freedom have entries set to '0', otherwise the entry is set to '1'. If we are imposing Dirichlet boundary conditions using only edge information in two dimensions (or face information in three dimensions) then care must be taken to ensure that all local contributions to the mask array in elements that only touch the Dirichlet boundary through a vertex are set. For example, setting the values of mask[e][i] in elements with edges that touch the Dirichlet boundary (i.e., $e = 1, 3$) in Fig. 4.20 will not enforce the single vertex touching the boundary in element 2 (highlighted by a box) to be zero. To ensure the mask array of this vertex point is set correctly we can use a global assembly and scatter operation using the bmap[e][i] array similar to that discussed in Section 4.2.1.1. In this case, we first initialise a global array, Gmask[i], to have a value of 1 for all entries and then perform the assembly

```
do e = 1, N_el
   do i = 0, n_b[e] - 1
      Gmask[bmap[e][i]] = Gmask[bmap[e][i]] × mask[e][i]
   continue
continue
```

The global array Gmask[i] will now contain entries of 1 only when all vertex points have a local mask which is 1. The local mask array can then be recovered through the scatter operation

[31] Manipulation of the global numbering scheme using elemental mapping arrays.

```
do e = 1, N_el
    do i = 0, n_b[e] - 1
        mask[e][i] = Gmask[bmap[e][i]]
    continue
continue
```

and mask[e][i] will then correspond to Fig. 4.20(b).

Finally, using the local mask array we can reorder the global numbering system using elemental information and another global array. We now initialise a global array Gbmap[i] to zero, and fill this array using the mask[e][i] array and by applying the following logic

```
let n_1 = n_2 = 1
do e = 1, N_el
    do i = 0, n_b[e] - 1
        if (Gbmap[bmap[e][i]] = 0)
            if (mask[e][i] = 1)
                Gbmap[bmap[e][i]] = n_1
                n_1 = n_1 + 1
            else
                Gbmap[bmap[e][i]] = n_2 + N_bslv
                n_2 = n_2 + 1
    continue
continue
```

where N_{bslv} is the number of unknown boundary degrees of freedom which typically has to be obtained as part of the loop and then added afterwards. The final form of the reordered global numbering scheme can be recovered into bmap[e][i] using a scatter operation of the form

```
do e = 1, N_el
    do i = 0, n_b[e] - 1
        bmap[e][i] = Gbmap[bmap[e][i]]
    continue
continue
```

The array bmap[e][i] now corresponds to the ordering in Fig. 4.20(d). If we make a global assembly with the new bmap[e][i] then we obtain a global ordering array of the form indicated in the right-hand side of Fig. 4.20, where the unknown degrees of freedom are listed first followed by the known Dirichlet values.

4.2.4.1 *Discrete lifting of the known solution*

We recall that, in a standard Galerkin implementation, Dirichlet boundary conditions can be enforced by 'lifting' a known solution satisfying these boundary conditions, see Section 2.2.1.3. Performing this operation leaves us with a homogeneous Dirichlet boundary problem where the same test and trial space can be applied to the problem. The separation of the global solution array $\hat{\boldsymbol{u}}$ into known (i.e., Dirichlet) and unknown boundary degrees of freedom provides a way of obtaining a discrete lifted boundary solution. If we denote the homogeneous unknown solution by $u^{\mathcal{H}}(\boldsymbol{x})$ and the known Dirichlet boundary conditions by $u^{\mathcal{D}}(\boldsymbol{x})$ then we can decompose the solution $u^{\delta}(\boldsymbol{x})$ into the form

$$u^{\delta}(\boldsymbol{x}) = u^{\mathcal{H}}(\boldsymbol{x}) + u^{\mathcal{D}}(\boldsymbol{x}) = \sum_{j}^{N^{\mathcal{H}}} \hat{u}_j^{\mathcal{H}} \Phi_j(\boldsymbol{x}) + \sum_{j=N^{\mathcal{H}}+1}^{N_{\mathrm{dof}}^{b}} \hat{u}_j^{\mathcal{D}} \Phi_j(\boldsymbol{x}),$$

where we recall that $\Phi_j(\boldsymbol{x})$ is the global expansion basis and $N^{\mathcal{H}}$ is defined as the number of global homogeneous boundary degrees of freedom ($N^{\mathcal{H}} = 17$ in the example of Fig. 4.20). We note that the important component of the definition is that the homogeneous solution has zero contribution from modes which are nonzero on Dirichlet boundaries. The homogeneous solution also contains the interior degrees of freedom which are defined to be zero on the boundaries. For the lifted solution the values of $\hat{u}_j^{\mathcal{D}}$ can be determined using a boundary transformation, as discussed in Section 4.3.2. In general, the lifted solution $u^{\mathcal{D}}(\boldsymbol{x})$ may also contain any pre-defined contribution associated with any homogeneous mode. This is typically possible for unsteady problems with steady boundary conditions where the previous solution that satisfies boundary conditions can be used as the lifted solution of the next time-step.

4.2.4.2 *Boundary matrix manipulation*

As a final example of the application of both the boundary numbering system and the lifted solution expressed in terms of expansion coefficients, we consider the matrix solution shown in Fig. 4.21. If we wish to directly invert a matrix problem $\boldsymbol{M}$ arising from our spectral/*hp* element formulation then we can apply the static condensation technique discussed in Section 4.2.3. By construction, this technique decouples the interior degrees of freedom, and for a symmetric matrix requires the inversion of a global boundary matrix of the form $\boldsymbol{M}_{\mathrm{SC}} = \boldsymbol{M}_b - \boldsymbol{M}_c \boldsymbol{M}_i^{-1} \boldsymbol{M}_c^{\top}$, where $\boldsymbol{M}_b$, $\boldsymbol{M}_i$, and $\boldsymbol{M}_c$ are submatrices of $\boldsymbol{M}$. For the standard Galerkin formulation $\boldsymbol{M}_{\mathrm{SC}}$ can be constructed from its elemental contributions $\boldsymbol{M}_{\mathrm{SC}}^{e}$ and by applying the global assembly technique over the boundary degrees of freedom, such that $\boldsymbol{M}_{\mathrm{SC}} = \boldsymbol{\mathcal{A}}_b^{\top} \underline{\boldsymbol{M}_{\mathrm{SC}}^{e}} \boldsymbol{\mathcal{A}}_b$, as discussed in Section 4.2.1.1. We note, however, that the matrix $\boldsymbol{M}_{\mathrm{SC}}$ constructed in this manner contains both the Dirichlet and unknown degrees of freedom. Strictly speaking, the matrix does not correspond to the Galerkin problem since it contains test (or weight) functions which are now zero on Dirichlet boundary conditions. The application of a boundary numbering scheme, where known Dirichlet values are listed after the unknown boundary degrees of freedom, allows us to manipulate the resulting matrix into the appropriate form.

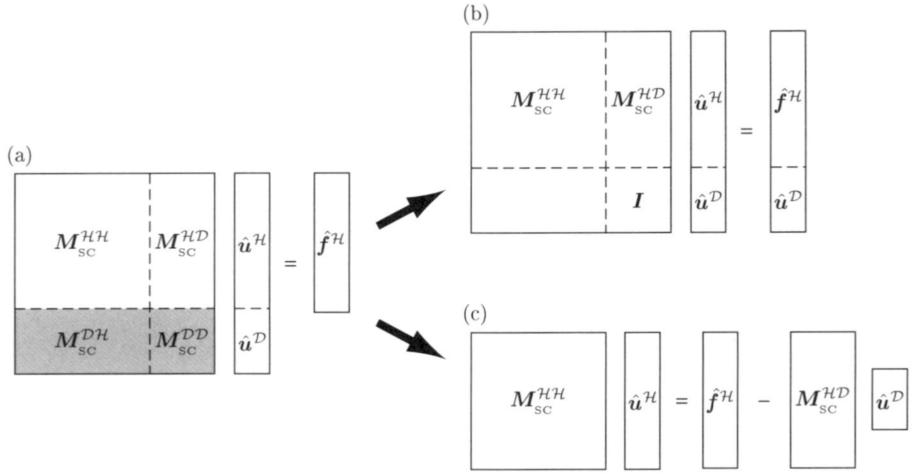

FIG. 4.21. Sorting the boundary vector $\hat{\boldsymbol{u}}$ into an unknown component $\hat{\boldsymbol{u}}^{\mathcal{H}}$ and a known (lifted) component $\hat{\boldsymbol{u}}^{\mathcal{D}}$, the full boundary matrix system $\boldsymbol{M}_{\mathrm{SC}}$ can be viewed as submatrices $\boldsymbol{M}^{\mathcal{HH}}_{\mathrm{SC}}$, $\boldsymbol{M}^{\mathcal{HD}}_{\mathrm{SC}}$, $\boldsymbol{M}^{\mathcal{DH}}_{\mathrm{SC}}$, and $\boldsymbol{M}^{\mathcal{DD}}_{\mathrm{SC}}$, as shown in (a). This can be solved by removing matrices $\boldsymbol{M}^{\mathcal{DH}}_{\mathrm{SC}}$ and $\boldsymbol{M}^{\mathcal{DD}}_{\mathrm{SC}}$ with the identity matrix as shown in (b) or, equivalently, by taking the product $\boldsymbol{M}^{\mathcal{HD}}_{\mathrm{SC}}\hat{\boldsymbol{u}}^{\mathcal{D}}$ to the right-hand side as shown in (c).

This process is highlighted in Fig. 4.21. The full boundary matrix $\boldsymbol{M}_{\mathrm{SC}} = \boldsymbol{\mathcal{A}}_b^{\top}\underline{\boldsymbol{M}}^{e}_{\mathrm{SC}}\boldsymbol{\mathcal{A}}_b$ is indicated in Fig. 4.21(a) where the numbering system allows us to view the matrix as a series of submatrices $\boldsymbol{M}^{\mathcal{HH}}_{\mathrm{SC}}$, $\boldsymbol{M}^{\mathcal{HD}}_{\mathrm{SC}}$, $\boldsymbol{M}^{\mathcal{DH}}_{\mathrm{SC}}$, and $\boldsymbol{M}^{\mathcal{DD}}_{\mathrm{SC}}$, corresponding to the homogeneous and Dirichlet solutions $\hat{\boldsymbol{u}}^{\mathcal{H}}$ and $\hat{\boldsymbol{u}}^{\mathcal{D}}$, respectively. The highlighted matrices $\boldsymbol{M}^{\mathcal{DH}}_{\mathrm{SC}}$ and $\boldsymbol{M}^{\mathcal{DD}}_{\mathrm{SC}}$ are not part of the Galerkin problem since they involve trial (or weight) functions which are not zero at Dirichlet boundary conditions. To enforce the Dirichlet boundary conditions we therefore have two choices. As shown in Fig. 4.21(b), we can replace the matrices $\boldsymbol{M}^{\mathcal{DH}}_{\mathrm{SC}}$ and $\boldsymbol{M}^{\mathcal{DD}}_{\mathrm{SC}}$ with a zero block and the identity matrix $\boldsymbol{I}$ on the diagonal entries as well as copy the vector $\hat{\boldsymbol{u}}^{\mathcal{D}}$ to the right-hand side. An equivalent technique is often applied in finite element methods where this approach is favoured since it is not necessary to reorder the degrees of freedom as a row containing a Dirichlet degree of freedom can be deleted and a unit value placed on the diagonal. There are, however, two potential drawbacks. The first is that the matrix is now not symmetric, even if the original matrix $\boldsymbol{M}_{\mathrm{SC}}$ was symmetric. The second is that the conditioning of the system can be significantly influenced by the introduction of diagonal terms if all other entries of the system are very small.

Alternatively, we can manipulate the matrix system as shown in Fig. 4.21(c). This system is the same as shown in Fig. 4.21(a,b) but the known Dirichlet boundary conditions $\hat{\boldsymbol{u}}^{\mathcal{D}}$ have now been taken to the right-hand side of the system. In this manipulation we have maintained the symmetry of the system

$\boldsymbol{M}_{\mathrm{sc}}^{\mathcal{HH}}$ by essentially lifting the known solution out of the problem. To apply this technique, however, a boundary numbering system is clearly necessary to construct a compact form of the matrix $\boldsymbol{M}_{\mathrm{sc}}^{\mathcal{HH}}$.

We close by noting that a more general lifted solution, $u^{\mathcal{D}}$, can be used involving all the boundary degrees of freedom. In this case a matrix contribution involving $\boldsymbol{M}_{\mathrm{sc}}^{\mathcal{HH}}$ will appear on the right-hand side of the problem. We remark, however, that right-hand-side matrix–vector products do not require the assembly of global matrix systems since they can (and should) be evaluated at an elemental level and then assembled.

4.3 Pre- and post-processing issues

In Section 4.1 we discussed the key elemental operations used in a spectral/*hp* element solver independent of the formulation, namely integration, differentiation, and basic elemental mappings. We then applied these operations in Section 4.2 within the context of global C^0 expansion for a classical Galerkin formulation by introducing the concept of global assembly. Although these two sections represent the operations behind the algorithms, practical implementation also requires various pre- and post-processing techniques, the discussion of which will conclude this chapter.

In this section we will discuss issues relevant to mesh generation and boundary representation, as well as particle tracking within a spectral/*hp* element computational setting. The first two topics, boundary condition discretisation and mesh generation, are important issues of any spectral/*hp* element formulation. The last topic, particle tracking, is a useful diagnostic tool in fluid mechanics as well as a key component in the strong semi-Lagrangian formulation, which is discussed in Chapters 6 and 8.

4.3.1 *Boundary condition discretisation*

Until now we have assumed that all boundary conditions have been specified in terms of the expansion coefficients (whether modal or nodal); however, for a general implementation this is not typically the case. In Section 4.1.3 we discussed how curved elements, which are typically due to a curved boundary of the solution domain, can be represented in terms of an iso-parametric mapping if the mapping of the edges, or faces in three dimensions, is provided. In both the case of the Dirichlet boundary condition and the curved boundary mapping we need to project the known function onto a discrete expansion basis which is also globally C^0 continuous.

In this section we will discuss how to generate C^0 approximations over the whole surface using local elemental information. Although we will focus on the Dirichlet boundary condition, an analogous approach can be used in the case of a curved surface mapping to define the elemental transformation.

4.3.2 *Elemental boundary transformation*

Given a Dirichlet boundary condition $g_{\mathcal{D}}(\boldsymbol{x})$, where $\boldsymbol{x} \in \partial\Omega$, we need a consistent method of approximating the boundary condition in terms of our discrete expansion. In two dimensions the boundary of the domain is simply a one-dimensional segment, but in three dimensions, depending on our spatial discretisation, the surface becomes a tessellation of triangles or rectangles, or even a combination of both shapes.

In general, we require that our discrete boundary approximation remains at least C^0 continuous. This condition could be ensured by formulating an approximation over the whole Dirichlet boundary. In two dimensions this would only involve a one-dimensional problem with multiple elemental segments. In a three-dimensional problem, however, this implies constructing a full two-dimensional system. It can be appreciated that for a general mesh this could become excessively complicated. A more desirable method is to perform a local projection within each element. However, if we performed an elemental Galerkin (L^2) projection of $g_{\mathcal{D}}$ onto the boundary modes within each individual element then we could not, in general, guarantee that our approximation would be C^0 continuous over the whole boundary.

We recall that elemental projections or forward transformations were previously discussed in Section 4.1.5.3. We note that the use of a collocation projection onto a set of nodal points which includes the boundary of the segment or face satisfied our requirement of a C^0 continuous approximation. An appropriate choice of collocation points is therefore the Gauss–Lobatto–Legendre quadrature points in a segment or rectangular region and any choice of the nodal non-tensorial distribution discussed in Section 3.3 on a triangular face. As we have seen previously, these points also have favourable interpolation properties from a Lebesgue constant point of view. Nevertheless, if a local Galerkin projection is desired then we need to modify the Galerkin projection to ensure C^0 continuity. In two dimensions this modified projection can be viewed as a collocation projection at the vertices, followed by an L^2 projection on all edges, with a final interior L^2 projection of the remaining function.

To illustrate the modified Galerkin projection we consider the case shown in Fig. 4.22. We wish to project a known boundary condition $g_{\mathcal{D}}$ onto a boundary of element 'e' lying on $\partial\Omega$. The discrete solution $u^\delta(x_1, x_2)$ along the edge can be written as

$$u^\delta(x_1, x_2) = \sum_{pq} \hat{u}^e_{pq}\phi^e_{pq}(\xi_1, -1) = \sum_{p=0}^{P_1} \hat{u}^e_{p0}\phi^e_{p0}(\xi_1, -1),$$

where the reduction of the summation is due to the boundary interior decomposition of the basis, and $x_1 = \chi^e_1(\xi_1, -1)$ and $x_2 = \chi^e_2(\xi_1, -1)$. If our expansion is of a modified modal type then $\phi_{p0}(\xi_1, -1) = \psi^a_p(\xi_1)\psi^a_0(-1) = \psi^a_p(\xi_1)$. Alternatively, if the expansion is of a nodal type then $\phi_{p0}(\xi_1, -1) = h_p(\xi_1)$.

For the modal expansion case we would like to determine $\hat{u}^e_{p0}$ such that

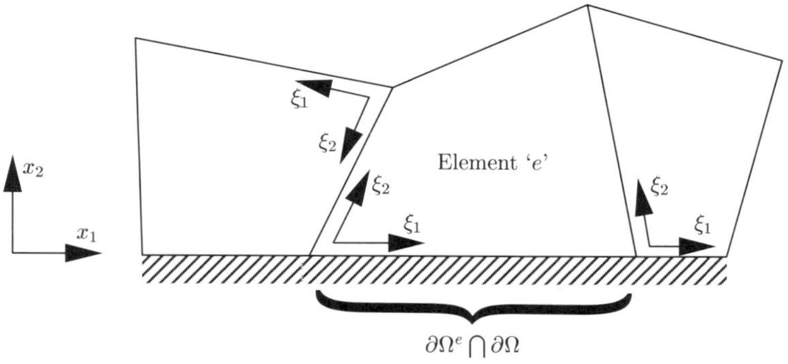

FIG. 4.22. Intersection of the boundary of the eth element with the domain boundary $\partial\Omega$. The orientation of the local coordinates is indicated by the (ξ_1, ξ_2) system.

$$\sum_{p=0}^{P_1} \hat{u}^e_{p0}\psi^a(\xi_1) \simeq g_{\mathcal{D}}(\chi^e_1(\xi_1, -1)) \,.$$

We can use the vertex–edge decomposition of the boundary modes to ensure that our approximation remains C^0 continuous over all elements. The vertex functions have, by definition, a unit value at the ends of an edge and all other boundary modes are zero at this point. C^0 continuity is, therefore, ensured if we set the vertex coefficients to

$$\begin{aligned} \hat{u}^e_{00} &= g_{\mathcal{D}}(\chi^e_1(-1, -1)) \,, \\ \hat{u}^e_{P_1 0} &= g_{\mathcal{D}}(\chi^e_1(1, 1)) \,. \end{aligned}$$

(Recall that $p = 0$ and $p = P_1$ refer to the vertex modes of the edge expansions $\psi^a(\xi)$ and $h_p(\xi)$.) We also note that $\chi^e_1(-1, -1)$, $\chi^e_1(1, -1)$ are simply the values of the vertex location (x_1, x_2). The remaining unknown coefficients at the boundary may now be written as

$$\sum_{p=1}^{P_1-1} \hat{u}^e_{p0}\psi^a(\xi_1) \simeq g_{\mathcal{D}}(\chi^e_1(\xi_1, -1)) - \hat{u}^e_{00}\psi^a_0(\xi_1) - \hat{u}^e_{P_1 0}\psi^a_{P_1}(\xi_1) \,.$$

Since the remaining modes do not contribute to the end-points we can solve for the remaining unknowns $\hat{u}^e_{p0}$ $(1 \leqslant p \leqslant P_1)$ without destroying the C^0 continuity. These coefficients can be found by setting up the following local Galerkin projection.

Find $\hat{u}^e_{p0}$ such that

$$\left(\phi_i, \sum_{p=1}^{P_1-1} \hat{u}^e_{p0}\phi_p\right) = (\phi_i, g_{\mathcal{D}} - \hat{u}^e_{00}\psi^a_0 - \hat{u}^e_{P_1 0}\psi^a_{P_1}) \,,$$

for all i $(0 \leqslant i \leqslant P_1 - 1)$.

Clearly, this involves constructing and inverting the one-dimensional mass matrix $\boldsymbol{M}^{1\mathrm{D}}[i][j] = (\phi_i(\xi_1), \phi_j(\xi_1))$ for $1 \leqslant i, j \leqslant P_1$. This Galerkin approximation minimises the error in the L^2 norm between the exact boundary condition and the edge approximation in the interior of the region (for a more detailed analysis, see Section 7.4 and also Babuška *et al.* [25]). In a discrete implementation of the projection we will be performing the collocation projection onto the quadrature points. However, if this collocation projection is of a sufficiently high order (i.e., the number of quadrature points is large enough) then the error in the collocation projection will be smaller than that of the L^2 projection. A directly analogous procedure can be followed for triangular regions as the boundary modes of the modified expansion are identical (see Section 3.2.3.3). Although we can also use the same technique when using the spectral element nodal expansion $h_p(\xi_1)$, we recall that if we evaluate the one-dimensional mass matrix $\boldsymbol{M}^{1\mathrm{D}}$ using a discrete inner product $(u, v)_\delta$ of the same order as $h_p(\xi_1)$ then the mass matrix is diagonal and we are essentially performing a collocation projection, see Sections 2.3.4.2 and 4.1.5.3. In this case the formulation is, therefore, identical to evaluating the boundary conditions at the nodal points.

For the three-dimensional case we follow an analogous procedure using three steps.

1. Set the expansion coefficients of the vertex modes to the value of $g_{\mathcal{D}}$ evaluated at the vertex points.
2. Subtract the vertex mode approximation from $g_{\mathcal{D}}$ and evaluate the coefficients of the edge modes using a local one-dimensional Galerkin projection within the interior of the edge.
3. Subtract the vertex and edge mode approximation from $g_{\mathcal{D}}$ and determine the face modes using a local two-dimensional Galerkin projection within the interior of the face.

The three-dimensional modified transformation for a triangular face is represented diagrammatically in Fig. 4.23. Using the collocation property of the vertex modes, we set the coefficient of the vertex mode to the physical value of the function evaluated at the vertex, as indicated in Fig. 4.23(a). Assuming that the boundary function is continuous, this ensures continuity of the vertex functions over the entire boundary. To calculate the remaining expansion coefficients we subtract the vertex mode approximation from the boundary function, as indicated in Fig. 4.23(b). We then locally project each of the edge functions onto the edge modes using a one-dimensional Galerkin approximation. As we have subtracted the vertex contribution, the function is zero at the end-points of an edge which is consistent with the shape of the edge modes. A possible source of error in evaluating the edge modes between two elements is in the numerical integration. However, if we use a symmetric quadrature rule, such as Gauss–Lobatto–Legendre quadrature, this error will be identical for any smooth function provided that the same quadrature order is used and so we maintain C^0 continuity. Finally, we subtract the edge and vertex modes approximation

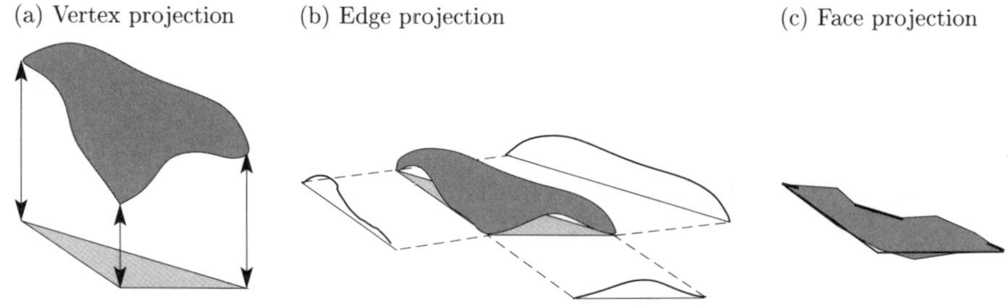

FIG. 4.23. Diagrammatic representation of the three-dimensional modified boundary transformation within a triangular face. Given an initial function, as shown in (a), the vertex coefficients are set to the value of the function at the vertices. This approximation is then subtracted, as shown in (b), permitting the edge contributions to be locally projected. Finally, the edge approximation is also subtracted and the remaining function is projected onto the face modes, as indicated in (c).

from the boundary function and project this onto the face modes, as shown in Fig. 4.23(c). This function will not, in general, be exactly zero along all edges since there is an error associated with the approximation of the edge functions. However, for a smooth boundary function the error will be consistent with the overall approximation.

4.3.3 *Mesh generation for spectral/hp element discretisation*

In this section we discuss issues concerning the generation of high-order or curvilinear meshes for spectral/*hp* element discretisations of complex geometries. The extension of standard mesh generation technology for spectral/*hp* element algorithms is a non-trivial exercise. Complications arise due to the conflicting requirements of generating coarse meshes for high-order polynomial approximations whilst maintaining good elemental properties in regions of high curvature. A potential problem is illustrated in Fig. 4.24 that shows the occurrence of invalid curvilinear spectral/*hp* elements in reconstructing an arterial bypass graft [434]. The straight-sided element discretisation does not include any invalid elements.

A necessary starting-point for discussing the issues involved in mesh generation is to define what makes the elements invalid. In the case shown in Fig. 4.24 the elements are invalid because the mapping between the physical region $\boldsymbol{x} \in \Omega^e$ and the standard region $\boldsymbol{\xi} \in \Omega_{\rm st}$ is not bijective. This is highlighted by the fact that the Jacobian of the mapping $x_i = \chi_i(\xi_1, \xi_2, \xi_3)$ for $i = 1, 2, 3$ is singular. A valid element can therefore be defined as an element where the Jacobian of the mapping is strictly positive, i.e.,

$$J^e(\boldsymbol{\xi}) = \left| \frac{\partial \chi_i^e}{\partial \xi_j} \right| > 0, \quad \forall\ \chi_i^e(\boldsymbol{\xi}) \in \Omega^e,\ \boldsymbol{\xi} \in \Omega_{\rm st}\,.$$

From a purely implementation point of view, differentiation in an arbitrary element, as discussed in Section 4.1.3.4, involves derivatives with respect to the local

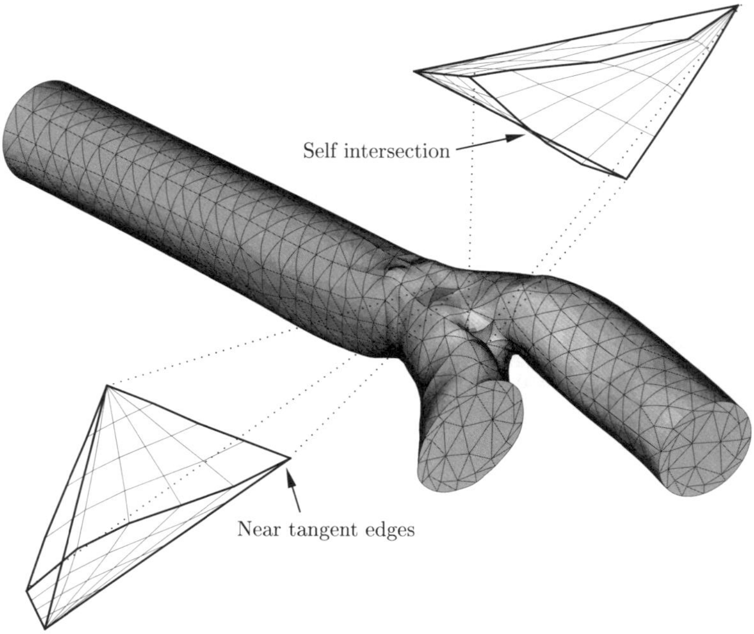

FIG. 4.24. Generation of high-order curvilinear elements from a coarse mesh. The deformation of straight-sided elements can lead to invalid elements being generated with singular mappings.

Cartesian coordinates divided by the Jacobian. A zero Jacobian therefore implies an infinite derivative, making the algorithm and element description invalid.

In general, we would like our mesh generation algorithm to take into account the possibility of generating invalid elements during the mesh construction process. However, such an approach is typically too complicated since it requires an algorithm to take into account all possible invalid generation scenarios. An alternative, more practical strategy is, therefore, to initially design a coarse mesh of straight-sided elements using as much information of the surface topology as possible. Subsequently, the coarse straight-sided mesh is deformed to conform to the curved boundary representation of the solution domain. Whilst a mesh of straight-sided elements may consist of purely valid elements, the deformation of these elements into curvilinear approximations can generate invalid elements, as shown in Fig. 4.24. Therefore, we need to employ strategies to minimise the generation of invalid elements such as curvature-driven mesh refinement, interior edge and face deformation, and the use of hybrid shape expansions.

In Section 4.3.4 we outline the basic concepts behind the geometry representation and mesh generation techniques for constructing a coarse mesh of straight-sided elements with vertices that conform to the geometry boundary. In Section 4.3.5 we then discuss how to deform the meshes by constructing local mappings to ensure boundary-conforming curvilinear elements in two and three dimensions.

4.3.4 *Global coarse meshing*

To understand the problem of curvilinear mesh generation we must first outline standard mesh generation techniques. We start by assuming that the geometry of the computational domain is defined through a *top-down* boundary representation. We will implicitly assume that the 'real' surface can be adequately represented by this boundary representation, which we subsequently treat as our exact definition of the solution domain boundary. As shown in Fig. 4.25, the concept of a top-down representation arises from the viewpoint that a volume is enclosed by a series of faces, which are themselves enclosed by a set of curves, which in turn are enclosed between two points. To ensure a complete boundary description each of these surfaces may only intersect one another along curves, and curves may only intersect one another at boundary points. Typically, these curves and surfaces are described using standard techniques of computer-aided design (CAD).

Many standard mesh generation techniques, such as advancing front, structured meshing, and Delaunay [462], follow a 'bottom-up' construction procedure. We will not discuss the specific details of the different types of mesh generation, but will simply outline the broad common strategy that all these techniques follow. For details on each of the different types of mesh generation techniques we

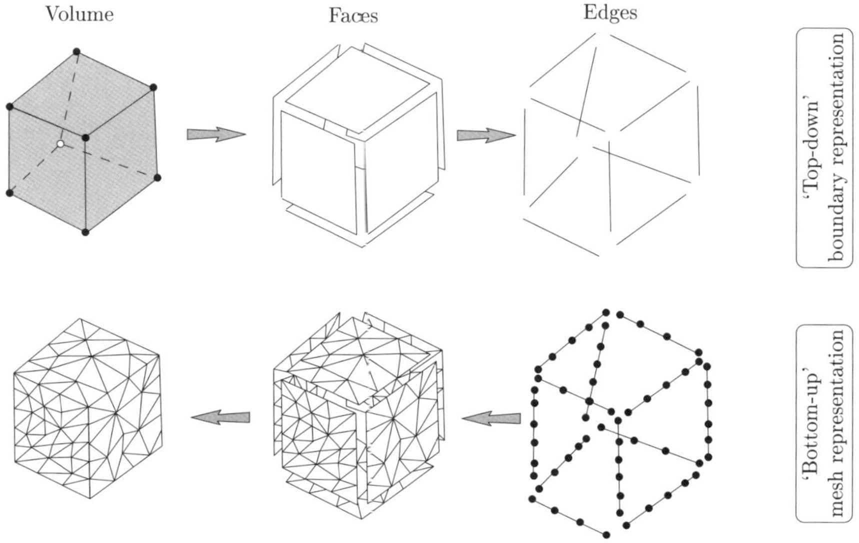

FIG. 4.25. Illustration of the 'top-down' boundary representation of a domain for use in a 'bottom-up' mesh generation strategy. (Courtesy of J. Peiró.)

point the reader to [462] and the references therein. As illustrated in Fig. 4.25, the 'bottom-up' approach initially discretises the edges of the boundary representation into discrete segments which conform to the points of the boundary representation. Every surface of the boundary representation is then bounded by a set of discretised edges, and so the next step of the generation is to develop a surface discretisation in terms of triangular or quadrilateral elements. Finally, the generation process is completed by constructing elements in the interior of the domain which comply to the face and edge definitions constructed in the previous steps. This way it is possible to construct a discretisation which conforms to the boundary requirements. To ensure a similar condition for curvilinear elements using spectral/*hp* element expansion we will follow a similar bottom-up strategy.

We note that different types of interior/volume discretisation are also possible. An example is shown in Fig. 4.24 where a *boundary-layer mesh* has been employed that produces a layer of high-aspect-ratio elements in a structured fashion adjacent to the geometry boundary. This type of discretisation can be very useful in viscous flows where boundary layers naturally occur as part of the physical solution. In this context, we are using some a priori information about the solution to dictate our meshing strategy. Although local refinement such as the boundary-layer meshing is useful from an approximation point of view, it can also produce more invalid curvilinear elements.

4.3.5 *High-order mesh generation*

We are now faced with the task of generating a curvilinear and boundary conforming mesh from a coarse mesh of straight elements. This coarse mesh contains vertices which conform to the points, curves, and faces of the boundary representation. The process of high-order mesh generation is to determine an elemental boundary transformation, in a bottom-up fashion. Having defined the boundary transformation, an elemental mapping from the standard region $\Omega_{\rm st}$ can be defined, as discussed in Section 4.1.3.2.

From our coarse mesh we know the location of the vertex points. The next stage is, therefore, to ensure that the edges of our element in two and three dimensions conform to the boundary representation, as discussed in Section 4.3.5.1. Subsequently, we can then define edges on the interior of the domain in two dimensions, as outlined in Section 4.3.5.2. In three dimensions, the interior edge evaluation is similar to the problem of determining edges lying within a boundary face. Finally, in three dimensions we also need to determine the interior face location. These three-dimensional construction issues are discussed in Section 4.3.5.3.

4.3.5.1 *Boundary-constrained edges in two dimensions*

The mappings involved in the definition of an edge constrained to a boundary segment are illustrated in Fig. 4.26. In general, we are seeking an elemental mapping $\chi_i(\xi_1, \xi_2)$ for $i = 1, 2$ which defines the transformation from the

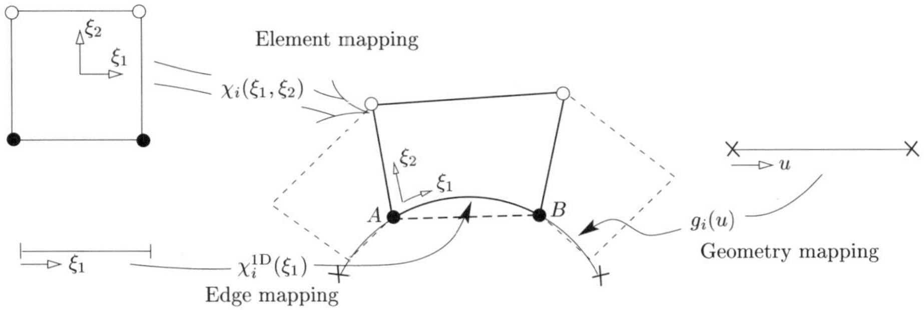

FIG. 4.26. Mappings involved in the generation of a curvilinear element. The mapping $\chi_i(\xi_1,\xi_2)$ defines the element to standard region transformation. This transformation requires the definition of the edge mapping $\chi_i^{1D}(\xi_1)$. Normally, a parametric mapping $g_i(u)$ will also exist which defines the geometry boundary.

boundary-conforming element to the standard region Ω_{st}. To define $\chi_i(\xi_1,\xi_2)$, however, we require the one-dimensional mapping $\chi_i^{1D}(\xi_1)$ which transforms the boundary-conforming edge of the physical element to one side of the standard region $-1 \leqslant \xi_1 \leqslant 1$. This mapping, combined with similar mappings for the other edges, would allow us to define $\chi_i(\xi_1,\xi_2)$, as discussed in Section 4.1.3.2 (note that $\chi_i^{1D}(\xi)$ was denoted as $f_i(\xi)$ in that section). To determine $\chi_i^{1D}(\xi)$ we need some information about the surface, which is typically defined in terms of another transformation $\boldsymbol{g}(u) = [g_1(u), g_2(u)]$. For example, if the boundary edge is a circle, then the mapping can be defined as $g_1(u) = R\cos u$, $g_2(u) = R\sin u$, where u is the azimuthal angle and R is the circle radius. Alternatively, $\boldsymbol{g}(u)$ might be expressed as a spline representation in terms of the local coordinate u. Typically, these geometrical definitions have been provided by the coarse mesh generator and are often C^1 continuous.

We next consider that our spectral/*hp* element approximation within an element is a polynomial of order P. For a consistent approximation of the geometry boundary we could also approximate the surface by a P-order polynomial expansion. Dey [133] and Dey *et al.* [135] argue that, to maintain optimal convergence of a P-order spectral/*hp* approximation to each integral, only a $P-1$ polynomial order approximation of the boundary is required.

Two constraints which we must impose on our polynomial approximation of $\chi_i^{1D}(\xi)$ arise because the vertex points common to adjacent edges ensure continuity of elements. To determine the mapping $\chi_i^{1D}(\xi)$ we, therefore, apply a collocation approximation which enforces the following conditions at the vertex points:

$$\boldsymbol{\chi}^{1D}(-1) = \boldsymbol{x}_A\,, \quad \boldsymbol{\chi}^{1D}(1) = \boldsymbol{x}_B\,,$$

where $\boldsymbol{x}_A$ and $\boldsymbol{x}_B$ are the coordinates of the end-points. For a P-order polynomial approximation we are now free to choose $P-1$ further conditions to define our mapping. These points can be determined by defining $P-1$ nodal points in

$-1 \leqslant \xi_1 \leqslant 1$ and $P-1$ corresponding points along the curvilinear edge. This point is illustrated in Fig. 4.27 where we show the discretisation of the same edge illustrated in Fig. 4.26. To define a $P = 4$ order Lagrange polynomial approximation for $\chi_i^{1D}(\xi_1)$ we require five nodal points. The first two points, denoted by the solid circles in Fig. 4.27, are provided by vertex locations. The three interior points, denoted by open circles, are to be determined. A reasonable choice for these collocation points in the standard region $-1 \leqslant \xi_1 \leqslant 1$ are the Gauss–Lobatto–Legendre integration points due to their favourable Lebesgue properties (see Section 3.3.1). We are, therefore, left with the task of determining the nodal points, $\boldsymbol{x}_i(\xi_1)$, along the curved boundary. There are a few possible methods we could apply.

An intuitive method for obtaining the surface collocation points is to initially determine the local Gaussian quadrature points along the straight line segment between the vertices. The desired collation points are then obtained by pushing these mapped quadrature points in the surface normal direction onto the geometry boundary. This technique is illustrated in Fig. 4.27(b). For a relatively well-behaved curved segment with curvature that does not vary significantly between the vertices, this produces reasonably distributed points, as shown in the same figure. Here, the cross points on the straight line segment represent the Gauss–Lobatto–Legendre points linearly mapped to the line A–B. These points are then pushed onto the surface in the normal direction. Problems arise with this technique when the curvature varies rapidly along the line segment. In this case the 'normal pushing' technique can produce a nodal distribution of points which

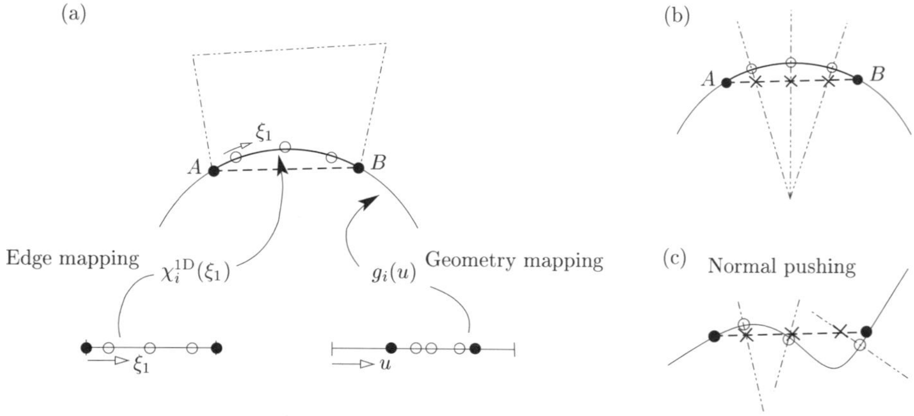

FIG. 4.27. (a) Defining the mapping, $\chi_i^{1D}(\xi_1)$, using a collocation projection between the standard region, $-1 \leqslant \xi_1 \leqslant 1$, and the geometry boundary segment A–B. (b) Ensuring that the vertices are at the end-points provides two collocation points (solid circles); however, we are free to define other collocation/nodal points (open circles) using normal pushing (as indicated in (b) and (c)) or throughout the surface parameter u space (as shown in (a)).

is highly irregular in spacing along the arc length of the curve, as illustrated in Fig. 4.27(c). A spacing along the arc length, which is significantly different from the spacing in the standard region (for example, the Gauss–Lobatto–Legendre integration points), will result in a highly nonlinear mapping $\chi_i^{1\mathrm{D}}(\xi_1)$. This can cause the Jacobian of the mapping to be singular and make the element invalid, thereby destroying the good convergence properties of the spectral/*hp* element approximation. We are, therefore, motivated to consider an 'optimal' transformation as one which minimises the distortion of the Jacobian of the mapping $\chi_i^{1\mathrm{D}}(\xi_1)$, $i = 1, 2$, as defined in Section 4.1.4. This criterion implies that the best mapping has a constant Jacobian, which is the case between a straight-sided element and the standard region under a linear mapping. A constant Jacobian mapping is also preferable in an approximation sense because a minimal quadrature order is required to ensure optimal integration of a polynomial integrand.

An alternative method of choosing the nodal points is to select points in the parametric space u which under the mapping $\boldsymbol{g}(u)$ fix the location of the nodal points on the geometry surface. The selection of the parametric nodal points u_k depends on the form of the mapping $\boldsymbol{x}_k = \boldsymbol{g}(u_k)$. An obvious initial distribution is to use the same parametric spacing in the u space as applied in the ξ_1 region. For many mappings this selection may be perfectly adequate and even optimal. For example, when the surface is defined as a circle, i.e., $g_1(u) = R\cos u$, $g_2(u) = R\sin u$, setting the collocation point in the u space can be shown to lead to a mapping $\chi_i^{1\mathrm{D}}(\xi_1)$ which has a constant Jacobian. However, for a more complex mapping $\boldsymbol{g}(u)$ this type of point selection can lead to a highly-distorted distribution of collocation points. In such a situation the mapping $\chi_i^{1\mathrm{D}}(\xi_1)$ can again become singular. A modification to this approach, proposed in [434], is to define a minimisation procedure to determine the parametric discrete points u_k that minimise the Jacobian of the mapping $\chi_i^{1\mathrm{D}}(\xi_1)$. In this work the definition of an optimal mapping was approximated as the linear spacing between two collocation points $\boldsymbol{x}_k = \boldsymbol{g}(u_k)$ on the curved geometry being the same as the spacing between the nodal points in the standard region. It is then possible to define a functional of the form

$$\mathcal{J}(u_1, \ldots, u_{P-1}) = \sum_{i=1}^{P-1} \frac{\|\boldsymbol{g}(u_{i+1}) - \boldsymbol{g}(u_i)\|^2}{\xi_{i+1} - \xi_i}, \tag{4.3.1}$$

where ξ_i, $i = 0, \ldots, P$, represent the Gauss–Lobatto–Legendre integration points. The minimisation of the function $\mathcal{J}$ provides the spacing of points u_k ($1 \leqslant k \leqslant P-1$) which gives a near-optimal spacing $\boldsymbol{x}_k = \boldsymbol{g}(u_k)$ in the sense of minimising the curve Jacobian. We note, however, that the nonlinearity of the mapping $\boldsymbol{g}(u)$ makes this a nonlinear optimisation problem.

Finally, we note that accounting for the surface curvature is an important prerequisite when constructing curvilinear elements. Mesh generation techniques exist (see Frey and George [167]) which take account of surface curvature as part of the mesh refinement strategy, and such an approach clearly has benefits when designing spectral/*hp* element meshes.

4.3.5.2 *Internal edges in two dimensions*

Having constructed a mapping for boundary-conforming edges, we require a definition of the edge mappings for all other edges interior to the solution domain. Since we know the vertex location of these internal edges from the coarse mesh description, and in the absence of any other information, using a linear mapping between the vertices for interior edges is the most obvious approach. Such an approach leads to the element shapes shown in Fig. 4.26 where only the bottom edge is deformed.

Applying linear mappings for interior edges is attractive since it simplifies the form of the elemental mapping of the interior elements which do not touch the solution domain boundary. This can be advantageous since the Jacobian of integral interior elements will not be influenced by any surface curvature. We note, however, that in cases where the structure of the solution is known the interior elements with curved boundaries may be advantageous from an approximation point of view. Nevertheless, a more important factor to consider is if an internal edge should be deformed to avoid the generation of invalid elements after surface edge deformation. This point is illustrated in Fig. 4.28 where we see the generation of a curvilinear mesh from a coarse mesh generation in a flower-type shape. Although this example is three-dimensional, the construction of elements within the planar surface of the surface mesh is identical to the generation of two-dimensional interior elements. Figure 4.28(a) shows a coarse straight-sided mesh, where a boundary-layer region has also been generated adjacent to the surface boundary. All elements in this mesh are valid. If we now deform the edges which touch the solution domain boundary then we observe in Fig. 4.28(b) that invalid elements are generated in regions where the surface is concave with respect to the local element (i.e., the curved edge reduces the local area of the straight-sided element). In general, these problems occur when the surface deformation is larger relative to the size of the element.

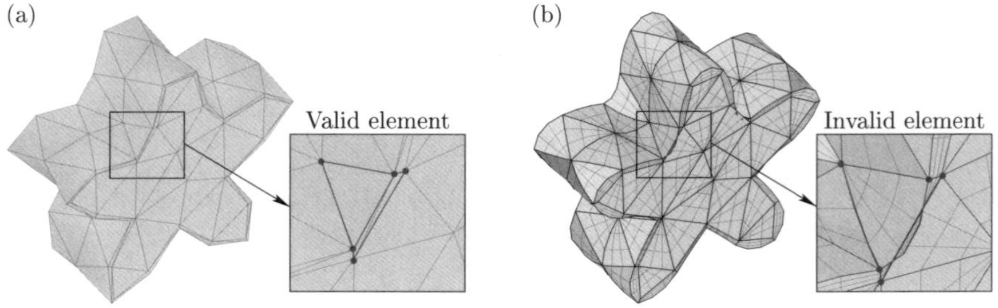

FIG. 4.28. Edge deformation of (a) a straight-sided coarse mesh leading to (b) invalid curvilinear meshes in regions of concave curvature relative to the element volume.

Invalid element detection

We are, therefore, faced with the problem of identifying invalid elements and applying a suitable modification to make them valid. One approach for detection is to numerically evaluate the Jacobian at a set of discrete points in the element and to identify if there is a sign change or ensure that the Jacobian is larger than zero by a suitable tolerance. Although such an approach cannot absolutely guarantee the validity of the element, since only a discrete set of points are evaluated, in practice, an inspection at quadrature points is normally sufficient. The approach can, however, be numerically quite expensive. Luo *et al.* [313] have proposed to use Bezier polynomials in their curvilinear meshing strategy. Bezier curves have a number of attractive features, including the property that the convex hull of the control points contains the Bezier curve. In [313] they used this property and the fact that the derivatives and products of Bezier curves can also be expressed as Bezier curves to determine the Jacobian as a Bezier form. It is then possible to determine if the Jacobian is greater than zero by ensuring that the expansion coefficients (control points) are greater than zero.

Eliminating invalid elements

Having identified the invalid elements, we are faced with the problem of deciding how to correct the invalid element. This point was addressed in the work of Dey [133] and Dey *et al.* [134] where they considered two approaches: edge swapping and interior edge deformation. Edge swapping has previously been used in optimising meshes [307], but in the curvilinear context can be applied if the edge swapping does not create any more invalid elements, otherwise an infinite swapping loop might be generated [134]. An example of effective edge swapping is shown in Fig. 4.29(a) where the triangle ABC is invalid due to the intersection of the curvilinear line A–B with the straight line A–C. In this case swapping edges A–C with B–D will make two valid triangles ABD and BDC. If edge swapping is not possible then the interior edge can be deformed. A strategy for quadratic-order edge deformation proposed by Dey *et al.* [134] is based on en-

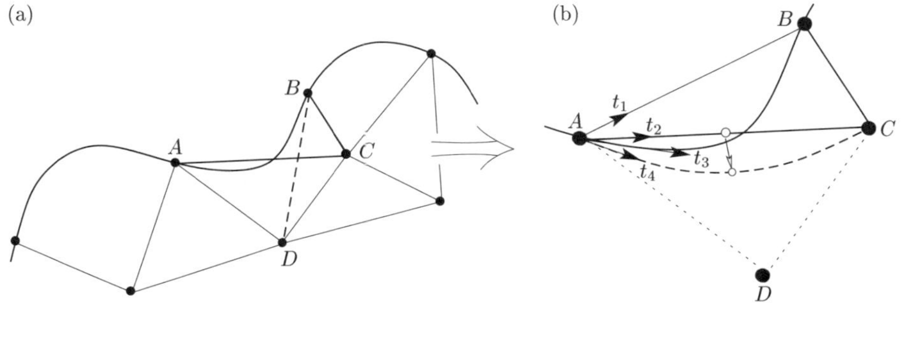

FIG. 4.29. Edge swapping and edge deformation to avoid the construction of invalid triangles.

suring that the normals of planes containing the straight edges have the same sign as the normals of the plane containing the curved edges. Therefore, in Fig. 4.29(b) we consider a case where edge AC is to be deformed by a quadratic fit and moving the midpoint in the normal direction of AC. We define $\boldsymbol{t}_1$ and $\boldsymbol{t}_2$ to be the tangent vector to the straight lines AB and AC, and $\boldsymbol{t}_3$ and $\boldsymbol{t}_4$ to be the tangent vector to the curved lines AB and AC. The criterion suggested by Dey *et al.* [134] is to move the point until the sign of the vector product $\boldsymbol{t}_4 \times \boldsymbol{t}_3$ is the same as that of $\boldsymbol{t}_2 \times \boldsymbol{t}_1$ or, equivalently,

$$(\boldsymbol{t}_2 \times \boldsymbol{t}_1) \cdot (\boldsymbol{t}_4 \times \boldsymbol{t}_3) > \epsilon \,,$$

where ϵ is a small constant.

Curvature-based refinement

We have observed that problems with invalid elements tend to arise due to excessive surface deformation relative to the size of the element. An alternative approach to eliminating invalid elements is to try to avoid their generation by controlling the coarse mesh element spacing relative to the surface deformation. Curvature-based refinement in which the mesh size is obtained as a function of the curvature has been proposed by several authors [167, 290] as a way to obtaining an accurate piecewise linear approximation of a curved surface. This type of approach has also been applied to curvilinear elements in [370] and uses a variation of the curvature-driven refinement of the coarse space mesh.

As shown in Fig. 4.30(a,b), in this technique a curve is locally approximated by a circle of radius R, the radius of curvature. We assume that the mesh spacing can be represented by a chord of length c in the circle and a spacing δ in the normal direction. In the modelling of viscous flows, the value of δ is usually prescribed to achieve a certain boundary-layer resolution. The value of c is, therefore, chosen to guarantee that the osculating circle representing the curve does not intersect the interior sides of the elements, i.e., $\theta \geqslant 90°$ for the triangular element. The value of c, which should be considered as a maximum mesh spacing, can now be obtained as a function of R and δ. Its value c_t for triangular elements is

$$c_t \leqslant R\sqrt{\frac{2\delta}{R+\delta}} \,. \tag{4.3.2}$$

The corresponding value c_q for quadrilateral elements is

$$c_q \leqslant \frac{2R\delta}{R+\delta}\sqrt{1+\frac{2R}{\delta}} \,, \tag{4.3.3}$$

where the displacement δ is assumed to be the same on either side of the rectangle. It is interesting to notice that, for a given δ, the quadratic element allows for a mesh spacing c_q which is about twice the value of the spacing c_t for the triangular element. If we apply the curvature-based refinement to the original

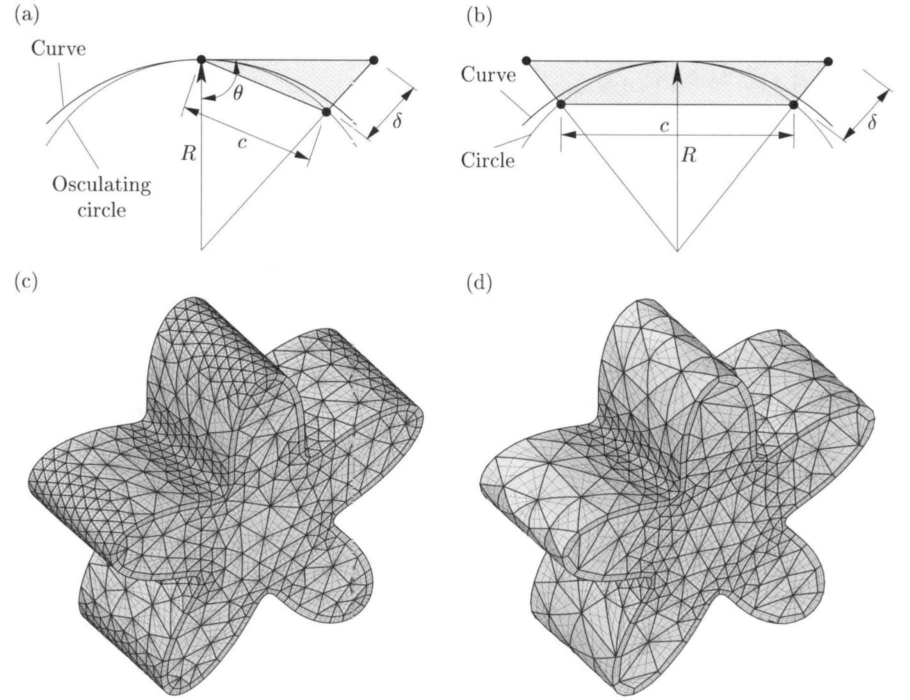

FIG. 4.30. Determination of the minimum spacing c based on the local radius of curvature R and the interior spacing δ for (a) triangular elements, and (b) quadrilateral elements. Application of curvature refinement to the problem of Fig. 4.28 using (c) indiscriminate curvature refinement, and (d) selective curvature refinement.

problem shown in Fig. 4.28 then we obtain a spacing which prevents the generation of invalid elements, as shown in Fig. 4.30(c,d). However, we note that indiscriminate application of the curvature criteria can generate too much refinement since invalid elements are only generated when the local curvature is concave with respect to the element volume. Selective refinement, where refinement is only applied in regions of the domain which are locally concave, provides a better mesh spacing, as shown in Fig. 4.30(d).

4.3.5.3 *Extension to three dimensions*

For many two-dimensional problems the techniques discussed in Sections 4.3.5.1 and 4.3.5.2 are often not necessary since in two dimensions meshes can often be 'hand crafted' to avoid invalid elements. However, in three dimensions this is not the case, partly due to the problems of even visualising computational meshes. It is in three dimensions that many of the more complicated meshing strategies are most beneficial. In this section we will outline how the concepts described in Sections 4.3.5.1 and 4.3.5.2 can be extended to three dimensions.

Boundary edges constrained to a surface

As we discussed in Section 4.3.5.1, when a CAD edge or surface is represented by an isometric mapping (i.e., those that preserve lengths) the procedure of defining the edge mapping $\chi_i^{1D}(\xi)$ based on the same distribution of collocation points in the u space as the ξ space (see Fig. 4.26) produces reasonably good quality elements. However, when the mapping is anisometric[1] this simple approach can produce highly-nonlinear mappings, leading to badly-shaped elements. A solution to avoid this problem is to apply an extension of the minimisation procedure similar to eqn (4.3.1). In a three-dimensional 'bottom-up' generation strategy the edges which lie on a boundary curve can be dealt with by optimising eqn (4.3.1). The next stage of the generation is to determine edges which lie within a boundary surface which is now represented by a two-dimensional parametric space $\boldsymbol{g}(\boldsymbol{u}) = [g_1(\boldsymbol{u}), g_2(\boldsymbol{u})]$, where $\boldsymbol{u} = [u, v]$. Equation (4.3.1), therefore, becomes a two-dimensional functional in terms of $[u, v]$ of the form

$$\mathcal{J}_e(\boldsymbol{u}_1, \dots, \boldsymbol{u}_{P-1}) = \sum_{i=1}^{P-1} \frac{\|\boldsymbol{g}(u_{i+1}, v_{i+1}) - \boldsymbol{g}(u_i, v_i)\|^2}{\xi_{i+1} - \xi_i}.$$

To determine the location of the points (u_i, v_i), for $i = 1, \dots, P-1$, we need to minimise $\mathcal{J}_e$, where once again ξ_i, $i = 0, \dots, P-1$, represents the nodal spacing in the standard region, for example, using Gauss–Lobatto–Legendre integration points. The minimisation of $\mathcal{J}_e$ provides a set of points, $\boldsymbol{u}_k$, which give a near-optimal spacing $\boldsymbol{x}_k = \boldsymbol{g}(u_k, v_k)$ in the sense of minimising the element surface Jacobian. The points $\boldsymbol{\xi}_k$ and $\boldsymbol{x}_k$ are then sufficient to generate a P-order polynomial approximation using a collocation projection. However, unlike the one-dimensional case, the edges are now geodesics of the surface. In [434] the minimisation procedure was also extended to determine the triangular face mapping for use in hybrid elements.

As a final illustration of the role of the mapping, we present the example shown in Fig. 4.31. We consider the generation of a tetrahedral spectral/*hp* mesh of fifth-order polynomials within a simple cubic computational domain $0 \leqslant x, y, z \leqslant 10$. The faces of the cube are located on tensor-product surfaces defined by an anisometric mapping shown in Fig. 4.31(a). The spacing varies linearly with a value of 0.1 at the boundary and 2.1 at the centre of the faces. The anisometry of the mapping is due to the fact that the unevenly-spaced network of lines depicted in Fig. 4.31(a) is obtained as the image of a network of coordinate lines $u =$ constant and $v =$ constant in the parametric plane which are uniformly spaced with $\Delta u = \Delta v = 1$. The curve definition of the edges representing the intersection between each of the faces was taken to be isometric.

A standard h-type unstructured mesh generation process was used to construct a coarse mesh with sixty-six elements. Although the p-type elements can easily be constructed on a planar surface by a linear interpolation between the

[1] Not isometric.

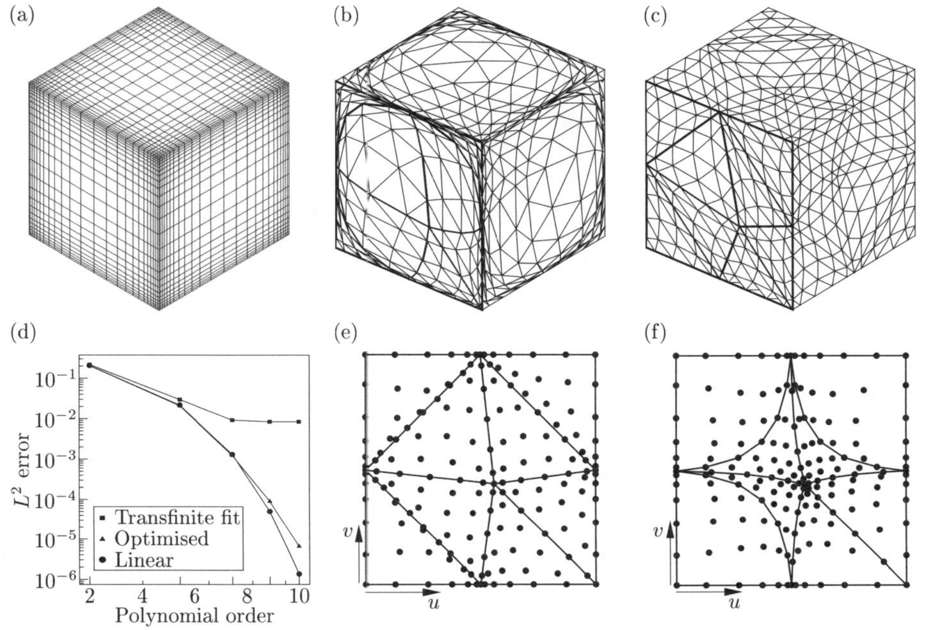

FIG. 4.31. Influence of surface and edge mappings on the spectral/*hp*-type mesh generation procedure. (a) Anisometric CAD description. (b) High-order mesh using a transfinite interpolation in the parametric space, as shown in (e). (d) High-order mesh using an optimised point placement in the parametric space, as shown in (f). (d) Comparison of errors in an elliptic problem using the different meshes.

vertices, we have chosen to reconstruct the surface elements using the parametric definition of the surface, as required for a non-planar surface. Therefore, in this geometry an optimal solution which would give elemental mappings with constant Jacobian is a linear distribution of points between the vertices.

Using a straightforward transfinite interpolation of points in the parametric space (u, v) results in the highly-distorted surface mesh shown in Fig. 4.31(b). In this figure we have connected all the interior edge and face nodal points using a triangular mesh and highlighted the element boundaries on one face with dark lines. The distribution of points in the parametric plane for the highlighted surface is shown in Fig. 4.31(e). If we now apply the optimisation procedure [434], then we obtain the surface mesh shown in Fig. 4.31(c) which produces a fit very close to a linear mapping between the vertices. A direct consequence of obtaining a good physical surface distribution is that the parametric distribution becomes very distorted, as shown in Fig. 4.31(f).

Finally, we compare the L^2 error to the solution $u(x, y, z) = \sin(0.2\pi x) \sin(0.2\pi y) \sin(0.2\pi z)$ of a Poisson equation to assess the influence of mesh distortion on the solution accuracy. Figure 4.31(d) compares the error of solutions obtained using computational meshes from a transfinite parametric fit (Fig. 4.31(b)) with those from an optimised parametric fit (Fig. 4.31(c)) and a standard linear fit between the physical vertices. The transfinite parametric fit leads to elements with a singular Jacobian, and so it is not surprising that the error quickly saturates at 1×10^{-2}. The optimised surface mapping, with a convergence tolerance in the parametric space of $\epsilon \propto 0.3/P^2$, follows the error of the linear fit up to a polynomial order of $P = 8$ where the error is of order 1×10^{-4}. For higher-order expansions the rate of convergence decreases when compared to the standard linear discretisation. However, increasing the convergence tolerance of the optimal iterative procedure reduces the saturation level.

Curvature-based refinement on a surface—hybrid meshing

The extension of the curvature-based refinement discussed in Section 4.3.5.2 to surfaces is relatively straightforward. The refinement criterion given by formulae (4.3.2) and (4.3.3) can be applied in the two principal directions of the surface, as illustrated in Fig. 4.32. The corresponding mesh spacings, c_1 and c_2, can then be calculated from the values of the principal curvatures $k_{1,2} = 1/R_{1,2}$ and the normal mesh spacing δ using formulae (4.3.2) and (4.3.3).

As noted previously, the minimum spacings c_1 and c_2 for a quadrilateral surface normal face are twice that of a triangular surface normal face. This leads us to the conclusion that *hexahedral* or *prismatic* elements adjacent to a curvilinear surface are less likely to generate invalid elements than tetrahedral elements. However, tetrahedral elements provide greater geometric flexibility in the interior of the domain. Therefore, a *hybrid* mesh containing a mix of these elements provides a good compromise between these two factors. Indeed, to generate boundary-layer meshes typically involves the extension in the surface normal direction of the surface discretisation. If the surface is discretised using triangles then the surface normal extension directly generates prismatic-shaped elements which can be subdivided into tetrahedrons if desired. However, direct use of the prismatic discretisation is favourable from an element validity point of view and

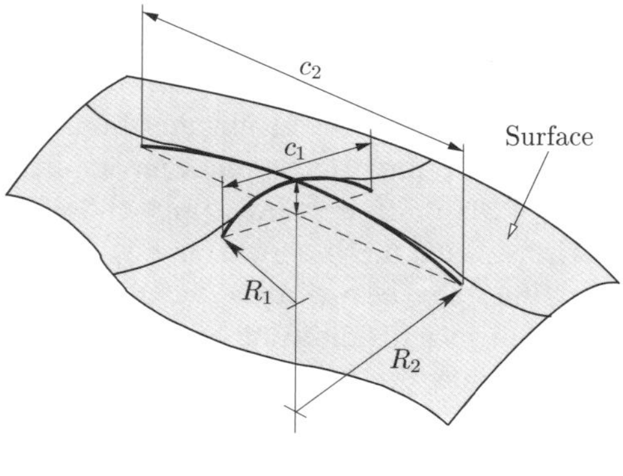

FIG. 4.32. Notation for curvature refinement in a surface.

local approximation. This type of strategy has been adopted in Figs 4.28 and 4.30, whereas a tetrahedral boundary layer was adopted in Fig. 4.24.

Interior element modifications

As discussed in the two-dimensional case, it is often attractive from an approximation point of view to maintain straight-sided elements when constructing three-dimensional elements. This arises simply as a consequence of the Jacobian of straight-sided elements requiring low-order polynomial approximations and so not affecting the quadrature order used in integral evaluations. Although surface optimisation and curvature-based refinement can reduce the potential for generation of invalid elements, it may still be necessary to deform interior edges and faces.

This point is illustrated in Fig. 4.33 due to Dey *et al.* [134] and Luo *et al.* [313]. Figure 4.33(a) shows an initial coarse straight-sided mesh of this complex domain representing an engineering component. Figure 4.33(b) shows the final high-order curvilinear mesh which reproduces the curved nature of the component. However, in order not to generate invalid elements the interior edges have to be curved, as shown in the close-up of the curvilinear mesh of the component in Fig. 4.33(c). Different strategies for interior modifications have been proposed in the work of Dey *et al.* [134, 135] and Luo *et al.* [313]. These strategies involved edge and face swapping and deletion, as well as interior edge and face deformation. The more recent work of this group has also involved the use of Bezier representation to produce a more automatic approach to invalid element

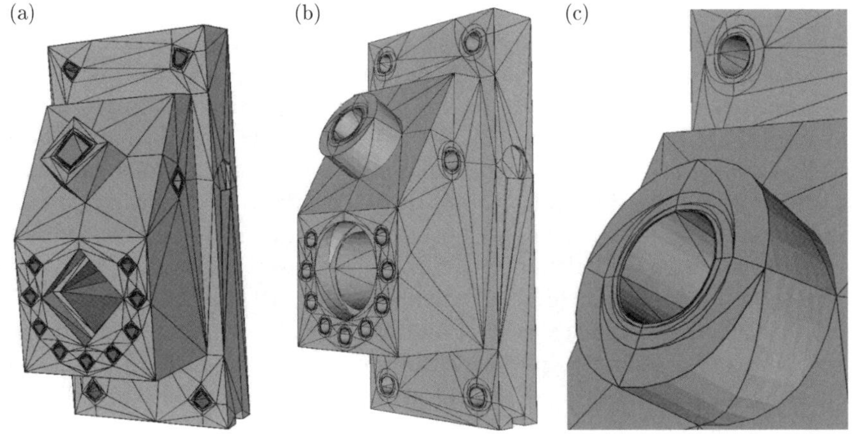

FIG. 4.33. High-order curvilinear mesh of an engineering component: (a) initial straight-sided coarse mesh, (b) high-order curved mesh conforming to curved geometry, and (c) close-up of a high-order mesh, demonstrating the interior element deformation required to prevent the generation of invalid elements. (Courtesy of M. Shephard.)

detection and modification [313].

4.3.6 *Particle tracking in spectral/hp element discretisations*

In the preceding section we discussed the formulation of spectral/hp element approximations for Eulerian descriptions of partial differential equations. These techniques will be applied to the advection equation in Chapter 6 and the Navier–Stokes equations in Chapters 8 and 10. Solving either the advection equation or the Navier–Stokes equations with a spectral/hp element formulation implies that we have a high-order polynomial approximation of the velocity field. A popular post-processing technique in fluid mechanics is to consider the streamlines, for steady flow, or the path lines, for unsteady flow. However, this requires being able to track particles over the high-order velocity field. One approach to implement particle tracking is to divide the macro element of the spectral/hp element discretisation into many small elements within which a linear approximation is applied. On this finer mesh a commercial particle tracker can be applied. For many post-processing requirements this may be satisfactory when high accuracy is not required. However, when developing algorithms such as the strong form of the semi-Lagrangian method discussed in Section 6.4, the error associated with such a linear approximation is inadequate.

The approximation of a spectral/hp element velocity field by piecewise linear polynomials on smaller elements can result in the inaccurate calculation of the trajectories, as shown in Fig. 4.34. Streamline integration is very sensitive to small changes in kinematics and therefore inconsistencies between the high-order and a linear velocity field approximation can result in kinematic changes that have a significant effect on the path lines in complex flow regions. Figure 4.34 compares the particle traces for a spectral/hp element velocity field calculated by the commercial package Tecplot [9] and the algorithm proposed in [109].

The focus of this section is, therefore, to review approaches to particle tracking using a consistent approximation of the velocity field. The problem of calculating flow lines for high-order polynomial element representations has received little attention to date. Two early examples of particle tracking within quadratic-order elements can be found in the literature of non-Newtonian flows [201, 444], where the strain history calculated along streamlines was used in the constitutive equations for the stress tensor. In Sun and Tanner [444] the time integration was applied in physical space on a triangular mesh, whereas in Goublomme *et al.* [201] time integration was performed in the parametric space of the standard region using unstructured quadrilateral meshes. Following [109], we will discuss both of these integration strategies in what follows.

4.3.6.1 *Runge–Kutta-based particle tracking*

The problem of finding the trajectory $\boldsymbol{x}(t)$ of a particle is formulated as a set of ordinary differential equations:

$$\frac{\mathrm{d}\boldsymbol{x}}{\mathrm{d}t} = \boldsymbol{u}\left(\boldsymbol{x}, t\right), \qquad (4.3.4)$$

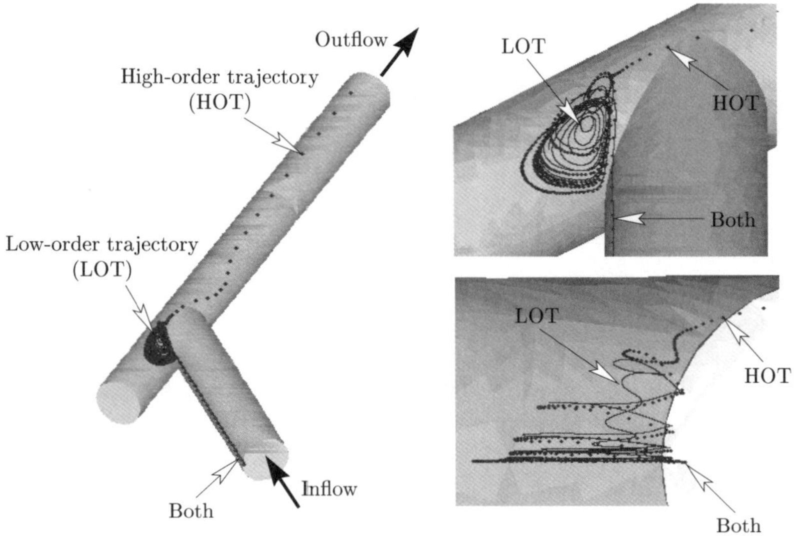

FIG. 4.34. Streamlines in a junction between two straight pipes. The high-order finite element mapping is based on a seventh-order polynomial expansion and the corresponding streamline is represented by the dotted line. The solid line is the streamline starting at the same point as calculated by a commercial package using linear interpolation on a subdivision of the high-order mesh. (Note that one high-order element was divided into forty-two linear elements.)

where $\boldsymbol{x}$ is the position in space and $\boldsymbol{u}$ is the velocity field. This simply states that the tangent to the trajectory curve at a point with coordinates $\boldsymbol{x}$ is parallel to the velocity $\boldsymbol{u}$ at that point. The initial condition for this problem amounts to specifying the position $\boldsymbol{x}_0$ of the particle at a specific time, $t = t_0$, such that

$$\boldsymbol{x}(t_0) = \boldsymbol{x}_0 .$$

The two main types of integration schemes for ordinary differential equations are multi-stage methods (mainly Runge–Kutta type) and multi-step methods (for example, Adams–Bashforth). Whilst multi-step methods are typically more efficient in problems where the velocity field is smooth, they require a start-up procedure and their time-step cannot be changed easily dynamically. In particle tracking, therefore, a multi-stage algorithm is attractive since it does not require a start-up procedure and the time-step can be altered dynamically.

The application of an s-stage explicit Runge–Kutta method to the general system of ordinary differential equations

$$\frac{\mathrm{d}\boldsymbol{y}}{\mathrm{d}t} = \boldsymbol{f}(\boldsymbol{y}, t),$$

with the initial conditions $\boldsymbol{y}(t_0) = \boldsymbol{y}_0$ results in the iteration

$$\boldsymbol{y}^{n+1} = \boldsymbol{y}^n + \Delta t \sum_{i=1}^{s} b_i \boldsymbol{f}_i \,, \tag{4.3.5}$$

where $\boldsymbol{y}^n$ denotes the value $\boldsymbol{y}(t^n)$,

$$\boldsymbol{f}_i = \boldsymbol{f}\left(\boldsymbol{y}^n + \Delta t \sum_{j=1}^{i-1} a_{ij} \boldsymbol{f}_j,\, t^n + c_i \Delta t\right), \tag{4.3.6}$$

s is the number of stages, and Δt is the time-step. The values b_i, c_i, and a_{ij} are the entries of the corresponding *Butcher array* [288]:

$$\begin{array}{c|ccc} c_1 & a_{11} & \cdots & a_{1s} \\ \vdots & \vdots & & \vdots \\ c_s & a_{s1} & \cdots & a_{ss} \\ \hline & b_1 & \cdots & b_s \end{array} \,.$$

The coefficients of this array define the particular scheme employed and some examples are given in Table 4.1.

The second term on the right-hand side of eqn (4.3.5) is a weighted average of the values $\boldsymbol{f}_i$ taken at each stage. Hence, if we set

$$\overline{\boldsymbol{f}} = \sum_{i=1}^{s} b_i \boldsymbol{f}_i$$

then a generic Runge–Kutta scheme, as represented by eqn (4.3.5), can be considered as an Euler scheme that marches in time using an averaged value of $\boldsymbol{f}$. The same consideration can be applied to each stage; hence the second term on the right-hand side of eqn (4.3.6) can be considered as an Euler step taken with average velocity

$$\hat{\boldsymbol{f}}_i = \sum_{j=1}^{i-1} a_{ij} \boldsymbol{f}_j \,.$$

This interpretation of the Runge–Kutta scheme is depicted in Fig. 4.35 for the three-stage scheme which shows how each stage of the Runge–Kutta integration can be envisaged as an Euler step in the direction of a suitably averaged velocity.

TABLE 4.1. Butcher arrays for a one-stage Euler method (RK1), two-stage improved Euler method (RK2), three-stage Kutta's formula (RK3), and four-stage classical Runge–Kutta (RK4) schemes.

RK1: $\begin{array}{c|c} 0 & 0 \\ \hline & 1 \end{array}$ RK2: $\begin{array}{c|cc} 0 & & \\ 1 & 1 & \\ \hline & \frac{1}{2} & \frac{1}{2} \end{array}$ RK3: $\begin{array}{c|ccc} 0 & & & \\ \frac{1}{2} & \frac{1}{2} & & \\ 1 & -1 & 2 & \\ \hline & \frac{1}{6} & \frac{2}{3} & \frac{1}{6} \end{array}$ RK4: $\begin{array}{c|cccc} 0 & & & & \\ \frac{1}{2} & \frac{1}{2} & & & \\ \frac{1}{2} & 0 & \frac{1}{2} & & \\ 1 & 0 & 0 & 1 & \\ \hline & \frac{1}{6} & \frac{1}{3} & \frac{1}{3} & \frac{1}{6} \end{array}$

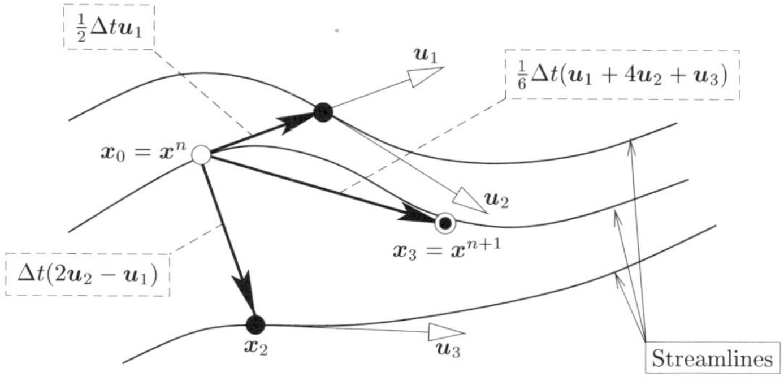

FIG. 4.35. Interpretation of the Runge–Kutta integration as a series of Euler steps using a suitably averaged velocity. The figure shows the steps for a three-stage scheme.

4.3.6.2 *Particle tracking for spectral/hp elements*

Two possible strategies for tracking particles depend on the space in which the time integration is performed [109]. The first strategy is to perform the time integration in the global physical space of the solution $\boldsymbol{x} \in \Omega$. Given the elemental definition of the flow field, this process involves searching for the element containing the point where the velocity is to be evaluated, followed by the interpolation of the velocity using the expansion basis within the element. The two computationally intensive operations to be performed are

(i) a nonlinear iterative procedure to find the local coordinates $\boldsymbol{\xi}_i$ in the parametric space from the Cartesian coordinates $\boldsymbol{x}_i$ in physical space, and

(ii) the interpolation of the velocity $\boldsymbol{u}$ at a point with parametric coordinates $\boldsymbol{\xi}_i$.

The second strategy performs the time integration in the parametric space of the standard element $\boldsymbol{\xi} \in \Omega_{\mathrm{st}}$. This approach involves advancing the particle within an element using a transformed velocity field in the parametric space until the particle reaches the element boundary. The approach is then continued in the neighbouring element sharing the boundary where the particle exits. The two main operations to be performed are

(i) the interpolation of the velocity $\boldsymbol{u}_{\xi}$ at a point with coordinates $\boldsymbol{\xi}_i$ in the parametric space, and

(ii) a nonlinear iterative procedure to find the intersection of a path line with an elemental boundary.

In the following sections we will provide an overview of the two approaches and also discuss ways of combining the methods.

Particle tracking in the physical space

We recall that the explicit Runge–Kutta scheme (4.3.5) applied to the particle trajectory equation (4.3.4) results in

$$\boldsymbol{x}^{n+1} = \boldsymbol{x}^n + \Delta t \sum_{i=1}^{s} b_i \boldsymbol{u}_i\,, \quad \boldsymbol{u}_i = \boldsymbol{u}\left(\boldsymbol{x}^n + \Delta t \sum_{j=1}^{i-1} a_{ij} \boldsymbol{u}_j,\, t^n + c_i \Delta t\right).$$

As mentioned previously, there are two computationally intensive steps. The first is to determine the inverse mapping $\boldsymbol{\xi}_i = \boldsymbol{\chi}^{-1}(\boldsymbol{x})$ and the second is to interpolate the velocity field to obtain u_i. In the above algorithm this is necessary for every substep which depends on the order of the scheme. However, before performing either of these operations it is necessary to know within which element the point x_i lies.

In the case of elements with linear mappings the inverse mapping $\boldsymbol{\xi} = \boldsymbol{\chi}^{-1}(\boldsymbol{x})$ is analytic. However, for curvilinear elements, or even bilinear and trilinear mapping of straight-sided quadrilaterals and hexahedral elements the computational expense is significantly higher. To appreciate why, we need to consider a way of determining the inverse mapping. One approach to determine the coordinates $\boldsymbol{\xi}_i$ such that $\boldsymbol{\chi}^e(\boldsymbol{\xi}_i) = \boldsymbol{x}_i$ is to formulate this as the problem of finding a zero of a function $\boldsymbol{F}(\boldsymbol{\xi}_i)$, where

$$\boldsymbol{F}(\boldsymbol{\xi}_i) = \boldsymbol{\chi}^e(\boldsymbol{\xi}_i) - \boldsymbol{x}_i\,. \tag{4.3.7}$$

The Newton–Raphson iteration [269] applied to eqn (4.3.7) can be written as

$$\boldsymbol{J}_e \cdot \left[\boldsymbol{\xi}_i^{k+1} - \boldsymbol{\xi}_i^k\right] = -\boldsymbol{F}(\boldsymbol{\xi}_i^k)\,, \quad \boldsymbol{J}_e = \frac{\partial \boldsymbol{F}}{\partial \boldsymbol{\xi}} = \frac{\partial \boldsymbol{\chi}^e}{\partial \boldsymbol{\xi}}\,,$$

where k represents an iteration counter, i denotes the discrete point, and $\boldsymbol{J}_e$ denotes the Jacobian of the mapping. In three dimensions the above procedure, per iteration, requires three interpolation (or backward transform) operations to determine $\boldsymbol{\chi}^e(\xi_i)$, and additionally nine interpolation operations to evaluate $\boldsymbol{J}_e^{-1}$ at $\boldsymbol{\xi}_i$. We note that $\boldsymbol{J}_e$ can be inverted analytically, as discussed in Section 4.1.3.4.

The potential cost of this evaluation of the inverse mapping puts a high cost on physical space particle tracking using high-order elements. The efficiency of the searching procedure to determine in which element a point, $\boldsymbol{x}_i$, lies, and therefore reduce the potential algorithm cost, can be improved by using appropriate data structures. If the number of elements in the mesh is large, tree structures [65] could be used to find the element containing the starting-point of a trajectory. In the data structuring of many spectral/*hp* element methods connectivity information between elements is stored and so can be used to identify adjacent elements. In [498] a more intensive check was adopted where the desired physical location, $\boldsymbol{x}_i$, was checked against the normal of every discrete coordinate point known along the element boundary at the quadrature points. This approach can guarantee whether a point lies in a straight-sided element, eliminating any erroneous searches and inverse iterations. However, it is not a sufficient test for a curvilinear element, and further checks may be necessary.

Particle tracking in the parametric space

An alternative method for particle tracking that avoids the iterative solution of eqn (4.3.7) is to use the parametric space description of the velocity. Rather than considering the rate of change of the position of a particle in physical space $\boldsymbol{x}$, we can transform the velocity field onto the standard region, $\boldsymbol{\xi} \in \Omega_{\mathrm{st}}$. We then obtain an equation representing the corresponding rate of change in time of the particle position in the parametric space within element e as

$$\frac{\mathrm{d}\boldsymbol{\xi}^e}{\mathrm{d}t} = \boldsymbol{u}_{\xi}^e(\boldsymbol{\xi}, t). \tag{4.3.8}$$

The transformed velocity $\boldsymbol{u}_{\xi}^e$ in the standard element can be evaluated in terms of the physical velocity $\boldsymbol{u}$ by applying the chain rule to each of its scalar components, so that

$$u_{\xi_i} = \frac{\mathrm{d}\xi_i}{\mathrm{d}t} = \frac{\partial \xi_i}{\partial x_1}\frac{\mathrm{d}x_1}{\mathrm{d}t} + \frac{\partial \xi_i}{\partial x_2}\frac{\mathrm{d}x_2}{\mathrm{d}t} + \frac{\partial \xi_i}{\partial x_3}\frac{\mathrm{d}x_3}{\mathrm{d}t}, \qquad i = 1, 2, 3,$$

which can be written in matrix form as

$$\boldsymbol{u}_{\xi}^e = \frac{\partial \boldsymbol{\xi}}{\partial \boldsymbol{\chi}^e}\boldsymbol{u}, \tag{4.3.9}$$

where we note that $\partial\boldsymbol{\xi}/\partial\boldsymbol{\chi}^e$ are the standard geometrical derivatives used to differentiate in an arbitrary region, and so are usually known at the quadrature points (see Section 4.1.3.4).

Omitting the index e for simplicity, the use of an s-stage Runge–Kutta method for the integration in time of eqn (4.3.8) leads to

$$\boldsymbol{\xi}^{n+1} = \boldsymbol{\xi}^n + \Delta t \sum_{i=1}^{s} b_i \boldsymbol{u}_{\xi_i}, \qquad \boldsymbol{u}_{\xi_i} = \boldsymbol{u}_{\xi}\left(\boldsymbol{\xi}^n + \Delta t \sum_{j=1}^{i-1} a_{ij}\boldsymbol{u}_{\xi_j}, t^n + c_i \Delta t\right). \tag{4.3.10}$$

We note, however, that eqn (4.3.9) can only be applied within the corresponding elemental region since the representation of parametric velocity, $\boldsymbol{u}_{\xi}^e$, and the geometric derivatives $\partial\boldsymbol{\xi}/\partial\boldsymbol{\chi}^e$ are only piecewise continuous. If we can determine the intersection of the particle trajectory with the boundary of the standard element, Ω_{st}, then the continuity of the local coordinates along a boundary can be used to advance the particle to the next adjacent element. Such an approach requires replacing Δt in eqn (4.3.10) with the time-step for the particle to reach the boundary $\Delta\tau$.

As noted in [109], the parametric approach has eliminated the nonlinear inverse mapping evaluation using a physical space approach. We are, however, now faced with the problem of determining the intersection of the trajectory with the boundary of the standard element. If we use an Euler scheme ($s = 1$), then the intersection with the boundary is easily determined since it is linearly dependent on Δt. For a multi-stage scheme ($s > 1$), we see, by inspection of eqn (4.3.10), that the point $\boldsymbol{\xi}^{n+1}$ is a nonlinear function of the time-step. This means that the calculation of the time-step $\Delta\tau$ required to move a particle exactly to the boundary is also a nonlinear problem. Although an iterative scheme can be devised to solve this problem, elemental mappings which contain singular factors are observed to be ill-conditioned, making this approach in curvilinear spectral/*hp* element methods unattractive [109].

Guided search approach to particle tracking

The previous sections have highlighted several problems in the implementation of the particle tracking algorithm in the physical and parametric spaces. The main weakness of the physical space approach is the need to solve the nonlinear inverse mapping problem. This deficiency can be overcome by using a time-integration scheme in the parametric space, but at the expense of requiring the solution of the nonlinear problem of finding the intersection of the trajectory with the elemental boundary. To overcome these problems a combined approach was proposed in [109] where the velocity is predominantly evaluated in physical space but utilises the parametric space. Such an approach eliminates the inverse mapping iteration at each substep, although it introduces an error associated with the variation of the Jacobian of the mapping.

The philosophy behind the *guided search* is to have a particle leaving an element in the parametric space, eliminating the need to resort to an iterative procedure to obtain $\boldsymbol{\xi}_i = \boldsymbol{\chi}^{-1}(\boldsymbol{x}_i)$. Since the Runge–Kutta scheme can be considered as a series of linear substeps, in this approach we take a series of linear substeps in the parametric space instead of the physical space. This approach is illustrated in Fig. 4.36 where we consider a step starting at point P in the physical space. A linear step in physical space $\Delta\boldsymbol{x} = \boldsymbol{v}\Delta t$ would take the particle to point Q. We then require the local parametric coordinate of point Q in order to proceed. In the guided search, the parametric point P' is advanced by a linear substep $\Delta\boldsymbol{\xi} = \boldsymbol{v}_{\xi}^{e_1}\Delta\tau_{e_1}$ based on the local parametric velocity $\boldsymbol{v}_{\xi}^{e_1}$. In general, the point will not remain within an element. The time taken for the point to meet a boundary of the parent element (point R' in Fig. 4.36) is $\Delta\tau_{e_1} \leqslant \Delta t$. Since the step is linear and the boundary is planar, the intersection can be evaluated analytically. To complete the guided search, a new parametric velocity $\boldsymbol{v}_{\xi}^{e_2}$ is then evaluated at point R' in element e_2. The point is then linearly advanced through element e_2 over a time $\Delta\tau_{e_2}$, which is evaluated as the time for the particle to reach S'. To complete the step an analogous procedure is followed in element e_3 using a new parametric velocity $\boldsymbol{v}_{\xi}^{e_3}$. The particle is then linearly advanced a time $\Delta\tau_{e_3}$ such that $\Delta t = \Delta\tau_{e_1} + \Delta\tau_{e_2} + \Delta\tau_{e_3}$.

This procedure significantly reduces the computation time required to trace a particle and overcomes the problems posed by the iterative solution of the nonlinear problems associated with the two previous particle tracking strategies.

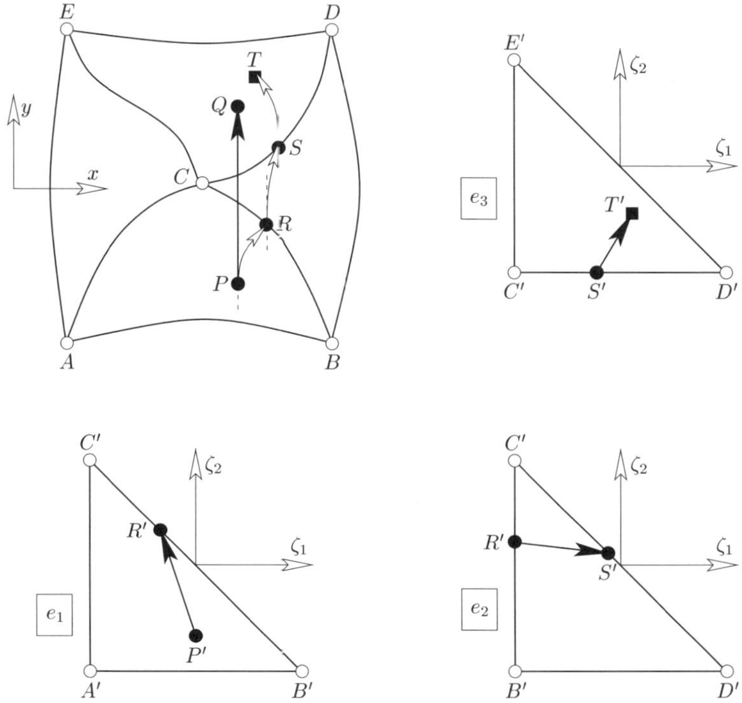

FIG. 4.36. Illustration of the guided search algorithm. The vector PQ represents the step in the physical space parallel to the global averaged velocity $\boldsymbol{v}_P$ evaluated at P. The straight segments $P'R'$, $R'S'$, and $S'T'$ are steps in the parametric space parallel to the transformed velocity $\boldsymbol{v}_\xi^e$ at point P. The local velocity $\boldsymbol{v}_\xi$ is evaluated at the points P', R', and S' for elements e_1, e_2, and e_3, respectively. The path $PRST$ is the image in physical space of the piecewise linear steps taken in the parametric space.

The computational saving is a consequence of the fact that each small substep of the above example only requires three interpolation operations to evaluate $\boldsymbol{v}_\xi$ at each boundary intersection. This should be compared to twelve interpolation operations required in the Newton–Raphson iteration. If many substeps are necessary due to multiple element-boundary crossing, then the guided search may still be expensive, but in general only one or two substeps are typically required. The guided search is exact when applied to elements with a constant Jacobian, but an error arises when the trajectory crosses elements with varying Jacobian. Often, high-order schemes use linear element mappings when dealing with straight-sided elements, and so varying Jacobians are usually associated with curvilinear elements.

4.3.6.3 *Examples of particle tracking schemes*

We now have a variety of possible strategies to handle particle tracking within high-order spatial representations and so we compare the following four approaches.

1. Particle tracking in the physical space evaluating the inverse mapping using a Newton–Raphson iteration as discussed in Section 4.3.6.2. We will denote this scheme as the *physical space* algorithm.
2. Particle tracking in the physical space using the guided search algorithm discussed in Section 4.3.6.2. We will denote this scheme as the *guided search* algorithm.
3. Particle tracking in the physical space using the guided search algorithm (see Section 4.3.6.2) and checking the error between the physical space advancement and the guided search. This allows the error introduced by curved elements to be monitored and requires an error tolerance ϵ above which the iterative technique to evaluate the inverse mapping is applied. We will refer to this scheme as the *guided search* (ϵ) algorithm.
4. Finally, a hybrid scheme where the particles are advanced in the parametric space, as discussed in Section 4.3.6.2, provided that they remain within the element during all substeps of the Runge–Kutta algorithm. If during a substep the particle leaves the elemental region then the physical space scheme using the error-checked guided search is applied. We will refer to this scheme as the *hybrid* algorithm.

In all of the above schemes care should be taken to consider the role of numerical precision on the criterion to determine whether a point lies within an element or when to say a point lies within an elemental face. For the interface between two curvilinear elements it is possible to get into an infinite loop if the velocity is almost tangent to the face which is only defined up to numerical round-off. Further discussion on intersection criteria can be found in [109].

In the following tests a range of schemes are considered, including Euler/RK1, RK2, RK3, and RK4, using the meshes shown in Fig. 4.37 which contain thirty-seven prismatic elements adjacent to the boundary and forty-six tetrahedral elements in the rest of the domain. The curvature of the surface is represented by positioning one of the triangular faces of the prismatic elements on the cylindrical surface, as shown in Fig. 4.37(a). The elemental boundary curvature can be removed to obtain a linear surface representation, as shown in Fig. 4.37(b). Within this domain an analytic unsteady solution, previously used in [122], was adopted of the form

$$u = -x\,, \quad v = -0.1y\,, \quad w = -20z\mathrm{e}^{-0.1t}\,,$$

which corresponds to a particle location at time t of

$$x(t) = x_0\mathrm{e}^{-t}\,, \quad y(t) = y_0\mathrm{e}^{-0.1t}\,, \quad z(t) = z_0\exp\left[200\left(\mathrm{e}^{-0.1t} - 1\right)\right],$$

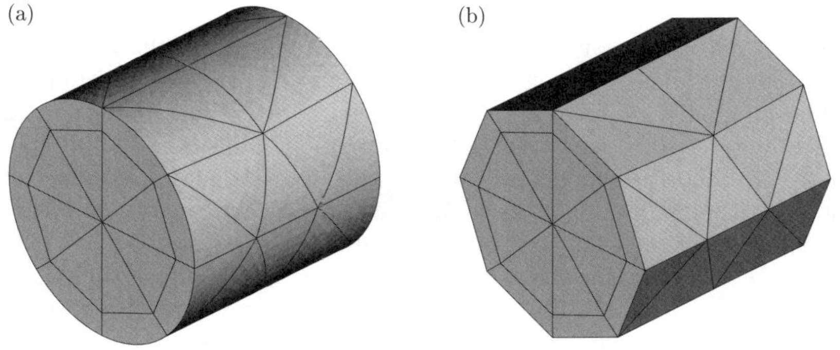

FIG. 4.37. Mixed prismatic and tetrahedral meshes using eighty-three elements within a cylindrical pipe: (a) curved elements, (b) straight-sided elements.

where x_0, y_0, z_0 are the initial coordinates of the particle. The starting-point was taken to be $x_0 = 0.5$, $y_0 = 0.25$, $z_0 = 0.35$, and since the solution of this system is relatively stiff a relatively short final time $T = 0.2$ was considered. Figure 4.38(a) shows a comparison of the convergence rate of the guided search algorithm with error checking, using a tolerance $\epsilon = 10^{-12}$, and the physical space scheme for all the Runge–Kutta schemes. The error in these tests is measured as the distance between the final location of the particle and the analytic solution relative to the exact value.

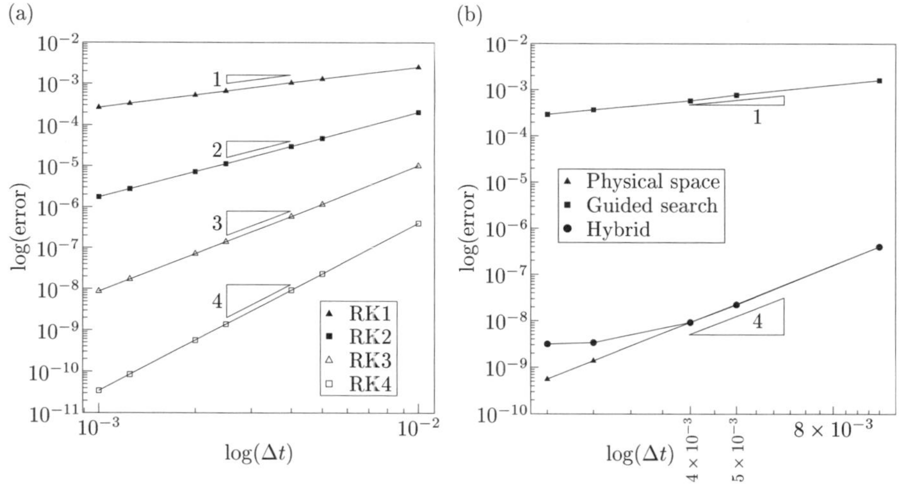

FIG. 4.38. (a) Temporal convergence for different Runge–Kutta schemes using an analytic solution with the physical space and guided search ($\epsilon = 10^{-12}$) algorithms. (b) Temporal convergence for the RK4 scheme using the physical space, guided search, and hybrid algorithms.

Figure 4.38(b) shows the convergence rate for three schemes using the RK4 time integration, where we observe that the guided search algorithm with no error checking only produces a linear convergence rate. Since the trajectory determined by the guided search is influenced by the nonlinear elemental mapping, the deterioration of convergence is to be expected. We note, however, that the hybrid algorithm maintains a fourth-order convergence rate until a level of 10^{-9}, where the error of the elemental mapping saturates the results.

To compare the relative merit of each scheme, Fig. 4.39 shows timings for two numerical experiments. In both cases, a circular ring of particles was released within the computational domain where the velocity was set to be the numerical solution to the Poiseuille flow. All the tests were performed using the RK4 scheme over 100 time-steps, with a time-step of $\Delta t = 0.0125$. In the first test, shown in Fig. 4.39(a), we consider a ring of diameter $0.45D$ chosen to guarantee that all particles remain within the tetrahedral mesh. Since all these elements have linear mappings the results indicate that there is practically no difference between the computational cost of the different algorithms for a fixed polynomial order. The scaling within this region is approximately $\mathcal{O}(P^{1.6})$ and is well below the asymptotic scaling value of $\mathcal{O}(P^3)$ expected when the interpolation of the velocity field dominates.

Releasing a ring of particles of a larger diameter, $0.9D$, produces a significant difference in the timings included in Fig. 4.39(b). These particles now travel within the curved prismatic region of the computational domain and are therefore more sensitive to the nonlinear mapping introduced by the deformation of

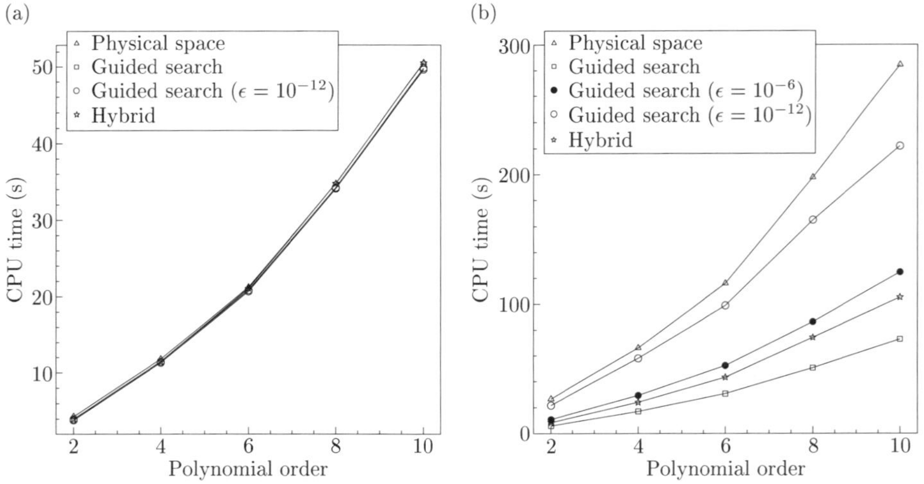

FIG. 4.39. (a) Time to march 100 particles on a circle of radius $0.45D$ through a mesh of tetrahedral elements. (b) Time to march 100 particles at a radius of $0.9D$ through a region discretised by prismatic elements.

the elements. In this example the physical space particle tracking is the most costly. It is approximately four times more expensive than the guided search algorithm without error checking. If we introduce error checking in the guided search algorithm, the cost depends on the error tolerance ϵ.

4.4 Exercises: implementation of a two-dimensional spectral/*hp* element solver for a global projection problem using a C^0 Galerkin formulation

To complement the discussion of local and global spectral/*hp* element operations ♦32
we propose the following series of exercises to help develop a two-dimensional spectral/*hp* element solver. In Section 3.5 we have already suggested some exercises to develop local elemental mass matrices for triangular and quadrilateral tensorial expansions. In this section we will build upon these exercises and develop a global C^0 continuous projection operator, as well as discuss how to enforce Dirichlet boundary conditions. In Section 5.6 we will then finally extend the multi-dimensional projection operator to a two-dimensional elliptic solver.

Some useful codes are also available on the web page

`http://www.nektar.info/2nd_edition`

1. Similar to Section 2.6, a good starting-point for the development of the solver is to set up routines to perform numerical integration in both the standard regions and a general element. We recall from Section 4.1.1 that for integration in the standard region we require the quadrature weights w_i, w_j and zeros z_{1i}, z_{2j}, which can either be generated as discussed in Appendix B.2 or using the code in the *Polylib* library from the web page. Having obtained the zeros and weights, and following Sections 4.1.1.1 and 4.1.1.2, try the following exercises.
   (a) Integrate $u(\xi_1, \xi_2) = [\xi_1]^6 \times [\xi_2]^6$ numerically as:
   $$\int_{-1}^{1}\int_{-1}^{1} u(\xi_1,\xi_2)\, d\xi_1\, d\xi_2 = \sum_{i=0}^{Q-1}\sum_{j=0}^{Q-1} w_i w_j\, u(\xi_{1i},\xi_{2j}) = \frac{4}{49},$$
   using Gauss–Lobatto–Legendre quadrature in both the ξ_1 and ξ_2 directions with $Q = 4$, 5, and 6.
   (b) Integrate $u(\xi_1, \xi_2) = [\xi_1]^6 \times [\xi_2]^6$ numerically as:
   $$\int_{-1}^{1}\int_{-1}^{-\xi_2} u(\xi_1,\xi_2)\, d\xi_1\, d\xi_2 = \int_{-1}^{1}\int_{-1}^{1} u(\eta_1,\eta_2)\frac{1-\eta_2}{2}\, d\eta_1\, d\eta_2$$
   $$= \sum_{i=0}^{Q-1}\sum_{j=0}^{Q-1} w_i w_j\, u(\eta_{1i},\eta_{2j})\frac{1-\eta_{2j}}{2} = \frac{2}{49},$$

[32] Exercises to help implement a two-dimensional spectral/*hp* element solver/projection operator.

where $\eta_1 = 2(1+\xi_1)/(1-\xi_2) - 1$, $\eta_2 = \xi_2$, and using Gauss–Lobatto–Legendre quadrature in the η_1 direction and Gauss–Radau–Legendre quadrature (including the point $\eta_2 = -1$) in the η_2 direction with $Q = 4$, 5, and 6. Note that the factor $(1-\eta_2)/2$ could also be incorporated into the integration by using Gauss–Radau–Jacobi–(1,0) quadrature, see Section 4.1.1.2.

(c) Consider a general straight-sided quadrilateral element, Ω^e, defined to have vertices $(x_1^A, x_2^A) = (0,0)$, $(x_1^B, x_2^B) = (1,0)$, $(x_1^C, x_2^C) = (2,1)$, and $(x_1^D, x_2^D) = (0,1)$. Numerically integrate $u(x_1,x_2) = [x_1]^6 \times [x_2]^6$, i.e.,

$$\int_{\Omega^e} u(x_1,x_2)\,\mathrm{d}x_1\,\mathrm{d}x_2 = \int_{\Omega_{\rm st}} u(\xi_1,\xi_2)|J|\,\mathrm{d}\xi_1\,\mathrm{d}\xi_2$$
$$= \sum_{i=0}^{Q-1}\sum_{j=0}^{Q-1} w_i w_j\, u(\xi_{1i},\xi_{2j})|J(\xi_{1i},\xi_{2j})|\,,$$

using Gauss–Lobatto–Legendre quadrature in both the ξ_1 and ξ_2 directions with $Q = 4$, 5, and 6. You will need to determine the local mapping $\chi_i(\xi_1,\xi_2)$, as discussed in Section 4.1.3, and calculate the Jacobian, as discussed in Section 4.1.3.3. For this case the Jacobian can either be evaluated analytically or numerically using the differentiation techniques discussed in Section 4.1.2.1. Also note that $u(\xi_1,\xi_2) = [\chi_1(\xi_1,\xi_2)]^6 \times [\chi_2(\xi_1,\xi_2)]^6$.

(d) Consider a general straight-sided triangular region, Ω^e, defined to have vertices $(x_1^A, x_2^A) = (1,0)$, $(x_1^B, x_2^B) = (2,1)$, and $(x_1^C, x_2^C) = (1,1)$. Numerically integrate $u(x_1,x_2) = [x_1]^6 \times [x_2]^6$, i.e.,

$$\int_{\Omega^e} u(x_1,x_2)\,\mathrm{d}x_1\,\mathrm{d}x_2 = \int_{\Omega_{\rm st}} u(\eta_1,\eta_2)\frac{1-\eta_2}{2}|J|\,\mathrm{d}\eta_1\,\mathrm{d}\eta_2$$
$$= \sum_{i=0}^{Q-1}\sum_{j=0}^{Q-1} w_i w_j\, u(\xi_{1i},\xi_{2j})\frac{1-\eta_{2j}}{2}|J(\xi_{1i},\xi_{2j})|\,,$$

using Gauss–Lobatto–Legendre quadrature in the ξ_1 direction and Gauss–Radau–Legendre quadrature in the ξ_2 direction with $Q = 4$, 5, and 6. As with the last exercise, you will need to determine the local mapping $\chi_i(\xi_1,\xi_2)$, as discussed in Section 4.1.3, and calculate the Jacobian, as discussed in Section 4.1.3.3.

(e) As a final exercise using the elemental domains defined in 1(c) and 1(d), calculate $\mathcal{I} = \int_{\Omega^e} \sin x_1 \sin x_2\,\mathrm{d}\boldsymbol{x}$ for $2 \leqslant Q \leqslant 8$ and plot the error, ε, between the exact and numerical integrals versus Q on semi-log axes. In using Gaussian quadrature we are essentially approximating the smooth function $\sin x \sin y$ with polynomials of order $Q-1$. The error, ε, will therefore be proportional to C^Q. Thus, plotting $\log\varepsilon$ versus Q should asymptotically give a straight line since taking the log of the error relationship we have $\log \varepsilon \propto Q \log C$.

2. Having developed routines for integration in a general elemental region, we continue our spectral/*hp* element construction by considering an elemental projection problem in two dimensions. Consider the projection problem $u^\delta(x_1, x_2) = f(x_1, x_2)$, where $f(x_1, x_2)$ is a known function, for example $f(x_1, x_2) = [x_1]^6 \times [x_2]^6$ or $f(x_1, x_2) = \sin x_1 \sin x_2$. We recall from Section 2.6 that projection problems are helpful since they do not require any boundary conditions to be imposed. Extending the formulation in Section 2.2, our Galerkin problem in the elemental region Ω^e can be stated as follows.

   Find $u^\delta \in \mathcal{X}^\delta$, such that

$$\int_{\Omega^e} v^\delta(\boldsymbol{x}) u^\delta(\boldsymbol{x})\, \mathrm{d}\boldsymbol{x} = \int_{\Omega^e} v^\delta f(\boldsymbol{x})\, \mathrm{d}\boldsymbol{x}\,, \quad \forall\, v^\delta \in \mathcal{V}^\delta\,,$$

   and for a Galerkin expansion we define the expansion space $\mathcal{X}^\delta$ to be the same as the test space $\mathcal{V}^\delta$.

   Defining a discrete expansion $\phi_{pq}(\boldsymbol{\xi})$ and accordingly a discrete solution $u^\delta(\boldsymbol{\xi}) = \sum_p \sum_q \hat{u}_{pq} \phi_{pq}(\boldsymbol{\xi})$ leads to the matrix problem (see also Section 4.1.5.3)

$$\boldsymbol{M}^e \hat{\mathbf{u}} = \left(\boldsymbol{B}^{e^\top} \boldsymbol{W}^e \boldsymbol{B}^e\right) \hat{\mathbf{u}} = \boldsymbol{B}^{e^\top} \boldsymbol{W}^e \boldsymbol{f}\,,$$

   where the above matrices and vectors were defined in Section 4.1.5.1.
   In Section 3.5 we previously discussed how to construct the elemental mass matrix in the standard elemental region Ω_{st}. In this exercise we need to construct the elemental matrix for a general-shaped subdomain. At the elemental level we could use any of the two-dimensional bases defined in Chapter 3. However, for use in the C^0 expansion we will focus on either the tensorial quadrilateral expansions $\phi_{pq}(\xi_1, \xi_2) = h_p(\xi_1) h_q(\xi_2)$ and $\phi_{pq}(\xi_1, \xi_2) = \psi_p^a(\xi_1)\psi_q^a(\xi_2)$ or the tensorial triangular expansion $\phi_{pq}(\xi_1, \xi_2) = \psi_p^a(\eta_1)\psi_{pq}^b(\eta_2)$. Consider the following tasks using one (or more) of these bases within either a straight-sided quadrilateral region with vertices $(x_1^A, x_2^A) = (0, 0)$, $(x_1^B, x_2^B) = (1, 0.5)$, $(x_1^C, x_2^C) = (2, 1)$, and $(x_1^D, x_2^D) = (0.5, 1)$ or a straight-sided triangular region with vertices $(x_1^A, x_2^A) = (0, 0)$, $(x_1^B, x_2^B) = (1, 0.5)$, and $(x_1^C, x_2^C) = (2, 1)$.

   (a) Construct the mass matrix $\boldsymbol{M}^e$ in the region Ω^e for $P_1 = P_2 = 8$ and $Q_1 = Q_2 = 10$, where

$$\boldsymbol{M}^e[m(pq)][n(rs)] = \int_{\Omega_{\mathrm{st}}} \phi_{pq}(\boldsymbol{\xi})\phi_{rs}(\boldsymbol{\xi})|J(\boldsymbol{\xi})|\, \mathrm{d}\boldsymbol{\xi}\,, \quad 0 \leqslant m, n \leqslant N_m\,.$$

   In the above equation we recall that $J(\boldsymbol{\xi})$ is the Jacobian of the mapping between the element Ω^e and the standard region Ω_{st}, and $m(pq)$ and $n(rs)$ represent mappings between the index pairs (p, q) and (r, s) with the unique indices m and n, as discussed in Section 4.1.5.1.

(b) Construct the right-hand-side vector $\hat{\boldsymbol{f}}$ $(= \boldsymbol{B}^\top \boldsymbol{W} \boldsymbol{f})$:

$$\hat{\boldsymbol{f}}[m(pq)] = \int_{\Omega_{\mathrm{st}}} \phi_{pq}(\boldsymbol{\xi}) f(\boldsymbol{\xi}) |J(\boldsymbol{\xi})| \, \mathrm{d}\boldsymbol{\xi} \,, \quad 0 \leqslant m \leqslant N_m \,,$$

where $f(\boldsymbol{\xi}) = [\chi_1(\xi_1, \xi_2)]^6 \times [\chi_2(\xi_1, \xi_2)]^6$, and use a uniform polynomial order $P = 8$ and uniform quadrature order $Q = 10$. Note that it is economical to evaluate the inner product operation $\boldsymbol{B}^\top \boldsymbol{W} \boldsymbol{f}$ by taking advantage of the tensorial nature of the expansion basis using the sum-factorisation technique discussed in Section 4.1.6.

(c) Using a matrix inversion technique (for example, the Cholesky factorisation and routines *dpptrf* and *dpptrs* in the *LAPACK* library [12]), invert the symmetric mass matrix $\boldsymbol{M}^e$ and solve for $\hat{\boldsymbol{u}} = [\boldsymbol{M}^e]^{-1} \hat{\boldsymbol{f}}$. Verify your solution by performing the backwards transformation $\boldsymbol{B}\hat{\boldsymbol{u}}$ (using sum-factorisation) to evaluate the solution at the quadrature points and verify the solution is equal to $u^\delta(\xi_{1i}, \xi_{2j}) = [\chi_1(\xi_{1i}, \xi_{2j})]^6 \times [\chi_2(\xi_{1i}, \xi_{2j})]^6$.

3. To solve the elemental projection problem of Exercise 2 using a non-tensorial nodal basis we can adopt a matrix construction since the sum-factorisation techniques are no longer applicable. As discussed in Section 4.1.5.1, the basis matrix for a Lagrange non-tensorial basis $\boldsymbol{B}[m][n] = L_n(\boldsymbol{\xi}_m)$ can be evaluated at the quadrature points $\boldsymbol{\xi}_m(ij) = [\eta_{1i}, \eta_{2j}]$ (see eqn (4.1.25)) as

$$\boldsymbol{B} = \boldsymbol{B}_{\mathcal{T}} \boldsymbol{B}_{\mathcal{N}}^{-1} \,,$$

where $\boldsymbol{B}_{\mathcal{N}}$ is a square matrix of the tensorial basis ϕ_{pq} evaluated at the nodal points $\boldsymbol{\zeta}_{n'}$:

$$\boldsymbol{B}_{\mathcal{N}}[n'][n(pq)] = \phi_{pq}(\boldsymbol{\zeta}_{n'}) \,,$$

and $\boldsymbol{B}_{\mathcal{T}}$ is a non-square matrix of the basis ϕ_{pq} evaluated at the quadrature points $\boldsymbol{\xi}_{m(ij)}$, i.e.,

$$\boldsymbol{B}_{\mathcal{T}}[m(ij)][n(pq)] = \phi_{pq}(\boldsymbol{\xi}_{m(ij)}) \,.$$

The elemental mass matrix can then be assembled in matrix form as $\boldsymbol{M}^e = \boldsymbol{B}^\top \boldsymbol{W} \boldsymbol{B}$, where $\boldsymbol{W}$ is defined in Section 4.1.5.1.

Solve the elemental projection problem in the triangular region defined in Exercise 2 using a Lagrange polynomial basis defined through the electrostatic points as given in Appendix D. Take $P = 8$ $(N_m = 45)$ and $f(\boldsymbol{\xi}) = [\chi_1(\xi_1, \xi_2)]^6 \times [\chi_2(\xi_1, \xi_2)]^6$, and verify that the expansion coefficients $\hat{\boldsymbol{u}} = [\boldsymbol{M}^e]^{-1} \hat{\boldsymbol{f}}$, where $\hat{\boldsymbol{f}} = \boldsymbol{B}^\top \boldsymbol{W} \boldsymbol{f}$, are the exact solution at the nodal/electrostatic points. Although any tensorial basis can be used, the properties of the orthogonal basis $\phi_{pq} = \widetilde{\psi}_p^a \widetilde{\psi}_{pq}^b$ are numerically advantageous.

4. The next step is to solve the projection problem for multiple elements. Consider the four-element quadrilateral or triangular mesh shown in Fig. 4.40. We want to construct the global projection for a $P = 6$ polynomial order expansion within this region. For either a quadrilateral or triangular expansion perform the following steps.

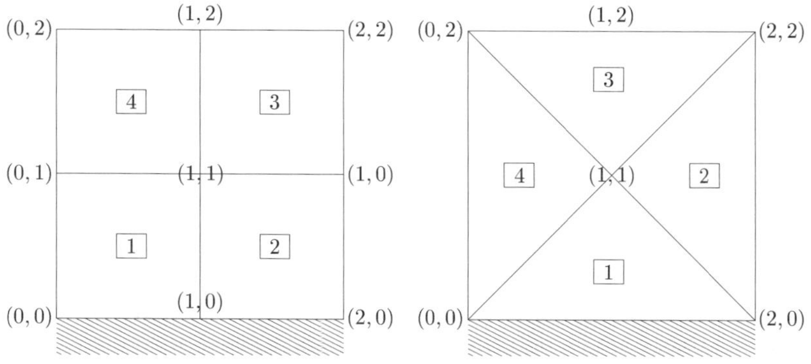

FIG. 4.40. Four element domains for global projection problems.

(a) Using the discussion in Sections 4.2.1 and 4.2.1.1, construct a boundary mapping array bmap[e][i] which maps the local elemental boundary degrees of freedom to the global boundary degrees of freedom. For a modal expansion you will also need to define the sign array sign[e][i].

(b) Construct the elemental mass matrices $\boldsymbol{M}^e$, $1 \leqslant e \leqslant 4$, for the elements in the mesh. In constructing the mass matrix use a local index ordering so that the boundary degrees are listed first and follow a similar ordering convention as the mapping array you have defined in (a). Note that, since the elements are straight-sided (and in the quadrilateral case they have similar angles to the standard region), the elemental mass matrices will be identical as each element has the same constant mapping Jacobian.

(c) Construct the global mass matrix $\boldsymbol{M} = \boldsymbol{\mathcal{A}}^\top \underline{\boldsymbol{M}^e} \boldsymbol{\mathcal{A}}$ as discussed in Section 4.2.2. Since we have only defined a boundary mapping array the interior modes can be ordered in continuous blocks looping over each element (see also Fig. 2.11). The action $\boldsymbol{\mathcal{A}}$ and $\boldsymbol{\mathcal{A}}^\top$ can be evaluated as shown in eqns (4.2.4)–(4.2.6).

(d) Construct the right-hand-side vector $\hat{\boldsymbol{f}}_g = \boldsymbol{\mathcal{A}}^\top \underline{\boldsymbol{B}^{e\top} \boldsymbol{W}^e \boldsymbol{f}^e}$, where $\boldsymbol{f}^e$ contains the function $f(x_1, x_2) = \cos x_1 \cos x_2$, for $(x_1, x_2) \in \Omega^e$, evaluated at the elemental quadrature points. A useful debugging case is to also consider the case where $f(x_1, x_2) = 1$.

(e) By inverting the global mass matrix solve the problem $\hat{\boldsymbol{u}}_g = \boldsymbol{M}^{-1} \hat{\boldsymbol{f}}_g$. This inversion can also be performed using the static condensation technique discussed in Section 4.2.3. To recover the local solution within each element we can use the assembly operator $\hat{\boldsymbol{u}}_l = \underline{\hat{\boldsymbol{u}}^e} = \boldsymbol{\mathcal{A}} \hat{\boldsymbol{u}}_g$. The local solution at the quadrature points can then be determined in each element from $\boldsymbol{u}^e = \boldsymbol{B} \hat{\boldsymbol{u}}^e$, as discussed in Section 4.1.5.2.

(f) Determine the L^2 error in the projection, $\varepsilon = \left[\int_\Omega (u^\delta - u^{\text{exact}})^2 \, d\boldsymbol{x}\right]^{1/2}$, for different polynomial orders $4 \leqslant P \leqslant 10$ and for different uniform discretisation in the x_1 and x_2 directions. Plot the error as a function of polynomial order and the square root of the number of elements (or element size h) to demonstrate the difference between h and p convergence for this smooth problem.

5. The last exercise is to impose Dirichlet boundary conditions on the solution. This will also be necessary in solving the elliptic problems suggested in Section 5.6. Dirichlet boundary conditions can be imposed using the lifting technique discussed in Sections 2.2.1.3 and 4.2.4.1. This requires linearly decomposing the solution into a known function, which satisfies the Dirichlet boundary conditions, and the remaining solution with homogeneous boundary conditions. At an implementation level this process involves renumbering the boundary mapping array bmap[e][i], as described in Section 4.2.4. We also require an elemental boundary transformation to determine the expansion coefficient of the degrees of freedom on the Dirichlet boundary, as discussed in Section 4.3.2. Consider the global projection problem of Exercise 4 with a Dirichlet boundary condition on the domain boundary $x_2 = 0$ of the domains shown in Fig. 4.40. The Dirichlet boundary condition is therefore $u(\partial\Omega_{\mathcal{D}}) = \cos x_1$. Solve the global projection problem of Exercise 4, constrained to satisfy the Dirichlet boundary condition in the following steps.
   (a) Determine the expansion coefficient of all boundary degrees of freedom using the elemental boundary transformation technique of Section 4.3.2.
   (b) Order the boundary mapping array bmap[e][i] so that the known (or lifted) boundary degrees of freedom are ordered after the unknown boundary degrees of freedom. This can either be performed 'by hand' or by implementing an automated procedure as discussed in Section 4.2.4.
   (c) Invert the global mass matrix using a static condensation technique where only the submatrix of the boundary matrix system corresponding to the unknown degrees of freedom is inverted, as shown in Section 4.2.4.2 and Fig. 4.21. Verify that your answer converges for different polynomial orders and numbers of elemental domains.

5

DIFFUSION EQUATION

This chapter is concerned with the discretisation of the parabolic-type diffusion equation

$$\frac{\partial u}{\partial t} = \alpha \nabla^2 u \,, \tag{5.0.1}$$

where α is the diffusivity.

Adopting the matrix and vector notation defined in Section 4.2.2, the standard Galerkin formulation of eqn (5.0.1) leads to a semi-discrete system of ordinary differential equations of the form

$$\boldsymbol{M}\frac{\mathrm{d}\hat{\boldsymbol{u}}_g}{\mathrm{d}t} = \alpha \boldsymbol{L}\hat{\boldsymbol{u}}_g \,, \tag{5.0.2}$$

where $\boldsymbol{M}$ and $\boldsymbol{L}$ are the mass and Laplacian matrices, respectively, which are here assumed to include the boundary condition contributions.

In discretising eqn (5.0.1) or eqn (5.0.2) in time we can choose to use either an explicit or implicit formulation. As will be discussed in Section 5.4.1, stability considerations mean that the explicit discretisation of eqn (5.0.2) requires a time-step restriction. This scales as the inverse of the growth rate of maximum eigenvalue of $\boldsymbol{M}^{-1}\boldsymbol{L}$. When applying a spectral/hp discretisation the largest eigenvalue of this operator grows as $\mathcal{O}(P^4)$. The severe time-step restriction at large polynomial orders motivates the use of an implicit discretisation in time. As an example, if we consider the implicit Euler-backward time discretisation over a time-step Δt, eqn (5.0.1) becomes

$$\frac{u^{n+1} - u^n}{\Delta t} = \alpha \nabla^2 u^{n+1} \,.$$

The above equation can be rearranged to obtain a Helmholtz equation for u^{n+1} such that

$$\nabla^2 u(\boldsymbol{x}) - \lambda\, u(\boldsymbol{x}) = f(\boldsymbol{x}) \,, \tag{5.0.3}$$

where $\lambda = 1/\alpha\, \Delta t$ is a *positive* constant and $f(\boldsymbol{x}) = -u^n(\boldsymbol{x})/\alpha\Delta t$. Once again, the equation is also supplemented with appropriate boundary conditions, which will typically be of Dirichlet or Neumann type. We assume that the minimum conditions of regularity are $u(\boldsymbol{x}) \in H^1$ and that $f(\boldsymbol{x}) \in L^2$.

In the above brief summary we have highlighted all the main steps required to discretise the diffusion equation. We start our more detailed discussion in Section 5.1 by considering the standard Galerkin formulation discretisation of the Helmholtz equation (5.0.3). In Section 5.2 we supplement this formulation with numerical examples demonstrating the convergence rate of the continuous Galerkin method in complex-geometry domains when the solution is sufficiently smooth.

In constructing eqn (5.0.3) we used an implicit Euler-backward discretisation in time; however, there is a range of choices of multi-stage and multi-step time-integration schemes. In Section 4.3.6 we considered the multi-stage Runge–Kutta schemes in the context of particle tracking. These schemes require multiple evaluations of the right-hand side of the ordinary differential equation and so can be quite costly for spectral/*hp* element schemes where these evaluations are relatively expensive. An alternative approach is to use a multi-step scheme which reuses right-hand-side evaluations from previous time-steps. In Section 5.3 we discuss different types of multi-step temporal discretisations which will also be applied to the advection equation in Chapter 6. Within this section we also consider the stability region of the multi-step schemes.

To determine the time-step restrictions associated with the explicit time integration of the diffusion equation, we need to understand the eigenspectra of the weak Laplacian matrix which is discussed in Section 5.4. In Section 5.4.1 we first consider the growth of the maximum eigenvalue of the Laplacian matrix $\boldsymbol{L}$ which directly relates to the time-step restriction of the full discretisation. Understanding of the eigenspectra also allows us to assess the cost of iteratively inverting the weak Laplacian matrix $\boldsymbol{L}$, which is related to the growth of the condition number of the matrix, and how it is affected by appropriate preconditioners. The growth rate of the condition number with polynomial order and number of elements of different preconditioned matrices is discussed in Section 5.4.2.

In the first four sections of this chapter we consider solutions which are well behaved in the sense that they were sufficiently smooth. However, elliptic problems, which can be seen as the steady state of the parabolic diffusion problem, may contain a non-smooth solution, for example, due to corner singularities. In Section 5.5 we analyse the convergence of the spectral/*hp* method for domains with corner singularities and suggest possible ways of restoring high accuracy.

5.1 Galerkin discretisation of the Helmholtz equation

♣[21] Considering a solution domain Ω with boundary $\partial\Omega$, we solve eqn (5.0.3) using a standard Galerkin construction. Equation (5.0.3) is expressed in a weak form by taking the inner product of eqn (5.0.3) with respect to a function $v(\boldsymbol{x})$ to obtain

$$(v, \nabla^2 u)_\Omega - \lambda(v, u)_\Omega = (v, f)_\Omega \,,$$

where $v(\boldsymbol{x}) \in H^1$ and is, by definition, zero on all Dirichlet boundary conditions. Since $\nabla^2 u = \nabla \cdot \nabla u$, we can apply the divergence theorem to the first term to obtain

[21] Discretisation of the Helmholtz equation.

$$(\nabla v, \nabla u)_\Omega + \lambda(v, u)_\Omega = \langle v, \nabla u \cdot n \rangle - (v, f)_\Omega \,, \tag{5.1.1}$$

where

$$\langle v, \nabla u \cdot n \rangle = \int_{\partial\Omega} v \nabla u \cdot \boldsymbol{n} \, \mathrm{d}\boldsymbol{x} \tag{5.1.2}$$

and $\boldsymbol{n}$ is the outward normal to the boundary $\partial\Omega$.

The flux contribution in eqn (5.1.1), explicitly defined by eqn (5.1.2), only makes a contribution on the Neumann boundary since, by definition, $v(\partial\Omega_D)$ is zero on all Dirichlet boundaries. Equation (5.1.2) therefore allows us to weakly enforce the Neumann boundary condition $\nabla u \cdot \boldsymbol{n} = g_{\mathcal{N}}$.

Dirichlet boundary conditions can be strongly imposed by *lifting* the known modes out of the system, as discussed in the one-dimensional context in Section 2.2.1.3 and in a discrete manner in Section 4.2.4.1. We recall that lifting a known solution is equivalent to decomposing the solution into an unknown homogeneous $u^{\mathcal{H}}(\boldsymbol{x})$ contribution and a known non-homogeneous $u^{\mathcal{D}}(\boldsymbol{x})$ contribution which satisfies the boundary condition $u(\partial\Omega_{\mathcal{D}}) = g_{\mathcal{D}}(\partial\Omega_{\mathcal{D}})$, that is,

$$u(\boldsymbol{x}) = u^{\mathcal{H}}(\boldsymbol{x}) + u^{\mathcal{D}}(\boldsymbol{x}) \,,$$

where

$$u^{\mathcal{H}}(\partial\Omega_{\mathcal{D}}) = 0 \,, \quad u^{\mathcal{D}}(\partial\Omega_{\mathcal{D}}) = u(\partial\Omega_{\mathcal{D}}) = g_{\mathcal{D}}(\partial\Omega_{\mathcal{D}}) \,.$$

If we substitute this into the initial definition of the Helmholtz equation (5.1.1) then we see that this is equivalent to solving the equation

$$\begin{aligned}(\nabla v, \nabla u^{\mathcal{H}})_\Omega + \lambda(v, u^{\mathcal{H}})_\Omega &= \langle v, g_{\mathcal{N}} \rangle - (v, f)_\Omega \\ &\quad - (\nabla v, \nabla u^{\mathcal{D}})_\Omega - \lambda(v, u^{\mathcal{D}})_\Omega \,.\end{aligned} \tag{5.1.3}$$

In the above problem the test and trial spaces for v and u can be identical as they are both defined to be zero on Dirichlet boundary conditions.

An alternative approach to imposing Dirichlet boundary conditions is to use a penalty method. In this formulation eqn (5.1.1) is augmented with a penalty term of the form

$$\tau \int_{\partial\Omega_{\mathcal{D}}} v \left(u(\partial\Omega) - g_{\mathcal{D}} \right) \mathrm{d}\boldsymbol{x} \,,$$

where τ, the penalty parameter, is a sufficiently large number to enforce the boundary constraint in a weak sense [247, 511].

Elemental contributions

Following the formulation of Chapter 4, we initially consider a single element and then construct the global system. In each element we approximate the solution by a polynomial expansion discretely represented by $\boldsymbol{u}^e$ ($= \boldsymbol{B}^e \hat{\boldsymbol{u}}^e$). Similarly, we evaluate the forcing function at the quadrature points and denote it by $\boldsymbol{f}^e$ (see Section 4.1.5.1 for further details). Taking the inner product with respect to the expansion basis (i.e., $v = \phi_{pqr}$), we can write the elemental contribution of eqn (5.1.1) in matrix form as

$$\boldsymbol{L}^e \hat{\boldsymbol{u}}^e + \lambda \boldsymbol{M}^e \hat{\boldsymbol{u}}^e = -(\boldsymbol{B}^e)^\top \boldsymbol{W}^e \boldsymbol{f}^e \,, \tag{5.1.4}$$

where we recall that

$$\begin{aligned}
\boldsymbol{M}^e &= (\boldsymbol{B}^e)^\top \boldsymbol{W}^e \boldsymbol{B}^e \,, \\
\boldsymbol{L}^e &= (\boldsymbol{D}^e_{x_1}\boldsymbol{B}^e)^\top \boldsymbol{W}^e \boldsymbol{D}^e_{x_1}\boldsymbol{B}^e + (\boldsymbol{D}^e_{x_2}\boldsymbol{B}^e)^\top \boldsymbol{W}^e \boldsymbol{D}^e_{x_2}\boldsymbol{B}^e + (\boldsymbol{D}^e_{x_3}\boldsymbol{B}^e)^\top \boldsymbol{W}^e \boldsymbol{D}^e_{x_3}\boldsymbol{B}^e \,,
\end{aligned}$$

$$\begin{aligned}
\boldsymbol{D}^e_{x_1} &= \boldsymbol{\Lambda}^e\left(\frac{\partial \xi_1}{\partial x_1}\right)\boldsymbol{D}^e_{\xi_1} + \boldsymbol{\Lambda}^e\left(\frac{\partial \xi_2}{\partial x_1}\right)\boldsymbol{D}^e_{\xi_2} + \boldsymbol{\Lambda}^e\left(\frac{\partial \xi_3}{\partial x_1}\right)\boldsymbol{D}^e_{\xi_3} \,, \\
\boldsymbol{D}^e_{x_2} &= \boldsymbol{\Lambda}^e\left(\frac{\partial \xi_1}{\partial x_2}\right)\boldsymbol{D}^e_{\xi_1} + \boldsymbol{\Lambda}^e\left(\frac{\partial \xi_2}{\partial x_2}\right)\boldsymbol{D}^e_{\xi_2} + \boldsymbol{\Lambda}^e\left(\frac{\partial \xi_3}{\partial x_2}\right)\boldsymbol{D}^e_{\xi_3} \,, \\
\boldsymbol{D}^e_{x_3} &= \boldsymbol{\Lambda}^e\left(\frac{\partial \xi_1}{\partial x_3}\right)\boldsymbol{D}^e_{\xi_1} + \boldsymbol{\Lambda}^e\left(\frac{\partial \xi_2}{\partial x_3}\right)\boldsymbol{D}^e_{\xi_2} + \boldsymbol{\Lambda}^e\left(\frac{\partial \xi_3}{\partial x_3}\right)\boldsymbol{D}^e_{\xi_3} \,.
\end{aligned}$$

The matrix $\boldsymbol{L}^e$ is the discrete representation of the weak form of the Laplacian $(\nabla v, \nabla u)_{\Omega^e}$. The elemental formulation of $\boldsymbol{L}^e$ given by eqn (5.1.4) ignores the contribution of boundary conditions, which will be handled subsequently. However, in general, the majority of elements may not touch the solution boundary $\partial\Omega$, and so eqn (5.1.4) provides the major contribution to the algebraic form. Figure 5.1 shows the structure of $\boldsymbol{L}^e$ in the standard straight-sided triangular and quadrilateral regions using modified modal expansion bases, $\phi_{pq} = \psi^a_p\,\psi^b_{pq}$

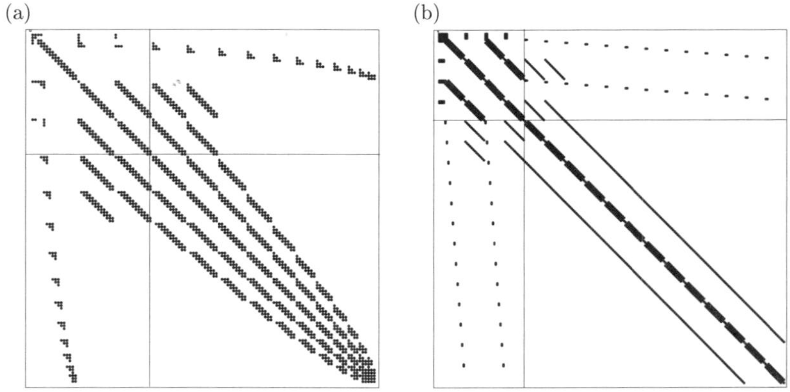

FIG. 5.1. Structure of the two-dimensional elemental Laplacian matrix at $P = 14$: (a) standard modal triangular expansion $\phi_{pq} = \psi^a_p\psi^b_{pq}$, and (b) standard modal quadrilateral expansion $\phi_{pq} = \psi^a_p\psi^a_q$. In both plots the boundary degrees of freedom were ordered first, followed by the interior degrees of freedom. The matrix is symmetric and we note that the banded structure of the interior–interior matrix is similar in form to the mass matrix shown in Fig. 3.18.

and $\phi_{pq} = \psi_p^a \psi_q^a$. The structure of the Laplacian matrix is very similar to the mass matrix for these types of bases (see Fig. 3.18). However, for the standard nodal quadrilateral expansion the matrix has an interior–interior system which is nearly of full rank, independent of the number of quadrature points used. Figure 5.2 shows the structure of the three-dimensional Laplacian matrix within a tetrahedral modal expansion as we increase the polynomial order from $P = 4$ to $P = 9$ and $P = 19$.

Global matrix assembly

Interpreting an underlined matrix as the block diagonal extension of this matrix, the discrete representation of eqn (5.1.4) over all elements has the form

$$[\underline{\boldsymbol{L}}^e + \lambda \underline{\boldsymbol{M}}^e]\, \hat{\boldsymbol{u}}_l = -(\underline{\boldsymbol{B}}^e)^\top \underline{\boldsymbol{W}}^e \boldsymbol{f}_l\,, \tag{5.1.5}$$

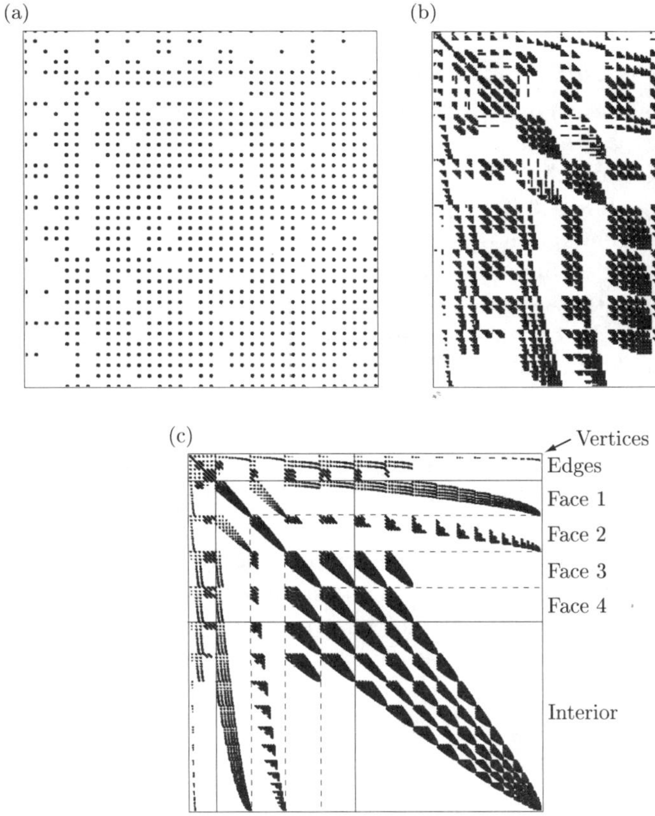

FIG. 5.2. Structure of the weak three-dimensional Laplacian matrix for expansion orders of (a) $P = 4$, (b) $P = 9$, and (c) $P = 19$.

where we recall from Section 4.2.2 that $\hat{\boldsymbol{u}}_l$ and $\boldsymbol{f}_l$ are the concatenation of $\hat{\boldsymbol{u}}^e$ and $\boldsymbol{f}^e$, that is, $\hat{\boldsymbol{u}}_l = \underline{\hat{\boldsymbol{u}}}^e$ and $\boldsymbol{f}_l = \underline{\boldsymbol{f}}^e$. For the solution to exist in H^1 and to enable the inversion of the left-hand matrix, we can enforce that the expansion is globally C^0 continuous. This is achieved by introducing the global expansion coefficients via the matrix operation $\boldsymbol{\mathcal{A}}$ (see Section 4.2.1). Expressing the local expansion coefficients in terms of the global expansion coefficients (that is, $\hat{\boldsymbol{u}}_l = \boldsymbol{\mathcal{A}}\hat{\boldsymbol{u}}_g$) and pre-multiplying eqn (5.1.5) by $\boldsymbol{\mathcal{A}}^\top$, we obtain

$$\boldsymbol{\mathcal{A}}^\top \left[\underline{\boldsymbol{L}}^e + \lambda \underline{\boldsymbol{M}}^e\right] \boldsymbol{\mathcal{A}}\hat{\boldsymbol{u}}_g = \boldsymbol{\Gamma} - \boldsymbol{\mathcal{A}}^\top (\underline{\boldsymbol{B}}^e)^\top \underline{\boldsymbol{W}}^e \boldsymbol{f}_l .$$

The above equation is the global description of the discrete Helmholtz equation, where we have introduced the vector $\boldsymbol{\Gamma}$ to represent the nonzero surface integral terms from eqn (5.1.1). To solve this system we therefore need to invert the weak Helmholtz matrix

$$\boldsymbol{H} = \boldsymbol{L} + \lambda \boldsymbol{M} = \boldsymbol{\mathcal{A}}^\top \left[\underline{\boldsymbol{L}}^e + \lambda \underline{\boldsymbol{M}}^e\right] \boldsymbol{\mathcal{A}},$$

subject to the imposition of boundary conditions.

Discrete boundary condition implementation

Although in general we can use any solution satisfying the boundary conditions as a lifted solution $u^{\mathcal{D}}(\boldsymbol{x})$, in practice it is convenient to transform any mode which touches a Dirichlet boundary using the boundary transformation described in Section 4.3.1. The lifted solution $u^{\mathcal{D}}(\boldsymbol{x})$ is then extended into the interior by the shape of the boundary modes in the interior of the element which touch the Dirichlet boundary. Once the expansion coefficients of modes touching the boundary are known, these degrees of freedom can be removed or *lifted* from the global matrix system by taking the appropriate components to the right-hand side, as discussed in Section 4.2.4.2. This is the discrete equivalent of eqn (5.1.3).

We note that the manipulation of the global matrix $\boldsymbol{H}$ to enforce Dirichlet boundary conditions only requires the consideration of the elemental boundary degrees of freedom since the interior degrees of freedom do not contribute to the expansion along any boundary. We also note that, although the discussion in Section 4.2.4.2 is motivated by the static condensation technique, the formulation can be used whether or not static condensation is applied.

To enforce the Neumann boundary condition we need to enforce the vector $\boldsymbol{\Gamma}$. We recall that the only contributions to the surface integral terms (represented by the vector $\boldsymbol{\Gamma}$) are from boundaries of a Neumann type. This term is of the form

$$\boldsymbol{\Gamma}^e[n(pqr)] = \int_{\partial\Omega^e \bigcap \partial\Omega_{\mathcal{N}}} \phi^e_{pqr} \nabla u \cdot \boldsymbol{n} \, \mathrm{d}\boldsymbol{x},$$
$$\boldsymbol{\Gamma} = \boldsymbol{\mathcal{A}}^\top \boldsymbol{\Gamma}^e,$$

where $n(pqr)$ represents the ordering of the three indices p, q, and r into a consecutive global list and $\partial\Omega_{\mathcal{N}}$ represents the boundary of Neumann conditions.

The majority of elements do not contribute to this term since they will not have an edge touching the boundary (i.e., $\partial\Omega^e \bigcap \partial\Omega = \emptyset$). However, on a boundary of the solution domain which has a Neumann condition we know the flux $\nabla u \cdot \boldsymbol{n} = g_{\mathcal{N}}$, and so we can discretely evaluate this integral by projecting the function onto the quadrature points (using a Galerkin or collocation projection) and using the surface Jacobian, as discussed in Section 4.1.4.

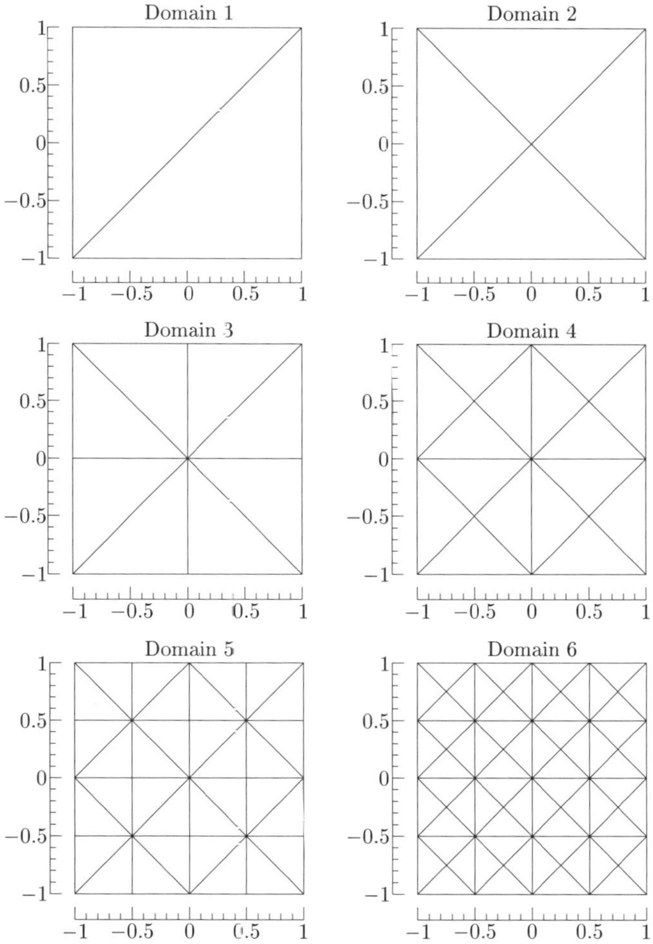

FIG. 5.3. Computational domains for the Helmholtz test problem. Each domain spans the region $-1 \leqslant x \leqslant 1$, $-1 \leqslant y \leqslant 1$ but is divided into $N_{\rm el} = 2, 4, 8, 16, 32,$ or 64 equal triangular elements.

5.2 Numerical examples

The initial test we shall consider is a two-dimensional Helmholtz problem with a coefficient of $\lambda = 1$, having a solution of the form

$$u(x_1, x_2) = \sin(\pi x_1)\cos(\pi x_2)\,.$$

The solution to this problem was obtained on a variety of domains using Dirichlet boundary conditions at different expansion orders. The computational domain spans the area $\{-1 \leqslant x_1 \leqslant 1, -1 \leqslant x_2 \leqslant 1\}$ and is subdivided into $N_{\text{el}} = 2$, 4, 8, 16, 32, or 64 equal triangular elements. We will refer to these as domains 1 through 6, as shown in Fig. 5.3. For each domain the Helmholtz problem was solved for a variety of expansion orders P and the H^1 error versus expansion order is plotted in Fig. 5.4. Spectral convergence is demonstrated by the fact that on this linear–log plot the convergence lines are slightly better than linear. The error is resolved to machine precision, which is of the order of 10^{-13}–10^{-14}. Finally, we consider a line of constant expansion order and convergence with the

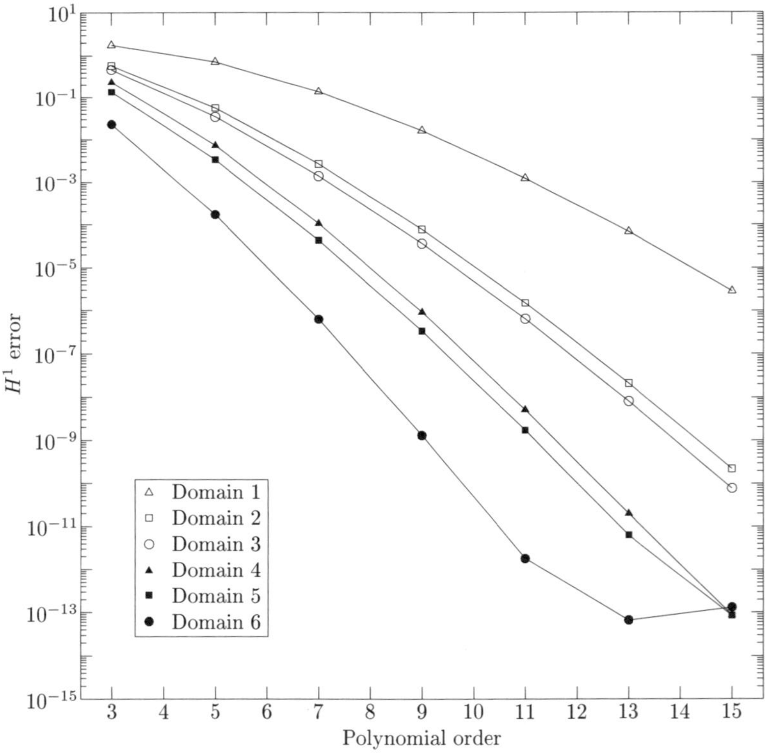

FIG. 5.4. Convergence plots in the H^1 norm for a Helmholtz problem with $\lambda = 1$ versus polynomial order P. The solution is obtained on different domains, shown in Fig. 5.3, which spanned the same area in x_1, x_2.

number of elements. The lines corresponding to domains 2 and 3, as well as 4 and 5, are closely spaced. This is not too surprising given that the resolution length along the $\pm 45°$ lines in domains 2 and 3 are the same as 4 and 5.

The second test we considered is one which involves a variety of triangular elements of different aspect ratios and orientations. Such a domain is shown in Fig. 5.5. The Dirichlet–Helmholtz problem with $\lambda = 1$ and $u(x_1, x_2) = \sin(\pi x_1)\cos(\pi x_2)$ was solved on this domain, and the L^∞ and H^1 errors were plotted with respect to expansion order, as can also be seen in Fig. 5.5. Spectral convergence is maintained in this more complicated domain even though we have used a random discretisation which is not a Delaunay triangulation.

The final two-dimensional result demonstrates the variable capabilities of the expansion basis. We consider the Helmholtz problem with a solution $u(x_1, x_2) = \exp(-2.5((x_1 - 1)^2 + (x_2 - 1)^2))$. As shown in Fig. 5.6(a), this solution is a boundary layer located at $x_1 = 1$, $x_2 = 1$ which decays radially away from this location. The problem was discretised using $N_{\text{el}} = 32$ uniform elemental regions, as shown in Fig. 5.6(b). If we solve this problem using Dirichlet boundary conditions and a fixed-order polynomial expansion of $P = 6$ within every element, then the H^1 error within all elemental regions is as shown in Fig. 5.6(c). The error in this plot clearly indicates the boundary-layer form of the solution and so the error is largest at $x_1 = 1$, $x_2 = 1$, decaying rapidly away from this point. Now, if we use a variable-order expansion where the order of the expansion is dictated by the polynomial order of the edge modes, as indicated in Fig. 5.6(b), then we obtain an H^1 elemental error distribution, as shown in Fig. 5.6(d). In this discretisation the order ranges from linear finite elements in the lower left-

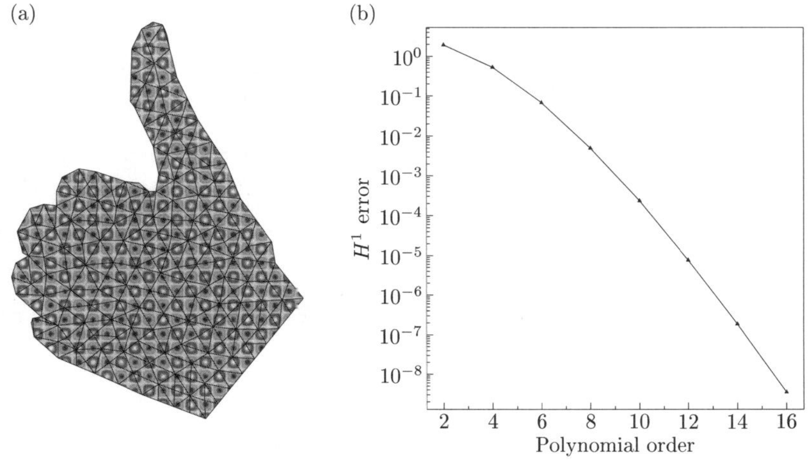

FIG. 5.5. For the complex domain shown in (a) a convergence test to the Helmholtz problem with solution $u(x_1, x_2) = \sin(\pi x_1)\cos(\pi x_2)$ was performed (b). Spectral convergence was obtained independently of the complexity of the discretisation.

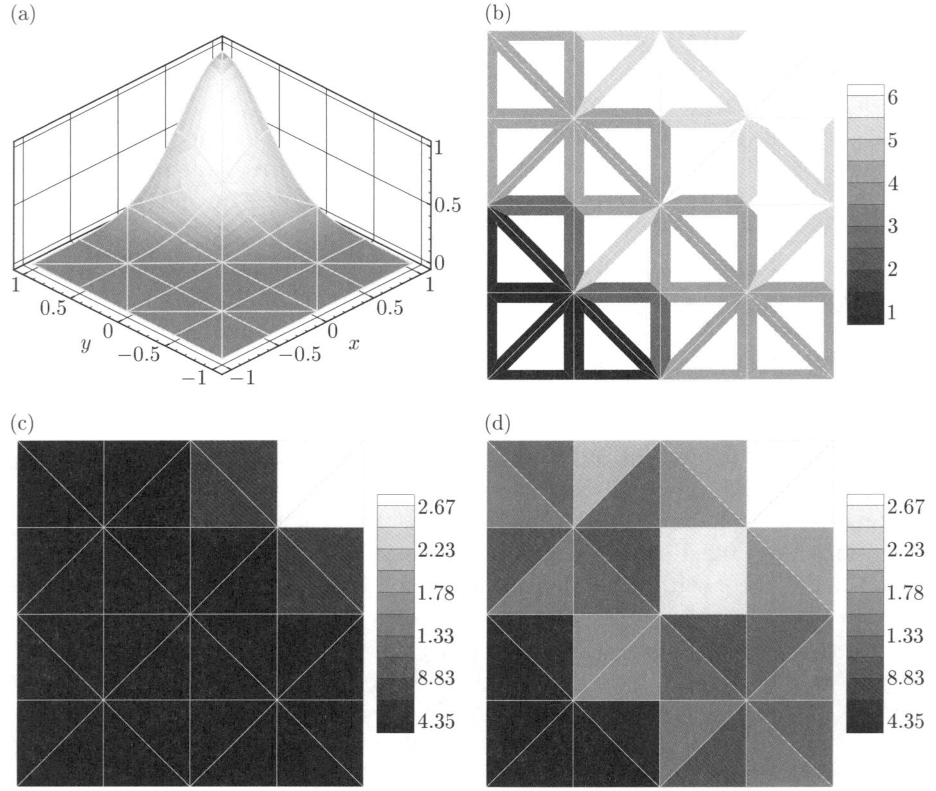

FIG. 5.6. A Helmholtz problem with a solution $u(x_1, x_2) = \exp(-2.5((x_1-1)^2 + (x_2-1)^2))$ as shown in (a) was discretised using $N_{\rm el} = 32$ elements, as shown in (b). Using a fixed polynomial order of $P = 6$ in all elements, the H^1 error within each element is shown in (c). Using a variable-order polynomial expansion where the order is also illustrated in (b), the H^1 error within each element is shown in (d). The error distribution due to the variable-polynomial-order expansion shown in (d) demonstrates a greater uniformity than the error distribution due to the case shown in (c).

hand corner building up to a $P = 6$ order expansion at the top right-hand corner. It can be appreciated that the local H^1 error is far more uniform and it never exceeds the largest H^1 error of the first calculation.

The result shown in Fig. 5.7 demonstrates fast convergence for three-dimensional curvilinear domains. In this figure we see the complex geometry of two circular pipes coalescing at an angle of 45°. The mesh was generated using the unstructured mesh generator of the *FELISA* package (Peiró *et al.* [371]) using $N_{\rm el} = 950$ tetrahedral elements. In a similar way to many automated mesh generators, the curved surface was represented using bi-cubic splines. Although

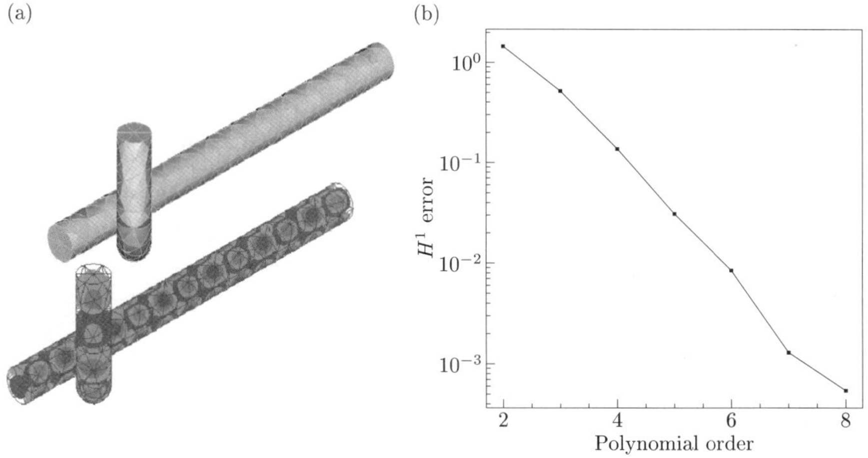

FIG. 5.7. (a) Solution of a Helmholtz problem with exact solution and boundary conditions given by $u = \sin(\pi x_1)\sin(\pi x_2)\sin(\pi x_3)$ in a bifurcated pipe discretised with $N_{\rm el} = 950$ elements. Exponential convergence is achieved, as shown in the error plots in (b).

the standard output from the generator typically only involves the vertices of linear elements, the bi-cubic spline input information of the surface was also used to reconstruct the necessary information within a curved face of the spectral/*hp* element.

Within this domain we consider the solution to a Helmholtz problem with a unit coefficient (that is, $\lambda = 1$) and a solution of the form

$$u(x_1, x_2, x_3) = \sin(\pi x_1)\sin(\pi x_2)\sin(\pi x_3)\,.$$

Figure 5.7 also shows the error in the H^1 norms as a function of expansion order, where we see that exponential convergence rate is, once again, evident from the form of the linear–log plot.

5.3 Temporal discretisation

We have seen that the standard Galerkin formulation of eqn (5.0.1) leads to a semi-discrete system of ordinary differential equations of the form

$$\boldsymbol{M}\frac{\mathrm{d}\hat{\boldsymbol{u}}_g}{\mathrm{d}t} = \alpha \boldsymbol{L}\hat{\boldsymbol{u}}_g \quad \Rightarrow \quad \frac{\mathrm{d}\hat{\boldsymbol{u}}_g}{\mathrm{d}t} = \alpha \boldsymbol{M}^{-1}\boldsymbol{L}\hat{\boldsymbol{u}}_g\,. \tag{5.3.1}$$

The complete discretisation of eqn (5.3.1) therefore involves the solution of a system of ordinary differential equations in time. We have previously discussed the *multi-stage* Runge–Kutta scheme in Section 4.3.6.1 in the context of particle tracking techniques. A similar approach can also be adopted when solving the system of ordinary differential equations (5.3.1). Alternatively, a *multi-step* scheme can be applied.

In the following section, we briefly review some standard *multi-step* schemes that are typically used for integrating systems of ordinary differential equations of the form

$$\frac{\mathrm{d}\boldsymbol{u}}{\mathrm{d}t} = \boldsymbol{F}(\boldsymbol{u})\,. \tag{5.3.2}$$

These schemes are governed by *Dahlquist's theorems* of stability that determine the accuracy J of the scheme. These theorems state the following.

- For an implicit s-step stable multi-step scheme the order J is less than $s+3$ or $s+2$ if s is even or odd, respectively.
- For an explicit scheme the order J is less than $s+1$. Further, no multi-step scheme of order greater than two is unconditionally stable.

We only consider a few discretisation schemes as many of the well-known schemes can be found in standard textbooks; for example, for general discretisations see Gear [179], and for spectral methods see Canuto *et al.* [86]. In general, we are interested in long-time integration for which *absolute* (i.e., asymptotic) stability of the discretisation is important as compared to weak stability or algebraic stability (see [179] and [199] for rigorous definitions of stability).

5.3.1 *Forward multi-step schemes*

The forward multi-step scheme which encompasses the Adams family of schemes can be expressed as

$$\frac{\boldsymbol{u}^{n+1} - \boldsymbol{u}^{n}}{\Delta t} = \sum_{q=0}^{J-1} \beta_q \boldsymbol{F}^{n+1-q}\,, \tag{5.3.3}$$

where for $\beta_0 = 0$ we obtain the explicit Adams–Bashforth family and for $\beta_0 \neq 0$ we obtain the implicit Adams–Moulton family.

The stability of the scheme can be found by considering the scalar linear problem where $\boldsymbol{F}(\boldsymbol{u}) = \lambda u$ $(\lambda \leqslant 0)$. For example, if we consider the first-order Adams–Bashforth scheme (which is equivalent to an Euler-forward scheme) where $J = 1$, $\beta_0 = 0$, and $\beta_1 = 1$, then when $\boldsymbol{F}(\boldsymbol{u}) = \lambda u$ $(\lambda \leqslant 0)$ eqn (5.3.3) can be written as

$$\frac{u^{n+1} - u^{n}}{\Delta t} = \lambda u^{n}$$

or

$$u^{n+1} = (1 + \Delta t \lambda) u^{n}\,. \tag{5.3.4}$$

From this example we observe that the scheme is stable if $|1+\Delta t\lambda| \leqslant 1$, which in the complex plane includes the region of a unit circle centred at $\lambda \Delta t = -1$. For higher-order schemes (i.e., $J > 1$) an extension of this analysis can be performed in terms of the eigenvalues of a matrix problem acting on a vector of the form $\boldsymbol{v}^{n+1} = [u^{n+1}, u^{n}, \ldots, u^{n-J}]^{\top}$, see [179].

In Table 5.1 we list the values of the coefficients β_q up to order $J = 3$, and in Fig. 5.8 we graphically show the stability regions for the explicit and implicit schemes. The diffusion operator $\boldsymbol{M}^{-1}\boldsymbol{L}$ typically has purely real eigenvalues, and so it can be appreciated that the lower-order explicit schemes when $J = 1$ will be the most stable. In Chapter 6 we will consider a similar analysis for the semi-discrete problem arising from the discretisation of a linear advection equation. In this case the matrix arising from the advection operator can have purely imaginary eigenvalues. It can therefore be seen that for the advection operator the third-order Adams–Bashforth scheme, which intersects the imaginary axis at approximately 0.723, is the most stable. Nevertheless, lower-order schemes, and particularly the second-order scheme, are often used in spectral discretisations of

TABLE 5.1. Weights for integrating the advection equation using Adams–Bashforth and Adams–Moulton algorithms.

	Coefficient	1st order	2nd order	3rd order
Adams–Bashforth	β_0	0	0	0
	β_1	1	3/2	23/12
	β_2	0	−1/2	−16/12
	β_3	0	0	5/12
Adams–Moulton	β_0	1	1/2	5/12
	β_1	0	1/2	8/12
	β_2	0	0	−1/12
	β_3	0	0	0

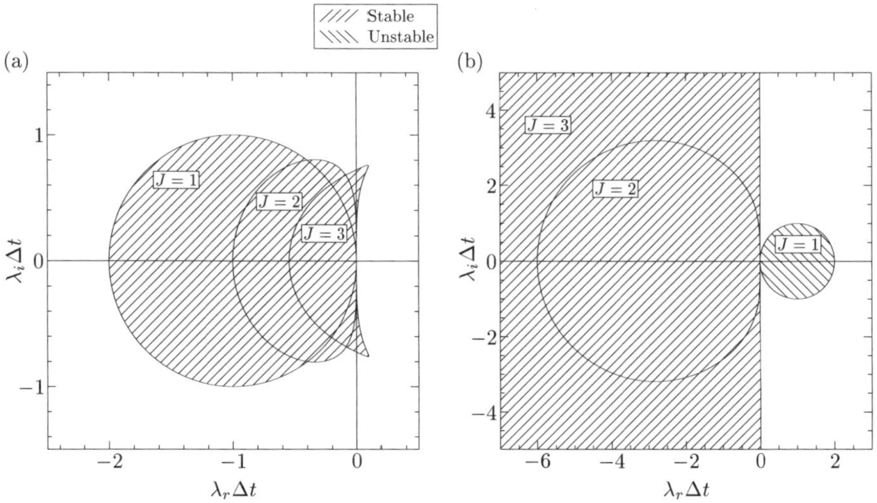

FIG. 5.8. Stability diagrams for (a) the Adams–Bashforth, and (b) the Adams–Moulton schemes.

the advection problem as boundary conditions may introduce eigenvalues with real and imaginary components.

The first-order and second-order Adams–Moulton schemes are unconditionally stable, with the stability region encompassing the entire imaginary axis. The third-order scheme is conditionally stable and touches the imaginary axis only in a small segment around the origin.

5.3.2 *Backward multi-step schemes*

Alternatively, we can use a backward multi-step scheme to discretise eqn (5.3.2) and obtain

$$\frac{\gamma_0 \boldsymbol{u}^{n+1} - \sum_{q=0}^{J-1} \alpha_q \boldsymbol{u}^{n-q}}{\Delta t} = \boldsymbol{F}^{n+1},$$

where for consistency $\gamma_0 = \sum_{q=0}^{J-1} \alpha_q$. This class of schemes is described as *stiffly stable* due to their stability properties, that is, they are stable in the left plane and accurate around the origin (see [179] for details). Explicit schemes can also be constructed by approximating the right-hand side of the implicit discretisation as

$$\boldsymbol{F}^{n+1} \approx \sum_{q=0}^{J-1} \beta_q \boldsymbol{F}^{n+1-q}.$$

The weights for both explicit (SE) and implicit (SI) schemes up to third-order are given in Table 5.2, with the implicit schemes obtained by setting $\beta_0 = 1$ and $\beta_q = 0,\ q > 0$. The corresponding stability regions are shown in Fig. 5.9. We see that the explicit schemes of second-order and above have stability regions similar to the Adams–Bashforth scheme, although somewhat larger. The third-order implicit scheme is slightly unstable around the origin, but otherwise its stability region encompasses the entire left plane. It is interesting to note that for the third-order stiffly-stable implicit scheme, a small amount of diffusion suppresses the instability region around the axis. In Chapter 8 we discuss the stiffly-stable schemes again in the context of Navier–Stokes discretisations.

TABLE 5.2. Weights for mixed explicit–implicit integration of advection–diffusion equations using stiffly-stable integrators (SE/SI scheme).

Coefficient	1st order	2nd order	3rd order
γ_0	1	3/2	11/6
α_0	1	2	3
α_1	0	−1/2	−3/2
α_2	0	0	1/3
β_0	1	2	3
β_1	0	−1	−3
β_2	0	0	1

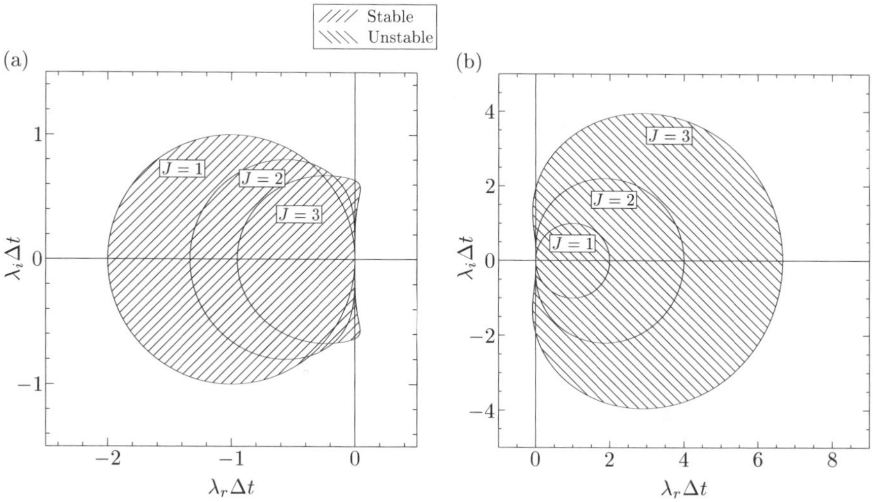

FIG. 5.9. Stability diagrams for (a) the explicit, and (b) the implicit stiffly-stable schemes.

5.4 Eigenspectra and iterative solution of the weak Laplacian operator

In this section we consider the iterative solution and the eigenspectrum of the weak Laplacian operator. This eigenspectrum is significant in the numerical discretisation of the diffusion equation in a number of ways. First, the growth of the largest eigenvalue of the Laplacian operator dictates how the time-step must be scaled in an explicit temporal treatment of the diffusion equation. Secondly, the conditioning of the matrix, which is related to the ratio of the maximum to minimum eigenvalue, informs us how many iterations are required to invert the matrix if an iterative inversion scheme, such as the preconditioned conjugate gradient technique, is applied. In the following section we discuss these two points in greater detail.

5.4.1 *Time-step restriction and maximum eigenvalue growth*

In Section 5.1 we have discussed how the spectral/*hp* element discretisation of the diffusion equation leads us to the semi-discrete system of ordinary differential equations of the form

$$\frac{\mathrm{d}\hat{\boldsymbol{u}}_g}{\mathrm{d}t} = \boldsymbol{M}^{-1}\boldsymbol{L}\hat{\boldsymbol{u}}_g\,, \tag{5.4.1}$$

where we have assumed that the matrix $\boldsymbol{M}^{-1}\boldsymbol{L}$ has been modified to take boundary conditions into consideration. If the matrix $\boldsymbol{M}^{-1}\boldsymbol{L}$ has distinct eigenvalues then we know that $\boldsymbol{M}^{-1}\boldsymbol{L} = \boldsymbol{Q}^{-1}\boldsymbol{\Lambda}\boldsymbol{Q}$, where $\boldsymbol{Q}$ is a matrix of eigenvectors and $\boldsymbol{\Lambda}$ is a diagonal matrix of eigenvalues of $\boldsymbol{M}^{-1}\boldsymbol{L}$. Therefore, if we pre-multiply eqn (5.4.1) by $\boldsymbol{Q}$ and let $\hat{\boldsymbol{v}}_g = \boldsymbol{Q}\hat{\boldsymbol{u}}_g$ then we obtain

$$\frac{d\hat{\boldsymbol{v}}_g}{dt} = \boldsymbol{\Lambda}\hat{\boldsymbol{v}}_g ,$$

which is a decoupled set of ordinary differential equations. As discussed in Section 5.3, the stability of a time-stepping scheme requires that $\lambda_i \Delta t$ (where $\lambda_i = \boldsymbol{\Lambda}[i,i]$) lies within the stability region of the time-integration scheme. For the case of the Galerkin formulation of the Laplacian matrix the eigenvalues of $\boldsymbol{M}^{-1}\boldsymbol{L}$ are real. We are, therefore, interested in the value of $\lambda_{\max} = \max_i \lambda_i$ to determine the stability requirements on Δt. For a C^0 modified modal expansion in a triangle the eigenspectrum of $\boldsymbol{M}^{-1}\boldsymbol{L}$ has been numerically evaluated for the case of a periodic bi-unit square domain formed by two triangles. In Fig. 5.10 we show the growth rate of the maximum eigenvalue as a function of polynomial order, and for this test case we observe that the growth rate $\lambda_{\max} \propto P^n$ is approximated by $n \approx 3.8$. Similarly, in Fig. 5.11 we see the growth rate of the maximum eigenvalue of the matrix $\boldsymbol{M}^{-1}\boldsymbol{L}$ as a function of polynomial order for the three-dimensional periodic cubic domain consisting of six tetrahedra. Considering the last three points, we find that the polynomial growth rate in this case $n \approx 3.7$. Therefore, we hypothesise that in both two and three dimensions the growth rate is bounded by P^4, which is consistent with the standard tensorial spectral methods for quadrilateral and hexahedral regions [86]. For many practical applications, however, a growth rate of $\mathcal{O}(P^4)$ is far too restrictive and so the diffusive terms must be handled implicitly.

5.4.2 *Iterative solution and preconditioners*

An implicit treatment of the diffusion terms naturally requires the inversion of the Helmholtz and/or the Laplacian matrix. Considering the memory available on most computers, in two dimensions direct solvers are tractable using LU

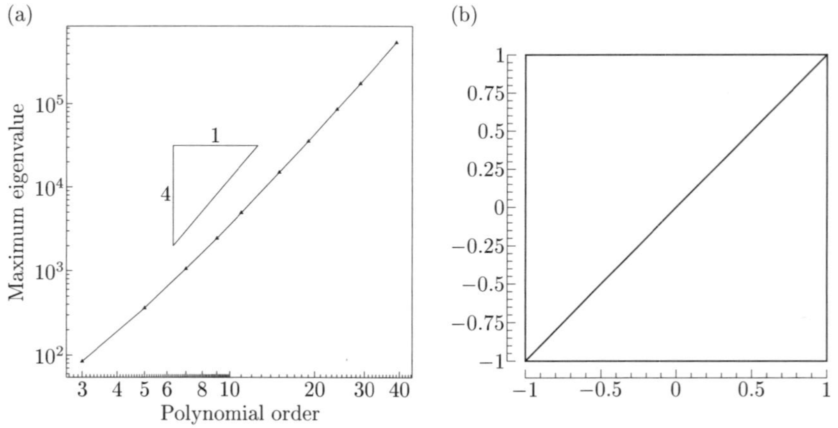

FIG. 5.10. (a) Plot of the maximum eigenvalue of the weak Laplacian matrix, $\boldsymbol{M}^{-1}\boldsymbol{L}$, as a function of polynomial order P for a bi-unit square periodic domain consisting of two triangles, as shown in (b).

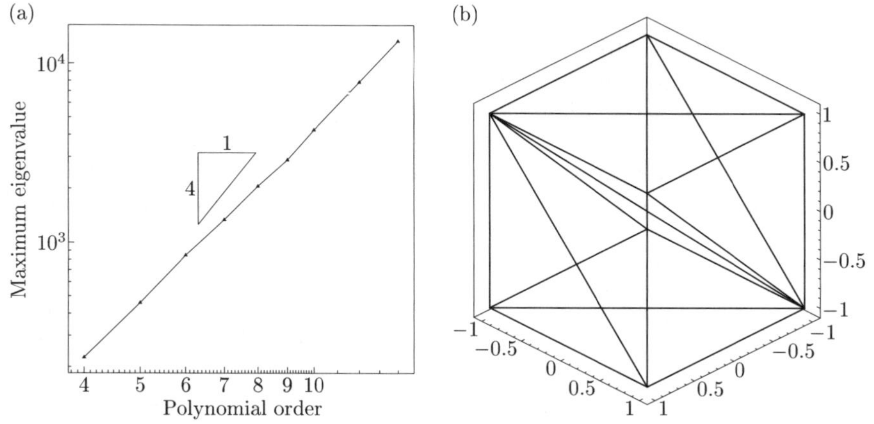

FIG. 5.11. (a) Growth rate of the matrix $\boldsymbol{M}^{-1}\boldsymbol{L}$ for the periodic cubic domain consisting of six tetrahedra shown in (b) as a function of polynomial order. The rate is bounded by P^4.

factorisation and its variant. The static condensation technique described in Section 4.2.3 decouples the interior from the boundary degrees of freedom, which helps to considerably reduce the memory requirement as compared to a full spare matrix inverse. Nevertheless, in three dimensions, when parallel implementation may also be required, iterative solutions become necessary.

One of the most commonly applied techniques in spectral/hp element methods, especially for symmetric positive-definite elliptic solvers, is the preconditioned conjugate gradient technique. For a fuller description of the technique, as well as a discussion on alternative iterative solution techniques, we refer the reader to [21, 129, 195]. Following [129], given a preconditioning matrix $\boldsymbol{P}$ to the matrix $\boldsymbol{A}$, the preconditioned conjugate gradient method for the solution of $\boldsymbol{Ax} = \boldsymbol{b}$ can be written as ♣22

compute $\boldsymbol{r}_0 = \boldsymbol{A}\boldsymbol{x}_0$, $\boldsymbol{z}_0 = \boldsymbol{P}^{-1}\boldsymbol{r}_0$, and $\boldsymbol{w}_0 = \boldsymbol{z}_0$
do $j = 0$, until convergence
$\alpha_j = (\boldsymbol{r}_j, \boldsymbol{z}_j)/(\boldsymbol{A}\boldsymbol{w}_j, \boldsymbol{w}_j)$
$\boldsymbol{x}_{j+1} = \boldsymbol{x}_j - \alpha_j \boldsymbol{w}_j$
$\boldsymbol{r}_{j+1} = \boldsymbol{r}_j - \alpha_j \boldsymbol{w}_j$
$\boldsymbol{z}_{j+1} = \boldsymbol{P}^{-1}\boldsymbol{r}_{j+1}$
$\beta_j = (\boldsymbol{r}_{j+1}, \boldsymbol{z}_{j+1})/(\boldsymbol{r}_j, \boldsymbol{z}_j)$
$\boldsymbol{w}_{j+1} = \boldsymbol{z}_{j+1} + \beta_j \boldsymbol{w}_j$
continue

22 Preconditioned conjugate gradient method.

The preconditioned conjugate gradient technique can be considered as semi-iterative since it is guaranteed to converge in a finite number of iterations, assuming exact arithmetic [195]. The convergence rate of the algorithm depends on the eigenspectra of the matrix $\boldsymbol{A}$ and the preconditioned matrix $\boldsymbol{P}^{-1}\boldsymbol{A}$. For the preconditioned conjugate gradient method the number of iterations, N_{iter}, to invert a matrix $\boldsymbol{A}$ scales with the square root of the condition number [129], that is,

$$N_{\text{iter}} \propto [\kappa_2(\boldsymbol{P}^{-1}\boldsymbol{A})]^{1/2}.$$

The L^2 condition number of a non-singular matrix $\boldsymbol{A}$ is defined as $\kappa_2(\boldsymbol{A}) = ||A||_2||A^{-1}||_2$, which for a symmetric positive-definite matrix is equivalent to the ratio of the largest to the smallest eigenvalue.

The choice of the preconditioner matrix $\boldsymbol{P}$ should be such that it is easy to invert and ideally closely approximates the eigenspectrum of $\boldsymbol{A}$. In the following section we briefly discuss results relating to the scaling of the condition number produced by different approaches to preconditioning. For the interested reader we refer to the book by Deville *et al.* [129] for *nodal basis* expansion preconditioning, the book by Smith *et al.* [439] which discuses more general domain decomposition and Schwarz methods, and finally the proceedings of the domain decomposition methods conferences [231].

5.4.2.1 *Condition number of the two-dimensional modal expansion Laplacian matrix*

The condition number of the *full* discrete two-dimensional Laplacian $\boldsymbol{L}$, based upon the modified C^0 basis introduced in Section 3.2.3, scales as

$$\kappa_2(\boldsymbol{L}) \propto N_{\text{el}}P^3.$$

The required number of iterations for a preconditioned conjugate gradient method can therefore be very high for large problems, and especially for high-order P. However, following the technique suggested in Section 4.2.3, we can consider the substructured or statically condensed problem. We recall this involved forming the Schur complement $\boldsymbol{L}_S$:

$$\boldsymbol{L}_S = \boldsymbol{L}_b - \boldsymbol{L}_c[\boldsymbol{L}_i]^{-1}\boldsymbol{L}_c^{\top},$$

where

$$\boldsymbol{L} = \begin{bmatrix} \boldsymbol{L}_b & \boldsymbol{L}_c \\ \boldsymbol{L}_c^{\top} & \boldsymbol{L}_i \end{bmatrix}.$$

Here $\boldsymbol{L}_b$ represents the boundary–boundary interaction of the matrix system, and $\boldsymbol{L}_c$ and $\boldsymbol{L}_i$ represent the boundary–interior and interior–interior part of the system, respectively. For a symmetric positive-definite system the condition number of the Schur complement matrix $\boldsymbol{L}_S$ can be no larger than the condition number of the complete system $\boldsymbol{L}$ [439].

Experimental results indicating the scaling of $\kappa_2(\boldsymbol{L}_S)$ have been obtained in the domains shown in Fig. 5.12. The relationship of the condition number with

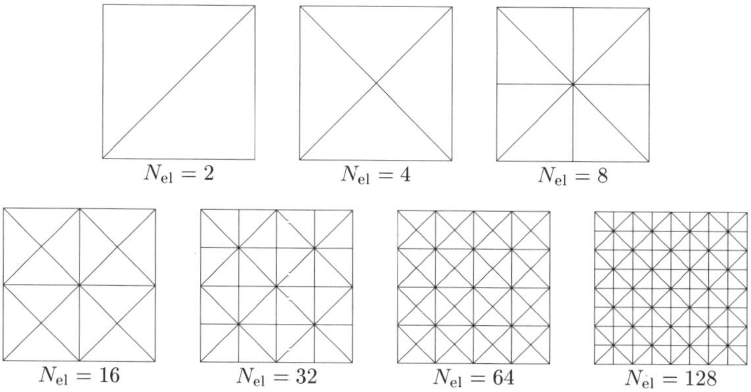

FIG. 5.12. Triangulations used in determining the eigenspectrum of the Laplacian operator.

polynomial order P is demonstrated in Fig. 5.13(a), and with the number of elements in Fig. 5.13(b). The variation of the maximum and minimum eigenvalue with respect to the expansion order is shown in Fig. 5.14. The maximum eigenvalue is independent of the order P, in accordance with the estimates in [89]. Also, the minimum eigenvalue varies as $\approx 1/P$, which is consistent with the theoretical upper bound estimate in [89] of $(\log P)/P$. For the range considered, these results also seem to indicate that the condition number grows at most linearly with the order P and slower than logarithmically with the number of elements N_{el}. However, experiments with a larger number of elemental do-

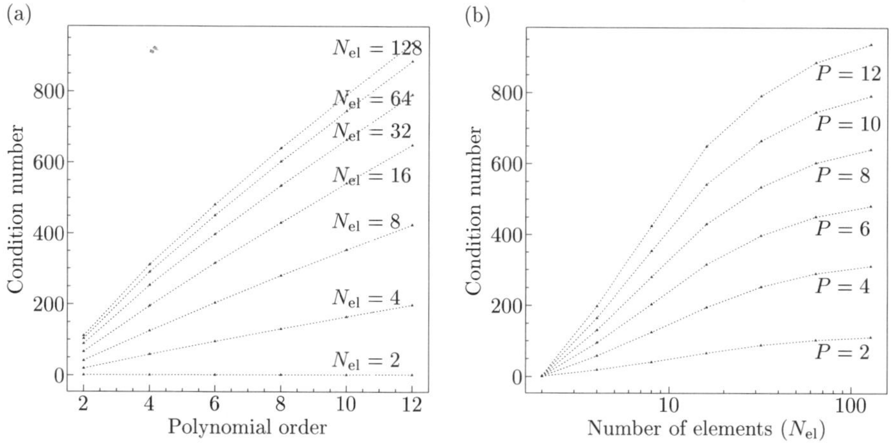

FIG. 5.13. Condition number variation of the Schur complement of the discrete Laplacian operator with respect to (a) the order, and (b) the number of elements.

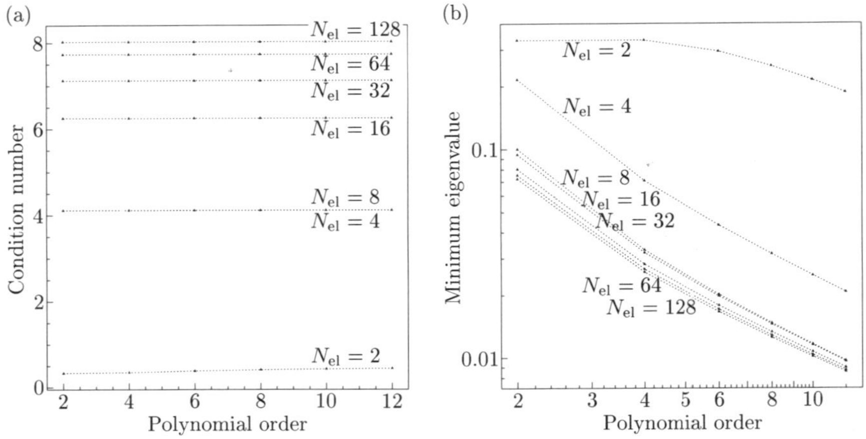

FIG. 5.14. (a) Maximum, and (b) minimum eigenvalue of the Schur complement of the discrete Laplacian operator with respect to the expansion order.

mains indicate that the condition number grows with N_{el} asymptotically, which is consistent with the theory for diagonal preconditioning.

To get a better indication of the asymptotic behaviour with the polynomial order, P, we consider the first case of two elemental regions for the polynomial range $8 \leqslant P \leqslant 48$, as shown in Fig. 5.15. In this test case we have imposed Dirichlet boundary conditions on all boundaries, and so when we statically condense the system we are only left with the interior edge system. In Fig. 5.15 we see the condition number for this problem plotted against the functions (a) $P \log P$, (b) P, and (c) $P/\log P$. As can be seen, the condition number clearly grows at a slower rate than $P \log P$ but faster than $P/\log P$, which is consistent with the

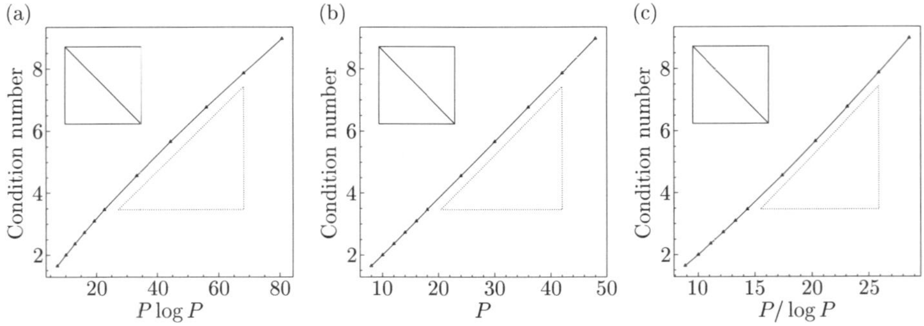

FIG. 5.15. Condition number of the Schur complement of the discrete Laplacian operator for the $N_{el} = 2$ element domain plotted as a function of (a) $P \log P$, (b) P, and (c) $P/\log P$.

results of [89]. Although not formally proven, the asymptotic rate is most likely to scale linearly with P. In the absence of any interior vertices, for example, a mesh of $1 \times N_{el}$ elements with Dirichlet boundary conditions, the above scaling for two elements still applies. However, if there is an interior vertex, known as a cross-point mesh, then it will scale as $(\log P)^2$.

5.4.2.2 *Preconditioned systems based on modal expansions*

The conditioning of the spectral/hp elliptic matrix systems with the number of elemental domains N_{el} when using a modal expansion is primarily contained within the linear vertex expansion modes. The hierarchical nature of the modal expansion means that the linear finite element matrix is contained within the higher-expansion-order matrix and can therefore be extracted by appropriate contraction of the full matrix. Using the vertex block submatrix as part of the preconditioner typically removes the number of elements scaling and therefore reduces the problem primarily to one of controlling the dependency with polynomial order P. Given the comparatively small amount of elements that are used in a spectral/hp element discretisation as compared to a classical linear finite element method, this coarse space submatrix can typically be directly inverted.

Diagonal preconditioning in two dimensions

For a two-dimensional mesh where the edges are diagonally preconditioned and the complete vertex block is applied, the condition number of the Schur complement, $\kappa_2(\boldsymbol{L}_S)$, can be bounded by

$$P \log P \leqslant \kappa_2(\boldsymbol{L}_S) \leqslant P(\log P)^3 .$$

Since the bounds are only realised for a large P, these estimates are often observed to be very conservative. Further, one would expect to observe a sharp upper bound of $P(\log P)^2$ based on the numerical behaviour shown in Fig. 5.15 for large P. These results may equally well be applied to the quadrilateral region which has similar edge support.

When diagonal edge preconditioning is applied as a preconditioner the absolute magnitude of the condition number is approximately one order of magnitude less compared with the unpreconditioned case. This is not, however, true for the modal basis constructed in [445] based on the integrated Legendre polynomials rather than the $P_p^{1,1}(x)$ Jacobi polynomials (see Section 2.3.3.3). The difference between these two formulations in the quadrilateral expansion is the presence of a factor of P that provides the appropriate scaling, and therefore the matrix demonstrates similar growth in the unpreconditioned case and the diagonal-preconditioned case.

Block diagonal preconditioning in two dimensions

If a block diagonal preconditioner is constructed based on blocks of edge–edge interactions as well as the vertex block, then the scaling changes substantially. Numerical experiments suggest a scaling of $(\log P)^2$, in agreement with the estimates reported in [24] for a similar construction. The condition number also seems to be independent of the number of elements for $N_{el} \geqslant 100$, again in agreement with the estimates in [24].

Low-energy preconditioning of the modal expansion in three dimensions

In three dimensions, it is more difficult to establish estimates of the condition number. In [367] a polylogarithmic bound for a wire basket preconditioner was found of the form

$$\kappa_2 \leqslant C(1 + \log P)^2 ,$$

which is independent of the number of elements. This estimate is valid for hexahedral elements, but a similar bound was obtained in [49] for tetrahedral elements. The main idea is to use a wire basket preconditioner that is based on a new, numerically-determined, set of vertex and edge basis functions of 'low energy'. In [429] the low-energy basis work of Bica [49] was implemented in a computationally efficient manner by using a symmetric reference element to design the low-energy functions. An example of the numerically-derived basis is shown in Fig. 5.16, where we show the standard linear finite element vertex mode in Fig. 5.16(a) and the numerically-determined low-energy vertex mode in Fig. 5.16(b). The low-energy functions have traces on the wire basket which decay much faster than the standard basis functions. The transformed Schur complement system using the low-energy basis has a much more diagonally-dominant structure, as shown in Fig. 5.16(c,d). In this figure we compare the Schur complement matrix, $\boldsymbol{S}_1$, using the standard closed-form expansion with the transformed Schur complement, $\boldsymbol{S}_2$, using the numerically-determined low-energy basis. The transformed Schur complement is, therefore, more amenable to block diagonal preconditioning. In [429], using regular-shaped elemental domains, a polylogarithmic condition number was numerically observed of the form $(1 + \log P)^2$, consistent with the findings in [49, 367].

An alternative approach that employs full orthogonalisation of each vertex function with respect to functions of its three faces, and each edge function with respect to functions of its two faces, was proposed in [322].

Scaling of the two-dimensional Helmholtz matrix

The efficiency by which we can invert the *Helmholtz* matrix depends on the *combined* spectrum of the Laplacian matrix and the mass matrix. The eigenspectrum of the Schur complement of the mass matrix has not been studied theoretically, but numerical experiments with triangular elements suggest a similar dependence to the Laplacian matrix with respect to the number of elements N_{el}. However, with respect to polynomial order P its condition number grows much faster. In particular, using the triangular modal basis with no preconditioning we observed a growth $\kappa_2(\boldsymbol{M}) \propto P^{5/2}$; with diagonal preconditioning $\kappa_2(\boldsymbol{M}) \propto P^{1.95}$; and with block diagonal preconditioning, as before, we obtained $\kappa_2(\boldsymbol{M}) \propto P^{1.6}$. These are probably polylogarithmic terms but we best-fitted experimental results to obtain these exponents.

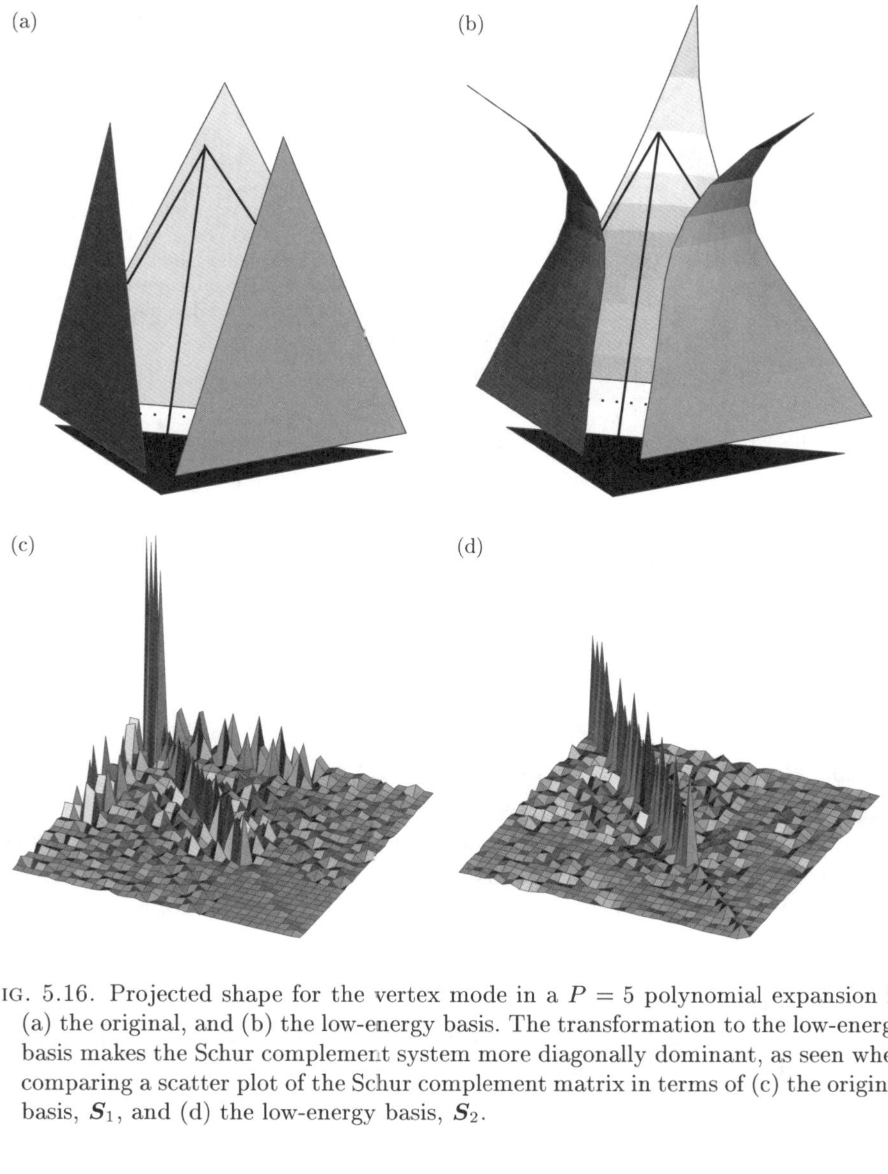

FIG. 5.16. Projected shape for the vertex mode in a $P = 5$ polynomial expansion in (a) the original, and (b) the low-energy basis. The transformation to the low-energy basis makes the Schur complement system more diagonally dominant, as seen when comparing a scatter plot of the Schur complement matrix in terms of (c) the original basis, $\boldsymbol{S}_1$, and (d) the low-energy basis, $\boldsymbol{S}_2$.

In summary, it is possible to apply preconditioning techniques for inverting the Helmholtz matrix similar to preconditioners for the Laplacian and the mass matrix. The effect on the convergence rate of a preconditioned conjugate gradient solver for the Helmholtz equation with constant $\lambda = 1$ is shown in Fig. 5.17, which also verifies the fast convergence for the Schur complement, in contrast to the full discrete Laplacian.

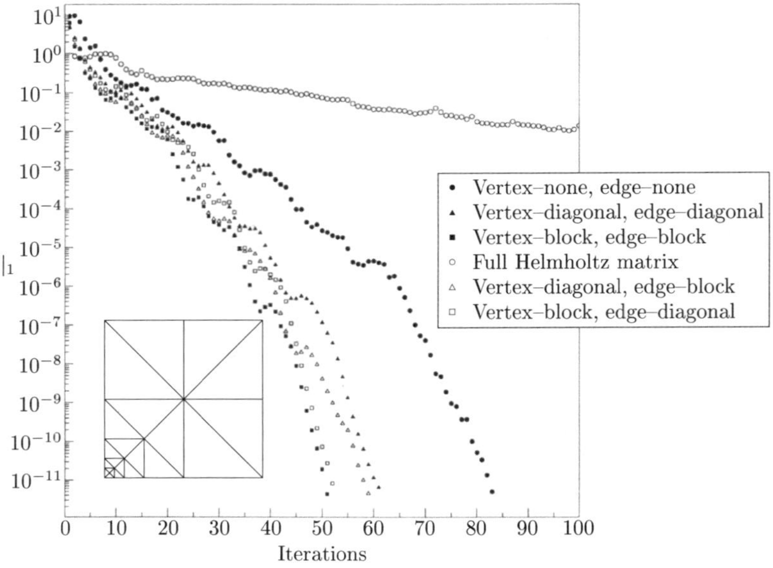

FIG. 5.17. Convergence rate of a preconditioned conjugate gradient solver for a Schur complement of the Helmholtz matrix using the indicated preconditioners. Also shown for comparison is the solution for the full Helmholtz matrix.

5.4.2.3 *Preconditioned systems based on nodal expansions*

As in Section 5.4.2.1, which relates to the hierarchical modal basis, the condition number of the Schur complement using a nodal expansion basis has been numerically observed to scale linearly with the order P. However, unlike the hierarchical basis, the nodal basis also grows linearly with the number of elements $N_{\rm el}$, even with a small number of elements. Preconditioning with the diagonal matrix does not change this scaling, although it substantially reduces the value of the condition number.

Numerical experiments with an incomplete Cholesky preconditioner on the Schur complement system indicated a scaling $\kappa_2 \propto P^{1/2}$, which does not change with the number of elements. The overall gain in computer time using the incomplete Cholesky preconditioner was a factor of 2, taking into account the overhead associated with the construction and inversion of the preconditioner.

Other more effective preconditioners can be used, such as the tridiagonal preconditioner in [92], which are independent of the expansion order; however, they are typically associated with a large computational overhead. Note that in [92] the condition number of the Schur complement was found to grow quadratically with the order P and only linearly, after preconditioning, with the diagonal. The difference from the linear scaling of the unpreconditioned Schur complement of the nodal basis is due to the different scaling of the basis employed, as discussed earlier.

When considering a Helmholtz problem with a large Helmholtz constant the diagonal mass matrix makes the system diagonally dominant and favourable to iterative methods based on the full matrix system rather than the Schur complement. One of the first people to advocate iterative solutions of collocation calculations was Orszag [352], who suggested that a finite difference scheme on the same collocation grid as the spectral method would result in a rapidly converging scheme. He further demonstrated that there was a spectral equivalence between the finite difference and spectral operators. This preconditioner concept was extended to include the finite element methods by Deville and Mund [131]. This was found to work even more effectively than the finite difference preconditioner. The finite element preconditioner has also been widely applied in iterative spectral element calculations, for example, see [132].

More recently, the overlapping Schwarz method has been adopted in the context of preconditioning of nodal spectral element methods [90, 161, 359]. In this approach local problems are solved on overlapping subdomains that contain a complete spectral element plus an overlap region that includes a few Gauss–Lobatto–Legendre collocation points in the adjacent elements. Using the spectral equivalence of a finite element method based on the Gauss–Lobatto–Legendre points, the finite element technique is applied to solve the overlapping subdomain problems. When combined with a coarse grid solver in a regular domain the overlapping Schwarz method has demonstrated convergence in a fixed number of iterations, independent of polynomial order and number of elements. A similar overlapping technique has also been applied to the Schur complement system using modal expansions in two and three dimensions by Pavarino and Warburton [368]. However, in the modal expansion case the overlap region is not obvious as there is no longer a collocation property to the expansion basis. In the work of Pavarino and Warburton [368] a whole element overlap was applied. This is relatively expensive and therefore reduces the efficiency of the technique as compared to its application in a nodal basis.

Finally, we note that multigrid has also been applied in nodal spectral elements preconditioning, see [129, 312] for an overview. The construction of both modal and nodal spectral/*hp* expansions suggests that using multigrid based upon successive reductions in polynomial order would be an effective preconditioner. Early work in this direction was undertaken by Rønquist and Patera [405]. Whilst the approach was shown to be effective in one dimension, in higher dimensions a degradation in performance was attributed to high-aspect-ratio cells. An alternative approach to using multigrid is to adopt a finite element (or finite difference) preconditioner at the Gauss–Lobatto–Legendre points and use classical multigrid to solve the finite element problem. Such an approach was adopted by Zang *et al.* [508]. A comprehensive review of these approaches and comparison with a recently suggested multigrid technique combined with the overlapping Schwarz approach is given by Lottes and Fischer [312]. In this paper they demonstrate that it is possible to achieve convergence rates with spectral element multigrid which are comparable to classic multigrid on regular grids.

5.5 Non-smooth domains

Exponential convergence is obtained with spectral/*hp* element methods if the solution is smooth, i.e., possessing a high degree of regularity. However, there are a number of cases for which the solution of a Helmholtz equation may be singular [205]. The specific cases that we examine in this section are solutions in a non-smooth domain. Other cases, not considered in this section, involve a smooth domain boundary but with a discontinuity in the boundary conditions, or in the specified data (for example, forcing). Here we assume that all the data, as well as the boundary conditions, are continuous and that singularities are due to corners in the domain. First derivatives are unbounded when the angle is reflexive or convex (as in Fig. 5.18), and second derivatives are unbounded when the angle is acute or obtuse. In this case, not only may the fast convergence of spectral/*hp* discretisation be destroyed, but also the numerical solution obtained (with any standard method) may be erroneous. An example of the latter case is the so-called drum problem [142] for computing eigenvalues. In general, theoretical results for vertex, edge, and combined vertex–edge singularities in three dimensions are more difficult to obtain, but work by Guo [216] has addressed these issues.

To proceed, we consider the domain shown in Fig. 5.18 with the corner located at the origin and with one side of the corner aligned along the x_1 axis, while the other is at an angle $\alpha\pi$, $0 < \alpha < 2$, in the counterclockwise direction. The solution in the local polar coordinates typically has the form

$$u(r, \theta) \propto r^{\beta} s(\theta)\chi(r) .$$

Here $s(\theta)$ is an analytic function and $\chi(r)$ is a smooth cut-off function, e.g.,

$$\chi(r) = \mathrm{e}^{-(r-R)^2} , \quad R \leqslant r \leqslant R^* ,$$

where, for $r > R^*$ and for $r < R$, $\chi(r)$ is equal to 0 and 1, respectively. In this case, the spectral element solution u^{δ} computed with order P satisfies

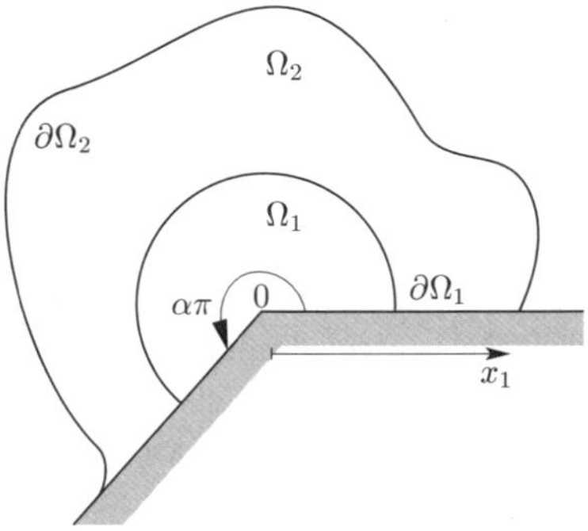

FIG. 5.18. Computational domain Ω with a single corner of angle $\theta = \alpha\pi$.

$$\|u - u^\delta\|_1 \leqslant CP^{-2\beta-\epsilon},$$

where $\epsilon > 0$ and the exponent β depends on the angle. For many problems $\beta = 1/\alpha$ (where $\theta = \alpha\pi$), and thus the convergence rate lies between $\mathcal{O}(P^{-1})$ and $\mathcal{O}(P^{-2})$.

In general, in spectral element h- and p-type methods for the Laplace operator in two-dimensional domains, the error bounds have the form

$$\|u - u^\delta\|_1 \leqslant Ch^n P^m, \tag{5.5.1}$$

where h denotes a characteristic length of the largest element and P denotes the polynomial degree of the test and trial functions. These bounds depend on the regularity of the exact solution (and therefore on the singularity exponents), the polynomial degree of the spectral elements, and the discretisation type, that is the h-version, p-version, or *hp* version; see [26, 27, 31] and references therein.

In Table 5.3 we summarise known results for the asymptotic rate of convergence for the two-dimensional solutions of Laplace's equation in the H^1 norm. In interpreting Table 5.3 we consider two types of convergence denoted as follows:

- algebraic rate: $\|u - u^\delta\|_1 \leqslant \dfrac{C}{P}$;
- exponential rate: $\|u - u^\delta\|_1 \leqslant \dfrac{C}{\exp(\gamma P^\kappa)}$.

In Table 5.3, 'Category A' refers to the solution u being analytic on each spectral element including the boundary; we say that u is a *smooth function.* Also, in 'Category B' the solution u is analytic on each spectral element, including the boundary, except at some points, namely the so-called *singular points.* The smoothness of u is characterised by $\beta = \min\{\tilde{\beta}_1\}$, so that $\tilde{\beta}_1$ is the smallest β over all singular points in the domain. Finally, 'Category C' includes solutions u

TABLE 5.3. Summary of convergence rates for the two-dimensional Laplace equation.

	Type of extension		
Category	h	p	hp
A	Algebraic $\kappa = P/2$	Exponential $\kappa \geqslant 1/2$	Exponential $\kappa \geqslant 1/2$
B	Algebraic (see Note 1) $\kappa = \frac{1}{2}\min(P, \beta)$	Algebraic $\kappa = \beta$	Exponential $\kappa \geqslant 1/3$
C	Algebraic $\kappa > 0$	Algebraic $\kappa > 0$	See Note 2

Note 1: Uniform or quasi-uniform meshes are assumed. The maximum possible value of κ obtainable with optimal (adaptively-determined) meshes is $P/2$.

Note 2: When the exact solution has an identifiable structure then nearly exponential convergence rates can be obtained with *hp* adaptive schemes.

that do not belong in either of the aforementioned categories. This happens in nonlinear problems or when the singular points or lines are not known a priori.

There are three main methods that allow us to recover, if not exponential, at least very fast convergence for most elliptic problems. They make use of the following.

1. Gradual h-refinement, usually using radical or geometrically-refined meshes (see [31]).
2. Conformal mappings to smooth the singularity.
3. Extra basis functions that contain the form of singularity referred to as *space enrichment methods*. These require knowledge of the eigenpair representation of the solution.

The first approach requires the application of an appropriate discretisation strategy and quasi-uniform meshing, but can be applied using the standard solution algorithm presented in Section 5.1. A geometric progression has been found to be effective with a ratio of $(\sqrt{2}-1)^2 \approx 0.17$, independently of the strength β of the singularity, see [211,445]. In practice a value of 0.15 is typically adopted. All the other approaches require explicit modification of the standard solution algorithm previously presented in Section 5.1. In the following we briefly review the main ideas of these methods with emphasis on the auxiliary mapping method.

5.5.1 *Laplace equation in two-dimensional domains*

The method of conformal or *auxiliary mapping* is relatively simple and very effective in practice. Let us start by considering Laplace's equation:

$$\nabla^2 u = 0\,,$$

with $u = 0$ on $\partial\Omega_1$ and $u = g(\boldsymbol{x})$ on $\partial\Omega_2$, see Fig. 5.18. In general, in the neighbourhood of the corner the solution can be expressed as [111,274]:

$$u(r,\theta) = \sum_{k=1}^{\infty}\sum_{j=1}^{J}\sum_{i=1}^{I} a_{kji} r^{\beta_k+j} (\ln r)^i s_{kji}(\theta)\,, \tag{5.5.2}$$

where I might be 1 for integer eigenvalues (and a right-hand-side term exists), and J differs from 0 if the boundaries intersecting at O are curvilinear arcs. The singular term due to the curvature (with $k = 1, j = 1$) may be more singular than the second term (with $k = 2, j = 0$).

Here, we concentrate on the most common cases for which eqn (5.5.2) is reduced to

$$u(r,\theta) = \sum_{k=0}^{\infty} a_k \varphi_k(r,\theta)\,,$$

where the coefficients a_k are determined by the boundary conditions, and the fundamental solutions in the case of homogeneous Dirichlet boundary conditions are

$$\varphi_k(r,\theta) = \begin{cases} r^{k/\alpha} \sin\left(\dfrac{k}{\alpha\theta}\right), & \dfrac{k}{\alpha} \text{ is not an integer}, \\ r^{k/\alpha}\left[\ln r \sin\left(\dfrac{k}{\alpha\theta}\right) + \theta\cos\left(\dfrac{k}{\alpha}\right)\right], & \dfrac{k}{\alpha} \text{ is an integer}. \end{cases}$$

For zero Neumann boundary conditions the sine term is replaced by $\cos(k/\alpha\theta)$, and for a problem with a Dirichlet condition on one side and Neumann on the other side, it is replaced by $\sin[(k/\alpha)(\theta/2)]$. In the following we assume that we have homogeneous Dirichlet boundary conditions along both sides and that the logarithmic term does not contribute to the solution. We introduce the mapping

$$z = \xi^\alpha, \quad \text{where } z = re^{i\theta}, \ \xi = \rho e^{i\theta},$$

which is shown graphically in Fig. 5.19. This mapping is conformal at all points except at the origin, which makes the transformed solution analytic in terms of the new variables, thus

$$u(r,\theta) = \sum_{k=0}^{\infty} a_k r^{k/\alpha} \sin\left(\frac{k}{\alpha}\theta\right) \quad \mapsto \quad u(\rho,\phi) = \sum_{k=0}^{\infty} a_k \rho^k \sin(k\phi).$$

For the spectral/*hp* element discretisation this method was first implemented in [29]. An alternative implementation was adopted in [364], where the domain close to the corner was patched to the rest of the domain using the iterative patching technique described in Section 7.2. The example of Fig. 5.20 is taken from [364] and corresponds to a spectral element solution. Without the mapping the leading-order term in the solution behaves as $r^{3/2}$ and the theoretically predicted convergence is $\mathcal{O}(P^{-6/4-\epsilon})$, which is close to the experimentally obtained rate of $\mathcal{O}(P^{-1.615})$. When the mapping is applied the error decays at an exponential rate which has a numerical fit of $\mathcal{O}(e^{-12.924P})$.

5.5.2 *Laplace equation in three-dimensional domains*

The solution of the Laplace equation in three dimensions, in the vicinity of singularities, can be decomposed into three different forms, depending whether

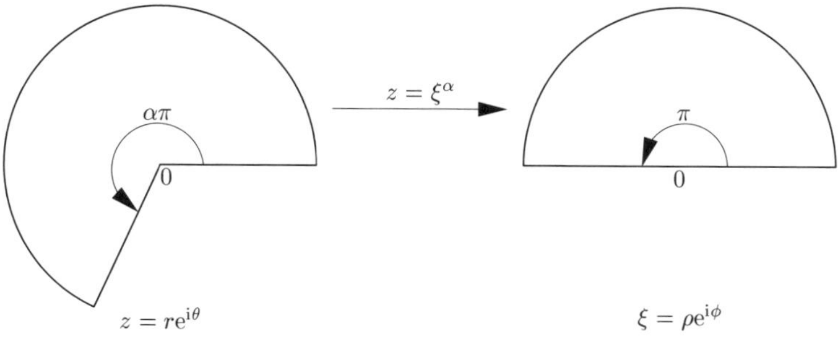

FIG. 5.19. Auxiliary mapping of the subdomain containing the singular corner of a domain with no singular corners.

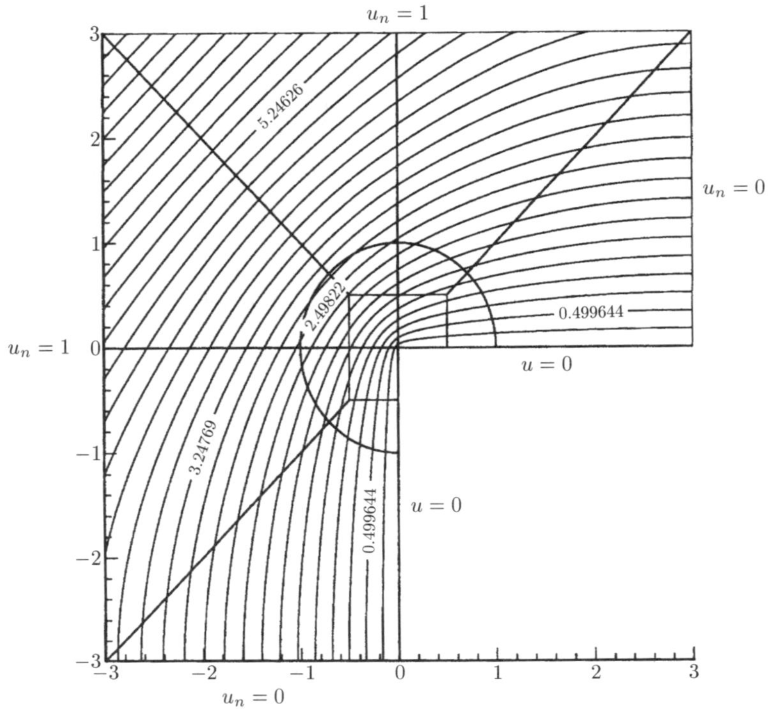

FIG. 5.20. Solution of Laplace's equation using the spectral element mesh shown. The circular subdomain is used in the auxiliary mapping. ([364].)

it is in the neighbourhood of an *edge*, a *vertex*, or an *intersection* of the edge and the vertex. Mathematical details on the decomposition can be found in [13, 33, 110, 124, 206, 291] and the references therein. A representative three-dimensional domain denoted by Ω, which contains typical three-dimensional singularities, is shown in Fig. 5.21. Vertex singularities arise in the neighbourhood of the vertices A_i, and edge singularities arise in the neighbourhood of the edges Λ_{ij}. In this problem only straight edges are considered; for curved edges see [110]. Close to the vertex–edge intersection, vertex–edge singularities arise. In the vicinity of edges or vertices of interest, it is assumed that homogeneous boundary conditions are applied for clarity and simplicity of presentation. We now provide a short outline of the singular solution decomposition in the neighbourhood of an edge, a vertex, or at a vertex–edge intersection; corresponding computational methods are presented in detail in Yosibash [501].

Edge singularities

Let us consider one of the edges denoted by Λ_{ij} connecting the vertices A_i and A_j. Moving away from the vertex a distance $\delta/2$, we create a cylindrical sector subdomain of radius $r = R$, with the edge Λ_{ij} as its axis. We denote this region

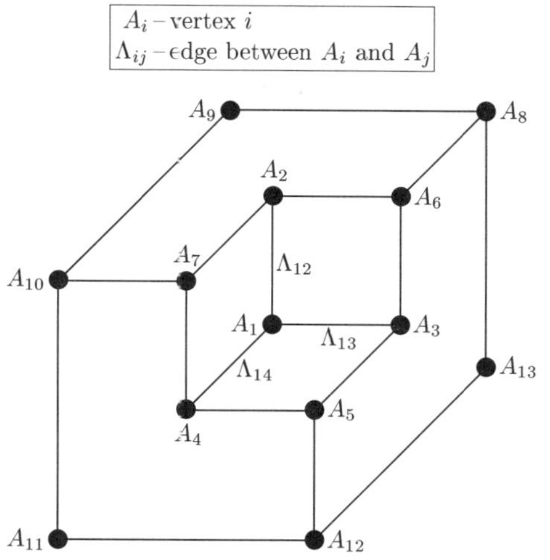

FIG. 5.21. Typical three-dimensional singularities.

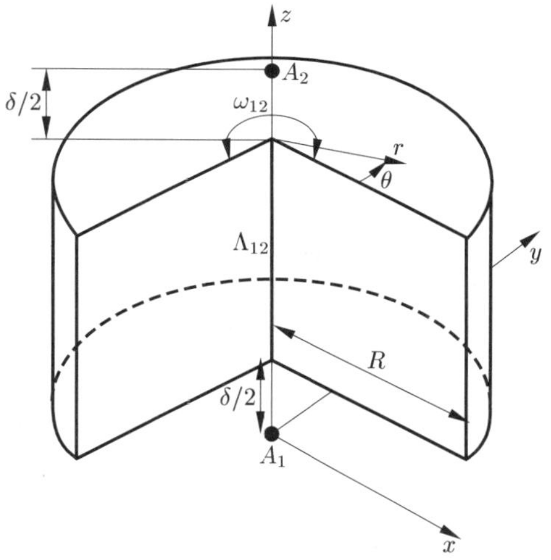

FIG. 5.22. The edge subdomain $\mathcal{E}_{\delta,R}(\Lambda_{12})$.

by $\mathcal{E}_{\delta,R}(\Lambda_{ij})$. This is shown in Fig. 5.22 for the edge Λ_{12}. We again emphasise that our attention is restricted to domains having straight edges. The solution in $\mathcal{E}_{\delta,R}$ can be decomposed as follows:

$$u(r,\theta,z) = \sum_{k=1}^{K}\sum_{\ell=0}^{L} a_{k\ell}(z)\, r^{\beta_k} (\ln r)^{\ell} s_{k\ell}(\theta) + v(r,\theta,z)\,, \tag{5.5.3}$$

where $L \geqslant 0$ is an integer which is zero unless β_k is an integer, $\beta_{k+1} \geqslant \beta_k$ are called edge eigenvalues, and the functions $a_{k\ell}(z)$ are analytic in z and are denoted by edge flux intensity functions. The functions $s_{k\ell}(\theta)$ are also analytic in θ and are referred to as edge eigenfunctions. The function $v(r,\theta,z)$ belongs to $H^s(\mathcal{E})$, the usual Sobolev space, where s can be as large as required and depends on K. We shall assume that β_k for $k \leqslant K$ are not integers, and that no 'crossing points' are of interest (see a detailed explanation in [110]). Therefore, eqn (5.5.3) becomes

$$u(r,\theta,z) = \sum_{k=1}^{K} a_k(z)\, r^{\beta_k} s_k(\theta) + v(r,\theta,z)\,. \tag{5.5.4}$$

Vertex singularities

A sphere of radius $\rho = \delta$, centred in the vertex A_{11} for example, is constructed and intersected with the domain Ω. Then, a cone having an opening angle $\theta = \sigma$ is constructed such that it intersects at A_{11}, and removed from the previously constructed subdomain, as shown in Fig. 5.23(a). The resulting vertex subdomain is denoted by $\mathcal{V}_\delta(A_{11})$, and the solution u can be decomposed in $\mathcal{V}_\delta(A_{11})$ using a spherical coordinate system as follows:

$$u(\rho,\phi,\theta) = \sum_{l=1}^{L} \sum_{q=0}^{Q} b_{lp}\, \rho^{\gamma_l} (\ln \rho)^q s_{lp}(\phi,\theta) + v(\rho,\phi,\theta)\,, \tag{5.5.5}$$

where $Q \geqslant 0$ is an integer which is zero unless γ_l is an integer, $\gamma_{l+1} \geqslant \gamma_l$ are referred to as vertex eigenvalues, and the functions $s_{lp}(\phi,\theta)$ are analytic in ϕ and θ *away from the edges* and are called vertex eigenfunctions; b_{lp} are denoted by vertex flux intensity factors. The function $v(r,\theta,z)$ belongs to $H^q(\mathcal{V})$, where

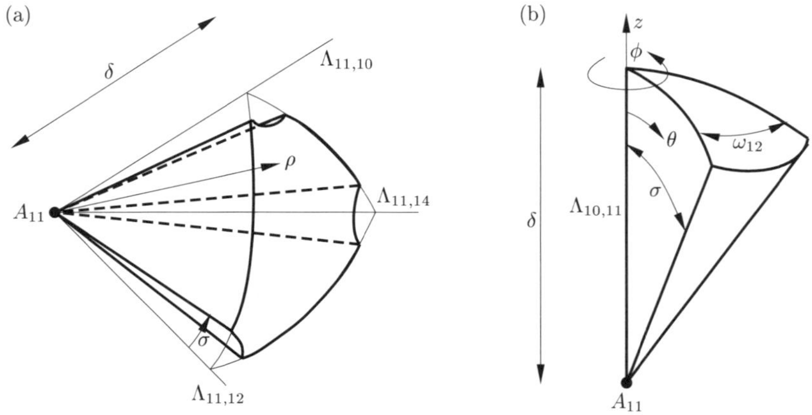

FIG. 5.23. (a) The vertex neighbourhood $\mathcal{V}_\delta(A_{11})$. (b) The vertex–edge neighbourhood $\mathcal{VE}_{\delta,R}(A_{11},\Lambda_{10,11})$.

q depends on L. We shall assume that γ_l for $l \leqslant L$ is not an integer; therefore, eqn (5.5.5) becomes

$$u(\rho, \phi, \theta) = \sum_{l=1}^{L} b_l \, \rho^{\gamma_l} s_l(\phi, \theta) + v(\rho, \phi, \theta) \,. \tag{5.5.6}$$

Vertex–edge singularities

The most complicated decomposition of the solution arises in the case of a vertex–edge intersection. For example, let us consider the neighbourhood where the edge $\Lambda_{10,11}$ approaches the vertex A_{11}. A spherical coordinate system is located in the vertex A_{11}, and a cone having an opening angle $\theta = \sigma$ with its vertex coinciding with A_{11} is constructed with $\Lambda_{10,11}$ being its centre axis. This cone is terminated by a ball-shaped basis having a radius $\rho = \delta$, as shown in Fig. 5.23(b). The resulting vertex–edge subdomain is denoted by $\mathcal{VE}_{\delta,R}(A_{11}, \Lambda_{10,11})$, and the solution u can be decomposed in $\mathcal{VE}_{\delta,R}(A_{11}, \Lambda_{10,11})$ as follows:

$$\begin{aligned} u(\rho, \phi, \theta) = \sum_{k=1}^{K} \sum_{q=0}^{Q} \left(\sum_{l=1}^{L} a_{kql} \, \rho^{\gamma_l} + m_{kq}(\rho) \right) (\sin\theta)^{\beta_k} \, [\ln(\sin\theta)]^q \, g_{kq}(\phi) \\ + \sum_{l=1}^{L} \sum_{p=0}^{P} c_{lp} \, \rho^{\gamma_l} (\ln\rho)^p h_{lp}(\phi, \theta) + v(\rho, \phi, \theta) \,, \end{aligned} \tag{5.5.7}$$

where the functions $m_{kq}(\rho)$ are analytic in ρ, and $g_{kq}(\phi)$ and $h_{lp}(\phi, \theta)$ are analytic in ϕ and θ. The function $v(r, \theta, z)$ belongs to $H^t(\mathcal{VE})$, where t is as large as required depending on L and K.

5.5.3 *Poisson equation*

For the Poisson equation

$$\nabla^2 u = f(\boldsymbol{x})$$

the situation is more complicated because the solution may not be analytic after the mapping. Typically, we decompose the solution into a homogeneous part $u^{\mathcal{H}}$, which has the singularity, and a particular part $u^{\mathcal{P}}$, which depends on the forcing. Complications arise due to the particular solution since, even if it is smooth in the original domain, it may be singular after the transformation. This can be seen by considering the transformed equation which has a singular forcing function

$$\nabla^2 u = \alpha^2 \rho^{2\alpha - 2} f(x(\rho, \xi), y(\rho, \phi)) \,.$$

Given the $\mathcal{O}(\rho^{-2\alpha})$ singularity, then the spectral element convergence is $\mathcal{O}(P^{-4\alpha-\epsilon})$. In order to enhance this convergence, we separate the two contributions so that we have an analytic contribution in the z plane and an analytic contribution in the ξ plane.

In the special case that the forcing is constant, that is, $f(\boldsymbol{x}) = C$, then it is straightforward to obtain the particular solution

$$u^{\mathcal{P}}(r,\theta) = \frac{Cr^2}{4}\left[1 - \frac{\cos(2\theta - \alpha\pi)}{\cos(\alpha\pi)}\right],$$

provided that α is not equal to 1/2 or 3/2. For the case when $\alpha = 1/2, 3/2$ we need to include logarithmic terms to obtain a particular solution of the form

$$u^{\mathcal{P}}(r,\theta) = \frac{Cr^2}{4\alpha\pi}[\alpha\pi + 2\ln r\sin(2\theta) + (2\theta - \alpha\pi)\cos(2\theta)].$$

The homogeneous solution written in terms of the fundamental solution is then $u^{\mathcal{H}}(r,\theta) = \sum_{k=0}^{\infty} c_k\varphi_k(r,\theta)$. Now if $u^{\mathcal{P}}$ vanishes on either side of the corner, then it follows that $u^{\mathcal{H}}$ satisfies Laplace's equation with the following boundary conditions on the outside boundary $\partial\Omega_2$:

$$u^{\mathcal{H}} = g(\theta) - u^{\mathcal{P}} \quad \text{on } \partial\Omega_2 .$$

Using the auxiliary mapping, $u^{\mathcal{H}}$ can then be obtained numerically with spectral accuracy as before.

In the case of *variable forcing*, progress can also be made if we assume that $f(\boldsymbol{x})$ is at least weakly regular, so that it can be represented as

$$f(r,\theta) = \sum_{k=0}^{\infty} f_k(r)\sin\left(\frac{k}{\alpha\theta}\right) = \sum_{k=0}^{\infty}\sum_{l=0}^{\infty} f_{kl}\, r^l \sin\left(\frac{k}{\alpha\theta}\right),$$

which implies that the particular part (due to the shifting theorem) can be written as

$$u^{\mathcal{P}}(r,\theta) = \sum_{k=0}^{\infty}\sum_{l=0}^{\infty} f_{kl}\varphi_{kl}(r,\theta),$$

where

$$\varphi_{kl}(r,\theta) = \begin{cases} \dfrac{\alpha^2}{\alpha^2(l+2)^2 - k^2}\, r^{l+2}\sin\left(\dfrac{k}{\alpha\theta}\right), & \dfrac{k}{\alpha} \neq l+2, \\ \dfrac{1}{2\alpha(k+1)} r^{k/\alpha}\ln r\sin\left(\dfrac{k}{\alpha\theta}\right), & \dfrac{k}{\alpha} = l+2. \end{cases}$$

The coefficients f_{kl} were computed in [364]. In the case that $f(\boldsymbol{x})$ is not known analytically, as, for example, in the pressure Poisson equation for the Navier–Stokes equations (see Chapter 8), then the right-hand side can still be approximated using a polynomial expansion and proceeding as above.

5.5.4 *Helmholtz equation*

For the Helmholtz equation

$$\nabla^2 u - \lambda u = f(\boldsymbol{x}), \quad \lambda > 0,$$

the conformal mapping is an effective way of enhancing convergence, although exponential convergence cannot be fully restored. The auxiliary mapping $z = \xi^\alpha$ converts the Helmholtz equation to

$$\nabla^2 u - \lambda \alpha^2 \rho^{2\alpha-2} u = \alpha^2 \rho^{2\alpha-2} f.$$

In terms of the original variables, the solution around the corner is

$$u(r,\theta) = \sum_{k=0}^{\infty} a_k I_{k/\alpha}(\sqrt{\lambda} r) \sin\left(\frac{k}{\alpha\theta}\right)$$

for k/α not an integer, where $I_m(z)$ is the modified Bessel function of the first kind. After application of the mapping, the solution has the form

$$u(\rho,\phi) = \sum_{k=0}^{\infty} a_k \rho^k \sin(k\theta) \sum_{j=0}^{\infty} c_j \rho^{2j\alpha},$$

with a leading singular term of order $\rho^{1+2\alpha}$. Therefore, the estimated convergence rate is $\mathcal{O}(P^{-2-4\alpha-\epsilon})$, which is algebraic but, in practice, adequately fast.

Similar to the Poisson equation, in three dimensions there are also three different types of singularities: the vertex, the edge, and the combined vertex–edge. Furthermore, solutions of elliptic problems are anisotropic in the neighbourhood of edges and vertex–edges. However, it is possible to obtain explicitly the form of such singularities [216], and thus the auxiliary mapping technique can be effectively used in three dimensions. The difference is that specific auxiliary mappings are required for each type of singularity; for example, vertex–edge mapping and edge mapping. This was obtained in the work of Lee *et al.* [291], who presented examples treating all three types of singularities. To deal with the singularities of these types, three auxiliary mappings and formulae for the transformed bilinear forms and linear functionals were developed. The effectiveness of this method was demonstrated on two polyhedra domains. The results with the mapping method were superior to results obtained with the p-version of the finite element method [445] or other low-order methods.

5.5.5 *Singular basis*

An alternative approach to using the auxiliary mapping is to use a set of supplementary basis functions which have the leading behaviour of the singularity in conjunction with the smooth basis $\Phi_k(\boldsymbol{x})$. For the Helmholtz equation the leading-order singular terms are

$$r^{1/\alpha}\,,\; r^{2/\alpha}\;(\alpha > 1/2)\,,\; r^{3/\alpha}\;(\alpha > 1)\,,\ldots\,,$$

which can be included into the expansion basis. However, we can do even better by supplementing the standard basis in the *mapped* domain. The transformed solution is then

$$u(\rho,\phi) = \sum_{k=1}^{\infty} a_k I_{k/\alpha}(\sqrt{\lambda}\rho^{\alpha})\sin(k\phi) = \sum_{k=1}^{\infty}\sum_{l=0}^{\infty} b_{kl}\rho^{k+2l\alpha}\sin(k\phi)\,,$$

and thus the leading singularities are weaker, i.e.,

$$\rho^{1+2\alpha}\,,\; \rho^{2+2\alpha}\;(\alpha > 1/2)\,,\; \rho^{1+4\alpha}\,,\;\ldots\,.$$

A comparison of the effect of adding supplementary basis terms in the original and the transformed domain is shown in Tables 5.4 and 5.5. It is clear that with one or two terms included in the transformed domain, a very fast convergence is obtained.

To implement this approach one can either explicitly subtract the leading singularity, as in [416], or discretise the Helmholtz equation with the augmented basis. In the latter case we obtain

$$\int_{\partial\Omega} \nabla v\cdot\nabla u \,\mathrm{d}\xi_1\,\mathrm{d}\xi_2 + \lambda\int_{\partial\Omega}\rho^{2\alpha-2} vu\,\mathrm{d}\xi_1\,\mathrm{d}\xi_2 = -\int_{\partial\Omega}\rho^{2\alpha-2} vf\,\mathrm{d}\xi_1\,\mathrm{d}\xi_2\,,$$

where

$$u(\xi_1,\xi_2) = \sum_{j=1}^{P_1} a_j\Phi_j(\xi_1,\xi_2) + \sum_{n=1}^{P_2} b_n\varphi_n(\xi_1,\xi_2)$$

and φ_n are the singular eigenfunctions. For good accuracy, the mixed inner products (Φ_i,φ_j) should be computed by using over-integration to deal with the

TABLE 5.4. Leading singular behaviour and convergence rate after removing the leading singular terms (before mapping) [364].

Number	Basis function	Leading singular term	Convergence rate
0	–	$r^{1/a}$	$P^{-2/a}$
1	$r^{1/a}\sin(1/\alpha\theta)$	$r^{2/a}$	$P^{-4/a}$
	$I_{1/a}(\sqrt{\lambda}r)\sin(1/\alpha\theta)$	$r^{2/a}$	$P^{-4/a}$
2	$\left\{r^{1/a}\sin(1/\alpha\theta)\,, r^{2/a}\sin(2/\alpha\theta)\right\}$	$r^{3/a}$	$P^{-6/a}$
	$\left\{I_{1/a}(\sqrt{\lambda}r)\sin(1/\alpha\theta), I_{2/a}(\sqrt{\lambda}r)\sin(2/\alpha\theta)\right\}$	$r^{3/a}$	$P^{-6/a}$

TABLE 5.5. Leading singular behaviour and convergence rate after removing the leading singular terms (after mapping) [364].

Number	Basis function	Leading singular term	Convergence rate
0	–	ρ^{1+a}	P^{-2-4a}
1	$\rho^{1-2a}\sin\phi$	ρ^{2+2a}	P^{-4-4a}
	$I_{1/a}(\sqrt{\lambda}\rho^{a})\sin\phi$	ρ^{2+2a}	P^{-4-4a}
2	$\left\{\rho^{+2/a}\sin\varphi, r^{2+2a}\sin(2\phi)\right\}$	ρ^{1+4a}	P^{2-8a}
	$\left\{I_{1/a}(\sqrt{\lambda}\rho^{a})\sin\varphi, I_{2/a}(\sqrt{\lambda}\rho^{a})\sin(2\phi)\right\}$	ρ^{3+2a}	P^{-6-4a}

non-polynomial behaviour of the singularity. In the limit the system becomes ill-conditioned as the augmented basis $\{\Phi_1, \Phi_2, \ldots, \varphi_1, \varphi_2, \ldots\}$ is nearly linearly dependent.

5.5.6 *Eigenpair representation: Steklov formulation*

A more recent method that treats singular solutions of both scalar and vector elliptic problems in the neighbourhood of corners was presented for two-dimensional domains by Yosibash and Szabó [446,502] and for three-dimensional domains by Yosibash in [500,501]. As we have already seen, such singular solutions are characterised by the form $u = r^{\beta}s(\theta)$ close to the corner.

Let us illustrate the basic characteristics of the solution to the Laplace problem over a two-dimensional domain as in Fig. 5.18. The Laplace equation over Ω_2, with Dirichlet boundary conditions on its boundary $\partial\Omega$, is cast in cylindrical coordinates as follows:

$$\nabla^2 u = \frac{\partial^2 u}{\partial r^2} + \frac{1}{r}\frac{\partial u}{\partial r} + \frac{1}{r^2}\frac{\partial^2 u}{\partial \theta^2} = 0 \quad \text{in } \Omega_2\,. \tag{5.5.8}$$

The solution in the vicinity of the singular point is sought by separation of variables.

Both positive and negative β_i values satisfy eqn (5.5.8). We denote by $s^{+}(\theta)$ the functions associated with the positive value of β, and by $s^{-}(\theta)$ the ones associated with the negative value of β. Although for the Laplace equation $s^{+}(\theta) \equiv s^{-}(\theta)$, for a general scalar elliptic equation this is not the case.

The restriction $\beta_i \geqslant 0$ is imposed because of 'physical' reasoning—since u at $r = 0$ should be finite—so that we deal with solutions belonging to the Sobolev space $H^1(\Omega_2)$. The negative values are nevertheless of interest for other mathematical manipulations and for describing the 'far field'. Thus, the solution to eqn (5.5.8) admits the expansion

$$u = \sum_{i=1}^{\infty} a_i r^{\beta_i} s_i^{+}(\theta)\,, \quad s_i^{+}(\theta) = \sin(\beta_i\theta)\,, \quad \text{with } \beta_i = \frac{i}{\alpha}\,, \tag{5.5.9}$$

where β_i and $s_i(\theta)$ denote *eigenpairs*, and these are determined uniformly by the geometry and boundary conditions in the neighbourhood of the singular point. We also denote as 'primal' eigenfunction and 'dual' eigenfunction the two functions corresponding to the same positive and negative eigenvalue: $s_i^+(\theta)$ and $s_i^-(\theta)$, respectively.

Notice that if $\beta_i < 1$ then the corresponding ith term in the expansion (5.5.9) for ∇u is unbounded as $r \to 0$. We say that u is singular at 0 if ∇u tends to infinity as $r \to 0$. The solution u in eqn (5.5.9) is therefore singular at 0 if $\alpha > 1$. The coefficients a_i depend on the boundary conditions away from the singular point as well as the forcing term in the Poisson equation.

Using the path-independent integral along an arc of radius R, centred at the singular point of interest (see [28, 60]), it can be shown that

$$\int_0^{\omega} s_i^+(\theta) s_j^-(\theta)\, \mathrm{d}\theta = 0\,, \quad i \neq j\,. \tag{5.5.10}$$

This means that the 'primal' and 'dual' eigenfunctions are orthogonal with respect to a path integral along an arc starting on one edge, intersecting at O, and terminating on the other.

For general singular points, analytical computation of eigenpairs is not practical, and numerical approximations are usually sought. One of the most robust and efficient methods for the computation of the eigenpairs in the vicinity of corners, and hence abrupt changes in material properties and boundary conditions, is the modified Steklov method, developed by Yosibash [501] and Yosibash and Szabó [502]. In the following sections we provide some details of this formulation.

5.5.6.1 *Formulation of the modified Steklov method*

Consider the Laplace equation in a two-dimensional domain in the vicinity of a singular point P. We may consider a zoomed subdomain close to P because the eigenpairs depend only on the geometry, material properties, and boundary conditions in the vicinity of P. Therefore, we define the subdomain Ω_R ('modified Steklov domain') as follows (see Fig. 5.24):

$$\Omega_R = \{\boldsymbol{x} \mid \boldsymbol{x} \in \Omega \cap \{R^* \leqslant r \leqslant R\}\}\,. \tag{5.5.11}$$

We have to solve over Ω_R the equation

$$\nabla^2 u = 0 \quad \text{in } \Omega_R\,, \tag{5.5.12}$$

subject to either Dirichlet or Neumann homogeneous boundary conditions on Γ_1 and Γ_2 (or a combination of these). We first consider homogeneous Neumann boundary conditions on Γ_1 and Γ_2 for simplicity of presentation. This problem is not well posed because boundary conditions have yet to be specified on the artificial circular boundaries Γ_{R^*} and Γ_R. To create these boundary conditions, let us assume that β and $s(\theta)$ are an eigenpair, satisfying eqn (5.5.12) and the

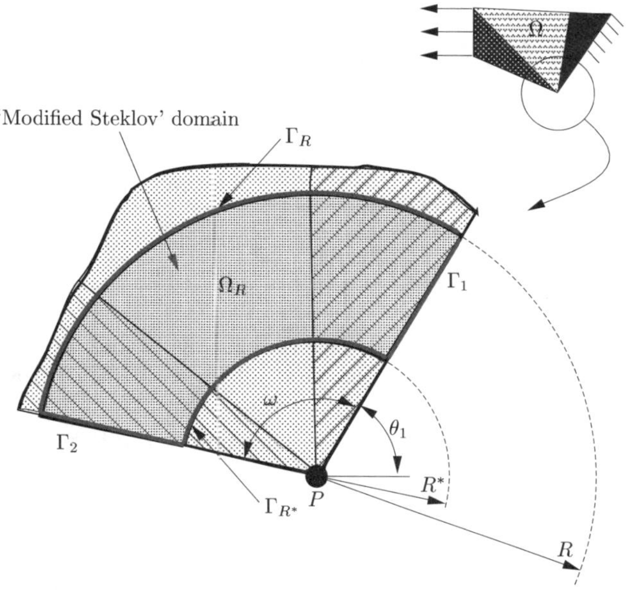

FIG. 5.24. The 'modified Steklov' domain and notations.

boundary conditions on Γ_1 and Γ_2. Thus, $u = r^\beta s(\theta)$ is a solution, and on Γ_R it satisfies

$$\frac{\partial u}{\partial n} = \frac{\partial u}{\partial r} = \beta r^{\beta-1} s(\theta)|_{r=R} = \frac{\beta}{R} u \quad \text{on } \Gamma_R\,. \tag{5.5.13}$$

A similar connection holds for Γ_{R^*}, so that we may now summarise the eigenproblem to be solved as follows:

$$\nabla^2 u = 0 \quad \text{in } \Omega_R\,, \tag{5.5.14}$$

$$\sum_{i=1}^{2} \partial_i u n_i = 0\,, \quad \theta = \theta_1, \theta_1 + \omega\,, \tag{5.5.15}$$

$$\frac{\partial u}{\partial r} = \frac{\beta}{R} u\,, \qquad r = R\,, \tag{5.5.16}$$

$$\frac{\partial u}{\partial r} = \frac{\beta}{R^*} u\,, \quad r = R^*\,. \tag{5.5.17}$$

Equations (5.5.14)–(5.5.17) form the 'generalised mixed problem of Steklov type'. Note that we have to find an eigenvalue, appearing here in the boundary condition, and an associated eigenfunction.

An analytic solution of the 'classical' Steklov eigenproblem (5.5.14)–(5.5.17) cannot be obtained in general. Thus the weak formulation has to be used, based on which the spectral element method will be applied. The associated weak form is

$$\begin{gathered}\text{seek } \beta \in \Re,\ 0 \neq u \in H^1(\Omega_R) \text{ such that} \\ \mathcal{B}(u,\chi) = \beta\left(\mathcal{M}_{R^*}(u,\chi) + \mathcal{M}_R(u,\chi)\right), \quad \forall\, \chi \in H^1(\Omega_R),\end{gathered} \tag{5.5.18}$$

where

$$\mathcal{B}(u,\chi) = \int_R^{R^*} \int_{\theta_1}^{\theta_1+\omega} \sum_{i=1}^{2} \partial_i \chi\, \partial_i u\, r\, \mathrm{d}r\, \mathrm{d}\theta\,, \tag{5.5.19}$$

$$\mathcal{M}_R(u,\chi) \stackrel{\text{def}}{=} \int_{\theta_1}^{\theta_1+\omega} [u\chi]_R\, \mathrm{d}\theta\,, \tag{5.5.20}$$

and $\mathcal{M}_{R^*}$ is exactly the same as in eqn (5.5.20), save that the expression is to be evaluated at R^*. One may observe that all forms in (5.5.18) are symmetric; thus only real eigenpairs are expected.

In the case of a general scalar elliptic equation of the form

$$\sum_{i,j=1,2} k_{ij}(\theta) \frac{\partial^2 u}{\partial x_i \partial x_j} = 0\,,$$

the 'modified Steklov weak eigenproblem' is more complicated, involving a non-symmetrical bilinear form $\mathcal{N}$:

$$\begin{gathered}\text{seek } \beta \in \Re,\ 0 \neq u \in H^1(\Omega_R) \text{ such that} \\ \mathcal{B}(u,\chi) - \left(\mathcal{N}_{R^*}(u,\chi) + \mathcal{N}_R(u,\chi)\right) = \beta\left(\mathcal{M}_{R^*}(u,\chi) + \mathcal{M}_R(u,\chi)\right), \\ \forall\, \chi \in H^1(\Omega_R),\end{gathered} \tag{5.5.21}$$

where

$$\mathcal{M}_R(u,\chi) \stackrel{\text{def}}{=} \int_{\theta_1}^{\theta_1+\omega} \left[k_{11}(\theta)\cos^2\theta + k_{12}(\theta)\sin(2\theta) + k_{22}(\theta)\sin^2\theta\right] [u\chi]_R\, \mathrm{d}\theta\,, \tag{5.5.22}$$

$$\mathcal{N}_R(u,\chi) \stackrel{\text{def}}{=} \int_{\theta_1}^{\theta_1+\omega} \left[(k_{22}(\theta) - k_{11}(\theta))\sin\theta\cos\theta + k_{12}(\theta)\cos(2\theta)\right] \left[\frac{\partial u}{\partial \theta}\chi\right]_R \mathrm{d}\theta\,, \tag{5.5.23}$$

and $\mathcal{M}_{R^*}$ and $\mathcal{N}_{R^*}$ are exactly the same as eqns (5.5.22) and (5.5.23), respectively, save that they are to be evaluated at R^*. Details can be found in [502].

Notice that the weak form does not exclude the existence of negative eigenpairs. This is because solutions of the form $r^{-\beta}s^-(\theta)$ do belong to the space $H^1(\Omega_R)$. Therefore, both positive and negative eigenpairs can be obtained. It is interesting to note that for the Laplace equation if β is an eigenvalue, with a corresponding eigenfunction called $s^+(\theta)$, then $-\beta$ is also an eigenvalue, with an associated eigenfunction $s^-(\theta) \equiv s^+(\theta)$.

The domain Ω_R does not include singular points, and hence no special refinement of the spectral element mesh is required. Furthermore, Ω_R is small in size so that very few spectral elements are needed in the discretisation.

5.5.6.2 *A benchmark problem*

Let Ω be the unit circle slit along the positive x axis, and denote by Γ_1 the upper face of the slit, by Γ_2 the lower face of the slit, and by Γ_R the circular portion of the boundary of Ω, see Fig. 5.25.

Consider the problem discussed in [28]:

$$\begin{gathered}\nabla^2 u = 0 \quad \text{in } \Omega\,, \\ u = 0 \quad \text{on } \Gamma_1\,, \qquad \frac{\partial u}{\partial \theta} = 0 \quad \text{on } \Gamma_2\,, \qquad \frac{\partial u}{\partial r} = y \quad \text{on } \Gamma_R\,.\end{gathered} \tag{5.5.24}$$

Then the solution to this problem, accurate up to the sixth significant digit, is

$$\begin{aligned}u(r,\theta) = -1.35812 r^{1/4} \sin\left(\frac{\theta}{4}\right) &+ 0.970087 r^{3/4} \sin\left(\frac{3\theta}{4}\right) \\ &+ 0.452707 r^{5/4} \sin\left(\frac{5\theta}{4}\right) + \mathcal{O}(r^{7/4})\,.\end{aligned} \tag{5.5.25}$$

Based on this problem, we first study the convergence of the eigenvalues computed by the modified Steklov method, and show that the value of R^* has minor influence on the accuracy of the results if chosen to be in the range $0.5 \leqslant R^* \leqslant 0.95$.

First, we compute the eigenpairs using three spectral elements having 120° each, with $R = 1$ and $R^* = 0.95$, as shown in Fig. 5.25. The convergence of the first three computed eigenvalues as the p-level over each element is increased from $P = 1$ up to $P = 6$ is demonstrated in Fig. 5.26. The absolute relative error in percentage is computed as

$$100 \times \text{abs}\left[\frac{(\beta_i)_n - (\beta_i)_e}{(\beta_i)_e}\right],$$

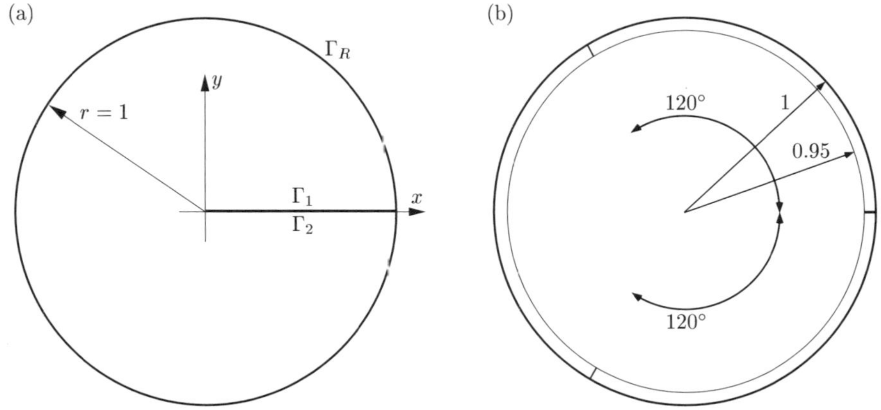

FIG. 5.25. Sketch for (a) the model problem, and (b) the spectral element mesh.

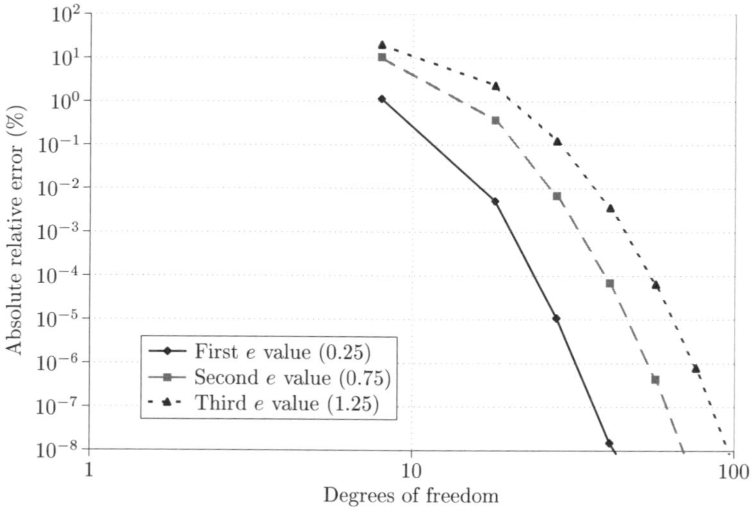

FIG. 5.26. Relative error (%) of the first three eigenpairs. (Courtesy of Z. Yosibash.)

where the subscripts n and e denote the numerical and exact values, respectively. This example demonstrates the efficiency and accuracy of the modified Steklov method in computing the eigenpairs. Eigenvalues have relative errors of less than 10^{-8}%, with less than 100 degrees of freedom.

5.5.7 *Singularities and Stokes flow*

Algorithms for the time-independent Stokes flow are presented in Chapter 8. Here we consider the particular case where the domain is of convex shape. In two-dimensional domains, as for the Helmholtz equation, the solution in the vicinity of any corner is singular, having similar characteristics. When applying the spectral/*hp* element method on uniform meshes, it is known that the convergence rate is determined by the strongest singularity (we noticed similar behaviour when addressing the Laplace equation), thus a high rate of convergence may not be obtained. Nevertheless, Schwab and Suri [418] showed that, by using *mixed hp* finite element methods in the classical framework of the velocity–pressure formulation, with properly chosen spaces, and applying a geometrical graded mesh, exponential convergence rates are regained:

$$\|\mathbf{u} - \mathbf{u}^\delta\|_{H^1} \leqslant C \exp(-bN_{\text{dof}}^{1/3}),$$
$$\|p - p^\delta\|_{L^2} \leqslant C \exp(-bN_{\text{dof}}^{1/3}),$$

where $\mathbf{u}$ and p denote the velocities and pressure. Schotzau and Schwab [413] also considered the *hp* version of least-squares (GLS) type for the Stokes equations in polygonal domains. Contrary to the standard Galerkin *hp* element methods, the least-squares approach admits equal-order interpolation in the velocity and the pressure, see Section 8.5. In conjunction with geometrically-refined meshes and linearly-increasing approximation orders, it was shown that the new method leads to exponential rates of convergence for solutions exhibiting singularities near corners. Several other methods, such as mixed discontinuous Galerkin approximations of the Stokes problem, were also analysed, and a priori error estimates for *hp* approximations on tensor product meshes were provided in [414].

Finally, Toselli and Schwab [469] considered the Stokes problem in a three-dimensional polyhedral domain discretised on hexahedral meshes anisotropically and non-quasi-uniformly refined towards faces, edges, and corners. The inf–sup constant of the discretised problem was shown to be independent of arbitrarily large aspect ratios. Their work generalises the available results for two-dimensional problems.

5.6 Exercises: implementation of a two-dimensional spectral/*hp* element solver for a Helmholtz problem using a C^0 Galerkin formulation

To complement the exercises of Sections 2.6, 3.5, and 4.4 we propose the following series of exercises to develop a two-dimensional spectral/*hp* element solver for the elliptic Helmholtz problem. In this section we will draw heavily from the exercises of Sections 3.5 and 4.4, with an additional focus on incorporating differential operations into the local and global operations. We also note that some useful codes are available on the web page ♦[33]

`http://www.nektar.info/2nd_edition`

From Section 5.1, we recall that the C^0 continuous Galerkin formulation of the Helmholtz problem can be expressed as eqn (5.1.3), i.e.,

$$(\nabla v, \nabla u^{\mathcal{H}})_\Omega + \lambda(v, u^{\mathcal{H}})_\Omega = \langle v, g_{\mathcal{N}}\rangle - (v, f)_\Omega - (\nabla v, \nabla u^{\mathcal{D}})_\Omega - \lambda(v, u^{\mathcal{D}})_\Omega\,,$$

where $g_{\mathcal{D}}$ and $g_{\mathcal{N}}$ are the Dirichlet and Neumann boundary conditions imposed on the boundary segments $\partial\Omega_{\mathcal{D}}$ and $\partial\Omega_{\mathcal{N}}$, respectively. To impose Dirichlet boundary conditions in the above, the solution $u(\boldsymbol{x})$ has also been decomposed into an unknown homogeneous component $u^{\mathcal{H}}(\boldsymbol{x})$ and a known inhomogeneous component $u^{\mathcal{D}}(\boldsymbol{x})$ such that

$$u(\boldsymbol{x}) = u^{\mathcal{H}}(\boldsymbol{x}) + u^{\mathcal{D}}(\boldsymbol{x})\,,$$

where

$$u^{\mathcal{H}}(\partial\Omega_{\mathcal{D}}) = 0\,, \quad u^{\mathcal{D}}(\partial\Omega_{\mathcal{D}}) = u(\partial\Omega_{\mathcal{D}}) = g_{\mathcal{D}}(\partial\Omega_{\mathcal{D}})\,.$$

1. A good starting-point for this exercise is to perform some differentiation exercises. Differentiation in two-dimensional elemental regions was introduced

[33] Exercises to help implement a two-dimensional spectral/*hp* element Helmholtz solver.

in Sections 4.1.2.1 and 4.1.3.4. When using tensorial-based expansions we can differentiate functions using a series of one-dimensional differentiation and applying the chain rule. As outlined in Section 4.1.2, we shall only consider differentiation in 'physical space' at the quadrature points, and so the first part of the exercise is to construct the one-dimensional derivative matrices through the quadrature zeros. This construction is discussed in Appendix C and routines are also available in the *Polylib* library on the web site. Having obtained a routine to calculate the differentiation matrix, $D[i][j]$, try the following tests.

(a) Differentiate the function $u(\xi_1, \xi_2) = [\xi_1]^7 \times [\xi_2]^6$ in the standard quadrilateral region using Gauss–Lobatto–Legendre quadrature points ξ_{i1}, ξ_{2j}, i.e.,

$$\frac{\partial u}{\partial \xi_1}(\xi_{1i}, \xi_{2j}) = [\xi_{2j}]^6 \sum_{k=0}^{Q_1-1} D[i][k]\ (\xi_{1k})^7 = 7[\xi_{1i}]^6 \times [\xi_{2j}]^6\,,$$

$$\frac{\partial u}{\partial \xi_2}(\xi_{1i}, \xi_{2j}) = [\xi_{1i}]^7 \sum_{k=0}^{Q_2-1} D[i][k]\ (\xi_{2k})^6 = [\xi_{1i}]^7 \times 6[\xi_{2j}]^5\,.$$

Use $Q_1 = Q_2 = 6$, 7, 8, and 9 and verify how many quadrature points are required to get an exact answer to numerical precision.

(b) Differentiate the function $u(\xi_1, \xi_2) = [\xi_1]^7 \times [\xi_2]^6$ in the standard triangular region using Gauss–Lobatto–Legendre quadrature points in the ξ_1 direction and Gauss–Radau–Legendre quadrature in the ξ_2 direction. This exercise is similar to Exercise 1(a), but we now must differentiate with respect to the local collapsed coordinate system (η_1, η_2) instead of the local Cartesian coordinates and apply the chain rule as previously given in eqn (4.1.4), i.e.,

$$\begin{pmatrix} \dfrac{\partial}{\partial \xi_1} \\ \dfrac{\partial}{\partial \xi_2} \end{pmatrix} = \begin{pmatrix} \dfrac{2}{1-\eta_2}\dfrac{\partial}{\partial \eta_1} \\ 2\dfrac{1+\eta_1}{1-\eta_2}\dfrac{\partial}{\partial \eta_1} + \dfrac{\partial}{\partial \eta_2} \end{pmatrix}.$$

Again use $Q_1 = Q_2 = 6$, 7, 8, and 9 quadrature points and verify how many quadrature points are required to get an exact answer to numerical precision.

(c) Consider a general straight-sided quadrilateral element, Ω^e, defined to have vertices $(x_1^A, x_2^A) = (0, 0)$, $(x_1^B, x_2^B) = (1, 0)$, $(x_1^C, x_2^C) = (2, 1)$, and $(x_1^D, x_2^D) = (0, 1)$. By defining a linear mapping $x_i = \chi_i(\boldsymbol{\xi})$ (see Section 2.3.1.2) and using the chain rule as shown in Section 4.1.3.4, evaluate $\partial u/\partial x_1$ and $\partial u/\partial x_2$ at the quadrature points in the element Ω^e when $u(x_1, x_2) = [x_1]^7 \times [x_2]^6$. Consider the error in the numerical evaluation of the partial derivatives when $Q_1 = Q_2 = 6$, 7, 8, and 9.

(d) Repeat Exercise 1(c) for the straight-sided triangular region, Ω^e, defined to have vertices $(x_1^A, x_2^A) = (1, 0)$, $(x_1^B, x_2^B) = (2, 1)$, and $(x_1^C, x_2^C) = (1, 1)$.

(e) For the quadrilateral region of Exercise 1(c) and the triangular region of Exercise 1(d) calculate

$$\mathcal{I} = -\int_{\Omega^e} \left[\frac{\mathrm{d}}{\mathrm{d}x_1} \cos(x_1 + x_2) + \frac{\mathrm{d}}{\mathrm{d}x_2} \cos(x_1 + x_2) \right] \mathrm{d}\boldsymbol{x}$$

for $2 \leqslant Q \leqslant 8$ and plot the error of this integral versus Q on semi-log axes. (Refer to Section 4.4 for exercises involving integration.)

2. The elemental Laplacian matrix $\boldsymbol{L}^e$ is defined as

$$\boldsymbol{L}^e[m(pq)][n(rs)] = \int_{\Omega^e} \nabla\phi_{pq} \cdot \nabla\phi_{rw} \, \mathrm{d}\boldsymbol{x} \,,$$

$$\boldsymbol{L}^e = (\boldsymbol{D}^e_{x_1}\boldsymbol{B}^e)^\top \boldsymbol{W}^e \boldsymbol{D}^e_{x_1}\boldsymbol{B}^e + (\boldsymbol{D}^e_{x_2}\boldsymbol{B}^e)^\top \boldsymbol{W}^e \boldsymbol{D}^e_{x_2}\boldsymbol{B}^e \,,$$

where

$$\boldsymbol{D}^e_{x_1} = \boldsymbol{\Lambda}^e \left(\frac{\partial \xi_1}{\partial x_1}\right) \boldsymbol{D}^e_{\xi_1} + \boldsymbol{\Lambda}^e \left(\frac{\partial \xi_2}{\partial x_1}\right) \boldsymbol{D}^e_{\xi_2} \,,$$

$$\boldsymbol{D}^e_{x_2} = \boldsymbol{\Lambda}^e \left(\frac{\partial \xi_1}{\partial x_2}\right) \boldsymbol{D}^e_{\xi_1} + \boldsymbol{\Lambda}^e \left(\frac{\partial \xi_2}{\partial x_2}\right) \boldsymbol{D}^e_{\xi_2} \,.$$

See Section 4.1.5.1 for further details on the matrix notation. In the following use the exercises of Section 3.5 with the differential techniques of Exercise 1 as a guide.

(a) Generate the elemental Laplacian matrix $\boldsymbol{L}^e$ in the quadrilateral standard region Ω_{st} using the modal expansion $\phi_{pq} = \psi^a_p \psi^a_q$ with a polynomial order $P_1 = P_2 = 14$ and exact numerical integration. Assemble the local matrix so that the local boundary degrees of freedom are listed first, followed by the interior degrees of freedom, and compare the structure of the matrix to Fig. 5.1(b).

(b) Generate the elemental Laplacian matrix $\boldsymbol{L}^e$ in the triangular standard region Ω_{st} using the modal expansion $\phi_{pq} = \psi^a_p \psi^b_{pq}$ within a polynomial order $P_1 = P_2 = 14$ and exact numerical integration. Once again, assemble the local matrix so that the local boundary degrees of freedom are listed first, followed by the interior degrees of freedom, and compare the structure of the matrix to Fig. 5.1(a). For this expansion the q index should run faster than the p index on the interior modes to obtain the structure shown in this figure.

3. By adding λ times the elemental mass matrix $\boldsymbol{M}^e$ (constructed in the exercises of Section 4.4) to the elemental Laplacian matrix $\boldsymbol{L}^e$, we can construct the elemental Helmholtz matrix $\boldsymbol{H}^e$, i.e.,

$$\boldsymbol{H}^e = \boldsymbol{L}^e + \lambda \boldsymbol{M}^e \,.$$

Although the local Laplacian matrix is not invertible, as there is a constant null space, the Helmholtz matrix can be inverted to solve an elemental problem.

(a) Solve the Helmholtz problem

$$\nabla^2 u - \lambda u = -(\lambda + \pi^2)\cos(\pi\xi_1)\cos(\pi\xi_2)$$

in the standard quadrilateral region $\Omega_{\rm st}$ with $\lambda = 1$. This problem can be written in weak form using matrix notation as

$$\boldsymbol{H}^e \hat{\boldsymbol{u}}^e = [\boldsymbol{L}^e + \lambda \boldsymbol{M}^e]\hat{\boldsymbol{u}}^e = \hat{\boldsymbol{f}}^e,$$

where

$$\hat{\boldsymbol{f}}^e[m(pq)] = \int_{\Omega_{\rm st}} \phi_{pq}(\lambda + \pi^2)\cos(\pi\xi_1)\cos(\pi\xi_2)\,\mathrm{d}\boldsymbol{\xi}.$$

Note the change in sign of the discrete and continuous problems due to the use of the divergence theorem in constructing the weak form of the Laplacian. The problem can be solved with either the modal expansion $\phi_{pq} = \psi_p^a\psi_q^a$ or the nodal expansion $\phi_{pq} = h_p h_q$, both of which are defined in Section 3.1. The analytic solution to this problem is $u(\xi_1, \xi_2) = \cos(\pi\xi_1)\cos(\pi\xi_2)$. By not explicitly imposing any boundary conditions in our solution we are essentially setting homogeneous Neumann boundary conditions (i.e., $g_{\mathcal{N}} = 0$) on all boundaries. For this problem in $\Omega_{\rm st} \in \{-1 \leqslant \xi_1, \xi_2 \leqslant 1\}$ the solution has been chosen to have zero Neumann boundary conditions. Verify that your solution converges exponentially with P by solving the problem for $2 \leqslant P \leqslant 10$ and plotting the log of error versus polynomial order.

(b) Solve the Helmholtz problem

$$\nabla^2 u - u = -(1 + \pi^2)\sin(\pi\xi_1)\sin(\pi\xi_2)$$

in the standard quadrilateral region. This problem has the analytic solution $u(\xi_1, \xi_2) = \sin(\pi\xi_1)\sin(\pi\xi_2)$, and so Neumann boundary conditions will be required; this requires augmenting the right-hand-side vector $\hat{\boldsymbol{f}}$ with the boundary integral $\langle v, g_{\mathcal{N}}\rangle$ over all edges. Verify that your solution is exponentially convergent.

(c) Solve the Helmholtz problem of Exercise 2(b) imposing a Dirichlet boundary condition along $\xi_2 = -1$. This requires introducing a known function, $u^{\mathcal{D}}$, which satisfies the Dirichlet boundary conditions. A computationally convenient choice is to perform an elemental boundary transformation, as discussed in Section 4.3.2, and then use these expansion coefficients, extended by zero for all other expansion coefficients, as the known function. You will then need to modify the right-hand-side terms by the additional contribution $(\nabla\phi_{pq}, \nabla u^{\mathcal{D}})_{\Omega^e} + \lambda(\phi_{pq}, u^{\mathcal{D}})_{\Omega^e}$. Verify that your solution is exponentially convergent.

(d) Repeat Exercises 2(b) and (c) for the triangular modal expansion $\phi_{pq} = \psi_p^a \psi_{pq}^b$.

4. Following Exercise 4 of Section 4.4, construct a global C^0 approximation to the Helmholtz equation $\nabla^2 u - \lambda u = f$ in the computational domains defined in Fig. 4.40. Setting $\lambda = 1$, consider the following problems.

   (a) Start by considering the forcing function $f(x_1, x_2) = -(1 + \pi^2)\cos(\pi x_1)\cos(\pi x_2)$, which has zero Neumann boundary conditions in this computational domain.

   (b) As previously, the next step is to consider a forcing function $f(x_1, x_2) = -(1 + \pi^2)\sin(\pi x_1)\sin(\pi x_2)$ with nonzero Neumann boundary conditions on all domain boundaries.

   (c) Similar to Exercise 5 of Section 4.4, introduce Dirichlet boundary conditions along the boundary $x_2 = 0$.

   (d) Verify your implementation by considering a range of different polynomial orders within the elements (i.e., $2 \leqslant P \leqslant 10$) and a range of uniform elemental refinements. Plot the error versus polynomial order and characteristic element size to observe exponential and algebraic convergence rates associated with the p- and h-type discretisations, respectively.

6

ADVECTION AND ADVECTION–DIFFUSION

To complement the discussion of Chapter 5 on the diffusion equation, in this chapter we analyse the advection and the advection–diffusion equations. A strong understanding of the numerical and physical properties of these equations enables us to extend these techniques to the more complicated Navier–Stokes equations. As model problems we shall consider the *linear* advection equation, with and without diffusion, as well as the viscous Burgers equation.

In Section 6.1 we start by providing a general discussion of the linear advection equation and the dispersion and diffusion errors associated with the discretisation of this equation. Subsequently, in Section 6.2 we discuss the standard Galerkin and discontinuous Galerkin formulations of a scalar hyperbolic conservation law, primarily based on the linear advection equation. In Section 6.2.1 we first present the formulation of the standard Galerkin method using globally-C^0-continuous basis functions. In Section 6.2.2 we then consider a discontinuous Galerkin formulation, which uses piecewise continuous basis functions which are globally in L^2. In these sections we also draw upon the previous general formulations presented in Chapter 4 and already adopted in Chapter 5.

Having introduced the discretisation techniques, we are then faced with the challenge of understanding the numerical stability of these schemes. This is important as the nonlinear advection operator associated with the Navier–Stokes equations in many spectral/*hp* formulations is typically treated explicitly in time. In Section 6.3 we therefore present numerical bounds for the eigenspectrum of the weak linear advection operator for both the Galerkin and discontinuous Galerkin formulations. These bounds determine the necessary time-step restrictions for the stability of the explicit temporal discretisation, i.e., the proper Courant–Friedrichs–Lewy (CFL) criterion.

We have previously considered the eigenspectrum of the weak Laplacian operator in Section 5.4, where we observed that the spectral radius grows as $\mathcal{O}(P^4)$. The rapid growth of this eigenspectrum typically means that the spectral/*hp* diffusion terms are treated implicitly in time, and therefore the time-step restrictions in an advection–diffusion system arise due to the explicit advection terms. In Section 6.4 we present a semi-Lagrangian formulation for advection–diffusion scalar equations, which allows explicit time integration with a time-step much larger than the one dictated by the CFL number in the standard Eulerian framework.

Finally, when we consider nonlinear advection terms such as those that arise in the viscous Burgers equation, steep gradients in the solution are possible.

In Section 6.5 we consider discontinuous solutions and address issues of monotonicity. We demonstrate how filtering, artificial viscosity, super-collocation, and upwind nodal distribution can be used to control high-frequency oscillations. The key issue here is how to maintain monotonicity and thus stability without sacrificing the exponential convergence of the method. These are very important issues for shock wave dynamics and we will also revisit them in Chapter 10.

6.1 Linear advection equation

To introduce the basic concepts associated with this chapter, we start by considering the one-dimensional scalar advection equation

$$\frac{\partial u}{\partial t} + V\frac{\partial u}{\partial x} = 0\,, \tag{6.1.1a}$$

$$u(x,0) = u_0(x)\,, \tag{6.1.1b}$$

where V is the constant advection velocity and $u_0(x)$ is the initial condition. We assume periodic boundary conditions in the interval $x \in [a,b]$. The exact solution is given by the propagation of the initial conditions with a velocity V, and so is given by $u(x,t) = u_0(x - Vt)$. Accordingly, the only extrema in the solution are those of the initial conditions and no new extrema develop in time. For the case of periodic boundary conditions the solution can be expressed in terms of Fourier components $\mathrm{e}^{\mathrm{i}(\omega t - kx)}$, where ω and k are the temporal and spatial wavenumbers, respectively. Substituting the Fourier component ($u(x,t) = \hat{u}\mathrm{e}^{\mathrm{i}(\omega t - kx)}$) into eqn (6.1.1a) leads to the relationship $\omega = kV$, which demonstrates that the solution corresponds to an exact phase speed $C_\phi \equiv \omega/k = V$, with each Fourier component $\mathrm{e}^{\mathrm{i}(\omega t - kx)}$ being advected by the constant velocity V.

6.1.1 *Dispersion and diffusion errors*

In general, for the discrete system the finite size of the resolution parameters $(\Delta x, \Delta t)$ introduces numerical dispersion, which results in each Fourier component of spatial wavenumber k travelling at a different phase speed. In this case the discrete phase speed C_ϕ^δ is therefore not constant. The specific form of the numerical dispersion depends on the particular discretisation scheme adopted, and so is another measure of the accuracy of the scheme. For an equidistant finite difference discretisation the equivalent differential equation to eqn (6.1.1a) that the approximate solution, u^δ, satisfies (sometimes referred to as the modified equation) has the general form

$$\frac{\partial u^\delta}{\partial t} + V\frac{\partial u^\delta}{\partial x} = F_0(\Delta x, \Delta t)\frac{\partial^p u^\delta}{\partial x^p} + F_1(\Delta x, \Delta t)\frac{\partial^{p+1} u^\delta}{\partial x^{p+1}}\,, \tag{6.1.2a}$$

where F_0 and F_1 are functions of Δx and Δt of appropriate order. If the leading error term has an even value of p then the error is dissipative and F_0 denotes the coefficient of numerical diffusion; if the leading error contains an odd value of p then the error is dispersive.

Finite difference method

As an example of diffusion and dispersion errors, in the Lax–Wendroff one-dimensional, second-order scheme

$$u^{n+1} = u^n - \frac{C}{2}(u_{j+1}^n - u_{j-1}^n) + \frac{C^2}{2}(u_{j+1}^n - 2u_j^n + u_{j-1}^n)\,, \tag{6.1.2b}$$

the corresponding leading error term contains a third derivative ($p = 3$) with $F_0 = -(V/6)\Delta x^2(1-C^2)$ and $F_1 = -(V/8)\Delta x^3 C(1-C^2)$, where $C = (V\Delta t)/\Delta x$ is the Courant–Friedrichs–Lewy (CFL) number. The dominant errors are therefore of dispersive type in this scheme, and they may lag or lead the exact waveform depending on the value of the Courant number C.

In contrast, in the one-dimensional Euler-forward/upwind difference scheme

$$u^{n+1} = u^n + C(u_j^n - u_{j-1}^n)\,, \tag{6.1.2c}$$

the corresponding leading error term contains an even derivative ($p = 2$), where $F_0 = (V\Delta x/2)(1 - C)$ is the coefficient of numerical diffusion.

The disastrous effects of numerical dispersion and diffusion are well documented in the literature. The effects of numerical dispersion, in particular, can be quantified by considering the advection of a single harmonic $e^{ik(C_\phi t - x)}$. To this end, we consider the discrete phase and group velocities as a function of spatial wavenumber for different central-differencing discretisations of eqn (6.1.1a), with formal accuracy ranging from 2 to 9. Figure 6.1 shows the normalised phase velocity $C_\phi^\delta/C^\phi = (\omega^\delta/k^\delta)/V$ and normalised group velocity $(\partial\omega^\delta/\partial k^\delta)/V$ as a function of spatial wavenumber normalised by $\pi/\Delta x$, the maximum wavenumber on a mesh of spacing Δx. The normalisation has been chosen so that the exact solution is given by a unit value in both plots for all wavenumbers. We observe

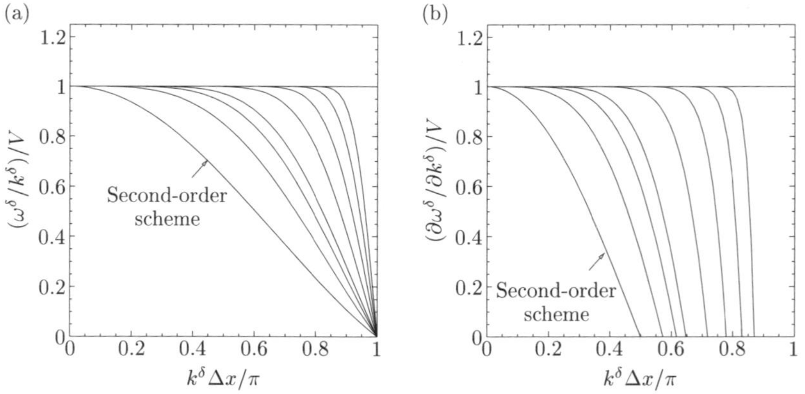

FIG. 6.1. (a) Phase, and (b) group velocity versus wavenumber for different orders of finite differencing. The lowest curve corresponds to second-order accuracy and the highest curve to ninth-order accuracy. (Giannakouros [185].)

that the correct phase and group velocity are approached as the order of the scheme increases. For low-order schemes the group velocity error (the deviation from a unit value) is very large, and at high wavenumbers the group velocity becomes negative.

For nonlinear advection problems or discretisations the numerical diffusion and dispersion can act in concert. For example, due to dispersion, different Fourier components will propagate at different phase speeds and may produce features which are not easily resolvable on a fixed mesh. This will, in turn, widen the spectrum by including higher components, which may eventually be damped due to dissipation. These considerations are very significant when discretising hyperbolic conservation laws, where monotonicity requires filtering or the use of limiters to handle artificial extrema. This issue is addressed later in this Chapter (see Section 6.5) and also in Chapter 10.

Spectral/hp element method

An interesting effect of the interaction between numerical dispersion and diffusion was demonstrated by Giraldo [188]. In particular, he performed a stability analysis of the linear advection equation (6.1.1a) by employing the standard spectral/*hp* element formulation and a semi-Lagrangian formulation. Details of these discretisations are presented in Sections 6.2.1 and 6.4. Giraldo showed that the application of the semi-Lagrangian technique to the spectral element discretisation introduces a proper dissipation mechanism required for high values of the Courant number.

In particular, for linear elements, i.e., $P = 1$, the dispersion errors are plotted in Fig. 6.2 for both types of discretisations. At this low value of polynomial order both schemes suffer from dispersion errors (isocontours with values less or greater than 1) but have different error distributions. Although not shown in this

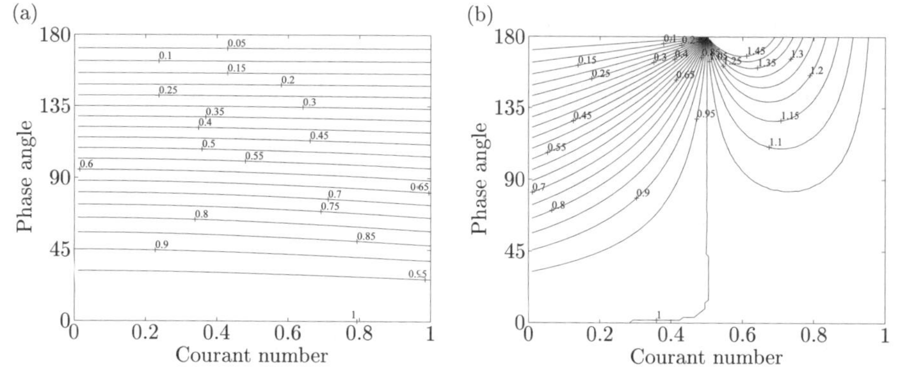

FIG. 6.2. Order $P = 1$. Isocontours of phase errors for different frequencies (vertical axis) and Courant number (horizontal axis). (a) Standard spectral element scheme, and (b) semi-Lagrangian scheme. (Courtesy of F. X. Giraldo.)

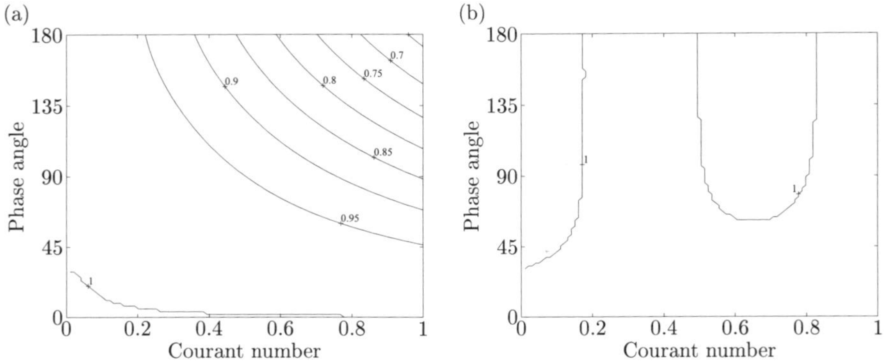

FIG. 6.3. Order $P = 4$. Isocontours of phase errors for different frequencies (vertical axis) and Courant number (horizontal axis). (a) Standard spectral element scheme, and (b) semi-Lagrangian scheme. (Courtesy of F. X. Giraldo.)

figure, there is no diffusion error for the standard spectral element method but the semi-Lagrangian scheme exhibits appreciable numerical diffusion, especially around Courant number $C = 0.5$ and at high frequencies. This is also the region where the dispersive errors are large as well, thus acting in concert with diffusion.

However, for order $P \geqslant 4$ all dispersion errors are eliminated in the semi-Lagrangian scheme but not in the standard spectral element scheme, as shown in Fig. 6.3. In particular, for $P \geqslant 4$ the standard scheme does not suffer any errors for values of Courant number $C \leqslant 1/4$, while for $1/4 \leqslant C \leqslant 1/2$ it exhibits errors for the high frequencies. For larger values of C, the standard scheme experiences lagging errors for the high frequencies and for a large part of the low frequencies. The semi-Lagrangian scheme exhibits neither numerical dispersion nor diffusion errors for $P \geqslant 4$. This behaviour of the semi-Lagrangian scheme is true even for Courant number $C \gg 1$, see [188].

6.2 Galerkin and discontinuous Galerkin discretisations

In general, we can consider a *nonlinear* hyperbolic conservation law which describes the advection of a conserved quantity u in a region Ω governed by the equation

$$\frac{\partial u}{\partial t} + \nabla \cdot \mathbf{F}(u) = 0\,. \qquad (6.2.1)$$

In eqn (6.2.1), $\mathbf{F}(u) = [f(u), g(u), h(u)]^\top$ is the *flux* vector which defines the transport of $u(\boldsymbol{x}, t)$. When $\mathbf{F}(u) = \mathbf{V}u$, where $\mathbf{V}$ is a divergence-free velocity (i.e., $\nabla \cdot \mathbf{V} = 0$), we recover the hyperbolic conservation law and eqn (6.2.1) reduces to the multi-dimensional linear advection equation.

6.2.1 *Galerkin discretisation of the linear advection equation*

For the Galerkin discretisation we consider the linear advection equation of the scalar quantity $u(\boldsymbol{x};t)$ in multiple dimensions, which we can write as

$$\frac{\partial u}{\partial t} + \mathbb{L}u = \frac{\partial u}{\partial t} + (\mathbf{V}\cdot\nabla)u = 0\,, \tag{6.2.2}$$

$$\mathbf{V} = [a(\boldsymbol{x}), b(\boldsymbol{x}), c(\boldsymbol{x})]^\top\,, \quad u(\boldsymbol{x};0) = u_0(\boldsymbol{x})\,.$$

We assume that the initial condition $u_0(\boldsymbol{x})$ is smooth (that is, C^0 continuous) and that the propagation velocity $\mathbf{V}$ is real and divergence-free (that is, $\nabla\cdot\mathbf{V} = 0$). Finally, we assume that the equation is supplemented with appropriate boundary conditions, consistent with the hyperbolic nature of this equation.

Weak formulation

To discretise eqn (6.2.2) using a standard Galerkin formulation we construct the weak integral form of eqn (6.2.2). Therefore, taking the inner product of eqn (6.2.2) with respect to the test function $v(\boldsymbol{x}) \in C^0$, we obtain

$$\left(v, \frac{\partial u}{\partial t}\right)_\Omega + (v, \mathbb{L}u)_\Omega = 0\,. \tag{6.2.3}$$

As illustrated in Fig. 6.4, we consider the solution in a finite domain Ω which is fixed in space and has a boundary $\partial\Omega$. This example assumes the use of triangular subdomains. Note, however, that a similar formulation would apply to a quadrilateral or hybrid discretisation. We divide the domain into N_{el} subdomains denoted by Ω^e, each of which has a local boundary $\partial\Omega^e$ as shown in Fig. 6.4. We shall assume that the initial condition exists within the approximation trial space $\mathcal{X}^\delta$. If this is not the case, then the initial conditions can be projected into the trial space using a Galerkin projection by inverting the global mass matrix, as illustrated later in this section. Alternatively, a collocation projection can be performed at a set of discrete quadrature points, which is computationally more convenient and maintains the C^0 continuity requirements (see Section 4.1.5.3). Similarly, the boundary conditions can be approximated, as explained in Section 4.3.2.

Matrix representation of the elemental problem

Similar to the construction of the Galerkin discretisation of the Helmholtz problem in Section 5.1, we initially consider a single element, Ω_e. Considering the solution to be restricted to a discrete trial space $u^\delta \in \mathcal{X}^\delta$ and considering the set of weight functions $v^\delta(\boldsymbol{x}) = \phi_{pqr}(\boldsymbol{x})$, eqn (6.2.3) becomes

$$\left(\phi_{pqr}, \frac{\partial u^\delta}{\partial t}\right)_{\Omega_e} + \left(\phi_{pqr}, \mathbb{L}u^\delta\right)_{\Omega_e} = 0\,, \quad \forall\,(p,q,r)\,. \tag{6.2.4}$$

Following the vector notation of Section 4.1.5.1 and since the discrete solution, u^δ, is a polynomial function, we can evaluate the solution at the quadrature

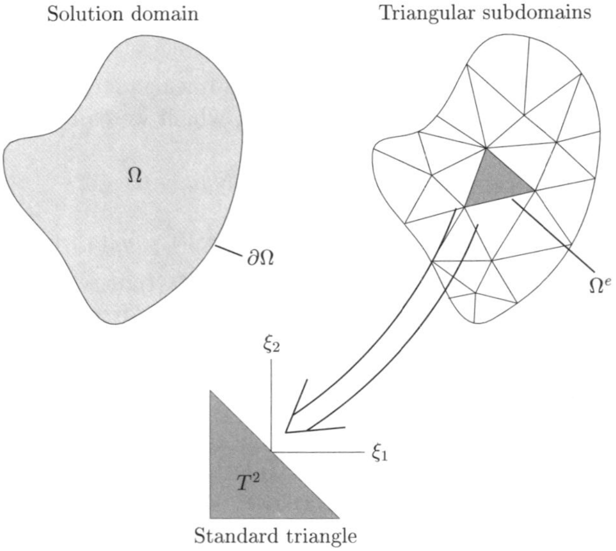

FIG. 6.4. In two dimensions the solution domain is divided into appropriate subdomains, Ω^e, which are then mapped to the standard region (triangle or quadrilateral). All operations are then performed in the local (ξ_1, ξ_2) space on the standard triangle or square.

points and denote it by the vector $\boldsymbol{u}^e$. Equation (6.2.4) can then be written in
♦34 matrix form as

$$(\boldsymbol{B}^e)^\top \boldsymbol{W}^e \frac{\partial \boldsymbol{u}^e}{\partial t} + (\boldsymbol{B}^e)^\top \boldsymbol{W}^e \mathbb{L} \boldsymbol{u}^e = 0 \,. \tag{6.2.5}$$

Finally, we recall that in the standard Galerkin approximation the same expansion space is used for both the trial and test bases, and so we can represent the solution vector in terms of the expansion coefficients as $\boldsymbol{u}^e = \boldsymbol{B}^e \hat{\boldsymbol{u}}^e(t)$. Inserting this relationship into eqn (6.2.5) and noting that

$$\frac{\partial \boldsymbol{u}^e}{\partial t} = \boldsymbol{B}^e \frac{\mathrm{d}\hat{\boldsymbol{u}}^e}{\mathrm{d}t} \,,$$

we obtain the Galerkin approximation to eqn (6.2.2) within an element in terms of the expansion coefficients $\hat{\boldsymbol{u}}^e$:

$$(\boldsymbol{B}^e)^\top \boldsymbol{W}^e \boldsymbol{B}^e \frac{\mathrm{d}\hat{\boldsymbol{u}}^e}{\mathrm{d}t} + (\boldsymbol{B}^e)^\top \boldsymbol{W}^e \mathbb{L} \boldsymbol{B}^e \hat{\boldsymbol{u}}^e = 0 \,. \tag{6.2.6}$$

This type of elemental formulation will also be used in the discontinuous Galerkin formulation discussed in Section 6.2.2. For the standard Galerkin formulation, if

[34] Elemental matrix construction of the multi-dimensional linear advection problem: standard Galerkin.

we were only considering a problem on a single element then eqn (6.2.6) could be discretised in time and Dirichlet boundary conditions could be imposed on the fully-discretised system by lifting a known solution out of the problem.

Matrix representation of the global problem

To construct the solution for multiple elements we must take the elemental contributions of eqn (6.2.6) and construct the global matrix problem. Adopting the matrix notation introduced in Section 4.2.2 where an underline matrix denotes a block diagonal extension to the local matrix, we can represent the local approximation to eqn (6.2.2) over all elements as

$$\underline{((\boldsymbol{B}^e)^\top \boldsymbol{W}^e \boldsymbol{B}^e)}\frac{\mathrm{d}\hat{\boldsymbol{u}}_l}{\mathrm{d}t} + \underline{((\boldsymbol{B}^e)^\top \boldsymbol{W}^e \mathbb{L} \boldsymbol{B}^e)}\hat{\boldsymbol{u}}_l = 0\,, \tag{6.2.7}$$

where we recall that $\hat{\boldsymbol{u}}_l$ denotes a concatenation of the vectors containing all the expansion coefficients from each element $\hat{\boldsymbol{u}}^e$ (that is, $\hat{\boldsymbol{u}}_l = \underline{\hat{\boldsymbol{u}}}^e$). In order to define the global expansion as a union of C^0 continuous expansions, we need to globally assemble eqn (6.2.7) to generate the global matrix system, as described in Section 4.2.1. In this global assembly process we recall that the local expansion coefficients, $\hat{\boldsymbol{u}}_l$, can be expressed in terms of global expansion coefficients, $\hat{\boldsymbol{u}}_g$, through the matrix assembly operation $\boldsymbol{\mathcal{A}}$ by

$$\hat{\boldsymbol{u}}_l = \boldsymbol{\mathcal{A}}\hat{\boldsymbol{u}}_g\,.$$

Inserting this relation into eqn (6.2.7) and pre-multiplying the entire equation by $\boldsymbol{\mathcal{A}}^\top$, we obtain the global matrix system

$$\boldsymbol{\mathcal{A}}^\top \underline{((\boldsymbol{B}^e)^\top \boldsymbol{W}^e \boldsymbol{B}^e)}\boldsymbol{\mathcal{A}}\frac{\mathrm{d}\hat{\boldsymbol{u}}_g}{\mathrm{d}t} + \boldsymbol{\mathcal{A}}^\top \underline{((\boldsymbol{B}^e)^\top \boldsymbol{W}^e \mathbb{L} \boldsymbol{B}^e)}\boldsymbol{\mathcal{A}}\hat{\boldsymbol{u}}_g = 0\,.$$

Depending on the imposition of appropriate boundary conditions, we invert the global mass matrix $\boldsymbol{M} = \boldsymbol{\mathcal{A}}^\top \underline{((\boldsymbol{B}^e)^\top \boldsymbol{W}^e \boldsymbol{B}^e)}\boldsymbol{\mathcal{A}}$ to obtain the semi-discrete system for the solution of the global expansion coefficients: ♦[35]

$$\frac{\mathrm{d}\hat{\boldsymbol{u}}_g}{\mathrm{d}t} = -\boldsymbol{M}^{-1}\boldsymbol{\mathcal{A}}^\top \underline{((\boldsymbol{B}^e)^\top \boldsymbol{W}^e \mathbb{L} \boldsymbol{B}^e)}\boldsymbol{\mathcal{A}}\hat{\boldsymbol{u}}_g\,. \tag{6.2.8}$$

This ordinary differential equation can then be discretised in time, as discussed in Section 5.3.

Boundary conditions

This problem may of course be solved with a range of appropriate boundary conditions. If the domain is periodic then the $\boldsymbol{\mathcal{A}}$ matrix would be constructed so that periodic edges are treated in the same fashion as elemental edges. If

[35] Global matrix construction of the multi-dimensional linear advection problem: standard Galerkin.

an inflow is specified then a Dirichlet boundary condition will be required on some boundaries. Initially, the boundary conditions at every time level can be evaluated by performing a boundary transformation in space (see Section 4.3.1) on any modes which have a nonzero component along the Dirichlet boundary. Similar to the Helmholtz equation formulation in Section 5.1, the Dirichlet modes are then lifted out of the global matrix system, as discussed in Section 4.2.4.2.

We recall that the lifting process can be interpreted as a special case of splitting the solution $u(\boldsymbol{x};t)$ into a homogeneous and boundary contribution, that is,

$$u(\boldsymbol{x};t) = u^{\mathcal{H}}(\boldsymbol{x};t) + u^{\mathcal{D}}(\boldsymbol{x};t)\,,$$

where

$$u^{\mathcal{H}}(\partial\Omega;t) = 0\,, \quad u^{\mathcal{D}}(\partial\Omega) = u(\partial\Omega)\,.$$

The above decomposition can be directly substituted into eqn (6.2.2), although this would lead to a right-hand side involving a time derivative of $u^{\mathcal{D}}(\boldsymbol{x})$. Since $u^{\mathcal{D}}(\partial\Omega)$ is a known function, theoretically, we know $\partial u^{\mathcal{D}}/\partial t$. However, in practice, this is inconvenient for a generalised formulation. Alternatively, we can temporally discretise $\partial u^{\mathcal{D}}/\partial t$ using the same scheme as we adopt for $\partial u^{\mathcal{H}}/\partial t$. It can be shown that such an approach is equivalent to temporally discretising eqn (6.2.2) and then lifting the Dirichlet boundary conditions from the system.

Elemental advection operator

To complete the discretisation we need to describe the form of the operator $\mathbb{L} \equiv \mathbf{V}\cdot\nabla$. The differential component of the operator acts on the expansion basis, which is in the polynomial space $\mathcal{P}_P(\Omega^e)$. It can therefore be evaluated exactly by differentiating the Lagrange polynomials representation through the quadrature points, as explained in Section 4.1.2. The operator in full is defined as

$$\mathbb{L} \equiv a(\boldsymbol{x})\frac{\partial}{\partial x_1} + b(\boldsymbol{x})\frac{\partial}{\partial x_2} + c(\boldsymbol{x})\frac{\partial}{\partial x_3}\,.$$

If we express the partial derivatives in x_i in terms of partial derivatives in ξ_i using the chain rule, we obtain

$$\mathbb{L} = \left(a\frac{\partial \xi_1}{\partial x_1} + b\frac{\partial \xi_1}{\partial x_2} + c\frac{\partial \xi_1}{\partial x_3}\right)\frac{\partial}{\partial \xi_1} + \left(a\frac{\partial \xi_2}{\partial x_1} + b\frac{\partial \xi_2}{\partial x_2} + c\frac{\partial \xi_2}{\partial x_3}\right)\frac{\partial}{\partial \xi_2}$$
$$+ \left(a\frac{\partial \xi_3}{\partial x_1} + b\frac{\partial \xi_3}{\partial x_2} + c\frac{\partial \xi_3}{\partial x_3}\right)\frac{\partial}{\partial \xi_3}\,,$$

and therefore the discrete version of the $\mathbb{L}$ operator acting at the quadrature points can be written as

$$\mathbb{L} = \boldsymbol{R}^e\boldsymbol{D}^e{}_{\xi_1} + \boldsymbol{S}^e\boldsymbol{D}^e{}_{\xi_2} + \boldsymbol{T}^e\boldsymbol{D}^e{}_{\xi_3}\,,$$

where

$$\boldsymbol{R}^e = \boldsymbol{\Lambda}^e\left(a\frac{\partial\xi_1}{\partial x_1} + b\frac{\partial\xi_1}{\partial x_2} + c\frac{\partial\xi_1}{\partial x_3}\right),$$
$$\boldsymbol{S}^e = \boldsymbol{\Lambda}^e\left(a\frac{\partial\xi_2}{\partial x_1} + b\frac{\partial\xi_2}{\partial x_2} + c\frac{\partial\xi_2}{\partial x_3}\right),$$
$$\boldsymbol{T}^e = \boldsymbol{\Lambda}^e\left(a\frac{\partial\xi_3}{\partial x_1} + b\frac{\partial\xi_3}{\partial x_2} + c\frac{\partial\xi_3}{\partial x_3}\right).$$

Both the differential matrices $\boldsymbol{D}^e{}_{\xi_1}$, $\boldsymbol{D}^e{}_{\xi_2}$, and $\boldsymbol{D}^e{}_{\xi_3}$ and the diagonal matrix operator $\boldsymbol{\Lambda}^e(\cdot)$ were defined in Section 4.2.2. For straight-sided triangular or tetrahedral elements (or hybrid elements with the same angles as the standard spaces) the geometric factors $\partial\xi_1/\partial x_1, \partial\xi_2/\partial x_1, \partial\xi_3/\partial x_1, \ldots$ are constants, and so if $a(\boldsymbol{x})$, $b(\boldsymbol{x})$, and $c(\boldsymbol{x})$ are also constants, then $\boldsymbol{R}^e$, $\boldsymbol{S}^e$, and $\boldsymbol{T}^e$ are simply scalars. Finally, even if $a(\boldsymbol{x})$, $b(\boldsymbol{x})$, and $c(\boldsymbol{x})$ are functions of $u(\boldsymbol{x})$ and so the operator becomes nonlinear, the action of the advection operator $\mathbb{L}u$ can still be evaluated through the matrix (or equivalent sum-factorisation) operations shown above.

6.2.1.1 *Numerical convergence*

Projection of initial conditions

As mentioned earlier, we expect the *initial conditions* to be expressed within the expansion space $\mathcal{X}^\delta$, although typically this will not be the case. Accordingly, we have to project the initial conditions. As mentioned previously, there are two types of projection we might consider: collocation and Galerkin. From an implementation point of view, it is computationally most convenient to perform a collocation-type projection of the initial conditions at the quadrature points. This is particularly convenient since all the operations can be treated locally, as discussed in Section 4.1.5.3. However, when using non-symmetric quadrature distributions, such as for the generalised tensor-based expansion in hybrid-shaped domains, this projection will not be symmetric even though the expansion bases are symmetric in the elemental region. Alternatively, a Galerkin projection can be performed which primarily involves the inversion of the global mass matrix.

As an example, the projection of the function $u(x_1, x_2) = \sin(\pi x_1)\sin(\pi x_2)$ is shown in Fig. 6.5(a) and the projection of the function $u(x_1, x_2, x_3) = \sin(\pi x_1)\sin(\pi x_2)\sin(\pi x_3)$ is shown in Fig. 6.5(b). In both figures we have considered periodic domains. Also shown in these figures is the convergence in the L^∞ and L^2 norms as a function of polynomial order. For the two-dimensional case the L^2 error is 10% at an expansion order of $P = 4$, whereas in the three-dimensional case the L^2 error is 10% at an expansion order of $P = 6$. Exponential convergence is only fully established at about $P = 6$ for the two-dimensional case and at about $P = 6$ to 10 in the three-dimensional case. This is not surprising if we note that all the elemental domains in both figures have edges that range over the whole solution region. If we assume that we need a cubic-type mode to capture a full period in one dimension, then for the two-dimensional function we would need a mode of the form $x_1^3 x_2^3$, which requires an expansion of polynomial

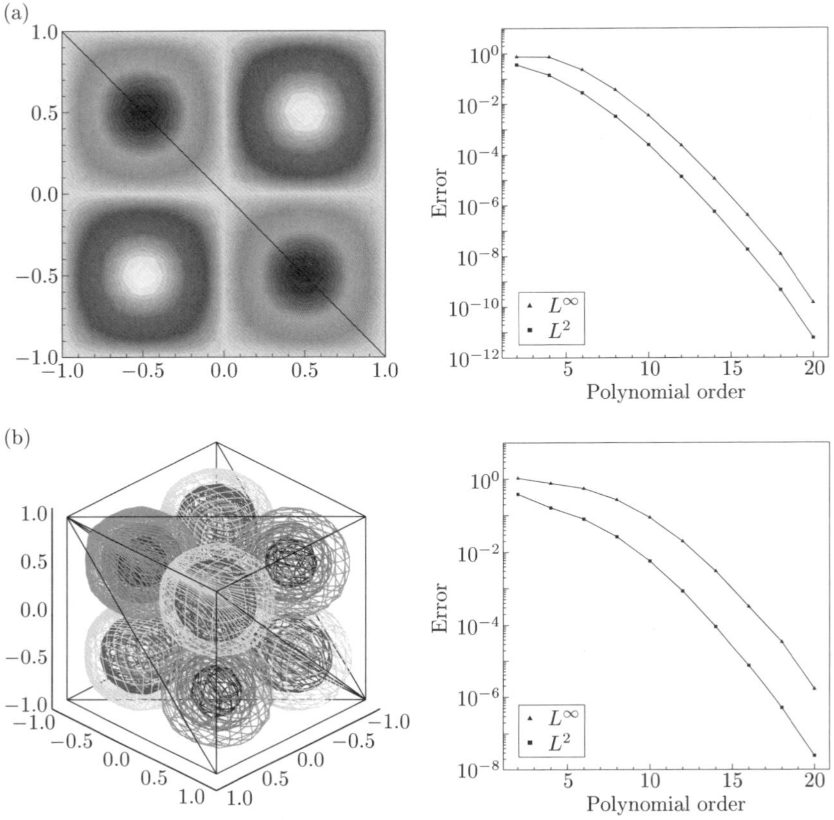

FIG. 6.5. Isocontours of projection functions in two and three dimensions as a function of polynomial order. (a) Errors in the two-dimensional projection of the function $u(x_1, x_2) = \sin(\pi x_1)\sin(\pi x_2)$ in a periodic domain. (b) Errors in the three-dimensional projection of the function $u(x_1, x_2, x_3) = \sin(\pi x_1)\sin(\pi x_2)\sin(\pi x_3)$ in a periodic domain.

order $P = 6$. Similarly, for the three-dimensional function we would need a mode of the form $x_1^3 x_2^3 x_3^3$, which requires an expansion of polynomial order $P = 9$ to be resolved.

Spatial convergence

To examine spatial convergence as a function of the polynomial order as well as the number of elements, we consider a problem with initial condition

$$u_0(\boldsymbol{x}) = \sin(\pi \cos(\pi x)),$$

used in [86], and apply a time-step, $\Delta t = 0.002$, which is small enough to eliminate time errors. The solution was propagated for one time period (that is, two time units) using a third-order Adams–Bashforth scheme. Four computational domains were considered, as shown in Fig. 6.6. Each domain spans the region $-1 \leqslant x \leqslant 1$, but is successively subdivided into equal lengths whilst maintaining the same aspect ratio. We can increase the expansion order on each domain to determine convergence with respect to expansion order, and consider the error on each subdomain at constant expansion order to ascertain the convergence with respect to the number of elements.

Figure 6.7(a) shows the L^∞ error versus expansion order for the four domains shown in Fig. 6.6. The L^∞ error is plotted on a logarithmic scale and we can therefore deduce spectral convergence from the approximately straight lines on these plots. The anomalous behaviour of the convergence rate in domains A and B at low expansion orders is due to the fact that the solution is under-resolved.

Figure 6.7(b) shows the L^∞ error versus the total number of degrees of freedom. We have used $N_{\rm el}P/2$ as the one-dimensional total degrees of freedom as this is the number of degrees of freedom along the bottom edge of the domain ($N_{\rm el}$ being the number of elements and P being the expansion order). This approximation is only relevant to the degrees of freedom in the x_1 direction. This is appropriate, in this case, as this is a one-dimensional problem. The data points connected with dashed lines are taken from Canuto *et al.* [86] and represent the L^∞ error of a Fourier–Galerkin solution, as well as a second-order and fourth-order finite difference solution, to this problem. By plotting the error in this

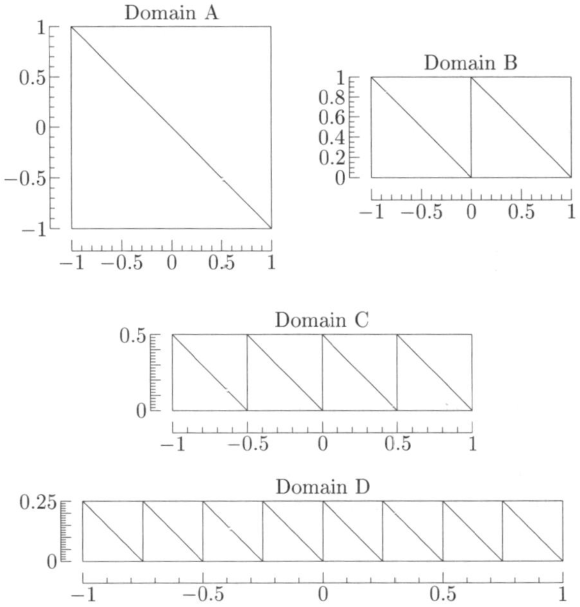

FIG. 6.6. Spatial accuracy tests are performed on four solution domains, as shown here. The solution is one-dimensional in x and each domain spans $-1 \leqslant x \leqslant 1$.

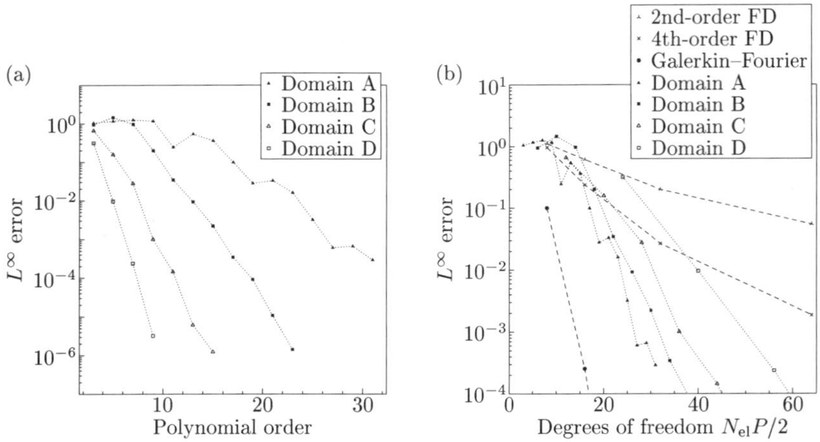

FIG. 6.7. (a) L^∞ error after one period for all domains shown in Fig. 6.6. (a) Error with respect to polynomial order P, and (b) error with respect to degrees of freedom. The points connected with dashed lines are results given in Canuto *et al.* [86] for Fourier–Galerkin and second-order and fourth-order finite difference solutions to this problem.

manner, we see that the simplest domain (domain A) has a faster convergence than the other domains solved with triangular spectral elements, although there is a larger operation count per degree of freedom for this solution. The convergence of this domain is still slower than the Fourier–Galerkin solution.

6.2.2 *Discontinuous Galerkin method*

We now consider the two-dimensional discontinuous Galerkin discretisation of the general hyperbolic equation

$$\frac{\partial u}{\partial t} + \nabla \cdot \mathbf{F}(u) = 0\,, \tag{6.2.9}$$

where we recall that $\mathbf{F}(u) = [f(u), g(u)]^\top$ is the two-dimensional *flux* vector which defines the transport of $u(\boldsymbol{x}, t)$. We note that the linear advection equation considered in the previous section is given by the choice $\mathbf{F}(u) = \mathbb{L}u = \mathbf{V} \cdot \nabla u$.

From the point of view of the hyperbolic conservation equation (6.2.9), we note that C^0 continuous function spaces applied in the standard Galerkin formulation are not the natural place to pose the problem. Mathematically, hyperbolic problems of this type tend to have solutions in spaces of bounded variation. In physical problems, the best one can hope for is that solutions will be piecewise smooth, that is, be smooth in regions separated by discontinuities (e.g., shocks).

This consideration suggests that a more appropriate formulation is one where the global trial space $\mathcal{X}^\delta$ can also contain discontinuous functions. Within the spectral/*hp* element framework, we therefore consider a space of piecewise continuous polynomial functions where the approximation space is defined as

$$\mathcal{X}^\delta = \{v \in L^2(\Omega) : v|\Omega^e \in \mathcal{P}_P(\Omega^e),\ \forall\, \Omega^e\}. \tag{6.2.10}$$

For elements with linear elemental mappings, $\mathcal{P}_P(\Omega^e)$ is the space of polynomials of order P defined in the element Ω^e. The space of test functions $\mathcal{V}^\delta$ is defined analogously.

To deal with this discontinuous approximation space requires a somewhat different approach to the standard Galerkin variational formulation. Initially, we follow the same elemental construction as the standard Galerkin formulation discussed in Section 6.2.1. Therefore, we approximate u by a polynomial expansion u^δ such that $u^\delta \in \mathcal{X}^\delta$, where $\mathcal{X}^\delta$ is a finite-dimensional subspace of the space of compactly-supported continuous functions. The elemental variational statement of the Galerkin formulation of (6.2.9) is obtained by multiplying by a test function $v^\delta = \phi_{pq}(\boldsymbol{x})$ and integrating over the elemental domain Ω^e, i.e.,

$$\int_{\Omega^e} v^\delta \frac{\partial u^\delta}{\partial t}\, \mathrm{d}\boldsymbol{x} + \int_{\Omega^e} v^\delta \nabla \cdot \mathbf{F}(u^\delta)\, \mathrm{d}\boldsymbol{x} = 0\,.$$

In contrast to the standard Galerkin formulation, we now apply the divergence theorem to the second advection term to obtain

$$\int_{\Omega^e} v^\delta \frac{\partial u^\delta}{\partial t}\, \mathrm{d}\boldsymbol{x} + \int_{\partial\Omega^e} v^\delta\, \mathbf{F}(u^\delta) \cdot \boldsymbol{n}\, \mathrm{d}s - \int_{\Omega^e} \nabla v^\delta \cdot \mathbf{F}(u^\delta)\, \mathrm{d}\boldsymbol{x} = 0\,. \tag{6.2.11}$$

The solution $u^\delta \in \mathcal{X}^\delta$ satisfies this equation for all $v^\delta \in \mathcal{V}^\delta$.

Equation (6.2.11) currently only involves local elemental operations. However, to keep the solution within a space of bounded variation and to allow information to propagate between elements requires some elemental coupling. This coupling arises as a result of the way the flux $\mathbf{F}(u^\delta)$ is evaluated in the second term of eqn (6.2.11). Boundary conditions within each elemental formulation given by eqn (6.2.11) are enforced through the flux $\mathbf{F}(u^\delta)$ that appears in the second term. As this flux is computed at the boundary between adjacent elements, where the solution may be discontinuous, we have two possible values of the discrete solution u^δ. If we denote these two values by u^δ_- (internal to the element) and u^δ_+ (external to the element), then the boundary flux is denoted by $\tilde{\boldsymbol{f}}^e(u^\delta_-, u^\delta_+)$.

To generate a stable scheme, upwinding considerations based on the natural propagation of information should dictate how this flux is computed. For example, if we consider the case of linear advection where $\mathbf{F}(u^\delta) = \mathbf{V}u^\delta$ then we could determine $\tilde{\boldsymbol{f}}^e(u^\delta_-, u^\delta_+)$ using an upwinded approach. If u^δ_- denotes the solution in element e and u^δ_+ denotes the solution in an adjacent element, then $\tilde{\boldsymbol{f}}^e(u^\delta_-, u^\delta_+)$ on the boundary of element e with outward normal $\boldsymbol{n}^e$ should be chosen so that

$$\tilde{\boldsymbol{f}}^e(u^\delta_-, u^\delta_+) = \begin{cases} \mathbf{V}u^\delta_-\,, & \mathbf{V}\cdot\boldsymbol{n}^e \geqslant 0\,, \\ \mathbf{V}v^\delta_+\,, & \mathbf{V}\cdot\boldsymbol{n}^e < 0\,. \end{cases}$$

For the more complicated case of a hyperbolic system of equations, a one-dimensional approximate Riemann solver can also be used to compute the value of the flux based on u_-^δ and u_+^δ, as discussed in Section 10.3.

Using the derivative and projection operators described in Section 4.1.5.1, it is straightforward to implement a numerical method for eqn (6.2.1) using formula (6.2.11), which can be rewritten (dropping the superscript δ) as

$$\partial_t(v, u)_{\Omega^e} + \langle v, \tilde{\boldsymbol{f}}(u_-, u_+)\rangle_{\partial\Omega_e} - (\nabla v, \mathbf{F}(u))_{\Omega^e} = 0\,, \tag{6.2.12}$$

where the elemental inner products $(\cdot,\cdot)_{\Omega^e}$ and $\langle\cdot,\cdot\rangle_{\Omega^e}$ are defined as follows:

$$\begin{aligned}
(a, b)_{\Omega^e} &\equiv \int_{\Omega^e} ab\,\mathrm{d}\boldsymbol{x}\,,\\
(\boldsymbol{a}, \boldsymbol{b})_{\Omega^e} &\equiv \int_{\Omega^e} \boldsymbol{a}\cdot\boldsymbol{b}\,\mathrm{d}\boldsymbol{x}\,,\\
\langle a, \boldsymbol{b}\rangle_{\partial\Omega^e} &\equiv \int_{\partial\Omega^e} a\boldsymbol{b}\cdot\boldsymbol{n}^e\,\mathrm{d}s\,.
\end{aligned}$$

We are now free to use any expansion basis, ϕ_{pq}, discussed in Chapter 3, whether it is modal or nodal or even fully orthogonal, and we can the write the discrete equation in matrix form as

$$\begin{aligned}
\frac{\mathrm{d}\hat{\boldsymbol{u}}^e}{\mathrm{d}t} = [\boldsymbol{M}^e]^{-1}\big[(\boldsymbol{D}^e{}_{x_1}\boldsymbol{B}^e)^\top \boldsymbol{W}^e\boldsymbol{\Lambda}^e(f(u))\\
+ (\boldsymbol{D}^e{}_{x_2}\boldsymbol{B}^e)^\top \boldsymbol{W}^e\boldsymbol{\Lambda}^e(g(u))\big] - [\boldsymbol{M}^e]^{-1}\boldsymbol{b}^e\,,
\end{aligned} \tag{6.2.13}$$

where $\boldsymbol{b}$ is a vector corresponding to the surface integral such that

$$\boldsymbol{b}^e[n(pq)] = \int_{\partial\Omega^e} \phi_{pq}\tilde{\boldsymbol{f}}\cdot\boldsymbol{n}^e\,\mathrm{d}s$$

and $n(pq)$ represents the mapping from the local basis indices p and q to a consecutive list. We note that eqn (6.2.13) only involves the inversion of the local elemental mass matrix, which is significantly cheaper to invert than the global mass matrix of the standard Galerkin discretisation. Further, $\boldsymbol{M}^e$ will be diagonal when using the orthogonal expansion $\phi_{pq}(\xi_1,\xi_2) = \widetilde{\psi}_p^a(\eta_1)\widetilde{\psi}_{pq}^b(\eta_2)$, making the inversion trivial. We also note that

$$\begin{aligned}
(\boldsymbol{D}^e{}_c\boldsymbol{B}^e)^\top \boldsymbol{W}^e\boldsymbol{\Lambda}^e(f) &= (\boldsymbol{B}^e)^\top \boldsymbol{D}^{e\top}_c \boldsymbol{W}^e\boldsymbol{\Lambda}^e(f)\\
&= ((\boldsymbol{B}^e)^\top \boldsymbol{W}^e)(\boldsymbol{W}^e)^{-1}\boldsymbol{D}^{e\top}_c \boldsymbol{W}^e\boldsymbol{\Lambda}^e(f)
\end{aligned}$$

for $c = x_1, x_2$. Therefore, if we evaluate the vector

$$(\boldsymbol{W}^e)^{-1}\boldsymbol{D}^{e\top}_c \boldsymbol{W}^e\boldsymbol{\Lambda}^e(f)$$

using the transposed derivative operator, then this term reduces to the standard inner product which we have seen in Section 4.1.6, and thus it may be efficiently evaluated using the sum-factorisation technique.

An *alternative* implementation of the discontinuous Galerkin method is to integrate the third term in eqn (6.2.11) by parts again. This formulation does not rely upon transposed derivative operators and so is more consistent with the operators used in the standard Galerkin formulation. The variational statement written for each element in this case is

$$\partial_t(v, u^\delta)_{\Omega^e} + \langle v, [\tilde{\boldsymbol{f}}(u_-^\delta, u_+^\delta) - \mathbf{F}(u^\delta)]\rangle_{\Omega^e} + (v, \nabla \cdot \mathbf{F}(u^\delta))_{\Omega^e} = 0\,, \qquad (6.2.14)$$

where $\tilde{\boldsymbol{f}}$ again denotes the numerical upwind flux and $\mathbf{F}(u^\delta)$ is the local elemental flux. This formulation has also been used by many authors, for example, Johnson [255].

When evaluating the surface integral vector $\boldsymbol{b}^e$ we find an extra consideration on the choice of *quadrature points*. Typically, we prefer to use the Gauss distribution of points when integrating along the edge rather than the Gauss–Radau or Gauss–Lobatto distributions. There are two reasons for this.

- First, we theoretically need a higher accuracy in the integration along the edges as a result of the interior integrals being evaluated over a volume h^2, and the edge integrals over h. In the error analysis, edge errors are multiplied by a larger constant (see [100] for a detailed account of truncation errors).
- Secondly, we want to avoid evaluation at the vertex points of the triangle to avoid the need to solve a multi-dimensional Riemann problem when considering the flux computation in nonlinear problems.

6.2.2.1 *Conservativity*

One of the main strengths of the discontinuous Galerkin formulation is that it automatically guarantees conservativity on an element-wise basis, similar to finite volume methods. In general, this property is difficult to satisfy in high-order discretisations. Another way of formulating conservative algorithms for spectral (collocation) discretisations is through the use of staggered grids, as we demonstrate in Section 10.1.

To further illustrate the discontinuous Galerkin formulation, we consider the one-dimensional version of eqn (6.2.9), which we put in weak form and integrate by parts to obtain

$$(v, \partial_t u) - (v_x, f(u)) + v f(u)|_{x_L}^{x_R} = 0\,, \qquad (6.2.15)$$

where $x \in [x_L, x_R]$, with x_L and x_R being the left and right boundaries, respectively, of a single element.

The treatment of the boundary terms is particularly important as it justifies the *conservativity property*. That is to say, the last term in eqn (6.2.15) expands to

$$v_R^- f_R^- - v_L^+ f_L^-\,,$$

which implies an *upwind* treatment (see the flux of the second term), and the test function v is evaluated inside the interval $[x_L, x_R]$. Note that f_L^- is a function of (u_L^-, u_L^+), and similarly for f_R^-. Integrating eqn (6.2.15) by parts again we obtain

$$(v, \partial_t u) + (v, f_x(u)) + v_R^- f_R^- - v_L^+ f_L^- - v_R^- f_R^- + v_L^+ f_L^+ , \tag{6.2.16}$$

which reduces to the form

$$(v, \partial_t u) + (v, f_x(u)) + v_L^+ (f_L^+ - f_L^-) . \tag{6.2.17}$$

This final equation represents the so-called weak imposition of boundary conditions through the jump term.

The analogy with the first-order upwind *finite volume* scheme can be established by considering the linear advection equation

$$\partial_t u + u_x = 0$$

in a periodic domain, $x \in [0, 2\pi]$, assuming that u is a polynomial of degree P in each cell. We denote the cell by $I_j = [x_{j-1/2}, x_{j+1/2}]$ and set $x_{1/2} = 0$ and $x_{N+1/2} = 2\pi$. We also denote the centre of the cell by x_j and assume an equidistant mesh with spacing Δx.

The variational statement for this one-dimensional problem then reads as follows.

Find u for all test functions v such that

$$\int_{I_j} u_t v \, dx - \int_{I_j} u v_x \, dx + \tilde{u}_{j+1/2} v_{j+1/2}^- - \tilde{u}_{j-1/2} v_{j-1/2}^+ = 0 , \tag{6.2.18}$$

where the numerical upwind flux is $\tilde{u}_{j\pm1/2} = u_{j\pm1/2}^-$ from upwind considerations.

By now choosing a trial basis of order P and inverting the local mass matrix of order $(P+1) \times (P+1)$, the above scheme can be written in block form as

$$\frac{d\boldsymbol{u}_j}{dt} + \frac{1}{\Delta x}(\boldsymbol{A}\boldsymbol{u}_j + \boldsymbol{B}\boldsymbol{u}_{j-1}) , \tag{6.2.19}$$

where $\boldsymbol{u}_j$ now denotes a vector of coefficients of length $P+1$. Also, $\boldsymbol{A}$ and $\boldsymbol{B}$ are $(P+1) \times (P+1)$ matrices with constant entries, and thus they can be pre-computed and stored at a pre-processing step.

In order to see the analogy with the upwind finite volume scheme, we set $P = 0$ in the above to obtain

$$\frac{du_j}{dt} + \frac{u_j - u_{j-1}}{\Delta x} = 0 , \tag{6.2.20}$$

which is the familiar upwind formula.

In summary, in order to achieve this result we have replaced the flux term by a single-valued numerical flux of the upwind type, and we have also replaced the test function at the boundaries by its values inside the cell at both ends. This is a general procedure for the discontinuous Galerkin method for advection.

6.3 Eigenspectra and time-step restriction of the weak advection operator

We have previously considered in Section 5.4 the eigenspectrum of the weak Laplacian operator. We recall that the eigenspectrum was important in determining the time-step restrictions of the explicit treatment of the diffusion operator. An analogous role is played by the weak advection operator for both the standard and discontinuous Galerkin formulations.

We recall that for the standard Galerkin formulation the semi-discrete form is given by eqn (6.2.8) and for the discontinuous Galerkin formulation the semi-discrete equation is given by eqn (6.2.13). In the standard Galerkin formulation and for a given time-stepping scheme, we recall that the temporal discretisation of the semi-discrete system will be stable if the eigenvalues of

$$\Delta t\, \boldsymbol{M}^{-1}\boldsymbol{\mathcal{A}}^{\top}\underline{((\boldsymbol{B}^e)^{\top}\boldsymbol{W}^e\mathbb{L}\boldsymbol{B}^e)}\boldsymbol{\mathcal{A}}$$

lie within the stability region of that temporal scheme. Here Δt is the time-step, the decay rate of which must be the same as the growth rate of the largest eigenvalue so that the spectrum remains within the stability region. An analogous result exists for the discontinuous Galerkin formulation.

To understand the form of the weak advection operation we can consider the matrix $(\boldsymbol{B}^e)^{\top}\boldsymbol{W}^e\mathbb{L}\boldsymbol{B}^e$ of the standard Galerkin formulation. This matrix represents the discrete form of the inner product of the expansion basis with the advection operator acting on the basis, that is, $(\phi_{pqr}, \mathbb{L}\phi_{lmn})_e$. If we assume that the propagation velocity $\mathbf{V}$ is divergence-free, then we can write

$$\mathbb{L}\phi_{pqr} = (\mathbf{V}\cdot\nabla)\phi_{pqr} = \nabla\cdot(\mathbf{V}\phi_{pqr})\,.$$

Since ϕ_{pqr} and ϕ_{lmn} are both scalar fields we can apply the vector identity

$$\phi\nabla\cdot(\psi\mathbf{V}) = \nabla\cdot(\phi\psi\mathbf{V}) - \nabla\phi\cdot(\psi\mathbf{V})\,,$$

and letting $\phi = \phi_{pqr}$ and $\psi = \phi_{lmn}$, we find

$$(\phi_{pqr}, \mathbb{L}\phi_{lmn})_e = -(\mathbb{L}\phi_{pqr}, \phi_{lmn})_e + \int_{\Omega^e}\nabla\cdot(\phi_{pqr}\mathbf{V}\phi_{lmn})\,\mathrm{d}\boldsymbol{x}\,.$$

Applying the divergence theorem to the last term, we arrive at

$$(\phi_{pqr}, \mathbb{L}\phi_{lmn})_e = -(\mathbb{L}\phi_{pqr}, \phi_{lmn})_e + \int_{\partial\Omega^e}\phi_{pqr}\phi_{lmn}\mathbf{V}\cdot\boldsymbol{n}\,\mathrm{d}\boldsymbol{x}\,,$$

where $\boldsymbol{n}$ is the outward normal along the boundaries of an element. This result shows that the operator is skew-symmetric if the surface integral, $\int_{\partial\Omega^e}\phi_{pqr}\phi_{lmn}\mathbf{V}\cdot\boldsymbol{n}\,\mathrm{d}\boldsymbol{x}$, is zero. If all terms are evaluated using consistent numerical integration then there is no approximation error in evaluating the operator $\mathbb{L}$, and thus the matrix representation of the discrete operator will also be skew-symmetric. The surface integral is zero if either of the expansion modes ϕ_{pqr} or

ϕ_{lmn} is an interior mode, since interior modes are zero along the boundaries by design in a C^0 expansion. In the standard Galerkin formulation we are explicitly imposing C^0 continuity. Therefore, if we assume a conforming discretisation then the surface integral along the interfaces between two elements will necessarily be equal and opposite. Therefore, when the system is globally assembled there will be zero net contribution from inter-elemental interfaces. The only remaining contribution from the surface integral term comes from the boundaries of the solution domain. If the boundary conditions are skew-symmetric (that is, periodic or zero Dirichlet) then the global operator is skew-symmetric.

As the global mass matrix $\boldsymbol{M}$ is symmetric and the weak advection operator $\boldsymbol{\mathcal{A}}^\top \underline{((\boldsymbol{B}^e)^\top \boldsymbol{W}^e \mathbb{L} \boldsymbol{B}^e)} \boldsymbol{\mathcal{A}}$ (with appropriate boundary conditions) is skew-symmetric, it can be shown that the eigenvalues of $\boldsymbol{M}^{-1} \boldsymbol{\mathcal{A}}^\top \underline{((\boldsymbol{B}^e)^\top \boldsymbol{W}^e \mathbb{L} \boldsymbol{B}^e)} \boldsymbol{\mathcal{A}}$ must be purely imaginary. This means that, for this case, we require a time-stepping scheme with a stability region encompassing the imaginary axis. For example, for stability for the third-order Adams–Bashforth scheme we should satisfy

$$\Delta t \cdot \lambda_{\max} \simeq 0.723\,,$$

where $\lambda_{\max}$ is the maximum eigenvalue. In discontinuous Galerkin methods the eigenvalues of the advection operator are also predominantly imaginary. However, depending on the type of numerical fluxes that are applied between the elemental boundaries, unresolved values also have a negative real part which leads to some numerical diffusion of these modes [245, 246, 426].

6.3.1 *Numerical evaluation of the time-step restriction*

♦[36] From the discussion in the previous section, we note that the evaluation of the convective time-step restriction is equivalent to bounding the largest eigenvalue of the weak advection operator pre-multiplied by the inverse of the mass matrix, i.e., $\boldsymbol{\mathcal{A}}^\top \underline{((\boldsymbol{B}^e)^\top \boldsymbol{W}^e \mathbb{L} \boldsymbol{B}^e)} \boldsymbol{\mathcal{A}}$. The implicit assumption is that the time eigenvalues are predominately imaginary and the stability region of the time-stepping scheme includes the imaginary axis.

In the next section we present some analytical and numerical results on the eigenspectrum of the advection operator in one, two, and three dimensions in the standard regions, $\Omega_{\rm st}$, for unit advection velocities. The results of this section will demonstrate that the growth of the maximum eigenvalue in the standard region, $\lambda^{\rm st}_{\max}$, can be bounded by

$$c_\lambda P^2 > \lambda^{\rm st}_{\max}\,, \tag{6.3.1}$$

where c_λ is a constant. However, for a general domain of characteristic length h with local advection velocity $\mathbf{V}$ we expect that the desired bound on the maximum eigenvalue of the full advection operator will be of the form

$$C(\mathbf{V}, h) P^2 > \lambda_{\max}\,.$$

[36] Numerical evaluation of the advection time-step restrictions.

The evaluation of $C(\mathbf{V}, h)$ is highly problem-dependent and so it is difficult to formalise it in an exact algebraic expression. Nevertheless, we now have a first condition that the time-step restriction for explicit evaluation of the advection operator is bounded by

$$\Delta t \leqslant \frac{\alpha_{\text{im}}}{C(\mathbf{V}, h)P^2}, \tag{6.3.2}$$

where α_{im} is the value obtained at the intersection of the stability region with the imaginary axis (i.e., $\alpha_{\text{im}} \simeq 0.723$ for third-order Adams–Bashforth).

To apply eqn (6.3.2) in a numerical algorithm we still require an estimate of the coefficient $C(\mathbf{V}, h)$. We note that an estimate of the constant c_λ in the local bound (6.3.1) can be obtained from numerical experiments and analytical results. The local elemental nature of this result makes it attractive from an implementation point of view. Therefore, we are motivated to relate every general elemental domain to the local domain and apply this relation rather than try to obtain $C(\mathbf{V}, h)$ directly.

For a general elemental region we have seen previously that we can map the elemental region Ω_e to the standard region Ω_{st}. We can also apply the same mapping to the velocity components within a general element to obtain a 'local velocity' in the standard region. Having obtained the local velocity, we can then scale the previous approximation of the maximum eigenvalue in the standard region which was based on a unit local velocity to approximate the maximum eigenvalue for a non-unit velocity in a general-shaped element. Denoting $\mathbf{V}^e(\boldsymbol{x}) = [u_1^e, u_2^e, u_3^e]$ as the velocity in element e and $\mathbf{V}^{\text{st}}(\boldsymbol{\xi}) = [u_1^{\text{st}}, u_2^{\text{st}}, u_3^{\text{st}}]$ as the velocity in the standard region, we can evaluate $\mathbf{V}^{\text{st}}(\boldsymbol{\xi})$ in terms of $\mathbf{V}^e(\boldsymbol{x})$ by applying the chain rule, i.e.,

$$\begin{aligned}
u_1^{\text{st}} &= u_1^e \frac{\partial \xi_1}{\partial x_1} + u_2^e \frac{\partial \xi_1}{\partial x_2} + u_3^e \frac{\partial \xi_1}{\partial x_3}, \\
u_2^{\text{st}} &= u_1^e \frac{\partial \xi_2}{\partial x_1} + u_2^e \frac{\partial \xi_2}{\partial x_2} + u_2^e \frac{\partial \xi_2}{\partial x_3}, \\
u_3^{\text{st}} &= u_1^e \frac{\partial \xi_3}{\partial x_1} + u_2^e \frac{\partial \xi_3}{\partial x_2} + u_2^e \frac{\partial \xi_3}{\partial x_3},
\end{aligned}$$

where we note that the geometric factors $\partial \xi_1 / \partial x_1$ are usually available since they are required to evaluate derivative operators (see Section 4.1.3.4). By considering the maximum local velocity $|\mathbf{V}^{\text{st}}|$ and using the bound on the local maximum eigenvalue of the form (6.3.1), we argue that an estimate of the global maximum eigenvalue, $\lambda_{\max}$, is of the form

$$\max_{e=1}^{N_{\text{el}}} |\mathbf{V}^{\text{st}}| \, c_\lambda \, P^2 \approx \lambda_{\max},$$

which is equivalent to approximating the constant $C(\mathbf{V}, h)$ by $C(\mathbf{V}, h) \approx c_\lambda \max |\mathbf{V}^{\text{st}}|$. In practice, we evaluate $\mathbf{V}^{\text{st}}$ at the quadrature points to determine the maximum value and typically choose a value of $c_\lambda = 0.2$.

In one dimension we note that the geometric factor which maps an element of size h to the standard segment is $\partial\xi_1/\partial x_1 = 2h$, and so the constant becomes

$$C(\mathbf{V}, h) \approx \max|V^e|\frac{c_\lambda P^2}{h}.$$

The term $c_\lambda P^2/h$ in the above expression can be considered as an approximation to the minimum mesh spacing Δx, and so it can be appreciated that $C(\mathbf{V}, h)\Delta t$ is directly analogous to the standard CFL definition in finite differences. Finally, although we have formulated the problem primarily from the point of view of the standard Galerkin formulation, the procedure is equally applicable to a discontinuous Galerkin formulation.

6.3.2 *Eigenspectrum of the weak advection operator in $\Omega_{\rm st}$*

6.3.2.1 *One dimension*

The eigenspectra of the first-order derivatives for a variety of global spectral discretisations in one dimension are reviewed in Canuto *et al.* [86]. In summary, for a Chebyshev–Gauss–Lobatto discretisation with Dirichlet boundary conditions, the eigenvalues have negative real parts and can be approximated by

$$\min|\lambda_k| \approx 0.46P\,, \quad \max|\lambda_k| \approx 0.089P^2\,.$$

For a similar Legendre discretisation with Dirichlet boundary conditions we have

$$\min|\lambda_k| \approx 0.66P\,, \quad \max|\lambda_k| \approx P\,.$$

Although for the scalar case the spectral radius for the Legendre expansion grows linearly with the order P, for hyperbolic systems the corresponding spectral radius grows as $\mathcal{O}(P^2)$. Even in the scalar case the effect of round-off error for $P \geqslant 64$ is the production of pseudo-eigenvalues which grow quadratically with P. We also note that when numerically evaluating the distributions of eigenspectra the result may be 'polluted' by spurious eigenvalues due to round-off, as pointed out in [471].

A detailed dispersion analysis of the one-dimensional spectral/hp element formulation for both the standard and discontinuous Galerkin formulations has been performed in [246, 426].

6.3.2.2 *Two dimensions*

Standard Galerkin eigenspectrum

If we consider a periodic solution domain split into two triangles as shown in Fig. 6.8, where one triangle is in the standard space, then we can determine the maximum eigenvalue for wave speeds $\mathbf{V}$ at different angles θ to the horizontal. The wave speed is assumed to have a unit magnitude (i.e., $|\mathbf{V}| = 1$).

We have only displayed eigenvalues for the range $-90^\circ \leqslant \theta \leqslant 90^\circ$ because a velocity with components $\mathbf{V} = [1/\sqrt{2}, 1/\sqrt{2}]$ will have the same maximum

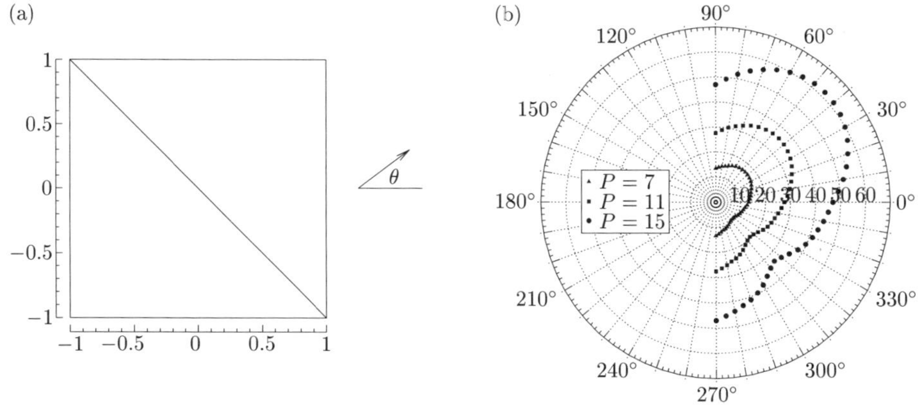

FIG. 6.8. For the periodic domain shown in (a) we consider a wave travelling at an angle θ to the horizontal with unit magnitude. For several wave orientations the maximum eigenvalue of the weak advection operator $\boldsymbol{M}^{-1}\boldsymbol{\mathcal{A}}^{\top}\underline{((\boldsymbol{B}^e)^{\top}\boldsymbol{W}^e\mathbb{L}\boldsymbol{B}^e)}\boldsymbol{\mathcal{A}}$ was determined and is shown in the polar plot in (b).

eigenvalue as a velocity with components $\mathbf{V} = [-1/\sqrt{2}, -1/\sqrt{2}]$. Therefore, the first quadrant is similar to the third quadrant, and the second quadrant is similar to the fourth. From the polar plot in Fig. 6.8 we see that the largest eigenvalue of the advection operator has a maximum value at $\theta = 45°$. Two other orientations were also considered where the upper triangle is rotated so that the collapsed vertex is pointing at the other corners, and *identical* distributions were found in each case.

A simpler example is a single domain, as shown in Fig. 6.9. In this example we use the standard triangle and, once again, propagate a wave of unit velocity at an angle θ to the horizontal. Since this domain is no longer periodic we have to impose boundary conditions. As indicated in Fig. 6.9, we apply homogeneous inflow boundary conditions on the horizontal and vertical boundaries and treat the sloping boundary as an outflow boundary where no condition is explicitly enforced. The imposition of boundary conditions means that the advection operator $\boldsymbol{M}^{-1}\boldsymbol{\mathcal{A}}^{\top}\underline{((\boldsymbol{B}^e)^{\top}\boldsymbol{W}^e\mathbb{L}\boldsymbol{B}^e)}\boldsymbol{\mathcal{A}}$ is no longer skew-symmetric. Nevertheless, we find that the largest eigenvalue lies close to the imaginary axis (see Fig. 6.10). These eigenvalues, for a range of wave velocities of unit magnitude at angles in the range $0° \leqslant \theta \leqslant 90°$, are also shown in Fig. 6.9. The distribution of eigenvalues is far more uniform over the θ range than in the periodic box test. It is also worth noting that for the lower expansion orders ($P = 7, 11$) the maximum eigenvalue is at $\theta = 45°$, whereas for the higher expansion orders ($P = 19, 23$) the maximum eigenvalues are at $\theta = 0°, 90°$.

This property is reflected in the spectra shown in Fig. 6.10. Here we see the complete spectra for a wave of unit magnitude propagating at $\theta = 45°$ to the horizontal for expansion orders of $P = 7$, 15, and 23. The lower expansion orders

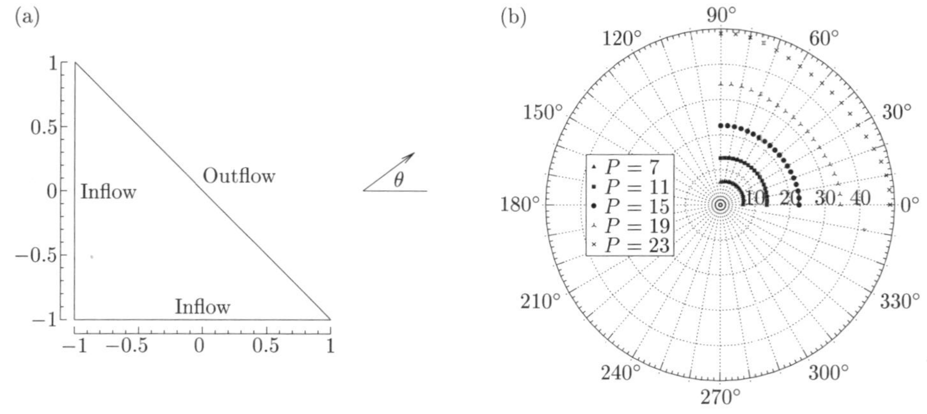

FIG. 6.9. For the domain shown in (a) we consider a wave travelling at an angle θ to the horizontal with unit magnitude. The maximum eigenvalue of the weak advection operator $\boldsymbol{M}^{-1}\boldsymbol{\mathcal{A}}^\top(\underline{(\boldsymbol{B}^e)^\top \boldsymbol{W}^e \mathbb{L}\boldsymbol{B}^e})\boldsymbol{\mathcal{A}}$ was determined for various wave orientations and is shown in the polar plot in (b).

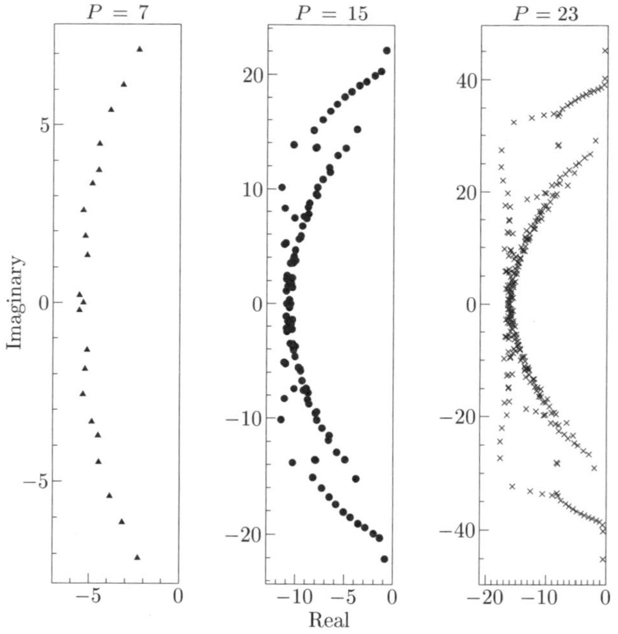

FIG. 6.10. The eigenspectra of the advection matrix system $\boldsymbol{M}^{-1}\boldsymbol{\mathcal{A}}^\top \underline{(\boldsymbol{B}^e)^\top \boldsymbol{W}^e \mathbb{L}\boldsymbol{B}^e}\boldsymbol{\mathcal{A}}$. The advection velocity $\mathbf{V}$ is at $\theta = 45°$. The spectra are for expansion orders of $P = 7$, 15, and 23.

($P = 7, 11$) have eigenvalues which lie close to a semicircular distribution in the left-hand plane, but the higher expansion order ($P = 23$) has a different, more elongated, distribution. We cannot, however, dismiss the possibility of round-off error at higher expansion orders due to some degree of non-normality of the matrices.

As already discussed, to determine the time-step restriction as a function of expansion order we need a bound for the maximum eigenvalue. For both of the previous examples, the maximum eigenvalue is plotted against expansion order in Fig. 6.11. This plot is on log–log axes and it is evident that for both examples the growth rate is bounded by P^2.

It is interesting to note that if a characteristic treatment is used for the flux instead of the centred-type flux imposed by the standard Galerkin construction, then the growth of the maximum eigenvalue is $\mathcal{O}(P)$. This linear growth is consistent with the spectral Legendre expansion. However, it is very sensitive to small perturbations due to either the boundary conditions or the interface treatment, and a quadratic growth is typically observed even for scalar problems.

The same trends in the growth of the spectral radius are evident in *hybrid discretisations* consisting of triangular and quadrilateral elements. This is shown in Fig. 6.12 which shows a polar plot of the spectral radius for different directions of the constant advection velocity **V** on three different meshes. The deformation of the triangles in mesh (ii) increases the spectral inhomogeneity, but overall a similar distribution is maintained for all three discretisations.

An *exact* fit for the spectral envelope of the Galerkin convective operator on

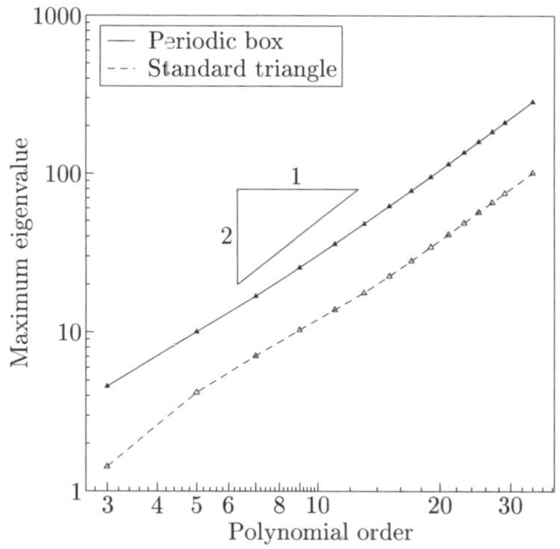

FIG. 6.11. Growth of the maximum eigenvalue $\lambda_{\max}$ with respect to polynomial order P. The 'periodic box' test case is shown in Fig. 6.8 and the 'standard triangle' is shown in Fig. 6.9. In both cases the growth rate is bounded by P^2.

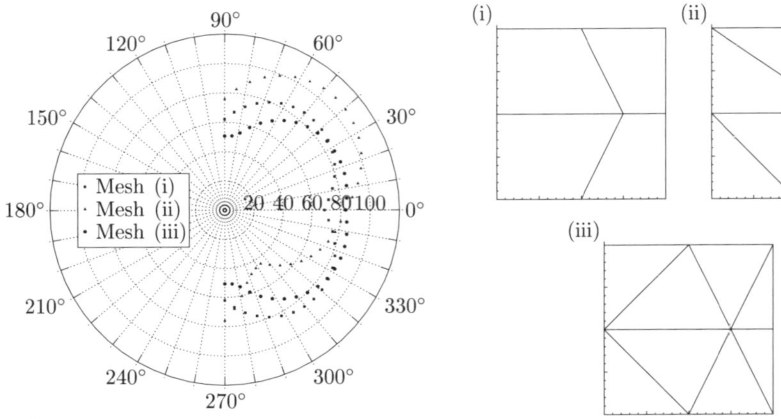

FIG. 6.12. Spectral radius of the advection operator for a hybrid two-dimensional discretisation and $P = 11$. The angle corresponds to the direction of a constant advection velocity. Periodic boundary conditions are assumed. ([483].)

a periodic square domain discretised with *regular quadrilaterals* was obtained in [483]. The fit depends upon the value of the maximum eigenvalue when $\theta = 0$, which is denoted by $\rho_\square(0)$ and is equivalent to the one-dimensional results in a segment. The fit has the following form:

$$\rho_\square(\theta) = \begin{cases} \sqrt{2}\rho_\square(0)\sin\left(\theta + \frac{\pi}{4}\right), & \theta > 0\,, \\ \sqrt{2}\rho_\square(0)\sin\left(\theta + \frac{3\pi}{4}\right), & \text{otherwise}\,. \end{cases}$$

Also, in [483] an *approximate* fit was obtained for the numerical spectral envelope of the Galerkin convective operator on a periodic square domain discretised with *regular triangles.* This is given by

$$\rho_\triangle(\theta) \approx \begin{cases} \sqrt{2}\left[\rho_\triangle(0)\sin\left(\theta + \frac{\pi}{4}\right) - \frac{\rho_\triangle(0)}{2}\sin(2\theta)\right], & \theta > 0\,, \\ \sqrt{2}\left[\rho_\triangle(0)\sin\left(\theta + \frac{3\pi}{4}\right) + (\rho_\triangle(0) - \rho_\square(0))\sin(2\theta)\right], & \text{otherwise}\,. \end{cases}$$

Although it is approximate, the above analytical formula provides a very close fit to the spectral radius that can be obtained more accurately numerically. Clearly, such formulae cannot take into consideration the skewness of the elements.

For *non-conforming* quadrilateral elements (see Chapter 7) the same quadratic growth of the spectral radius was documented in numerical experiments in [229]. In addition, for skew-symmetric boundary conditions a non-conforming discretisation results in purely imaginary eigenvalues, as shown in Fig. 6.13. The computed eigenvalues for the advection operator are plotted in a conforming mesh with resolution $N_{\text{el}} = 1$, $P = 10$, and in a non-conforming mesh with resolution $N_{\text{el}} = 3$, $P = 6$; in both cases purely imaginary eigenvalues are obtained. In a similar investigation, however, in [404] it was reported that eigenvalues with positive real parts may be obtained if a non-conforming discretisation is used in the case where the advection velocity is in the direction from low to high resolution. The interface condition in this case (upwind or centred flux) was not identified. As previously mentioned, this is important as it will ultimately decide the particular form of the eigenspectrum.

Discontinuous Galerkin eigenspectrum

For a single triangle where inflow/outflow conditions are prescribed (see Fig. 6.9(a)) the discontinuous Galerkin discretisation shares the property of the one-dimensional Legendre spectral methods in that the maximum eigenvalue magnitude grows linearly as the polynomial order P increases. Similar to the one-dimensional case, the high degree of non-normality in the matrix equations implies that, in practice, the time-step in a numerical scheme is inversely proportional to P^2, and not P as the von Neumann stability analysis predicts [86]. In fact, this linear growth of eigenvalue magnitude is destroyed as the problem is perturbed, as shown in computational experiments [310].

The same is true for multi-element discretisations, as can be seen by considering the case of two triangles as shown in Fig. 6.8(a). On this mesh, we have applied an *upwind flux* for $\tilde{\boldsymbol{f}}$ in eqn (6.2.12), with inflow conditions on the bottom and left boundaries, and outflow conditions on the top and right boundaries. With an upwind flux, the computation on each individual triangle is very close to the case of a single triangle, and the eigenvalue spectrum looks the same (see Fig. 6.14). However, if $\tilde{\boldsymbol{f}}$ is changed to be a *centred flux*, that

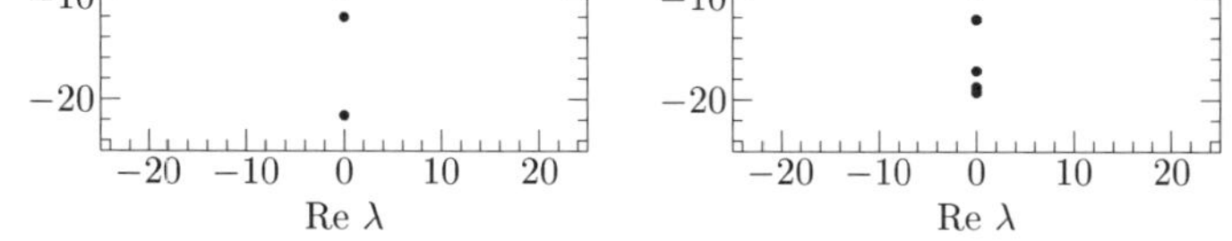

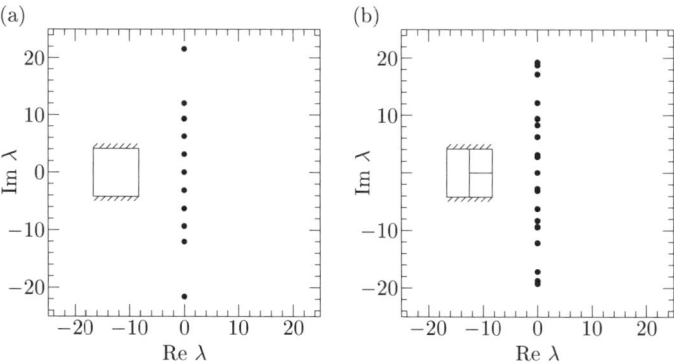

FIG. 6.13. Eigenvalue distribution for (a) conforming ($P = 10$), and (b) non-conforming ($P = 6$) discretisations. ([229].)

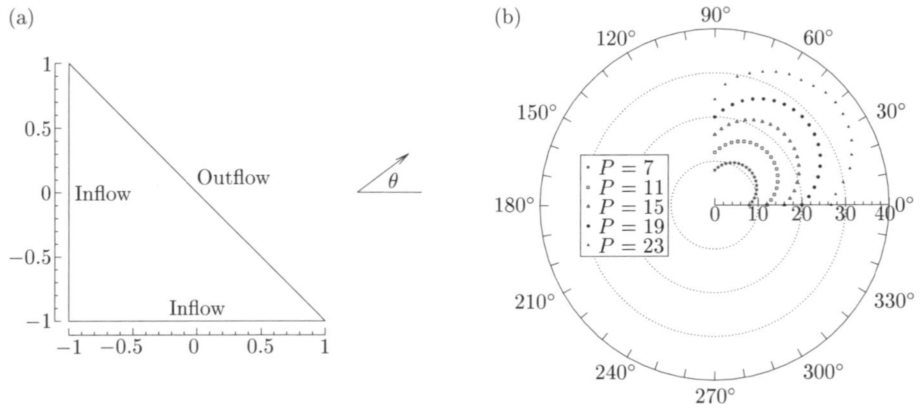

FIG. 6.14. For the single triangle shown in (a), the maximum eigenvalue magnitude for the linear advection operator was determined and is shown in (b), for $0 \leqslant \theta \leqslant 90°$.

is, $\tilde{\boldsymbol{f}}(u_-^\delta, u_+^\delta) = \boldsymbol{f}(u_-^\delta)/2 + \boldsymbol{f}(u_+^\delta)/2$, then the eigenvalue magnitudes will grow as P^2. If the problem is changed to be periodic on all four sides, or even just periodic in one direction, then the eigenvalue magnitudes will also grow as P^2. Figure 6.15 demonstrates this P^2 dependence for a two-triangle domain with *periodic* boundary conditions.

6.3.2.3 *Three dimensions*

We have already seen that the standard tensor product nature of the quadrilateral expansion means that in non-deformed elements the spectral radius can be

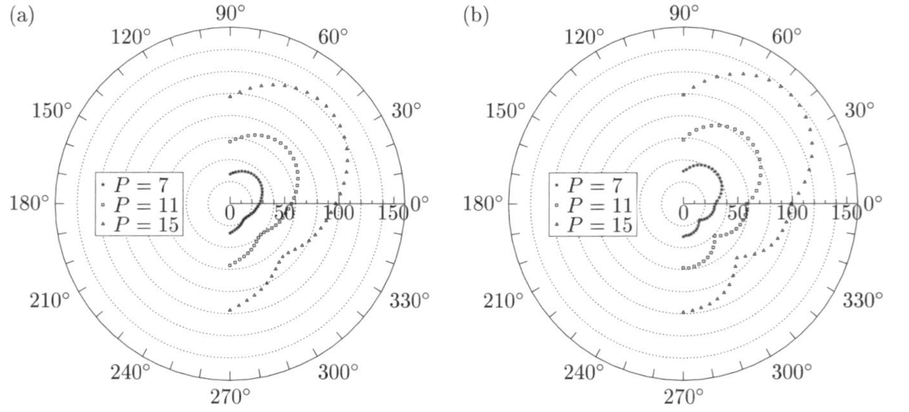

FIG. 6.15. Maximum eigenvalue magnitude for the linear advection operator on a two-triangle periodic box (see Fig. 6.8). (a) Centred flux, and (b) upwind flux for $0 \leqslant \theta \leqslant 90°$.

related to the one-dimensional spectral radius. It is therefore not unreasonable to assume that a similar extension exists in three dimensions. However, for the tetrahedral element, when we apply a generalised tensorial expansion the introduction of another principal function means that the connection to the triangular results are not so obvious. In this final expansion we present numerical results relating to the spectral radius in the standard tetrahedral domain for the C^0 continuous expansion.

In order to determine the maximum eigenvalue of the matrix $\boldsymbol{M}^{-1}\boldsymbol{\mathcal{A}}^{\top}\underline{((\boldsymbol{B}^e)^{\top}\boldsymbol{W}^e\mathbb{L}\boldsymbol{B}^e)}\boldsymbol{\mathcal{A}}$ in three dimensions, we choose a periodic domain formed from six tetrahedra, as shown in Fig. 6.16. The domain spans the space $\{-1 \leqslant x_1, x_2, x_3 \leqslant 1\}$ and, as indicated by the circles in the plot, the collapsed vertex of all tetrahedral elements is placed at the point E $(1, -1, 1)$. The triangles in this figure indicate the location of the base degenerate vertices (see Section 4.2.1.1). Since the advection operator $\mathbb{L}$ is a function of the propagation velocity $\mathbf{V}$, the matrix $\boldsymbol{M}^{-1}\boldsymbol{\mathcal{A}}^{\top}\underline{((\boldsymbol{B}^e)^{\top}\boldsymbol{W}^e\mathbb{L}\boldsymbol{B}^e)}\boldsymbol{\mathcal{A}}$ must also be a function of $\mathbf{V}$. The variation of maximum eigenvalues as a function of the propagation velocity is shown in Fig. 6.16. In constructing this plot we have assumed that the propagation velocity has unit magnitude and is orientated in the direction given by a vector connecting the origin to a point on the surface of the hemisphere. It is only necessary to determine this range of propagation velocities since propagation in the $[1/\sqrt{3}, 1/\sqrt{3}, 1/\sqrt{3}]$ direction forms a matrix which is the negative of the matrix due to propagation in the $[-1/\sqrt{3}, -1/\sqrt{3}, -1/\sqrt{3}]$ direction, and so they will have the same maximum eigenvalues. Since we have restricted the propagation velocity to unit magnitude, it is possible to describe any vector by

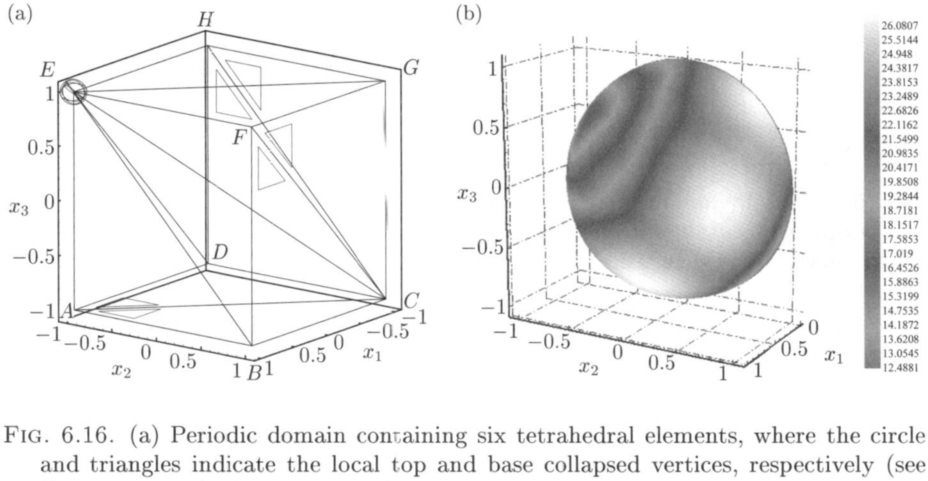

FIG. 6.16. (a) Periodic domain containing six tetrahedral elements, where the circle and triangles indicate the local top and base collapsed vertices, respectively (see Section 4.2.1.1). For a range of propagation velocities at $P = 9$ the maximum eigenvalue of the matrix $\boldsymbol{M}^{-1}\boldsymbol{\mathcal{A}}^{\top}\underline{((\boldsymbol{B}^e)^{\top}\boldsymbol{W}^e\mathbb{L}\boldsymbol{B}^e)}\boldsymbol{\mathcal{A}}$ was calculated and is shown in (b). The propagation velocities have unit magnitude and are parallel to the vectors connecting the origin in (b) to a point on the hemisphere.

two spherical angles ϕ and θ, as shown in Fig. 6.17. In this figure we also see the absolute maximum eigenvalues as shown in Fig. 6.16, parameterised with the spherical angles ϕ and θ. We have reversed the direction of the θ axis to make the plot consistent with Fig. 6.16.

The values of ϕ and θ at the nearest calculated position to the extrema shown in Fig. 6.17 are given in Table 6.1. The position and form of the extrema can be attributed to the structure of the tetrahedral domain shown in Fig. 6.16. For example, the global minimum (extremum 1) corresponds to a unit propagation velocity in the $[1/\sqrt{3}, -1/\sqrt{3}, 1/\sqrt{3}]$ direction, which is parallel to the diagonal bisector of the whole domain running from position C to E in Fig. 6.16. The triangular structure of this minimum is consistent with the velocity directions aligned with the bisectors of the faces of the domain. These bisectors run in the directions from W to E, C to F, and G to E, which correspond to the spherical angles $\phi = -180°$ and $\theta = 45°$, $\phi = 0°$ and $\theta = 45°$, and $\phi = -45°$ and $\theta = 180°$, respectively. Figure 6.17 demonstrates that these angles make a

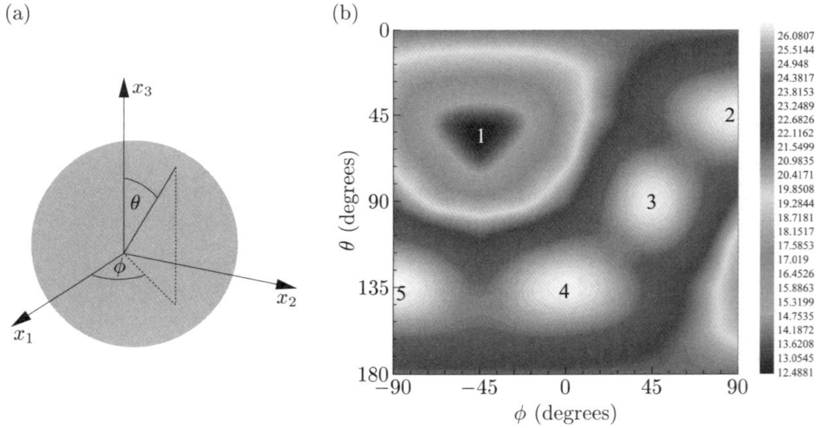

FIG. 6.17. (a) Definition of spherical angles ϕ and θ, and (b) plot of the absolute maximum eigenvalues for a unit propagation vector. This is the same distribution as shown in Fig. 6.16.

TABLE 6.1. Table of position and type of turning point in the absolute maximum eigenvalue distribution shown in Fig. 6.17.

	Type	ϕ (degrees)	θ (degrees)	Eigenvalue
1	Minimum	−45.3	56.2	12.35
2	Maximum	90.0	45.3	26.65
3	Maximum	45.3	90.0	26.65
4	Maximum	0.0	135.2	26.65
5	Maximum	−90.0	135.2	26.65

triangle of similar orientation to the region surrounding the minimum extremum at position 1.

Let us now consider vectors which bisect the faces in the other directions, that is, from A to F, G to W, and D to W, which correspond to the spherical angles $\phi = 90°$ and $\theta = 45°$, $\phi = 0°$ and $\theta = 135°$, and $\phi = 45°$ and $\theta = 90°$, respectively. We see from Table 6.1 that these angles correspond to the maximum extrema 2, 4, and 3. Maximum 5 corresponds to a propagation velocity of equal and opposite direction to maximum 2. Finally, the local minima between the maxima 2–3, 3–4, and 4–5 can be attributed to the three remaining diagonal bisectors of the complete domain, that is, from D to F, H to W, and G to A. We see that propagation velocities aligned with the element edges lead to minima in the eigenvalue distribution, whereas maxima correspond to propagation in the orthogonal direction to those that lead to the minima. This type of behaviour is consistent with the two-dimensional expansion shown in Section 3.2.3.

The growth rate for three propagation directions is shown in Fig. 6.18. The most critical directions are those corresponding to the maximum extrema of the eigenvalues in Fig. 6.17. We see that the growth rate in the $\phi = 45°$, $\theta = 90°$ direction is asymptotically faster than in the other sampled directions, but is still bounded by a slope of 2. If we consider the last three points of these curves then we find that the slopes are 1.88, 1.79, and 1.82 for the propagation velocities of $\phi = 45°$ and $\theta = 90°$ (D–B), $\phi = 45°$ and $\theta = 56.2°$ (D–F), and $\phi = -45°$ and $\theta = 56.2°$ (C–E), respectively. Once again, the growth rate is bounded by P^2.

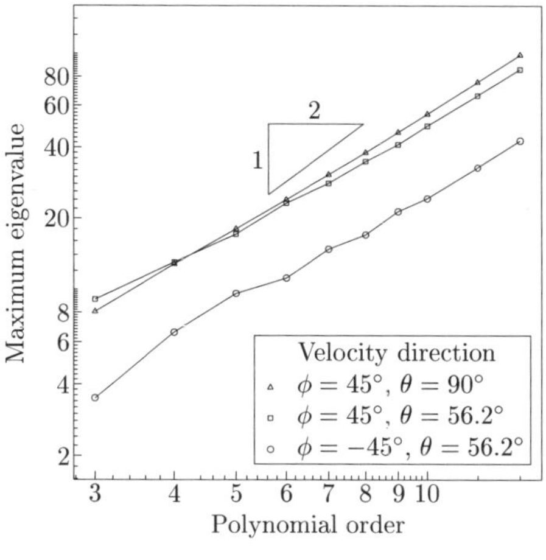

FIG. 6.18. The growth of the maximum eigenvalue of the matrix $\boldsymbol{M}^{-1}\boldsymbol{\mathcal{A}}^{\top}((\boldsymbol{B}^e)^{\top}\boldsymbol{W}^e\mathbb{L}\boldsymbol{B}^e)\boldsymbol{\mathcal{A}}$ as a function of polynomial order for propagation velocities of magnitude $|\mathbf{V}| = \sqrt{3}$, with directions of $\phi = 45°$ and $\theta = 90°$ (D–B), $\phi = 45°$ and $\theta = 56.2°$ (D–F), and $\phi = -45°$ and $\theta = 56.2°$ (C–E).

6.4 Semi-Lagrangian formulation for advection–diffusion

Having already studied the continuous and discontinuous formulations for the advection equation and their corresponding eigenspectra, we can appreciate the difficulties arising in integrating this equation explicitly. In particular, in explicit integrations the size of time-step required for spectral solutions of advection equations is dictated by the CFL number, which scales as P^2 as already mentioned. The same is true for advection–diffusion equations, assuming that the advection is treated explicitly and the diffusion implicitly (i.e., a semi-implicit scheme). This is true for discretisations in the Eulerian framework. Progress can be made by employing semi-Lagrangian (SL) temporal discretisation, which could increase significantly the maximum allowable time-step while maintaining the efficiency of symmetric solvers. We have discussed in the introduction that the semi-Lagrangian scheme completely eliminates dispersion and diffusion errors for the linear advection equation if the spectral order $P \geqslant 4$, which is a very desirable property.

The semi-Lagrangian approach has long been used in meteorology for numerical weather prediction, where the use of a large time-step is essential for efficiency [308]. This method was introduced at the beginning of the 1980s [378, 396], and the basic idea is to discretise the Lagrangian derivative of the solution in time instead of the Eulerian derivative. As we shall see in the following example, this requires the solution at the foot of the characteristic from each discrete mesh point. The solution at the characteristic foot can either be determined using backward particle tracking or, equivalently, by solving an auxiliary advection equation. The first version of the method is often referred to as the *strong* scheme, while the second is referred to as the *auxiliary* scheme.

To demonstrate these two approaches we consider the scalar advection–diffusion equation

$$\frac{\partial \phi}{\partial t} + \boldsymbol{a} \cdot \nabla \phi = \nu \nabla^2 \phi \,, \tag{6.4.1}$$

where $\boldsymbol{a}(\boldsymbol{x}, t)$ is an advection field. As mentioned previously, under any explicit discretisation the time-step restriction associated with the diffusion operator, $\nu \nabla^2 \phi$, is greater than that for the advection operator, $\boldsymbol{a} \cdot \nabla \phi$. For this reason it is often desirable to treat the diffusion operator implicitly, whilst, in general, we would like to consider nonlinear convection, and so the advection terms are typically handled explicitly. This type of strategy for a one-dimensional finite difference scheme can be symbolically represented on a discrete x–t diagram, as shown in Fig. 6.19. In this figure the black squares represent the implicit contribution from a centred approximation to the diffusion operator (i.e., $\nabla \phi|_i^{n+1} = (1/\Delta x^2)(\phi_{i+1}^{n+1} - 2\phi_i^{n+1} + \phi_{i-1}^{n+1})$, whilst the black circles represent the contributions from an upwinded first-order approximation to the advection operator assuming $|\boldsymbol{a}| > 0$ (i.e., $\boldsymbol{a} \cdot \nabla \phi_i^n = (\boldsymbol{a}/\Delta x)(\phi_i^n - \phi_{i-1}^n)$).

Equation (6.4.1) can equivalently be written in Lagrangian form as

$$\frac{\mathrm{D}\phi}{\mathrm{D}t} = \nu \nabla^2 \phi \,, \quad \text{with } \frac{\mathrm{D}}{\mathrm{D}t} = \frac{\partial}{\partial t} + \boldsymbol{a} \cdot \nabla \,, \tag{6.4.2}$$

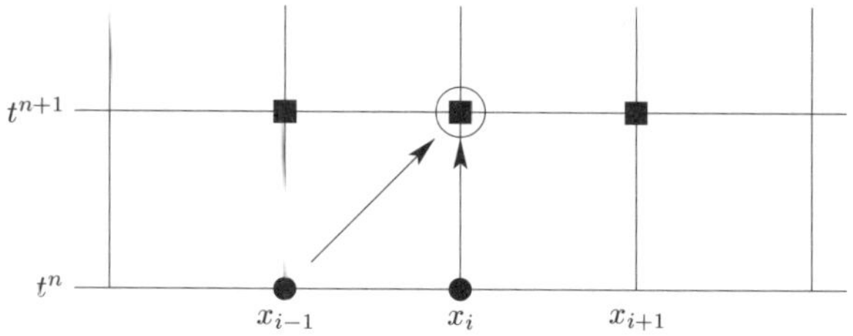

FIG. 6.19. Schematic diagram of the standard finite difference approximation to the advection–diffusion equations using an implicit diffusion and explicit advection discretisation.

where $\mathrm{D}/\mathrm{D}t$ is the Lagrangian or material derivative moving with the advection velocity, i.e., $\mathrm{D}\boldsymbol{x}/\mathrm{D}t = \boldsymbol{a}$. Following the semi-Lagrangian approach, we can discretise eqn (6.4.2) in time at point x_i using a first-order implicit scheme to obtain

$$\frac{\phi_i^{n+1} - \phi_d^n}{\Delta t} = \nu \nabla^2 \phi_i^{n+1}, \tag{6.4.3}$$

where $\phi_d^n = \phi^n(\boldsymbol{x}_d, t^n)$ and $\boldsymbol{x}_d$ is the so-called *departure* point.

Since the material derivative is evaluated along the characteristics, we can determine the departure point $\boldsymbol{x}_d$ by solving the characteristic equation $\mathrm{D}\boldsymbol{x}/\mathrm{D}t = \boldsymbol{a}$ backward in time from $t^{n+1} \geqslant t \geqslant t^n$ using the initial conditions $\boldsymbol{x}(t^{n+1}) = \boldsymbol{x}_i$. This approach is schematically represented in Fig. 6.20(a), where we show the one-dimensional discretisation on a discrete x–t diagram. Again, the implicit discretisation of the diffusion term is represented by the black squares. However, in the strong semi-Lagrangian approach we track particles backwards along the characteristic, $\mathrm{D}\boldsymbol{x}/\mathrm{D}t = \boldsymbol{a}$, of the hyperbolic advection operator to determine point $\boldsymbol{x}_d$. Subsequently, $\phi(\boldsymbol{x}_d)$ is evaluated and the approximation to the La-

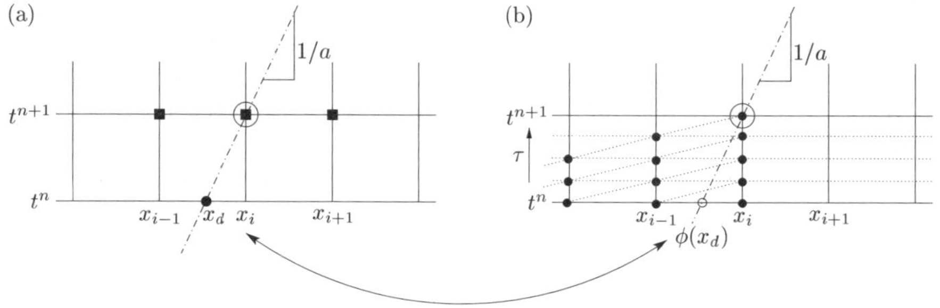

FIG. 6.20. (a) Schematic of the strong semi-Lagrangian approximation where the particle is backward transformed from x_i^{n+1} to determine x_d and $\phi(x_d)$. (b) Schematic representation of the auxiliary semi-Lagrangian approximation where a hyperbolic advection equation is advanced using a smaller time-step to determine $\tilde{\phi}(\boldsymbol{x}_i, t^{n+1})$.

grangian time derivative can then be determined. The complexity of backward particle tracking depends on the form of $\boldsymbol{a}(\boldsymbol{x},t)$. If $\boldsymbol{a}(\boldsymbol{x},t)$ is time independent then the calculation is relatively straightforward on a well-behaved discretisation. However, if the transport velocity is time dependent then different time-integration strategies can be adopted, as discussed in the next section. We also note that this scheme requires an interpolation operation to evaluate $\phi(\boldsymbol{x}_d, t^n)$.

In the auxiliary semi-Lagrangian approach [320] the starting-point is similar to the strong semi-Lagrangian technique; however, instead of backward particle tracking we solve for the departure solution ϕ_d^n directly. This can be achieved by solving the advection part of the problem independently in its Eulerian form and with a smaller time-step. To this end, we introduce an intermediate solution $\tilde{\phi}(\boldsymbol{x},\tau)$ and solve the problem

$$\frac{\mathrm{D}\tilde{\phi}}{\mathrm{D}\tau} = \frac{\partial\tilde{\phi}}{\partial\tau} + \boldsymbol{a}\cdot\nabla\tilde{\phi} = 0\,, \quad t^n \leqslant \tau \leqslant t^{n+1}\,, \tag{6.4.4}$$

with initial conditions $\tilde{\phi}(\boldsymbol{x}_i, t^n) = \phi(\boldsymbol{x}_i, t^n)$. Since this is a strictly hyperbolic equation then along the characteristic the solution is constant, and so $\phi(\boldsymbol{x}_d, t^n) = \tilde{\phi}(\boldsymbol{x}_i, t^{n+1})$. The auxiliary semi-Lagrangian method is shown schematically in Fig. 6.20(b), where the solution at the departure point is determined by the discretisation of eqn (6.4.4) using a smaller time-step. The explicit solution of eqn (6.4.4) means that from stability considerations the time-step is restricted by a CFL condition. Nevertheless, provided that the solution of eqn (6.4.4) can be determined more efficiently than the implicit diffusion operator, then eqn (6.4.4) can be discretised with a time-step near to the CFL limit. The implicit diffusion operator is solved less frequently in time, thereby reducing the cost of the overall algorithm for a fixed integration time. Accordingly, it is the ratio of the cost of the explicit advection term to the implicit diffusion operator which ultimately limits the possible speed-up. We also note that the time accuracy is governed by the larger time-step applied on the implicit diffusion operator.

In the following, we will present both versions in the context of spectral/*hp* element discretisations. While both approaches can be thought of as discrete local versions of the method of characteristics, implementation details are important and result in different attributes of the strong and auxiliary form. From the theoretical standpoint, the strong form requires regularity in the trajectory to locate the departure point $\boldsymbol{x}_d$ but not for the corresponding value $\phi(\boldsymbol{x}_d)$, whereas the opposite is true for the auxiliary forms.

6.4.1 *Strong semi-Lagrangian method*

We recall that the strong method involves backward time integration of a characteristic equation to find the departure point due to the advection velocity from an Eulerian grid point. The solution variable at the departure point is then obtained by interpolation. Unlike purely Lagrangian methods, there is no mesh deformation because the *arrival points* of any trajectory are defined to be the grid points. However, there may be significant interpolation costs to obtain the solution values at the 'departure points'.

We consider the advection–diffusion equation (6.4.1) and use a semi-implicit scheme for time discretisation, i.e., we employ a second-order Adams–Bashforth scheme for the advection term and a Crank–Nicolson scheme for the diffusion term:

$$\frac{\phi^{n+1} - \phi^n}{\Delta t} + \boldsymbol{a} \cdot \nabla \left(\frac{3\phi^n - \phi^{n-1}}{2} \right) = \nu \nabla^2 \left(\frac{\phi^{n+1} + \phi^n}{2} \right). \tag{6.4.5a}$$

The Lagrangian form of eqn (6.4.1) is

$$\frac{\mathrm{D}\phi}{\mathrm{D}t} = \nu \nabla^2 \phi \,, \tag{6.4.5b}$$

$$\frac{\mathrm{D}\boldsymbol{x}}{\mathrm{D}t} = \boldsymbol{a}(\boldsymbol{x}, t) \,. \tag{6.4.5c}$$

The idea of a *purely* Lagrangian approach is to solve eqn (6.4.5b) along the characteristic lines (6.4.5c). This leads to an effective decoupling of the advection and diffusion terms and an unconditionally stable scheme. However, as the fluid particles move along, they may result in a heavily distorted and irregular mesh. Hence, expensive remeshing is required between time-steps. In the strong semi-Lagrangian approach, the computational mesh is *fixed*. At each time-step, a discrete set of particles arriving at the grid points is tracked backwards over a single time-step along its characteristic line up to its departure points. The solution value at the departure points is then obtained by interpolation. The second-order Crank–Nicolson semi-Lagrangian scheme is of the form

$$\frac{\phi^{n+1} - \phi^n_d}{\Delta t} = \nu \nabla^2 \left(\frac{\phi^{n+1} + \phi^n_d}{2} \right), \tag{6.4.6a}$$

$$\frac{\mathrm{D}\boldsymbol{x}}{\mathrm{D}t} = \boldsymbol{u}(\boldsymbol{x}, t) \,, \quad \boldsymbol{x}^{n+1} = \boldsymbol{x}(t^{n+1}) = \boldsymbol{x}_a \,. \tag{6.4.6b}$$

Here ϕ^n_d denotes the value of ϕ at the *departure* points $\boldsymbol{x}_d$ at time-level n, and $\boldsymbol{x}_a$ is the position of the *arrival* points which are the grid points. The characteristic equation (6.4.6b) is solved backward, i.e., we solve for the departure point at time-level n, $\boldsymbol{x}^n_d$, with the initial condition $\boldsymbol{x}^{n+1} = \boldsymbol{x}_a$. The departure points do not coincide with the grid points, and thus a search–interpolation procedure is needed. Also, the overall accuracy and efficiency of the semi-Lagrangian scheme depends critically on both the accuracy of the backward integration of eqn (6.4.6b) as well as the accuracy of the interpolation method. In the following we discuss how these algorithms can be implemented.

The semi-Lagrangian method depends strongly on the spatial discretisation. Specifically, its accuracy is particularly sensitive to the method of backward integration of the characteristic equation as well as the interpolation scheme to evaluate the solution at the departure points. This has been shown by Falcone and Ferretti [157], who conducted a rigorous analysis of the stability and convergence properties of semi-Lagrangian schemes. Typically, the backward integration is performed by employing second-order schemes (i.e., the mid-point rule), explicitly or implicitly. Also, a fourth-order Runge–Kutta method can be employed, but the results do not improve over the second-order schemes if low spatial resolution is used. In fact, the simplest semi-Lagrangian scheme with linear interpolation is equivalent to the classical first-order upwinding scheme, which is excessively dissipative [396]. A popular and effective choice for interpolation methods (in space) is to use cubic splines [8].

An intriguing finding is that the error of semi-Lagrangian schemes in solving advection–diffusion equations *decreases* as the time-step increases in a certain range of parameters, and this has initially led to some erroneous justifications. The error analysis in [157] showed that the overall error of the semi-Lagrangian method is indeed *not* monotonic with respect to the time-step Δt, and, in particular, it has the form

$$\mathcal{O}\left(\Delta t^k + \frac{\Delta x^{P+1}}{\Delta t}\right),$$

where k refers to the order of the backward time integration and P refers to the interpolation order; similar conclusions had been reached earlier in [327].

Backward integration

We solve eqn (6.4.6b) for one single time-step in order to obtain $\boldsymbol{x}_d = \boldsymbol{x}(t^n)$ by the explicit second-order mid-point rule:

$$\hat{\boldsymbol{x}} = \boldsymbol{x}_a - \frac{\Delta t}{2}\boldsymbol{a}(\boldsymbol{x}_a, t^n), \tag{6.4.7a}$$

$$\boldsymbol{x}_d = \boldsymbol{x}_a - \Delta t\, \boldsymbol{a}\left(\hat{\boldsymbol{x}}, t^n + \frac{\Delta t}{2}\right). \tag{6.4.7b}$$

By defining

$$\alpha = \boldsymbol{x}_a - \boldsymbol{x}_d,$$

we can rewrite the *explicit* mid-point rule as

$$\boldsymbol{\alpha} = \Delta t\, \boldsymbol{a}\left(\boldsymbol{x}_a - \frac{\Delta t}{2}\boldsymbol{a}(\boldsymbol{x}_a, t^n), t^n + \frac{\Delta t}{2}\right). \tag{6.4.7c}$$

Similarly, we can employ *implicit* integration for eqn (6.4.7a) by setting

$$\hat{\boldsymbol{x}} = \boldsymbol{x}_a - \frac{\Delta t}{2}\boldsymbol{a}\left(\hat{\boldsymbol{x}}, t^n + \frac{\Delta t}{2}\right),$$

to arrive at the implicit mid-point rule

$$\alpha = \Delta t\, \boldsymbol{a}\left(\boldsymbol{x}_a - \frac{\boldsymbol{\alpha}}{2}, t^n + \frac{\Delta t}{2}\right). \tag{6.4.7d}$$

This is the backward integration algorithm used in most semi-Lagrangian schemes. Although the explicit and implicit schemes are formally of second-order, some accuracy improvement can be obtained for the implicit method. Equation (6.4.7d) has to be solved iteratively, but numerical experiments show that only a few iterations are needed for convergence (typically around five). This result is valid if a high-order predictor is employed in the iterative process [440]. For an advection–diffusion equation with the velocity field known analytically, this additional cost is negligible. However, for a velocity field known only in numerical form, the iteration is costly because each substep requires a search–interpolation procedure. Numerical results obtained in [498] show that the two methods give almost identical results, but for more general problems, especially for Navier–Stokes equations, the explicit method (eqn (6.4.7c)) is preferred. To enhance the accuracy further, higher-order time-integration methods can be used.

In general, the departure points do not coincide with the grid points and so a particle tracing procedure must be used to determine $\boldsymbol{x}_d$, as discussed in Section 4.3.6. If the elements are straight sided then the particle tracking can be efficiently evaluated, especially if inter-element connectivity information is maintained in the numerical algorithm. Since there are normally fewer larger elements in spectral/*hp* element methods as compared to *h*-type discretisation, it is not unusual to maintain information about inter-element connectivity.

6.4.2 *Auxiliary semi-Lagrangian method*

Similar to the strong semi-Lagrangian formulation, in the auxiliary semi-Lagrangian formulation of the advection–diffusion equation (6.4.1) we once again solve the Lagrangian form of the equation, i.e.,

$$\frac{\mathrm{D}\phi}{\mathrm{D}t} = \nu \nabla^2 \phi \,.$$

This equation can be discretised using a Crank–Nicolson scheme as given by eqn (6.4.6a), which requires the value of $\phi^n_d = \phi^n(\boldsymbol{x}_d)$ for every grid point. In contrast to the strong semi-Lagrangian formulation where we solved the equation $\mathrm{D}\boldsymbol{x}/\mathrm{D}t = u(\boldsymbol{x}, t)$ backwards in time from a collocation point to determine $\boldsymbol{x}_d$ and subsequently ϕ^n_d, in the auxiliary semi-Lagrangian formulation we solve the equation

$$\frac{\mathrm{D}\tilde{\phi}}{\mathrm{D}\tau} = \frac{\partial \tilde{\phi}}{\partial \tau} + \boldsymbol{a} \cdot \nabla \tilde{\phi} = 0 \,, \quad t^n \leqslant \tau \leqslant t^{n+1} \,, \tag{6.4.8}$$

where $\tilde{\phi}(\boldsymbol{x}, t^n) = \phi(\boldsymbol{x}, t^n)$. As illustrated in Fig. 6.20, the strictly hyperbolic nature of this problem implies that for a collocation point $\boldsymbol{x}_i$ we know that $\phi^n_d = \phi(\boldsymbol{x}_i, t^{n+1})$, which allows us to complete the discretisation.

In the auxiliary semi-Lagrangian approach we do not need to interpolate the solution field to determine ϕ^n_d, but instead we can apply standard Eulerian techniques to solve eqn (6.4.8). Nevertheless, the auxiliary semi-Lagrangian technique is still limited by an explicit time-step restriction which potentially could be quite severe, and this may make the solution of eqn (6.4.4) relatively costly. To make this technique tractable, it requires an efficient technique to solve eqn

(6.4.8), which has implications for the type of spatial discretisation adopted. It also requires an estimate of the time-step restriction for the advection problem, as discussed in Section 6.3.

When a classical spectral element continuous Galerkin discretisation is adopted, we recall that the choice of the nodal points at the Gauss–Lobatto–Legendre quadrature points leads to a diagonal mass matrix, provided that the quadrature order is one degree higher than the polynomial order (see Section 2.3.4.2). Therefore, an explicit solution of eqn (6.4.8), which simply requires the inverse of the global mass matrix, is particularly efficient when compared to the implicit solution of the diffusion operator. This approach has been adopted by a variety of researchers, e.g., see [320, 335].

We recall, however, that the choice of a C^0 continuous modal spectral/*hp* element applied to a continuous Galerkin formulation (as defined in Section 3.2.3) does not produce a diagonal mass matrix. Therefore, solving eqn (6.4.8) using a continuous Galerkin formulation and a modal expansion is no more efficient than implicitly solving the diffusion operator, $\nabla^2\phi$, and so there is no particular advantage in using the auxiliary semi-Lagrangian. However, we have seen in Section 6.2.2 that the explicit formulation of eqn (6.4.8) using a discontinuous Galerkin formulation can be solved very efficiently since it leads to a block diagonal matrix (or even a purely diagonal matrix depending on the expansion basis) which can be efficiently inverted. An auxiliary semi-Lagrangian formulation using a discontinuous Galerkin formulation for the advection problem was proposed in [427] and has also been applied to the incompressible Navier–Stokes equations, as discussed in Section 8.7.2.

Operator–integration-factor splitting

In the previous sections we motivated the formulation of the auxiliary semi-Lagrangian formulation using the hyperbolic characteristic nature of the equation. However, a more mathematically general construction can be considered in terms of an operator–integration-factor splitting approach, as suggested by Maday *et al.* [320].

We start by considering a semi-discrete formulation of the advection–diffusion equation (6.4.1) of the form

$$\boldsymbol{M}\frac{\mathrm{d}\hat{\boldsymbol{\phi}}(t)}{\mathrm{d}t} + \boldsymbol{A}\hat{\boldsymbol{\phi}}(t) = \nu\boldsymbol{L}\hat{\boldsymbol{\phi}}(t)\,, \tag{6.4.9}$$

where under a Galerkin formulation we understand the matrix operators $\boldsymbol{M}$, $\boldsymbol{A}$, and $\boldsymbol{L}$ to be defined as

$$\boldsymbol{M}[i][j] = (\Phi_i, \Phi_j)_\Omega\,, \quad \boldsymbol{A}[i][j] = (\Phi_i, \boldsymbol{a}\cdot\nabla\Phi_j)_\Omega\,, \quad \boldsymbol{L}[i][j] = (\nabla\Phi_i\cdot\nabla\Phi_j)_\Omega$$

and we recall that $\hat{\phi}(t)$ denotes the time-dependent expansion coefficients. Without loss of generality, we shall assume that $\boldsymbol{M} = \boldsymbol{I}$; for example, for a nodal spectral element expansion basis (alternatively, we could have multiplied the equation through by $\boldsymbol{M}^{-1}$).

We now introduce an integration factor $\mathcal{Q}(t)$ and multiply eqn (6.4.9) by this factor to obtain

$$\mathcal{Q}(t)\frac{\mathrm{d}\hat{\boldsymbol{\phi}}(t)}{\mathrm{d}t} - \mathcal{Q}(t)\boldsymbol{A}\hat{\boldsymbol{\phi}}(t) = \nu\mathcal{Q}(t)\boldsymbol{L}\hat{\boldsymbol{\phi}}(t)\,. \tag{6.4.10a}$$

We choose the integration factor $\mathcal{Q}(t)$ to satisfy

$$\frac{\mathrm{d}\mathcal{Q}\hat{\boldsymbol{\phi}}}{\mathrm{d}t} = \mathcal{Q}\frac{\mathrm{d}\hat{\boldsymbol{\phi}}}{\mathrm{d}t} + \mathcal{Q}\boldsymbol{A}\hat{\boldsymbol{\phi}}\,, \quad \text{which implies } \hat{\boldsymbol{\phi}}\frac{\mathrm{d}\mathcal{Q}}{\mathrm{d}t} = \mathcal{Q}\boldsymbol{A}\hat{\boldsymbol{\phi}}\,, \tag{6.4.10b}$$

with the further constraint that at time $t = t^{n+1}$ we have

$$\mathcal{Q}(t^{n+1}) = \boldsymbol{I}\,. \tag{6.4.10c}$$

Note that, in general, an arbitrary time $t = t^{\star}$ could have been chosen to set this constraint. Condition (6.4.10b) implies that we can recast eqn (6.4.10a) as

$$\frac{\mathrm{d}\mathcal{Q}\hat{\boldsymbol{\phi}}}{\mathrm{d}t} = \nu\mathcal{Q}\boldsymbol{L}\hat{\boldsymbol{\phi}}\,. \tag{6.4.11}$$

To illustrate how the operator–integration-factor method directly relates to the previous description of the auxiliary semi-Lagrangian formulation, it is convenient to continue our discussion by time-discretising eqn (6.4.10c) with a backward time-difference formula. We note, however, that this is not necessary and a more generalised, although more involved, formulation is possible, as discussed in [320]. Discretising eqn (6.4.11) with a backward Euler scheme and noting that $(\mathcal{Q}\hat{\boldsymbol{\phi}})^{n+1} = \hat{\boldsymbol{\phi}}^{n+1}$, from condition (6.4.10c) we obtain

$$\frac{\hat{\boldsymbol{\phi}}^{n+1} - (\mathcal{Q}\hat{\boldsymbol{\phi}})^{n}}{\Delta t} = \nu\boldsymbol{L}\hat{\boldsymbol{\phi}}^{n+1}\,. \tag{6.4.12}$$

We note the similarity of this equation with eqn (6.4.3) and observe that the term $(\mathcal{Q}\hat{\boldsymbol{\phi}})^{n}$ plays the same role as ϕ_d^n. However, without making this assumption we require a method to evaluate $(\mathcal{Q}\hat{\boldsymbol{\phi}})^{n}$. The last stage of the operator–integration-factor method is to evaluate this term without the need to explicitly evaluate the integration factor $\mathcal{Q}$. This can be achieved by solving an auxiliary equation of the form

$$\frac{\mathrm{d}\tilde{\boldsymbol{\phi}}}{\mathrm{d}t} + \boldsymbol{A}\tilde{\boldsymbol{\phi}} = 0\,, \quad \text{subject to } \tilde{\boldsymbol{\phi}}(t^n) = \hat{\boldsymbol{\phi}}(t^n)\,. \tag{6.4.13}$$

If we multiply eqn (6.4.13) by $\mathcal{Q}$ and combine it with eqn (6.4.10b) (substituting $\tilde{\boldsymbol{\phi}}$ for $\hat{\boldsymbol{\phi}}$) then we can show that

$$\frac{\mathrm{d}\mathcal{Q}\tilde{\boldsymbol{\phi}}}{\mathrm{d}t} = 0 \quad \text{or} \quad \mathcal{Q}\tilde{\boldsymbol{\phi}} = C\,.$$

Since, by definition, $\tilde{\boldsymbol{\phi}}(t^n) = \hat{\boldsymbol{\phi}}(t^n)$, we can determine the constant as $C = \mathcal{Q}(t^n)\tilde{\boldsymbol{\phi}}(t^n)$ and so

$$\mathcal{Q}(t)\tilde{\phi}(t) = \mathcal{Q}(t^n)\tilde{\phi}(t^n)\,. \tag{6.4.14}$$

The right-hand side of eqn (6.4.14) is the required quantity in eqn (6.4.11). Recalling that $\mathcal{Q}(t^{n+1}) = \boldsymbol{I}$, we can evaluate eqn (6.4.14) at t^{n+1} to arrive at

$$\mathcal{Q}(t^{n+1})\tilde{\phi}(t^{n+1}) = \tilde{\phi}(t^{n+1}) = \mathcal{Q}(t^n)\tilde{\phi}(t^n)\,.$$

Therefore, by integrating eqn (6.4.13) from t^n to t^{n+1} to determine $\tilde{\phi}(t^{n+1})$, we can evaluate $\mathcal{Q}(t^n)\tilde{\phi}(t^n)$, which is the unknown term in eqn (6.4.11). This process allows us to determine $\mathcal{Q}(t^n)\tilde{\phi}(t^n)$ without evaluating the integration factor $\mathcal{Q}$ and is the direct analogue to determining ϕ_d^n in the previous auxiliary formulation.

6.4.3 *Convergence, efficiency, and stability of semi-Lagrangian schemes*

Convergence

Adopting the standard benchmark applied in [498], we consider the advection and diffusion of a Gaussian cone with a transport velocity field of

$$u = +x_2\,, \qquad v = -x_1$$

and initial condition

$$\phi(x_1, x_2, 0) = \exp\left\{-\frac{1}{2\lambda^2}\left[(x_1 - x_1^0)^2 + (x_2 - x_2^0)^2\right]\right\}.$$

The exact solution to this problem is

$$\phi(x_1, x_2, t) = \frac{\lambda^2}{\lambda^2 + 2\nu t}\exp\left\{-\frac{\hat{x}_1^2 + \hat{x}_2^2}{2(\lambda^2 + 2\nu t)}\right\},$$

where

$$\hat{x}_1 = x_1 - x_1^0\cos t - x_2^0\sin t\,, \quad \hat{x}_2 = x_1 + x_1^0\sin t - x_2^0\cos t\,.$$

Fixing the constants as $\lambda = \frac{1}{8}$ and $(x_1^0, x_2^0) = (-\frac{1}{2}, 0)$, we discretise the problem with a mesh consisting of 10×10 quadrilateral elements in the region $-1 \leqslant x_1, x_2 \leqslant 1$. The solution is then time-integrated for *one revolution*, corresponding to a final time of $t = 2\pi$. In the following we define the Courant number and the non-dimensional diffusion coefficient as

$$C = \max\left(\frac{|\boldsymbol{a}|\Delta t}{\Delta s}\right) \quad \text{and} \quad D = \max\left(\frac{\nu\Delta t}{\Delta s^2}\right),$$

respectively, where $|\boldsymbol{a}| = \sqrt{a_1^2 + a_2^2}$ and $\Delta s = \sqrt{(\Delta x_1)^2 + (\Delta x_2)^2}$.

In Fig. 6.21 we plot the L^2 error from an Eulerian method (Adams–Bashforth combined with Crank–Nicolson, ABCN) and the auxiliary semi-Lagrangian spectral/*hp* element (SLSE) method with fixed $C = 0.5$ and $D = 0.01$. The backward integration is the explicit mid-point rule (eqn (6.4.7c)), denoted here as the RK2

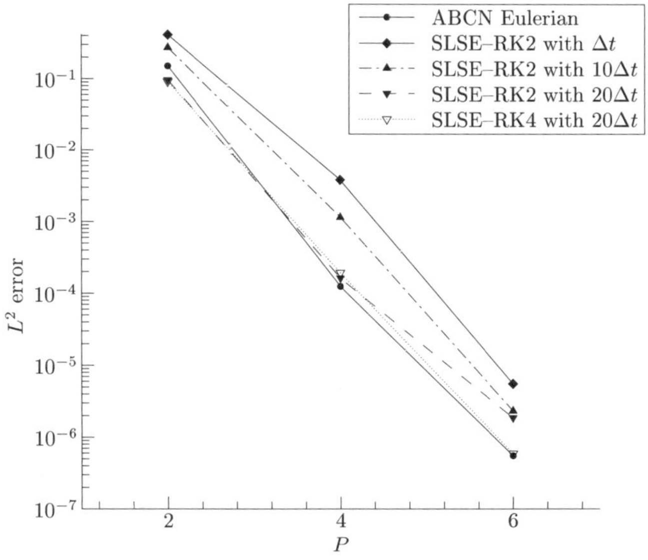

FIG. 6.21. Spatial convergence of Eulerian and strong semi-Lagrangian methods at large Δt for a Gaussian-cone problem.

method. The spectral order varies from $P = 2$ to 6. We observe on this log–linear plot that spectral convergence is achieved for both methods. The SLSE method gives relatively larger error at lower polynomial order P, but it quickly reaches the $\mathcal{O}(\Delta t^2)$ temporal error limit at $P = 6$. Results with $10\Delta t$ and $20\Delta t$, which correspond to CFL numbers of 5 and 10, are also plotted. We observe that, as the time-step increases, the error is reduced, matching the error of the Eulerian scheme but at a time-step size *twenty* times larger. Also, a further improvement with the fourth-order Runge–Kutta method (RK4) is obtained at $20\Delta t$ with polynomial order $P = 6$.

The error of the strong semi-Lagrangian method consists of two parts: the error of the backward integration $\mathcal{O}(\Delta t^{k+1})$ and the error from interpolation $\mathcal{O}(\Delta x^{P+1})$, where k is the order of the integration method and P is the order of the polynomial basis. Therefore, the overall accuracy of the semi-Lagrangian method is

$$\frac{\boldsymbol{a}^{n+1} - \boldsymbol{a}_d^n}{\Delta t} = \frac{\mathrm{d}\boldsymbol{a}}{\mathrm{d}t} + \mathcal{O}\left(\Delta t^k + \frac{\mathcal{O}(\Delta x^{P+1})}{\Delta t}\right). \tag{6.4.15}$$

A rigorous derivation of the above expression can be found in [157]. Equation (6.4.15) shows that the error is not monotonic with respect to Δt. When the polynomial order P is small, the interpolation error dominates; as Δt increases, the overall error decreases. It can also be appreciated that when the first term $\mathcal{O}(\Delta t^k)$ is subdominant, further increasing k will not improve the overall accuracy, which is why there is no improvement with a fourth-order Runge–Kutta method over the second-order methods for low-order discretisations. On the other hand, when the spatial error is subdominant at high P, increasing Δt increases the first error term in eqn (6.4.15) and thus the overall error is larger. In this case, a higher-order backward integration method (higher k, e.g., Runge–Kutta of fourth-order) reduces the dominant first term and improves the solution.

To further study the structure of the error of both the strong and auxiliary semi-Lagrangian methods, we test the SLSE method at different time-steps and different spectral orders. We set the viscosity to a small value, $\nu = 4.6 \times 10^{-6}$, in order to emphasise the effect of the advection. The range for Δt is 0.01 to 0.05, which corresponds to a CFL number of 5 to 25, for $P = 10$. In Fig. 6.22(a) we plot results obtained with second-order backward integration ($k = 2$) on log–log axes for the strong semi-Lagrangian method. We make the following observations.

- $P = 4$
  The interpolation error is relatively large, and thus the second error term in eqn (6.4.15) dominates. As Δt increases, the overall accuracy improves almost monotonically up to a large Δt when the first error term becomes significant.
- $P = 6$
  The interpolation error is smaller and the first error term in eqn (6.4.15), $\mathcal{O}(\Delta t^2)$, is comparable with the second term. As Δt increases, the error starts to decrease first. The $\mathcal{O}(\Delta t^2)$ term then becomes dominant and the overall error starts to increase. At this intermediate spatial resolution there is clearly a competition between the two error terms, resulting in the minimum error around $\Delta t \approx 0.024$.
- $P = 8$
  The interpolation error is sufficiently small and thus the $\mathcal{O}(\Delta t^2)$ term domi-

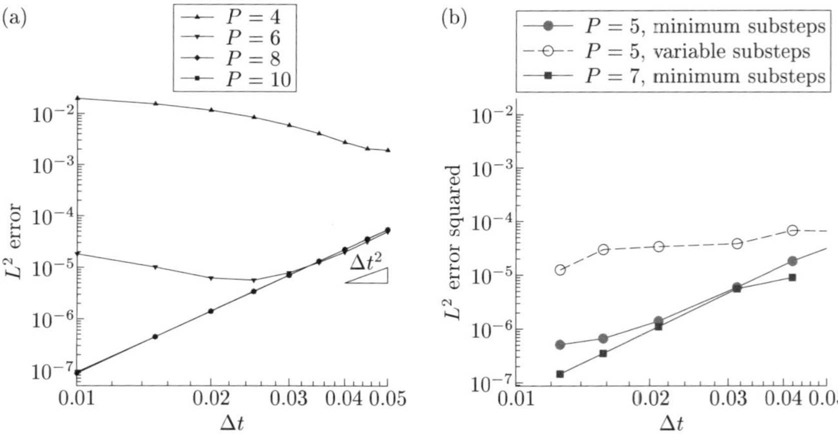

FIG. 6.22. Error dependence upon Δt with second-order integration for (a) strong semi-Lagrangian methods, and (b) the auxiliary semi-Lagrangian method.

nates. The overall error then grows at an algebraic second-order rate.

- $P = 10$
  The result is identical to that of $P = 8$, because the dominant error is the $\mathcal{O}(\Delta t^2)$ term, which does not depend on P.

A similar behaviour can be obtained with fourth-order Runge–Kutta backward integration ($k = 4$). It is worth noticing that in this case with high-order interpolation polynomials, $P = 8$ and $P = 10$, the interpolation error is very small and so is the backward integration error ($\mathcal{O}(\Delta t^4)$). The competition between the two error terms results in a *plateau*, and the overall accuracy is preserved over a large range of time-steps.

In Fig. 6.22(b) we plot results obtained with second-order integration on log–log axes for the auxiliary semi-Lagrangian method. We make the following observations.

- $P = 5, 7$, *minimum substeps*
  In this case the minimum substeps used in the solution of the advection problem was set to 10. Due to the small viscosity applied in this test the error is dominated by the substepping evaluation, and so as Δt is increased the error also increases monotonically. At lower Δt the spatial error starts to dominate with $P = 5$, and so there is deviation of the $P = 5$ and $P = 7$ errors.
- $P = 5$, *variable substeps*
  In this case we have used a CFL estimator to determine the number of substeps required for a specified Δt. At larger Δt more substeps are therefore used. Due to the dominance of the error in the advection step for this problem, we observe a *plateau* of this error for a range of Δt values.

Efficiency

Both the strong and auxiliary semi-Lagrangian methods require more computational cost than the Eulerian counterpart on a *per-time-step* consideration. In this strong semi-Lagrangian method we require backward particle tracking from every quadrature point and then interpolation of the polynomial approximation at non-quadrature points. Backward particle tracking can be expensive in deformed elements and is more intricate to parallelise. In contrast, the auxiliary semi-Lagrangian method uses 'more standard' implementation from the Eulerian point of view, and so can benefit from existing concepts for parallelisation. However, the cost of performing each full time-step Δt in the auxiliary approach is dependent upon the number of substeps that are necessary to maintain the CFL stability of the advection problem (eqn (6.4.8)). Therefore, if the cost of solving one step of the advection problem is the same as the cost of inverting the implicit diffusion operator in the previous advection–diffusion example then there would not appear to be any benefit in using the auxiliary semi-Lagrangian method, independent of the size of Δt that would be achieved. The ratio of the computational cost of the advection step to the diffusion step is therefore very important. However, for the strong semi-Lagrangian method there is a distinct advantage to using larger time-steps, provided that stability restrictions are not violated. This is because the cost of the backward particle tracking in the strong semi-Lagrangian method is relatively independent of Δt. It would therefore seem that for intermediate time-steps the auxiliary semi-Lagrangian method is preferable, whereas for large time-steps the strong semi-Lagrangian method will prevail.

For the advection–diffusion equations, particularly in non-deformed elements where a fast particle tracking algorithm can be applied, the computational cost of strong SLSE can be of the same order as the Eulerian spectral/*hp* element method. In Table 6.2 we compare the computational cost of the two approaches for the aforementioned Gaussian-cone problem. We see that the SLSE method is less than twice more expensive than the Eulerian method, and as the spectral order increases the two costs are comparable. However, with much larger allowable CFL numbers, the total CPU time required for the SLSE method to reach a certain time-level is significantly less than that of the Eulerian method. The overall speed-up list in the table is obtained at a CFL number of 20. Therefore, at least one order of magnitude in speed-up is achieved.

Stability

We now turn our attention to the ***stability*** of the strong semi-Lagrangian method, which has been studied by Falcone and Ferretti [157]. In particular, L^∞ stability holds for any low-order interpolation, i.e., piecewise linear or bilinear approximation. In the general linear high-order case, unconditional stability with respect to time-step holds for fixed spatial discretisations for any $\Delta t > 0$. However, this is not true with respect to any order of interpolation, say order P, unless monotone interpolating schemes are used for the departure point.

With respect to the L^2 stability, using von Neumann analysis, Falcone and Ferretti [157] showed that, for equidistant interpolations of order higher than second, instabilities may arise on a fixed grid. However, this can be overcome by employing smaller 'sliding' stencils of grid points surrounding the departure point $\boldsymbol{x}_d$. This is what is actually done in practice. For non-equidistant grids with Lagrangian or Hermite interpolations, L^2 stability holds, although a rigorous theory is incomplete.

Another cause of instability may be due to the *intersection* of the approximate trajectories. A sufficient condition to avoid this has been derived in [440] and

TABLE 6.2. Comparison of the computational cost of the Eulerian method (SE) and the strong semi-Lagrangian method (SLSE).

Polynomial order P	SLSE–RK4 (sec/step)	ABCN Eulerian (sec/step)	Ratio (SLSE/ABCN)	Overall speed-up (SLSE/ABCN)
4	0.15	0.08	1.88	10.6
6	0.39	0.21	1.86	10.8
8	0.80	0.51	1.57	12.7
10	1.49	1.19	1.25	16.0

states that

$$\Delta t < |\boldsymbol{J}_a^{-1}|, \tag{6.4.16}$$

involving the Jacobian with respect to the velocity field **a**, which in general varies in space and time.

Finally, *boundary conditions* should be carefully treated as the tracking procedure may search for points outside the domain due to the large time-step taken. In [157], a modified algorithm is proposed where the grid points around the boundary are treated with a time-step $\delta t < \Delta t$. This clearly complicates the implementation.

6.5 Wiggles and high order: stabilisation techniques

So far we have mostly dealt with smooth solutions of infinite regularity for which exponential convergence rates are realised. However, there are many hyperbolic problems which lead to discontinuous solutions or solutions of limited regularity. Also, in advection-dominated flows at marginal resolutions, wiggles appear that may render the computation unstable. The Gibbs phenomenon associated with high-order schemes in the presence of discontinuities causes loss of monotonicity in the solution of hyperbolic conservation laws. Appropriate reconstruction procedures have been developed by Gottlieb and his collaborators based on Gegenbauer polynomials to eliminate the loss of accuracy associated with the Gibbs phenomenon [200]. However, this is typically done at a post-processing stage, and the issue of monotonicity loss still remains for time-dependent problems.

Godunov [194] first proved that there are no linear second-order or higher-order schemes which guarantee monotonicity. In addition to these difficulties, it seems from the finite volume formulation literature (see [349] and references therein) that there is a conflict between the conservation property and monotonicity. For example, enforcing conservation can cause a positive physical quantity (for example, density) to become negative. We will revisit this issue again in Chapter 10.

The straightforward incorporation of artificial dissipation to control oscillation is not usually appropriate for high-order methods. However, high-frequency oscillations of the numerical solution can be controlled by filters, mode-dependent artificial viscosity, or by over-integration, as we describe in the next few sections. In physical space, we can also introduce an *upwind grid* to satisfy the differential equations leading to superconsistent discretisations. This is also presented in the last section of this chapter.

A fundamental result on the application of spectral methods for the scalar advection equation with non-smooth solution was obtained by Gottlieb *et al.* [198]. They proved that, for a Chebyshev collocation scheme, the limit solution of the nonlinear advection equation

$$\frac{\partial u}{\partial t} + \frac{\partial f(u)}{\partial x} = 0 \tag{6.5.1}$$

is a weak solution, that is, it satisfies the integral form

$$\iint \left(u\frac{\partial \psi}{\partial t} - f(u)\frac{\partial \psi}{\partial x} \right) \mathrm{d}x\, \mathrm{d}t = \int u(x,0)\psi(x,0)\, \mathrm{d}x\,,$$

(for smooth and homogeneous functions ψ), and that it satisfies the jump condition

$$[f(u)] = f(u_u) - f(u_d) = 0\,,$$

where brackets denote the jump in flux over the shock based on the upstream, u_u, and downstream, u_d, conditions. This implies that shocks are propagated with the correct speed for a viscous limiting solution. In addition, the transition zone is resolved within one grid point. Therefore, accurate position of the shock and correct shock speed are inherent in spectral methods, despite the high-frequency oscillations present in spectral solutions for discontinuous problems. Such a claim is consistent with the argument made by Lax [289] that information is contained in the oscillations associated with high-order schemes, and that high-order schemes retain more information than low-order schemes.

Convergence of spectral methods for nonlinear conservation laws was first proved by Tadmor [449], who showed that the inviscid Burgers equation converges to the unique entropy solution, that is, the spectral solution satisfies the entropy condition

$$\frac{\partial}{\partial t}\left(\frac{u^2(x,t)}{2}\right) + \frac{\partial}{\partial x}\left(\frac{u^3(x,t)}{3}\right) \leqslant 0\,. \tag{6.5.2}$$

The key idea behind this proof is the augmentation of the spectral approximation with a spectrally-vanishing viscosity kernel, which controls the higher modes in the spectrum and guarantees spectral convergence, independent of whether or not the solution is smooth. In a subsequent paper [321], the L^∞ stability bound was derived, and in [448] the proof was extended to nonlinear systems of conservation laws.

Earlier in this chapter we considered scalar hyperbolic equations with *smooth initial conditions* and argued that spectral/*hp* element discretisations can be very effective in obtaining accurate numerical solutions. We return again to the linear advection equation (6.1.1a), subject to a *discontinuous* initial condition of the form shown in Fig. 6.23 in a periodic domain. The spatial discretisation is the standard piecewise spectral approximation based on Chebyshev collocation. Three elements of equal size are used in the discretisation, with sixty Chebyshev–Gauss–Lobatto points per element. For the time integration, a second-order Adams–Bashforth scheme was used for a total of 750 000 time-steps ($\Delta t = 10^{-3}$). In other words, the waveform was advected seventy-five times around the (periodic) domain. There are two important features characteristic of this numerical solution:

- first, large amplitude oscillations appear everywhere in the domain, creating new extrema; and

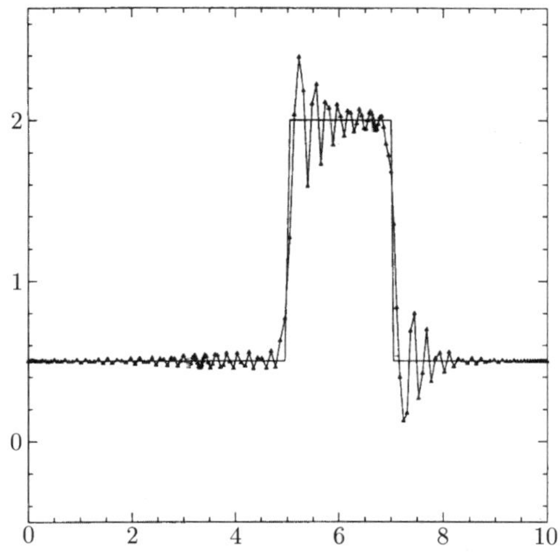

FIG. 6.23. Linear advection of a square wave using spectral element discretisation ($N_{\rm el} = 3, P = 59$). Periodic boundary conditions are imposed, while the time integration proceeds for 750 000 time-steps of size $\Delta t = 10^{-3}$. ([187].)

- second, the numerical solution has been convected with the *correct phase speed*, even after this very large number of time-steps. This feature is indicative of the very small dispersion error of high-order methods, as discussed earlier in this chapter.

The formal accuracy of spatial discretisation is also important in resolving contact discontinuities. This has been demonstrated by Harten [218], who considered eqn (6.1.1a) with the leading term in the truncation error as follows:

$$\frac{\partial u}{\partial t} + V\frac{\partial u}{\partial x} = \alpha\frac{\Delta x^{p+1}}{\Delta t}\frac{\partial^{p+1} u}{\partial x^{p+1}}\,. \tag{6.5.3a}$$

We assume that a discontinuity exists in the initial condition $u_0(x)$ so that

$$u_0(x) = \begin{cases} u_L\,, & x \leqslant 0\,, \\ u_R\,, & x > 0\,. \end{cases} \tag{6.5.3b}$$

After integration for N_s time-steps (total time $N_s\Delta t$), the discontinuity has spread and its width is δ_p. Introducing the transformation $(x,t) \to (y,t) \equiv (x - Vt, t)$, eqn (6.5.3a) reduces to

$$\frac{\partial u}{\partial t} = \alpha\frac{\Delta x^{p+1}}{\Delta t}\frac{\partial^{p+1} u}{\partial y^{p+1}}\,, \tag{6.5.3c}$$

which has a similarity solution with the similarity variable defined as $\eta = y/\delta_p$, where $\delta_p = (\alpha\Delta x^{p+1}N_s)^{1/(p+1)}$. The transition layer becomes narrower as the spatial accuracy (order p) of the scheme increases, that is, $\delta_p \propto N_s^{1/p+1}$. This is especially important for a large number of time-steps. To make this proof more rigorous, the very early times where the layer is still very thin have to be carefully treated, see Harten [218]. Also, this result is based on exact time integration, but it demonstrates how the *formal spatial* accuracy of the scheme can play a significant role in long-time integration.

6.5.1 *Filters and relaxation*

The Gibbs phenomenon can potentially be treated by filtering the high-frequency oscillations either in physical or modal space, and both approaches have been used in the past. In modal space, for example, the coefficients $\hat{u}$ of a spectral expansion for u are multiplied by the filter $\sigma(\theta)$, where $\theta \in [-\pi, \pi]$. The filter is an infinitely-differentiable function, and it should be equal to unity around the origin ($\theta = 0$) so that it does not change the mean value of the filtered function. It is also zero outside the range of θ (that is, for $|\theta| \geqslant \pi$). A particular class of filters corresponding to different orders p was constructed by Vandeven [478] in the form

$$\sigma_p(x) = 1 - \frac{(2p-1)!}{(p-1)!^2} \int_0^x [t(1-t)]^{p-1}\, dt\,. \tag{6.5.4}$$

A plot of this filter for different values of p is shown in Fig. 6.24. For $p = 1$ we obtain $\sigma_1(x) = 1 - x$, which is the classical Fejer's filter. For $p = 3$ the Vandeven filter is very similar to the classical raised cosine filter, which is defined as

$$\sigma_{\mathrm{RC}}(\theta) = \frac{1}{2}(1 + \cos\theta)\,.$$

In physical space and on an equidistant grid the raised cosine filter transforms the u_j value to u_j^*:

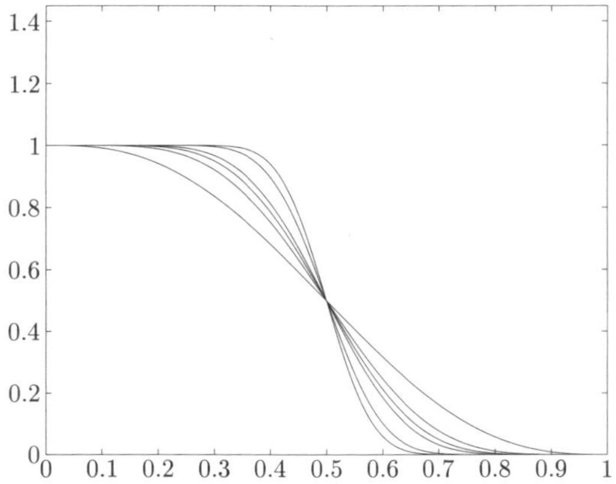

FIG. 6.24. Vandeven filter for different orders ($p = 3$, 6, 8, 10, 20, and 30).

$$u_j^* = \frac{u_{j-1} + 2u_j + u_{j+1}}{4} = u_j + \frac{\Delta x^2}{4} \frac{u_{j-1} - 2u_j + u_{j+1}}{\Delta x^2},$$

which has the form of a second-order artificial viscosity term.

For $p = 8$, the Vandeven filter is similar to the sharpened raised cosine filter

$$\sigma_{\mathrm{SRC}}(\theta) = \sigma_{\mathrm{RC}}^4 \left(35 - 84\sigma_{\mathrm{RC}} + 70\sigma_{\mathrm{RC}}^2 - 20\sigma_{\mathrm{RC}}^3\right).$$

Kopriva [277] applied the sharpened raised cosine filter to the advection equation (using Fourier collocation) with the following initial discontinuous condition:

$$u(x) = \frac{1}{1 + \cos^2(5\pi x/4)}, \quad 0 \leqslant \frac{x}{\pi} \leqslant 2. \tag{6.5.5}$$

The numerical solution, filtered with the sharpened raised cosine, was of very high accuracy. However, as it was demonstrated, the convergence rate was not exponential. Vandeven [478] has shown that a pre-specified accuracy (controlled directly by the order of the filter p) can be recovered away from the discontinuity. By appropriately selecting the order p in relation to the number of Chebyshev modes (e.g., $p \propto P^{1+\epsilon/4}$, where $\epsilon > 0$), exponential accuracy can be achieved away from the discontinuity. Vandeven developed his filter for Chebyshev discretisations, but numerical experiments indicate that the filter is equally effective for Legendre discretisations.

In Fig. 6.25 we plot the pointwise error for the function given in eqn (6.5.5) used in Kopriva's experiments. A Chebyshev representation with 128 modes and with a filter of order $p = 3$, 10, 20, and 30 were used. We see that high accuracy is recovered away from the discontinuity, but there are still large boundary errors. Also, the solution in the neighbourhood of the discontinuity is still misrepresented, with the unfortunate occurrence that the width of the affected area increases with the filter order. The latter is certainly an unwanted result in

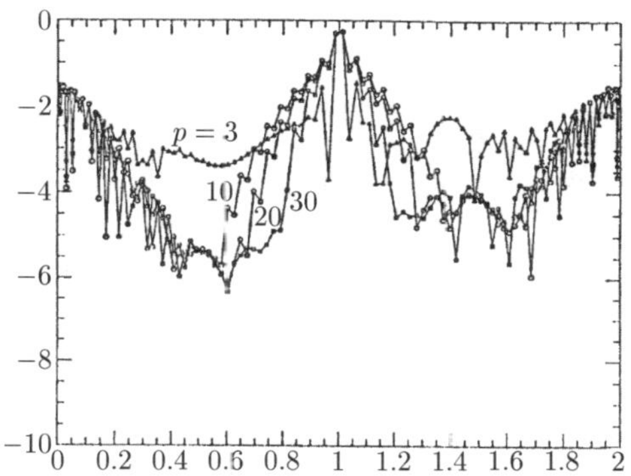

FIG. 6.25. Pointwise error for the function described in eqn (6.5.5) filtered with the Vandeven filter and order $p = 3$, 10, 20, and 30; 128 Chebyshev modes were used. (Courtesy of I. G. Giannakouros.)

simulating phenomena where most of the physics stems from the discontinuous interface, as, for example, is the case of compressible flow in the presence of a shock wave as discussed in Chapter 10.

The Vandeven filter is effective in recovering exponential accuracy for several cases we have tested in approximating discontinuous functions. However, its use in the discretisation of the linear advection equation gives no improvement at all. To demonstrate this we plot in Fig. 6.26 the results of a spectral element (collocation projection) simulation of a square wave, linearly advected to the right for various values of the filter order p. In this test the filter is applied here at selected time intervals: at the last time-step (Fig. 6.26(b)); every time-step (Fig. 6.26(c)); and every 100 time-steps (Fig. 6.26(d)). For comparison, we also plot the unfiltered solution in Fig. 6.26(a). The results are qualitatively different. In the third case the waveform is 'smoothed' globally; in other cases major errors are confined around the discontinuities and the boundary points.

One of the problems of filtering is that it often violates the boundary conditions of the problem. This issue has been addressed by Boyd [72], who proposed

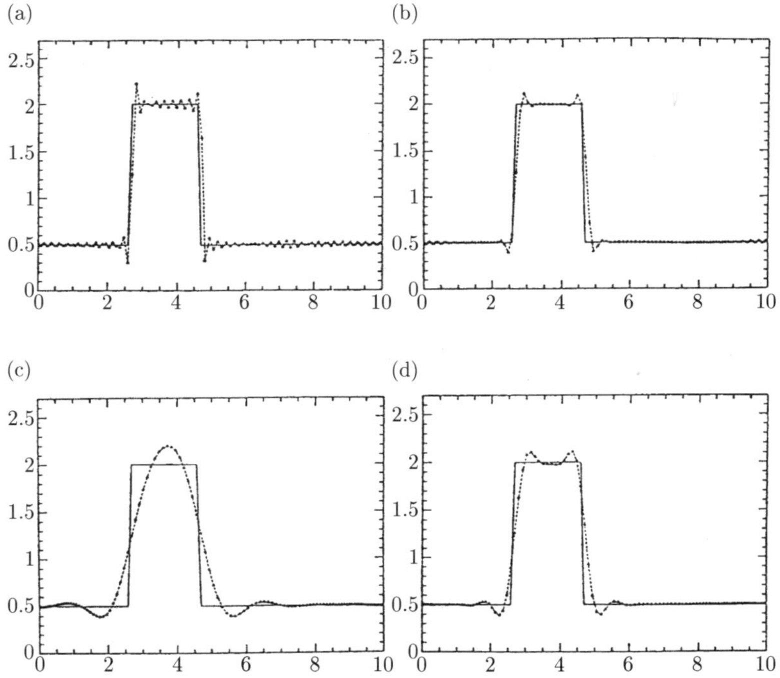

FIG. 6.26. Linear advection of a square wave (a) unfiltered solution, and Vandeven filter applied at (b) the last time-step, (c) every time-step, and (d) every 100 time-steps. The solution is advected for 4800 time-steps with a time-step $\Delta t = 1.6 \times 10^{-3}$ (128 collocation points). (Courtesy of I. G. Giannakouros.)

a simple but effective modification to filtering so that the filtered solution satisfies the same boundary conditions as the original solution. The key idea is to rewrite the numerical solution in terms of new basis functions $\phi(x)$, which individually satisfy homogeneous boundary conditions. The inhomogeneous part can be represented by the linear modes, as we have already seen in Section 4.3.2.

Two special sets of basis functions proposed in [72] are written in terms of Chebyshev $T_k(x)$ and Legendre polynomials $L_k(x)$ as follows:

$$\phi_k(x) = T_{k+2}(x) - T_k(x) \quad \text{or} \quad \phi_k(x) = L_{k+2}(x) - L_k(x) . \tag{6.5.6}$$

Clearly, $\phi_k(\pm 1) = 0$ for $x \in [-1, 1]$. This basis is appropriate for Dirichlet boundary conditions, but for Neumann or Robin boundary conditions appropriate modifications are required.

To clarify this concept, let us consider the Legendre expansion

$$u(x) = \sum_{k=0}^{P} \hat{u}_k L_k(x) = ax + b + \sum_{k=0}^{P-2} \hat{U}_k \phi_k(x) ,$$

where $\phi_k(x)$ is given in eqn (6.5.6). By equating the Pth-order and $(P-1)$th-order terms, we obtain $\hat{U}_{P-2} = \hat{u}_P$ and $\hat{U}_{P-3} = \hat{u}_{P-1}$. Subsequently, we can compute

$$\hat{U}_{k-2} = \hat{U}_k + \hat{u}_k , \quad k = P-2, P-4, \dots ,$$

for the even coefficient (assuming P is even), and a similar expression can be derived for the odd coefficients.

If we now apply a *sharp cut-off* filter F_P to simply cancel the last $(P-2)$th mode of the homogeneous component in the expansion of $u(x)$, then we obtain the following filtered expansion:

$$\begin{aligned} F_P u(x) &= u(x) - \hat{U}_{P-2}\phi_{P-2} \\ &= u(x) - \hat{u}_P[L_P(x) - L_{P-2}(x)] \\ &= \sum_{k=0}^{P-3} \hat{u}_k L_k(x) + (\hat{u}_{P-2} + \hat{u}_P) L_{P-2}(x) + \hat{u}_{P-1} L_{P-1}(x) . \end{aligned} \tag{6.5.7}$$

This expression is, in fact, identical to a filtering procedure proposed by Fischer and Mullen in physical space for the Gauss–Lobatto–Legendre (GLL) collocation points [162]. They proposed to compute the solution $u(x)$ on a GLL grid with $P+1$ nodes and subsequently to transfer the solution to a GLL grid with P nodes. The appropriate projector operator can be defined as

$$\mathbb{P}_{P-1} \equiv \mathcal{I}_{P-1}^{P} \cdot \mathcal{I}_{P}^{P-1}$$

in terms of the interpolator operators on the right-hand side. In [362] it was shown that this inter-grid transfer is equivalent to the sharp cut-off filter given by eqn (6.5.7), i.e.,

$$\mathbb{P}_{P-1} \equiv F_P \,.$$

Better results of this filtering procedure have been achieved in [162] by applying a relaxation to the above filtering, thus defining a new filter operator of the form

$$F_\alpha \equiv \alpha \mathbb{P}_{P-1} + (1-\alpha)\mathcal{I}_P^P \,,$$

where the last operator in the above equation is the identity operator. Clearly, $0 \leqslant \alpha \leqslant 1$, but the recommended range for advection–diffusion problems is $0.05 \leqslant \alpha \leqslant 0.3$. Fischer and Mullen have also recognised that the suppressed mode is the $(P-2)$th mode in the basis suggested by Boyd [72], i.e., the mode

$$L_P(x) - L_{P-2} = \frac{2P-1}{P(P-1)}(1-x^2)L'_{P-1}(x) \,,$$

where the above equality is based on the properties of the Legendre polynomials (see Appendix A, noting $L_p(x) = P_p^{0,0}(x)$).

From the physical point of view, it seems that the interlacing of the nodal basis on the Pth-order GLL grid with the $(P-1)$th-order GLL grid leads to dampening of high-frequency oscillations that are responsible for wiggles and eventual blow-ups. However, as pointed out in [362], in contrast to the usual filters, this type of inter-grid interpolation, which effectively modifies the $(P-2)$th mode, does not imply a dissipation of 'energy'. In particular, for the Legendre polynomials, we obtain the following L^2 weighted norm:

$$||u||_w^2 - ||F_P u||_w^2 = \hat{u}_P^2 ||L_P||_w^2 + \hat{u}_{P-2}^2 ||L_{P-2}||^2 - (\hat{u}_P + \hat{u}_{P-2})^2 ||L_{P-2}||_w^2 \,,$$

which can be positive or negative depending on the flow problem. This is also true for the Chebyshev polynomials. It is conceivable that at marginal resolutions and large values of the Péclet (or Reynolds) number a long-term instability may arise if the above difference is negative.

6.5.2 *Spectral vanishing viscosity (SVV)*

Artificial viscosity has been used in many discretisation methods to suppress wiggles associated with high wavenumbers. Hyperviscosity, in particular, has been found to be very effective in simulations with Fourier methods. A refined idea that is based on a second-order diffusion (convolution) operator was proposed by Tadmor [449]. He introduced the concept of spectral vanishing viscosity (SVV) using the inviscid Burgers equation

$$\frac{\partial}{\partial t}u(x,t) + \frac{\partial}{\partial x}\left(\frac{u^2(x,t)}{2}\right) = 0 \,, \tag{6.5.8}$$

subject to proper initial and boundary conditions. The distinct feature of solutions to this problem is that spontaneous jump discontinuities (shock waves) may develop, and hence *a class* of weak solutions can be admitted. Within this class, there are many possible solutions, and in order to single out a physically relevant one an additional entropy condition is applied of the form

$$\frac{\partial}{\partial t}\left(\frac{u^2(x,t)}{2}\right)+\frac{\partial}{\partial x}\left(\frac{u^3(x,t)}{3}\right)\leqslant 0\,. \tag{6.5.9}$$

In practical applications, spectral methods are often augmented with smoothing procedures in order to reduce the Gibbs oscillations [137] associated with discontinuities arising at the domain boundaries or due to under-resolution. However, when considering nonlinear problems, convergence of the Fourier method, for example, may fail despite additional smoothing of the solution. Tadmor [449] introduced the spectral vanishing viscosity method, which adds a small amount of controlled dissipation that satisfies the entropy condition, yet retains spectral accuracy. It is based on viscosity solutions of nonlinear Hamilton–Jacobi equations, which have been studied systematically in [116]. The viscosity solution for the Burgers equation has the form

$$\frac{\partial}{\partial t}u(x,t)+\frac{\partial}{\partial x}\left(\frac{u^2(x,t)}{2}\right)=\epsilon\frac{\partial}{\partial x}\left[F_\epsilon\frac{\partial u}{\partial x}\right], \tag{6.5.10}$$

where ϵ ($\to 0$) is a viscosity amplitude and F_ϵ is a viscosity kernel, which may be nonlinear and, in general, a function of x. Convergence may then be established by compensated compactness estimates combined with entropy dissipation arguments [449]. To respect spectral accuracy, the SVV method makes use of viscous regularisation, and eqn (6.5.10) may be rewritten in discrete form (retaining P modes) as

$$\frac{\partial}{\partial t}u_P(x,t)+\frac{\partial}{\partial x}\left[\mathbb{P}_P\left(\frac{u^2(x,t)}{2}\right)\right]=\epsilon\frac{\partial}{\partial x}\left[F_P*\frac{\partial u_P}{\partial x}\right], \tag{6.5.11}$$

where the star ($*$) denotes convolution and $\mathbb{P}_P$ is a projection operator. Here F_P is a (possibly nonlinear) viscosity kernel, which is only activated for high wavenumbers. In Fourier space, this kind of spectral viscosity can be efficiently implemented as multiplication of the Fourier coefficients of u_P with the Fourier coefficients of the kernel F_P, i.e.,

$$\epsilon\frac{\partial}{\partial x}\left[F_P*\frac{\partial u_P}{\partial x}\right]=-\epsilon\sum_{M_{\mathrm{SVV}}\leqslant|k|\leqslant P}k^2\hat{F}_k(t)\hat{u}_k(t)\mathrm{e}^{\mathrm{i}kx}\,,$$

where k is the wavenumber, P is the number of Fourier modes, and M_{SVV} is the wavenumber above which the spectral vanishing viscosity is activated.

Originally, Tadmor [449] adopted

$$\hat{F}_k=\begin{cases}0\,, & |k|\leqslant M_{\mathrm{SVV}}\,,\\ 1\,, & |k|>M_{\mathrm{SVV}}\,,\end{cases} \tag{6.5.12}$$

with $\epsilon M_{\mathrm{SVV}}\sim 0.25$ based on the consideration of minimising the total variation of the numerical solution. In subsequent work, however, a smooth kernel was used, since it was found that the C^∞ smoothness of $\hat{F}_k$ improves the resolution of the SVV method. For Legendre pseudo-spectral methods, Maday *et al.* [318] used $\epsilon \approx P^{-1}$, activated for modes $k > M_{\mathrm{SVV}} \approx 5\sqrt{P}$, with

$$\hat{F}_k = \exp\left\{ -\frac{(k-P)^2}{(k-M_{\mathrm{SVV}})^2} \right\}, \quad k > M_{\mathrm{SVV}}\,. \tag{6.5.13}$$

Karamanos and Karniadakis [259] made the first extension of the spectral vanishing viscosity concept to spectral/*hp* element methods. In [259], the general form of the SVV operation as presented by Tadmor is maintained; however, polynomial filtering is used to mimic the convolution operator in Tadmor's formulation. In Kirby [270] several improvements were implemented, both for the continuous Galerkin and the discontinuous Galerkin formulations. The implementation for the latter is straightforward as a fully-orthogonal L^2 basis can be employed. However, for the continuous Galerkin formulation the C^0 basis is semi-orthogonal, and thus some correction is required. We present some of the implementation details in the following.

We start by considering the Burgers equation written in a strong form with the SVV term added to the right-hand side, i.e.,

$$\frac{\partial u}{\partial t} + \frac{1}{2}\frac{\partial u^2}{\partial x} = \epsilon \frac{\partial}{\partial x}\left(F \frac{\partial u}{\partial x} \right). \tag{6.5.14}$$

If we examine the weak form of the SVV term only, and ignore boundary terms and the leading coefficient, we have the following basic form of the SVV operator:

$$\left(\frac{\partial v}{\partial x}, F\frac{\partial u}{\partial x} \right), \tag{6.5.15}$$

where v is a test function taken from the space of one-dimensional C^0 principal functions $\{\psi_k\}$, and $u = \sum_k \hat{u}_k \psi_k$. In the derivation below, we will assume that all discrete summations are from $1, \ldots, P$. In the notation above and henceforth, $(\cdot,\cdot)$ denotes the L^2 inner product, and it is assumed that the continuous and discrete inner products are interchangeable given sufficient quadrature order.

Let $\boldsymbol{T}$ be a matrix which transforms the modal coefficients $\hat{\boldsymbol{u}}$ related to basis functions $\{\psi_k\}$ to $\tilde{\boldsymbol{u}}$, which are related to functions $\{\widetilde{\psi}_k\}$, where $\{\psi_k\}$ is our C^0 basis used for the continuous Galerkin formulation and where $\{\widetilde{\psi}_k\}$ is an orthonormal basis which spans the same space as $\{\psi_k\}$. Let $\boldsymbol{F}$ be a diagonal matrix which acts as a filtering function (the diagonal entries of which are given by eqn (6.5.13)) in modal space.

In the notation above, we have that $\tilde{\boldsymbol{u}} = \boldsymbol{T}\hat{\boldsymbol{u}}$. Our goal is to filter the coefficients $\tilde{u}$ instead of filtering the coefficients $\hat{u}$. Hence we want to transform (by the matrix $\boldsymbol{T}$) to the orthogonal space, filter, and then transform back. This operation is accomplished as follows:

$$\hat{\hat{\boldsymbol{u}}} = \boldsymbol{T}^{-1}\boldsymbol{F}\boldsymbol{T}\hat{\boldsymbol{u}}\,. \tag{6.5.16}$$

We can rewrite this expression as $\hat{\hat{\boldsymbol{u}}} = \boldsymbol{\Theta}\hat{\boldsymbol{u}}$, where $\boldsymbol{\Theta} = \boldsymbol{T}^{-1}\boldsymbol{F}\boldsymbol{T}$. In discrete form eqn (6.5.15) using matrix notation is as follows:

$$\boldsymbol{S}^{\top}\boldsymbol{T}^{-1}\boldsymbol{F}\boldsymbol{T}\boldsymbol{M}^{-1}\boldsymbol{S}\hat{\boldsymbol{u}}\,,$$

where $\boldsymbol{S}_{ij} = (\phi_i, \partial\phi_j/\partial x)$ and $\boldsymbol{M}_{ij} = (\phi_i, \phi_j)$. In the above equation, it can be shown that $\boldsymbol{T}^{-1} = \boldsymbol{M}^{-1}\boldsymbol{T}^{\top}$, and hence we have that the discrete form of the SVV operator for the continuous Galerkin method is given by the following expression:

$$\boldsymbol{S}^{\top}\boldsymbol{M}^{-1}\boldsymbol{T}^{\top}\boldsymbol{F}\boldsymbol{T}\boldsymbol{M}^{-1}\boldsymbol{S}\hat{\boldsymbol{u}} = \left(\boldsymbol{M}^{-1}\boldsymbol{S}\right)^{\top}\boldsymbol{T}^{\top}\boldsymbol{F}\boldsymbol{T}\boldsymbol{M}^{-1}\boldsymbol{S}\hat{\boldsymbol{u}}\,.$$

The discrete SVV operator is symmetric and semi-positive definite [270].

Next we test this implementation of SVV in the context of the inviscid Burgers equation. In Fig. 6.27 we plot the solution of the inviscid Burgers equation at time $T = 0.5$, with and without SVV. Five equally-spaced elements spanning $[-1, 1]$ were used, each element containing sixteen modes. In this example, a wave cut-off of $M_{\mathrm{SVV}} = 8$ and amplitude of $\epsilon = 1/16$ were applied. We note that in the standard formulation if an element interface is aligned with the discontinuity then the wiggles in the numerical solution are significantly suppressed.

Observe that the SVV has two positive effects: in the region of the discontinuity (the discontinuity is centred at the origin), SVV greatly minimises the

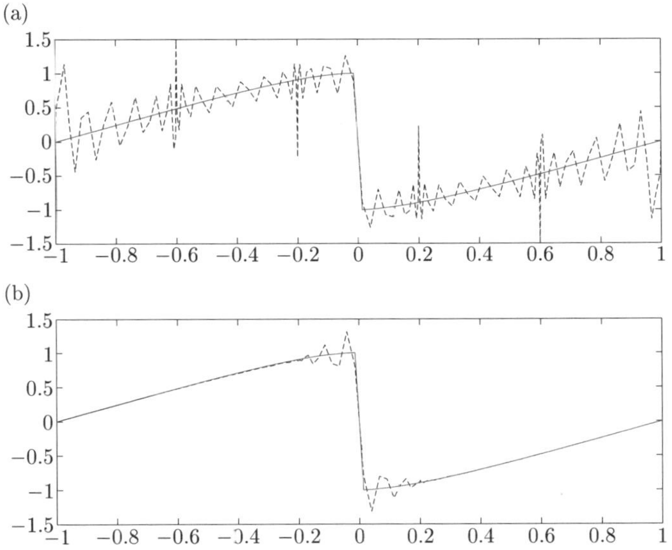

FIG. 6.27. Solution of the inviscid Burgers equation at time $T = 0.5$ using continuous Galerkin (a) without, and (b) with SVV. Five equally-spaced elements spanning $[-1, 1]$ were used, each of which contained sixteen modes. ([270].)

variation of the solution; away from the discontinuity, the 'wiggles' in the solution have been removed. This feature of SVV is consistent with the results shown for Fourier methods in [449] and for continuous Galerkin spectral element methods in [259]. The SVV result away from the discontinuity looks very smooth. This can be attributed to the fact that, for the one-dimensional continuous Galerkin method, only the linear modes directly transmit information between elements (all higher-order modes are bubble modes interior to the element and zero on the elemental boundary). Hence information is only propagated through the linear modes to elements away from the sharp gradient, and SVV immediately dampens the energy introduced to the higher modes through the nonlinear terms.

Similar results can be obtained with the *discontinuous Galerkin method* using the above formulation. A slightly modified formulation was developed in [217] with the SVV of the form

$$\mathrm{SVV}_2 = \epsilon F \frac{\partial^2}{\partial x^2} F u \,, \tag{6.5.17}$$

where F is a polynomial filter, and the second derivative operations are given by the appropriate numerical discontinuous Galerkin derivatives. This modified form applies the polynomial filter twice, instead of filtering only the intermediary-stage first derivative; the SVV of eqn (6.5.17) filters both the initial solution and the resulting second derivative. Results obtained in Kirby [270] with the modified form (6.5.17) show that it is less dissipative than the original form (6.5.15).

6.5.3 *Over-integration of the viscous Burgers equation*

We now examine how over-integration (introduced earlier in Section 2.4.1.2) can be employed to deal with the small scales (i.e., high frequencies) generated by the nonlinear terms. In order to test the integration of the nonlinear terms, we solve the viscous Burgers equation

$$\frac{\partial u}{\partial t} + \frac{1}{2}\frac{\partial u^2}{\partial x} = \nu \frac{\partial^2 u}{\partial x^2} \tag{6.5.18}$$

with initial condition $u(x) = -\sin(\pi x)$ using the *continuous Galerkin* method for $\nu = 10^{-5}$; Gauss–Lobatto–Legendre points are used for each element. The simulation parameters are presented in Table 6.3; five equal spectral elements were used, each of which contained $P = 16$ modes in its polynomial expansion. The number of elements and the element spacing were chosen so that the steep gradient in the solution occurs in the middle of an element and *not* at an element interface. This choice forces the *middle* element to accommodate the steep gradient, involving most of the modes in its elemental polynomial expansion, hence making it a prime candidate for aliasing problems. The rationale for this example is to test whether the solution of a system with a small amount of viscosity is sufficient to overcome the 'numerical crimes' committed by under-integrating the nonlinear terms, and hence render the solution stable.

TABLE 6.3. Parameters used for the results presented in Fig. 6.28.

Parameters	Values
Method	Continuous Galerkin
Number of elements	5
Modes	16
Time integrator	2nd-order Adams–Bashforth
Final time T	0.5
Δt	0.0001

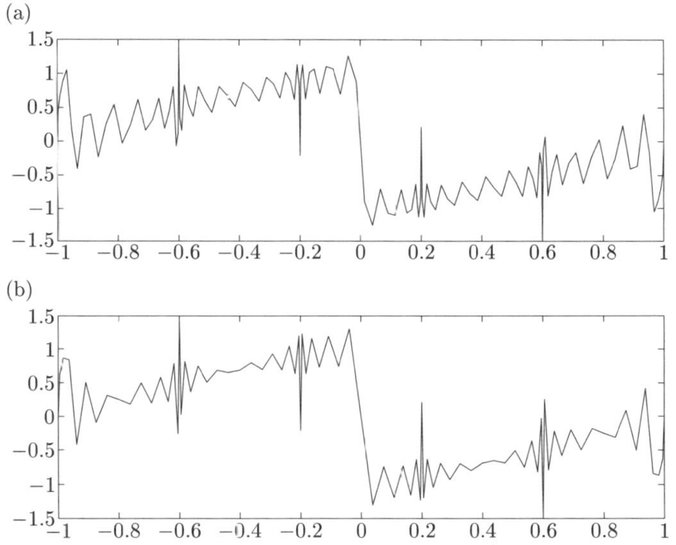

FIG. 6.28. Solution of the viscous Burgers equation with $\nu = 10^{-5}$ evaluated at $T = 0.5$. In (a), $Q = 24$ quadrature points are used for integrating both the advection and diffusion terms; and in (b), $Q = 24$ quadrature points are used for the advection term, and only $Q = 17$ points are used for the diffusion terms. ([270].)

If $Q = P + 1 = 17$ quadrature points are used for integrating both the advection and diffusion terms, then the solution is unstable. If, however, $Q = 3P/2 = 24$ quadrature points are used, as in Fig. 6.28(a), then the solution is stable. To verify that indeed over-integration is only necessary for the advection (nonlinear) terms and not the diffusion terms (which, because they are linear, should only require $P + 1$ points for exact integration), we also computed the solution when $Q = 24$ points were used for integrating the advection term and $Q = 17$ points for integrating the diffusion term. The result is presented in Fig. 6.28(b). The difference in the L^∞ norm of the solution is negligible.

To further extend our understanding of the problems which may arise due to under-integration, we also solved the viscous Burgers equation with $\nu = 10^{-2}/\pi$

using the continuous Galerkin method. This problem was used as a test problem in [37] for evaluating the effectiveness of different spatial discretisations. As is pointed out in that paper, for $\nu = 10^{-2}/\pi$ the solution develops into a saw-tooth wave centred at the origin around the time $t = 1/\pi$ (an approximation which comes from the inviscid theory for Burgers equation), and the maximum gradient of the solution occurs at approximately $t = 0.5$. As in [37], the Cole transformation is used to obtain the exact solution to eqn (6.5.18):

$$u(x,t) = -\frac{\int_{-\infty}^{\infty} \sin(\pi(x-\eta)) f(x-\eta) \exp(-\eta^2/4\nu t)\,\mathrm{d}\eta}{\int_{-\infty}^{\infty} f(x-\eta) \exp(-\eta^2/4\nu t)\,\mathrm{d}\eta}, \tag{6.5.19}$$

where

$$f(y) = \exp\left[-\cos\left(\frac{\pi y}{2\pi\nu}\right)\right]. \tag{6.5.20}$$

In the current study, the convolution given in eqn (6.5.19) was computed numerically using Gauss–Hermite integration with up to thirty terms.

Our rationale in this example is to see how much viscosity is necessary to stabilise the solution when under-integration is involved, and to compare solutions against an exact solution. The following three cases were examined (and are plotted in Fig. 6.29).

1. Case A: $Q = 17$ quadrature points were used for integrating both the advection and diffusion terms (Fig. 6.29(a)).
2. Case B: $Q = 24$ quadrature points were used for integrating both the advection and diffusion terms (Fig. 6.29(b)).
3. Case C: $Q = 24$ points were used for integrating the advection term and $Q = 17$ points were used for integrating the diffusion term (Fig. 6.29(c)).

In Table 6.4 we present the error in the discrete L^∞ norm for the three cases presented in Fig. 6.29. Observe that both case B and case C show marked improvement over case A. The discrepancy between case B and case C can be explained by the fact that the discrete (pointwise) L^∞ norm is taken over the number of points used for the diffusion operator. Hence, in case B more (and different) points were used for the diffusion operator than in both case A and case C. The overall trend, and hence the point of the test, was not affected however.

As is evident by these examples, given sufficiently large viscosity, the solution can be rendered stable. However, since we have an interest in the numerical simulation of high Reynolds number flows, special care must be taken when integrating the nonlinear terms as the viscosity may be insufficient in rendering the solution stable.

6.5.4 *Superconsistent collocation for advection–diffusion*

So far we have dealt primarily with stabilisation techniques in modal space, although filtering and SVV in physical space is also possible [137, 318]. We now present another idea due to Funaro [170], which enhances the stability of

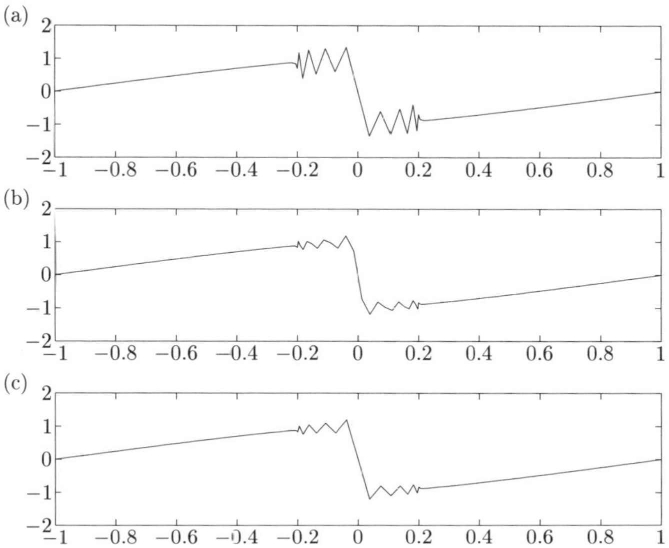

FIG. 6.29. Solution of the viscous Burgers equation with $\nu = 10^{-2}/\pi$ evaluated at $T = 0.5$. In (a), $Q = 17$ quadrature points are used for integrating both the advection and diffusion terms; in (b), $Q = 24$ quadrature points are used for integrating both the advection and diffusion terms; and in (c), $Q = 24$ quadrature points are used for the advection term, and only $Q = 17$ points are used for the diffusion terms. ([270].)

TABLE 6.4. Error in the discrete L^∞ norm for the three cases A, B, and C.

Case	Error in the discrete L^∞ norm
A	0.5090
B	0.2729
C	0.2038

advection–diffusion systems. To illustrate the key idea we study only a single one-dimensional element, but multi-dimensional extensions can be found in Funaro's book [170].

Let us consider the steady advection–diffusion equation

$$V\frac{\partial u}{\partial x} = \nu\frac{\partial^2 u}{\partial x^2}, \quad x \in [-1, 1],$$
$$u(-1) = u(1) = 0,$$

where V is constant and ν is the viscosity. The exact solution is

$$U_{\mathrm{ex}}(x) = \frac{\mathrm{e}^{x\mathrm{Pe}} - \mathrm{e}^{-\mathrm{Pe}}}{\mathrm{e}^{\mathrm{Pe}} - \mathrm{e}^{-\mathrm{Pe}}},$$

where $\mathrm{Pe} = V \cdot l/\nu$ is the *Péclet number* and l is a characteristic length. For very large values of Pe a steep boundary layer develops at $x = 1$. A standard spectral collocation method based on Legendre polynomials enforces

$$V\frac{\partial v_P}{\partial x}(\xi_i) = \nu\frac{\partial^2 v_P}{\partial x^2}(\xi_i)\,,$$

where v_P is the numerical solution,

$$v_P(x) = \sum_{j=1}^{P} v_P(\xi_j^h) h_j^P(x)\,,$$

and $h_j^P(\xi_i^P) = \delta_{ij}$ is the Lagrangian interpolant.

When P is sufficiently large, e.g., of order $\mathcal{O}(\sqrt{\mathrm{Pe}})$, then we can accurately resolve the boundary layer and monotonicity is preserved. However, for lower resolutions (i.e., smaller P) large wiggles appear, monotonicity is not preserved, and eventual blow-up may occur.

This situation can be alleviated by choosing another set of points for which the residual of the differential equation is zero. This new set of points depends on the operator we are considering. In this particular case we have that

$$\mathbb{L}u \equiv \mathrm{Pe}\frac{\partial u}{\partial x} - \frac{\partial^2 u}{\partial x^2} = 0\,.$$

So, Funaro [170] suggested that in addition to the *representation* grid ξ_i^P that defines the Lagrangian interpolant, we can determine a new grid so that

$$\mathbb{L}_d u_d|_{\zeta_d} = 0\,,$$

where ζ_d defines the new (unknown) collocation points, and $\mathbb{L}_d$ and u_d are the discrete operator and numerical solution, respectively.

In order to determine how good the discretisation is in terms of *consistency*, we can estimate the dimension of the space of functions v such that

$$\mathbb{L}v|_{\zeta_d} - \mathbb{L}_d v = 0\,. \qquad (6.5.21)$$

The key idea is to increase the dimension of that space. Consistency requires that this dimension be equal to the number of grid points on the representation grid in the collocation method. *Superconsistency* implies that this dimension is even bigger. For example, if

$$v = \sum_{j=1}^{P} h_i v_j$$

then the proper space for superconsistency is determined by

$$\{h_1, h_2, \ldots, h_P, \phi_d\}\,,$$

where ϕ_d is an extra function which is not a linear combination of $h_1, \ldots, h_P$.

Funaro [171] suggested the following procedure for determining ϕ_d. First we write

$$\mathbb{L}_d v = \sum_{j=1}^{P} v_j (\mathbb{L}_d h_j) = \sum_{j=1}^{P} v_j \left(\mathbb{L} h_j|_{\zeta_d} \right) = \mathbb{L} v|_{\zeta_d} , \tag{6.5.22}$$

and then we require that

$$\phi_d|_{\xi_j} = 0 , \quad j = 1, \ldots, P .$$

Now, the ith component of the discrete operation $\mathbb{L}_d$ applied to ϕ_d will be zero, i.e.,

$$(\mathbb{L}_d \phi_d)_i = \sum_{j=1}^{P} \phi_d|_{\xi_i} \left(\mathbb{L} h_j|_{\zeta_i} \right) = 0 ,$$

so if we set $v = \phi_d$ in eqn (6.5.21) then we obtain

$$\mathbb{L}\phi_d|_{\zeta_i} = 0 . \tag{6.5.23}$$

The above is the required equation in determining the *new collocation grid* ζ_i.

Returning to the operator

$$\mathbb{L} \equiv \mathrm{Pe}\frac{\partial}{\partial x} - \frac{\partial^2}{\partial x^2} ,$$

we can choose $\phi_d = (1 - x^2)L_P$, where L_P is a Legendre polynomial, so that it vanishes at all Gauss–Lobatto–Legendre points. Applying eqn (6.5.23) we require that

$$\mathrm{Pe}\,\frac{\partial \phi_d}{\partial x}(\zeta_i) - \frac{\partial^2 \phi_d}{\partial x^2}(\zeta_i) = 0 , \quad i = 1, \ldots, P-1$$

or

$$\mathrm{Pe}[(1 - x^2)L'_P]' - [(1 - x^2)L'_P]'' = 0 .$$

We can simplify this equation by using the definition of the Legendre polynomial in the Sturm–Liouville problem, i.e.,

$$[(1 - x^2)L'_P(x)]' = -P(P+1)L_P(x) ,$$

and thus we can determine the new collocation points from

$$P(P+1)[\mathrm{Pe}\, L_P(\zeta_i) - L'_P(\zeta_i)] = 0 , \quad i = 1, \ldots, P-1 . \tag{6.5.24}$$

A plot of the distribution of the locations of zero, i.e., ζ_i, is shown in Fig. 6.30. We have used $P = 8$ points for clarity and varied the Péclet number Pe (vertical axis). For Pe $= 0$ the ζ_i points coincide with the Gauss–Lobatto points ξ_i, but for increasing values of Pe the points move upwind and for Pe $\to \infty$ they coincide with the Gauss points z_i (where $L_P(z_i) = 0$). This can also be seen from the following approximate formulae given in [169].

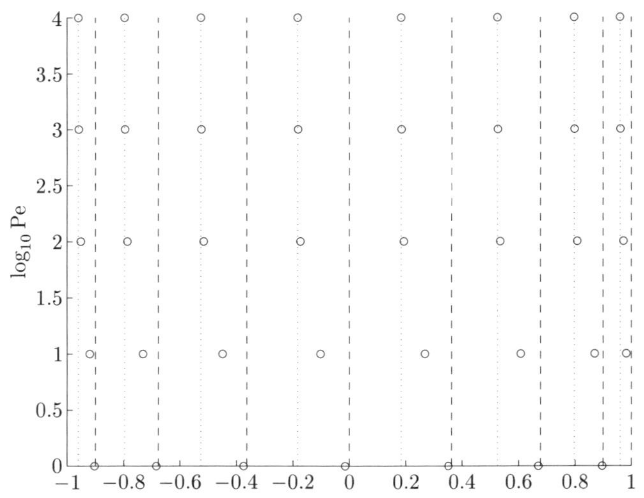

FIG. 6.30. Upwind collocation grid (circles) in the domain $[-1,1]$ for different values of the Péclet number corresponding to the one-dimensional advection–diffusion operator. Also shown are the Gauss–Legendre points (dotted line) and the Gauss–Lobatto–Legendre (dashed line).

Upwind points—new collocation grid:

$$\zeta_j \approx -\cos\left[\frac{2j-1/2}{2P+1}\pi + \frac{2}{2P+1}\cos^{-1}\left(\frac{2}{\sqrt{4+(2P+1)^2/\mathrm{Pe}^2}}\right)\right],$$

$$j = 1,\ldots,P-1\,.$$

Gauss points:

$$z_j \approx -\cos\left(\frac{2j-1/2}{2P+1}\pi\right), \quad j = 1,\ldots,P\,.$$

Gauss–Lobatto points:

$$\xi_j \approx -\cos\left(\frac{2j+1/2}{2P+1}\pi\right), \quad j = 1,\ldots,P-1\,.$$

Numerical experiments in [169] have shown that stability is indeed enhanced significantly by employing the new points, even for severely under-resolved cases. Unfortunately, there is no rigorous theory to justify the extra stability. The accuracy is improved as well, but the formal convergence rate of the method remains the same. Results in two dimensions and extensions to spectral elements can be found in [170].

7

NON-CONFORMING ELEMENTS

In many flow simulations there is often a need to refine the mesh locally; for example, to resolve the geometric singularity in a flow over a backwards-facing step. In time-dependent flows the refinement or coarsening process may need to be performed repeatedly as flow structures are convected through the computational domain. It would, therefore, be desirable to contain the mesh refinement or coarsening locally as needed and not propagate these mesh changes globally. Local refinement or coarsening in *hp* methods can also be achieved by h-type mesh adaptation or by changing the order P^e of each element to accommodate resolution requirements. Such features of selective local refinement and automatic mesh adaptation are particularly crucial in increasing the computational efficiency of direct (DNS) and large-eddy simulations (LES) of turbulent flows (see Chapter 9). In either h- or p-refinement (or coarsening) a *geometric* and/or *functional* incompatibility is produced.

The formulation presented so far for multi-dimensional space (see Chapters 4 and 5) has dealt with conformity of elements where vertices of adjoining elements coincide, and correspondingly a C^0 functional condition is satisfied at the elemental interfaces. In this chapter, we consider again *second-order* spatial operators, as in Chapter 5, but will allow for non-conforming discretisations. That is, we will no longer require that the vertices of adjoining elements coincide. Instead, we will develop a framework that allows for arbitrary connections between elements.

In the first part of the chapter we introduce two formulations that employ geometrically non-conforming elements but which maintain the C^0 continuity of the global polynomial expansion. We refer to them as the *iterative patching* and the *constrained approximation*. In the second part, we introduce the *mortar element* and the *discontinuous Galerkin method* for second-order elliptic and parabolic problems. In these cases the C^0 continuity is no longer imposed and new weak forms of the problem are developed. Some interesting possibilities exist with the discontinuous Galerkin method for second-order elliptic and parabolic problems, and so we pay particular attention to this formulation.

Examples and notation

To illustrate the differences in discretisation flexibility, we revisit the example for flow past a half-cylinder, which was first presented in Chapter 1. The mesh for the conforming discretisation is shown in Fig. 7.1(b). The total number of elements shown in this conforming discretisation is $N_{\rm el} = 276$. Alternatively, ig-

(a)

(b)

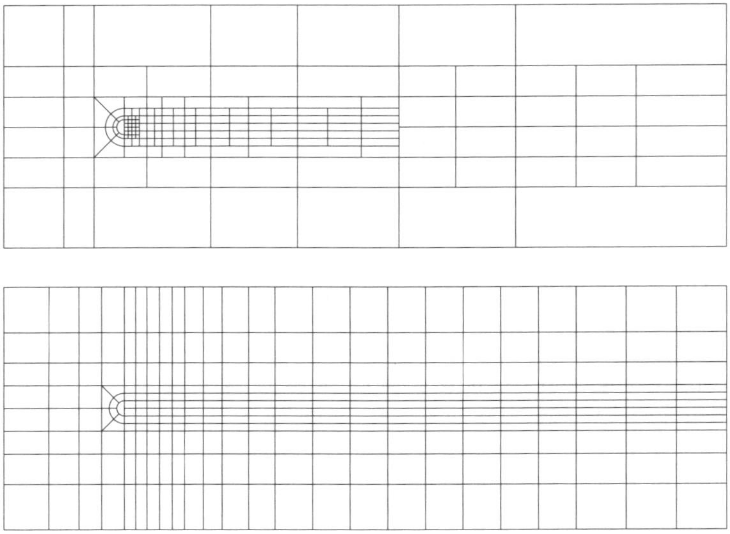

FIG. 7.1. (a) Non-conforming mesh, and (b) conforming mesh for flow past a half-cylinder. ([229].)

noring geometric compatibility between elements, an intuitively better elemental decomposition is achieved in Fig. 7.1(a). In this mesh more elements of smaller size have been added in the near-wake, whereas larger elements have been placed on the side boundaries where the flow is expected to be more uniform. The total number of elements in this case is now $N_{\text{el}} = 176$. The differences in the quality of the corresponding two solutions are shown in Fig. 1.5 in terms of vorticity contours (obtained as the curl of the computed velocity field). The noisy solution on the conforming mesh is due to the under-resolution of the smaller-scale structures in the near-wake. The corresponding computational time for the non-conforming simulation is smaller if proper solution solvers are employed, as discussed in [228] where more details for this problem can be found.

The mesh shown in Fig. 7.1(a) is an example of a non-conforming discretisation, and by that we refer to the *geometric* incompatibility. We also use the same term to refer to *functional* incompatibility between adjacent elements, where a different number of edge degrees of freedom are used on either side of two adjoining elements. Alternatively, one element may be split into two or more elements, as in the examples of Fig. 7.2. Some of the numerical issues associated with either case are similar and we address them both under a unified formulation. To simplify the presentation we consider a model problem as shown in Fig. 7.2(a), where the domain Ω is decomposed into $N_{\text{el}} = 4$ non-overlapping subdomains, that is,

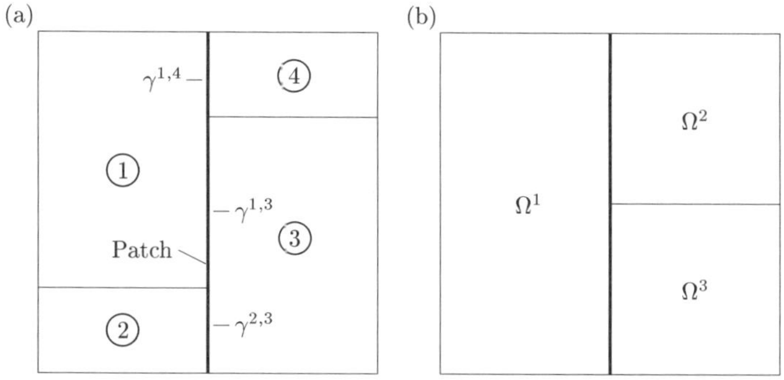

FIG. 7.2. Non-conforming mesh for the model problem: (a) general, and (b) simple splitting.

$$\Omega = \cup_{e=1}^{N_{\rm el}} \Omega^e\,,$$
$$\Omega^{e_1} \cap \Omega^{e_2} = \emptyset\,, \quad 1 \leqslant e_1 < e_2 \leqslant N_{\rm el}\,.$$

To complete the notation we also introduce the elemental segments as intersections of adjacent elemental boundaries, that is, $\gamma^{k,l} = \partial\Omega^k \cap \partial\Omega^l$, where the interpretation is shown schematically in Fig. 7.2; we will refer to these segments as *mortars* γ^m, where $m(k,l)$ denotes a unique index of the pair (k,l). We also define the *patch* as a set that includes all mortars in a non-conforming contiguous interface. For example, in Fig. 7.2(a) we only have one patch but we have three elemental interfaces. Finally, for a quadrilateral mesh the 'skeleton' is defined by $\mathcal{S} = \cup_{e=1}^{N_{\rm el}} \cup_{j=1}^{4} \Gamma^{e,j}$, where $\Gamma^{e,j}$ is the edge j of element e and $N_{\rm el}$ is the total number of elements. For a triangular element the summation over edges is appropriately modified.

7.1 Interface conditions and implementation

Next we revisit the elliptic problem first considered in Chapter 2 in order to examine the imposed elemental interface conditions. In particular, we want to investigate what is the loss of accuracy (if any) of various ways of imposing the weak C^0 continuity condition and examine if there is any gain in accuracy by restoring a C^1 condition. In its strong form, the second-order elliptic problem requires continuity of function values as well as first derivatives. For the sake of simplicity, we consider a discretisation on only two subdomains (elements) Ω^1 and Ω^2, as shown in Fig. 7.3.

We consider three forms of the discrete system corresponding to a Helmholtz equation with homogeneous data:

$$\nabla^2 u - \lambda^2 u + f(\boldsymbol{x}) = 0 \quad \text{in } \Omega\,, \tag{7.1.1a}$$
$$u = 0 \quad \text{on } \partial\Omega\,. \tag{7.1.1b}$$

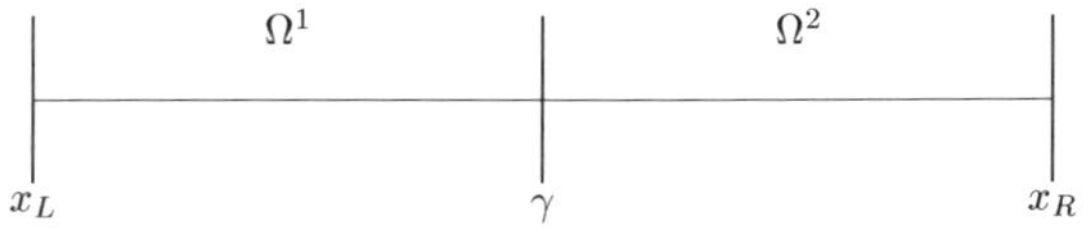

FIG. 7.3. Domain partitioning in a one-dimensional example.

Consider the one-dimensional version of eqn (7.1.1a) as first discussed in Chapter 2. We recall the one-dimensional bilinear operator for the Helmholtz equation

$$a(u,v) = \left(\frac{\partial u}{\partial x}, \frac{\partial v}{\partial x}\right) + \lambda^2 (u,v) .$$

The first formulation (method A) is the standard Galerkin formulation presented in Chapter 2 and may be written in the following form.

Method A

$$\mathop{\mathcal{A}}_{e=1}^{N_{\mathrm{el}}} a(v^e, u^e) = \mathop{\mathcal{A}}_{e=1}^{N_{\mathrm{el}}} (v^e, f) \quad \text{in } \Omega^e , \tag{7.1.2}$$

where u^e (with $e = 1, 2$) is the numerical solution and v^e are the test functions. The C^0 continuity condition is enforced through the global assembly operator $\mathcal{A}_{e=1}^{N_{\mathrm{el}}}$ described in Section 4.2.1.

The next formulation is of mixed type and combines method A with a weak C^1 condition.

Method B

$$a(v^1, u^1) - \left(v^1, \frac{\partial u^2}{\partial x}\right) = (v^1, f) \quad \text{in } \Omega^1 , \tag{7.1.3a}$$

$$- \left(v^2, \frac{\partial u^1}{\partial x}\right) + a(v^2, u^2) = (v^2, f) \quad \text{in } \Omega^2 . \tag{7.1.3b}$$

This system enforces C^0 continuity and, in addition, it enforces explicitly the C^1 condition on each element in the variational setting. The final formulation, in addition to the C^0 condition, also enforces the C^1 condition explicitly on each element but in a pointwise collocation manner.

Method C

$$a(v^1, u^1) + a(v^2, u^2) = (v^1, f) + (v^2, f) \quad \text{in } \Omega^1 \cup \Omega^2 , \tag{7.1.4a}$$

$$\left.\frac{\partial u^1}{\partial x}\right|_{\gamma} = \left.\frac{\partial u^2}{\partial x}\right|_{\gamma} . \tag{7.1.4b}$$

For all three formulations the test functions $v \in \mathcal{V} \subset H^1(\Omega)$ satisfy homogeneous boundary conditions, as defined in Chapter 2. Implementation of the boundary condition at the interface point γ for methods B and C can be obtained directly by formulating the discrete equations as a coupled system or iteratively, as we will see in the next section.

These three schemes were applied to solve the one-dimensional Helmholtz equation version of eqn (7.1.1a) with $\lambda^2 = 1$. The right-hand side is chosen such that the exact solution is given by $u = e^{-x^2} \cos(7\pi x/2)$ and appropriate forcing over the domain $-1 \leqslant x \leqslant 1$. The domain is split into two elements of equal size. Figure 7.4 shows convergence of the discrete solution with respect to the total degrees of freedom in the mesh, which is refined by increasing the polynomial order.

A solution obtained on a single domain is shown for reference as it represents the corresponding global spectral solution using Legendre polynomials $L_p(\xi)$; as expected, it achieves the best convergence rate. The three multi-domain approximations have the same asymptotic convergence rate, but method A is optimal and gives a slightly better approximation for a given mesh than either the mixed formulation or the collocation method. Of these latter two, the mixed formulation is both more accurate and more economical because of the weaker coupling.

The results of these numerical experiments are consistent with the theoretical

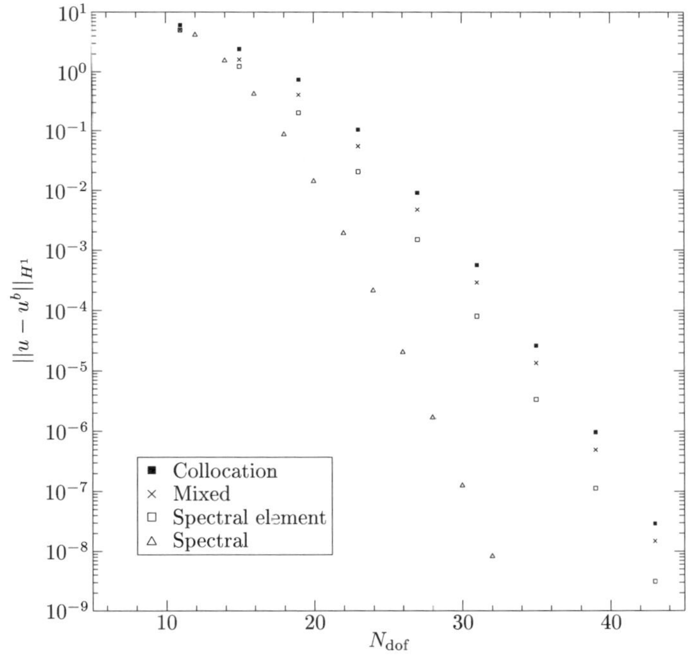

FIG. 7.4. Convergence in the H^1 norm of methods A (spectral element), B (mixed), and C (collocation). Also included for reference are results from a global (Legendre) spectral method. ([229].)

results of Bernardi *et al.* [44] in coupling spectral and finite element methods for the Laplace equation. It was shown that the derived error estimates are best for the *integral matching* constraint and not for the pointwise matching. We will revisit this issue in Section 7.4.

7.2 Iterative patching

Implementation of the formulations of methods B and C of the previous section can be obtained using efficient iterative techniques in the spirit of the alternating Schwarz algorithm for spectral methods [85]. The original Schwarz algorithm employs overlapping domains, but other versions [305] employ non-overlapping subdomains. Here we present a version first proposed by Funaro *et al.* [173] and modified in [227].

7.2.1 *One-dimensional discretisation*

Let us consider the domain shown in Fig. 7.3 where we designate arbitrarily the master domain as $\Omega^1 \equiv \Omega^m$ with length l_m and the slave domain as $\Omega^2 \equiv \Omega^s$ of length l_s. Let us also denote by γ the interface point between the two subdomains. The problem we want to solve in the entire domain $\Omega = \Omega^m \cup \Omega^s$ is as follows:

$$-u_{xx} + \lambda^2 u = f(x)\,, \tag{7.2.1a}$$

$$u(x_L) = u(x_R) = 0\,. \tag{7.2.1b}$$

To this end, we set up an iteration scheme that involves a Dirichlet and a Neumann problem assigned to the master and slave domains, respectively. In the following the iteration number is denoted by n.

Master domain: Dirichlet problem

$$-(u_{xx})_m^n + \lambda^2 u_m^n = f(x) \quad \text{on } \Omega^m\,, \tag{7.2.2a}$$

$$u(x_L) = 0\,, \tag{7.2.2b}$$

$$u_m^n(\gamma) = u_s^{n-1}(\gamma)\,. \tag{7.2.2c}$$

Slave domain: Neumann problem

$$-(u_{xx})_s^n + \lambda^2 u_s^n = f(x) \quad \text{on } \Omega^s\,, \tag{7.2.2d}$$

$$u(x_R) = 0\,, \tag{7.2.2e}$$

$$(u_x)_s^n(\gamma) = (u_x)_m^k(\gamma)\,. \tag{7.2.2f}$$

In eqn (7.2.2f) the index k could take two values depending on whether there is a serial or parallel implementation. These values are: (1) *serial* version: $k = n$; and (2) *parallel* version: $k = n - 1$. The convergence properties of the two versions are somewhat different, as we will see next. The following simple theorem holds for this scheme.

Convergence theorem 7.2.1 *The iterative procedure (7.2.2a)–(7.2.2f) converges if and only if $l_m/l_s > 1$.*

This is easily demonstrated by analysing the errors at iteration 'n' in the master ε_m^n and slave ε_s^n domains, which are defined as follows:

$$\varepsilon_m^n(x) = u_m^n(x) - u_m^{n-1}(x),$$
$$\varepsilon_s^n(x) = u_s^n(x) - u_s^{n-1}(x).$$

The errors satisfy the homogeneous equation corresponding to eqns (7.2.2a) and (7.2.2d), and therefore the corresponding solutions are linear functions in x. On the slave side, we obtain at the interface point

$$\varepsilon_s^n(\gamma) = -G(\lambda)\,\varepsilon_s^{k-1}(\gamma), \tag{7.2.3}$$

where the growth factor $G(\lambda)$ is defined as

$$G(\lambda) = \begin{cases} \dfrac{l_s}{l_m}, & \lambda = 0, \\ \dfrac{\tanh(\lambda l_s)}{\tanh(\lambda l_m)}, & \text{otherwise}, \end{cases}$$

and λ^2 is a constant in eqn (7.2.1a). For convergence, we require that the growth factor be less than unity for $l_m/l_s > 1$, for either case. In the *parallel* version the convergence rate is half of the convergence rate of the serial case as $Z_P^2 = Z_S = G(\lambda)$, where we define $Z \equiv \varepsilon^n/\varepsilon^{n-1}$. Here we assume that, if the solution at the interface converges as the number of iterations increases, then the same is true for the solution in the entire domain. For a rigorous proof of this assertion we refer the reader to Marini and Quarteroni [323].

This dependence on the relative size of the domains is a serious constraint for practical applications. It is associated with the Dirichlet (essential) boundary condition imposed at the interface γ. To circumvent this difficulty, Funaro *et al.* [173] proposed the following procedure to relax this interface boundary condition:

$$u_m^n(\gamma) = \theta u_s^{n-1}(\gamma) + (1-\theta)u_m^{n-1}(\gamma), \tag{7.2.4}$$

where $\theta \in (0, 1]$ (the semi-open interval) is the relaxation parameter; for $\theta = 1$ we obtain the original unrelaxed iteration scheme. The corresponding error equation at the interface on the master side can now be obtained for both the serial and parallel versions as before by incorporating eqn (7.2.4). Here, we write the final error equation on the master side in terms of the convergence rate as

$$\text{serial:} \quad (Z_m)_S^{n+1} = -\theta G(\lambda) + (1-\theta), \tag{7.2.5a}$$
$$\text{parallel:} \quad (Z_m)_P^2 - (\theta - 1)(Z_m)_P + G(\lambda)\theta = 0, \tag{7.2.5b}$$

where we recall that the subscript m refers to the master domain.

Therefore, we see that convergence (i.e., $|Z| < 1$) for the *serial* algorithm is guaranteed for any size of domains as long as the relaxation parameter θ is chosen so that

$$0 < \theta < \Theta_S = \frac{2}{1 + G(\lambda)}\,. \tag{7.2.6}$$

For the *parallel* algorithm we see from eqn (7.2.5b) that, for $\theta \in (0, 1]$, we always obtain two (positive) Z roots smaller than unity, and therefore convergence is guaranteed for any under-relaxed case (that is, $\theta < 1$).

For both the serial and parallel iteration schemes there is an optimum value of the relaxation parameter that makes the error identically zero in a finite number of iterations. In particular, for the serial case we can use eqn (7.2.5a) to set $\varepsilon_m^{n+1} = 0$ by taking the relaxation parameter to be

$$(\theta_S)_{\text{opt}} = \frac{\Theta_S}{2}\,, \quad n \geqslant 2\,, \tag{7.2.7}$$

which implies that the serial iterative scheme converges after two steps. For the parallel scheme the same type of analysis using eqn (7.2.5b) yields

$$(\theta_P)_{\text{opt}} = \frac{\varepsilon_m^{n-1}(\gamma)}{\varepsilon_m^{n-1}(\gamma) + G(\lambda)\varepsilon_m^{n-2}(\gamma)}\,, \quad n \geqslant 3\,, \tag{7.2.8}$$

which implies that the parallel iterative scheme converges after three steps. From this formula it is obvious that the optimum value for the parallel scheme is changing in each iteration, unlike the serial algorithm which uses a fixed $(\theta_S)_{\text{opt}}$. However, a similar dynamic relaxation parameter can also be chosen for the serial scheme by taking $\theta_S^2 = \theta_{\text{opt}}$ during the second iteration, that is,

$$\theta_S^n = \frac{u_m^n(\gamma) - u_m^{n-1}(\gamma)}{(u_m^n(\gamma) - u_m^{n-1}(\gamma)) - (u_s^n(\gamma) - u_s^{n-1}(\gamma))}\,, \quad n \geqslant 2\,. \tag{7.2.9}$$

In this case, the exact solution is obtained after three steps, independently of the choice of the initial guess for θ_S^1. This choice of θ_S^n corresponds to the minimisation of the interface error $|u_m^{n+1}(\gamma) - u(\gamma)|$ and, in fact, it zeros the corresponding error, that is, $\varepsilon_m^{n+1}(\gamma) = 0$, for $n \geqslant 2$.

7.2.2 *Two-dimensional discretisation*

In the one-dimensional iteration procedure the dynamic relaxation parameter is chosen so that it minimises the distance $||S^n(\theta) - S^{n-1}(\theta)||$, where the sequence S^n is defined as the relaxed Dirichlet boundary condition at the interface, that is,

$$S^n(\theta) = \theta u_s^n(\gamma) + (1 - \theta)u_m^n(\gamma)\,. \tag{7.2.10}$$

This interpretation given by Funaro *et al.* [173] is useful because it extends readily to multiple dimensions, where an analytical formula for an optimum relaxation parameter is not available. More specifically, in two dimensions we now impose the interface conditions on γ as follows:

$$u_m^n = \theta u_s^{n-1} + (1-\theta)\mathcal{I}_m^s u_m^{n-1}\,, \tag{7.2.11a}$$

$$\boldsymbol{n}\cdot\nabla u_s^n = -\mathcal{I}_s^m\,\boldsymbol{n}\cdot\nabla u_m^k\,, \tag{7.2.11b}$$

where $\boldsymbol{n}$ is the outwards unit normal and $\mathcal{I}$ is a general interpolation operator between the two sides of the interface; for example, $\mathcal{I}_s^m$ interpolates from the slave to the master domain. Here again, $k = n$ for the serial case and $k = n-1$ for the parallel case. We examine, first, convergence with the fixed relaxation parameter and subsequently we consider the dynamic relaxation procedure.

Convergence theorem 7.2.2. (Serial algorithm) *The iterative procedure for two dimensions with the interface conditions of eqns (7.2.11a) and (7.2.11b) and $k = n$ converges, irrespective of the original guess $u_m^0(\gamma)$ for a fixed relaxation parameter θ, if and only if*

$$0 < \theta < \Theta(\lambda_q) = \frac{2}{1 + G(\lambda_q)}\,, \tag{7.2.12}$$

where

$$\lambda_q^2 = \lambda^2 + \frac{q^2\pi^2}{4}\,, \quad q \geqslant 1\,. \tag{7.2.13}$$

Convergence theorem 7.2.3. (Parallel algorithm) *The iterative procedure for two dimensions with the interface conditions of eqns (7.2.11a) and (7.2.11b) and $k = n-1$ converges, irrespective of the original guess $u_m^0(\gamma)$ for a fixed relaxation parameter θ, if and only if*

$$\theta \in (0, 1]\,.$$

Therefore, the convergence properties of the two-dimensional algorithm are similar to the one-dimensional algorithm, with the incorporation of the frequency components q. For the special case where the two subdomains are identical ($l_m = l_s$), the serial version of the algorithm converges exactly after two steps if $\theta = \frac{1}{2}$. The proof for stability of the serial version in the multi-dimensional case is based on eigenfunction analysis and separation of variables [173].

The *dynamic* relaxation parameter is now chosen by minimising the distance $||S^n(\theta) - S^{n-1}(\theta)||^2$, where $S^n = u_m^{n+1}$ is obtained from eqns (7.2.11a) and (7.2.11b). This gives the following formula:

$$\theta^n = \frac{(\varepsilon_m^n, \varepsilon_m^n - \varepsilon_s^n)}{||\varepsilon_m^n - \varepsilon_s^n||^2}\,, \quad n \geqslant 2\,. \tag{7.2.14}$$

If any of the errors $\varepsilon_m^{n+1}(\gamma)$ or $\varepsilon_s^{n+1}(\gamma)$ exceeds a specified tolerance, then a new relaxation parameter is computed using θ^n and a new iteration begins. However, unlike the one-dimensional case, this choice of θ does not minimise the true interface error $||u_m^{n+1}(\gamma) - u(\gamma)||_{L^2}$. If the relaxation parameter is chosen dynamically from eqn (7.2.14) then convergence is guaranteed for both the serial and parallel algorithms. There is a possibility that some values of the dynamic parameter may violate the strict stability conditions, but numerical experiments [227] show that even in these cases convergence is obtained. As in the fixed relaxation case for equal subdomains, for $\theta^n = \frac{1}{2}$ the serial algorithm terminates after three steps, while the parallel algorithm terminates a step later. However, unlike the one-dimensional case, in general, there is no optimum parameter.

7.2.3 *Variational formulation*

We first consider the conforming discretisation, where we define the approximation space $\mathcal{X} \subset H^1(\Omega)$ and the test functions in spaces $\mathcal{V}^e$ associated with the subdomains/elements as follows:

$$\mathcal{V}^e = \left\{v \in H^1(\Omega^e),\ v|_\gamma = 0\right\},$$

and thus $\mathcal{V}_0^e \subset H_0^1(\Omega^e)$. We also define the space of continuous functions on the patch as

$$\mathcal{W} = \left\{w|_\gamma : w \in H^1(\Omega)\right\}$$

and the elemental bilinear form $a^e(\cdot,\cdot)$ corresponding to eqn (7.1.1a), that is,

$$a^e(v,u) = \int_{\Omega^e} [\nabla v\, \nabla u + \lambda^2 u\, v]\, \mathrm{d}\boldsymbol{x}\,.$$

We define the subdomain scalar product $(\cdot,\cdot)_e$ similarly. We can now write the variational form of eqn (7.1.1a) for the master and slave domains connected by a single segment γ as

$$a^m(v, u_m^{n+1}) = (v,f)_m\,, \quad \forall\, v \in \mathcal{V}_0^m\,, \tag{7.2.15a}$$

$$u_m^{n+1}|_\gamma = S^n(\theta)\,, \tag{7.2.15b}$$

$$a^s(v, u_s^{n+1}) = (v,f)_s - a^m(R^m v|_\gamma, u_m^k) + (R^m v|_\gamma, f)_m\,, \quad \forall\, v \in \mathcal{V}_0^s\,, \tag{7.2.15c}$$

where we define the extension of $w \in \mathcal{W}$ to Ω^m as follows.

Find $R^m w \in \mathcal{V}^m$:

$$a^m(R^m w, v) = 0\,, \quad \forall\, v \in \mathcal{V}_0^m\,, \tag{7.2.15d}$$

$$R^m w = w \quad \text{on } \gamma\,. \tag{7.2.15e}$$

Based on this definition of solution extensions, Marini and Quarteroni [323] derived an optimum value for the relaxation parameter, which can be used with fixed iteration or dynamic iteration as before. Then eqn (7.2.10) is used to update the interface value $S^n(\theta)$. In particular, they proposed the following formula for convergence acceleration:

$$\theta_{\text{opt}} = \frac{\tau + 1}{\sigma^2\tau + \tau + 2}\,, \tag{7.2.16}$$

where the parameters σ and τ are defined from

$$\sigma = \sup\left\{\frac{||R^m w||_1^2}{||R^s w||_2^2},\ w \in \mathcal{W}\right\},$$

$$\tau = \sup\left\{\frac{||R^s w||_2^2}{||R^m w||_1^2},\ w \in \mathcal{W}\right\}.$$

In the dynamic relaxation procedure, the parameter θ^n is chosen so that it converges to θ_{opt} obtained from eqn (7.2.16) for an arbitrary initial guess θ^1. A specific algorithm that involves θ^n is presented in Marini and Quarteroni [323].

The correct construction of the interpolation operators $\mathcal{I}_m^s$ and $\mathcal{I}_s^m$ in eqns (7.2.11a) and (7.2.11b) is crucial in obtaining exponential convergence. For the Dirichlet boundary conditions the interpolation operator should have the form $\mathcal{I}_s^m = h^{P_s}(\xi^m)$, where $h^P(\xi)$ denotes the Lagrangian interpolation polynomial of order P, and ξ^m are the coordinates of the points on the interface γ on the master side. As an example, let us consider two elements, only connected at the interface γ, but corresponding to different orders, P_m and P_s. To honour the C^0 condition in this case (and thus avoid any consistency errors due to the non-conformity in spaces) we have to impose that the order P_m on the master side be larger than the order P_s on the slave side. This is consistent with the aforementioned choice of the interpolation operator for Dirichlet data. However, if $P_m < P_s$ then convergence is much slower.

In general, however, for a non-conforming geometric discretisation space conformity is not guaranteed, and thus the C^0 condition cannot be honoured unless the method of constrained approximation is followed, as discussed in the next section. This implies that there is a consistency error due to non-conformity of the approximation space.

The strength of this approach lies in the ease with which the patching condition can be incorporated into an existing code as nothing more than a 'special' boundary condition. Also, the interpolation operator can be constructed for a wide class of discretisations, including hybrid spectral element–finite difference schemes; for further details see [227].

One drawback associated with the patching conditions (7.2.11a) and (7.2.11b) is the increased complexity of the coupling (*element*wise versus *point*wise) that makes it very difficult to implement other than as an iterative procedure. Also, to guarantee convergence a patch must terminate on an external boundary. While this is always possible, it does place some restrictions on mesh construction. Finally, because each iteration requires the solution of a Helmholtz equation on every subdomain, the method can be computationally expensive unless the solution phase is properly balanced by computing the solution updates in parallel.

7.2.4 *Interpretation of the relaxation procedure*

Let us now try to interpret the relaxation procedure outlined in the previous section in terms of more standard relaxation schemes; we follow here, again, the work of Marini and Quarteroni [323]. To this end, we set up the solution algorithm in matrix form in terms of the unknowns $\boldsymbol{U}_m$, $\boldsymbol{U}_s$, and $\boldsymbol{U}_\gamma$ on the master, slave, and interface domains, respectively. The corresponding indices are m, s, and γ and $\boldsymbol{F}$ denotes the known forcing:

$$\begin{bmatrix} \boldsymbol{A}_{mm} & 0 & \boldsymbol{A}_{m\gamma} \\ 0 & \boldsymbol{A}_{ss} & \boldsymbol{A}_{s\gamma} \\ \boldsymbol{A}_{m\gamma}^\top & \boldsymbol{A}_{s\gamma}^\top & \boldsymbol{A}_{\gamma\gamma} \end{bmatrix} \begin{bmatrix} \boldsymbol{U}_m \\ \boldsymbol{U}_s \\ \boldsymbol{U}_\gamma \end{bmatrix} = \begin{bmatrix} \boldsymbol{F}_m \\ \boldsymbol{F}_s \\ \boldsymbol{F}_\gamma \end{bmatrix},$$

where the interface matrix $\boldsymbol{A}_{\gamma\gamma}$ can be split into two contributions due to the master and slave domains, that is, $\boldsymbol{A}_{\gamma\gamma} = \boldsymbol{A}_{\gamma\gamma}^m + \boldsymbol{A}_{\gamma\gamma}^s$. In matrix form, the Dirichlet problem on the master side is obtained from

$$\boldsymbol{U}_m^{n+1} = \boldsymbol{A}_{mm}^{-1}(\boldsymbol{F}_m - \boldsymbol{A}_{m\gamma}\boldsymbol{U}_\gamma^n),$$

assuming that $\boldsymbol{U}_\gamma^n$ is known from iteration n. Eliminating the master unknowns, we obtain the subsystem

$$\begin{bmatrix} \boldsymbol{A}_{ss} & \boldsymbol{A}_{s\gamma} \\ \boldsymbol{A}_{s\gamma}^\top & \boldsymbol{A}_{\gamma\gamma} \end{bmatrix} \begin{bmatrix} \boldsymbol{U}_s \\ \boldsymbol{U}_\gamma^{n+1/2} \end{bmatrix} = \begin{bmatrix} \boldsymbol{F}_s \\ \boldsymbol{F}_\gamma - \boldsymbol{A}_{m\gamma}^\top \boldsymbol{U}_m^k - \boldsymbol{A}_{\gamma\gamma}^m \boldsymbol{U}_\gamma^n \end{bmatrix},$$

from which we can obtain intermediate values for the array $\boldsymbol{U}_\gamma^{n+1/2}$:

$$\boldsymbol{S}_s\boldsymbol{U}_\gamma^{n+1/2} = \boldsymbol{F}^* - \boldsymbol{S}_m\boldsymbol{U}_\gamma^n,$$

where $\boldsymbol{F}^*$ refers to the rearranged right-hand side. Here $\boldsymbol{S}_m$ and $\boldsymbol{S}_s$ are matrices resulting from condensing the 2×2 system and are defined as

$$\boldsymbol{S}_m = \boldsymbol{A}_{\gamma\gamma}^m - \boldsymbol{A}_{m\gamma}^\top|\boldsymbol{A}_{mm}^{-1}\boldsymbol{A}_{m\gamma},$$

and similarly for $\boldsymbol{S}_s$. Using the relaxation presented in eqn (7.2.10), we can now update the interface values at the current iteration level, $\boldsymbol{U}_\gamma^{n+1} = \theta^n\boldsymbol{U}_\gamma^{n+1/2} + (1-\theta^n)\boldsymbol{U}_\gamma^n$, which after substitution can be rewritten as

$$\boldsymbol{U}_\gamma^{n+1} = \theta^n[\boldsymbol{S}_s^{-1}(\boldsymbol{F}^* - \boldsymbol{S}_m\boldsymbol{U}_\gamma^n)] + (1-\theta^n)\boldsymbol{U}_\gamma^n. \tag{7.2.17}$$

We can contrast this with the form $(\boldsymbol{S}_m + \boldsymbol{S}_s)\boldsymbol{U}_\gamma = \boldsymbol{F}^*$ or, equivalently, $\boldsymbol{U}_\gamma^{m+1} = \boldsymbol{S}_s^{-1}(\boldsymbol{F}^* - \boldsymbol{S}_m\boldsymbol{U}_\gamma^n)$, resulting from the original 3×3 system. The comparison suggests that the iterative or Zanolli patching can be thought of as an under-relaxation method or as a Richardson method with $\boldsymbol{S}_s$ as preconditioner.

Similarly, by rewriting eqn (7.2.17) as

$$\boldsymbol{S}_s(\boldsymbol{U}_\gamma^{n+1} - \boldsymbol{U}_\gamma^n) = \theta^n(\boldsymbol{F}^* - \boldsymbol{S}\boldsymbol{U}_\gamma^n),$$

we also see the analogy with a conjugate gradient preconditioner for elliptic solvers, originally proposed by Bjorstad and Widlund [52].

7.3 Constrained approximation

The constrained approximation was introduced by Oden and his associates [127, 346], and [390] to deal with geometrically non-conforming discretisations introduced by refinement. The main idea is to maintain C^0 continuity across elemental interfaces by modifying the unconstrained basis functions appropriately. In other words, the approximation space is a constrained space which is still a subset of $H^1(\Omega)$ for second-order elliptic problems.

To illustrate this approach we consider a domain decomposed into three elements, with e_1 being the largest element, as shown in Fig. 7.5. This type of mesh is called irregular, with index of regularity one as just one node is 'floating' on one of the sides of the largest element. An element e_1 which is split into three smaller elements will correspond to an irregularity index of two, and so on. Meshes with a high index of irregularity are difficult to handle, and so are meshes such as the one shown in Fig. 7.2(a). Nevertheless, a great deal of flexibility is achieved, even for the two-to-one splitting rule as shown in Fig. 7.5.

For simplicity, let us consider bi-quadratic elements, that is, $N_{\rm eof} = 9$ for all the three elements, and denote the common side of element e_1 by ABC, corresponding to the nodal degrees of freedom on that side. The corresponding nodal degrees of freedom on the element are ADB and BEC for elements e_2 and e_3, respectively. We can enforce continuity in this case at the non-coinciding degrees of freedom D and E by simply setting

$$u^\delta(D) = \alpha_D u^\delta(A) + \beta_D u^\delta(B) + \gamma_D u^\delta(C)\,, \tag{7.3.1a}$$

$$u^\delta(E) = \alpha_E u^\delta(A) + \beta_E u^\delta(B) + \gamma_E u^\delta(C)\,, \tag{7.3.1b}$$

where α, β, and γ are appropriate weights. These coefficients can be considered as contributions to the assembly matrix $\boldsymbol{\mathcal{A}}$ described in Section 4.2.1, where previously we have only considered unit entries into this matrix. The number of global degrees of freedom corresponds to the expansion along the master element,

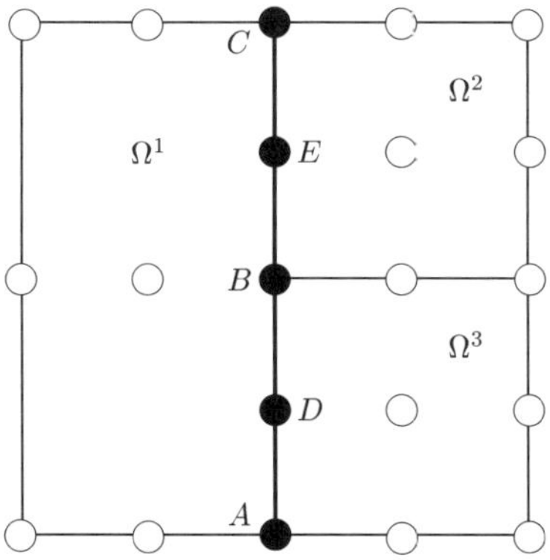

FIG. 7.5. Irregular mesh of index one.

which for the case shown in Fig. 7.5 is three. For higher-order approximations, which are typical in *hp* methods, a more systematic procedure can be established which can also handle different polynomial orders on adjacent elements.

To construct the constrained approximation space $\mathcal{X}^\delta$ we can separate the global (that is, over all elements) degrees of freedom into active N^a_{dof} and constrained N^c_{dof}. In the example above $N^a_{\text{dof}} = 19$ and $N^c_{\text{dof}} = 2$. The constrained degrees of freedom can be thought of as linear combinations of active degrees of freedom, and correspondingly their functional can be written as linear combinations of the functionals of active nodes. Following Demkowicz *et al.* [127], we assume that each constrained degree of freedom v^δ_i, where $i \in N^c_{\text{dof}}$, is influenced by a set of nodes $I(i) \subset N^a_{\text{dof}}$; we also introduce its 'inverse' set $J(i) = \{j \in N^c_{\text{dof}} | i \in I(j)\}$. The linear dependence is expressed as

$$v^\delta_i = \sum_{j \in I(i)} r_{ij} v^\delta_j \,, \quad \forall\, i \in N^c_{\text{dof}} \,. \tag{7.3.2}$$

The space of constrained approximations can then be defined as the subspace of the unconstrained approximation space $\mathcal{X}^\delta$ whose elements satisfy eqn (7.3.2) and are chosen in such a way that they are continuous functions. The corresponding basis functions can be easily obtained from the basis functions of the unconstrained approximation ϕ. We start by expressing $u^\delta \in \mathcal{X}^\delta$ in terms of ϕ, that is,

$$u^\delta = \sum_{i \in N^a_{\text{dof}}} u_i \phi_i + \sum_{j \in N^c_{\text{dof}}} u_j \phi_j = \sum_{i \in N^a_{\text{dof}}} u_i \phi_i + \sum_{j \in N^c_{\text{dof}}} \sum_{k \in I(j)} r_{jk} u_k \phi_j \,.$$

Using the definition of the inverse set $J(i) \subset N^c_{\text{dof}}$, we can rewrite the above expression as

$$u^\delta = \sum_{i \in N^a_{\text{dof}}} u_i \left[\phi_i + \sum_{j \in J(i)} r_{ji} \phi_j \right] = \sum_{i \in N^a_{\text{dof}}} u_i \tilde{\phi}_j \,,$$

and therefore we obtain for the constrained approximation basis:

$$\tilde{\phi}_i = \phi_i + \sum_{j \in J(i)} r_{ji} \phi_j \,, \quad \forall\, i \in N^a_{\text{dof}} \,.$$

Based on this equation, the modified right-hand-side vector and the stiffness matrix can be efficiently constructed using appropriate data structures and data management schemes, as demonstrated in [127, 346, 390]. The specific values of the elements r_{ij} are computed by matching degrees of freedom between a one-dimensional edge which is split into two smaller edges, as shown in Fig. 7.5.

7.4 Mortar patching

The method that we describe in this section was first introduced by Patera and his associates, see [10, 324], who coined the term 'mortar element method' because the discretisation introduces a set of functions that mortar the brick-like elements together; see also [46]. The method is general and allows for the coupling of variational discretisations of different types in non-overlapping domains, that is, the non-conformity may be due to geometry, approximation spaces, or both.

7.4.1 *Projection and non-conforming spaces*

In the mortar patching method, interface conditions are imposed variationally through an L^2-minimisation condition. We therefore introduce a new approximation space $\mathcal{X}^\delta$ to facilitate such a projection, which in the case of conformity in geometry and polynomial order along elemental interfaces reduces to a subspace of $H^1(\Omega)$. However, for non-conforming discretisations this is not, in general, true any more. The problem is therefore to estimate and bound the consistency error, which measures the deviation of $\mathcal{X}^\delta$ from the space of continuous functions $H^1(\Omega)$.

Associated with this method is a new space that contains the mortar functions, which are continuous polynomials on the spectral element skeleton $\mathcal{S}$ containing all mortar segments, that is,

$$\mathcal{W}^\delta = \left\{ w \mid w \in C^0(\mathcal{S}) \text{ and } w_{|\gamma^m} \in \mathcal{P}_P(\gamma^m) \right\}. \tag{7.4.1}$$

The mortar functions are also homogeneous on the domain boundary, that is, $w_{|\partial\Omega} = 0$. The discrete approximation space $\mathcal{X}^\delta$ is the space of functions v^δ which satisfy the essential boundary conditions, and whose restriction $(v^\delta)_{|\Omega^e}$ belongs to a space which is a subset of $H^1(\Omega^e)$. In addition, the functions v^δ satisfy the following integral and vertex conditions, respectively:

$$\int_{\Gamma^{e,j}} [(v^\delta)_{|\Omega^e} - w^m]\, \psi \, \mathrm{d}s = 0\,, \quad \forall\, \psi \in \mathcal{P}_{P-2}(\Gamma^{e,j})\,, \tag{7.4.2a}$$

$$[(v^\delta)_{|\Gamma^{e,j}} - w^m]_{|s=0,\bar{\gamma}^m} = 0\,, \tag{7.4.2b}$$

where $s = 0, \bar{\gamma}^m$ are the end-points of the mortar γ^m. The superscript m corresponds to the element/edge pair (e, j). Note that, since we use the two vertex conditions at the end-points of the mortar, we need to reduce by 2 the order of the test functions $\psi(s)$. By requiring eqn (7.4.2a) to hold for every $\psi \in \mathcal{P}_{P-2}(\Gamma^{e,j})$, we effectively solve for an L^2-minimisation of the jump in function values across the patch. Combined with the C^0 condition associated with the conforming regions, eqn (7.4.2b) ensures that the mortar functions are C^0 along a patch. Consistent with this vertex condition, the test functions should satisfy $\psi(0) = \psi(\bar{\gamma}^m) = 0$. The combination of this L^2 projection expressed by the integral constraint as well as the continuity at the end-points of a mortar is effectively equivalent to an H^1 projection. This interpretation is discussed in [324] and it is based on the theoretical proofs in [44].

The elements of $\mathcal{X}^\delta$ are not in $H^1(\Omega)$ but are locally in $H^1(\Omega^e)$, and therefore we need to introduce local ('broken') norms as follows:

$$H^*(\Omega) = \{v \mid v \in L^2(\Omega),\ v_{|\Omega^e} \in H^1(\Omega^e)\}\,, \tag{7.4.3a}$$

$$\forall\, v \in H^*(\Omega)\,, \quad ||v||_*^2 = \sum_{e=1}^{N_{\rm el}} ||v_{|\Omega^e}||^2_{H^1(\Omega^e)}\,. \tag{7.4.3b}$$

We also define the bilinear form

$$(u,v)_* = \sum_{e=1}^{N_{\rm el}} \int_{\Omega^e} \nabla u_{|\Omega^e} \nabla v_{|\Omega^e}\, \mathrm{d}\boldsymbol{x}\,, \tag{7.4.4}$$

for any functions $u, v \in H^1(\Omega)$. This local inner product satisfies the inequality $(v^\delta, v^\delta)_* \geqslant C||v^\delta||_*^2$, where C is a constant. The proof is given in [47, Proposition A1]. Having defined such a scalar product, we can then naturally define a projection operator $\mathbb{P}^\delta$ from $H^1(\Omega)$ onto $\mathcal{X}^\delta$, $\forall\, v \in H^1(\Omega)$, as follows:

$$\mathbb{P}^\delta v \in \mathcal{X}^\delta\,, \quad (v - \mathbb{P}^\delta v, u^\delta)_* = 0\,, \quad \forall\, u^\delta \in \mathcal{X}^\delta\,. \tag{7.4.5}$$

The projection error is bounded from above by the regularity of the local (elemental) restriction of the solution, as proved by Anagnostou *et al.* [11] for elements of general shape. The following theorem then holds.

Projection theorem 7.4.1 *Assuming that the function u belongs to $H^1(\Omega)$ and that its restriction $u_{|\Omega^e}$ belongs to $H^{\sigma_e}(\Omega^e)$ (with $\sigma_e \geqslant 1$), then*

$$||u - \mathbb{P}^\delta u||_* \leqslant C \sum_{e=1}^{N_{\rm el}} P^{1-\sigma_e} ||u_{|\Omega^e}||_{H^{\sigma_e}(\Omega^e)}\,. \tag{7.4.6}$$

The L^2 error estimate is similar, with the term $P^{-\sigma_e}$ replacing $P^{1-\sigma_e}$ in the inequality.

Similarly, theoretical results for approximating smooth functions also make use of local norms, as shown by Anagnostou *et al.* [11].

Approximation theorem 7.4.2 *Let us assume that the function $u \in H^1(\Omega)$ and that its restriction $u_{|\Omega^e} \in H^{\sigma_e}(\Omega^e)$ (with $\sigma_e > \frac{3}{2}$), then there exist $u^\delta \in \mathcal{X}^\delta$ such that*

$$||u - u^\delta||_* \leqslant C \sum_{e=1}^{N_{\rm el}} P^{1-\sigma_e} ||u_{|\Omega^e}||_{H^{\sigma_e}(\Omega^k)}. \tag{7.4.7}$$

Therefore, the approximation error depends only on the local regularity of the exact solution on each subdomain Ω^e.

7.4.2 *The discrete second-order problem*

The Poisson equation considered in eqn (7.1.1a) (with $\lambda = 0$) can be solved using the decomposition of the domain Ω into many non-overlapping non-conforming elements Ω^e. In variational form we have the following statement.

Find $u^\delta \in \mathcal{X}^\delta$ such that, $\forall\ v^\delta \in \mathcal{X}^\delta$,

$$\sum_{e=1}^{N_{\rm el}} \int_{\Omega^e} \nabla u^\delta \nabla v^\delta \, \mathrm{d}\boldsymbol{x} = \sum_{e=1}^{N_{\rm el}} \int_{\Omega^e} f(\boldsymbol{x}) v^\delta(\boldsymbol{x}) \, \mathrm{d}\boldsymbol{x} \, .$$

Next we perform a numerical quadrature of the integrals over Ω^e by Gauss–Lobatto (GL) quadrature, and therefore the variational statement is now as follows.

Find $u^\delta \in \mathcal{X}^\delta$ such that, $\forall\ v^\delta \in \mathcal{X}^\delta$,

$$a^\delta(u^\delta, v^\delta) \equiv \sum_{e=1}^{N_{\rm el}} (\nabla u^\delta, \nabla v^\delta)^\delta_{\Omega^e} = \sum_{e=1}^{N_{\rm el}} (f(\boldsymbol{x}), v^\delta(\boldsymbol{x}))^\delta_{\Omega^e} \, , \tag{7.4.8}$$

and such that

$$(u, v)^\delta_{\Omega^e} = \sum_{\rm GL} u(\boldsymbol{x}_i)\, v(\boldsymbol{x}_i)\, \rho_i J_i \, ,$$

where ρ_i denotes the two- or three-dimensional quadrature weights and J_i^e is the Jacobian of the mapping from Ω^e to the standard region evaluated at $\boldsymbol{x}_i$.

The well-posedness of the problem described in eqn (7.4.8) depends on the ellipticity of the discrete bilinear form $a^\delta(\cdot,\cdot)$ on $\mathcal{X}^\delta$, that is,

$$\forall\ v^\delta \in \mathcal{X}^\delta \, , \quad \sum_{e=1}^{N_{\rm el}} |v^\delta|^2_{H^1(\Omega^e)} \geqslant \alpha ||v^\delta||^2_* \, ,$$

for a positive constant α.

Similarly, the continuity condition holds on $\mathcal{X}^\delta$, that is,

$$\forall\ u^\delta, v^\delta \in \mathcal{X}^\delta \, , \quad a^\delta(u^\delta, v^\delta) \leqslant \gamma ||u^\delta||_* \, ||v^\delta||_* \, ,$$

and, by an extension of the Lax–Milgram theorem, we conclude that the problem described in eqn (7.4.8) is well posed and its solution is unique, that is,

$$||u^\delta||_* \leqslant \frac{\gamma}{\alpha} ||f||_{C^0(\Omega)} . \tag{7.4.9}$$

A rigorous proof is quite complicated and for the interested reader the details can be found in [47]. Here we provide a sketch of a proof for uniqueness of the solution u^δ and the associated mortar function w. Uniqueness of the mortar function is not necessary for uniqueness in the discrete solution, but it guarantees that no spurious modes will create any numerical difficulties in solving the combined discrete problem. We consider the simple non-conforming model problem of Fig.

7.2. Let us define $v^\delta = u_1^\delta - u_2^\delta$ as the difference of two possible solutions. From ellipticity (in the broken norm) we obtain

$$0 = (\nabla v^\delta, \nabla v^\delta)_* = \sum_{e=1}^{N_{\text{el}}} \int_{\Omega^e} (\nabla v^\delta_{|\Omega^e})^2 \, \mathrm{d}\boldsymbol{x} \,, \tag{7.4.10}$$

and therefore v^δ is piecewise constant. Moreover, $v^\delta \in \mathcal{X}^\delta$ is zero on the boundary $\partial\Omega$ and it satisfies the vertex continuity condition. These two constraints guarantee that $v^\delta \equiv 0$ everywhere, and therefore the two solutions u_1^δ and u_2^δ are identical.

To prove the uniqueness of w, we again recall that the conditions at the vertices, where the mortar function is identical to the solution, and thus we can write

$$(w - v^\delta_{|\gamma^m})(s) = (1 - s^2) g(s) \,,$$

where s is a normalised coordinate along the mortar segment and $g(s) \in \mathcal{P}_{P-2}(s)$. Integrating this equation over the mortar and using the orthogonality relation (integral condition), we obtain that $\int_{\gamma^m} (1-s^2) g^2(s) \, \mathrm{d}s = 0$, and therefore $g(s) \equiv 0$. We conclude that w coincides exactly with v^δ on the mortar segment and, since u^δ is unique, so is w.

Error estimate

The error estimate for the discrete problem of eqn (7.4.8) can be decomposed into three contributions according to Strang's second lemma [443]: the best fit (approximation error, ε^a), the quadrature error ε^Q, and the consistency error ε^c (due to non-conforming approximation), which are, respectively,

$$||u - u^\delta||_* \leqslant \inf_{v^\delta \in \mathcal{X}^\delta} ||u - v^\delta||_* \tag{7.4.11a}$$

$$+ \inf_{v^\delta \in \mathcal{X}^\delta} \sup_{z^\delta \in \mathcal{X}^\delta} \frac{a(v^\delta, z^\delta) - a^\delta(v^\delta, z^\delta)}{||z^\delta||_*} \tag{7.4.11b}$$

$$+ \sup_{z^\delta \in \mathcal{X}^\delta} \frac{\int_\Omega f(\boldsymbol{x}) z^\delta \, \mathrm{d}\boldsymbol{x} - \sum_{e=1}^{N_{\text{el}}} \sum_{\text{GL}} f_{|\Omega^e} z^\delta_{|\Omega^e}}{||z^\delta||_*} \tag{7.4.11c}$$

$$+ \sup_{z^\delta \in \mathcal{X}^\delta} \sum_{e=1}^{N_{\text{el}}} \sum_{l=e+1}^{N_{\text{el}}} \frac{\int_{\gamma^{\text{el}}} \partial u / \partial n [z^\delta] \, \mathrm{d}s}{||z^\delta||_*} \,,$$

where the bracket denotes jump-across interfaces. The best fit error is given by the approximation theorem stated in the previous section. The quadrature error is standard in spectral discretisations and an estimate is given in [87]:

$$\varepsilon^Q \leqslant C \sum_{e=1}^{N_{\text{el}}} \left[P^{-\rho_e} ||f_{|\Omega^e}||_{H^{\rho_e}(\Omega^e)} + P^{1-\sigma_e} ||u_{|\Omega^e}||_{H^{\sigma_e}(\Omega^e)} \right] ,$$

where $u \in H^{\sigma_e}(\Omega^e)$ and $f_{|\Omega^e} \in H^{\rho_e}(\Omega^e)$. The bound for the consistency error is more involved but it is of the same order as the approximation error, see Bernardi *et al.* [47]. This result depends strongly on the integral matching condition. To appreciate this let us instead consider another choice. The often-used pointwise matching across a non-conforming interface (or patch) would result in a very slow conditional convergence rate. The differences between integral and pointwise matching were systematically investigated in Bernardi *et al.* [44], where a finite element triangulation was coupled to a spectral discretisation through an interface γ. We can use the results of Bernardi *et al.* [44] in the current context by considering two adjacent spectral elements with the polynomial interpolation $P^- < P^+$ along the interface γ from the left and right, respectively. In the pointwise matching case, we obtain for the consistency error ε_p^c

$$\varepsilon_p^c \leqslant C\,(P^+)^{-1/2+\alpha}||u||_*\,, \quad 0<\alpha<\frac{1}{2}\,,$$

whereas in the integral matching condition we obtain for the consistency error ε_i^c as

$$\varepsilon_i^c \leqslant C\,[(P^-)^{1-\sigma} + (P^+)^{1-\sigma}]||u||_*\,,$$

where σ measures the regularity of the solution, that is, $u \in H^\sigma$. We see that the integral matching condition indeed leads to an optimum estimate of spectrally small error, whereas the pointwise matching leads to fixed-order slow convergence.

In summary, the total error bound of eqn (7.4.11a)–(7.4.11c) then becomes

$$||u-u^\delta||_* \leqslant C\sum_{e=1}^{N_{\rm el}}\left[P^{1-\sigma_e}||u_{|\Omega^e}||_{H^{\sigma_e}(\Omega^e)} + P^{-\rho_e}||f_{|\Omega^e}||_{H^{\rho_e}(\Omega^e)}\right], \qquad (7.4.12)$$

which is a similar error bound to that in the standard conforming spectral element case (see Chapter 2).

Boundary conditions

The mortar projection also provides an optimal way of imposing Dirichlet boundary conditions. Typically, boundary data are interpolated using standard Lagrangian interpolation. In that case, the error is

$$||u-\mathcal{I}^\delta u||_{L^2(\partial\Omega)} \leqslant C\,P^{1/2-\sigma}||u||_{H^\sigma(\partial\Omega)}\,,$$

from which we can deduce that the corresponding $H^{1/2}$-norm error estimate is bounded by $P^{3/2-\sigma}$, which is an order worse due to the derivatives involved. However, using the mortar projection operator $\mathbb{P}^\delta$, we obtain

$$||u-\mathbb{P}^\delta u||_{H^{1/2}(\partial\Omega)} \leqslant C\,P^{1/2-\sigma}||u||_{H^\sigma(\partial\Omega)}\,,$$

and therefore we gain one order of approximation using the mortar projection. This can be especially important in practice if the boundary data used are discontinuous, as in the case of the classical driven-lid cavity flow (see Fig. 1.3). A numerical example to demonstrate this effect was presented by Mavriplis [324]. We can also interpret this result as the mortar projection (that is, combined vertex condition and L^2 minimisation), which is equivalent to an H^1 projection for which a similar error bound is valid, ensuring optimality of the formulation.

7.4.3 *Implementation*

To incorporate the patching condition into the discrete equations, a basis is required for the mortars and test functions [10,324]. The mortar w^m is simply u^δ restricted to the edge of an element, so it can be expanded using one-dimensional Lagrangian interpolants as

$$u^{e,l} = u^m = \sum_{i=0}^{P} u_i h_i(s)\,,$$

while the test function ψ becomes

$$\psi = \sum_{i=1}^{P-1} \psi_i g_i^{P-2}(s)\,, \quad \text{where } g_i^{P-2}(s) = (-1)^{P-i}\frac{L'_P(s)}{\xi_i - s}\,,$$

and ξ_i denotes the local Cartesian nodal points. Inserting this set of basis functions into eqn (7.4.2a) for the integral matching condition, we obtain the algebraic form

$$\sum_{j=1}^{P-1} \boldsymbol{M}^m[i][j]\,\boldsymbol{u}^{e,l}[j] = \sum_m \sum_{j=0}^{P} \boldsymbol{P}^m[i][j]\,\boldsymbol{w}^m[j]\,, \tag{7.4.13}$$

where the mortar mass matrix, $\boldsymbol{M}^m[i][j] \to \int_{\Gamma^{e,l}} v_i \psi_j\, ds$, and the projection matrix, $\boldsymbol{P}^m[i][j] \to \int_{\Gamma^{e,l}} w_i|_{\gamma^m}\psi_j\, ds$, are obtained from the following:

$$\boldsymbol{M}^m[i][j] = \frac{|\Gamma^{e,l}|}{2}(-1)^{P-i}[-L''_P(\xi_i)]\rho_i\delta_{ij}\,, \tag{7.4.14a}$$

$$\boldsymbol{P}^m[i][j] = \frac{|\gamma^m|}{2} g_i^{P-2}(\xi_j)\rho_j - \frac{|\Gamma^{e,l}|}{2} g_i^{P-2}(-1)\rho_0\delta_{0j}\,, \tag{7.4.14b}$$

with ρ_i being the quadrature weights (note that previously we have used w_i). The alternating sign term in $\boldsymbol{M}^m[i][j]$ ensures that the elements of the mass matrix are positive. Also, the above expression for the projection operator $\boldsymbol{P}^m[i][j]$ is only valid if $|\gamma^m| = |\Gamma^{e,l}|$; if that is not the case, then the difference $|\Gamma^{e,l}| - |\gamma^m|$ has to be included in eqn (7.4.14b) and appropriate modifications be made (see [10]).

Solving for the unknown velocities, we invert eqn (7.4.13):

$$\boldsymbol{u}^{e,l} = (\boldsymbol{M}^m)^{-1}\sum_m \boldsymbol{P}^m\boldsymbol{w}^m = \sum_m \boldsymbol{Z}^m\boldsymbol{w}^m\,, \tag{7.4.15}$$

where the summation over m indicates piecewise integration of eqn (7.4.2a) over all the mortar patches. For a master element $\boldsymbol{Z}^m$ is the identity matrix, and hence $\boldsymbol{w}^m$ can be replaced in eqn (7.4.15) by the solution along the corresponding master edge. For a slaved element, let $\tilde{\boldsymbol{u}}_b$ represent the values of the local degrees of freedom along the boundary (conforming edges plus mortars) and $\boldsymbol{u}_i$ represent the values on the interior of the element. Figure 7.6 shows how the basis for a non-conforming element is 'expanded' to include the boundary points of the master elements. After assembling the individual projection matrices $\boldsymbol{Z}^m$ into a single matrix, that is,

$$\tilde{\boldsymbol{Z}} = \mathop{\mathcal{A}}_{m} \boldsymbol{Z}^m ,$$

the vector of nodal coefficients for a non-conforming element can be related to the standard (conforming) coefficients through the matrix equation

$$\boldsymbol{u}^e = \begin{bmatrix} \boldsymbol{u}_b \\ \boldsymbol{u}_i \end{bmatrix} = \begin{bmatrix} \tilde{\boldsymbol{Z}} & \\ & \boldsymbol{I} \end{bmatrix} \begin{bmatrix} \tilde{\boldsymbol{u}}_b \\ \boldsymbol{u}_i \end{bmatrix} = \hat{\boldsymbol{Z}}^e \tilde{\boldsymbol{u}}^e . \tag{7.4.16}$$

The above equation provides the principal relation required to form the elemental matrix system, and the key to the simplicity of this approach introduced in [229]. If the local system for a standard element is given by

$$\boldsymbol{H}^e \boldsymbol{u}^e = \boldsymbol{M}^e \boldsymbol{f}^e ,$$

then the corresponding system for a patched element is

$$(\hat{\boldsymbol{Z}}^e)^\top \boldsymbol{H}^e \hat{\boldsymbol{Z}}^e \tilde{\boldsymbol{u}}^e = (\hat{\boldsymbol{Z}}^e)^\top \boldsymbol{M}^e \boldsymbol{f}^e .$$

The conforming and non-conforming formulations can be freely mixed within the computational domain, and the global matrices are again formed by summing the elemental matrices:

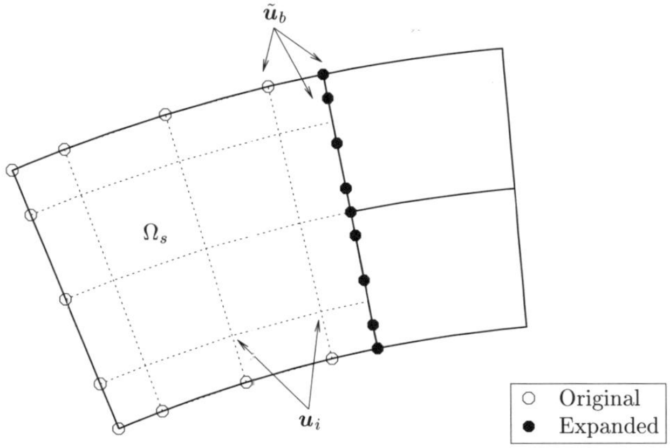

FIG. 7.6. Expanded basis of a non-conforming element. The solid circles represent nodal points linked to the master elements and related to the original boundary points of Ω_s through the projection matrix.

$$\boldsymbol{H} = \boldsymbol{\mathcal{A}}^\top \underline{(\hat{\boldsymbol{Z}}^e)^\top \boldsymbol{H}^e \hat{\boldsymbol{Z}}^e} \boldsymbol{\mathcal{A}},$$
$$\boldsymbol{f} = \boldsymbol{\mathcal{A}}^\top \underline{(\hat{\boldsymbol{Z}}^e)^\top \boldsymbol{M}^e}\, \underline{\boldsymbol{f}^e},$$

where $\hat{\boldsymbol{Z}}^e$ is the identity matrix for all but the slaved elements, and therefore corresponding operations can be simplified. The assembled matrix $\boldsymbol{H}$ is still symmetric and positive definite, with a slightly larger bandwidth than a purely conforming matrix. The larger bandwidth arises because the coupling between elements now extends beyond nearest neighbours to each element connected by the patch.

7.4.4 *Condition number of the Laplacian*

In non-conforming discretisations the success of the entire refinement process depends on the conditioning of the assembled stiffness matrix. For two-dimensional *conforming* spectral/*hp* discretisations the condition number of the full Laplacian matrix for the *nodal basis* scales as

$$\kappa_2(\boldsymbol{L}) \propto \mathcal{O}(N_{\mathrm{el}} P^3)\,. \tag{7.4.18}$$

This estimate assumes that all elements have the same order P.

The derivation of a theoretical estimate for the non-conforming Laplacian is more involved. Instead, we summarise the numerical results obtained in [229] for the condition number of the global Laplacian matrix $\boldsymbol{L}$. A model problem for refinement is used where half of the domain is halved successively, while the other part remains the same. The condition number for corresponding fully-conforming discretisations is included in the computation for reference. Using a forward

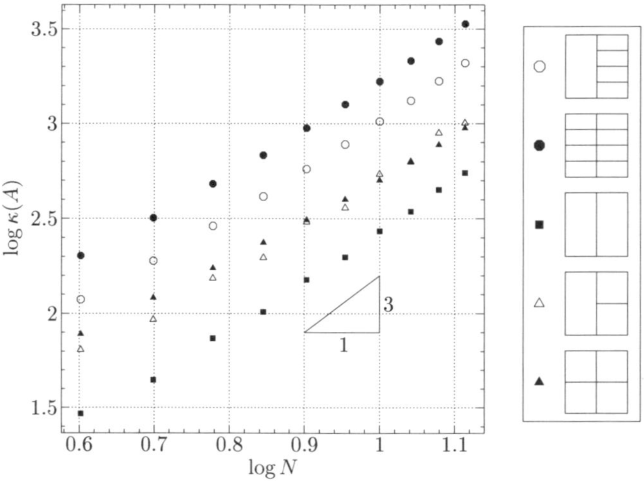

FIG. 7.7. Condition number of the discrete Laplace matrix for a family of conforming and non-conforming spectral element meshes ([229]).

and inverse power method, condition numbers were computed for a series of meshes that combine elements of various sizes [229]. Figure 7.7 shows a family of curves for these domains, each one exhibiting a growth rate of $\mathcal{O}(P^3)$. Comparing domains with similarly-sized elements, we see that $\boldsymbol{L}$ contains a blend of the eigenvalues from the largest and smallest, but overall adheres to the estimate given in eqn (7.4.18). This test also shows that local refinement to resolve a particular flow feature produces a matrix with a lower condition number than the one that would correspond to a refinement which was propagated throughout.

In summary, the global discrete Laplacian matrix which contains both conforming as well as non-conforming patches does not exhibit any spurious eigenvalues, and therefore similar strategies can be used to precondition both conforming and non-conforming Laplacian matrices.

7.5 Discontinuous Galerkin method (DGM)

Although the original application of most discontinuous Galerkin methods (DGM) was in solving hyperbolic problems, more recent work has led to formulations for parabolic and elliptic problems [102]. For example, works such as [38], in which the viscous compressible Navier–Stokes equations were solved, required that a discontinuous Galerkin formulation be extended beyond the hyperbolic advection terms to the viscous terms of the Navier–Stokes equations. Concurrently, both in [105] and [41, 345] other discontinuous Galerkin formulations for parabolic and elliptic problems were proposed. In an effort to classify all the efforts made toward the use of discontinuous Galerkin methods for elliptic problems, Arnold *et al.*, first in [16] and then more fully in [17], published a unified analysis of discontinuous Galerkin methods for elliptic problems.

In the following, we first present the two main formulations for the one-dimensional diffusion equation and then we present the unified formulation of Arnold *et al.* for multiple dimensions. We compare most of the known versions of DGM for diffusion, examining accuracy, stability, computational complexity, as well as ease of implementation in the context of spectral/*hp* methods.

7.5.1 *An inconsistent formulation*

Let us consider the one-dimensional diffusion equation

$$u_t = u_{xx} \tag{7.5.1}$$

with periodic boundary conditions in order to illustrate different formulations of the discontinuous Galerkin method for diffusion. To this end, we follow the weak formulation of the advection, eqn (6.2.17), where we replace u by u_x to obtain

$$\int_{\Omega^j} u_t v \,\mathrm{d}x + \int_{\Omega^j} u_x v_x \,\mathrm{d}x - (\tilde{u}_x)_R v_R^- + (\tilde{u}_x)_L v_L^+ = 0\,, \tag{7.5.2}$$

in the one-dimensional element denoted by Ω^j. Here, $j \pm 1/2$ denote the end-points of the element. This form is discussed in [510] and was also proposed in [185] in an attempt to obtain a conservative formulation for the diffusion equation. Specifically, the issue of the value of the flux $(\tilde{u}_x)_{L,R}$ was considered in [185] as follows. Assuming that we start with an initial condition

$$u(x,0) = u^- + H(x)(u^+ - u^-)\,,$$

where $H(x)$ denotes a Heaviside function, then the solution of the diffusion equation can be obtained in terms of the error function, i.e.,

$$u(x,t) = \frac{1}{2}(u^+ + u^-) + \frac{1}{2}(u^+ - u^-)\,\mathrm{erf}\left(\frac{x}{\sqrt{4t}}\right).$$

Here u^+ and u^- denote constant states right and left, respectively, of the initial discontinuity placed at $x = 0$. We note that in the above at $x = 0$

$$u(0,t) = \frac{1}{2}(u^+ + u^-)\,, \quad \forall\, t > 0\,.$$

Thus, a natural choice for the numerical flux is to average the left and right values at the ends of an element, i.e.,

$$(\tilde{u}_x)_R = \frac{1}{2}\left[(u_x)_R^- + (u_x)_R^+\right]. \tag{7.5.3}$$

Substitution of this flux into eqn (7.5.2) completes the numerical scheme. In matrix form we obtain for an equidistant mesh

$$\frac{\mathrm{d}\boldsymbol{u}_j}{\mathrm{d}t} + \frac{1}{\Delta x^2}[\boldsymbol{A}\boldsymbol{u}_{j-1} + \boldsymbol{B}\boldsymbol{u}_j + \boldsymbol{C}\boldsymbol{u}_{j+1}] = 0\,, \tag{7.5.4}$$

where $\boldsymbol{u}_j$ is a vector of length $P+1$ and $\boldsymbol{A}$ and $\boldsymbol{B}$ are matrices of size $(P+1)\times(P+1)$, assuming that a polynomial trial basis of degree P is used in all elements. However, as pointed out in [510], the above discontinuous Galerkin formulation is *inconsistent* although weakly stable. Indeed, for $P=0$ the above statement gives

$$\frac{\mathrm{d}\boldsymbol{u}_j}{\mathrm{d}t} = 0\,,$$

which is obviously wrong.

The consequences of such inconsistency are demonstrated in Fig. 7.8, where a numerical solution is obtained corresponding to the initial condition $u(x,0) = \sin x$, $x \in [0, 2\pi]$. We observe that for a fixed polynomial order the method does not converge upon elemental refinement. For a fixed number of elements (forty evenly-spaced elements), upon p-refinement, we initially observe what appears to be convergence. As the polynomial order is increased, however, the solution starts to diverge from the true solution. This inconsistency was also rigorously documented by Zhang and Shu [510] using Fourier analysis.

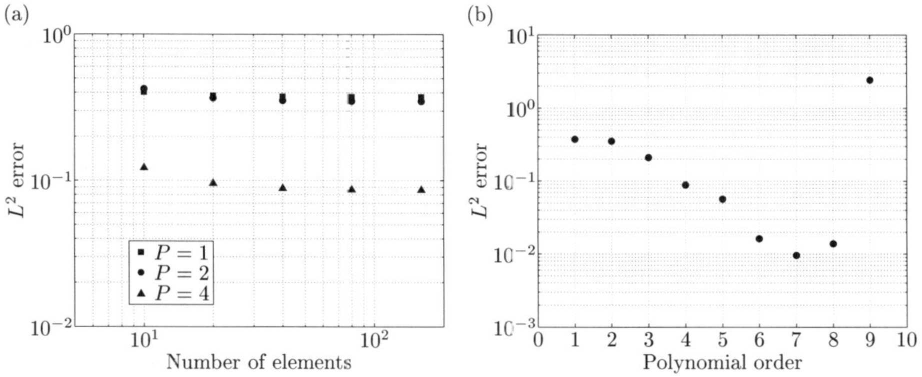

FIG. 7.8. Convergence study of the inconsistent formulation based upon the model problem evaluated at time $T = 0.7$. In (a) we present the L^2 error versus the number of evenly-spaced elements, having polynomial orders $P = 1$ (squares), $P = 2$ (circles), and $P = 4$ (triangles). In (b) we present the L^2 error versus polynomial order for a mesh consisting of forty evenly-spaced elements. (Courtesy of R. M. Kirby.)

7.5.2 *Local discontinuous Galerkin method (LDG)*

We can modify the above inconsistent formulation by introducing the auxiliary flux variable

$$q = u_x$$

and rewrite the diffusion equation as a first-order system, i.e.,

$$\begin{aligned} u_t - q_x &= 0\,, \\ q - u_x &= 0\,. \end{aligned} \tag{7.5.5}$$

To obtain the weak form we test the above with test functions v and w (in the same space as u and q):

$$\int_{\Omega^j} u_t v \, \mathrm{d}x + \int_{\Omega^j} q v_x \, \mathrm{d}x - \tilde{q}_R v_R^- + \tilde{q}_L v_L^+ = 0\,, \tag{7.5.6}$$

$$\int_{\Omega^j} q w \, \mathrm{d}x + \int_{\Omega^j} u w_x \, \mathrm{d}x - \tilde{u}_R w_R^- + \tilde{u}_L w_L^+ = 0\,, \tag{7.5.7}$$

where we have taken the test functions evaluated inside the cell at both ends. With regards to boundary terms for $\tilde{u}_{L,R}$ and $\tilde{q}_{L,R}$, we can use the arithmetic mean as before, see also [38]. Another suggestion was made by Cockburn and Shu [105], as follows:

$$\tilde{u}_R = u_R^+ \quad \text{and} \quad \tilde{q}_R = q_R^- \tag{7.5.8}$$

or

$$\tilde{u}_R = u_R^- \quad \text{and} \quad \tilde{q}_R = q_R^+\,. \tag{7.5.9}$$

That is, we *alternate* the left and right values of the numerical fluxes in u and q. This choice turns out to be more effective both in terms of accuracy as well as in computational efficiency, as we will discuss later.

7.5.3 *The Baumann–Oden discontinuous Galerkin method*

A different way of correcting the inconsistency of the original DGM was proposed by Baumann and Oden, see [40,41,345]. In this version more terms are included in the weak statement, allowing jumps in the original variable at the ends of the cell. The weak statement is now defined to be

$$\int_{\Omega^j} u_t v \, dx + \int_{\Omega^j} u_x v_x \, dx - (\tilde{u}_x)_R v_R^- + (\tilde{u}_x)_L v_L^+ \\ - \frac{1}{2}(v_x)_R^-(u_R^+ - u_R^-) - \frac{1}{2}(v_x)_L^+(u_L^+ - u_L^-) = 0\,, \tag{7.5.10}$$

where the flux is taken as the arithmetic mean, i.e.,

$$(\tilde{u}_x)_R = \frac{1}{2}[(u_x)_R^- + (u_x)_R^+]\,.$$

The main difference with the LDG formulation is that the Baumann–Oden discrete system is non-symmetric. This fundamentally changes the eigenstructure of the (symmetric) diffusion operator and has significant consequences on the computational complexity and temporal behaviour, as we will examine later.

We note that for $P = 0$ the Baumann–Oden scheme degenerates again to

$$\frac{du_j}{dt} = 0\,, \tag{7.5.11}$$

which is inconsistent. However, for $P \geqslant 1$ this scheme is stable and consistent. Stability results have been derived theoretically for $P > 1$. For $P = 1$ we also note that numerical results show that this scheme is stable [23].

7.5.4 *A unified formulation*

We now present a mathematical framework for relating the different discontinuous Galerkin approaches for elliptic problems adopted from [17]. A comparison of the different DGM approaches in the context of spectral/*hp* element methods has been performed by Kirby [270]. Similar studies have been presented by Castillo [91] and Shu [436], but for lower polynomial order P and non-spectral bases. In the following, we present some of the developments in these studies.

We consider the second-order elliptic model with Dirichlet boundary conditions (homogeneous, for simplicity) in two dimensions

$$-\nabla^2 u = f \quad \text{in } \Omega\,, \tag{7.5.12a}$$

$$u = 0 \quad \text{on } \partial\Omega\,, \tag{7.5.12b}$$

and introduce the auxiliary variable $\boldsymbol{q} = \nabla u$, as in the one-dimensional formulation, to obtain the first-order system, i.e.,

$$\boldsymbol{q} = \nabla u \quad \text{in } \Omega\,, \tag{7.5.13a}$$
$$-\nabla \cdot \boldsymbol{q} = f \quad \text{in } \Omega\,, \tag{7.5.13b}$$
$$u = 0 \quad \text{on } \partial\Omega\,. \tag{7.5.13c}$$

In order to discretise the above system we define two appropriate spaces on a tessellation $\mathcal{T}^\delta = \{\Omega^e\}$ of the domain Ω as follows:

$$\mathcal{V}^\delta := \{v \in L^2(\Omega) : v|_{\Omega^e} \in \mathcal{P}(\Omega^e),\ \forall\, e \in \mathcal{T}^\delta\}\,,$$
$$\mathcal{W}^\delta := \{\boldsymbol{w} \in [L^2(\Omega)]^2 : \boldsymbol{w}|_{\Omega^e} \in \mathcal{W}(\Omega^e),\ \forall\, e \in \mathcal{T}^\delta\}\,,$$

where e refers to an element number, $\mathcal{P}(\Omega^e)$ is the space of polynomial functions of degree at most $P \geqslant 1$ on Ω^e, and $\mathcal{W}^\delta(\Omega^e) = [\mathcal{P}(\Omega^e)]^2$. The discrete problem can be stated as follows.

Find $u^\delta \in \mathcal{V}^\delta$ and $\boldsymbol{q}^\delta \in \mathcal{W}^\delta$ such that, for all $e \in \mathcal{T}^\delta$,

$$\int_{\Omega^e} \boldsymbol{q}^\delta \cdot \boldsymbol{w}\,\mathrm{d}x = -\int_{\Omega^e} u^\delta \nabla^\delta \cdot \boldsymbol{w}\,\mathrm{d}x + \int_{\partial\Omega^e} \tilde{u}_{\gamma^e}\boldsymbol{n}_e \cdot \boldsymbol{w}\,\mathrm{d}s\,, \tag{7.5.14}$$
$$\int_{\Omega^e} \boldsymbol{q}^\delta \cdot \nabla v\,\mathrm{d}x = \int_{\Omega^e} f\, v\,\mathrm{d}x + \int_{\partial\Omega^e} v\tilde{\boldsymbol{q}}_{\gamma^e} \cdot \boldsymbol{n}_e\,\mathrm{d}s\,, \tag{7.5.15}$$

where the numerical fluxes $\tilde{\boldsymbol{q}}_{\gamma^e}$ and $\tilde{u}_{\gamma^e}$ are approximations to the traces of $\boldsymbol{q} = \nabla u$ and u, respectively, i.e., approximations on the interface edge of $\partial\Omega^e$ of element e.

Given this general unified formulation of the discrete problem, the two remaining choices which determine exactly which DG methodology is used are the choices of the numerical fluxes $\tilde{\boldsymbol{q}}_{\gamma^e}$ and $\tilde{u}_{\gamma^e}$. In Table 7.1 we present some representative formulations. The operator $\{\cdot\}$ denotes averaging across the interface, while $[[\cdot]]$ denotes the jump difference across the interface. Specifically, assuming two neighbouring elements e_1 and e_2 sharing an edge $\partial\Omega^{e_1,e_2}$, with corresponding normals $\boldsymbol{n}_{e_1}$ and $\boldsymbol{n}_{e_2}$, we define the following:

$$\{v\} \equiv \frac{1}{2}(v_{\gamma^{e_1}} + v_{\gamma^{e_2}}) \quad \text{and} \quad [[v]] \equiv v_{\gamma^{e_1}}\boldsymbol{n}_{e^1} + v_{\gamma^{e_2}}\boldsymbol{n}_{e^2}\,,$$

and similar definitions are used for the flux $\boldsymbol{q}$.

The interior penalty (IP) method is the oldest of these versions [15, 140, 491]; it is a symmetric DG method with provable error estimates. The Bassi–Rebay (BR) method is conceptually the simplest and easiest to implement [38]. The local discontinuous Galerkin (LDG) method employs a different flux for $\beta \neq 0$ and a penalty term similar to the IP scheme. The original Baumann–Oden (BO) method [41, 345] does not include a penalty term and it is non-symmetric due to the jump terms in the flux of the variable u. However, a new version of the Baumann–Oden method includes a penalty term on the interior boundaries only [401]. Finally, the non-symmetric interior penalty Galerkin method (NIPG) is a

TABLE 7.1. Proposed discontinuous Galerkin methodologies for elliptic problems and the flux choices they represent.

Method	$\tilde{u}_\gamma$	$\tilde{\boldsymbol{q}}_\gamma$
Bassi–Rebay (BR) [38]	$\{u^\delta\}$	$\{\boldsymbol{q}^\delta\}$
Interior penalty (IP) [140]	$\{u^\delta\}$	$\{\nabla^\delta u^\delta\} - \eta h^{-1}([[u^\delta]])$
Brezzi *et al.* (BA) [77]	$\{u^\delta\}$	$\{\boldsymbol{q}\} - \alpha_r([[u^\delta]])$
LDG [105]	$\{u^\delta\} - \beta \cdot [[u^\delta]]$	$\{\boldsymbol{q}^\delta\} + \beta \cdot [[\boldsymbol{q}^\delta]] - \eta h^{-1}([[u^\delta]])$
Baumann–Oden (BO) [41]	$\{u^\delta\} + \boldsymbol{n}_e \cdot [[u^\delta]]$	$\{\nabla^\delta u^\delta\}$
NIPG-j [393]	$\{u^\delta\} + \boldsymbol{n}_e \cdot [[u^\delta]]$	$\{\nabla^\delta u^\delta\} - \eta h^{-j}([[u^\delta]])$

modification of the Baumann–Oden approach, with two versions for $j = 1$ and $j = 3$ [393, 394].

In the following, we examine some of the properties of the different DG methods related to the accuracy, stability, and computational complexity in implementing a particular version. We follow the work of [270], considering primarily the Bassi–Rebay, LDG, and Baumann–Oden versions. By an algebraic manipulation, we can rewrite eqn (7.5.15) to eliminate the auxiliary variable $\boldsymbol{q}$ from the formulation (taking into account the proper flux when manipulating the variable out of the expression). This elimination is possible because the numerical flux $\hat{u}_\gamma$ depends only on the variable u. We illustrate this elimination by rewriting the above system in matrix form:

$$\boldsymbol{M}\boldsymbol{Q} + \boldsymbol{D}_1\boldsymbol{U} = \boldsymbol{F}\,,$$
$$\boldsymbol{D}_2\boldsymbol{Q} + \boldsymbol{C}\boldsymbol{U} = \boldsymbol{G}\,,$$

where $\boldsymbol{M}$ is the (block diagonal) mass matrix, $\boldsymbol{D}_1$ and $\boldsymbol{D}_2$ correspond to the gradient and divergence operators, $\boldsymbol{C}$ corresponds to the stabilisation terms, and $\boldsymbol{U}$ and $\boldsymbol{Q}$ are the vectors of unknowns for the variables u and $\boldsymbol{q}$, respectively. Elimination of the auxiliary variable leads to

$$(\boldsymbol{C} - \boldsymbol{D}_2\boldsymbol{M}^{-1}\boldsymbol{D}_1)\boldsymbol{U} = \boldsymbol{G} - \boldsymbol{D}_2\boldsymbol{M}^{-1}\boldsymbol{F}\,.$$

Following this procedure for the one-dimensional parabolic problem of eqn (7.5.1) in the interval $[0, 1]$ with periodic boundary conditions, this manipulation leads to the following systems for Bassi–Rebay, LDG, and Baumann–Oden, respectively:

$$\frac{\mathrm{d}\hat{\boldsymbol{U}}_j}{\mathrm{d}t} = \boldsymbol{A}_{-2}^{\mathrm{BR}}\hat{\boldsymbol{U}}_{j-2} + \boldsymbol{A}_{-1}^{\mathrm{BR}}\hat{\boldsymbol{U}}_{j-1} + \boldsymbol{A}_{0}^{\mathrm{BR}}\hat{\boldsymbol{U}}_{j} + \boldsymbol{A}_{1}^{\mathrm{BR}}\hat{\boldsymbol{U}}_{j+1} + \boldsymbol{A}_{2}^{\mathrm{BR}}\hat{\boldsymbol{U}}_{j+2}\,, \tag{7.5.16a}$$

$$\frac{\mathrm{d}\hat{\boldsymbol{U}}_j}{\mathrm{d}t} = \boldsymbol{A}_{-1}^{\mathrm{LDG}}\hat{\boldsymbol{U}}_{j-1} + \boldsymbol{A}_{0}^{\mathrm{LDG}}\hat{\boldsymbol{U}}_{j} + \boldsymbol{A}_{1}^{\mathrm{LDG}}\hat{\boldsymbol{U}}_{j+1}\,, \tag{7.5.16b}$$

$$\frac{\mathrm{d}\hat{\boldsymbol{U}}_j}{\mathrm{d}t} = \boldsymbol{A}_{-1}^{\mathrm{BO}}\hat{\boldsymbol{U}}_{j-1} + \boldsymbol{A}_{0}^{\mathrm{BO}}\hat{\boldsymbol{U}}_{j} + \boldsymbol{A}_{1}^{\mathrm{BO}}\hat{\boldsymbol{U}}_{j+1}\,, \tag{7.5.16c}$$

where $\hat{\boldsymbol{U}}$ denotes a vector of the modal coefficients of the polynomial expansion on an element j, and the matrices $\boldsymbol{A}_k$ are formulated based upon the choice of the numerical fluxes $\tilde{\boldsymbol{q}}_\gamma$ and $\tilde{u}_\gamma$. The subscript k on each matrix $\boldsymbol{A}_k$ denotes the offset from the current element j for which the solution is being sought. The particular LDG stencil above corresponds to a choice of $\beta = 1/2$, as in the work of [436]. A different choice of the β parameter may lead to a wider stencil for LDG.

In the following, we will examine a variety of factors such as the stencil width, eigenspectrum, h-convergence properties, and p-convergence properties of the different numerical fluxes to explore the differences between the presented DGM formulations.

7.5.5 *Compactness of the stencil*

The first observation that can be made immediately upon examination of eqns (7.5.16a)–(7.5.16c) is that both LDG and Baumann–Oden have shorter stencils than Bassi–Rebay. The LDG and Baumann–Oden schemes require information only from nearest-neighbouring elements, hence producing a three-element stencil, while the Bassi–Rebay scheme requires a five-element stencil. In the two-dimensional case, as shown in Fig. 7.9, LDG and Baumann–Oden require only local elemental communication, which for triangular meshes corresponds to only a four-element stencil. For Bassi–Rebay, however, information from as many as ten elements may be required for the computation of the solution on a single element.

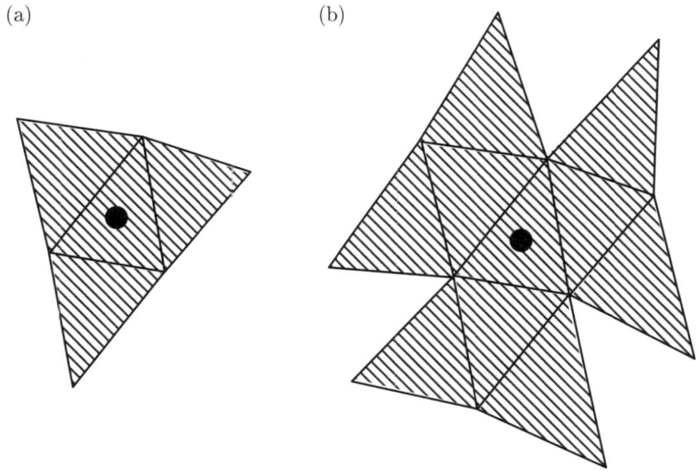

FIG. 7.9. (a) LDG and Baumann–Oden stencils, and (b) a Bassi–Rebay stencil. The element containing the black dot denotes the element on which the solution is being computed, and shaded areas denote the elements from which information is required for completing that computation.

This fact concerning the different methods is important when considering parallel communication costs; depending on the way in which the communications are implemented, the Bassi–Rebay method requires twice the communication as both the LDG and Baumann–Oden methods.

7.5.6 *Eigenspectrum*

We consider again the one-dimensional parabolic equation (7.5.1), which upon discretisation we write in matrix form

$$\frac{\mathrm{d}\hat{\boldsymbol{U}}_g}{\mathrm{d}t} = \boldsymbol{A}\hat{\boldsymbol{U}}_g\,, \tag{7.5.17}$$

where $\hat{\boldsymbol{U}}_g$ denotes the concatenation of modal coefficients of each element. Hence, given N_{el} elements, each having $M = P + 1$ modal coefficients, the size of $\hat{\boldsymbol{U}}_g$ is $N_{\mathrm{el}} \times M$, and $\boldsymbol{A}$ is a size($\hat{\boldsymbol{U}}_g$) $\times$ size($\hat{\boldsymbol{U}}_g$) square matrix. We now examine the eigenvalues of the operator $\boldsymbol{A}$ for the three different fluxes for the case in which ten equally-spaced elements are used. Eigenspectra of $\boldsymbol{A}$ for one to nine modes per element, $P = 0$ to $P = 8$, respectively, were computed in [270] for the Bassi–Rebay (BR), LDG, and Baumann–Oden (BO) fluxes. Here, we present typical plots for seven ($P = 6$), eight ($P = 7$), and nine ($P = 8$) modes in Fig. 7.10.

Examination of the eigenspectrum leads to the following observations.

- Both the Bassi–Rebay and LDG fluxes are purely diffusive (all eigenvalues lie on the negative real axis) up to machine precision. This is consistent with the fact that for both Bassi–Rebay and LDG the matrix $\boldsymbol{A}$ is real and symmetric. The use of Baumann–Oden fluxes leads to a real non-symmetric matrix $\boldsymbol{A}$, which is evident by eigenvalues which have nonzero imaginary components.
- When using only one mode ($P = 0$, Godunov scheme), the Baumann–Oden version reduces to an inconsistent scheme, which corresponds to the eigenvalues lying at $(0, 0)$. For modes greater than or equal to two (i.e., $P \geqslant 1$), the Baumann–Oden flux leads to a consistent scheme, in accordance with the numerical and theoretical findings in [23].
- The Baumann–Oden flux, although dispersive, does not show significant dispersion for a low number of modes ($P = 2, 3$). It is conceivable that this fact can be exploited to create a symmetric preconditioner which would accelerate the convergence of implicit methods which use Baumann–Oden fluxes.
- Comparison of the eigenspectrum of the operator formed using LDG fluxes and the operator formed using Bassi–Rebay fluxes shows that LDG requires a more stringent time-step if an explicit time-stepping method is used. In one dimension, this fact can be rationalised by examining the width of the stencil that is created (as is done in [436]) by the two methods considered. As pointed out earlier, for LDG the width of the stencil is three, whereas for Bassi–Rebay the width of the stencil is five. Hence, the effective spacing Δx for the LDG operator is smaller than that of the Bassi–Rebay operator. This fact requires

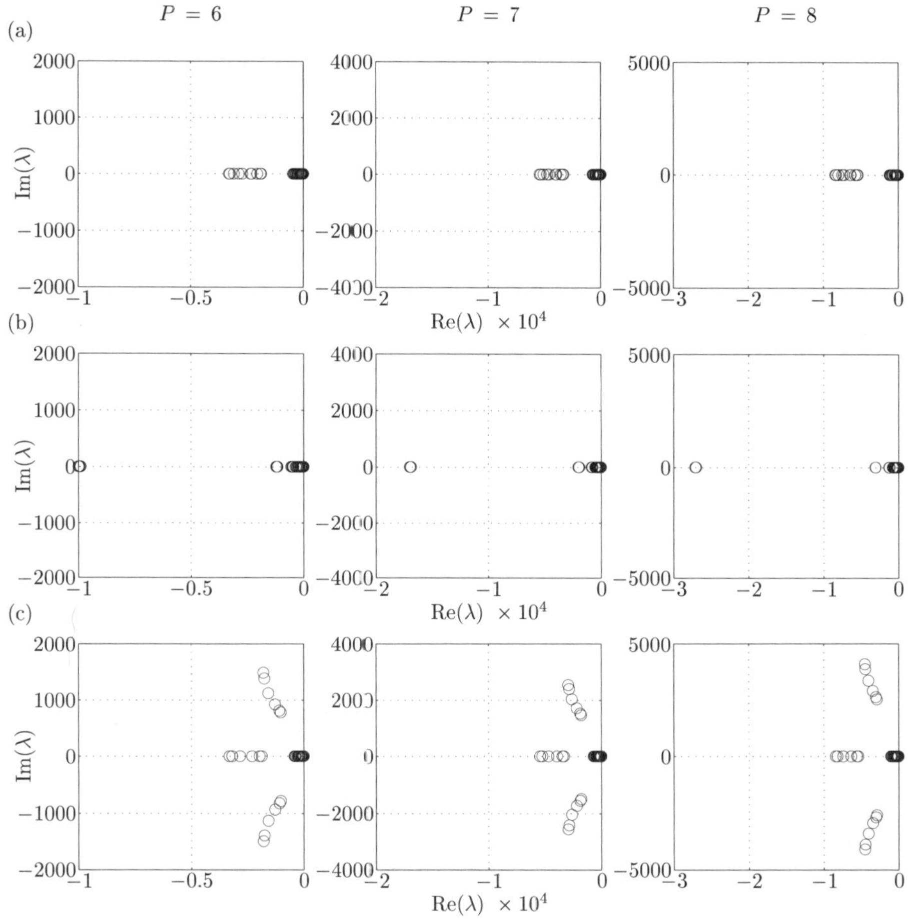

FIG. 7.10. Eigenvalues of the matrix $\boldsymbol{A}$ using (a) Bassi–Rebay fluxes, (b) LDG fluxes, and (c) Baumann–Oden fluxes. Ten elements are used in all cases. The ordinate is the imaginary axis, and the abscissa is the real axis. ([270].)

that, for explicit time integration, the time-step of the LDG method will be smaller than that of the Bassi–Rebay method so that the diffusion number limit can be maintained.

In Fig. 7.11 we compare the maximum absolute eigenvalue versus polynomial order for the three consistent schemes. An $N_{\text{el}} = 40$ evenly-spaced elemental mesh was used. All three flux choices exhibit $\mathcal{O}(P^4)$ growth. Observe that LDG has the largest absolute eigenvalue, implying that LDG will be the most restrictive when applying an explicit time-stepping algorithm. Bassi–Rebay is less restrictive than LDG (as stated previously, about three times less restrictive), and Baumann–Oden is the least restrictive based upon maximum eigenvalue

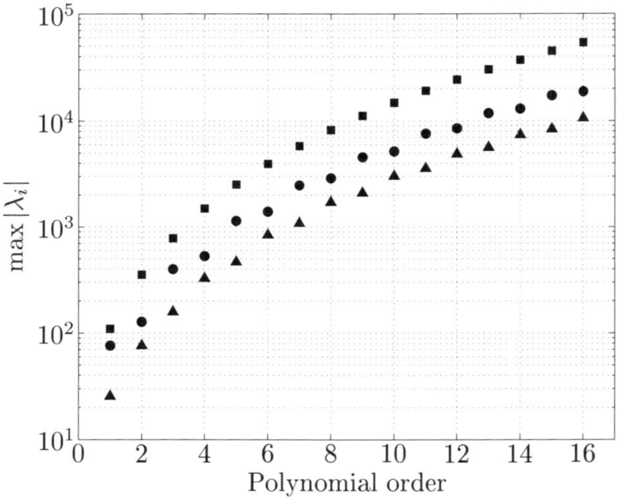

FIG. 7.11. Modulus of the maximum eigenvalues versus the number of modes per element for Bassi–Rebay (circles), LDG (squares), and Baumann–Oden (triangles). ([270].)

magnitude. For Baumann–Oden, however, we must remember that the explicit time-stepping scheme must contain a large region of the complex half-plane to encompass the dispersive eigenvalues of the Baumann–Oden spatial operator.

With respect to the dependence of the condition number of the stiffness matrix $\boldsymbol{A}$, Castillo [91] has provided both analytical and numerical results. Specifically, analytical results for the IP and LDG methods show an h^{-2} dependence on the element size h. For the Baumann–Oden and the NIPG-1 methods, numerical results show an h^{-2} dependence, in contrast with the NIPG-3 method that shows h^{-4} dependence. The latter gives an optimal rate of convergence in the L^2 norm for the original variable u, but it is obviously very inefficient for time-dependent computations.

7.5.7 *Convergence rate*

We now consider the convergence properties of the various DGM formulations both in terms of h-refinement as well as p-refinement.

With respect to the Bassi–Rebay flux, the convergence is $\mathcal{O}(h^P)$ for the odd-order polynomials and $\mathcal{O}(h^{P+1})$ for the even-order polynomials. This is typically referred to as a ***sub-optimal*** convergence rate. The LDG method with the alternating pair of fluxes, as discussed in Section 7.5.2, leads to a uniform $\mathcal{O}(h^{P+1})$ convergence rate for all polynomial orders P. Similarly, the Baumann–Oden method as well as the NIPG-1 exhibit a sub-optimal convergence rate. However, the trend is reversed compared to the Bassi–Rebay method, and thus the convergence rate is $\mathcal{O}(h^P)$ for the even-order polynomials and $\mathcal{O}(h^{P+1})$ for the odd-order polynomials. The NIPG-3 method leads to the optimal rate $\mathcal{O}(h^{P+1})$ in the L^2 norm for the original variable u, but not for the auxiliary variable.

For the Helmholtz equation

$$-\nabla^2 u + cu = f\,, \tag{7.5.18}$$

with mixed Dirichlet/Neumann conditions, Prudhomme *et al.* [387] have obtained the following theorem.

Theorem 7.5.1. (Error estimate of Baumann–Oden DGM) *Let $u \in H^1(\Omega) \cap H^s(\mathcal{P}^\delta)$ (where $s \geqslant 2$) and u^δ be the corresponding weak solution with spectral order $P \geqslant 2$. Then, the numerical error $\epsilon \equiv u - u^\delta$ is bounded as follows:*

$$||\epsilon||_* \leqslant C \sum_{e\in\mathcal{P}^\delta} \frac{h^{\mu-1}}{P_e^{s-5/2}} ||u||_{s,e}\,, \tag{7.5.19}$$

where $\mu = \min(P_e + 1, s)$. Also, the broken norm is defined here as

$$||\epsilon||_*^2 \equiv \sum_{e\in\mathcal{P}^\delta} ||\epsilon|\,^2_{*,e} = \sum_{e\in\mathcal{P}^\delta} \int_e (|\nabla\epsilon|^2 + c|\epsilon|^2)\,dx\,,$$

i.e., summing contributions over all elements numbered by e.

This result is valid for all values of the Helmholtz constant $c \geqslant 0$. For $c \neq 0$, another estimate was also obtained by Prudhomme *et al.* following the approach of [244], as follows:

$$||e||_* \leqslant C \sum_{e\in\mathcal{P}^\delta} \frac{h^{\mu-2}}{P_e^{s-3/2}} ||u||_{s,e}\,. \tag{7.5.20}$$

This estimate is sub-optimal in h but consistent with numerical experiments, e.g., in [23, 40], showing a $P_e^{s-3/2}$ convergence rate in polynomial order.

A stabilised version of the Baumann–Oden method proposed in [401] improves the h-convergence to optimal, but only in the H^1 norm. Similar results were obtained by Houston *et al.* [244] for a stabilised method. In the modification introduced in [394], the so-called NIPG-3 method, the resulting convergence rate is optimal for the variable u (in the L^2 norm) but not for the gradient. This improvement in accuracy comes, however, at a cost of adding a penalty term $\mathcal{O}(h^{-3})$, which leads to a condition number that scales as $\mathcal{O}(h^{-4})$, instead of the typical $\mathcal{O}(h^{-2})$ with the original Baumann–Oden method, as we have already seen.

7.5.8 *Examples and comparisons*

We present three examples to investigate further the convergence of DGM. The first example verifies the non-uniform h-convergence. The second example considers the solution to the viscous Burgers equation, and, finally, the third example considers the role of significant mesh distortion in the solution of the diffusion equation.

h-convergence

The h-convergence of the Bassi–Rebay, LDG, and Baumann–Oden DGM versions was studied numerically in [270] by obtaining solutions of the one-dimensional diffusion equation (7.5.1) in the interval $[0, 1]$ with periodic boundary conditions. An initial condition of $u(x) = \sin(2\pi x)$ was used, and the error was examined at time $t = 0.1$. To determine the convergence rate for each flux choice, the number of modes per element was fixed (both $P = 1$ and $P = 2$ polynomial order), while successive doubling of the number of elements was employed in the spatial discretisation. A plot of the L^2 error versus the number of elements is presented in Fig. 7.12.

We observe that the Baumann–Oden method achieves $\mathcal{O}(h^{P+1})$ when P is odd and $\mathcal{O}(h^P)$ when P is even. In contrast, the Bassi–Rebay achieves $\mathcal{O}(h^P)$ when P is odd and $\mathcal{O}(h^{P+1})$ when P is even. The LDG achieves method $\mathcal{O}(h^{P+1})$ for both odd and even P. In this example, for an even polynomial order, $P = 2$, the Bassi–Rebay and LDG versions have the same convergence rate and have similar absolute errors. In general, the LDG method with the *alternating flux* shows superior performance to both the Bassi–Rebay and the Baumann–Oden versions in that it has a consistent convergence rate, not dependent on the parity

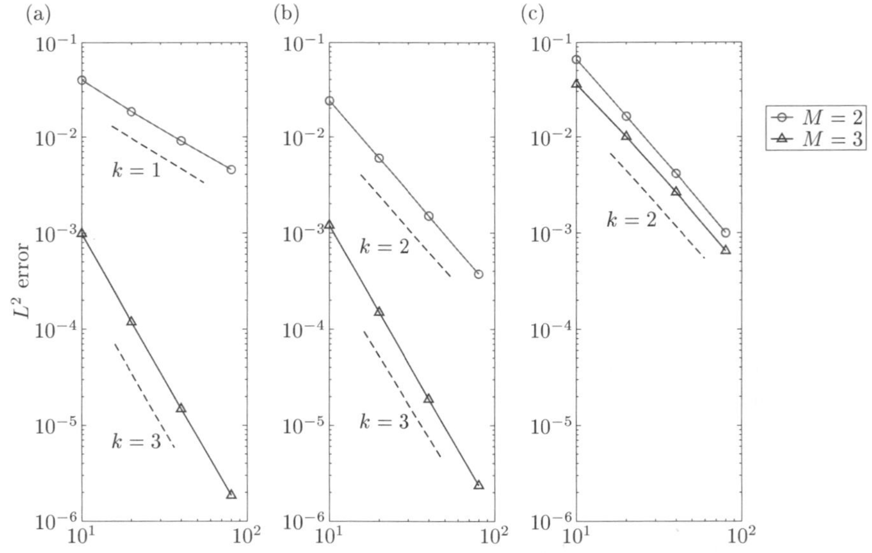

FIG. 7.12. Comparison of the rate of convergence for (a) Bassi–Rebay, (b) LDG, and (c) Baumann–Oden fluxes for linear ($P = 1$, $M = 2$) and quadratic ($P = 2$, $M = 3$) modes per element. The L^2 error is given on the ordinate, and the number of elements is given on the abscissa. Here k denotes the magnitude of the slope. ([270].)

of the polynomial order employed. However, for other flux choices the convergence of LDG is sub-optimal and similar to the Bassi–Rebay method.

Burgers equation

We compare the three main DGM formulations for diffusion by examining a benchmark solution of the viscous Burgers equation

$$\frac{\partial u}{\partial t} + \frac{1}{2}\frac{\partial u^2}{\partial x} = \nu \frac{\partial^2 u}{\partial x^2}, \tag{7.5.21}$$

where $\nu = 10^{-2}/\pi$. This problem was used as a benchmark problem in [37] for evaluating the effectiveness of different spatial discretisations. For $\nu = 10^{-2}/\pi$ the solution develops into a sawtooth wave centred at the origin around the time $t = 1/\pi$ (an approximation which comes from the inviscid theory for Burgers equation), and the maximum gradient of the solution occurs at approximately $t = 0.5$. For the advection operator the conservative form was employed, and upwinding was used in all three methods. The following results are adopted from the work of Kirby [270].

To evaluate the three different fluxes, we examined two quantities: the L^2 error and the magnitude of the slope of the approximate solution at the origin. The slope at the origin is a sensitive quantity and hence provides a good measure of the accuracy of the numerical methods. The analytical solution admits a maximum magnitude of the slope equal to 152.00516 at exactly $t = 0.51047356$; in the example considered we compare the three methods at $t = 0.5105$. In order to calculate the derivative of the approximate solution, the auxiliary variable $\boldsymbol{q}^\delta$, which approximates the first derivative, was used. This measure takes into account, not only the elemental derivative, but also the jump terms.

In Fig. 7.13 we compare the rate of convergence of the L^2 error for (a) Bassi–Rebay, (b) LDG, and (c) Baumann–Oden fluxes evaluated at time $t = 3/\pi$. Six equal elements were used, and the polynomial order was varied from $P = 3$ to $P = 7$. In Table 7.2 we compare the L^2 error and the value of $|\partial u/\partial x(0)|$. In this

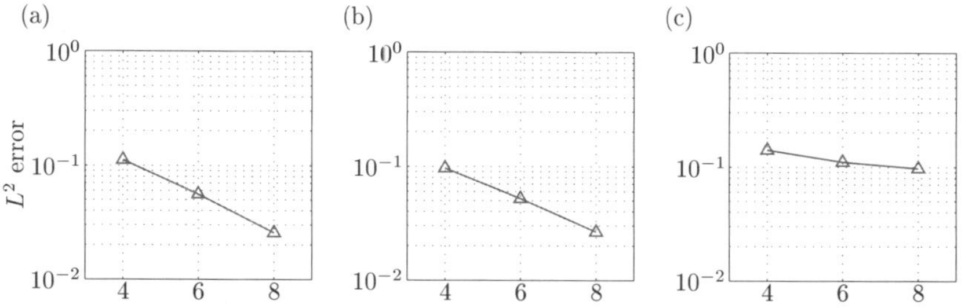

FIG. 7.13. Comparison of the rate of convergence for (a) the Bassi–Rebay, (b) LDG, and (c) Baumann–Oden fluxes for the viscous Burgers equation with $\nu = 10^{-2}/\pi$. Six equal elements were used. The L^2 error is given on the ordinate, and the number of modes per element $P + 1$ is given on the abscissa. ([270].)

TABLE 7.2. L^2 error and the value of $|\partial u/\partial x(0)|$ for the Bassi–Rebay, LDG, and Baumann–Oden fluxes for the viscous Burgers equation.

Flux term	L^2 error	$\|\partial u/\partial x(0)\|$
Bassi–Rebay	0.0135916	$\boldsymbol{q}^{\mathrm{BR}} = 150.171$
LDG	0.0103299	$\boldsymbol{q}^{\mathrm{LDG}} = 167.946$
Baumann–Oden	0.028627	$\boldsymbol{q}^{\mathrm{BO}} = 150.858$
Exact value	–	152.00516

test four equal elements were used, within which an expansion order of $P = 15$ was applied.

Both Bassi–Rebay and LDG show similar p-convergence properties for this problem, and both perform better than the Baumann–Oden flux. When examining the quantity $|\partial u/\partial x(0)|$, both Bassi–Rebay and Baumann–Oden are superior to LDG. This may be due, however, to the symmetries in the problem and the averaging nature of the Bassi–Rebay and Baumann–Oden fluxes. We recall that the LDG flux takes information from only one side during the intermediate stage (the computation of $\boldsymbol{q}^{\mathrm{LDG}}$) and uses information from the other direction in the final computation. This alternating feature of LDG may act as a form of prediction and correction so that the final computation of the second derivative yields an accurate solution. From the theoretical point of view, the gradient is approximated very accurately in the LDG method and, in fact, some superconvergence results have been reported on Cartesian grids in [101]. The following inequality establishes a relation between $\boldsymbol{q}^\delta$ and the gradient ∇u^δ in the L^2 norm:

$$\sum_e ||\boldsymbol{q}^\delta - \nabla u^\delta||_e^2 \leqslant C \int_{\Gamma_i} \frac{1}{h}[[u^\delta]]^2 \, \mathrm{d}\boldsymbol{x} \, ,$$

where Γ_i denotes the set of all interior edges.

Element distortion

Our second test demonstrates how convergence is affected by element distortion. We solve the parabolic equation $u_t = \nabla^2 u$ for an analytical solution of the form

$$u = \mathrm{e}^{\pi^2 t/12} \sin\left(\frac{\pi x}{6}\right) \sin\left(\frac{\pi y}{6}\right) \sin\left(\frac{\pi z}{6}\right),$$

with exact boundary conditions prescribed at all boundaries. The integration is for 1000 time-steps with $\Delta t = 10^{-5}$ to eliminate any temporal errors. We consider four different meshes consisting of twelve tetrahedra, as shown in Fig. 7.14. All tetrahedra share a common vertex at the centre of the box, which we move as shown to cause distortion of the tetrahedra. In the domain D elements with aspect ratio of 20 are obtained. The error plot of Fig. 7.14 shows that there is an effect of distortion, but exponential convergence is maintained.

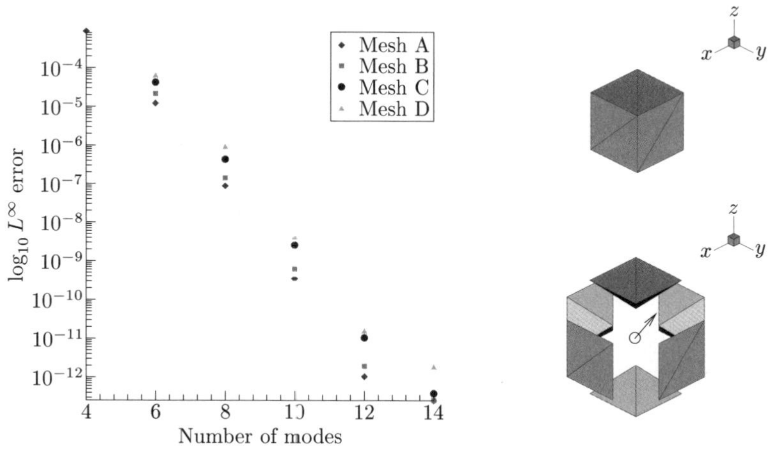

FIG. 7.14. L^∞ error of a three-dimensional parabolic problem as a function of the number of modes. Domain D has elements with aspect ratio of 20. ([309].)

7.5.9 *Stabilisation*

Starting with the interior penalty (IP) method, most discontinuous Galerkin methods, except the original Baumann–Oden and the Bassi–Rebay versions, employ a penalty term on the numerical flux of the form

$$\alpha_j = \eta h^{-1} [[u^\delta]] ,$$

with the free parameter η specifying, in some sense, how 'weak' or 'strong' the element interface condition is enforced. In the Brezzi *et al.* formulation [77], an alternative jump term was presented using the lift operator r_γ given by

$$\int_\Omega r_\gamma(\phi) \cdot \boldsymbol{w} \, \mathrm{d}x = - \int_\gamma \phi \cdot \{\boldsymbol{w}\} \, \mathrm{d}s , \quad \forall \, \boldsymbol{w} \in \mathcal{W}^\delta , \; \phi \in [L^1(\gamma)]^2 ,$$

where γ denotes an edge within the tessellation (which may be owned by only one element within the tessellation or may be shared by two elements), and where

$$\alpha_r(\phi) = -\eta r_\gamma(\phi) . \tag{7.5.22}$$

The free parameter η allows this stabilising factor to be tuned for the problem being solved. As with the IP and LDG, this stabilising factor is subtracted from the standard Bassi–Rebay numerical flux approximation $\tilde{\boldsymbol{q}}^\delta_\gamma = \{\boldsymbol{q}^\delta\}$, yielding a modified numerical flux $\tilde{\boldsymbol{q}}_\gamma = \{\boldsymbol{q}^\delta\} - \alpha_r([[u^\delta]])$.

Stabilisation of the Baumann–Oden method has been attempted in order to improve the accuracy of the method, as we have already presented. The NIPG-j versions of Riviere *et al.* [393, 394] are stabilisations of the Baumann–Oden method. The version with $j = 1$ corresponds to a stabilisation term similar to the interior penalty (IP) and LDG methods. It maintains a condition number for the stiffness matrix of $\mathcal{O}(h^{-2})$ but does not restore the loss of accuracy for even-degree polynomials. In contrast, the version NIPG-3 ($j = 3$) improves the accuracy for u only (in the L^2 norm), but the corresponding condition number is $\mathcal{O}(h^{-4})$ [91]. Both versions maintain a constant condition number as the stabilisation factor $\eta \to 0$ for $P \geqslant 2$, while for $P = 1$ the condition number grows but it is bounded. For large values of $\eta \gg 1$, the condition number scales linearly with η.

The stabilisation term introduced in [401] is of $\mathcal{O}(h/P^2)$ and is applied to the interior element interfaces only (i.e., the domain boundaries are excluded) in order to maintain conservativity. This stabilisation term establishes optimal convergence for all polynomials orders (i.e., odd or even) in the H^1 norm, but not in the L^2 norm. In addition, it was rigorously proved that the case of $P = 1$ converges sub-optimally at order $\frac{1}{2}$, although numerical experiments show a slope of order one.

The IP method converges only for very large values of the stabilisation factor η. In particular, it has to be above a constant value that depends on the regularity of the mesh as well as the polynomial order P [15]. For a mesh-dependent positive constant η_0, the condition number $\kappa(\eta)$ varies linearly with η for $\eta \gg 1$ and inversely proportionally with $\eta - \eta_0$, i.e., as $\mathcal{O}(1/(\eta - \eta_0))$ if $\eta - \eta_0 \ll 1$. The values of η_0 for $P = 1$ and $P = 2$ are 2.866 and 6.902, respectively, see [91]. Unlike the IP, the LDG method is stable for any positive value of η and, qualitatively, it behaves as the IP method for $\eta_0 = 0$.

Finally, we note that the theoretical and numerical results obtained by Castillo [91] show that the same accuracy is obtained for all DGM versions (including IP) for very large values of the stabilisation parameters.

7.5.10 *Discontinuous Galerkin versus mixed formulation*

In this section we compare the more classical mixed formulation with the discontinuous Galerkin formulation presented previously. To this end, we formulate a *mixed Galerkin* method involving two sets of test functions, one set in $L^2(\Omega)$ and the other one in $H^1(\Omega)$. We again consider the model problem for the scalar field $u(\boldsymbol{x})$:

$$-\nabla^2 u = f \qquad \text{in } \Omega\,, \tag{7.5.23a}$$

$$u = g(\boldsymbol{x}) \quad \text{on } \partial\Omega\,, \tag{7.5.23b}$$

and introduce, as before, the flux variable

$$\boldsymbol{q} = \nabla u\,,$$

with $\boldsymbol{q}(\boldsymbol{x}) \in \mathbf{H}(\text{div};\Omega)$, where we define the new functional space as

$$\mathbf{H}(\text{div};\Omega) = \{\boldsymbol{w} \in L^2(\Omega),\ \nabla \cdot \boldsymbol{w} \in L^2(\Omega)\}\,,$$

which is a Hilbert space equipped with the appropriate norm, that is,

$$||\boldsymbol{w}||_{\mathbf{H}(\mathrm{div};\Omega)} = \{||\boldsymbol{w}||^2 + ||\nabla \cdot \boldsymbol{w}||^2\}^{1/2} .$$

Upon substitution of the flux variable into the above system, we obtain

$$\boldsymbol{q} = \nabla u \quad \text{in } \Omega , \tag{7.5.24a}$$

$$\nabla \cdot \boldsymbol{q} = f \quad \text{in } \Omega , \tag{7.5.24b}$$

$$u = g(\boldsymbol{x}) \quad \text{on } \partial\Omega . \tag{7.5.24c}$$

The variational form can now be derived by testing eqn (7.5.24a) with functions $\boldsymbol{w} \in \mathbf{H}(\mathrm{div};\Omega)$. Correspondingly, we test eqn (7.5.24b) against functions $v \in L^2(\Omega)$, and subsequently integrate by parts. The Dirichlet variational problem corresponding to eqns (7.5.24a)–(7.5.24c) is then stated as follows.

Find $(\boldsymbol{q}, u) \in \mathbf{H}(\mathrm{div};\Omega) \times L^2(\Omega)$ such that

$$(\boldsymbol{w}, \boldsymbol{q}) = -(\nabla \cdot \boldsymbol{w}, u) + \langle g, \boldsymbol{w} \cdot \boldsymbol{n} \rangle_{\partial\Omega} , \quad \forall\, \boldsymbol{w} \in \mathbf{H}(\mathrm{div};\Omega) , \tag{7.5.25a}$$

$$(v, \nabla \cdot \boldsymbol{q}) = -(v, f) , \quad \forall\, v \in L^2(\Omega) , \tag{7.5.25b}$$

where $\boldsymbol{n}$ is the outwards unit normal; parentheses denote standard inner products over the *entire domain* Ω, and angular brackets denote inner products over the domain boundary $\partial\Omega$. This is a coupled system for $\boldsymbol{q}$ and u.

Next we define appropriate polynomial spaces for $\boldsymbol{w}, v$, and the unknowns $\boldsymbol{q}$ and u to guarantee stability of the approximation. This problem will be treated in greater detail when considering the Stokes problem (see Chapter 8), which is often treated using a mixed formulation [319]. Following similar arguments, we choose the polynomial space for $\boldsymbol{q}$ and $\boldsymbol{w}$ to be $\mathcal{P}_P(\Omega^e)$, and the polynomial space for u and v to be $\mathcal{P}_{P-1}(\Omega^e)$. For the numerical quadrature the same Gauss–Lobatto–Jacobi (or Gauss–Radau–Jacobi) points can be used for the flux variable and the velocity.

An example of the stability of the approximation with this choice of polynomial spaces is shown in Fig. 7.15, where we integrate the one-dimensional parabolic equation (7.5.1) by marching explicitly in time. The initial condition corresponds to a step function in the interval $[0, 1]$. The proper choice of polynomial space should produce the correct solution. A polynomial representation of equal order, however, produces unphysical oscillations which do not decay in time.

The *mixed formulation* produces exponential convergence of the error. To demonstrate this, we obtain the numerical solution to the heat equation $u_t = \nabla^2 u$ in a two-dimensional domain of a channel with a semicircular bump, shown in Fig. 7.16(b). The exact solution $u = \sin x \sin y\, \mathrm{e}^{-2t}$ was used and Dirichlet boundary conditions were applied on all boundaries. The final integration time was $t = 1.0$ and a small time-step of $\Delta t = 10^{-5}$ (using a third-order explicit scheme) was chosen to avoid any temporal discretisation errors. The plot of the L^∞ error is shown in Fig. 7.16(a) as a function of the expansion order for a fixed number of elements. We note that exponential convergence is demonstrated for

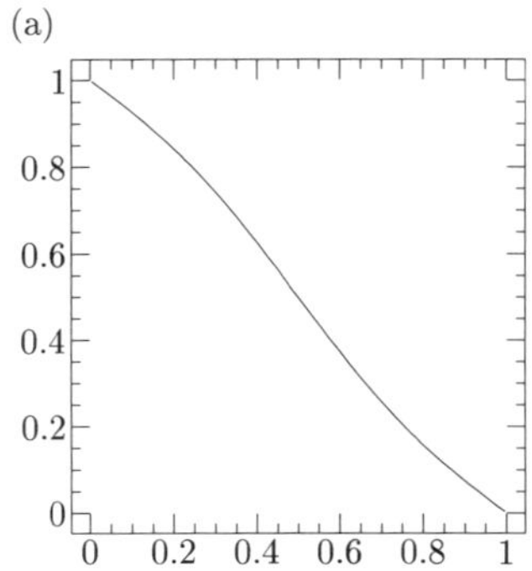

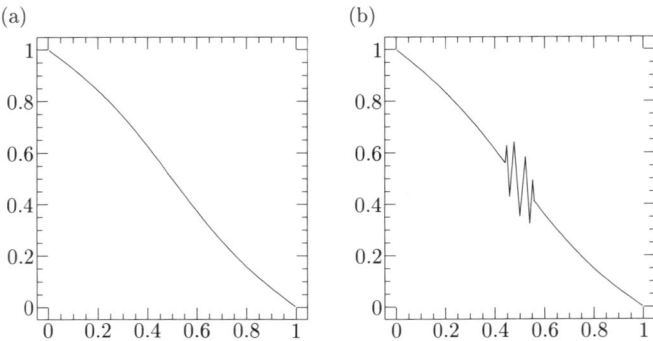

FIG. 7.15. Integration of the parabolic equation with discontinuous initial data for 50 000 time-steps, with $\Delta t = 10^{-6}$. The solution in the interval $x \in [0, 1]$ at $t = 0.05$ is plotted. (a) A stable approximation is obtained with polynomial orders $P = 6$ and 5 for the flux variable and the solution, respectively, while oscillations appear with equal polynomial order $P = 6$ (b). ([309].)

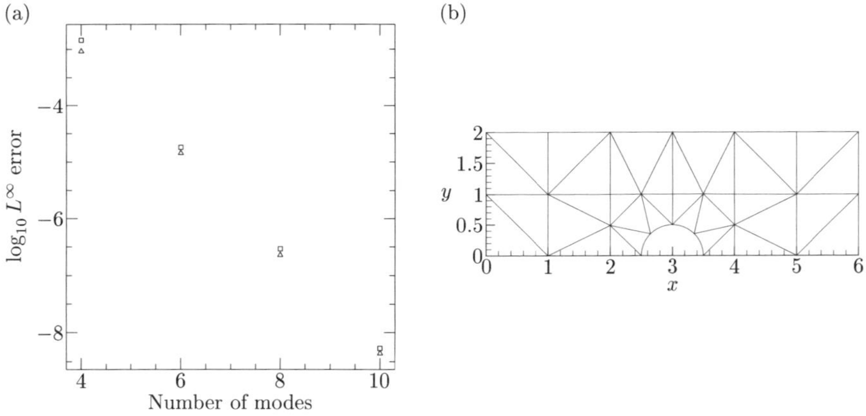

FIG. 7.16. (a) Convergence of the mixed ($\triangle$) and discontinuous ($\square$) Galerkin method for an analytical solution. The computational domain is the curvilinear region shown in (b). The parabolic equation is integrated up to time $t = 1$ with a time-step of $\Delta t = 10^{-5}$. ([309].)

this curvilinear geometry both for the mixed as well as the discontinuous Galerkin formulation (employing the Bassi–Rebay flux), as shown in Fig. 7.16(a).

As a final point, we note that for the explicit-in-time discontinuous Galerkin formulation, and unlike the mixed Galerkin formulation, an *equal-order* polynomial interpolation for the solution and the flux variable leads to a stable approximation [309].

7.5.11 *Which DGM version to use?*

The flexibility in choosing the numerical flux in DGM from either physical or mathematical considerations has led to several different formulations. Here we summarise the most popular features of the main versions associated with stability, accuracy, conditioning, and computational complexity.

- The Baumann–Oden method is inconsistent for $P = 0$, in contrast to the LDG and Bassi–Rebay methods, which are consistent and stable.
- The condition number of the stiffness matrix for the IP, Bassi–Rebay, LDG, Baumann–Oden, and NIPG-1 scales as $\mathcal{O}(h^{-2})$, whereas for the NIPG-3 it scales as $\mathcal{O}(h^{-4})$. In general, for a stabilisation term of the order $\mathcal{O}(h^{\alpha})$ the condition number of the corresponding stiffness matrix scales as $\mathcal{O}(h^{\alpha+1})$.
- The LDG method is more stable than the IP method since its stabilisation factor does not depend on the polynomial order or the mesh regularity. For very large values of the stabilisation factor, the stability of IP and LDG is the same.
- The maximum eigenvalue of the inverse of the mass matrix multiplied by the Laplacian matrix, for all DG methods, scales as $\mathcal{O}(P^4)$, similar to standard Galerkin methods, see Chapter 5. The LDG method has larger absolute values compared to the Bassi–Rebay or Baumann–Oden methods.
- The LDG method requires a more stringent time-step if an explicit time-stepping is used to advance the time-dependent parabolic equation.
- The LDG method with alternating fluxes achieves a uniform convergence rate of $\mathcal{O}(h^{P+1})$ for any polynomial order P. The Bassi–Rebay method achieves an $\mathcal{O}(h^P)$ rate for P odd and an $\mathcal{O}(h^{P+1})$ rate for P even. In contrast, the Baumann–Oden and NIPG-1 methods achieve an $\mathcal{O}(h^{P+1})$ rate for P odd and an $\mathcal{O}(h^P)$ rate for P even. The NIPG-3 version, and similarly the stabilised version of Baumann–Oden, converge uniformly in the L^2 and H^1 norms, respectively.
- For small values of the stabilisation factor η, the LDG method is more accurate than the IP or NIPG-j methods. However, for very large values of η all methods obtain the same accuracy.
- The IP, LDG, and Bassi–Rebay methods correspond to symmetric discretisations, unlike the Baumann–Oden method and its stabilised extensions. Therefore, the iterative solution of the corresponding linear systems is more efficient for the symmetric versions. However, the Bassi–Rebay method has a larger stencil and thus it requires more memory and operation counts, and it is less amenable to parallelism.

8

ALGORITHMS FOR INCOMPRESSIBLE FLOWS

In this chapter we discuss algorithms for incompressible flows appropriate for direct (DNS) and large-eddy (LES) simulations in complex-geometry domains. We are primarily interested in those formulations which are extendable to three dimensions and, therefore, we will focus on the primitive variables, velocity–vorticity, and the gauge formulations. Other formulations for general discretisations can be found in [215] and in [389]; a streamfunction–vorticity formulation of the unstructured spectral/*hp* element method can be found in [412].

This chapter is structured as follows. In Section 8.1 we discuss the variational formulation of the Stokes problem, which is the foundation of the following Navier–Stokes formulations where we will typically treat the nonlinear terms explicitly. In Section 8.2 we present coupled algorithms for the incompressible Navier–Stokes equations in terms of the primitive variables, that is, velocity–pressure $(\mathbf{v}, p)$. We then discuss in Section 8.3 several ways of splitting the Stokes operator so that we derive a more efficient approach to solving the Navier–Stokes equations. In Section 8.4 we present a formulation in terms of velocity–vorticity $(\mathbf{v}, \boldsymbol{\omega})$, and discuss two different ways of implementing this approach. Section 8.5 presents an alternative formulation, based on a least-squares projection, where vorticity is used as an auxiliary variable. In Section 8.6 we briefly discuss the gauge method in terms of two new variables, a gauge and auxiliary vector variable. Finally, in Section 8.7 we discuss the discretisation of the nonlinear terms in the Navier–Stokes formulation both in space as well as time using a semi-Lagrangian formulation.

8.1 Variational formulation

Let us consider an incompressible, isothermal flow with constant properties and homogeneous boundary conditions. The governing equations are the Navier–Stokes coupled to a velocity divergence-free constraint in a domain Ω. In terms of the primitive variables $(\mathbf{v}, p)$, they are written as

$$\frac{\partial \mathbf{v}}{\partial t} + \mathbf{v} \cdot \nabla \mathbf{v} = -\nabla p + \nu \nabla^2 \mathbf{v}\,, \tag{8.1.1a}$$

$$\nabla \cdot \mathbf{v} = 0\,, \tag{8.1.1b}$$

$$\mathbf{v} = 0 \quad \text{on } \partial\Omega\,, \tag{8.1.1c}$$

where $p(\boldsymbol{x}, t)$ is the pressure and ν is the kinematic viscosity. For simplicity, we assume that the density is $\rho = 1$.

To write these equations in variational form, we need to define appropriate functional spaces for the velocity and pressure fields. These are dictated by the highest spatial derivatives involved, and thus we consider the corresponding steady Stokes problem [191]:

$$-\nu\nabla^2\mathbf{v} + \nabla p = \boldsymbol{f}\,, \tag{8.1.2a}$$
$$-\nabla\cdot\mathbf{v} = 0\,, \tag{8.1.2b}$$
$$\mathbf{v} = 0 \quad \text{on } \partial\Omega\,, \tag{8.1.2c}$$

where $\boldsymbol{f}$ is a body force which may include all contributions resulting from an explicit (in time) treatment of the unsteady equations. The appropriate spaces are then defined as

$$H_0^1(\Omega) = \{w \in H^1(\Omega) \mid w = 0 \text{ on } \partial\Omega\}\,,$$
$$L_0^2(\Omega) = \left\{q \in L^2(\Omega) \;\middle|\; \int_\Omega q \,\mathrm{d}x = 0\right\}.$$

The zero Dirichlet nature of the space H_0^1 is consistent with a Galerkin formulation where nonzero Dirichlet boundary conditions have been 'lifted' from the problem. We also observe that the zero subscript of the space L_0^2 denotes a zero mean rather than a zero Dirichlet constraint on the boundary of the domain. Based on these definitions, the variational form of the Stokes problem (eqns (8.1.2a)–(8.1.2c)) can then be written as

$$(\nabla\mathbf{w}, \nu\nabla\mathbf{v}) - (\nabla\cdot\mathbf{w}, p) = (\mathbf{w}, \boldsymbol{f})\,, \quad \forall\, \mathbf{w} \in \mathcal{U} \equiv H_0^1(\Omega)^3\,, \tag{8.1.3a}$$
$$-(q, \nabla\cdot\mathbf{v}) = 0\,, \quad \forall\, q \in \mathcal{M} \equiv L_0^2(\Omega)\,, \tag{8.1.3b}$$

where $\mathcal{U}$ and $\mathcal{M}$ are the appropriate spaces for the velocity and pressure.

The discrete spaces for the velocity and pressure $(\mathcal{U}^\delta, \mathcal{M}^\delta)$ can be formed as subsets of the spaces defined above using polynomials of degree $\leqslant P$ for a subdomain Ω^e. However, the velocity and the pressure cannot be approximated independently due to the form of the coupling between them [22, 76]. A compatibility condition known as the inf–sup condition has to be satisfied by the discrete spaces for stability and uniqueness of the discrete solution. To this end, the pressure space $\mathcal{M}$ has to be restricted to exclude any parasitic (spurious) modes, defined by the condition

$$(\mathbf{w}, \nabla p^\star) = 0\,.$$

If the entire space $\mathcal{M}^\delta$ is employed then any pair $[\mathbf{v}^\delta, p^\delta + p^\star]$ is also a solution of the discrete version of the Stokes problem defined in eqns (8.1.3a) and (8.1.3b). To illustrate how the inf–sup condition is used to determine the appropriate pressure space $\mathcal{M}^\delta \subset \mathcal{M}$, we define the forms

$$a(\mathbf{w}, \mathbf{v}) = \nu(\nabla\mathbf{w}, \nabla\mathbf{v})\,,$$
$$b(\mathbf{w}, q) = -(\nabla\cdot\mathbf{w}, q)\,,$$

$$c(\mathbf{w}, \mathbf{u}, \mathbf{v}) = \int_\Omega \mathbf{w} \cdot (\mathbf{u} \cdot \nabla \mathbf{v})\, \mathrm{d}\boldsymbol{x}\,,$$

where the last term $c(\mathbf{w}, \mathbf{u}, \mathbf{v})$ is used in the Navier–Stokes formulation only. The discrete Stokes problem can be written in a compact form for the solution pair $[\mathbf{v}^\delta, p^\delta]$ as

$$a(\mathbf{w}, \mathbf{v}^\delta) + b(\mathbf{w}, p^\delta) = (\mathbf{w}, \boldsymbol{f})\,, \quad \forall\, \mathbf{w} \in \mathcal{U}^\delta\,,$$
$$b(\mathbf{v}^\delta, q) = 0\,, \quad \forall\, q \in \mathcal{M}^\delta\,,$$

where we assume that all inner products are computed exactly using appropriate Gauss quadratures. This problem has a unique solution if there exists a constant $\beta_\delta > 0$ such that

$$\sup_{\mathbf{v} \in \mathcal{U}^\delta} \frac{b(\mathbf{v}, q)}{||\mathbf{v}||_\mathcal{U}} \geqslant \beta_\delta ||q||_\mathcal{M}\,, \quad \forall\, q \in \mathcal{M}^\delta \quad \text{inf–sup condition}\,. \tag{8.1.4}$$

This condition excludes parasitic modes since they will make the left-hand side zero in the above equation. The uniqueness of the solution is then guaranteed by the estimates

$$||\mathbf{v}^\delta||_\mathcal{U} \leqslant \frac{\gamma}{\alpha_\delta} ||\boldsymbol{f}||_{\mathcal{U}'}\,, \quad ||p^\delta||_Q \leqslant \frac{1}{\beta_\delta}\left(1 + \frac{\gamma^2}{\alpha_\delta}\right) ||\boldsymbol{f}||_{\mathcal{U}'}\,,$$

where the constant γ is related to the continuity property of the bilinear form $a(\mathbf{w}, \mathbf{v})$, that is,

$$|a(\mathbf{w}, \mathbf{v})| \leqslant \gamma ||\mathbf{w}||_\mathcal{U} ||\mathbf{v}||_\mathcal{U}\,,$$

and α_δ is related to the ellipticity property of the same operator, that is,

$$a(\mathbf{w}, \mathbf{w}) \geqslant \alpha_\delta ||\mathbf{w}||^2_\mathcal{U}\,.$$

Here $\mathcal{U}'$ is the dual space of $\mathcal{U}$. The uniqueness of the approximation is guaranteed if these constants (including β_δ) are independent of the discretisation parameter $\delta = (N_{\mathrm{el}}, P)$, where we recall that N_{el} is the number of spectral elements Ω^e and P is the polynomial order. Convergence results and details can be found in [191] and in [86, 417].

Both finite element and spectral discretisations suffer from the presence of parasitic modes, as do spectral/*hp* element methods. The choice of the basis is not important, although purely Fourier methods do not have any spurious modes because of the lack of boundaries. Some of the theoretical results may be slightly different for Chebyshev or Legendre discretisations [86], but there is no theory yet specifically for general Jacobi polynomials that we deal with in this book. A straightforward choice for the restricted pressure space is the space formed by taking $\mathcal{M}^\delta$ as the subset of polynomials orthogonal to the space $\mathcal{M}$ but excluding the parasitic modes. However, such a space is not a good choice from the practical point of view as it leads to non-optimum values of β_δ. The standard approach is to reduce the polynomial space for the pressure representation accordingly (Bernardi and Maday [45]).

An example for spectral element methods, [319], is to determine the pressure space as

$$\mathcal{M}^{\delta_*} = \mathcal{M} \cap [\mathcal{P}_{P-2}(\Omega)]^3 ,$$

where $\delta_* = (N_{\text{el}}, P-2)$, that is, the pressure is approximated with a lower-degree polynomial. For the velocity space we have

$$\mathcal{U}^{\delta} = \mathcal{U} \cap [\mathcal{P}_P(\Omega)]^3 ,$$

where $\delta = (N_{\text{el}}, P)$. The theory for this so-called $\mathcal{P}_P(\Omega^e)/\mathcal{P}_{P-2}(\Omega^e)$ formulation has been presented in [45, 319]. A particular implementation using a staggered grid consisting of Gauss–Legendre and Gauss–Lobatto–Legendre collocation points for pressure and velocity, respectively, can be found in [402]. However, in that implementation the constant β_δ is not independent of the discretisation, and in three dimensions it scales as $1/P$ for a single element. This choice produces optimum error bounds for velocity but not for pressure. For a fixed number of elements N_{el} with this discretisation, we have

$$||\mathbf{v} - \mathbf{v}^{\delta}||_1 \leqslant C\{P^{1-m}(||\mathbf{v}||_m + ||p||_{m-1}) + P^{-\sigma}||\boldsymbol{f}||_{\sigma}\} ,$$
$$||p - p^{\delta}||_0 \leqslant C\{P^{2-m}(||\mathbf{v}||_m + ||p||_{m-1}) + P^{1-\sigma}||\boldsymbol{f}||_{\sigma}\} ,$$

where

$$\mathbf{v} \in H^m(\Omega)^3 , \quad p \in H^{m-1}(\Omega) ,$$

and

$$\boldsymbol{f} \in H^{\sigma}(\Omega)^3 .$$

The staggered grid has the same difficulty with implementation in three dimensions as the MAC method for finite differences [395]. The modal (rather than nodal) version of the spectral element method may be an easier approach in implementing this formulation. Some authors have used a single grid for discretisation, approximating the pressure in $\mathcal{P}(\Omega)$ with $\delta = (N_{\text{el}}, P)$. For example, in [376] a singular-value decomposition of the pressure operator was used to filter the spurious modes, as applied in other finite element methods. A domain decomposition approach which solves the Schur complement problems on the boundary of elemental regions using modal spectral/*hp* element methods has also been investigated in [5, 297]. This approach uses a single grid, but still applies different polynomial spaces for the pressure and velocities. Another approach is to use stabilisation methods of the form that Brezzi and Hughes introduced in finite element methods [78, 166]. This, however, introduces extra free parameters, which can be undesirable in direct and large-eddy simulation studies. Application of bubble functions in stabilising spectral discretisations (e.g., see [84]) shows that the extra pressure modes are suppressed and a unique solution is obtained.

The previous discussion has focused on the coupled Stokes system. An alternative way of solving the Navier–Stokes equations is to decouple or *split* the pressure and velocity fields in the Stokes problem. An elliptic consistent Poisson equation for pressure is derived which, along with the correct pressure boundary conditions, leads to unique solutions. This approach does not usually require the use of staggered grids for high-order elements and is therefore quite efficient, especially for high Reynolds numbers for which $\nu\Delta t$ is small. In the classical splitting methods the velocity error scales linearly with $\nu\Delta t$, but in the modern versions this error is of high order and of the form $\mathcal{O}((\nu\Delta t)^J)$, where $J \geqslant 2$ is the time-integration order. For larger values of $\nu\Delta t$, for example, in creeping flows, the splitting error may dominate, and thus a coupled approach may be more appropriate.

8.2 Coupled methods for primitive variables

In this section we discuss two approaches for discretising the primitive variable Stokes problem using a coupled velocity and pressure system. We start in Section 8.2.1 by discussing the Uzawa formulation, and then in Section 8.2.2 we discuss a substructuring approach to solving the system.

8.2.1 *The Uzawa algorithm*

We can write the discrete *steady Stokes* problem in matrix form by introducing the Laplacian matrix $\boldsymbol{L}$ and the mass matrix $\boldsymbol{M}$ as follows:

$$\nu \boldsymbol{L}\mathbf{v} - \boldsymbol{D}^\top \mathbf{p} = \boldsymbol{M}\boldsymbol{f}\,, \tag{8.2.1a}$$

$$-\boldsymbol{D}\mathbf{v} = 0\,, \tag{8.2.1b}$$

where the matrix $\boldsymbol{D} = (\boldsymbol{D}_{x_1}, \boldsymbol{D}_{x_2}, \boldsymbol{D}_{x_3})$ is the discrete gradient operator based on the derivative matrices defined in Chapter 4, and $\top$ denotes the transpose. In the system (8.2.1a) and (8.2.1b) we have dropped the superscript δ denoting the finite dimension space used in Section 8.1, for notational convenience. It is assumed that all boundary conditions are included in the matrix operators. The Uzawa algorithm provides a *saddle decomposition* of the Stokes problem, see [454]. We begin by solving eqn (8.2.1a) for the velocity:

$$\mathbf{v} = \frac{1}{\nu}(\boldsymbol{L}^{-1}\boldsymbol{D}^\top \mathbf{p} + \boldsymbol{L}^{-1}\boldsymbol{M}\boldsymbol{f})\,,$$

and we substitute into the continuity equation to obtain

$$-\boldsymbol{D}\mathbf{v} = -\frac{1}{\nu}(\boldsymbol{D}\boldsymbol{L}^{-1}\boldsymbol{D}^\top \mathbf{p} + \boldsymbol{D}\boldsymbol{L}^{-1}\boldsymbol{M}\boldsymbol{f}) = 0\,.$$

The Stokes problem (8.2.1a) and (8.2.1b) can then be recast into the equivalent form

$$\nu \boldsymbol{L}\mathbf{v} = \boldsymbol{D}^\top \mathbf{p} + \boldsymbol{M}\boldsymbol{f}\,, \tag{8.2.2a}$$

$$\boldsymbol{S}_0 \mathbf{p} = -\boldsymbol{D}\boldsymbol{L}^{-1}\boldsymbol{M}\boldsymbol{f}\,, \tag{8.2.2b}$$

where we define the pressure matrix

$$\boldsymbol{S}_0 \equiv \boldsymbol{D}\boldsymbol{L}^{-1}\boldsymbol{D}^\top ,$$

and we note that $\boldsymbol{L}$ and $\boldsymbol{S}_0$ are positive semi-definite symmetric matrices. The saddle decomposition corresponds to a decoupling of the original Stokes problem into two symmetric positive-definite forms. We also note that $\boldsymbol{S}_0$ will not be invertible, due to a rank deficiency, if the pressure space dimension of $\boldsymbol{D}$ is larger than the dimension of $\boldsymbol{L}$, which corresponds to the velocity space. This necessary constraint can be seen as a discrete motivation for the inf–sup condition.

In the solution process, we first solve for the pressure and subsequently for the velocity. No boundary conditions are needed for the pressure as we assume that pressure is discontinuous across elemental interfaces and is not defined on any boundaries. For example, the $\mathcal{P}_P(\Omega^e)/\mathcal{P}_{P-2}(\Omega^e)$ formulation discussed in the previous section, with the pressure defined on the $P-2$ Gauss points and the velocity defined on the P Gauss–Lobatto points, fulfils such a requirement. Alternatively, a $(P-2)$-order modal expansion could have been adopted for the pressure and a P modal expansion for the velocity using the same quadrature points, which consistently integrate the richer velocity space. The important point here is that, although the Uzawa algorithm decouples the pressure and velocity, it does not require any new discretisation and is simply a formal rearrangement of the original Stokes problem. The pressure matrix $\boldsymbol{S}_0$ is a full matrix due to the presence of the inverse Laplacian, which reduces the efficiency of an iterative solver to invert this system. However, $\boldsymbol{S}_0$ is very well conditioned, as can be argued heuristically by writing

$$\boldsymbol{S}_0 \mapsto \nabla \cdot (\nabla^2)^{-1} \nabla \approx \frac{\nabla \cdot \nabla}{\nabla^2} \approx 1 .$$

This implies that the matrix $\boldsymbol{S}_0$ is very close to the mass matrix $\boldsymbol{M}$, which is the discrete analogue of the identity matrix. This point has motivated the use of a preconditioned conjugate gradient algorithm (PCG) in [402] with an inner–outer iteration procedure. Similar to the scalar definition in Section 5.4.2, the *outer iteration* is

$$
\begin{aligned}
&\text{compute } \boldsymbol{p}_0;\ \boldsymbol{r}_0 = \boldsymbol{D}\boldsymbol{L}^{-1}\boldsymbol{M}\boldsymbol{f} + \boldsymbol{S}_0\boldsymbol{p}_0;\ \boldsymbol{q}_0 = \boldsymbol{M}_p^{-1}\boldsymbol{r}_0;\ \boldsymbol{p}_0 = \boldsymbol{q}_0\\
&\quad \text{do } m = 0, \text{ until convergence}\\
&\qquad \alpha_m \quad\ = \boldsymbol{q}_m^\top \boldsymbol{r}_m / \boldsymbol{p}_m^\top \boldsymbol{S}_0 \boldsymbol{p}_m\\
&\qquad \mathbf{v}_{m+1} = \mathbf{v}_m + \alpha_m \boldsymbol{p}_m\\
&\qquad \boldsymbol{r}_{m+1} = \boldsymbol{r}_m - \alpha_m \boldsymbol{S}_0 \boldsymbol{p}_m\\
&\qquad \boldsymbol{q}_{m+1} = \boldsymbol{M}_p^{-1} \boldsymbol{r}_{m+1}\\
&\qquad \beta_m \quad\ = \boldsymbol{q}_{m+1}^\top \boldsymbol{r}_{m+1} / \boldsymbol{q}_m^\top \boldsymbol{r}_m\\
&\qquad \boldsymbol{p}_{m+1} = \boldsymbol{q}_{m+1} + \beta_m \boldsymbol{p}_m\\
&\quad \text{continue}
\end{aligned}
$$

Here $\boldsymbol{p}$, $\boldsymbol{q}$, and $\boldsymbol{r}$ are vectors and α_m and β_m are scalars. The mass matrix $\boldsymbol{M}_p$ associated with the pressure grid discretisation has been used as a preconditioner.

Note that the residual $\boldsymbol{r}_m$ corresponds to the discrete-divergence $-\boldsymbol{D}\mathbf{v}$, and thus the specified convergence tolerance reflects *directly* the degree to which the velocity field is divergence-free in a weak sense.

The computationally expensive component of the above algorithm is the matrix–vector evaluation of $\boldsymbol{S}_0\boldsymbol{p}$, which can be evaluated with an *inner iteration* to invert $\boldsymbol{L}$ based on decomposing $\boldsymbol{S}_0$ into its separate components as

$$\begin{aligned}\boldsymbol{y} &= \boldsymbol{D}^\top \boldsymbol{p}\,,\\ \boldsymbol{L}\boldsymbol{z} &= \boldsymbol{y}\,,\\ \boldsymbol{S}_0\boldsymbol{p} &= \boldsymbol{D}\boldsymbol{z}\,,\end{aligned}$$

where $\boldsymbol{y}$ and $\boldsymbol{z}$ are intermediate vectors. The number of full PCG iterations depends on the condition number of the matrix $\boldsymbol{M}_p^{-1}\boldsymbol{S}_0$, the spectrum of which was analysed by Maday *et al.* [317]. For periodic problems the condition number is exactly one, and for semi-periodic problems the entire spectrum of eigenvalues is one, except for the two eigenvalues associated with the two non-periodic boundary conditions for which

$$\lambda_0 = \frac{1}{2} - \frac{k}{\sinh(2k)}\,, \qquad \lambda_1 = \frac{1}{2} + \frac{k}{\sinh(2k)}\,,$$

where k is the wavenumber. For small and moderate k the numerical results of Rønquist [402] agree with the theory; however, for large k the discrete condition number seems to be higher than that predicted theoretically due to the finite resolution of the spectral element discretisation.

It was documented in [317] that the condition number of the pressure matrix relates to the inf–sup constant β_δ and is proportional to $(1/\beta_\delta)^2$; thus the number of conjugate gradient (CG) iterations scales as $\mathcal{O}((\beta_\delta)^{-1})$. Given that in three dimensions $\beta_\delta \approx \mathcal{O}(P^{-1})$ for the discrete spectral element Legendre discretisation introduced in [402], it is clear that for multi-dimensional problems there is some dependence of the number of CG iterations on the polynomial order of the problem. Numerical experiments presented in [402] were for the two-dimensional case for which $\beta_\delta \approx \mathcal{O}(P^{-1/2})$, and thus no such obvious dependence was observed, especially as relatively small values of spectral order ($P \approx 10$) were employed.

We now consider the *unsteady Stokes* equation and apply an implicit Euler scheme to discretise in time as follows:

$$\boldsymbol{M}\frac{\mathbf{v}^{n+1} - \mathbf{v}^n}{\Delta t} + \nu\boldsymbol{L}\mathbf{v}^{n+1} - \boldsymbol{D}^\top\mathbf{p}^{n+1} = \boldsymbol{M}\boldsymbol{f}^{n+1}\,, \tag{8.2.3a}$$

$$-\boldsymbol{D}\mathbf{v}^{n+1} = 0\,. \tag{8.2.3b}$$

This system is *absolutely* stable, as can be shown using energy arguments. We consider only the *homogeneous* part of eqn (8.2.3a) and multiply by $\mathbf{v}^{n+1}$:

$$\mathbf{v}^{n+1^\top} \cdot \left(\frac{\boldsymbol{M}}{\Delta t} + \nu\boldsymbol{L}\right)\mathbf{v}^{n+1} = \mathbf{v}^{n+1^\top} \cdot \frac{\boldsymbol{M}}{\Delta t}\mathbf{v}^n + \mathbf{v}^{n+1^\top} \cdot \boldsymbol{D}^\top\mathbf{p}^{n+1}\,.$$

The last term is zero since, from eqn (8.2.3b), we observe that

$$\mathbf{v}^{n+1^\top} \cdot \boldsymbol{D}^\top \mathbf{p}^{n+1} = \mathbf{p}^{n+1^\top} \boldsymbol{D}\mathbf{v}^{n+1} = 0\,,$$

and therefore we obtain

$$\begin{aligned} \mathbf{v}^{n+1^\top} \cdot \frac{\boldsymbol{M}}{\Delta t}\mathbf{v}^{n+1} &< \mathbf{v}^{n+1^\top} \cdot \left(\frac{\boldsymbol{M}}{\Delta t} + \nu \boldsymbol{L}\right) \mathbf{v}^{n+1} \\ &= \mathbf{v}^{n+1^\top} \cdot \frac{\boldsymbol{M}}{\Delta t}\mathbf{v}^{n} \leqslant \left(\mathbf{v}^{n+1^\top} \cdot \frac{\boldsymbol{M}}{\Delta t}\mathbf{v}^{n+1}\right)^{1/2} \left(\mathbf{v}^{n^\top} \cdot \frac{\boldsymbol{M}}{\Delta t}\mathbf{v}^{n}\right)^{1/2}, \end{aligned}$$

which implies

$$\mathbf{v}^{n+1^\top} \cdot \frac{\boldsymbol{M}}{\Delta t}\mathbf{v}^{n+1} < \mathbf{v}^{n^\top} \cdot \frac{\boldsymbol{M}}{\Delta t}\mathbf{v}^{n}\,,$$

and thus absolute stability is established. The first inequality is valid due to the fact that $\boldsymbol{L}$ is positive definite and ν is a positive constant. The last inequality is due to the Cauchy–Schwartz inequality since $\boldsymbol{M}$ is also positive definite.

We can use the Uzawa algorithm, as before, to decouple the pressure and velocity fields:

$$\boldsymbol{H}\mathbf{v}^{n+1} - \boldsymbol{D}^\top \mathbf{p}^{n+1} = \boldsymbol{M}\bar{\boldsymbol{f}}^{n+1}\,, \tag{8.2.4a}$$

$$\boldsymbol{S}_t \mathbf{p}^{n+1} = -\boldsymbol{D}\boldsymbol{H}^{-1}\boldsymbol{M}\bar{\boldsymbol{f}}^{n+1}\,, \tag{8.2.4b}$$

where

$$\bar{\boldsymbol{f}}^{n+1} = \boldsymbol{f}^n + \frac{\boldsymbol{M}\mathbf{v}^n}{\Delta t}$$

and the Helmholtz operator and *unsteady* pressure operator $\boldsymbol{S}_t$ are defined as

$$\boldsymbol{H} \equiv \frac{\boldsymbol{M}}{\Delta t} + \nu \boldsymbol{L}\,, \quad \boldsymbol{S}_t \equiv \boldsymbol{D}\boldsymbol{H}^{-1}\boldsymbol{D}^\top\,.$$

Similar to the steady Stokes case, the matrix $\boldsymbol{S}_t$ is a full matrix due to presence of the inverse Helmholtz operator. However, unlike the matrix $\boldsymbol{S}_0$, the 'unsteady' matrix $\boldsymbol{S}_t$ is not well conditioned due to the unsteady term which dominates for $\Delta t \to 0$, in which case

$$\boldsymbol{S}_t \to (\Delta t)^{-1}\boldsymbol{E}\,,$$

with

$$\boldsymbol{E} \equiv \boldsymbol{D}\boldsymbol{M}^{-1}\boldsymbol{D}^\top\,.$$

That is, for small time-steps the pressure matrix approaches a pseudo-Laplacian $\boldsymbol{E}$, which is poorly conditioned. This pseudo-Laplacian matrix also appears if the unsteady Stokes problem is discretised in time using an explicit Euler scheme. In that case, we have

$$\begin{aligned} -\boldsymbol{E}p^{n+1} &= \boldsymbol{D}\boldsymbol{M}^{-1}\boldsymbol{f}^n\,, \\ \boldsymbol{M}\mathbf{v}^{n+1} &= \boldsymbol{D}^\top p^{n+1} + \boldsymbol{f}^n\,, \end{aligned}$$

where $\boldsymbol{f}^n$ has been redefined to include all explicit contributions. In the next section we discuss approaches to inverting the unsteady pressure matrix $\boldsymbol{S}_t$ efficiently.

8.2.1.1 *Pressure preconditioners*

Typically, in the solution process for the incompressible Navier–Stokes equations, the most computationally intensive part is the pressure solver. For relatively large values of $\nu\Delta t$ the *inverse pressure mass matrix* is a good preconditioner, but at small values of $\nu\Delta t$ it is no longer effective. In this case, new preconditioners have to be devised so that the coupled methods studied in this section can be competitive with the splitting methods, especially at high Reynolds numbers. The problem in inverting the pressure matrix operator is related to the lack of continuity across interfaces. This contrasts with the velocity operator, which is relatively easier to invert.

A key idea in obtaining effective pressure preconditioners was put forward by Cahouet and Chabard [81] based on the following operator identity:

$$-\nabla \cdot (\Delta t^{-1}\boldsymbol{I} - \nu\nabla^2)^{-1}\nabla = (\nu\boldsymbol{I}^{-1} - \Delta t^{-1}(\nabla^2)^{-1})^{-1},$$

where $\boldsymbol{I}$ is the identity operator. Although this identity is not exactly true for the corresponding discrete operators, it provides a good starting-point for the construction of appropriate preconditioners. For example,

$$\boldsymbol{Q}^{-1} = \nu\boldsymbol{M}_p^{-1} + \Delta t^{-1}(\boldsymbol{D}\boldsymbol{M}^{-1}\boldsymbol{D}^\top)^{-1} = \nu\boldsymbol{M}_p^{-1} + \Delta t^{-1}\boldsymbol{E}^{-1} \tag{8.2.5}$$

is a good preconditioner as $\boldsymbol{Q}^{-1}\boldsymbol{S}_t \approx \boldsymbol{I}$. In fact, for a purely Fourier discretisation, the spectrum of $\boldsymbol{S}_t^{-1}$ is identical to that of $\boldsymbol{Q}^{-1}$.

In spectral element methods, the inversion of the pseudo-Laplacian ($\boldsymbol{E}^{-1}$) is required in order to construct the preconditioner. This, in itself, is a difficult task as $\boldsymbol{E}$ is ill-conditioned and therefore inversion requires significant computational work. To this end, Rønquist [403] has proposed solving the pressure equation in two steps, first solving for a *coarse-grid* pressure on the mesh macro-skeleton consisting of constant pressure in each element, and subsequently solving a *fine-grid* correction for pressure variations within the element. Denoting the two pressure fields by $\mathbf{p}_c$ and $\mathbf{p}_f$, respectively, we solve the following pressure system:

$$\boldsymbol{S}_t(\boldsymbol{\mathcal{I}}\mathbf{p}_c + \mathbf{p}_f) = -\boldsymbol{D}\boldsymbol{H}^{-1}\boldsymbol{M}\boldsymbol{f}^{n+1},$$

where $\boldsymbol{\mathcal{I}}$ corresponds to an interpolation operator that maps the constant pressure levels to the standard (fine) pressure grid. To obtain a separate equation for $\mathbf{p}_c$, we pre-multiply by $\boldsymbol{\mathcal{I}}^\top$ and denote by $\boldsymbol{S}_t^c \equiv \boldsymbol{\mathcal{I}}^\top\boldsymbol{S}_t\boldsymbol{\mathcal{I}}$ and $\boldsymbol{S}_t^f \equiv \boldsymbol{G}_t\boldsymbol{S}_t$ (where $\boldsymbol{G}_t \equiv \boldsymbol{I} - \boldsymbol{S}_t\boldsymbol{\mathcal{I}}(\boldsymbol{S}_t^c)^{-1}\boldsymbol{\mathcal{I}}^\top$) the coarse-grid and fine-grid pressure operators, respectively, to arrive at

$$\boldsymbol{S}_t^f\mathbf{p}_f = -\boldsymbol{G}_t\boldsymbol{D}\boldsymbol{H}^{-1}\boldsymbol{M}\boldsymbol{f}^{n+1}, \tag{8.2.6a}$$

$$\boldsymbol{S}_0^c\mathbf{p}_c = \boldsymbol{\mathcal{I}}^\top(-\boldsymbol{D}\boldsymbol{H}^{-1}\boldsymbol{M}\boldsymbol{f}^{n+1} - \boldsymbol{S}_t\mathbf{p}_f). \tag{8.2.6b}$$

This system is still coupled but the fine-grid solution, which can be efficiently preconditioned by the elemental matrix $\boldsymbol{S}_t^e$, can be solved first. $\boldsymbol{S}_t^e$ is constructed based on individual elements (decoupled) and zero Dirichlet conditions incorporated in $\boldsymbol{H}$ at all boundaries. A standard preconditioned conjugate gradient (PCG) method can then be used for the solver in eqn (8.2.6a). A direct solver can be employed for solving eqn (8.2.6b) as it involves a relatively small number of degrees of freedom, even in three dimensions.

We have now transferred the problem to inverting the *local* $\boldsymbol{S}_t^e$ matrices instead of the pseudo-Laplacian $\boldsymbol{E}$, but these are still difficult to invert. Again, adopting the idea of Cahouet and Chabard [81], we can employ the spectrally equivalent *local* matrix

$$[(\nu \boldsymbol{M}_p^{-1} + \Delta t^{-1} \boldsymbol{E}^{-1})^{-1}]^e ,$$

so that we only need to invert the *local* pseudo-Laplacian on each element. This approach was adopted in Rønquist [403] using a direct method, which is costly, especially in three dimensions. However, other cost-effective approaches based on fast diagonalisation techniques have been proposed that work well both for simple and deformed geometries [115]. An even more effective approach, suggested by Couzy and Deville [115], is to use the coarse/fine-grid decomposition directly on the pseudo-Laplacian $\boldsymbol{E} \mapsto (\boldsymbol{E}^c, \boldsymbol{E}^f)$ to invert it and then use a PCG iteration on the original $\boldsymbol{S}_t$ pressure matrix. This is equivalent to using the preconditioner of eqn (8.2.5) directly. The local pseudo-Laplacian $\boldsymbol{E}^e$ can also be used as the preconditioner here with mixed Dirichlet/Neumann boundary conditions at interfaces [114].

The preceding discussion is appropriate for coupled methods on staggered grids. However, if a single grid is used then other more straightforward approaches are possible. For example, in [463] a finite element preconditioner for the Poisson equation was proposed, where linear elements are constructed based on a Gauss–Lobatto–Legendre mesh. It was shown that the convergence rate was independent of the spectral order P. The linear finite element operator is a banded matrix with twenty-seven diagonals in three dimensions, and thus standard inversion algorithms for banded matrices (e.g., a skyline solver) can be used to invert the preconditioner in a pre-processing stage. This approach can also be very effective, especially for the splitting methods that we discuss in Section 8.3, where a Poisson equation for pressure is usually involved.

8.2.2 *Substructured Stokes system*

An alternative approach to solving the coupled Stokes system using the Uzawa approach is to apply the static condensation or substructuring technique presented for the scalar elliptic equation in Section 4.2.3. The formulation and preconditioning of this approach has previously been investigated by LeTallec and Patra [297] and Ainsworth and Sherwin [5, 428].

We start by partitioning the velocity space, $\mathcal{U}^\delta$, into two sets consisting of elemental interior modes, denoted by a subscript i, and elemental boundary modes, denoted by a subscript b. We also decompose the pressure space, $\mathcal{M}^\delta$, into two sets consisting of the average (constant) pressure modes within each element, denoted by a subscript a, and the remaining interior pressure modes, denoted by a subscript 'ι'. We note that, similar to the Uzawa algorithm, the pressure space is discontinuous and so the mean average pressure mode will typically consist of a single mode, and thus all other expansion modes are treated as interior modes. The discrete Stokes system can then be written as

$$\left[\begin{array}{cc|cc} \nu\boldsymbol{L}_{bb} & \boldsymbol{D}_{ab}^{\top} & \nu\boldsymbol{L}_{bi} & \boldsymbol{D}_{\iota b}^{\top} \\ \boldsymbol{D}_{ab} & & \boldsymbol{D}_{ai} & \\ \hline \nu\boldsymbol{L}_{ib} & \boldsymbol{D}_{ai}^{\top} & \nu\boldsymbol{L}_{ii} & \boldsymbol{D}_{i\iota}^{\top} \\ \boldsymbol{D}_{\iota b} & & \boldsymbol{D}_{\iota i} & \end{array}\right] \left[\begin{array}{c} \boldsymbol{v}_b \\ \boldsymbol{p}_a \\ \hline \boldsymbol{v}_i \\ \boldsymbol{p}_\iota \end{array}\right] = \left[\begin{array}{c} \boldsymbol{f}_b \\ \boldsymbol{g}_a \\ \hline \boldsymbol{f}_i \\ \boldsymbol{g}_\iota \end{array}\right], \tag{8.2.7}$$

where $\boldsymbol{L}$ and $\boldsymbol{D}$ denote the weak Laplacian and derivative matrices, and the subscripts correspond to the subspaces defined earlier. Further, $\boldsymbol{f}$ and $\boldsymbol{g}$ represent the right-hand terms of the momentum and divergence equations after boundary conditions have been imposed. Note that $\boldsymbol{g}$ may be nonzero due to the lifting of a known solution to impose a Dirichlet boundary condition. In general, we do not have to consider the interior velocity and pressure modes as being restricted to a single elemental domain since we could also choose a subdomain defined as a cluster of adjacent elements. Finally, we note that the pressure and velocity spaces must satisfy the inf–sup condition. When using nodal spectral element methods, the obvious space is to use $\mathcal{P}_P/\mathcal{P}_{P-2}$ for the velocity and pressure. However, for a modal expansion, different combinations of spaces are possible, as discussed by Stenberg and Suri [441].

The interior unknown in the system (8.2.7) can be eliminated using static condensation, yielding a boundary system of the form

$$\boldsymbol{S}\begin{bmatrix} \boldsymbol{v}_b \\ \boldsymbol{p}_a \end{bmatrix} = \begin{bmatrix} \boldsymbol{f}_b \\ \boldsymbol{g}_a \end{bmatrix} - \begin{bmatrix} \nu\boldsymbol{L}_{bi} & \boldsymbol{D}_{\iota b}^{\top} \\ \boldsymbol{D}_{ai} & \end{bmatrix} \boldsymbol{T} \begin{bmatrix} \boldsymbol{f}_i \\ \boldsymbol{g}_\iota \end{bmatrix}, \tag{8.2.8}$$

where

$$\boldsymbol{T} = \begin{bmatrix} \nu\boldsymbol{L}_{ii} & \boldsymbol{D}_{i\iota}^{\top} \\ \boldsymbol{D}_{\iota i} & \end{bmatrix}^{-1}$$

and

$$\boldsymbol{S} = \begin{bmatrix} \boldsymbol{S}_{bb} & \boldsymbol{S}_{ab}^{\top} \\ \boldsymbol{S}_{ab} & \boldsymbol{S}_{aa} \end{bmatrix} = \begin{bmatrix} \nu\boldsymbol{L}_{bb} & \boldsymbol{D}_{ab}^{\top} \\ \boldsymbol{D}_{ab} & \end{bmatrix} - \begin{bmatrix} \nu\boldsymbol{L}_{bi} & \boldsymbol{D}_{\iota b}^{\top} \\ \boldsymbol{D}_{ai} & \end{bmatrix} \boldsymbol{T} \begin{bmatrix} \nu\boldsymbol{L}_{ib} & \boldsymbol{D}_{ai}^{\top} \\ \boldsymbol{D}_{\iota b} & \end{bmatrix} \tag{8.2.9}$$

is the Schur complement. Once the boundary unknowns have been determined, the interior unknowns can be obtained by back-substitution such that

$$\begin{bmatrix} \boldsymbol{v}_i \\ \boldsymbol{p}_\iota \end{bmatrix} = \boldsymbol{T}\begin{bmatrix} \boldsymbol{f}_b \\ \boldsymbol{g}_a \end{bmatrix} - \begin{bmatrix} \nu\boldsymbol{L}_{bi} & \boldsymbol{D}_{\iota b}^{\top} \\ \boldsymbol{D}_{ai} & \end{bmatrix} \boldsymbol{T} \begin{bmatrix} \boldsymbol{v}_b \\ \boldsymbol{p}_a \end{bmatrix}. \tag{8.2.10}$$

The reduction of the full system to the boundary system given by eqn (8.2.8) is the classical substructuring or static condensation procedure. The main advantage, which may not be immediately transparent from the above presentation, is that the matrix $\boldsymbol{T}$ has a block *diagonal structure* based around each of its elemental (or subdomain) components. The block diagonal structure reflects that there is no direct coupling between interior modes on the distinct elements. This structure can be exploited in so far as both the matrices $\boldsymbol{T}$ and $\boldsymbol{S}$ can be constructed at an elemental level and then globally assembled if necessary. This is possible since

$$\boldsymbol{T} = \underline{\boldsymbol{T}}^e = \underline{\begin{bmatrix} \nu \boldsymbol{L}^e_{ii} & \boldsymbol{D}^{e\;\top}_{\iota i} \\ \boldsymbol{D}^e_{\iota i} & \end{bmatrix}^{-1}},$$

where the superscripts denote the elemental contribution and we recall that the underline implies a block diagonal extension of the matrices. The Schur complement $\boldsymbol{S}$ can therefore be written as

$$\boldsymbol{S} = \boldsymbol{\mathcal{A}}_b^\top \underline{\boldsymbol{S}}^e \boldsymbol{\mathcal{A}}_b\,,$$

where

$$\boldsymbol{S}^e = \begin{bmatrix} \nu \boldsymbol{L}^e_{bb} & \boldsymbol{D}^{e\;\top}_{ab} \\ \boldsymbol{D}^e_{ab} & \end{bmatrix} - \begin{bmatrix} \nu \boldsymbol{L}^e_{bi} & \boldsymbol{D}^{e\;\top}_{\iota b} \\ \boldsymbol{D}^e_{ai} & \end{bmatrix} \boldsymbol{T}^e \begin{bmatrix} \nu \boldsymbol{L}^e_{ib} & \boldsymbol{D}^{e\;\top}_{ai} \\ \boldsymbol{D}^e_{\iota b} & \end{bmatrix}$$

and $\boldsymbol{\mathcal{A}}_b^\top$ denotes the local to global assembly operation of the elemental boundary degrees of freedom for the velocity–pressure system, analogous to the definition discussed in Section 4.2.1.1. The matrix $\boldsymbol{T}^e$ represents a coupled system between the pressure and velocity interior degrees of freedom. However, for a Newtonian fluid the matrix $\boldsymbol{L}^e_{ii}$ is a block diagonal system containing the Laplacian matrix operating on the components of the velocity. We can make further use of the structure in this system by applying the static condensation concept at an elemental level to evaluate the operation of $\boldsymbol{T}$. We also note that the problem

$$\begin{bmatrix} \boldsymbol{v}^e_i \\ \boldsymbol{p}^e_\iota \end{bmatrix} = \boldsymbol{T}^e \begin{bmatrix} \boldsymbol{f}^e_b \\ \boldsymbol{g}^e_a \end{bmatrix} \tag{8.2.11}$$

may be reformulated as

$$\begin{bmatrix} \nu \boldsymbol{L}^e_{ii} & \boldsymbol{D}^{e\;\top}_{\iota i} \\ & \boldsymbol{D}^e_{ii}[\nu \boldsymbol{L}^e_{ii}]^{-1} \boldsymbol{D}^{e\;\top}_{ii} \end{bmatrix} \begin{bmatrix} \boldsymbol{v}^e_i \\ \boldsymbol{p}^e_\iota \end{bmatrix} = \begin{bmatrix} \boldsymbol{f}^e_b \\ \boldsymbol{D}^e_{ii}[\nu \boldsymbol{L}^e_{ii}]^{-1} \boldsymbol{f}^e_b - \boldsymbol{g}_a \end{bmatrix}, \tag{8.2.12}$$

where the second row block is initially solved to determine $\boldsymbol{p}^e_\iota$, and then $\boldsymbol{v}^e_i$ is evaluated by solving the first row of eqn (8.2.12). We note that this approach only involves dealing with the positive-definite matrices $\boldsymbol{L}^e_{ii}$ and $\boldsymbol{D}^e_{\iota i}[\nu \boldsymbol{L}^e_{ii}]^{-1} \boldsymbol{D}^{e\;\top}_{\iota i}$.

Similar to the Uzawa algorithm in Section 8.2.1, the introduction of the unsteady term leads to the Laplacian matrix $\nu \boldsymbol{L}$, with a Helmholtz operator $\boldsymbol{H} = \nu \boldsymbol{L} + \boldsymbol{M}/\Delta t$. The substructuring leads to a global boundary system which is decoupled from the interior pressure and velocity modes. We are, however, left with the problem of inverting $\boldsymbol{S}$, which may need to be performed iteratively, particularly in three dimensions.

8.2.2.1 *Preconditioned iterative solution of the Schur complement system*

The interface problem of the Schur complement system $\boldsymbol{S}$ defined in eqn (8.2.9) has one negative eigenvalue for each element or subdomain of the domain decomposition. Therefore, the conjugate gradient algorithm is not directly applicable owing to the indefiniteness of the matrix. However, it is still possible to reformulate the problem on a subspace so that the conjugate gradient method is applicable.

A number of preconditioners for the Schur complement system have been investigated both numerically and theoretically in [5, 297]. In [5] two preconditioners were considered, both of which used an additive Schwarz preconditioner to approximate the $\boldsymbol{S}_{bb}$ submatrix of $\boldsymbol{S}$. The additive Schwarz velocity preconditioner, denoted by $\boldsymbol{C}_{bb}$, approximates $\boldsymbol{S}_{bb}$ using a coarse space block corresponding to the vertex modes, together with the block diagonal component corresponding to each subdomain edge within the domain.

As discussed in [5], for this type of preconditioner the matrix $\boldsymbol{S}_{bb}$ can be bounded from above and below by the preconditioner $\boldsymbol{C}_{bb}$ by an expression of the form

$$\beta_\delta^2 \left[1 + \log\left(\frac{HP}{h}\right)\right]^{-1} \boldsymbol{C}_{bb} \leqslant \boldsymbol{S}_{bb} \leqslant \min\left[\frac{H}{h}, \beta_\delta^{-2}(1 + \log^2 P)\right] \boldsymbol{C}_{bb}\,,$$

where all constants are independent of H, P, and h. In the above, β_δ is the inf–sup constant defined by eqn (8.1.4), h is the characteristic size of an element, and H is the characteristic size of a subdomain, which, for example, can be a group of neighbouring elements. In the case where the coarse space is taken to be the element vertices we have $H/h = 1$.

Block diagonal preconditioner for the Stokes operator

A preconditioner for the full Schur complement system $\boldsymbol{S}$ has been proposed in [5] of the form

$$\boldsymbol{C}_{\text{block}} = \begin{bmatrix} \boldsymbol{C}_{bb} & 0 \\ 0 & \boldsymbol{M}_{aa} \end{bmatrix},$$

where $\boldsymbol{M}_{aa}$ denotes the mass matrix formed from the constant pressure modes on the elements or subdomains. In [5] the preconditioner $\boldsymbol{C}_{\text{block}}$ was used with a preconditioned conjugate residual method. In the same paper it was also shown that the largest absolute eigenvalue of the preconditioned matrix is of order $\min(H/h, \beta_\delta^{-2}(1+\log^2 P))$ and the smallest absolute eigenvalue is of order $\beta_\delta^2(1+\log(HP/h))^{-1}$.

Preconditioned conjugate gradient solution of the Stokes operator

An alternative to the preconditioned conjugate gradient residual method is to apply the preconditioned conjugate gradient method to a subspace where $\boldsymbol{S}$ is positive definite, as formulated by LeTallec and Patra [297]. To construct the appropriate subspace we start by decomposing the solution vector into two components such that

$$\begin{bmatrix} \boldsymbol{v}_b \\ \boldsymbol{p}_a \end{bmatrix} = \begin{bmatrix} \boldsymbol{v}_b^1 \\ 0 \end{bmatrix} + \begin{bmatrix} \boldsymbol{v}_b^0 \\ \boldsymbol{p}_a \end{bmatrix},$$

where $\boldsymbol{v}_b^1$ satisfies

$$\boldsymbol{D}_{ab}\boldsymbol{v}_b^1 = \boldsymbol{g}_a^\star = \boldsymbol{g}_a + \boldsymbol{D}_{ai}\boldsymbol{f}_i\,. \tag{8.2.13}$$

Although eqn (8.2.13) is over-determined since the number of boundary velocity degrees of freedom is far larger than the space of average pressure degrees of freedom, one possibility is a least-squares-type approach of the form

$$\boldsymbol{D}_{ab}^\top \boldsymbol{D}_{ab}\boldsymbol{v}_b^1 = \boldsymbol{D}_{ab}^\top \boldsymbol{g}_a^\star\,.$$

The matrix $\boldsymbol{D}_{ab}^\top \boldsymbol{D}_{ab}$ has rank equal to the number of elements in the mesh, which is not too large to be directly inverted in a spectral/hp element discretisation. The positive-definite Schur complement system amenable to the preconditioned conjugate gradient method is then

$$\boldsymbol{S}\begin{bmatrix} \boldsymbol{v}_b^0 \\ \boldsymbol{p}_a \end{bmatrix} = \begin{bmatrix} \boldsymbol{f}_b^\star \\ 0 \end{bmatrix} - \boldsymbol{S}\left[\boldsymbol{v}_b^1\right],$$

where

$$\boldsymbol{f}_b^\star = \boldsymbol{f}_b - \left[\nu \boldsymbol{L}_{bi}\,\boldsymbol{D}_{\iota b}^\top\right]\boldsymbol{T}\begin{bmatrix} \boldsymbol{f}_i \\ \boldsymbol{g}_\iota \end{bmatrix}.$$

In [5] the conjugate gradient method was applied to a system with a preconditioner of the form

$$\boldsymbol{C}_{\text{sys}} = \begin{bmatrix} \boldsymbol{C}_{bb} & \boldsymbol{D}_{ab}^\top \\ \boldsymbol{D}_{ab} & 0 \end{bmatrix}.$$

This preconditioner was solved in an analogous manner to the approach shown in eqns (8.2.11) and (8.2.12), and therefore requires the assembly and inversion of a matrix of rank equal to the number of elements or subdomains. However, in [5] the condition number of $\boldsymbol{C}_{\text{sys}}^{-1}\boldsymbol{S}$ was theoretically determined to satisfy

$$\kappa(\boldsymbol{C}_{\text{sys}}^{-1}\boldsymbol{S}) \leqslant \beta_\delta^{-2}\min\left(\frac{H}{h}, \beta_\delta^{-2}(1+\log^2 P)\right)\left(1+\log^2\left(\frac{HP}{h}\right)\right).$$

Numerical experiments in [5] demonstrated that the preconditioned conjugate gradient method with preconditioner $\boldsymbol{C}_{\text{sys}}$ required fewer iterations than the preconditioned residual method with preconditioner $\boldsymbol{C}_{\text{block}}$ [5]. However, when taking into account the additional expense of inverting $\boldsymbol{C}_{\text{sys}}$ as compared to $\boldsymbol{C}_{\text{block}}$, the two methods have a similar overall cost.

8.3 Splitting methods for primitive variables

An important aspect of the numerical solution of the incompressible Navier–Stokes equations is that we can solve for the velocity field without approximating the pressure field. We can do this by defining an appropriate projection operator $\mathbb{P}_D$. To this end, we decompose the space $L^2(\Omega)$ into two parts, namely a divergence-free contribution and an irrotational contribution, as follows:

$$\mathcal{D}(\Omega) = \{\mathbf{u},\, \nabla \cdot \mathbf{u} = 0 \text{ in } \Omega,\, \mathbf{u} \cdot \boldsymbol{n} = 0 \text{ on } \partial\Omega\}\,, \quad \mathcal{G}(\Omega) = \{\mathbf{w},\, \mathbf{w} = \nabla\phi\}\,,$$

so that functions from the two spaces are by definition orthogonal, that is,

$$\int_\Omega \mathbf{u}\mathbf{w}\, \mathrm{d}s = \int_\Omega \mathbf{u}\nabla\phi\, \mathrm{d}s = -\int_\Omega \nabla \cdot \mathbf{u}\phi\, \mathrm{d}s + \int_{\partial\Omega} \phi\mathbf{u} \cdot \boldsymbol{n}\, \mathrm{d}\sigma = 0\,.$$

With this construction we recall that for a general vector, $\boldsymbol{q}$, the Helmholtz decomposition applies, that is, $\boldsymbol{q} = \mathbf{u} + \mathbf{w}$. The operator $\mathbb{P}_D$ is then defined as

$$\mathbb{P}_D : L^2(\Omega) \mapsto \mathcal{D}(\Omega)\,,$$

and therefore $\mathbf{u} = \mathbb{P}_D\boldsymbol{q}$. Using this projection, the Navier–Stokes equations (8.1.1a)–(8.1.1c) can be written as

$$\frac{\partial \mathbf{v}}{\partial t} = \mathbb{P}_D[\nu\nabla^2\mathbf{v} - \mathbf{v} \cdot \nabla\mathbf{v}]\,, \tag{8.3.1}$$

which is an evolution equation for velocity, independent of pressure. Note that, by construction, we have $\mathbb{P}_D\partial\mathbf{v}/\partial t = \partial\mathbf{v}/\partial t$ since $\nabla \cdot \mathbf{v} = 0$.

In the Helmholtz decomposition of the vector $\boldsymbol{q}$ it is usually easier to determine the irrotational part $\mathbf{w}$ rather than the divergence-free part $\mathbf{u}$. This is achieved by introducing the rotational ϕ and solving the Neumann elliptic problem

$$\nabla^2\phi = \nabla \cdot \boldsymbol{q}\,, \quad \frac{\partial\phi}{\partial n} = \boldsymbol{q} \cdot \boldsymbol{n}\,,$$

with the compatibility condition for the Neumann problem automatically satisfied. Then $\mathbf{w} = \nabla\phi$ and therefore $u = q - w$.

In summary, the projection method, which is the basis of splitting or fractional methods, approximates the evolution eqn (8.3.1) and not the equation of momentum conservation (see Temam [454] for a more extensive discussion).

8.3.1 *First-order schemes*

In the projection method, as introduced by Chorin [97] and Temam [453], the incompressibility constraint is initially ignored and the evolution equation is solved for an intermediate velocity field $\hat{\mathbf{v}}$. This is subsequently corrected by projecting $\hat{\mathbf{v}}$ onto a divergence-free space. More specifically, in the first substep we have

$$\frac{\hat{\mathbf{v}} - \mathbf{v}^n}{\Delta t} + (\mathbf{v}^{n+1/2} \cdot \nabla)\mathbf{v}^{n+1/2} = \nu \nabla^2 \hat{\mathbf{v}} \,, \tag{8.3.2a}$$

$$\hat{\mathbf{v}} = 0 \quad \text{on } \partial\Omega \,, \tag{8.3.2b}$$

and in the second substep we enforce a *pressure correction*:

$$\frac{\mathbf{v}^{n+1} - \hat{\mathbf{v}}}{\Delta t} = -\nabla p^{n+1} \,, \tag{8.3.2c}$$

$$\nabla \cdot \mathbf{v}^{n+1} = 0 \,, \tag{8.3.2d}$$

$$\mathbf{v}^{n+1} \cdot \boldsymbol{n} = 0 \,. \tag{8.3.2e}$$

For clarity in the following exposition we assume homogeneous Dirichlet boundary conditions for the velocity. Note that the nonlinear terms can be evaluated in an explicit form at some intermediate time-level denoted by $(n + 1/2)\Delta t$. The quantity p that enforces the incompressibility in the second substep is not the exact pressure p_e, but it satisfies the following equations:

$$\nabla^2 p^{n-1} = \nabla \cdot \left(\frac{\hat{\mathbf{v}}}{\Delta t} \right) \quad \text{in } \Omega \,, \tag{8.3.3a}$$

$$\frac{\partial p^{n+1}}{\partial n} = 0 \quad \text{on } \partial\Omega \,. \tag{8.3.3b}$$

A *dual splitting* or *velocity-correction* scheme that reverses the order of the above substeps is also possible and was introduced by Orszag and collaborators, see, for example, [354, 366]. Here, in the first substep a divergence-free field is obtained from

$$\frac{\hat{\mathbf{v}} - \mathbf{v}^n}{\Delta t} + (\mathbf{v}^{n+1/2} \cdot \nabla)\mathbf{v}^{n+1/2} = -\nabla p^{n+1} \,, \tag{8.3.4a}$$

$$\nabla \cdot \hat{\mathbf{v}} = 0 \,, \tag{8.3.4b}$$

$$\hat{\mathbf{v}} \cdot \boldsymbol{n} = 0 \quad \text{on } \partial\Omega \,, \tag{8.3.4c}$$

and, in the second substep, a *velocity correction* is enforced:

$$\frac{\mathbf{v}^{n+1} - \hat{\mathbf{v}}}{\Delta t} = \nu \nabla^2 \mathbf{v}^{n+1} \,, \tag{8.3.4d}$$

$$\mathbf{v}^{n+1} = 0 \quad \text{on } \partial\Omega \,. \tag{8.3.4e}$$

We note that in this scheme eqn (8.3.4a) is typically solved by taking the divergence of this equation to obtain a pressure Poisson equation. The weak form of the pressure Poisson equation with no additional boundary conditions implies that a similar pressure boundary condition to the first scheme is enforced, i.e., it satisfies the homogeneous Neumann condition of eqn (8.3.3b). On the other hand, the *exact* pressure $p_e(\boldsymbol{x}, t)$ satisfies

$$\nabla^2 p_e = -\nabla \cdot [(\mathbf{v} \cdot \nabla)\mathbf{v}] \quad \text{in } \Omega\,, \tag{8.3.5a}$$

$$\frac{\partial p_e}{\partial n} = \boldsymbol{n} \cdot \nu \nabla^2 \mathbf{v} \quad \text{on } \partial\Omega\,, \tag{8.3.5b}$$

that is, the pressure satisfies the non-homogeneous boundary condition given by eqn (8.3.5b), which is different from the boundary conditions imposed in the two previously discussed first-order splitting schemes that satisfy eqn (8.3.3b). This implies that p is an approximation to p_e and approaches the exact pressure in a weak sense, that is, it is little more than a distribution (see [455]). We will see in the following that imposition of the correct pressure boundary condition is crucial in obtaining high-order accuracy of the semi-discrete Navier–Stokes system.

The Neumann boundary condition, eqn (8.3.5b), is derived from the momentum equation projected in the normal direction. A tangential pressure boundary condition can also be obtained, which is equivalent to the Neumann condition in the case of smooth, divergence-free *initial* data. However, if this condition is violated then only the Neumann condition leads to the correct enforcement of the incompressibility constraint at *all* times [203].

The first theoretical result regarding first-order schemes was reported in Shen [420]. It was shown using energy methods that velocity is approximated with first-order accuracy in the discrete L^2 norm, i.e., $\mathcal{O}(\Delta t)$, and that pressure is approximated with only half-order accuracy, i.e., $\mathcal{O}(\sqrt{\Delta t})$. If, however, a better pressure correction is used (see below) then the pressure is also approximated with first-order accuracy in Δt. Error estimates in the maximum norm, L^∞, were derived by E and Liu [148]. For smooth and compatible (pressure) initial conditions, the discrete velocity $\mathbf{v}_{\Delta t}$ and discrete pressure $p_{\Delta t}$ satisfy

$$||\mathbf{v} - \mathbf{v}_{\Delta t}||_\infty + \Delta t^{1/2} ||p - p_{\Delta t}||_\infty \leqslant C \Delta t\,.$$

This is a general error estimate for the original splitting scheme, and it is irrespective of the formal order of the time integrator employed to obtain the semi-discrete system of the incompressible Navier–Stokes equations.

8.3.1.1 *LU factorisation*

An interesting interpretation of the splitting method was given by Perot [373], who used an LU factorisation of the spatially discretised system

$$\begin{pmatrix} \boldsymbol{H} & -\boldsymbol{D}^\top \\ -\boldsymbol{D} & 0 \end{pmatrix} \begin{pmatrix} \mathbf{v}^{n+1} \\ p^{n+1} \end{pmatrix} = \begin{pmatrix} \boldsymbol{M}\boldsymbol{f}^{n+1} \\ 0 \end{pmatrix}, \tag{8.3.6}$$

where $\boldsymbol{D}$ is the discrete gradient operator defined in the previous section, and

$$\boldsymbol{H} \equiv \frac{\boldsymbol{M}}{\Delta t} + \nu \boldsymbol{L}\,.$$

Now, if we define $\boldsymbol{Q} = \Delta t \boldsymbol{M}^{-1}$ then we observe that

$$\boldsymbol{HQ} = \left(\frac{1}{\Delta t}\boldsymbol{M} + \nu \boldsymbol{L}\right)(\Delta t \boldsymbol{M}^{-1}) = \boldsymbol{I} + \nu \Delta t \boldsymbol{L} \boldsymbol{M}^{-1}. \tag{8.3.7}$$

Therefore, the following approximation for the gradient term in system (8.3.6) with $\mathcal{O}(\nu \Delta t)$ can be made:

$$\begin{pmatrix} \boldsymbol{H} & -\boldsymbol{HQD}^\top \\ -\boldsymbol{D} & 0 \end{pmatrix} \begin{pmatrix} \mathbf{v}^{n+1} \\ p^{n+1} \end{pmatrix} = \begin{pmatrix} \boldsymbol{M f}^{n+1} \\ 0 \end{pmatrix}. \tag{8.3.8}$$

Note that, if $\boldsymbol{Q} = \boldsymbol{H}^{-1}$ then we recover the Uzawa algorithm. We can now apply block LU factorisation to approximation (8.3.8) to obtain

$$\begin{pmatrix} \boldsymbol{H} & 0 \\ -\boldsymbol{D} & -\boldsymbol{DQD}^\top \end{pmatrix} \begin{pmatrix} \hat{\mathbf{v}} \\ p^{n+1} \end{pmatrix} = \begin{pmatrix} \boldsymbol{M f}^{n+1} \\ 0 \end{pmatrix}$$

and

$$\begin{pmatrix} \boldsymbol{I} & -\boldsymbol{QD}^\top \\ 0 & \boldsymbol{I} \end{pmatrix} \begin{pmatrix} \mathbf{v}^{n+1} \\ p^{n+1} \end{pmatrix} = \begin{pmatrix} \hat{\mathbf{v}} \\ p^{n+1} \end{pmatrix}.$$

This represents the two substeps of the splitting scheme. Substituting $\boldsymbol{Q} = \Delta t \boldsymbol{M}^{-1}$ leads to the splitting scheme presented previously. Note that no boundary conditions are needed for $\hat{\mathbf{v}}$ and p^{n+1}. This is accomplished by incorporating the velocity boundary conditions into the discrete operators during spatial discretisation before any splitting takes place. Note also that the error in the classical splitting is $\mathcal{O}(\nu \Delta t \nabla^2 p^{n+1}) \simeq \mathcal{O}(\nu \Delta t \boldsymbol{L} p^{n+1})$, which is equivalent to the leading error in the approximation of the LU factorisation (see eqn (8.3.8)) when $\boldsymbol{M} \to \boldsymbol{I}$, which is, for example, the case in a mass lumped nodal basis.

The fully-discrete Navier–Stokes system is not positive definite, but a small modification of the incompressibility constraint can make it positive definite. For example, the modified system has the block form

$$\begin{pmatrix} \boldsymbol{H} & -\boldsymbol{D}^\top \\ -\boldsymbol{D} & -\epsilon \boldsymbol{I} \end{pmatrix} \begin{pmatrix} \mathbf{v}^{n+1} \\ p^{n+1} \end{pmatrix} = \begin{pmatrix} \boldsymbol{M f}^{n+1} \\ 0 \end{pmatrix},$$

which is also the starting-point of many stabilisation methods proposed for the solution of incompressible Navier–Stokes equations.

8.3.1.2 *Penalty formulation*

Another method of eliminating the pressure from the momentum equation is to enforce the incompressibility constraint through a *penalty method* as follows:

$$\begin{aligned} \frac{\mathbf{v}^{n+1} - \mathbf{v}^n}{\Delta t} + (\mathbf{v}^{n+1/2} \cdot \nabla)\mathbf{v}^{n+1/2} + \frac{1}{2}(\nabla \cdot \mathbf{v}^{n+1/2})\mathbf{v}^{n+1/2} \\ = \frac{1}{2}\nu \nabla^2(\mathbf{v}^n + \mathbf{v}^{n+1}) + \tau \nabla(\nabla \cdot \mathbf{v}^{n+1}), \end{aligned}$$

where the extra divergence term on the left-hand side was added by Temam [452] for enhanced stability. This form, although written using a Crank–Nicolson scheme, is not second-order accurate. In particular, its accuracy is dictated by the penalty parameter τ, so that incompressibility is enforced with accuracy $\mathcal{O}(1/\tau)$. This accuracy is good if the penalty parameter takes large values; however, for very large values of τ the resulting system becomes stiff and difficult to invert.

A modification of the penalty method was introduced by Boffi and Funaro [64], who have applied it to mono-domain spectral methods. To illustrate the principal idea, we consider the Stokes problem (eqn (8.1.2a)) and assume that non-homogeneous boundary conditions sustain the flow. The proposed equivalent system of equations is

$$-\nu\nabla^2\mathbf{v} + C\frac{\nabla b}{b}\nabla\cdot\mathbf{v} + \nabla p = 0\,,$$
$$\nabla\cdot(b\nabla p) = \nu\nabla b\cdot\nabla^2\mathbf{v}\,,$$

where $b(\boldsymbol{x})$ is a bubble function which is zero on the boundary $\partial\Omega$, similar to the interior modes described in Chapter 3. The use of this bubble function eliminates the need for boundary pressure as it makes the elliptic pressure operator singular at the boundary. The constant C should be positive and bounded from above ($0 < C < C_0$) for a finite C_0 in order to prove the equivalence of the new system with the Stokes problem. The penalty parameter in this case is the term $C\nabla b/b$, which is finite in the interior of the domain.

8.3.2 *High-order schemes*

There have been several attempts to modify the original first-order projection method in order to derive higher-order approximations. In general, these approaches try to do the following:

1. modify the boundary condition for the intermediate field;
2. formulate a pressure-correction scheme;
3. enforce the correct pressure boundary condition; and
4. generalise the block LU decomposition.

A key observation is that the use of the correct Neumann boundary condition for the pressure written in *rotational form* is crucial in achieving high-order accuracy for the semi-discrete Navier–Stokes system. The use of the rotational form was first advocated by Orszag *et al.* [353] and subsequently documented by Karniadakis *et al.* [264].

8.3.2.1 *Inhomogeneous boundary condition*

The first approach using inhomogeneous boundary conditions was originally adopted in the work of Fortin *et al.* [165] and later used by Kim and Moin [267] in conjunction with a finite difference scheme on a staggered grid. A second-order $\mathcal{O}(\Delta t^2)$ velocity approximation is obtained from

$$\frac{\hat{\mathbf{v}} - \mathbf{v}^n}{\Delta t} + (\mathbf{v}^{n+1/2} \cdot \nabla)\mathbf{v}^{n+1/2} = \frac{1}{2}\nu\nabla^2(\hat{\mathbf{v}} + \mathbf{v}^n)\,, \tag{8.3.9a}$$

$$\hat{\mathbf{v}} = \Delta t\, \nabla p^n \quad \text{on } \partial\Omega\,, \tag{8.3.9b}$$

and in a second substep incompressibility is enforced through the pressure equation (see eqns (8.3.2c)–(8.3.2e)) to obtain p^{n+1}. The inhomogeneous boundary condition for $\hat{\mathbf{v}}$ (eqn (8.3.9b)) cancels the $\mathcal{O}(\Delta t)$ error term, so the slip velocity at the boundary is $\mathcal{O}(\Delta t^2)$. However, enforcing a nonzero *normal* velocity for $\hat{\mathbf{v}}$ leads to weak instability, which is especially 'dangerous' in long-time integration with spectral discretisations [507]. Normal mode analysis performed by Orszag *et al.* [353] has revealed the form of this weak instability mode. A better boundary condition is produced by enforcing the inhomogeneous boundary condition for the *tangential* component of $\hat{\mathbf{v}}$ only, while enforcing $\hat{\mathbf{v}} \cdot \boldsymbol{n} = 0$. The stability analysis presented by E and Liu [148] adopted this type of tangential boundary condition.

For a *third-order scheme*, the intermediate tangential velocity at the boundary should satisfy

$$\hat{\mathbf{v}} \cdot \boldsymbol{t} = \Delta t \frac{\partial p^n}{\partial x} + \Delta t^2 \frac{\partial^2 p^n}{\partial t \partial x}\,,$$

which can be approximated as

$$\hat{\mathbf{v}} \cdot \boldsymbol{t} = \Delta t \left(2\frac{\partial p^n}{\partial x} - \frac{\partial p^{n-1}}{\partial x} \right).$$

Note that, in the modified scheme, the exact pressure boundary condition is still not satisfied since

$$\frac{\partial p^{n+1}}{\partial n} = \frac{\partial p^n}{\partial n} = \cdots = \frac{\partial p^0}{\partial n}\,,$$

so again p differs from the exact p_e by a factor $\mathcal{O}(\nu \Delta t \nabla^2 p)$. This creates a normal artificial boundary layer of the form $(\nu\Delta t)^{1/2}$, as was shown in the analysis of E and Liu [148]. More specifically, the following error bound was established for smooth initial data for the discrete solution $\mathbf{v}_{\Delta t}$:

$$||\mathbf{v} - \mathbf{v}_{\Delta t}||_{L^\infty} + \Delta t||p - p_{\Delta t}||_{L^\infty} \leqslant C\Delta t^2\,.$$

Thus, the numerical pressure $p_{\Delta t}$ is $\mathcal{O}(\Delta t)$ accurate, but for general initial data it is only $\mathcal{O}(\Delta t^{1/2})$ accurate at the boundary. In the interior, at distances greater than $(\nu\Delta t)^{1/2}$, second-order accuracy for the numerical pressure $p_{\Delta t}$ is also established, that is,

$$\max |p - p_{\Delta t}| \leqslant C\Delta t^2\,.$$

An overall better scheme is presented next which has been used with great success in spectral element formulations.

8.3.2.2 *Standard pressure-correction formulation*

The standard pressure-correction formulation, first proposed by Goda [193], is essentially a predictor-corrector scheme of the classical first-order splitting scheme, and leads to second-order accuracy in the velocity. In this modification, the pressure term from the previous time-step is retained in the first substep:

$$\frac{\hat{\mathbf{v}} - \mathbf{v}^n}{\Delta t} + (\mathbf{v}^{n+1/2} \cdot \nabla)\mathbf{v}^{n+1/2} + \nabla p^n = \frac{1}{2}\nu\nabla^2(\hat{\mathbf{v}} + \mathbf{v}^n),$$
$$\hat{\mathbf{v}} = 0 \quad \text{on } \partial\Omega,$$

and a correction is included in the second substep:

$$\begin{aligned} \frac{\mathbf{v}^{n+1} - \hat{\mathbf{v}}}{\Delta t} &= -\nabla(\delta p)^{n+1}, \\ \nabla \cdot \mathbf{v}^{n+1} &= 0, \\ \mathbf{v}^{n+1} \cdot \boldsymbol{n} &= 0 \quad \text{on } \partial\Omega, \end{aligned} \tag{8.3.10}$$

where $(\delta p)^{n+1} = p^{n+1} - p^n$. This form was proposed by Bell *et al.* [42] and a normal mode analysis was performed by E and Liu [149]. The scheme is based on the original form proposed by Van Kan [477], where a somewhat different discrete time-evolution equation was employed.

From the second substep (8.3.10) of this scheme, we note that on the boundary $\hat{\mathbf{v}} \cdot \boldsymbol{n} = \mathbf{v}^{n+1} \cdot \boldsymbol{n} = 0$ and so

$$\frac{\partial p^{n+1}}{\partial n} = \frac{\partial p^n}{\partial n} = \cdots = \frac{\partial p^0}{\partial n}.$$

Therefore, the field p does not satisfy a pressure boundary condition consistent with the original momentum equation. This artificial Neumann boundary condition on the pressure introduces a normal boundary layer, just as in the previous scheme, and this limits the accuracy of the pressure to first order, i.e., $\mathcal{O}(\Delta t)$. We refer to this scheme as *standard* to distinguish it from a modified one which we study in the next section; the latter satisfies the correct pressure boundary condition.

An interesting aspect of the standard pressure-correction scheme is that it leads to errors that have a different structure to those of the splitting scheme with the inhomogeneous velocity boundary condition of Section 8.3.2.1. Specifically, in the inhomogeneous velocity boundary condition case, a spurious mode with a normal boundary layer is developed. In contrast, in the standard pressure-correction scheme an *oscillatory mode* is present, according to the analysis by E and Liu [149].

To demonstrate the differences between the two schemes we consider *Stokes flow* in a channel, with the homogeneous flow (x direction) represented by a Fourier mode corresponding to a wavenumber k. Without a driving force the motion decays in proportion to $e^{\sigma t}$, and the normal modes of the flow can be obtained exactly. For example, from the exact analysis the pressure *symmetric mode* is

$$\tilde{p}_e(y) = \frac{\sigma}{k} \cos\mu \frac{\sinh(ky)}{\cosh k},$$

where $\sigma = -k^2 - \mu^2$ is the decay rate and the dispersion relation

$$\mu \tan\mu + k \tanh k = 0$$

defines the value of μ given the mode number k.

The normal mode analysis applied to the semi-discrete pressure-correction scheme of Bell *et al.* [42] leads to the following approximation for the pressure mode:

$$\tilde{p}(y) = \frac{4\tilde{\sigma} + 2\tilde{\sigma}^2 \Delta t}{(2 + \tilde{\sigma}\Delta t)^2} \cos\tilde{\mu} \left(\frac{\sinh(ky)}{k \cosh k} - \frac{\sin(\lambda y)}{\lambda \cos\lambda} \right),$$

where $\tilde{\sigma} = -k^2 - \tilde{\mu}^2$, and we also define $\kappa = (2+\tilde{\sigma}\Delta t)/(2-\tilde{\sigma}\Delta t)$. The approximate dispersion relation is

$$\tilde{\mu} \tan\tilde{\mu} + k \tanh k = k \left(\frac{2\tilde{\sigma}\Delta t}{2 + \tilde{\sigma}\Delta t} \right)^2 \left(\tanh k - \frac{k}{\lambda} \tanh\lambda \right),$$

where $\lambda = [2\kappa/((1-\kappa)\Delta t) - k^2]^{1/2}$ corresponds to the spurious mode, which is oscillatory in nature.

Figure 8.1 shows the two spurious modes induced, respectively, by a boundary layer in the first modified scheme with inhomogeneous boundary conditions, and by the oscillation in the pressure-correction scheme. It is interesting to notice that for a finite difference discretisation the intrinsic numerical diffusion may

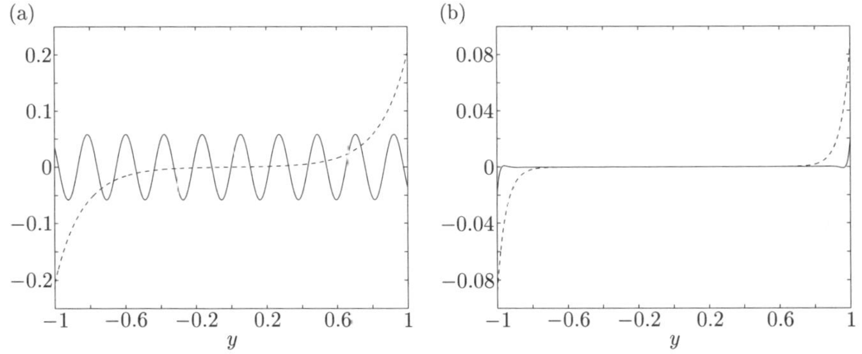

FIG. 8.1. Error in the pressure modes across a channel ($y \in [-1, 1]$) for the splitting scheme modified by the intermediate velocity (boundary layer) and by a pressure correction (oscillation). (a) Continuous results, and (b) the discrete result after spatial discretisation with second-order finite differences. (Courtesy of W. E and J.-G. Liu.)

damp these high-frequency oscillations, as shown in Fig. 8.1(b). However, these oscillations are not eliminated in spectral discretisations. The first results of the implementation of the pressure-correction scheme in conjunction with a spectral element discretisation were presented by Couzy [114]. It was reported that no steady-state results were exactly achieved, presumably because of the instability that the spatial oscillation introduced.

An alternative approach to deriving error estimates using the energy method was developed by Shen [422]. It was assumed that $d\mathbf{v}/dt$ is not necessarily smooth at $t = 0$ (e.g., an impulsive flow start), in which case the accuracy in Δt is only first order (in the maximum norm) for both velocity and pressure.

8.3.2.3 *Rotational pressure-correction formulation*

A more accurate and robust pressure-correction scheme was proposed by Timmermans *et al.* [464] for spectral element discretisations. It is based on the aforementioned standard pressure-correction formulation but it includes a *divergence correction.* A second-order stiffly-stable time integrator (backwards differentiation) was employed by Timmermans *et al.* [464], as we present here.

The first step is the same as in the standard pressure-correction formulation, i.e.,

$$\frac{3\hat{\mathbf{v}} - 4\mathbf{v}^n + \mathbf{v}^{n-1}}{2\Delta t} + \nabla p^n = \nu \nabla^2 \hat{\mathbf{v}} + f^{n+1}\,,$$
$$\hat{\mathbf{v}} = 0 \quad \text{on } \partial\Omega\,,$$

where the last term, f^{n+1}, in the first equation represents the nonlinear contributions combined with other external forces. The proposed divergence correction is included in the second substep:

$$\begin{aligned}\frac{3\mathbf{v}^{n+1} - 3\hat{\mathbf{v}}}{2\Delta t} &= -\nabla[(\delta p)^{n+1} + \nu\nabla\cdot\hat{\mathbf{v}}]\,,\\ \nabla\cdot\mathbf{v}^{n+1} &= 0\,,\\ \mathbf{v}^{n+1}\cdot\boldsymbol{n} &= 0 \quad \text{on } \partial\Omega\,.\end{aligned}$$

By defining the quantity $q^{n+1} \equiv (\delta p)^{n+1} + \nu\nabla\cdot\hat{\mathbf{v}}$, we can solve the Poisson equation

$$\begin{aligned}\nabla^2 q^{n+1} &= \frac{3}{2\Delta t}\nabla\cdot\hat{\mathbf{v}}\,,\\ \frac{\partial q^{n+1}}{\partial n} &= 0 \quad \text{on } \partial\Omega\end{aligned}$$

to obtain the corrected pressure $p^{n+1} = q^{n+1} + p^n - \nu\nabla\cdot\hat{\mathbf{v}}$. It was shown by Timmermans *et al.* [464] that the homogeneous Neumann boundary condition on q^{n+1} is a consistent boundary condition, and thus the solvability of the elliptic equation is guaranteed.

In order to understand the reported improved accuracy and enhanced stability (especially in simulations with singular solutions, see [333]) we derive the corresponding boundary condition that the pressure p^{n+1} satisfies. To this end, we combine the two substeps to obtain

$$\frac{3\mathbf{v}^{n+1} - 4\mathbf{v}^n + \mathbf{v}^{n-1}}{2\Delta t} = -\nabla p^{n+1} - \nu\nabla \times \nabla \times \hat{\mathbf{v}} + f^{n+1}\,,$$
$$\nabla \cdot \mathbf{v}^{n+1} = 0\,,$$
$$\mathbf{v}^{n+1} \cdot \boldsymbol{n} = 0 \quad \text{on } \partial\Omega\,,$$

where we made use of the identity

$$\nabla^2 \hat{\mathbf{v}} = \nabla(\nabla \cdot \hat{\mathbf{v}}) - \nabla \times \nabla \times \hat{\mathbf{v}}\,.$$

We can further eliminate the $\hat{\mathbf{v}}$ field in the above equation by noticing that

$$\nabla \times \nabla \times \hat{\mathbf{v}} = \nabla \times \nabla \times \mathbf{v}^{n+1}$$

from the second substep of the splitting scheme. By taking the inner product of the combined scheme with the unit normal, we arrive at the following pressure boundary condition:

$$\frac{\partial p^{n+1}}{\partial n} = [f^{n+1} - \nu\nabla \times \nabla \times \mathbf{v}^{n+1}] \cdot \boldsymbol{n}$$

on the boundary $\partial\Omega$, which is the correct pressure boundary condition. The significance of the rotational form of the boundary condition was first reported in Orszag *et al.* [353] and documented by Karniadakis *et al.* [264]; it is discussed in detail in the next section.

Having eliminated the splitting errors due to the pressure boundary conditions, the only remaining source of inaccuracy is the inexact *tangential* velocity boundary condition. A rigorous analysis of the error estimates for the rotational pressure-correction scheme was conducted by Guermond and Shen [209]. It was found that the velocity is $\mathcal{O}(\Delta t^2)$ while the pressure is $\mathcal{O}(\Delta t^{3/2})$, that is, we achieve an increase of $\mathcal{O}(\sqrt{\Delta t})$ in pressure compared with the standard pressure-correction formulation. Moreover, for smooth domains a second-order accuracy in the pressure can be achieved. This is supported by the normal mode analysis of Brown *et al.* [79] for a periodic channel, as well as the numerical examples of Guermond and Shen [208]. The implications of the divergence-correction term have been discussed by Minev and Gresho [333] in the context of singular benchmark solutions; this simple correction leads to a very robust pressure solver.

8.3.2.4 *Consistent pressure boundary condition*

The importance of the correct pressure boundary condition and its effect on the accuracy of the splitting scheme was first recognised in work by Orszag *et al.* [353] and Karniadakis *et al.* [264]. In particular, in these works the *dual splitting scheme* (eqns (8.3.4a)–(8.3.4e)) was employed along with the correct pressure boundary condition

$$\frac{\partial p^{n+1}}{\partial n} = -\nu(\nabla \times \boldsymbol{\omega})^{n+1} \cdot \boldsymbol{n} \quad \text{on } \partial\Omega\,. \tag{8.3.11}$$

This rotational form of the boundary condition for the pressure is equivalent to the Laplacian form of the boundary condition in eqn (8.3.5b), but, unlike the latter, it satisfies the compatibility condition and it also reinforces the incompressibility condition since $\nabla^2\mathbf{v} = \nabla(\nabla \cdot \mathbf{v}) - \nabla \times \boldsymbol{\omega}$, see also Karniadakis *et al.* [264]. In addition, it leads to a stable approximation as the boundary divergence is directly controlled by the time-step, see Petersson [375]. Specifically, it was shown rigorously by Leriche and Labrosse [294] that *ellipticity* is lost if the Laplacian form of the pressure boundary condition is employed. Therefore, instabilities may arise in such a Stokes solver, irrespective of the time-integration scheme.

To illustrate the differences between the rotational (eqn (8.3.11)) and Laplacian (eqn (8.3.5b)) form of the pressure boundary condition, we consider the exact boundary condition at time-step $(n+1)\Delta t$, that is,

$$\frac{\partial p^{n+1}}{\partial n} = \nu\left(\frac{\partial Q^{n+1}}{\partial n} - \omega_s^{n+1}\right),$$

where we have introduced $\omega_s = \nabla \times \boldsymbol{\omega} \cdot \boldsymbol{n}$ and $Q = \nabla \cdot \mathbf{v}$. We can now expand ω_s in a Taylor series to obtain

$$\frac{\partial Q^{n+1}}{\partial n} = \frac{1}{\nu}\frac{\partial p^{n+1}}{\partial n} + \omega_s^n + \Delta t \frac{\partial \omega_s^n}{\partial t} + \dots .$$

Inserting the Laplacian form (eqn (8.3.5b)) in the above equation, we obtain

$$\frac{\partial Q^{n+1}}{\partial n} \propto \frac{\partial Q^n}{\partial n} + \Delta t \frac{\partial \omega_s^n}{\partial t},$$

which shows an accumulation of divergence flux at the boundary every time-step, and therefore the possibility for instability. In contrast, if the rotational form in eqn (8.3.11) is used then we obtain

$$\frac{\partial Q^{n+1}}{\partial n} \propto \Delta t \frac{\partial \omega_s^n}{\partial t},$$

and therefore the magnitude of the *boundary divergence flux* is controlled directly by the time-step.

In order to decouple the pressure and velocity at the boundary and also in order to reduce the *boundary divergence errors*, we can use an explicit multi-step approximation to represent the right-hand side in eqn (8.3.11):

$$\frac{\partial p^{n+1}}{\partial n} = -\nu \boldsymbol{n} \cdot \sum_{q=0}^{J_p-1} \beta_q (\nabla \times \boldsymbol{\omega})^{n-q}\,, \tag{8.3.12}$$

where J_p is the number of previous steps from which information is used. Following the above argument, we find that

$$\frac{\partial Q^{n+1}}{\partial n} \propto (\Delta t)^{J_p}\,,$$

and therefore the boundary divergence flux can be made arbitrarily small by controlling the time-step Δt. Note that, for the inviscid pressure boundary condition

$$\frac{\partial p^{n+1}}{\partial n} = 0\,,$$

the boundary divergence flux is $\mathcal{O}(1)$, independent of the size of the time-step Δt!

To relate the boundary divergence to the overall accuracy of the velocity field, we consider the equation that the divergence $Q \equiv Q^{n+1}$ satisfies, that is,

$$\frac{Q}{\Delta t} - \gamma_0 \nu \nabla^2 Q = 0\,,$$

where we set the right-hand side to zero since the pressure satisfies a consistent Poisson equation, and the divergence at previous time-steps $(Q^n, Q^{n-1}, \ldots)$ is assumed to be zero; γ_0 is a coefficient due to the stiffly-stable time discretisation, see Section 8.3.2.5. It is clear, therefore, that there exists a numerical boundary layer of thickness $\delta = \sqrt{\gamma_0 \nu \Delta t}$, so that $Q = Q_w \mathrm{e}^{-s/\delta}$, and thus the boundary divergence is $Q_w = -\delta(\partial Q/\partial n)_w$. (Here s is a general coordinate normal to the boundary.) Similarly, from order-of-magnitude analysis we have $Q_w = \mathcal{O}(\partial v/\partial n)$, and therefore

$$v \propto Q_w \delta \propto \left(\frac{\partial Q}{\partial n}\right)_w \gamma_0 \nu \Delta t\,.$$

This relation shows that the time-differencing error of the velocity field is an order smaller in Δt than the corresponding error in the boundary divergence. For the inviscid-type pressure boundary condition, however, we are limited to first-order accuracy since the boundary divergence flux is $\mathcal{O}(1)$. In general, we obtain

$$v \propto (\Delta t)^{J_p+1}$$

if a high-order time-integration scheme is used to advance the velocity field. In numerical experiments, it was found in Karniadakis *et al.* [264] that with $J_p = 2$ we obtain a third-order accurate velocity field. Note that the boundary divergence scales as

$$Q_w \propto \sqrt{\nu}(\Delta t)^{J_p}\,.$$

The above heuristic arguments have been documented and confirmed by numerical results in Karniadakis *et al.* [264]. To demonstrate the effect of the incorrect inviscid pressure boundary condition versus the correct rotational form in the boundary condition, we consider a decaying Stokes channel flow subject to compatible initial conditions [130], similar to the ones analysed in Section 8.3.2.2. In Fig. 8.2 we plot the divergence of the velocity field across the channel. It is seen that the incorporation of the rotational form of the pressure boundary condition almost eliminates the artificial boundary layer. In Fig. 8.3 we plot the divergence for different multi-step integration levels. High-order treatment produces smaller boundary divergence errors, which is consistent with the aforementioned arguments.

8.3.2.5 *Stiffly-stable time integrators*

To extend the temporal accuracy to orders higher than second, appropriate *stable* time integrators have to be used. As was demonstrated by Karniadakis *et al.* [264], the Adams–Moulton schemes of order higher than second lead to severe time-step limitations. An alternative approach is to use stiffly-stable schemes, which are accurate for all components around the origin in the stability diagram and absolutely stable away from the origin in the left imaginary plane, see Section 5.3. For an advection–diffusion equation, the standard practice is to treat the advection terms explicitly and the diffusion terms implicitly. The stability diagram for a third-order scheme is shown in Fig. 8.4 at various values of the diffusion coefficient. The stability region is considerably larger than a corresponding Adams–Bashforth/Adams–Moulton third-order scheme.

The original stiffly-stable scheme is implicit but we can use consistent-order interpolation for the advection terms to obtain a mixed explicit/implicit scheme, which we denote as SE/SI. To this end, a backwards differentiation is first applied and the nonlinear terms are extrapolated with consistent order in time.

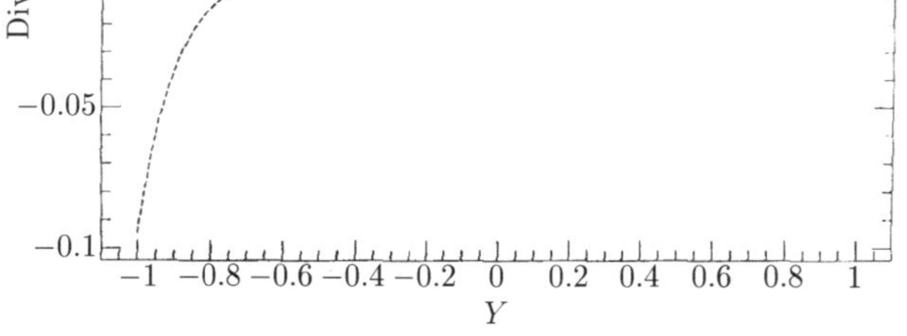

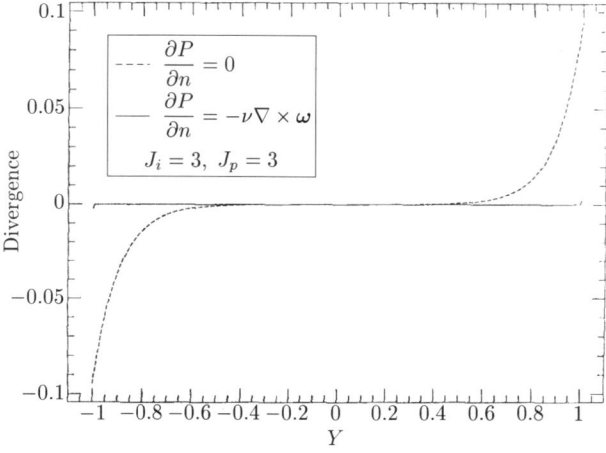

FIG. 8.2. Divergence of the velocity field across the channel for a Stokes flow, with $\Delta t = 10^{-2}$. The spatial discretisation is based on $N_{\rm el} = 20$ spectral elements of order $P = 10$ that eliminates any spatial errors. ([466].)

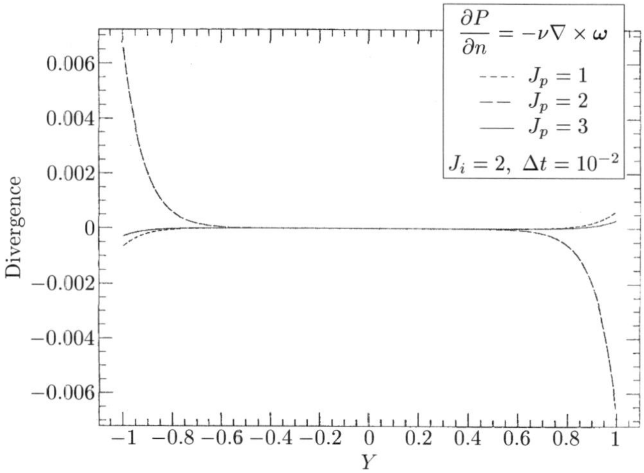

FIG. 8.3. Effect of explicit time-integration order in the pressure boundary condition upon the accuracy of the solution for a Stokes channel flow. ([466].)

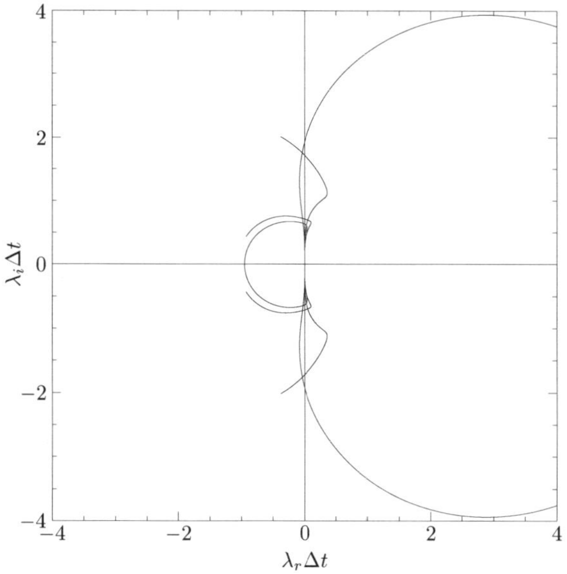

FIG. 8.4. Stability diagram for a third-order explicit/implicit stiffly-stable scheme (SE3/SI3) for an advection–diffusion equation. The region of *absolute stability* is outside the closed curves. The cell Reynolds numbers are 1000 (closed curve on the right plane), 2.5, 0.5, and 0.001.

For the mixed explicit/implicit treatment of advection and viscous terms, respectively, the overall scheme exhibits identical accuracy to the integration rule used, with no splitting error present in the steady-state case. These schemes were first proposed by Karniadakis *et al.* [264] and later analysed by Couzy [114], who found them to be more accurate than other mixed schemes. For example, a third-order backwards differentiation (SI3) for diffusion combined with a fourth-order Runge–Kutta scheme for advection is less accurate than a third-order stiffly-stable (SE3/SI3) scheme. It was also found that an SE3/SI1 scheme is less stable than an SE3/SI3 scheme, unlike the multi-step Adams-type schemes whose stability regions 'shrinks' as the order increases (see Gear [179]). For the SE3/SI3 the residual error is proportional to

$$\Delta t^3 \left[\frac{\mathrm{d}^3}{\mathrm{d}t^3}(\mathbf{v}\cdot\nabla) + 3\nu\frac{\mathrm{d}^3}{\mathrm{d}t^3}\nabla^2\mathbf{v} \right].$$

Therefore, for stationary problems no splitting errors remain in splitting the nonlinear from the linear contributions in advection–diffusion equations, unlike other mixed schemes analysed by Couzy [114].

8.3.2.6 *High-order dual splitting scheme*

To implement high order in the *dual splitting scheme* for the Navier–Stokes equations, we assume an order of integration J_e for the advection terms, J_i for the diffusion terms, and J_p for the pressure boundary condition (eqn (8.3.12)). The semi-discrete system can then be written as

$$\frac{\hat{\mathbf{v}} - \sum_{q=0}^{J_i-1}\alpha_q \mathbf{v}^{n-q}}{\Delta t} = -\sum_{q=0}^{J_e-1}\beta_q[(\mathbf{v}\cdot\nabla)\mathbf{v}]^{n-q}, \tag{8.3.13a}$$

$$\frac{\hat{\hat{\mathbf{v}}} - \hat{\mathbf{v}}}{\Delta t} = -\nabla p^{n+1}, \tag{8.3.13b}$$

$$\frac{\gamma_0\mathbf{v}^{n+1} - \hat{\hat{\mathbf{v}}}}{\Delta t} = \nu\nabla^2\mathbf{v}^{n+1}, \tag{8.3.13c}$$

where the integration weights are given in Table 5.2. The second equation for the pressure and the third equation for the viscous correction can be recast into a Poisson and a Helmholtz equation, respectively, which can be discretised as discussed in Chapters 5 and 7. For the pressure field, assuming the intermediate field $\hat{\hat{\mathbf{v}}}$ is divergence-free, we obtain

$$\nabla^2 p^{n+1} = \nabla\cdot\left(\frac{\hat{\mathbf{v}}}{\Delta t}\right), \tag{8.3.14}$$

which is solved with consistent boundary conditions

$$\frac{\partial p^{n+1}}{\partial n} = -\left[\frac{\partial \mathbf{v}}{\partial t}^{n+1} + \nu\sum_{q=0}^{J_p-1}\beta_q(\nabla\times\boldsymbol{\omega})^{n-q} + \sum_{q=0}^{J_e-1}\beta_q[(\mathbf{v}\cdot\nabla)\mathbf{v}]^{n-q}\right]\cdot\boldsymbol{n}.$$

These boundary conditions are analogous to the pressure boundary conditions given in eqn (8.3.12), but we have now included the unsteady boundary velocity and nonlinear advection terms. For the viscous correction, we obtain

$$\left(\nabla^2 - \frac{\gamma_0}{\nu\Delta t}\right)\mathbf{v}^{n+1} = -\left(\frac{\hat{\hat{\mathbf{v}}}}{\nu\Delta t}\right), \tag{8.3.15}$$

with Dirichlet velocity boundary conditions.

Discretising scheme (8.3.13a)–(8.3.13c) in space using a Galerkin projection introduces a mass matrix in eqn (8.3.13a) in an analogous fashion to the advection equation discussed in Chapter 6. For the *nodal* spectral element expansion this does not represent any significant numerical cost since the mass matrix is diagonal under an appropriate choice of quadrature points. For a *modal* expansion the inversion of a mass matrix represents a similar cost to solving the pressure Poisson equation or a single Helmholtz system. However, we can analytically combine eqns (8.3.13a) and (8.3.13b) before discretising in space to obtain a two-step scheme of the form

$$\frac{\hat{\hat{\mathbf{v}}} - \sum_{q=0}^{J_i-1}\alpha_q\mathbf{v}^{n-q}}{\Delta t} = -\sum_{q=0}^{J_e-1}\beta_q[(\mathbf{v}\cdot\nabla)\mathbf{v}]^{n-q} - \nabla p^{n+1}, \tag{8.3.16a}$$

$$\frac{\gamma_0\mathbf{v}^{n+1} - \hat{\hat{\mathbf{v}}}}{\Delta t} = \nu\nabla^2\mathbf{v}^{n+1}. \tag{8.3.16b}$$

Once again, we can take the divergence of eqn (8.3.16a) to obtain a Poisson equation for pressure which is equivalent to eqn (8.3.14) at the semi-discrete level.

Leriche *et al.* [295] have also presented a variant of the above scheme which does not directly require splitting the time operator. Their scheme has a directly analogous structure to eqns (8.3.16a) and (8.3.16b) (or, equivalently, (8.3.13a)–(8.3.13c)), but when constructing the pressure Poisson equation (8.3.14) divergence of $\hat{\hat{\mathbf{v}}}$ and $\mathbf{v}^{n-q}$ is enforced so that the pressure Poisson equation becomes

$$\nabla^2 p^{n+1} = \nabla\cdot\left(\sum_{q=0}^{J_e-1}\beta_q[(\mathbf{v}\cdot\nabla)\mathbf{v}]^{n-q}\right). \tag{8.3.17}$$

We now comment on the ***stability*** of the explicit (in time) pressure boundary condition. Numerical evidence and eigenvalue analysis by Leriche and Labrosse [294] have shown that the Stokes solver is unconditionally stable for the first-order and second-order schemes, i.e., $J_p = 1, 2$. However, for the third-order scheme ($J_p = 3$) (also for higher order, $J_p > 3$) the splitting scheme for the unsteady Stokes operator is only conditionally stable. Specifically, for global spectral methods the maximum eigenvalue of the unsteady Stokes operator scales as

$\mathcal{O}(P^4)$. For specific cases analysed by Leriche and Labrosse [294] with Chebyshev collocation the onset of instability for a third-order scheme was at

$$\nu\Delta t \approx 10^{-4} ,$$

so for a high Reynolds number flow simulation, i.e., Re $\geqslant$ 100, the time-step for stability is $\mathcal{O}(10^{-2})$, which is a reasonable value in order to have acceptable temporal accuracy. This implies that, in practice, conditional stability (for a third-order scheme) is not a real problem.

We also note that extrapolation of the pressure boundary condition with, say $J_p = 2$, leads to a velocity accuracy of $\mathcal{O}(\Delta t^{J_p+1}) = \mathcal{O}(\Delta t^3)$. Therefore, the highest order of the unconditionally stable splitting scheme for the velocity is third. For the non-splitting scheme using the pressure Poisson equations (8.3.17) only second-order accuracy has been reported [295].

8.3.2.7 *Rotational velocity-correction formulation*

The dual splitting scheme described by eqns (8.3.4a)–(8.3.4e) can be recast, by analogy with the pressure-correction scheme, in a velocity-correction formulation, as suggested by Guermond and Shen [209]. To this end, let us rewrite the first two substeps using second-order backwards differentiation as follows:

$$\frac{3\hat{\mathbf{v}} - 4\mathbf{v}^n + \mathbf{v}^{n-1}}{2\Delta t} - \nu\nabla^2\mathbf{v}^n = -\nabla p^{n+1} + \boldsymbol{f}^{n+1} , \tag{8.3.18a}$$

$$\nabla \cdot \hat{\mathbf{v}} = 0 , \tag{8.3.18b}$$

$$\hat{\mathbf{v}} \cdot \boldsymbol{n} = 0 \quad \text{on } \partial\Omega , \tag{8.3.18c}$$

where $\boldsymbol{f}^{n+1}$ represents collectively nonlinear contributions and other external forces. In the second substep we enforce a *velocity correction* in the form

$$\frac{3\mathbf{v}^{n+1} - 3\hat{\mathbf{v}}}{2\Delta t} = \nu\nabla^2(\mathbf{v}^{n+1} - \mathbf{v}^n) , \tag{8.3.18d}$$

$$\mathbf{v}^{n+1} = 0 \quad \text{on } \partial\Omega . \tag{8.3.18e}$$

In this form, the first step uses an explicit treatment for the viscous term (predictor), while the second substep uses an implicit treatment (corrector). It can therefore be thought of as the *dual* of the pressure-correction scheme [209].

We now examine what type of boundary condition the pressure satisfies given the eqns (8.3.18a)–(8.3.18e). To this end, we observe from eqn (8.3.18d) that $\nabla^2(\mathbf{v}^{n+1} - \mathbf{v}^n) \cdot \boldsymbol{n} = 0$, which implies that

$$\nabla^2\mathbf{v}^{n+1} \cdot \boldsymbol{n} = \nabla^2\mathbf{v}^n \cdot \boldsymbol{n} = \cdots = \nabla^2\mathbf{v}^0 \cdot \boldsymbol{n} .$$

Using the above equation, we obtain the following Neumann pressure boundary condition from eqn (8.3.18a):

$$\frac{\partial p^{n+1}}{\partial n} = [\boldsymbol{f}^{n+1} + \nu\nabla^2\mathbf{v}^n] \cdot \boldsymbol{n} = [\boldsymbol{f}^{n+1} + \nu\nabla^2\mathbf{v}^0] \cdot \boldsymbol{n} ,$$

which is clearly incorrect, and will therefore limit the accuracy of the splitting scheme as we have seen before.

In order to obtain a high-order splitting scheme we resort again to the rotational form of the viscous terms, now inserting it directly in the equations, as we have done for the pressure boundary condition earlier. That is, we replace $\nabla^2 \mathbf{v}^n$ by $-\nabla \times \nabla \times \mathbf{v}^n$ in both substeps, thereby imposing indirectly the divergence-free constraint $\nabla \cdot \mathbf{v}^n = 0$. The modified velocity-correction scheme in *rotational form* is then ([209])

$$\frac{3\hat{\mathbf{v}} - 4\mathbf{v}^n + \mathbf{v}^{n-1}}{2\Delta t} + \nu \nabla \times \nabla \times \mathbf{v}^n = -\nabla p^{n+1} + \boldsymbol{f}^{n+1}, \tag{8.3.19a}$$

$$\nabla \cdot \hat{\mathbf{v}} = 0, \tag{8.3.19b}$$

$$\hat{\mathbf{v}} \cdot \boldsymbol{n} = 0 \quad \text{on } \partial\Omega \tag{8.3.19c}$$

and

$$\frac{3\mathbf{v}^{n+1} - 3\hat{\mathbf{v}}}{2\Delta t} = \nu \nabla^2 \mathbf{v}^{n+1} + \nu \nabla \times \nabla \times \mathbf{v}^n, \tag{8.3.19d}$$

$$\mathbf{v}^{n+1} = 0 \quad \text{on } \partial\Omega. \tag{8.3.19e}$$

As before, we can derive the corresponding Neumann pressure boundary condition on the boundary $\partial\Omega$:

$$\frac{\partial p^{n+1}}{\partial n} = [\boldsymbol{f}^{n+1} - \nu \nabla \times \nabla \times \mathbf{v}^n] \cdot \boldsymbol{n} = [\boldsymbol{f}^{n+1} + \nu \nabla^2 \mathbf{v}^{n+1}] \cdot \boldsymbol{n},$$

which is the correct pressure boundary condition, first introduced by Orszag *et al.* [353] (see also [264]). Note that the last equality is justified by the second substep (eqns (8.3.19d) and (8.3.19e)).

For Galerkin spectral element discretisations which satisfy a C^0 condition at the element interfaces the above velocity-correction formulation is not convenient because of the presence of the second derivative. To this end, following Guermond and Shen [209], we can form the difference between the expressions generated by eqns (8.3.19a)–(8.3.19c) at time-steps $n\Delta t$ and $(n+1)\Delta t$ and also incorporate the second substep evaluated at time $n\Delta t$. The reformulated scheme for the computation of pressure is then

$$\frac{3\hat{\mathbf{v}} - 7\mathbf{v}^n + 5\mathbf{v}^{n-1} - \mathbf{v}^{n-2}}{2\Delta t} + \nabla q^{n+1} = \boldsymbol{f}^{n+1} - \boldsymbol{f}^n, \tag{8.3.20a}$$

$$\nabla \cdot \hat{\mathbf{v}} = 0, \tag{8.3.20b}$$

$$\hat{\mathbf{v}} \cdot \boldsymbol{n} = 0 \quad \text{on } \partial\Omega, \tag{8.3.20c}$$

where $q^{n+1} \equiv p^{n+1} - p^n$ is an auxiliary unknown variable. The above formulation can be handled with standard conforming C^0 spectral elements.

With regards to accuracy, it was shown by Guermond and Shen [209] that the velocity is $\mathcal{O}(\Delta t^2)$ while the pressure is $\mathcal{O}(\Delta t^{3/2})$ (in the maximum norm), similar to the rotational pressure-correction formulation. In contrast, the standard velocity-correction formulation (8.3.18a)–(8.3.18e) leads to first-order accuracy in the pressure and second-order accuracy in the velocity.

8.3.2.8 *A fully-consistent second-order splitting scheme*

Neither the pressure-correction nor the velocity-correction schemes have optimum accuracy for the pressure (and thus the vorticity) field. A new class of truly consistent splitting schemes for incompressible flows was proposed by Guermond and Shen [207]. It is based on the pressure-correction scheme and leads to second-order accuracy for both the velocity and the pressure fields.

The key idea of the formulation proposed by Guermond and Shen [207] is to compute the pressure from a Galerkin projection of the momentum equation tested against *gradients* of a function $q \in H^1(\Omega)$. The projected momentum equation takes the form

$$\int_\Omega \nabla q \cdot \nabla p \, \mathrm{d}\boldsymbol{x} = \int_\Omega \nabla q \cdot (\boldsymbol{f} + \nu \nabla^2 \mathbf{v}) \, \mathrm{d}\boldsymbol{x} \,, \quad \forall \, q \in H^1(\Omega) \,, \tag{8.3.21a}$$

where $\boldsymbol{f}$ contains the nonlinear contributions. Based on the pressure-correction scheme, the following system gives the semi-discrete equations:

$$\frac{3\mathbf{v}^{n+1} - 4\mathbf{v}^n + \mathbf{v}^{n-1}}{2\Delta t} + \nabla(2p^n - p^{n-1}) = \nu \nabla^2 \mathbf{v}^{n+1} + \boldsymbol{f}^{n+1} \,, \tag{8.3.22}$$

$$\mathbf{v}^{n+1} = 0 \quad \text{on } \partial\Omega \,, \tag{8.3.23}$$

$$(\nabla q, \nabla p^{n+1}) = (\nabla q, \boldsymbol{f}^{n+1} - \nu \nabla \times \nabla \times \mathbf{v}^{n+1}) \,, \quad \forall \, q \in H^1(\Omega) \,, \tag{8.3.24}$$

where the rotational form has also been employed. For C^0-based discretisations a different formulation has been proposed in order to avoid computing explicitly $\nabla \times \nabla \times \mathbf{v}^{n+1}$ [209]. To this end, we take the inner product of the first equation above with ∇q and subtract the result from the second equation. This leads to the computationally friendly form

$$\frac{3\mathbf{v}^{n+1} - 4\mathbf{v}^n + \mathbf{v}^{n-1}}{2\Delta t} + \nabla(2p^n - p^{n-1}) = \nu \nabla^2 \mathbf{v}^{n+1} + \boldsymbol{f}^{n+1} \,, \tag{8.3.25a}$$

$$\mathbf{v}^{n+1} = 0 \quad \text{on } \partial\Omega \,, \tag{8.3.25b}$$

$$(\nabla q, \nabla \psi^{n+1}) = \left(\nabla q, \frac{3\mathbf{v}^{n+1} - 4\mathbf{v}^n + \mathbf{v}^{n-1}}{2\Delta t} \right) , \quad \forall \, q \in H^1(\Omega) \,, \tag{8.3.25c}$$

$$p^{n+1} = \psi^{n+1} + 2p^n - p^{n-1} - \nu \nabla \cdot \mathbf{v}^{n+1} \,. \tag{8.3.25d}$$

The theoretical proof by Guermond and Shen [207] is for the first-order scheme, but numerical examples show consistently second-order convergence in Δt for both the velocity and the pressure field. It was also found in the numerical experiments of Guermond and Shen [207] that the inf–sup condition is not required for the system (8.3.22)–(8.3.24) but it is required for the system (8.3.25a)–(8.3.25d). This, in turn, implies that equal-order interpolation is possible for the first system but not for the second; see further discussion in Section 8.3.3.

8.3.2.9 *Generalised block LU decomposition*

The first-order LU approximation presented in Section 8.3.1.1 can be improved by constructing a better approximation to $\boldsymbol{H}^{-1}$, as suggested by Dukowicz and

Dvinsky [145] and Perot [373]. Here, we follow the analysis of Perot [373]. A second-order approximation of $\boldsymbol{H}$ is obtained by using a Taylor series expansion up to second order as follows:

$$\boldsymbol{Q} = \Delta t \boldsymbol{M}^{-1} - \Delta t^2 \nu \boldsymbol{M}^{-1} \boldsymbol{L} \boldsymbol{M}^{-1} .$$

This leads to the following splitting error of order $\mathcal{O}(\Delta t^2)$ measured by the difference

$$-\boldsymbol{D}^\top p^{n+1} - (-\boldsymbol{H}\boldsymbol{Q}\boldsymbol{D}^\top p^{n+1}) = (\nu \Delta t)^2 (\boldsymbol{L}\boldsymbol{M}^{-1})^2 \boldsymbol{D}^\top p^{n+1} .$$

However, as pointed out in Couzy [114], this choice for the matrix $\boldsymbol{Q}$ is not good in practice, as positive-definiteness may be lost for large values of $\nu\Delta t$. A third-order projection is better, corresponding to

$$\boldsymbol{Q} = \Delta t \boldsymbol{M}^{-1} - \Delta t^2 \nu \boldsymbol{M}^{-1} \boldsymbol{L} \boldsymbol{M}^{-1} + \Delta t^3 \nu^2 (\boldsymbol{M}^{-1}\boldsymbol{L})^2 \boldsymbol{M}^{-1} ,$$

which is always positive definite with a splitting error of third order, that is,

$$-\boldsymbol{D}^\top p^{n+1} - (-\boldsymbol{H}\boldsymbol{Q}\boldsymbol{D}^\top p^{n+1}) = (\nu \Delta t)^3 (\boldsymbol{L}\boldsymbol{M}^{-1})^3 \boldsymbol{D}^\top p^{n+1} .$$

Similarly, high-order approximations can be constructed by retaining more terms in the Taylor series for $\boldsymbol{Q}$, but these are very involved and require several matrix-inversion and matrix-multiplication operations.

Using the LU decomposition, the previously discussed splitting methods can be reformulated. For example, Couzy [114] has rewritten the pressure-correction method as

$$\begin{pmatrix} \boldsymbol{H} & -\boldsymbol{H}\boldsymbol{Q}\boldsymbol{D}^\top \\ -\boldsymbol{D} & 0 \end{pmatrix} \begin{pmatrix} \mathbf{v}^{n+1} \\ p^{n+1} - p^n \end{pmatrix} = \begin{pmatrix} \boldsymbol{M}\boldsymbol{f}^{n+1} + \boldsymbol{D}^\top p^n \\ 0 \end{pmatrix} ,$$

with $\boldsymbol{Q} = \Delta t \boldsymbol{M}^{-1}$, which, although a first-order approximation to $\boldsymbol{H}^{-1}$, leads to a second-order overall scheme. Moreover, the splitting error is proportional to $\Delta t \times (p^{n+1} - p^n) \propto \mathcal{O}(\Delta t^2)$ since $p^{n+1} = p^n + \mathcal{O}(\Delta t)$. In this case, the splitting error vanishes for steady-state problems, unlike the original splitting methods.

8.3.3 *The inf–sup condition*

One of the main questions in the full discretisation of incompressible Navier–Stokes equations is the order of interpolation of the velocity and pressure fields. It was found by Karniadakis *et al.* [264] that an equal-order polynomial interpolation leads to stable simulations. This was also verified for the rotational pressure-correction formulation for spectral element discretisations of Timmermans *et al.* [464]. In general, it has been observed that in splitting schemes the inf–sup condition between the velocity and the pressure approximation spaces is not required. However, it has been shown that, for the pressure-correction schemes, (see Auteri *et al.* [20]), if the inf–sup condition is violated then the accuracy in pressure is sub-optimal.

Another interesting interpretation of this has been given by Minev [332]. It was shown that the consistent pressure boundary condition used in the rotational velocity-correction formulation of Karniadakis *et al.* [264] leads to a continuity equation in the weak sense, enhanced by the residual of the momentum equation times a small factor. This is, in fact, the classical stabilisation scheme of Franca *et al.* [166], which does not require the inf–sup condition.

8.3.4 *Comparisons and recommendations*

There is now complete theory that proves rigorously the stability and convergence of almost all versions of splitting schemes. For high Reynolds number flows, splitting methods are more efficient computationally and competitive in accuracy compared to the more expensive coupled methods. For relatively large values of $\nu\Delta t$, however, coupled methods are more accurate. In the following we consider a particular Navier–Stokes solution used by Couzy [114] to evaluate the different ways of treating the pressure in Navier–Stokes calculations. The two-dimensional time-dependent solution is

$$\mathbf{v}(\boldsymbol{x}, t) = \left[-\cos\left(\frac{\pi x_1}{2}\right)\sin\left(\frac{\pi x_2}{2}\right), \sin\left(\frac{\pi x_1}{2}\right)\cos\left(\frac{\pi x_2}{2}\right)\right]^\top \sin(\pi t)$$

on a multiply-connected domain consisting of $N_{\text{el}} = 8$ elements ($P = 5$) in $[0, 1] \times [0, 1]$, with a hole in the middle $[0.4, 0.6] \times [0.4, 0.6]$. The initial condition is zero velocity and pressure, and the solution is integrated using a third-order scheme until time $t = 0.75$. The Reynolds number is $\text{Re} = 10$.

The results from the calculations of Couzy are shown in Table 8.1. Here LU1 refers to a first-order LU decomposition, which corresponds to the classical splitting scheme, and LU3 refers to the third-order scheme; PC corresponds to the pressure-correction scheme, and LU3–PC corresponds to a third-order pressure-correction scheme constructed using an LU decomposition, as explained in the previous section for a second-order scheme. The results are presented in terms

TABLE 8.1. Number of correct decimal digits ($-\log(\text{max error})$) for the pressure (ϵ_p, first column) and streamwise velocity (ϵ_u, second column) ($t = 0.75$, $\text{Re} = 10$). LU1 refers to an LU decomposition of first order, PC to a pressure correction of second order, LU3 the third-order scheme, LU3–PC a combined LU decomposition with pressure correction leading to third order, and UZ3 to Uzawa of third order. All data are from Couzy [114].

Method	$\Delta t = \frac{1}{40}$ (ϵ_p; ϵ_u)	$\Delta t = \frac{1}{80}$ (ϵ_p; ϵ_u)	$\Delta t = \frac{1}{160}$ (ϵ_p; ϵ_u)	$\Delta t = \frac{1}{320}$ (ϵ_p; ϵ_u)
LU1	0.66; 1.55	0.85; 1.81	1.07; 2.09	1.31; 2.38
PC	1.71; 2.17	2.06; 3.22	2.64; 3.82	3.18; 4.42
LU3	0.75; 0.71	1.14; 1.55	1.68; 2.53	2.45; 3.48
LU3–PC	1.66; 1.72	2.11; 3.07	3.23; 4.30	4.29; 5.36
UZ3	3.03; 4.82	3.92; 5.63	4.82; 5.82	5.42; 5.71

of the number of correct decimal digits; since $-\log 0.5 \approx 0.301$, by halving the time-step size, a gain of 0.3 digits corresponds to order 1, a gain of 0.6 digits corresponds to order 2, and a gain of 0.9 corresponds to order 3. The accuracy in pressure is less than the accuracy in velocity, in agreement with the analysis of the boundary divergence in the previous section. This example can be used as a benchmark for developing accurate pressure solvers.

In summary we have the following.

- Splitting methods are cost-effective algorithms and are particularly accurate in simulating high Reynolds number flows.
- Both pressure-correction and velocity-correction formulations lead to second-order accuracy in velocity, i.e., $\mathcal{O}(\Delta t^2)$, and $\mathcal{O}(\Delta t^{3/2})$ in pressure if the rotational version of the schemes is employed.
- The velocity-correction scheme can be extended to any high order, but it is conditionally stable for third or higher order due to its explicit treatment of the pressure boundary condition.
- The pressure-correction scheme is stable for up to second order and unstable for fourth order or higher. Third-order schemes have been developed and claimed stable, e.g., see Heinrichs [223], but a weak instability that affects long-time integration has been reported by Shen [421].
- An appropriate variational formulation of the projection step, i.e., testing against gradients, leads to a full second-order accuracy for the pressure field as well.
- Time integration based on backwards differentiation (stiffly-stable schemes) provides enhanced stability.
- The inf–sup condition is not required for properly implemented splitting schemes, and thus equal-order interpolation can be used for the velocity and pressure, unlike the coupled formulations.

8.4 Velocity–vorticity formulation

There are several general advantages of the vorticity–velocity formulation over other formulations, as well as specific advantages in the context of the spectral/*hp* element formulation. Among the advantages are the following.

1. The elimination of pressure, which leads to a simple formulation for a diffusion operator rather than the Stokes operator.
2. The divergence-free condition is imposed implicitly.
3. The formulation works in both two and three dimensions.
4. All non-inertial effects arising from the rotation and translation of the reference frame enter into the solution of the problem through the initial and boundary conditions, and thus no additional computational work is required to evaluate non-inertial terms.

5. Vorticity is a more relevant physical variable in vortex-dominated flows.

In the spectral/*hp* element context the vorticity–velocity formulation produces a vorticity field that is C^0 continuous across elemental interfaces, unlike the velocity–pressure formulation where continuity of vorticity across elemental interfaces is achieved only upon convergence. In addition, spectrally accurate vorticity boundary conditions are imposed, a crucial element of the formulation for the two implementations that we will present here.

The vorticity form of the incompressible Navier–Stokes equations is complicated by the lack of vorticity boundary conditions and by the incompressibility constraint on the velocity, that is, $\nabla \cdot \mathbf{v} = 0$. In the velocity–vorticity $(\mathbf{v}, \boldsymbol{\omega})$ formulation presented here, the incompressibility constraint will be replaced by a Poisson equation in $\mathbf{v}$. In addition, the definition of vorticity, $\boldsymbol{\omega} = \nabla \times \mathbf{v}$, on the domain boundary $\partial\Omega$ will provide a necessary constraint on the vorticity in order to prove *equivalence* with the standard velocity–vorticity formulation. Essentially, the constraint will act as a vorticity boundary condition. The system of governing equations and constraints in three dimensions is

$$\frac{\partial \boldsymbol{\omega}}{\partial t} + \nabla \times (\boldsymbol{\omega} \times \mathbf{v}) = -\nu \nabla \times \nabla \times \boldsymbol{\omega} \quad \text{in } \Omega\,, \tag{8.4.1a}$$

$$\nabla^2 \mathbf{v} = -\nabla \times \boldsymbol{\omega} \quad \text{in } \Omega\,, \tag{8.4.1b}$$

$$\boldsymbol{\omega} = \nabla \times \mathbf{v} \quad \text{on } \partial\Omega\,, \tag{8.4.1c}$$

$$\mathbf{v} = 0 \quad \text{on } \partial\Omega\,, \tag{8.4.1d}$$

$$\boldsymbol{\omega} = \nabla \times \mathbf{v} \quad \text{at } t = 0 \quad \text{in } \Omega\,, \tag{8.4.1e}$$

$$\int_{\partial\Omega} \mathbf{v} \cdot \boldsymbol{n}\, \mathrm{d}\boldsymbol{x} = 0 \quad \text{or} \quad \nabla \cdot \mathbf{v}(\boldsymbol{x}_0) = 0\,, \quad \text{where } \boldsymbol{x}_0 \in \partial\Omega\,. \tag{8.4.1f}$$

The difference from the original velocity–vorticity system in its standard form (see Chapter 1) is that the vorticity definition is enforced only at the boundary of the domain rather than the entire domain.

The proof of equivalence can be established in three steps. The first step proves that $\boldsymbol{\omega}$ in this $(\mathbf{v}, \boldsymbol{\omega})$ formulation is divergence-free. The second step shows that the definition of vorticity is satisfied everywhere in the domain if it is satisfied at the boundary. Finally, the third step proves that the velocity is divergence-free. Equivalence is then complete because the vorticity transport equation is solved and $\boldsymbol{\omega}$ and $\mathbf{v}$ are divergence-free [123, 213].

Lemma 8.4.1 *The* $(\mathbf{v}, \boldsymbol{\omega})$ *equations (8.4.1a)–(8.4.1f) satisfy the condition* $\nabla \cdot \boldsymbol{\omega} = 0$ *in* Ω *for all times if* $\nabla \cdot \boldsymbol{\omega} = 0$ *in* Ω *at* $t = 0$.

Proof Consider the vorticity transport equation, eqn (8.4.1a), the divergence of which yields

$$\frac{\partial Q}{\partial t} = 0\,, \quad Q \equiv \nabla \cdot \boldsymbol{\omega} \quad \text{in } \Omega\,.$$

Therefore, Q is identically zero assuming that, at $t = 0$, $Q(x, 0) = 0$. □

Lemma 8.4.2 *Necessary and sufficient conditions for* $\boldsymbol{\omega} = \nabla \times \mathbf{v}$ *in* Ω *are*

$$\boldsymbol{\omega} = \nabla \times \mathbf{v} \quad \text{in } \partial\Omega$$

and

$$\nabla \cdot \boldsymbol{\omega} = 0 \quad \text{in } \Omega\,.$$

Proof Consider the vector identity $\nabla^2 \boldsymbol{q} = \nabla(\nabla \cdot \boldsymbol{q}) - \nabla \times (\nabla \times \boldsymbol{q})$ and eqn (8.4.1b). Equating the right-hand sides of these two equations gives

$$\nabla \times (\boldsymbol{\omega} - \nabla \times \mathbf{v}) = -\nabla(\nabla \cdot \mathbf{v})\,.$$

Taking the *curl* of this equation gives

$$\nabla \times [\nabla \times (\boldsymbol{\omega} - \nabla \times \mathbf{v})] = -\nabla \times \nabla(\nabla \cdot \mathbf{v})\,,$$

where the right-hand side is zero because derivatives of $\nabla \cdot \mathbf{v}$ are interchangeable. Applying the previous vector identity with $\boldsymbol{q} = \boldsymbol{\omega} - \nabla \mathbf{v}$ to the left-hand side leads to

$$\nabla[\nabla \cdot (\boldsymbol{\omega} - \nabla \times \mathbf{v})] - \nabla^2(\boldsymbol{\omega} - \nabla \times \mathbf{v}) = 0\,,$$

where the first term drops out because $\nabla \cdot \boldsymbol{\omega} = 0$ in Ω from Lemma 8.4.1 and $\nabla \cdot (\nabla \times \mathbf{v}) = 0$ by definition. Therefore,

$$\nabla^2(\boldsymbol{\omega} - \nabla \times \mathbf{v}) = 0\,.$$

This elliptic equation has homogeneous boundary conditions because of the validity of the vorticity definition at the boundary, and thus $\boldsymbol{\omega} = \nabla \times \mathbf{v}$ everywhere in Ω. □

Lemma 8.4.3 *A necessary and sufficient condition for* $\nabla \cdot \mathbf{v} = 0$ *in* Ω *to be satisfied is*

$$\boldsymbol{\omega} = \nabla \times \mathbf{v} \quad \text{in } \Omega$$

and

$$\int_{\partial\Omega} \mathbf{v} \cdot \boldsymbol{n}\, \mathrm{d}\boldsymbol{x} = 0$$

or

$$\nabla \cdot \mathbf{v} = 0 \quad \text{at one point on } \partial\Omega\,.$$

Proof As in Lemma 8.4.2, we equate the vector identity and eqn (8.4.1b), giving

$$\nabla(\nabla \cdot \mathbf{v}) - \nabla \times (\nabla \times \mathbf{v}) = -\nabla \times \boldsymbol{\omega}\,.$$

But $\boldsymbol{\omega} = \nabla \times \mathbf{v}$ in Ω from Lemma 8.4.2, so the equation reduces to

$$\nabla(\nabla \cdot \mathbf{v}) = 0 \quad \text{in } \Omega\,.$$

Integrating the above equation leads us to $\nabla \cdot \mathbf{v} = C$, where this constant is independent of space. To eliminate the integration constant we can use either of the constraints in eqn (8.4.1f). □

The pertinent step in proving equivalence for the $(\mathbf{v}, \boldsymbol{\omega})$ formulation of the incompressible Navier–Stokes equations is enforcing $\boldsymbol{\omega} = \nabla \times \mathbf{v}$ on $\partial\Omega$. This important result from the equivalence statement provides a linear coupling between the vorticity and velocity on $\partial\Omega$ and guarantees that eqns (8.4.1a)–(8.4.1f) generate the correct vorticity.

For *multiply-connected domains*, such as in flow past a circular cylinder, for uniqueness of the solution an extra constraint must be imposed. On the cylinder surface we apply the momentum equation, that is,

$$0 = -\nabla p - \nu \nabla \times \boldsymbol{\omega}$$

and, after integrating along the cylinder surface, we obtain

$$\int_0^{2\pi} \nabla p|_{r=R} (r \, \mathrm{d}\theta) = - \int_0^{2\pi} \nu (\nabla \times \boldsymbol{\omega})|_{r=R} R \, \mathrm{d}\theta \, .$$

Therefore, the following condition should be satisfied on the cylinder surface:

$$\int_0^{2\pi} (\nabla \times \boldsymbol{\omega})|_{r=R} R \, \mathrm{d}\theta = 0 \, .$$

8.4.1 *Semi-discrete equations*

We will use a semi-implicit time-integration scheme to decouple the velocity and vorticity equations and thus linearise the vorticity transport equation. The following scheme was proposed by Trujillo [472]. First, the advection terms are discretised:

$$\frac{\boldsymbol{\omega}^{\star} - \boldsymbol{\omega}^n}{\Delta t} = \sum_{q=0}^{J-1} \beta_q [\nabla \times (\mathbf{v} \times \boldsymbol{\omega})]^{n-q} \, ,$$

where β_q are appropriate weights.

The diffusion terms are treated implicitly using the θ method (a stiffly-stable scheme could also be used). For $\theta = \frac{1}{2}$ we recover the Crank–Nicolson scheme and for $\theta = 1$ the first-order implicit Euler scheme:

$$\frac{\boldsymbol{\omega}^{n+1} - \boldsymbol{\omega}^{\star}}{\Delta t} = \nu [\theta \nabla^2 \boldsymbol{\omega}^{n+1} + (1 - \theta) \nabla^2 \boldsymbol{\omega}^n] \, .$$

The resulting equivalent semi-discrete system of equations is

$$\begin{aligned}
\left(\frac{1}{\theta \nu \Delta t} - \nabla^2 \right) \boldsymbol{\omega}^{n+1} &= f && \text{in } \Omega \, , \\
\nabla^2 \mathbf{v}^{n+1} &= -\nabla \times \boldsymbol{\omega}^{n+1} && \text{in } \Omega \, , \\
\nabla \times \mathbf{v}^{n+1} &= \boldsymbol{\omega}^{n+1} && \text{on } \partial\Omega \, , \\
\mathbf{v}^{n+1} &= 0 && \text{on } \partial\Omega \, , \\
\int_{\partial\Omega} \mathbf{v}^{n+1} \cdot \boldsymbol{n} \, \mathrm{d}\boldsymbol{x} &= 0 && \text{on } \partial\Omega \, .
\end{aligned}$$

The vorticity boundary conditions are treated consistently through the enforcement of the vorticity definition at the boundary. The Crank–Nicolson time integrator, $\theta = \frac{1}{2}$, generates high-frequency odd–even modes that are not damped out, while the implicit Euler time integrator, $\theta = 1$, heavily damps the high frequencies. The odd–even mode generated by the Crank–Nicolson time integrator for high frequencies can lead to numerical instability, as reported in Deville *et al.* [130]. However, the growth rate of the instability is very slow, so the Crank–Nicolson time integrator can be used for relatively short-time integration.

8.4.2 *Influence matrix implementation*

The influence matrix technique to impose boundary conditions is based on the discrete Green's function method, which in turn depends on the linearity of the semi-discrete equations. The equivalent semi-discrete system of equations can be split into the homogeneous contribution, $\hat{\mathbf{v}}, \hat{\boldsymbol{\omega}}$, and the particular contribution, $\tilde{\mathbf{v}}, \tilde{\omega}$ as follows:

$$\mathbf{v} = \tilde{\mathbf{v}} + \sum_{k=1}^{N_b} \lambda_k \hat{\mathbf{v}}_k \tag{8.4.3}$$

and

$$\omega = \tilde{\omega} + \sum_{k=1}^{N_b} \lambda_k \hat{\omega}_k \,, \tag{8.4.4}$$

where the values of the λ_ks are determined by enforcing the definition of vorticity on the boundary, and N_b is the number of degrees of freedom on the physical boundary. In this section we present this implementation for the two-dimensional case only, since computer memory requirements make this approach not so attractive in practice for three dimensions.

The discrete *homogeneous* problem is

$$\begin{aligned}
\left(\frac{1}{\theta\nu\Delta t} - \nabla^2\right)\hat{\omega}_k^{n+1} &= 0 && \text{in } \Omega\,,\\
\hat{\omega}_k(\gamma_j) &= \delta_{kj}\,, && \forall\ \gamma_j \in \partial\Omega\,,\\
\nabla^2 \hat{\mathbf{v}}_k^{n+1} &= -\nabla \times (\hat{\omega}_k^{n+1}\hat{\boldsymbol{z}}) && \text{in } \Omega\,,\\
\hat{\mathbf{v}}^{n+1} &= 0 && \text{on } \partial\Omega\,,
\end{aligned}$$

where $\hat{\omega}_k^{n+1}$ is the kth discrete Green's function and $\hat{\boldsymbol{z}}$ is the unit vector in the z direction.

The *particular* problem is

$$\begin{aligned}
\left(\frac{1}{\theta\nu\Delta t} - \nabla^2\right)\tilde{\omega}^{n+1} &= f && \text{in } \Omega\,,\\
\tilde{\omega}(\gamma_j) &= \tilde{\omega}_{\partial\Omega} && \text{on } \partial\Omega\,,\\
\nabla^2 \tilde{\mathbf{v}}^{n+1} &= -\nabla \times \tilde{\omega}^{n+1}\hat{\boldsymbol{z}} && \text{in } \Omega\,,\\
\tilde{\mathbf{v}}^{n+1} &= 0 && \text{on } \partial\Omega\,,
\end{aligned}$$

where the forcing function, f, includes primarily the influence of the advection terms and $\tilde{\omega}_{\partial\Omega}$ is an arbitrary smooth distribution of vorticity on the boundary. The influence matrix is constructed from the definition of vorticity on the boundary $\partial\Omega$ as follows:

$$(\tilde{\eta} - \tilde{\omega})(\gamma_j) + \sum_{k=1}^{N_b} \lambda_k (\hat{\eta}_k - \hat{\omega}_k)(\gamma_j) = 0, \quad \forall\, j = 1, 2, \ldots, N_b, \quad \gamma_j \in N_b,$$

where

$$\tilde{\eta} = (\nabla \times \tilde{\mathbf{v}}) \cdot \hat{\boldsymbol{z}}$$

and

$$\hat{\eta}_k = (\nabla \times \hat{\mathbf{v}}_k) \cdot \hat{\boldsymbol{z}}.$$

The influence matrix problem is then

$$\boldsymbol{A\lambda} = \boldsymbol{f},$$

where

$$\boldsymbol{A}[j][k] = (\hat{\eta}_k - \hat{\omega}_k)(\gamma_j)$$

and

$$\boldsymbol{f}[j] = -(\tilde{\eta} - \tilde{\omega})(\gamma_j).$$

The solution $\boldsymbol{\lambda} = \boldsymbol{A}^{-1}\boldsymbol{f}$ added to the arbitrary smooth vorticity on the boundary, $\tilde{\omega}_{\partial\Omega}$, provides the vorticity on the boundary, $\omega_{\partial\Omega}$, that is,

$$(\omega_{\partial\Omega})_k = (\tilde{\omega}_{\partial\Omega})_k + \lambda_k, \quad \forall\, k = 1, 2, \ldots, N_b. \tag{8.4.7}$$

We note that $\boldsymbol{A}$ only depends on the homogeneous solution which can be generated once and stored.

Therefore, in order to determine the correct boundary vorticity it is not necessary to store the discrete Green's functions $(\hat{\mathbf{v}}_k, \hat{\omega}_k)$. However, in order to construct the entire solution without storing the Green's functions it is necessary to resolve a system similar to the *particular* system, but with the correct boundary vorticity obtained from eqn (8.4.7).

The accurate solution of the Helmholtz equations requires very high resolution, especially as $\nu \to 0$. This was demonstrated by Trujillo and Karniadakis [473], who provided several examples with instabilities arising due to under-resolution of this numerical boundary layer, proportional to $\sqrt{\nu}$. A possible remedy is offered by Clercx [99], who proposed to compute an additional set of complementary solutions to eqns (8.4.3) and (8.4.4) as

$$\mathbf{v} = \tilde{\mathbf{v}} + \sum_{k=1}^{N_b} \lambda_k \hat{\mathbf{v}}_k + \sum_{l=1}^{L} \mu_l \hat{\hat{u}}_l \tag{8.4.8}$$

and

$$\omega = \tilde{\omega} + \sum_{k=1}^{N_b} \lambda_k \hat{\omega}_k + \sum_{l=1}^{L} \mu_l \hat{\hat{\omega}}_l \,. \qquad (8.4.9)$$

The additional fields satisfy the same set of governing equations with homogeneous boundary conditions for both the velocity and vorticity, and a modified right-hand side in the Helmholtz equation for the vorticity, see [99] for details.

8.4.3 *Penalty method implementation*

A more general method of imposing the vorticity definition on the boundary, according to the formulation in eqns (8.4.1a)–(8.4.1f) is to incorporate the boundary constraint eqn (8.4.1c) into the vorticity evolution equation (8.4.1a) using the penalty method. This is justified as the governing equation holds arbitrarily close to the boundary and there is a natural transition to the boundary condition. Unlike the penalty method used in the primitive formulation to enforce the incompressibility constraint everywhere in the domain, here only the *domain boundary* is involved in the penalty procedure, thus avoiding the stiffness typically associated with such constrained systems. This approach will be considered again in Chapter 10 to deal with interface boundary conditions for compressible flows. The vorticity transport equation then becomes

$$\frac{\partial \boldsymbol{\omega}}{\partial t} + \nabla \times (\boldsymbol{\omega} \times \mathbf{v}) = -\nu \nabla \times \nabla \times \boldsymbol{\omega} - \tau(\boldsymbol{\omega} - \nabla \times \mathbf{v})|_{\partial\Omega} \quad \text{in } \Omega \,, \qquad (8.4.10)$$

where τ is a large positive parameter that enforces the boundary constraint. The range of this penalty parameter is chosen so that stability of the discrete scheme is guaranteed; it is a function of the spectral order P and the Reynolds number. The additional constraint for multiply-connected domains discussed previously can also be implemented by the penalty method by adding the constraint to the right-hand side of eqn (8.4.10), corresponding to another penalty parameter τ_1. For stability of the approximation, it is necessary to select $\tau_1 < \tau$, as has been verified in numerical experiments [472]. A possible time discretisation is the following semi-implicit scheme:

$$\frac{\boldsymbol{\omega}^{n+1} - \boldsymbol{\omega}^n}{\Delta t} + \nabla \times (\boldsymbol{\omega}^n \times \mathbf{v}^n) = -\nu \nabla \times \nabla \times \boldsymbol{\omega}^{n+1} - \tau(\boldsymbol{\omega}^{n+1} - \nabla \times \mathbf{v}^n) \,.$$

This particular scheme is $\mathcal{O}(\Delta t)$ in accuracy, but higher time discretisations can be constructed. Note that for $\tau \to \infty$ the scheme is unstable and therefore an explicit treatment of the boundary constraint alone does not lead to stable approximations.

8.4.4 *Spatial discretisation*

Unlike the variational formulation of the Navier–Stokes equations in terms of primitive variables for which there is a rigorous theoretical framework [191], there is no variational formulation in terms of $(\mathbf{v}, \boldsymbol{\omega})$ which has yielded to rigorous analysis. To examine the underlying difficulties, we consider the steady linear equations rewritten in the form

$$-\nabla^2\mathbf{v} = \nabla \times \boldsymbol{\omega}\,, \tag{8.4.11a}$$

$$-\nu\nabla \times \nabla \times \boldsymbol{\omega} = \nabla \times \boldsymbol{f}\,, \tag{8.4.11b}$$

$$\boldsymbol{\omega} = \nabla \times \mathbf{v} \quad \text{on } \partial\Omega\,, \tag{8.4.11c}$$

$$\mathbf{v} = 0 \qquad \text{on } \partial\Omega\,, \tag{8.4.11d}$$

where $\boldsymbol{f} \in L^2(\Omega)$ is a driving force. For the sake of simplicity we assume only a simply-connected domain Ω. A variational approach proposed by Ruas [406] assumes that $\mathbf{v} \in H^1(\Omega)$ and the vorticity space is chosen so as to require the weakest possible regularity for well-posedness, that is,

$$\boldsymbol{\omega} \in \mathcal{X}(\Omega)\,, \quad \mathcal{X}(\Omega) = \{\chi \mid \chi \in L^2(\Omega),\ \nabla^2\chi \in H^{-1}(\Omega)\}\,.$$

The variational statement for eqns (8.4.11a)–(8.4.11d) is as follows.

Find $(\mathbf{v}, \boldsymbol{\omega}) \in H^1(\Omega) \times \mathcal{X}(\Omega)$ such that

$$(\mathbf{w}, \mathbf{v})_1 = (\nabla \times \mathbf{w}, \boldsymbol{\omega})\,, \quad \forall\, \mathbf{w} \in H^1(\Omega)\,,$$
$$\nu\langle \phi, -\nabla^2\boldsymbol{\omega}\rangle_1 = (\phi, \nabla \times \boldsymbol{f})\,, \quad \forall\, \phi \in H^1(\Omega)\,,$$

where $(\cdot,\cdot)$ is the standard L^2 inner product, $(\cdot,\cdot)_1$ denotes the product in H^1, and $\langle\cdot,\cdot\rangle_1$ denotes the duality product between H^{-1} and H^1.

We have not included the boundary conditions in the variational statement, the incorporation of which will require the selection of appropriate subspaces of $H^1(\Omega)$ and $\mathcal{X}(\Omega)$. Although it was proven by Ruas [406] that this formulation was equivalent to the Stokes problem in primitive variables, this formal setting suggests the use of mixed finite element methods (see Chapter 5), which complicates the discretisation. No stable numerical results have been reported adopting this formulation.

An alternative approach is to consider both velocity and pressure in $H^1(\Omega)$, as was adopted by Gunzburger *et al.* [214], in which case the variational statement has the following form.

Find $(\mathbf{v}, \boldsymbol{\omega}) \in H^1(\Omega) \times H^1(\Omega)$ such that

$$(\mathbf{v}, \mathbf{w})_1 = (\nabla \times \mathbf{w}, \boldsymbol{\omega})\,, \quad \forall\, \mathbf{w} \in H^1(\Omega)\,,$$
$$\nu(\boldsymbol{\omega}, \phi)_1 = (\phi, \nabla \times \boldsymbol{f})\,, \quad \forall\, \phi \in H^1(\Omega)\,.$$

The difficulty here arises from the boundary condition $\boldsymbol{\omega} = \nabla \times \mathbf{v}$ on $\partial\Omega$. Since $\mathbf{v} \in H^1(\Omega)$, we have that $\nabla \times \mathbf{v} \in L^2(\Omega)$, and thus we cannot restrict its components at the boundary to be equal to $\boldsymbol{\omega} \in H^1(\Omega)$. The proper formulation would have been to require $\mathbf{v} \in H^2(\Omega)$ and $\boldsymbol{\omega} \in H^1(\partial\Omega)$. This additional regularity for the velocity and vorticity fields has not been proven, and, in addition, it increases the discretisation complexity as it requires C^1 elements. These seemingly conflicting statements have precluded any rigorous analyses of the variational problem, including the derivation of error estimates.

Nevertheless, progress can be made by imposing the vorticity boundary condition, eqn (8.4.11c), indirectly, that is, by projection. In the continuous case

$$\int_{\partial\Omega} \boldsymbol{\phi} \cdot \boldsymbol{\omega} \, \mathrm{d}\boldsymbol{x} = \int_{\partial\Omega} \boldsymbol{\phi} \cdot \nabla \times \mathbf{v} \, \mathrm{d}\boldsymbol{x}, \quad \forall \, \phi \in H^1(\partial\Omega),$$

but this projection is still not properly defined as the right-hand side of the above equation does not make sense (see [214]) since $\mathbf{v} \in H^1(\Omega)$ and $\phi \in H^{1/2}(\Omega) \equiv H^1(\partial\Omega)$. However, with the discrete approximations for $\mathbf{v}$ and ϕ both being piecewise continuous polynomials, the projection is well defined. In certain low-order discretisations (see, for example, [210]) the vorticity interpolation is of lower order for polynomial order compatibility, but this is not necessary in high-order approximations for which equal-order approximations were found to be stable and more accurate.

The above formulation is not optimal (see [214]). This is manifested in the loss of accuracy for the vorticity in numerical experiments with quadratic elements. However, if refinement close to the boundary is performed, almost full accuracy is recovered. Numerical experiments with spectral elements which can provide high resolution at the boundary (Trujillo [472]) suggest that high-order approximations do not require unequal interpolation for stability.

8.5 Least-squares method

In this section we present another approach to dealing with the Stokes operator using the least-squares method. A significant difference between the coupled (i.e., mixed) method and the least-squares method is that in the latter no special compatibility between the functional spaces is required to ensure uniqueness. We follow here the work of Proot [384] and also Proot and Gerritsma [385], who have extended the least-squares method developed for finite element discretisations to spectral elements [62, 63, 253]. The main objective is to provide a spectral element formulation that circumvents the inf–sup condition for ellipticity, thus employing equal-order interpolation. This is similar to what is possible with appropriately-formulated splitting schemes; however, in the least-squares formulation the Stokes operator remains unsplit.

8.5.1 *Formulation*

In particular, the velocity–vorticity–pressure formulation was adopted in [384], which in two dimensions has the form

$$\nabla p - \nu \nabla \times \boldsymbol{\omega} = \boldsymbol{f} \quad \text{in } \Omega, \tag{8.5.1a}$$

$$\boldsymbol{\omega} - \nabla \times \mathbf{v} = 0 \quad \text{in } \Omega, \tag{8.5.1b}$$

$$\nabla \cdot \mathbf{v} = 0 \quad \text{in } \Omega. \tag{8.5.1c}$$

This system involves four equations and four unknowns $(\mathbf{v}, \omega, p)$ in two dimensions, but seven equations in three dimensions. For a C^0 continuous approximation, the least-squares formulation is normally applied to a first-order system, and this is primarily the reason for adopting the above representation of the Stokes problem.

Uniqueness of the solution (expressed in terms of the H^1-coercivity condition) depends on the boundary conditions. An appropriate set of boundary conditions is listed in Table 8.2; it involves combinations of normal and tangential velocity values as well as pressure and vorticity values. The justification of the compatibility of these unusual boundary conditions with the differential equations is provided by the ADN (Agmon–Douglis–Nirenberg) theory for first-order elliptic systems [2]. This theory shows that the coercivity estimates for the Stokes operator are not unique, but they depend on the specific boundary condition. Correspondingly, the rate of h-convergence of the least-squares formulation depends strongly on the boundary conditions involved. Specifically, all boundary conditions of Table 8.2 are Dirichlet since weak (Neumann) boundary conditions lead to loss of H^1-coercivity. We note also that a unique aspect of the least-squares formulation is that it can solve over-determined systems, e.g., extra boundary conditions, as long as they are consistent. This, however, may result in a sub-optimal convergence rate.

In the least-squares formulation the first step is to establish a priori estimates and, subsequently, to obtain a quadratic least-squares functional. Upon minimisation of the functional, the weak formulation is obtained. An appropriate a priori estimate for the velocity–vorticity–pressure formulation is

$$||\mathbf{v}||_1 + ||\boldsymbol{\omega}||_1 + ||p||_1 \leqslant C(||\nabla p + \nu \nabla \times \boldsymbol{\omega}||_0 + ||\nabla \cdot \mathbf{v}||_0 + ||\boldsymbol{\omega} - \nabla \times \mathbf{v}||_0) . \quad (8.5.2)$$

This estimate shows that the formulation is fully H^1-coercive and thus standard C^0 spectral/*hp* elements of equal order can be employed in the discretisation, see [253, 384]. The corresponding quadratic functional is

$$\mathcal{I}(u) = \frac{1}{2}(||\nabla p + \nu \nabla \times \boldsymbol{\omega} - \boldsymbol{f}||_0^2 + ||\nabla \cdot \mathbf{v}||_0^2 + ||\boldsymbol{\omega} - \nabla \times \mathbf{v}||_0^2) , \quad (8.5.3)$$

which upon minimisation gives the weak statement of the problem cast in the standard C^0 setting. In [63] a *weighted* functional was introduced of the form

$$\mathcal{I}(u) = \frac{1}{2}(||\nabla p + \nu \nabla \times \boldsymbol{\omega} - \boldsymbol{f}||_0^2 + h^{-2}||\nabla \cdot \mathbf{v}||_0^2 + h^{-2}||\boldsymbol{\omega} - \nabla \times \mathbf{v}||_0^2) , \quad (8.5.4)$$

TABLE 8.2. Boundary conditions for the least-squares formulation.

BC1: symmetry	BC2: inflow	BC3: outflow	BC4: outflow	BC5: wall, outflow	BC6: outflow
$\boldsymbol{n} \cdot \mathbf{v} = 0$	$\boldsymbol{n} \cdot \mathbf{v} = 0$	$\boldsymbol{n} \times \mathbf{v} = 0$	$\boldsymbol{n} \times \mathbf{v} = 0$	$\boldsymbol{n} \cdot \mathbf{v} = 0$	$\omega = 0$
$\omega = 0$	$p = 0$	$p = 0$	$\omega = 0$	$\boldsymbol{n} \times \mathbf{v} = 0$	$p = 0$

which is not strictly coercive, but a standard spectral/*hp* element discretisation can still be applied. The parameter $h > 0$ represents the size of the element and should be scaled with the order of the polynomial P in spectral elements. For example, Proot used in two dimensions $h \propto A_e/P^2$, where A_e is the area of the element e. For this functional, a sub-optimal convergence rate is obtained if an equal-order interpolation is involved. However, in [63] it was shown that if the pressure and vorticity are approximated with one order less than the velocity then optimal h-convergence is achieved.

8.5.2 *Performance*

We report here some results obtained by Proot [384] on the model Stokes problem:

$$u(x, y) = -\sin(2\pi x)\cos(2\pi y), \quad v(x, y) = \cos(2\pi x)\sin(2\pi y),$$

and

$$p(x, y) = \sin(\pi x)\sin(\pi y).$$

This solution satisfies the steady Stokes equations for the two-dimensional forcing

$$f_x = \pi\cos(\pi x)\sin(\pi y) - 8\pi^2\sin(2\pi x)\cos(2\pi y)$$

and

$$f_y = \pi\sin(\pi x)\cos(\pi y) + 8\pi^2\cos(2\pi x)\sin(2\pi y).$$

Proot solved this model problem for all the boundary conditions listed in Table 8.2 and also for an over-determined case for which all variables are specified at the boundary. The discretisation involved a nodal spectral element method with Legendre polynomials. In addition, comparisons were made against the Uzawa $\mathcal{P}_P/\mathcal{P}_{P-2}$ formulation with Gauss–Lobatto–Legendre nodal points. A summary of the results in terms of the h-convergence rate is shown in Table 8.3 for various boundary conditions. Here 'OD' refers to the over-determined case, while BC5a and BC5b correspond to the wall/outflow boundary condition with the standard and weighted functional, respectively. Clearly, the weighted functional leads to one order lower in the accuracy of the vorticity and pressure due to the loss of strict coerciveness [63]. Similarly, for the over-determined case a sub-optimal h-convergence is obtained. On the other hand, the same convergence rate is obtained for the velocity field for all conditions.

TABLE 8.3. Effect of boundary conditions on the h-convergence rate of the least-squares method. Values of the h-convergence rate are shown ($P = 4$).

	H^0 error				H^1 error			
Function	BC2	OD	BC5a	BC5b	BC2	OD	BC5a	BC5b
u	5.02	5.02	5.09	5.01	4.00	4.00	4.04	4.00
v	5.02	5.02	5.09	5.01	4.00	4.00	4.04	4.00
ω	4.99	4.99	6.10	4.42	3.99	3.99	4.60	3.52
p	5.86	5.86	6.81	5.69	4.91	4.91	5.72	4.67

Tests for p-convergence also included in [384] revealed that exponential accuracy is obtained for all variables. In particular, it was found that prescribing the pressure on the boundary (the fully-H^1-coercive formulation) leads to a pressure accuracy which is half an order better than the accuracy in velocity—an unusual result. In contrast, when the boundary pressure is not prescribed then the velocity accuracy is one order better than the vorticity and pressure accuracy, as expected. Also, the over-determined case yields results similar to the inflow condition (best case), implying indirectly that this formulation also behaves as fully H^1-coercive.

A comparison of the h-convergence versus the p-convergence for the least-squares spectral element method is given in Fig. 8.5. The test problem is still the aforementioned Stokes model problem. It shows that the combination of least-squares and spectral discretisation leads to very accurate solutions at a substantially lower number of degrees of freedom. Finally, a comparison of the least-squares formulation with the Galerkin $\mathcal{P}_P/\mathcal{P}_{P-2}$ method is shown in Table 8.4. Both the standard (BC5a) and the weighted formulations (BC5b) are included in the comparison. While the h-convergence rate for velocity is the same

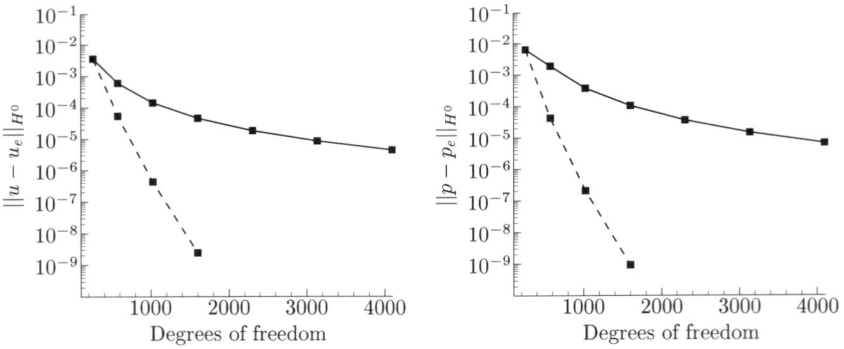

FIG. 8.5. Comparison of the h-convergence (solid lines) and p-convergence (dashed lines) rates for a model Stokes problem using the least-squares spectral element formulation. (Courtesy of Michael Proot.)

TABLE 8.4. Comparison between the least-squares spectral element method (LSQ–SEM) and the Galerkin formulation for the model Stokes problem ($P = 4$).

Variable	LSQ–SEM BC5a	LSQ–SEM BC5b	Galerkin
u	5.09	5.01	4.91
v	5.09	5.01	4.91
ω	6.10	4.42	3.97
p	6.81	5.69	3.12

for all formulations, the pressure and vorticity are substantially less accurate for the Galerkin method. With respect to vorticity, this is expected since in the Galerkin formulation it is derived from the velocity field by differentiation, whereas in the least-squares method it is computed directly. With regards to p-convergence, the vorticity is again predicted more accurately by the least-squares method, while the pressure is predicted more accurately by the Galerkin method.

In summary, it is interesting to note that the p-convergence rates of the least-squares spectral element formulation do not seem to be affected by the choice of the boundary conditions, in contrast to the h-convergence rates.

With regards to *computational complexity*, the least-squares method leads to a symmetric positive-definite system which can be solved very effectively with substructuring following the techniques of Section 4.2.3. In the least-squares finite element literature, the Jacobi conjugate gradient method is typically found to be very effective. However, numerical tests in [384] have shown that for spectral order $P \geqslant 4$ the Schur complement method outperforms the Jacobi conjugate gradient by a factor of approximately twenty.

There are some open issues related to the least-squares formulation in the context of Navier–Stokes equations, which have also been addressed in [384]. For example, the 'loss of mass', i.e., violation of the incompressibility constraint, in internal flows was studied in detail. To this end, a modified functional can be introduced that incorporates the divergence-free condition as a constraint to deal with this. However, this leads to complications and the loss of coercivity, which, in turn, leads to a requirement for different spaces between the variables, as in the standard Galerkin formulation. Alternatively, a special weighting of the divergence of the velocity field can be pursued in the form

$$\mathcal{I}(u) = \frac{1}{2}(||\nabla p + \nu \nabla \times \boldsymbol{\omega} - \boldsymbol{f}||_0^2 + W||\nabla \cdot \mathbf{v}||_0^2 + ||\boldsymbol{\omega} - \nabla \times \mathbf{v}||_0^2),$$

where W is the weighting parameter; values of $W \geqslant 10$ were suggested in [125] in order to reduce the mass loss to acceptable levels. However, as pointed out by Proot, this weighting may have an adverse effect on the accuracy of the momentum conservation. In general, the use of weighted or constrained least-squares formulations seem to take away the beneficial attributes of the standard formulation developed by Proot, namely the simple equal-order discretisation and the fast scalable positive-definite algebraic system.

8.6 The gauge method

An interesting method that avoids splitting and which is similar to the vorticity–velocity formulation is the gauge method, introduced by E and Liu [150, 151]. The starting-point is to replace pressure by a gauge variable ϕ, and introduce an auxiliary field defined as

$$\mathbf{a} = \mathbf{v} - \nabla\phi\,. \tag{8.6.1}$$

If we now substitute eqn (8.6.1) into the incompressible Navier–Stokes equations (eqn (8.1.1a)), then we obtain an equation for the new field:

$$\mathbf{a}_t + (\mathbf{v} \cdot \nabla)\mathbf{v} = \nu \nabla^2 \mathbf{a}\,, \tag{8.6.2}$$

where we have also set

$$\nabla^2 \phi = -\nabla \cdot \mathbf{a}\,. \tag{8.6.3}$$

The pressure can be computed from the gauge function

$$p = \phi_t - \nu \nabla^2 \phi\,. \tag{8.6.4}$$

To close the system we need to specify consistent boundary conditions for the gauge function and the auxiliary field. By choosing arbitrarily

$$\frac{\partial \phi}{\partial \boldsymbol{n}} = 0$$

and using the definition (eqn (8.6.1)), we obtain that

$$\mathbf{a} \cdot \boldsymbol{n} = 0\,, \quad \mathbf{a} \cdot \boldsymbol{t} = -\frac{\partial \phi}{\partial \boldsymbol{t}}\,,$$

where $\boldsymbol{n}$ and $\boldsymbol{t}$ are the unit vectors in the normal and tangential directions to the domain boundary, respectively.

From the computational point of view, this approach involves equivalent complexity (in terms of the number of elliptic solves and fields) as the velocity–pressure formulation, and less complexity than the velocity–vorticity formulation. On the other hand, it does not involve any splitting errors and it avoids solving directly for the pressure, which is a computationally expensive component.

8.7 Discretisation of nonlinear terms

8.7.1 *Spatial discretisation*

In spectral/*hp* element methods the quadratic nonlinear terms in the Navier–Stokes equations are commonly computed in the physical space, even when a modal basis is employed for the discretisation. For the nodal basis all variables can be interpreted as being the physical value at the nodal points of the basis. Therefore, the nonlinear products can be obtained using a collocation projection at these nodal points, which are usually (at least for the Legendre method) also the quadrature points. For a modal expansion, the velocity field can typically be transformed into physical space and, subsequently, the nonlinear products are obtained at all quadrature points in a collocation fashion (see also Section 2.4.1.2). Another transform is then performed to bring the results back to modal space.

As discussed in Chapter 1, the form in which we write the nonlinear terms, that is, in convective (flux), skew-symmetric, or rotation form is important. In spectral DNS of boundary layers and channel flows, the rotation form is usually preferred over the convective form as it semi-conserves energy (in the inviscid limit) which, in general, makes it more stable, especially for the long-time integration required in DNS. In addition, it is more economical as it requires the evaluation of only six derivatives, whereas the convective form requires nine derivative evaluations. The skew-symmetric form is typically found to be more 'forgiving' in aliasing errors in under-resolved simulations of homogeneous turbulence compared to the rotation form. This, however, requires the evaluation of eighteen derivatives.

There is no strong evidence yet with spectral/*hp* element DNS to conclusively suggest one form or the other, although there is some consensus that the convective form is quite accurate and leads to stable discretisations. A comparison of the different forms (convective, flux, and skew-symmetric) was performed by Couzy [114] for a constant advection velocity as well as for a spatially-varying divergence-free velocity $(u, v) = (-\sin x_2 \cos x_1, \sin x_1 \cos x_2)$. The discretisation was based on a nodal Gauss–Lobatto–Legendre basis. The result was that, for the constant advection velocity, all forms were the same in that they produced identical eigenspectra with all imaginary eigenvalues. However, for the variable advection velocity, only the skew-symmetric form gave imaginary eigenvalues, with the convective and conservative form producing complex eigenvalues with positive real parts. For purely convection equations, these spurious positive values may lead to instabilities if explicit time-stepping is used. However, for Navier–Stokes computations at modest Reynolds number no such instabilities have been observed, presumably due to the stabilising role of the viscous terms.

The choice of which form to use may also be dictated by the specific spectral/*hp* discretisation employed. For example, a conforming discretisation leads to different stability properties than a non-conforming one. Such behaviour was encountered by Rønquist [404] in unsteady calculations for flow past a cylinder using a nodal Gauss–Lobatto–Legendre discretisation. It was observed that, for a mixed time-stepping scheme (treating the nonlinear terms explicitly and the viscous terms implicitly), the skew-symmetric form honours the CFL time-step restriction. In contrast, the convective form leads to instability whenever a non-conforming discretisation (of the mortar type) is used with the flow entering from a lower-order element into a higher-order element. This again was explained in terms of the eigenspectrum, as complex eigenvalues with large spurious positive parts may be excited that render the solution unstable. This problem is usually eliminated if, instead of a mortar projection in adjacent elements with different degrees of freedom on the edge, an interpolation is performed similar to the constrained approximation (see Section 7.3), where the low-order polynomial dictates the accuracy.

Although the skew-symmetric form is usually considered to be the most accurate, problems may also be encountered with this form for Dirichlet and inflow/outflow conditions. To illustrate this point, we report an experiment presented by Rønquist [404] for the Kovasznay flow discussed in Section 9.1.3 in a domain $\Omega = [-0.5, 1] \times [-0.5, 1.5]$, discretised with $N_{\rm el} = 6$ equal quadrilaterals with matching and non-matching spectral order across the interfaces. In Fig.

8.6(a) the relative error as a function of the polynomial degree P for a fixed number of elements is shown using the convective form, and in Fig. 8.6(b) it is shown for the skew-symmetric form. Although in both cases exponential convergence is achieved, the accuracy is superior with the more economical convective form. The reason for this is that the skew-symmetric form requires a more exact representation of the boundary conditions, which for this problem were specified as piecewise continuous polynomials at all boundaries.

For staggered grids employing the $\mathcal{P}_P/\mathcal{P}_{P-2}$ formulation, Wilhem and Kleiser [492] observed numerical instabilities in Navier–Stokes simulations for different treatments of the advection terms. Specifically, they found that stability is obtained for the *convective* formulation but not for the conservative or skew-symmetric formulations. Also, a modified rotational form, where the $\frac{1}{2}\nabla|\mathbf{v}|^2$ term is explicitly computed, was also found to be stable. It was shown that the instability is due to the staggered-grid discretisation, and the fact that divergence errors may grow exponentially at the Gauss–Lobatto–Legendre (GLL) velocity points given that the continuity equation is discretised on the Gauss–Lobatto (GL) pressure points. The linearised Navier–Stokes equations also exhibited the same problem. Numerical experiments revealed that only with sufficient viscous damping a stable solution was possible, irrespective of the time-integration scheme.

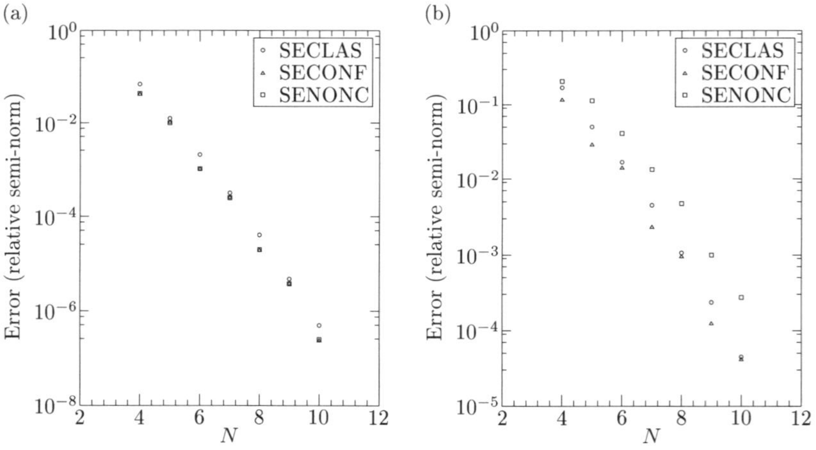

FIG. 8.6. Convergence for the Kovasznay problem. (a) The convective form of the nonlinear terms has been used. SECLAS refers to a standard Gauss–Lobatto–Legendre spectral element discretisation with fixed order P; SECONF is still a conforming mesh but with half of the elements of polynomial order P and the other half with order $P+3$; SENONC refers to a non-conforming discretisation with polynomial order fixed at P. (b) The skew-symmetric form of the nonlinear terms has been used. In both (a) and (b) $N = P + 1$. (Courtesy of E. M. Rønquist.)

Finally, errors may be caused by insufficient quadrature used in the spectral/*hp* element discretisation of the nonlinear terms, especially in complex-geometry flows. These errors can be eliminated effectively by employing over-integration, i.e., integrating the nonlinear terms in the variational statement with higher-order quadrature than the one employed for the linear contributions, e.g., pressure and viscous terms. This is discussed further in the next chapter (see Section 9.3.2) where an example of a turbulent flow simulation is presented.

8.7.2 *Temporal discretisation: semi-Lagrangian method*

We have already discussed (see Section 8.3.2.5) how to discretise explicitly the nonlinear advection terms independently of the treatment of the Stokes operator, the latter being treated implicitly. While these semi-implicit schemes are efficient for low-to-moderate-size flow problems, for high-resolution turbulent simulations (see Chapter 9) the CFL constraint may be too severe. This simply implies that we may be unnecessarily over-resolving the smallest physical time-scale. This is primarily due to the scaling of the maximum eigenvalue of the advection operator in spectral/*hp* element methods, see Section 6.3 and also [498].

In this section, we apply the semi-Lagrangian spectral element (SLSE) method introduced in Section 6.4 to incompressible Navier–Stokes equations, following Xiu and Karniadakis [498] and Xiu *et al.* [499]. We consider the Navier–Stokes equations in Lagrangian form:

$$\frac{d\mathbf{v}}{dt} = -\nabla p + \nu \nabla^2 \mathbf{v}\,, \tag{8.7.1a}$$

$$\nabla \cdot \mathbf{v} = 0\,, \tag{8.7.1b}$$

and present a second-order scheme based on backwards differentiation. We note here that, in the numerical experiments of [498], a second-order scheme employing Crank–Nicolson integration for the viscous terms exhibited numerical instabilities after a long-time integration.

A second-order time-discretisation based on ***stiffly-stable*** integration is

$$\frac{\frac{3}{2}\mathbf{v}^{n+1} - 2\mathbf{v}_d^n - \frac{1}{2}\mathbf{v}_d^{n-1}}{\Delta t} = (-\nabla p + \nu \nabla^2 \mathbf{v})^{n+1}\,, \tag{8.7.2}$$

where $\mathbf{v}_d^n$ is the velocity $\mathbf{v}$ at the departure point $\mathbf{x}_d^n$ at time-level t^n, and $\mathbf{v}_d^{n-1}$ is the velocity at the departure point $\mathbf{x}_d^{n-1}$ at time-level t^{n-1}. The departure point $\mathbf{x}_d^n$ is obtained by solving the characteristic equation

$$\frac{d\mathbf{x}}{dt} = \mathbf{v}^{n-1/2}(\mathbf{x}, t)\,, \quad \mathbf{x}(t^{n+1}) = \mathbf{x}_a\,, \tag{8.7.3}$$

where $\mathbf{x}_a$ is the position vector of the arrival points, which coincide with the grid points. The velocity at $t^{n+1/2}$ is approximated by the second-order extrapolation

$$\mathbf{v}^{n-1/2} = \frac{3}{2}\mathbf{v}^n - \frac{1}{2}\mathbf{v}^{n-1} \tag{8.7.4}$$

over one single time-level Δt, while the point $\mathbf{x}_d^{n-1}$ is obtained by solving

$$\frac{\mathrm{d}\mathbf{x}}{\mathrm{d}t} = \mathbf{v}^n(\mathbf{x}, t)\,, \quad \mathbf{x}(t^{n+1}) = \mathbf{x}_a \tag{8.7.5}$$

over two time-levels $2\Delta t$. By using the above characteristic equations, the resulting scheme (8.7.2) is second-order accurate in time. Alternatively, an auxiliary Eulerian equation can be solved to obtain $\mathbf{v}_d^n$ and $\mathbf{v}_d^{n-1}$ directly [320, 427, 499], as discussed in Section 6.4.2.

A three-step dual-type splitting scheme can then be applied to solve system (8.7.2), i.e.,

$$\frac{\hat{\mathbf{v}} - 2\mathbf{v}_d^n + \frac{1}{2}\mathbf{v}_d^{n-1}}{\Delta t} = 0\,, \tag{8.7.6a}$$

$$\frac{\hat{\hat{\mathbf{v}}} - \hat{\mathbf{v}}}{\Delta t} = -\nabla p^{n+1}\,, \tag{8.7.6b}$$

$$\frac{\frac{3}{2}\mathbf{v}^{n+1} - \hat{\hat{\mathbf{v}}}}{\Delta t} = \nu \nabla^2 \mathbf{v}^{n+1}\,. \tag{8.7.6c}$$

The Taylor vortex

We consider the Taylor vortex problem, an exact solution to the unsteady Navier–Stokes equations (see Section 9.1), in order to quantify the error in the semi-Lagrangian spectral element (SLSE) method. We denote the backward particle tracking approach as the strong SLSE and the Eulerian evaluation of the velocity at the departure point as the auxiliary SLSE. The computational domain is a square defined by the coordinates $\left[-\frac{1}{2}\pi, \frac{1}{2}\pi\right]$ in each direction. A mesh consisting of 2×2 quadrilateral elements is used, and the spectral order ranges as $P = 8$, 10, 12, and 14. The Reynolds number is fixed at 10^6, and the L^∞ norm is used to examine the error. Note that, despite the high Reynolds number, this is a relatively simple flow to resolve spatially and also it decays exponentially in time.

In Fig. 8.7 we plot the dependence of the error upon the spectral polynomial order. The L^∞ error of the velocity is measured at $t = 2\pi$, i.e., after one revolution of the flow. We see in this semi-log plot that spectral convergence is achieved by both Eulerian and semi-Lagrangian spectral/*hp* element methods. The time-step here is chosen by fixing the CFL number at 0.6. In Fig. 8.8 we plot the dependence of the error upon the size of the time-step for both the strong and auxiliary SLSE methods. Note here that the accuracy of the strong SLSE method is again dictated by the term $\mathcal{O}(\Delta t^2 + \Delta x^{P+1}/\Delta t)$, as we discussed in Section 6.4. At low P the interpolation error dominates, and increasing Δt decreases the overall error. When the interpolation error is small at $P = 12$, the Δt^2 term dominates and a further increase in Δt increases the overall error. The $P = 8$ curve shows the competition between these two terms. In this plot, the largest time-step, 0.03, corresponds to the CFL number being about 4. We emphasise that it is the size of Δt and *not* the CFL number that restricts the

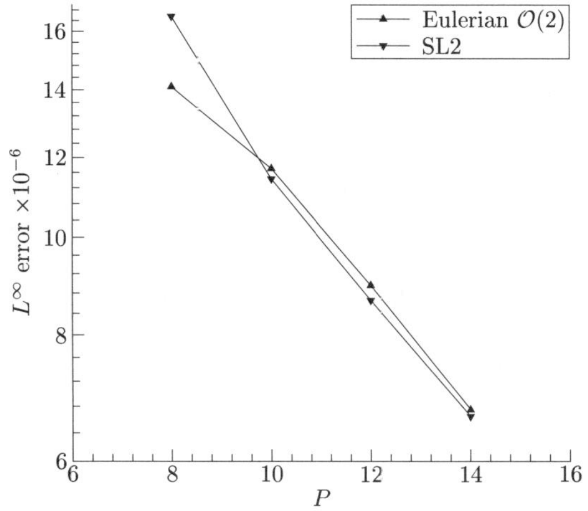

FIG. 8.7. Spectral convergence for Eulerian SE and strong SLSE methods.

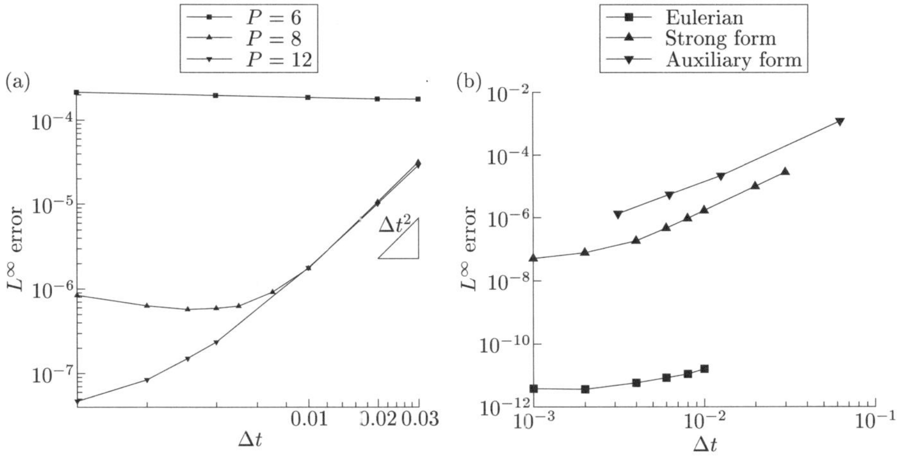

FIG. 8.8. Error dependence on Δt. (a) Strong SLSE results, and (b) a comparison of strong and auxiliary SLSE for $P = 12$.

use of the semi-Lagrangian method. In other words, for the SLSE method, the restriction on the size of time-step is solely due to accuracy considerations, but not due to stability.

In the Navier–Stokes implementation, another contribution to the error that needs to be examined is associated with the splitting procedure, as presented earlier. For the second-order strong SLSE method used here, the splitting error scales as $\mathcal{O}(\nu \Delta t^2)$. This error is only visible for relatively viscous flows. We use two extreme cases to examine this error. One is at $\nu = 1$ (Re = 1) where the splitting error is of the same order as the temporal error; the other is at $\nu = 10^{-6}$ (Re = 10^6) where the splitting error is negligible. In Fig. 8.9 we show the results for these two cases, and a comparison with the second-order Eulerian method for exactly the same spatial discretisation. We see that at Re = 1 the $\mathcal{O}(\Delta t^2)$ error dominates at very small values of the time-step ($\Delta t \sim 10^{-3}$). However, at Re = 10^6 the $\mathcal{O}(\Delta t^2)$ error starts to dominate only at $\Delta t \sim 10^{-2}$. A comparison of the two responses suggests that the $\mathcal{O}(\Delta t^2)$ error observed at Re = 1 is the splitting error, and it is comparable in that case to the second-order Runge–Kutta error associated with the solution of the departure points. The second-order Eulerian method generates the exact same error in this low Reynolds number case and reinforces the above statement. It should be noted that the error from the semi-Lagrangian part, i.e., $\mathcal{O}(\Delta t^2 + \Delta x^{P+1}/\Delta t)$, contains another $\mathcal{O}(\Delta t^2)$ term. In advection-dominated flows the Reynolds number is greater than one, and thus the splitting error is always subdominant.

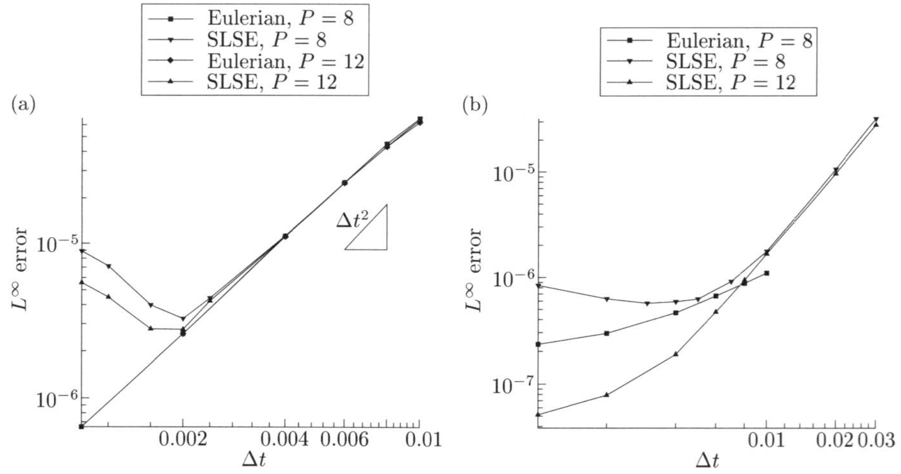

FIG. 8.9. Error dependence on Δt. (a) Re = 1 ($\nu = 1$), and (b) Re = 10^6 ($\nu = 10^{-6}$).

9

INCOMPRESSIBLE FLOW SIMULATIONS: VERIFICATION AND VALIDATION

In this chapter we present several examples of incompressible flow simulations. First, in Section 9.1 we consider a series of relatively simple laminar flows which have exact solutions and thus can be considered as benchmark problems. These solutions are useful for verification purposes, i.e., to ensure that the spectral/*hp* element discretisation applied and the associated software is implemented correctly. In Section 9.2 we consider the BiGlobal instability of small-amplitude disturbances superimposed on two-dimensional flow states. This type of analysis can aid in understanding the onset of three-dimensional, or transition to, turbulence. In Section 9.3, we discuss numerical issues in direct numerical simulation (DNS), including diagnostics for under-resolution. Specifically, we include a case study of turbulent channel flow which has been extensively simulated with global spectral methods. Subsequently, in Section 9.4, we summarise the main issues in large-eddy simulations (LES) with focus on high-order discretisation. The results here are compared with available experimental data and serve to validate the LES models. We conclude the chapter by considering a computing mode that involves resolution steering. To this end, we present two adaptive procedures based on error estimates as well as other heuristics, and a parallel paradigm suitable for dynamic *p*-refinement.

9.1 Exact Navier–Stokes solutions

In this section we consider several examples of the exact solution of the incompressible Navier–Stokes equations. These solutions are therefore suitable for code validation and convergence studies. Computations are also presented, illustrating both quadrilateral and triangular spectral/*hp* element discretisation. The formulation in terms of primitive variables is also compared against the velocity–vorticity formulation presented in the previous chapter. ♦[37]

9.1.1 *Moffatt eddies*

The first example is of viscous flow near a sharp corner at zero Reynolds number (Stokes flow), which was first presented in Chapter 1. We consider a wedge which has an aspect ratio of 2 : 1 as shown in Fig. 1.3, which means that the apex angle is approximately 28.1°. The wedge was discretised into $N_{\rm el} = 30$ triangular elements, while an expansion order of $P = 8$, 11, 14, and 17 was used

[37] Debugging cases for the incompressible Navier–Stokes equations.

to determine the steady-state solution using the implicit Euler scheme with a time-step of $\Delta t = 10^{-4}$. The flow was driven by a parabolic forcing velocity along the top of the wedge, which has a maximum value of one. Having allowed the solution to reach a steady state, we see that at the highest resolution considered ($P = 17$) we can resolve nine eddy structures, as indicated by the streamline plot in Fig. 1.3.

Using a similarity solution, Moffatt [334] derived an asymptotic result for the strength and location of an 'infinite' number of eddies in Stokes flow near sharp corners. The relations are dependent on the wedge angle; for the wedge angle 28.1° it is predicted that the strength of each eddy should asymptotically (i.e., away from the forced top section) be about 406 times weaker than the previous eddy. Moffatt's measure of the 'intensity' of consecutive eddies was the ratio of the local maximum transverse velocity along the centre-line. If we take a profile along the centre-line of our solution and plot the transverse (x_1 direction) velocity as a function of perpendicular distance from the top of the wedge, then we obtain the distributions shown in Fig. 1.4 for the four resolutions $P = 8$, 11, 14, and 17 (note that we have used the *logarithm* of the absolute velocity on the x_2 axis in Fig. 1.4 as the eddies decay so rapidly).

At the centre of an eddy we expect the transverse velocity to be zero, as indicated by the spikes in Fig. 1.4. After each spike there are local maxima which are tabulated in Table 9.1. Using these values we can determine the ratio of maximum velocities in order to evaluate Moffatt's eddy 'intensity'. These values are also recorded in Table 9.1. We note from Fig. 1.4 that at all resolutions the first four eddies have been resolved to the accuracy of the plot. The $P = 8$ run has resolved the fifth to the seventh eddies, although they still appear to be converging to their final values. Both the $P = 11$ and $P = 14$ runs resolve the fifth to the eighth eddies and they appear to be converged to the accuracy of the plot. The ninth eddy is not resolved by the $P = 11$ run, but it is quite well resolved by the $P = 14$ run. The $P = 17$ run adequately resolves the ninth eddy but there is no definite appearance of the tenth eddy. This is perhaps understandable since there are now three eddies in the last two elements; to capture further eddies would require h-type refinement in these two elements.

A more informative measure of convergence is shown in Table 9.1. As mentioned, the first four eddies were resolved by all expansion orders. This result is substantiated by the fact that the measured values for the first three eddies are within 1.5% of each other. The fourth eddy was not completely captured by the $P = 8$ case, and this is shown by the ratio between the third and the fourth eddies at this resolution. All other runs have captured the eddies very accurately and there is a definite convergence to a ratio value of 406.72. The eighth and the ninth eddies are only captured by the $P = 14$ and $P = 17$ runs and the ratio has not quite converged at $P = 17$, although it is within 1.0%.

TABLE 9.1. Values of the maximum transverse velocity along the centre-line of Stokes solution in a wedge, as shown in Fig. 1.3, using expansion orders $P = 8$, 11, 14, and 17. The highest value corresponds to the largest eddy near to the top of the wedge. Also shown is the ratio of the relative velocity as a measure of eddy 'intensity'. For this case Moffatt predicted an asymptotic ratio of 406.

	Expansion order							
Eddy	$P = 8$		$P = 11$		$P = 14$		$P = 17$	
	Maximum velocity	Ratio	Maximum velocity	Ratio	Maximum velocity	Ratio	Maximum velocity	Ratio
1	-2.0122×10^{-1}		-2.0099×10^{-1}		-2.0101×10^{-1}		-2.0101×10^{-1}	
		401.26		400.28		400.31		400.31
2	5.0147×10^{-4}		5.0213×10^{-4}		5.0213×10^{-4}		5.0213×10^{-4}	
		398.35		406.63		406.72		406.72
3	-1.2588×10^{-6}		-1.2349×10^{-6}		-1.2346×10^{-6}		-1.2346×10^{-6}	
		458.10		406.71		406.72		406.72
4	2.7480×10^{-9}		3.0362×10^{-9}		3.0354×10^{-9}		3.0354×10^{-9}	
		370.72		406.72		406.72		406.72
5	-7.4125×10^{-12}		-7.4650×10^{-12}		-7.4630×10^{-12}		-7.4630×10^{-12}	
		127.26		406.74		406.72		406.72
6	5.8248×10^{-14}		1.8353×10^{-14}		1.8349×10^{-14}		1.8349×10^{-14}	
		286.76		406.74		406.72		406.72
7	-2.0312×10^{-16}		-4.5123×10^{-17}		-4.5114×10^{-17}		-4.5114×10^{-17}	
				414.33		406.75		406.72
8	unresolved		1.0890×10^{-19}		1.1091×10^{-19}		1.1092×10^{-19}	
						398.75		402.25
9	unresolved		unresolved		-2.7819×10^{-22}		-2.7575×10^{-22}	

9.1.2 *Wannier flow*

The second example is an exact solution to the Stokes equations, but for a relatively complicated flow with curvilinear boundaries. It was originally derived by Wannier [482] for creeping flow past a rotating circular cylinder next to a moving wall. The solution depends only on the cylinder radius, r, its rate of rotation, ω, the distance from the centre of the cylinder to the moving wall, d, and the velocity of the wall, U. We assume that the centre of the cylinder is at $(x_1, x_2) = (0, 0)$, and for convenience we define $s^2 = d^2 - r^2$ and $\Gamma = (d + s)/(d - s)$, and the constants

$$a_0 = \frac{U}{\ln \Gamma}, \qquad a_1 = -d\left(a_0 + \frac{1}{2}\frac{r^2\omega}{s}\right),$$
$$a_2 = 2(d+s)\left(a_0 + \frac{1}{2}\frac{r^2\omega}{s}\right), \quad a_3 = 2(d-s)\left(a_0 + \frac{1}{2}\frac{r^2\omega}{s}\right).$$

Next, we introduce the following functions that depend on the position (x_1, x_2):

$$Y_1(x_2) = x_2 + d, \qquad Y_2(x_2) = 2Y_1(x_2),$$
$$K_1(x_1, x_2) = x_1^2 + (s + Y_1(x_2))^2, \quad K_2(x_1, x_2) = x_1^2 + (s - Y_1(x_2))^2.$$

In terms of these quantities, the solution can be written as

$$\begin{aligned} u(x_1, x_2) &= U - 2(a_1 + a_0 Y_1)\left[\frac{s + Y_1}{K_1} + \frac{s - Y_1}{K_2}\right] - a_0 \ln\left(\frac{K_1}{K_2}\right) \\ &\quad - \frac{a_2}{K_1}\left[s + Y_2 - \frac{(s+Y_1)^2 Y_2}{K_1}\right] - \frac{a_3}{K_2}\left[s - Y_2 + \frac{(s-Y_1)^2 Y_2}{K_2}\right], \\ v(x_1, x_2) &= \frac{2x_1}{K_1 K_2}(a_1 + a_0 Y_1)(K_2 - K_1) - \frac{x_1 a_2 (s + Y_1) Y_2}{K_1^2} - \frac{x_1 a_3 (s - Y_1) Y_2}{K_2^2}. \end{aligned}$$

This problem was solved using triangular and non-conforming quadrilateral elements. The corresponding domains are shown in Fig. 9.1 along with streamlines.

The H^1 error of the solution at steady state for various expansion orders are shown in Fig. 9.2. In both cases we achieve spectral convergence for this flow solution, as indicated by the near linear slope of the lines. The solution is very sensitive to the mesh resolution near the cylinder, and the finer resolution of the triangular mesh around the cylinder is necessary to achieve the same convergence rate as in the non-conforming quadrilateral mesh, which is coarser in this region but which has a richer polynomial space.

9.1.3 *Kovasznay flow*

In 1948, Kovasznay solved the problem of steady, laminar flow behind a two-dimensional grid [283]. This exact solution to the Navier–Stokes equations is given by

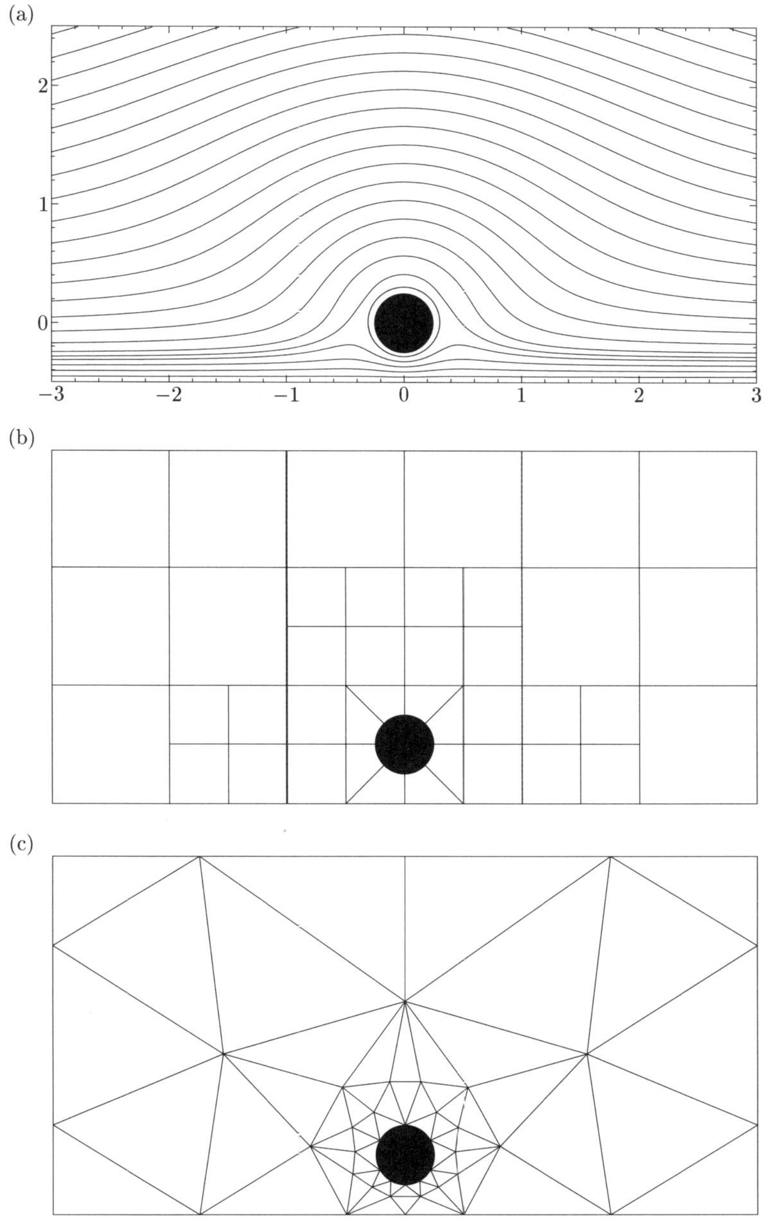

FIG. 9.1. Creeping flow past a rotating circular cylinder near a moving wall (Wannier flow): (a) streamlines of the exact solution corresponding to the parameters $r = 0.25$, $d = 0.5$, $U = 1$, and $\omega = 2$; the computational domain was discretised using (b) $N_{\rm el} = 40$ quadrilateral spectral elements; and (c) $N_{\rm el} = 65$ triangular spectral elements.

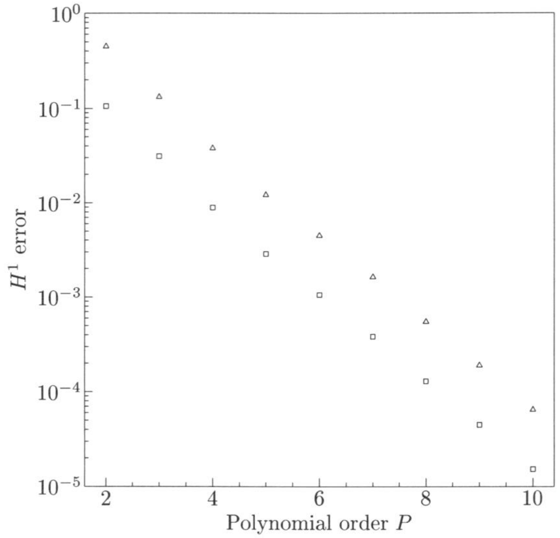

FIG. 9.2. Convergence of the velocity field in the H^1 norm to the exact solution for Wannier flow shown in Fig. 9.1: (□) quadrilateral and (△) triangular spectral elements. The error is calculated relative to the H^1 norm of unity over the domain.

$$u = 1 - e^{\lambda x_1}\cos(2\pi x_2)\,, \quad v = \frac{\lambda}{2\pi} e^{\lambda x_1}\sin(2\pi x_2)\,, \quad p = \frac{1}{2}(1 - e^{2\lambda x_1})\,,$$

where $\lambda = 1/2\nu - [(1/4\nu^2) + 4\pi^2]^{1/2}$. The solution looks similar to the low-speed flow of a viscous fluid past an array of cylinders. Figure 9.3 shows streamlines of the steady solution for Reynolds number $\mathrm{Re} = 1/\nu = 40$. The exact solution was used to compute Dirichlet boundary conditions, and the Navier–Stokes equations were integrated to obtain a steady-state solution on the interior of the domain. Knowing the exact solution, we can calculate the convergence rate with respect to expansion order and this is shown in Fig. 9.4. In this figure, we plot the H^1 error as a function of the expansion order for different approximations corresponding to nodal quadrilateral spectral elements and modal triangular spectral elements. Exponential convergence of the error is obtained in all cases.

The convergence of this problem using the vorticity–velocity formulation with the penalty method (see Section 8.4) is shown in Fig. 9.5. In Fig. 9.5(a) we consider convergence as a function of the penalty parameter, τ, used to enforce boundary conditions in this method, as discussed in Section 8.4.3. From this log–log plot we observe that a linear convergence is obtained with respect to the inverse penalty parameter. We recall, that unlike other penalty methods where the solution is penalised everywhere in the domain, in this formulation only the boundary is restricted. This avoids any potential problems with stiffness in the solution process, and thus very large values of the penalty parameter can be applied. In Fig. 9.5(b) we plot the error of the Kovasznay solution on the same

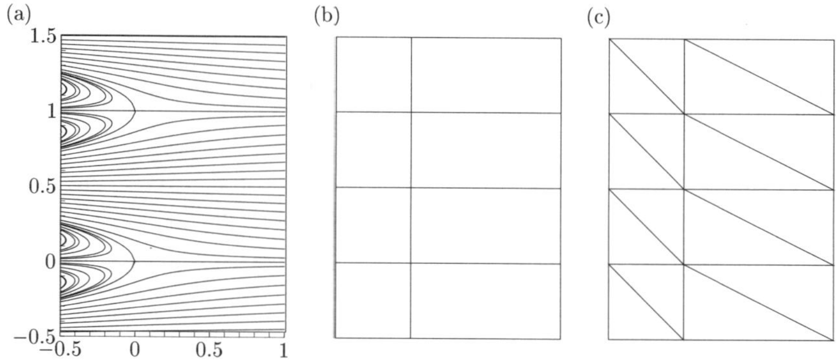

FIG. 9.3. (a) Streamlines of the exact solution ($\nu = 1/40$); the computational domain was discretised using (b) $N_{\mathrm{el}} = 8$ quadrilateral spectral elements, and (c) $N_{\mathrm{el}} = 16$ triangular spectral elements.

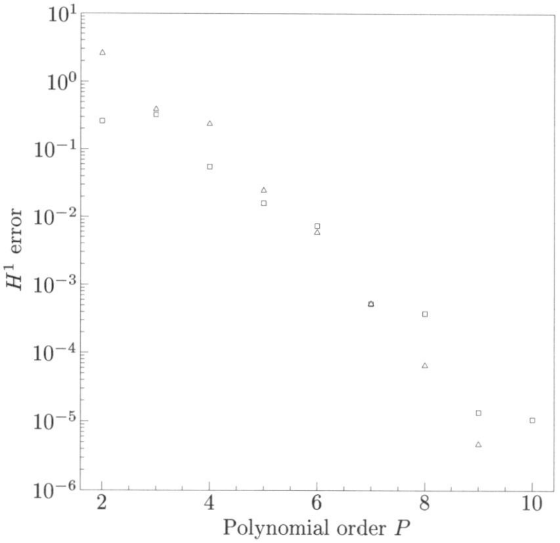

FIG. 9.4. Convergence of the velocity field in the H^1 norm to the exact solution for Kovasznay flow, as shown in Fig. 9.3; squares (□) denote nodal quadrilaterals, and triangles (△) denote modal triangular elements. The error is calculated relative to the H^1 norm of unity over the domain.

mesh as Fig. 9.5(a) using a very large penalty parameter. For a sufficiently high penalty parameter, exponential convergence is obtained for both velocity and vorticity. For smaller values of τ the error is limited by the value of $1/\tau$.

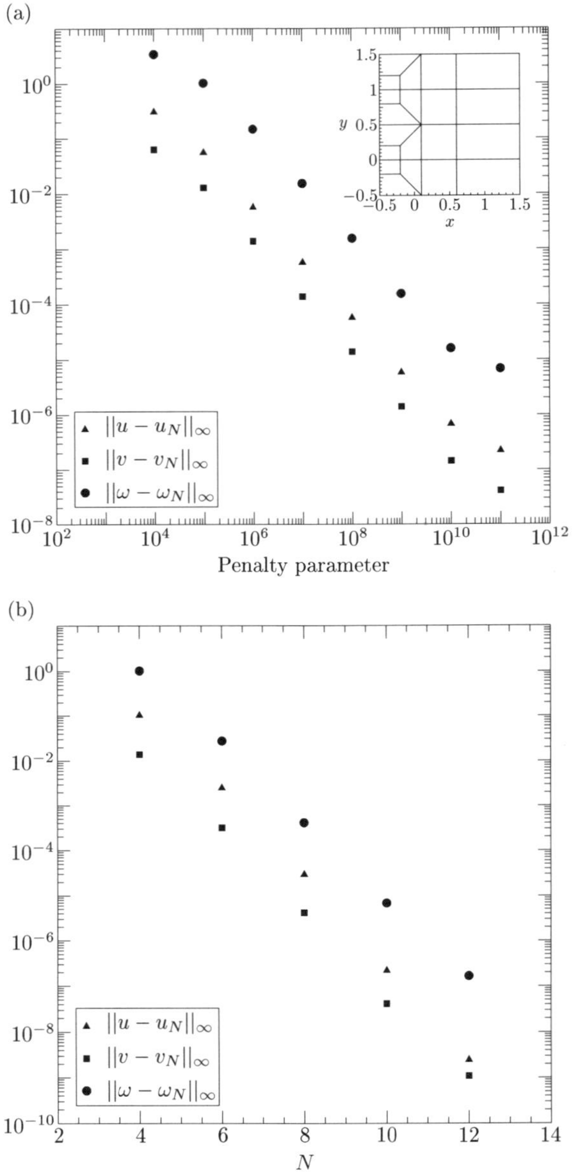

FIG. 9.5. Convergence of the velocity and vorticity field in the L^∞ norm to the exact Kovasznay solution using the velocity–vorticity formulation. (a) Convergence as a function of the penalty parameter. (b) Convergence as a function of polynomial order using a large penalty parameter. ($N = P - 1$.) ([472].)

9.1.4 *Triangular duct flow*

The fully-developed flow in the x_3 direction within a triangular duct of equilateral-triangular cross-section has an exact solution of the form

$$u(x_1, x_2, x_3) = 0\,, \quad v(x_1, x_2, x_3) = 0\,, \quad w(x_1, x_2, x_3) = \left(x_2 - \frac{\sqrt{3}}{2}\right)(3x_1^2 - x_2^2)\,.$$

Here we have assumed that the cross-section lies in the x_1–x_2 plane and the flow is in the x_3 direction. Note that the equilateral cross-section is the only triangular form to have a polynomial solution which satisfies the fully-developed equations. The symmetry of this cross-section causes the Laplacian of the solution to be constant (i.e., $\nabla^2 w = -2\sqrt{3}$), which is necessary to balance the pressure term in the fully-developed w momentum equation. This flow can be exactly represented with an expansion of order $P = 3$ or higher. This was verified in numerical experiments in [433] using four tetrahedra; spectral convergence for this three-dimensional flow is easily established.

9.1.5 *The Taylor vortex*

So far, the exact solutions presented were for time-independent solutions. The Taylor vortex is a relatively simple flow but it is time-dependent, and thus it provides a good test-bed for evaluating the time-accuracy of a Navier–Stokes solver. It has the form

$$\begin{aligned} u &= -\cos x_1 \sin x_2\, \mathrm{e}^{-2t/\mathrm{Re}}\,, \\ v &= \sin x_1 \cos x_2\, \mathrm{e}^{-2t/\mathrm{Re}}\,, \\ p &= -\frac{1}{4}\left(\cos(2x_1) + \cos(2x_2)\right) \mathrm{e}^{-4t/\mathrm{Re}}\,. \end{aligned}$$

For a square domain defined by the coordinates $\left[-\frac{1}{2}\pi, \frac{1}{2}\pi\right]$ in each direction, either periodic or Dirichlet boundary conditions can be prescribed. An arbitrary domain can also be employed if appropriate Dirichlet boundary conditions are imposed, given by the exact solution.

9.2 BiGlobal stability analysis of complex flows

BiGlobal stability analysis is concerned with the evolution of small-amplitude disturbances superimposed upon a basic flow state. Making the assumption that the flow field can be decomposed into a basic flow plus a small linear perturbation, this decomposition can be substituted into the Navier–Stokes equations to obtain a governing equation for the disturbance. For a sufficiently small disturbance it can be argued that the quadratic (or higher) terms in the disturbance are negligible, and so we obtain a linear system of equations. Obtaining the dominant eigenvalues and eigenmodes of the linear disturbance equations can be an aid to understanding the onset of three-dimensionality or the transition to turbulence. However, numerical solution of the fully three-dimensional (global) linear eigenvalue problems can be computationally demanding and so may not offer any significant advantages over performing a direct numerical simulation.

To perform a linear stability analysis of flows varying in more than one dimension a number of simplifications have typically been adopted. The most radical simplification is to consider a basic state which only depends on one spatial coordinate and where the other two spatial directions are taken as homogeneous [141]. This type of assumption allows the use of normal-mode analysis in the two homogeneous directions, where the perturbation fields are assumed to have a periodic structure in these directions. This construction has been widely employed in the last century to study, for example, the instability of boundary layers. In this case 'parallel-flow' approximations are typically employed, where the basic flow velocity component in the wall-normal direction as well as boundary-layer growth effects are neglected.

The predictions of classical linear stability theory, based on one-dimensional flow states, have been met with mixed success. One of the most notable verifications of the theory was the experimental identification by Schubauer and Skramstad [415] of the instability waves on a flat plate boundary layer, postulated by Tollmien [465]. However, one of the most notable failures of the theory is its inability to predict instability of Hagen–Poiseuille flow in a pipe at all Reynolds numbers. Instability and transition to turbulence in this flow have been known at least since the celebrated experiments of Reynolds in 1883 [392]. This discrepancy is all the more alarming, given that the basic flow is truly parallel. However, Trefethen *et al.* [470] have demonstrated that consideration of the pseudospectra of the linearised operator may be a significant factor in understanding the transition to turbulence.

Between the two extremes of basic flows dependent upon one and three spatial directions, we can consider steady or time-periodic basic flows dependent upon two spatial directions and impose three-dimensional disturbances that are periodic in the third, homogeneous spatial direction. This non-parallel (BiGlobal) linear instability analysis discussed in this section represents the natural extension of the classic linear stability theory eigenvalue problem of Tollmien [465]. In BiGlobal stability analysis the classic ordinary differential Orr–Sommerfeld and Squire equations, which result from a one-dimensional basic flow assumption, are replaced with a system of partial differential equations. The link with classical analyses may, in some cases, be obtained as a limiting case of solutions to the partial-derivative eigenvalue problem, as shown in [457, 459].

BiGlobal stability analysis is numerically demanding, and so the low dissipation and dispersion errors of high-order discretisations, such as the spectral/*hp* element method, are very attractive. For example, a physical instability may be suppressed in a numerical method with excessive artificial viscosity, or it may be triggered prematurely by numerical dispersion. Further, the geometric flexibility of the spectral/*hp* element method offers the ability to simulate the linearised stability of geometrically complex flows not previously possible in the classical approach [458]. For example, in Fig. 9.6 we observe the Floquet modes from linearised instability around the von Karman shedding frequency [36,56].

9.2.1 *Formulation of the linearised eigenproblem*

Adopting the formulations in [53,225,474], we start by denoting the incompressible Navier–Stokes equations as

$$\partial_t \boldsymbol{u} = -\boldsymbol{u} \cdot \nabla \boldsymbol{u} - \nabla p + \nu \nabla^2 \boldsymbol{u}\,, \tag{9.2.1a}$$

$$\nabla \cdot \boldsymbol{u} = 0\,. \tag{9.2.1b}$$

For a linearised stability analysis we decompose the solution, $\boldsymbol{u}$, into a 'base flow', $\boldsymbol{U}$, and a perturbation field, $\boldsymbol{u}'$, such that

$$\boldsymbol{u} = \boldsymbol{U} + \boldsymbol{u}'\,. \tag{9.2.2}$$

Similarly, we can consider the pressure as $p = \bar{P} + p'$, where $\bar{P}$ corresponds to the base flow and p' corresponds to the perturbation. The velocity decomposition (9.2.2) implies that

$$\boldsymbol{u} \cdot \nabla \boldsymbol{u} = \boldsymbol{U} \cdot \nabla \boldsymbol{U} + \boldsymbol{U} \cdot \nabla \boldsymbol{u}' + \boldsymbol{u}' \cdot \nabla \boldsymbol{U} + \boldsymbol{u}' \cdot \nabla \boldsymbol{u}'\,,$$

and so substitution of eqn (9.2.2) into eqn (9.2.1a) leads to

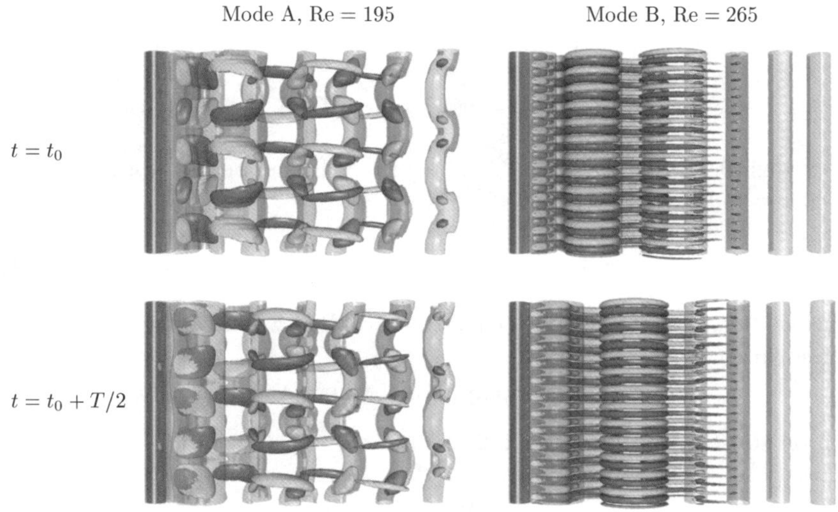

FIG. 9.6. Vorticity isosurfaces for the synchronous wake modes of the circular cylinder, shown for a $10D$ spanwise domain extent, and viewed from the cross-flow direction. Translucent isosurfaces are for the spanwise vorticity component, and solid surfaces are for the streamwise component. (Courtesy of H. Blackburn.)

$$\begin{aligned}\partial_t \boldsymbol{u}' = & -\boldsymbol{U}\cdot\nabla\boldsymbol{u}' - \boldsymbol{u}'\cdot\nabla\boldsymbol{U} - \nabla p' + \nu\nabla^2\boldsymbol{u}' \\ & + (-\partial_t\boldsymbol{U} - \boldsymbol{U}\cdot\nabla\boldsymbol{U} - \nabla\bar{P} + \nu\nabla^2\boldsymbol{U}) - \boldsymbol{u}'\cdot\nabla\boldsymbol{u}' .\end{aligned}$$

The second row of the above equation is identically eqn (9.2.1a) expressed in terms of the base flow velocity $\boldsymbol{U}$ and associated pressure $\bar{P}$. Therefore, if $(\boldsymbol{U}, \bar{P})$ satisfies the Navier–Stokes equations then this term will be identically zero. If we now consider a sufficiently small velocity $\boldsymbol{u}'$ then the quadratic term $\boldsymbol{u}'\cdot\nabla\boldsymbol{u}'$ will be significantly smaller than all the remaining linear terms and so it can be neglected. Enforcing these two assumptions, we obtain the *linearised* Navier–Stokes equations

$$\partial_t \boldsymbol{u}' = -\boldsymbol{U}\cdot\nabla\boldsymbol{u}' - \boldsymbol{u}'\cdot\nabla\boldsymbol{U} - \nabla p' + \nu\nabla^2\boldsymbol{u}' , \tag{9.2.3a}$$

$$\nabla\cdot\boldsymbol{u}' = 0 . \tag{9.2.3b}$$

We note that eqns (9.2.3a) and (9.2.3b) are identical in form to eqns (9.2.1a) and (9.2.1b) with the nonlinear advection term $\boldsymbol{u}\cdot\nabla\boldsymbol{u}$ replaced by the linearised advection terms $\boldsymbol{U}\cdot\nabla\boldsymbol{u}' + \boldsymbol{u}'\cdot\nabla\boldsymbol{U}$. Therefore, the solution techniques for solving the Navier–Stokes equations discussed in Chapter 8 can equally well be applied to the linearised Navier–Stokes equations.

In an incompressible flow, pressure can also be considered as a constraint enforcing the divergence-free condition on the velocity field. Mathematically, it can be eliminated from the problem through a pressure Poisson equation. Following Tuckerman and Barkley [474], we can construct a linearised system, independent of p', by taking the divergence of eqn (9.2.3a) and imposing the divergence condition (9.2.3b) to obtain the pressure Poisson equation

$$\nabla^2 p' = \nabla\cdot(\boldsymbol{U}\cdot\nabla\boldsymbol{u}' + \boldsymbol{u}'\cdot\nabla\boldsymbol{U}) = (\boldsymbol{U}\cdot\nabla + (\nabla\boldsymbol{U})^\top)\boldsymbol{u}' ,$$

and so eqns (9.2.3a) and (9.2.3b) can be written in an operator form without the explicit dependence on p' as

$$\partial_t \boldsymbol{u}' = \boldsymbol{A}(\boldsymbol{U})\boldsymbol{u}' = -(\boldsymbol{I} - \nabla\nabla^{-2}\nabla\cdot)(\boldsymbol{U}\cdot\nabla + (\nabla\boldsymbol{U})^\top)\boldsymbol{u}' + \nu\nabla^2\boldsymbol{u}' . \tag{9.2.4}$$

Equation (9.2.4) reduces to an eigenvalue problem when we take $\boldsymbol{u}'(\boldsymbol{x}, t)$ to be of the form

$$\boldsymbol{u}'(\boldsymbol{x}, t) = \boldsymbol{q}'(\boldsymbol{x})\exp(\lambda t) , \quad \text{where } \lambda = \sigma + \mathrm{i}\omega ,$$

which on substitution into eqn (9.2.4) leads to

$$\boldsymbol{A}\boldsymbol{q}' = \lambda\boldsymbol{q}' , \tag{9.2.5}$$

where we recall that $\boldsymbol{A} = \boldsymbol{A}(\boldsymbol{U})$ is a function of the base flow velocity $\boldsymbol{U}$. In practice, we do not need to construct the discrete matrix representation of $\boldsymbol{A}$ since we can equivalently use the standard time-stepping methods to integrate eqns (9.2.3a) and (9.2.3b) directly.

In BiGlobal stability analysis we consider the case where the base flow $\boldsymbol{U}$ is represented by a two-dimensional function in space, which can also be time periodic, i.e.,

$$\boldsymbol{U}(\boldsymbol{x};t) = \begin{bmatrix} u_1(x_1,x_2;t) \\ u_2(x_1,x_2;t) \\ u_3(x_1,x_2;t) \end{bmatrix}.$$

The perturbation velocity, $\boldsymbol{u}'$, is then expressed in a similar form, but with the dependence on the third homogeneous direction incorporated through the Fourier mode:

$$\boldsymbol{u}'(\boldsymbol{x};t) = \begin{bmatrix} \hat{u}_1'(x_1,x_2;t)\mathrm{e}^{\mathrm{i}\beta x_3} \\ \hat{u}_2'(x_1,x_2;t)\mathrm{e}^{\mathrm{i}\beta x_3} \\ \hat{u}_3'(x_1,x_2;t)\mathrm{e}^{\mathrm{i}\beta x_3} \end{bmatrix},$$

where $\hat{u}_1'$, $\hat{u}_2'$, and $\hat{u}_3'$ are complex. In the most general case the base flow can involve all three components of flow within the two-dimensional field and the matrix operator $\boldsymbol{A}$ contains entries for both the real and imaginary parts of every component of the perturbation velocity. However, we note that the base flow $\boldsymbol{U}$ may also have a different structure which modifies the structure of the operator $\boldsymbol{A}$. For example, in the case of a completely two-dimensional base flow where $u_3(x_1,x_2) = 0$, the system $\boldsymbol{A}$ can be reduced to a problem containing only the real parts of the perturbation velocities $\hat{u}_1'$ and $\hat{u}_2'$, and the imaginary part of the perturbation velocity $\hat{u}_3'$. This essentially halves the number of degrees of freedom in the system. However, if the eigenvalues in this case are complex then the full system must be considered to recover the correct (travelling wave) eigenmodes.

9.2.2 *Iterative solution of the eigenproblem*

We are now faced with the challenge of solving the eigenvalue problem (9.2.5). To determine the linearised stability of the base flow $\boldsymbol{U}$ it is sufficient to know whether any eigenvalues of $\boldsymbol{A}$ have a positive real part. Additional information about the leading parts of the spectrum as well as the corresponding eigenvectors may also be helpful. We are, therefore, faced with the challenge of determining complete information about a few of the leading eigenvalues or pairs. We also assume that the systems being considered are too large to construct the discrete representation of $\boldsymbol{A}$, and compute all eigenvalues and eigenvectors via the QR algorithm.

9.2.2.1 *Power method*

Following [225, 474], a basic technique to compute a single real eigenvalue is the power method. In this approach the action of the system $\boldsymbol{A}$ is repeatedly applied to an arbitrary initial vector $\boldsymbol{u}_0$ to produce the sequence of vectors $\boldsymbol{u}_n = \boldsymbol{A}^n\boldsymbol{u}_0$. This sequence approaches the dominant eigenvector, and the sequence of Rayleigh quotients

$$\lambda_n = \frac{\boldsymbol{u}_n^\top \boldsymbol{A}\boldsymbol{u}_n}{\boldsymbol{u}_n^\top \boldsymbol{u}_n}$$

converges to the corresponding eigenvalue.

9.2.2.2 *Arnoldi/block power method*

The power method only permits us to identify a single eigenvalue, so in order to calculate a block of eigenvalues we require a further modification by applying the Arnoldi method or one of its variations [18,408]. In this approach we initially construct the sequence $\boldsymbol{u}_0, \boldsymbol{A}\boldsymbol{u}_0, \ldots, \boldsymbol{A}^{K-1}\boldsymbol{u}_0$ whose span defines a Krylov space, where K is the number of eigenvalues sought (e.g., $K \approx 8$). The vectors are then orthonormalised, for example, using a modified Gram–Schmidt process, to form a basis $\boldsymbol{v}_1, \boldsymbol{v}_2, \ldots, \boldsymbol{v}_K$ which spans the Krylov space. We next define a matrix $\boldsymbol{V}^\top[k] \equiv \boldsymbol{v}_k$ and the $K \times K$ Hessenberg matrix $\boldsymbol{H} \equiv \boldsymbol{V}^\top \boldsymbol{A}\boldsymbol{V}$, which is a k-dimensional generalisation of the Rayleigh quotient. When $\boldsymbol{H}$ is diagonalised, its eigenvalues approximate K eigenvalues of $\boldsymbol{A}$. Further, $\boldsymbol{V}$ times the eigenvectors of $\boldsymbol{H}$ approximate the K dominant eigenvectors of $\boldsymbol{A}$. Repeated application of the Arnoldi technique leads to a block power method to determine the K leading-order eigenvalues and eigenvectors.

The dominant eigenvalues of $\boldsymbol{A}$ are not of particular interest since, in general, the eigenvalues of largest magnitude correspond to the most quickly damped modes of the system. We are instead interested in the leading eigenvalues with the largest real part, and near the onset of instability the real component may be close to zero. Two approaches to identifying these eigenvalues are the exponential power method and the inverse power method.

Exponential power method

The general solution of the linear system (9.2.4) and, equivalently, the solution to eqns (9.2.3a) and (9.2.3b) is

$$\boldsymbol{u}'(t_0 + \tau) = \exp(\tau \boldsymbol{A})\boldsymbol{u}'(t_0)\,.$$

The dominant eigenvalues of the matrix $\boldsymbol{B} = \exp(\tau \boldsymbol{A})$ correspond to the leading eigenvalues of $\boldsymbol{A}$. If λ is an eigenvalue of $\boldsymbol{A}$, then the corresponding eigenvalue of $\exp(\tau \boldsymbol{A})$ is $\Gamma = \exp(\lambda\tau)$, while the eigenvectors are the same. We can therefore use the discrete solution of the linearised Navier–Stokes equations (9.2.3a) and (9.2.3b) over a time interval τ as the inner operation to the Arnoldi iteration.

Inverse power method

In the most straightforward case we are interested in obtaining the eigenvalues nearest to zero. An alternative method to achieve this is to apply the inverse power method. This technique is analogous to the block power method but repeatedly applies the action of $\boldsymbol{A}^{-1}$ instead of $\boldsymbol{A}$. In this method an efficient technique to invert the matrix $\boldsymbol{A}$ is therefore required.

9.2.3 *Floquet analysis*

Often we are interested in the stability of periodic orbits of the basic flow states. The exponential power method can also be applied to the case where the base flow is time periodic. In this case the basic flow $\boldsymbol{U}(t)$ is time periodic over an interval T such that $\boldsymbol{U}(t+T) = \boldsymbol{U}(t)$. When the base flow is T-periodic the linearised operator $\boldsymbol{A}(\boldsymbol{U}(t))$ is also time dependent and T-periodic. In this case the general solution of the linear system (9.2.4) and, equivalently, the solution to eqns (9.2.3a) and (9.2.3b) becomes

$$\boldsymbol{u}'(t_0+\tau) = \exp\left(\int_{t_0}^{t_0+\tau} \boldsymbol{A}(\boldsymbol{U}(t'))\,\mathrm{d}t'\right)\boldsymbol{u}'(t_0)\,.$$

The stability of this system is now determined by the eigenvalues of the evolution (monodromy) operator over the periodic time T defined as

$$\boldsymbol{C} \equiv \exp\left(\int_{t_0}^{t_0+T} \boldsymbol{A}(\boldsymbol{U}(t'))\,\mathrm{d}t'\right).$$

The operator $\boldsymbol{C}$ takes a perturbation $\boldsymbol{u}'(t_0)$ and evolves it around the periodic orbit to give the perturbation at time $t_0 + T$. In Floquet stability analysis [53, 225, 474] we determine the eigenvalues of the operator $\boldsymbol{C}$. As discussed previously, the dominant eigenvalues of $\boldsymbol{C}$ can be determined using the exponential power method, where the action of $\boldsymbol{C}$ is incorporated in an Arnoldi iteration. We emphasise that, in practice, the action of $\boldsymbol{C}$ on the perturbation is computed by time integrating the system (9.2.3a) and (9.2.3b) over the period T. This necessarily requires that the base flow $\boldsymbol{U}(t)$ is known at every time-step. Since $\boldsymbol{U}(t)$ is time periodic, a natural interpolation technique is to represent $\boldsymbol{U}(t)$ in terms of a Fourier series in time. This only requires the storage of sufficient Fourier modes to be consistent with the accuracy of the rest of the computation.

The eigenvalues μ of $\boldsymbol{C}$ are known as Floquet multipliers and can also be expressed as

$$\mu = \exp(\sigma T)\,,$$

where σ is known as the Floquet exponent. In general, both σ and μ can be complex and a nonzero imaginary component often indicates travelling wave solutions in three-dimensional flows. Any perturbation $\boldsymbol{u}'(t)$ can be written as a sum of components:

$$\boldsymbol{u}'(t) = \sum_n \tilde{\boldsymbol{q}}_n(t)\exp(\sigma_n t)\,,$$

where $\tilde{\boldsymbol{q}}_n(t)$ are the T-periodic Floquet eigenmodes of $\boldsymbol{C}(t)$.

9.2.4 *Applications of BiGlobal stability*

Results obtained using BiGlobal stability analysis have arisen in many areas of fluid mechanics, following the pace of hardware and algorithmic developments. The first spectral element simulations for three-dimensional stability of complex-geometry flows with one homogeneous direction were presented in [262]. Morzynski and Thiele [337] were among the first to consider the instability of laminar flow behind a circular cylinder using finite element techniques. However, one of the first applications of spectral/*hp* element methods to this type of analysis was by Barkley and Henderson [36], who addressed the Floquet instability of laminar flow behind a circular cylinder and also classified the nonlinear characteristics of the bifurcation to three-dimensional flow. In Fig. 9.6 we show a visualisation of the two Floquet modes known as mode 'A' and 'B' [493], which are the first linearised three-dimensional unstable modes that appear on the well-known two-dimensional von Karman shedding mode. The visualisations due to Blackburn *et al.* [56] were obtained from direct numerical simulation with restricted spanwise periodic length, at Reynolds numbers slightly above the onset for the two modes. The original analysis wavelength and spatial structure was observed from a BiGlobal stability computation [36]. Barkley *et al.* [35] and Henderson [225] have also applied this type of analysis to the case of flow over a backwards facing step.

Robichaux *et al.* [399] and Sheard *et al.* [419] have investigated the instability of flow past square and circular cylinders using spectral/*hp* element discretisation. In these papers, one-dimensional power methods were adopted and synchronous modes, similar in form to the work of Barkley and Henderson, were observed. These investigations also suggested the existence of subharmonic modes for these geometries but, as discussed by Blackburn and Lopez [54], the eigenmethod employed in this work did not permit the occurrence of complex eigenvalues which correspond to quasi-periodic modes that are non-synchronous to the base flow. Another classical flow, the lid-driven cavity, has been considered using a global spectral Chebyshev method by Theofilis [456], and these results were also compared with a spectral/*hp* element method and other experiments and calculations in [458]. The case of a periodically-lid-driven cavity was investigated by Blackburn and Lopez [55], and Blackburn [53] has also applied the spectral/*hp* element BiGlobal stability analysis to steady and periodic flows in an oscillatory asymmetric swirling flow. Finally, the spectral element method has been applied to a BiGlobal stability analysis of non-Newtonian flow in a duct in [159], as well as constricted (stenotic) channels and pipes in [380].

Application of classical spectral methods to BiGlobal stability analysis also include the work of Lin and Malik [304], who consider the swept (Hiemenz) attachment-line boundary-layer flow. This problem was also investigated by Theofilis *et al.* [460], who demonstrated that the three-dimensional BiGlobal linear eigenmodes of this flow may be modelled and recovered as solutions of a sequence of one-dimensional eigenvalue problems of the Orr–Sommerfeld class. Hein and Theofilis [221] have also applied BiGlobal stability analysis to models of trailing-vortex systems using a classical spectral discretisation.

Ehrenstein [153] also used a classical spectral method to solve the BiGlobal eigenvalue problem in a channel, one wall of which was modelling a riblet geometry, and showed that the primary instability of this channel flow was enhanced when compared with the classic plane Poiseuille flow. Wintergerste and Kleiser [497] have studied the stability of nonlinearly-generated cross-flow vortices during the late transitional stages of boundary-layer flow.

9.3 Direct numerical simulations—DNS

The simulation of turbulent flows by accurately solving the Navier–Stokes equations has been one of the greatest successes of CFD, and it is due mainly to simultaneous advances in both spectral methods and computational resources since the 1970s. In 1970, Howard Emmons [154] reviewed the possibilities for numerical modelling of fluid dynamics and concluded that, 'The problem of turbulent flows is still the big holdout. This straightforward calculation of turbulent flows—necessarily three-dimensional and non-steady—requires a number of numerical operations too great for the foreseeable future.' However, within a year of the publication of that article, the field of direct numerical simulation (DNS) of turbulence was initiated by achieving accurate simulations of wind-tunnel flows at moderate Reynolds numbers by Orszag and Patterson [355].

In the last few decades the field has developed significantly. DNS of turbulence is now regularly performed in simple geometries based primarily on global spectral methods. For example, in a turbulent channel flow with smooth walls Chebyshev polynomials are used in the inhomogeneous directions and Fourier expansions in the homogeneous (streamwise/spanwise) directions. The first DNS of turbulent flow in a more complex geometry was based on spectral element discretisation [98]. It involved turbulent flow over riblets, as shown in Fig. 9.7. In the homogeneous streamwise direction one-dimensional Fourier expansions

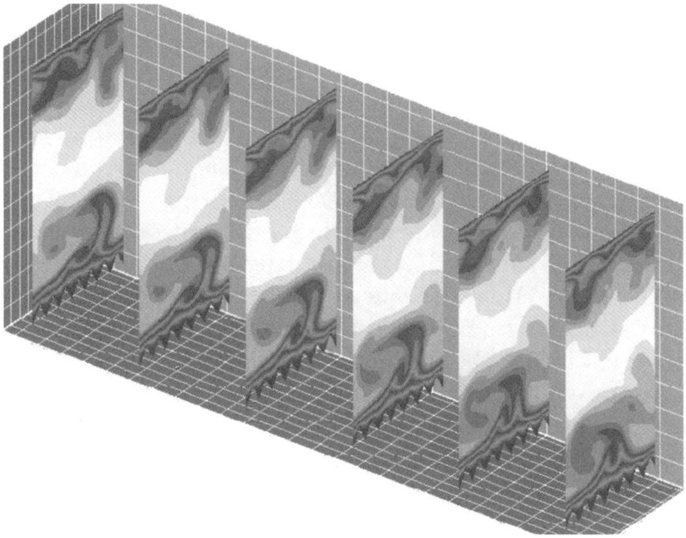

FIG. 9.7. Direct numerical simulation of turbulent flow in a channel with riblets on the lower wall. Plotted are instantaneous streamwise velocity contours at cross-flow planes. ([118].)

were employed, while nodal quadrilateral spectral elements were used in the cross-flow planes. At lower Reynolds numbers fully three-dimensional flows using spectral/*hp* methods have been applied in complex geometries of interest to biomedical [311, 361, 430, 435] and offshore engineering [330, 331] applications.

In this section, we discuss some of the practical aspects and implementation issues of DNS using spectral/*hp* element methods. We start by providing simple diagnostics for under-resolved simulations for transitional and turbulent flows. We then present a couple of different methods in stabilising and improving the quality of the results in marginally resolved simulations. The background material for these stabilisation methods is presented in Section 6.5.

9.3.1 *Under-resolution and diagnostics*

Spectral and spectral/*hp* element methods behave, in general, differently than low-order methods in under-resolved simulations. For example, there are many more wiggles, and therefore it is easier to detect suspicious simulations before resorting to more rigorous error estimation. In addition, finite difference methods have a numerical viscosity due to low-resolution discretisation, which effectively lowers the nominal Reynolds number of the simulated flow; this is not necessarily true in spectral/*hp* discretisations. The presence of wiggles can therefore be the source of erroneous instabilities in under-resolved flow simulations.

Specifically, the resulting velocity profiles may not be monotonic and thus they may be susceptible to inviscid-type instabilities, which in turn promote a transition from steady to unsteady flow or a transition to higher-order bifurcations and turbulence. For open unsteady flows, the amplitude of the oscillation in an under-resolved simulation is usually over-predicted. Similarly, in under-resolved simulations of wall-bounded flows using spectral/*hp* methods the effective Reynolds number may appear, in some cases, higher than its nominal value. An example of this behaviour for a global spectral method was presented by Henningson *et al.* [230], in the study of bypass transition in a boundary flow. Using a Chebyshev discretisation in the inhomogeneous direction and Fourier expansions in the other two directions, they demonstrated that with thirty-three Chebyshev modes their simulation showed that a wave structure was present which initiated a secondary instability. However, when twice the number of Chebyshev modes was used, this structure changed and the instability vanished completely. This result is plotted in Fig. 9.8, showing both the erroneous and the correct structure. Simulations with even higher resolution confirmed this explanation.

A similar example of such behaviour for spectral/*hp* element methods is the simulation of flow over a backwards-facing step, first presented in [256]. For the resolution shown in Fig. 9.9(a) the flow is unsteady, as is evident in the plot that shows time history in Fig. 9.10. However, if higher resolution is used, as shown in Fig. 9.9(b), then the flow is steady (see Fig. 9.10); this is true even at higher Reynolds numbers up to about $\mathrm{Re} \approx 2500$. Interestingly, the results of the under-resolved simulation are not totally irrelevant. For example, the two frequencies present in the unsteady case are the natural frequencies of the

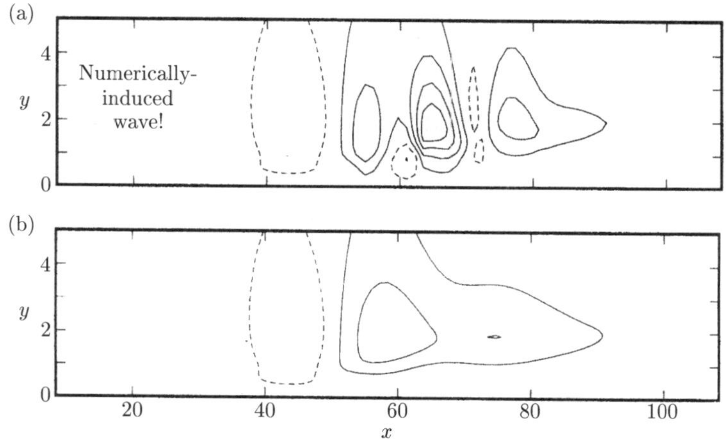

FIG. 9.8. Contours of normal velocity for describing the evolution of a finite-amplitude disturbance in a boundary-layer flow with normal resolution: (a) thirty-three Chebyshev modes, and (b) sixty-five Chebyshev modes. (Courtesy of D. Henningson.)

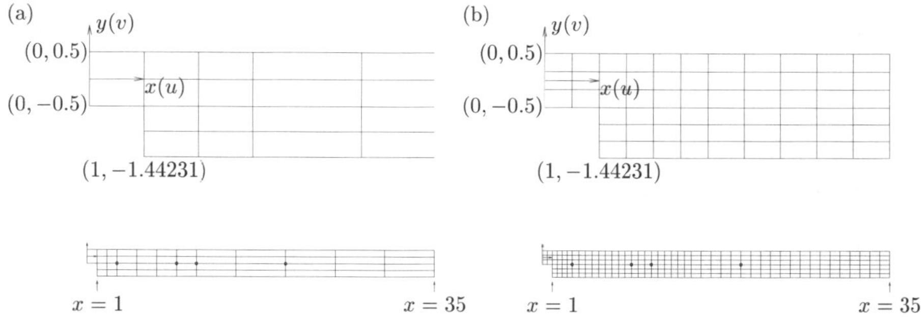

FIG. 9.9. Meshes for flow over a backwards-facing step at Re = 700. The Reynolds number is defined as $\text{Re} = 2\dot{Q}/\nu$, where $\dot{Q}$ is the flow rate. (a) Low-resolution mesh, and (b) high-resolution mesh.

shear-layer instability at the step corner and the Tollmien–Schlichting waves in the downstream channel portion. These modes are excited either by background noise, for example, some small turbulence level at the inflow, or spontaneously at a higher Reynolds number. Since no absolutely quiet wind tunnels exist, the results of the under-resolved 'noisy' simulation, in this case, match the results of the experiment [257].

We now turn to DNS of turbulent flows and discuss how under-resolved simulations behave in order to attempt to derive a set of empirical rules for users of spectral/*hp* element methods. As there is currently not sufficient experience with truly complex geometries, we will consider turbulent flow in a channel with

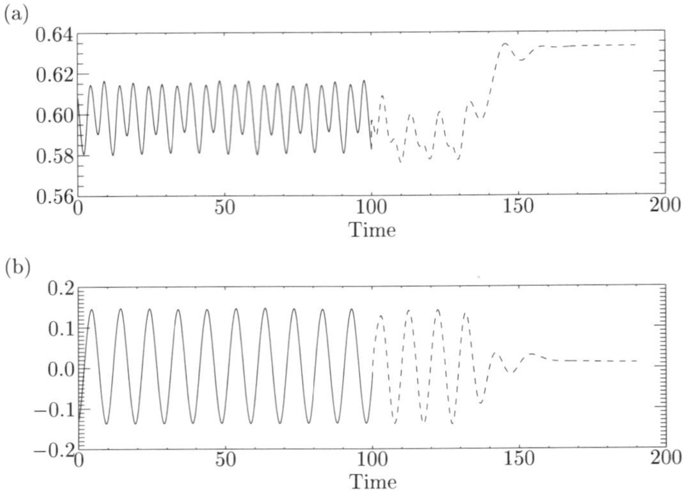

FIG. 9.10. Time history at the third point (shown in the mesh) in the step flow at Re = 700: (a) u velocity, and (b) v velocity at $x = 20$, $y = -0.5$. The solid line corresponds to the low-resolution simulation and the dashed line corresponds to the high-resolution simulation, which used as a restart the low-resolution simulation field. (Courtesy of L. Kaiktsis.)

flat walls. This flow has been simulated extensively with spectral [268, 314] and high-order finite difference methods [391]. The results agree very well with experimental data, which makes this flow a standard benchmark for resolution tests in DNS.

In order to test the spectral/*hp* element method we consider a mixed representation, where two-dimensional spectral elements are used in one plane and Fourier expansions in one of the homogeneous directions, as also discussed in Section 3.4.2 [261]. In this context, field variables are represented as

$$\mathbf{v}(x_1, x_2, x_3, t) = \sum_{m=0}^{M-1} \mathbf{v}_m(x_2, x_3, t) \mathrm{e}^{\mathrm{i}\beta m x_1} ,$$

where β is the x_1 direction wavenumber defined as $\beta = 2\pi/L_{x_1}$, and L_{x_1} is the length of the computational domain in the x_1 direction. We now take the Fourier transform of the Navier–Stokes equations to get the coefficient equation for each mode m of the expansion:

$$\frac{\partial \mathbf{v}_m}{\partial t} = -\tilde{\nabla}_m p_m + \nu \tilde{\nabla}_m^2 \mathbf{v}_m + [\widehat{\mathbf{v} \cdot \nabla \mathbf{v}}]_m \quad \text{in } \Omega_m , \quad m = 0, \ldots, M-1 ,$$

where the 'hat' denotes the Fourier transform and the equations are solved on each two-dimensional slice Ω_m. We also define

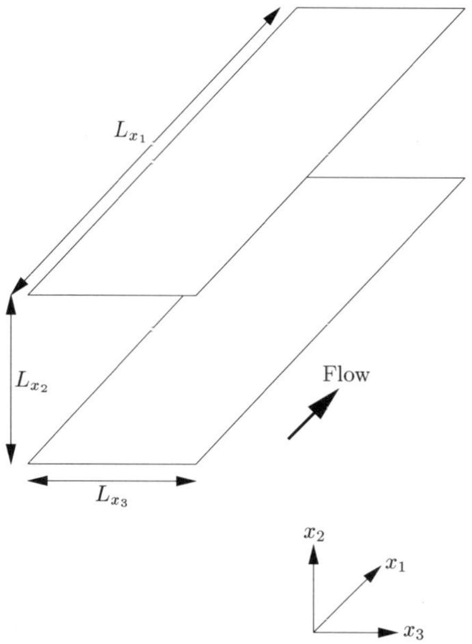

FIG. 9.11. Channel geometry used in the resolution study.

$$\tilde{\nabla}_m = \left(\frac{\partial}{\partial x_2}, \frac{\partial}{\partial x_3}, im\beta\right), \quad \tilde{\nabla}^2_m = \left(\frac{\partial^2}{\partial {x_2}^2} + \frac{\partial^2}{\partial {x_3}^2} - \beta^2 m^2\right).$$

The domain and coordinate system are shown in Fig. 9.11. The velocity components in the x_1, x_2, and x_3 directions are u, v, and w, respectively. The flow is driven by a constant flow rate Q:

$$Q = \int_{A_{cs}} u\,\mathrm{d}A = \frac{2}{3} A_{cs} U_l\,,$$

where $A_{cs} = L_{x_2} L_{x_3}$ is the cross-sectional area and U_l is the centre-line velocity of a laminar parabolic profile with the same volumetric flux. In general, the results of under-resolution in one or more directions are manifested in the mean properties as well as first-order or higher-order statistics of the flow. As an example, in Fig. 9.12 we plot the streamwise root-mean-square value at $\mathrm{Re}_\tau = 140$ based on wall-shear velocity and the channel half-width. These results are taken from [117], where more details can be found. It is seen that if the flow is under-resolved in the spanwise direction then u_{rms} is over-predicted, whereas if it is under-resolved in the normal direction then u_{rms} is under-predicted. Two practical guides were established in [117]. *First, the discretisation across the channel (x_2 direction) should mimic a Chebyshev distribution of points; and second, there should be at least ten collocation points (in x_2) in the first nine wall units.* ♦38

38 Good practice guidelines for DNS of channels.

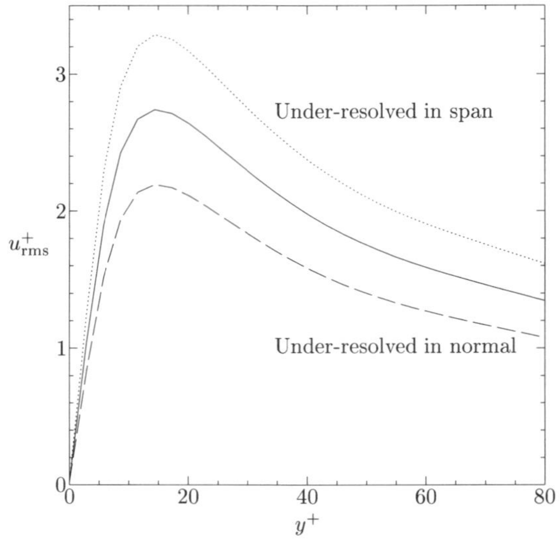

FIG. 9.12. Streamwise turbulence intensity $u_{\rm rms}$ at $\mathrm{Re}_\tau = 140$. The solid line denotes a well-resolved simulation from Kim *et al.* [268].

In order to help the reader in developing spectral/*hp* element codes for DNS of turbulence we now consider specific aspects of a systematic resolution study undertaken in [118]. The Reynolds number, Re, for these simulations is 5000, based on U_l and the channel half-width, $\delta = L_{x_2}/2$. This value corresponds to Reynolds number $\mathrm{Re}_\tau = 200$. The flow is periodic in the spanwise and streamwise directions, and no-slip conditions are used at the channel walls. The dimensions of the channel are $L_{x_1} = 5.61$, $L_{x_2} = 2.0$, and $L_{x_3} = 2.0$ when normalised by the channel half-width, δ. We define a wall unit length scale as $l^+ = lu_\tau/\nu$, so that the dimensions in wall units are $L^+_{x_1} \approx 1000$, $L^+_{x_2} \approx 400$, and $L^+_{x_3} \approx 400$. This geometry can be compared with the standard minimal flow unit to sustain turbulence reported in [254] of $L^+_{x_1} \approx 200 - 500$ and $L^+_{x_3} \approx 100$. The initial conditions can be obtained by starting with a parabolic velocity profile and some small noise in the first few Fourier modes (see [117]).

For a time-step of $\Delta t = 0.005$ or below, the results are essentially independent of the time-step for second-order time integration. The time-step restriction is usually imposed by the CFL requirement in mixed explicit–implicit integration. Since the spatial resolution is very fine, this time-step is also small enough to resolve the smallest time-scales in the flow. Care should be exercised when one uses fully-implicit time-stepping algorithms. Complete temporal resolution tests can be found in [118].

From resolution tests in the *streamwise direction* (in this case the Fourier direction), it was determined that the $N_{x_1} = 64$ (or, equivalently, $M = 32$) resolution was adequate to predict the turbulent statistics at Reynolds number

$\mathcal{O}(5000)$ (based on centre-line velocity and half-channel height). The statistical results from $N_{x_1} = 32$, $N_{x_1} = 64$, and $N_{x_1} = 128$ were similar, with the smallest differences between the $N_{x_1} = 64$ and $N_{x_1} = 128$ cases. The baseline case for this study used a mesh of $N_{\rm el} = 120$ elements, 11×11 quadrature in each element ($P = 10$th-order polynomials), and thirty-two Fourier modes (sixty-four physical planes in the streamwise direction). The x_2–x_3 projection of the mesh is shown in Fig. 9.13. A summary of the cases studied, as well as a comparison with previous channel simulations, is shown in Table 9.2.

From Table 9.2 we see that the Re_τ of the simulations appears to 'level out' at N_{x_1}-64. The simulation at N_{x_1}-16 greatly over-predicts the Re_τ of the channel flow; this over-prediction is also apparent from analysis of other zero-order

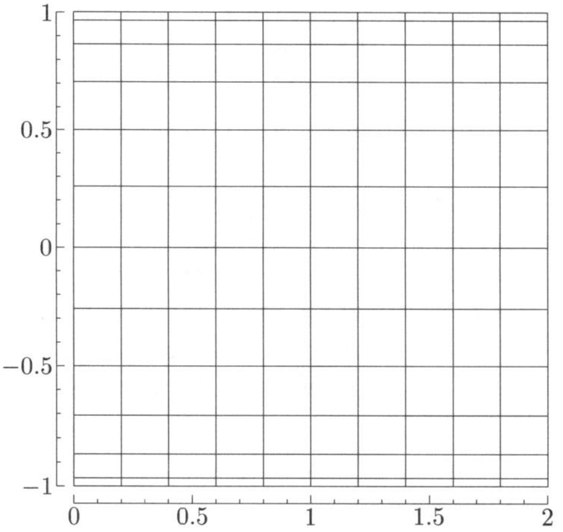

FIG. 9.13. Spectral element mesh of the two-dimensional solution domain, Ω_m, used for baseline simulation of turbulent flow in a channel; the flow is perpendicular to Ω_m.

TABLE 9.2. Numerical simulation parameters from simulation tests in [118] of streamwise resolution and previous numerical studies. All length-scales are normalised by the channel half-height δ.

Case	L_{x_1}	N_{x_1}	Δ_{x_1}	Re_τ
N_{x_1}-16	5.61	16	0.3506	222
N_{x_1}-32	5.61	32	0.1753	209
N_{x_1}-64 (baseline)	5.61	64	0.0877	207
N_{x_1}-128	5.61	128	0.0438	206.4
Chu and Karniadakis [98]	5.0	16	0.3125	131
Kim *et al.* [268]	4π	192	0.0654	180
Choi *et al.* [95]	π	16	0.1944	180

statistical data, such as the pressure drop in the channel. Results from the pressure drop data plotted in global coordinates, [118], show that the frequency of the data was less influenced by the streamwise resolution than the mean of the signal. The mean streamwise velocity was also affected by the streamwise resolution. Figure 9.14 shows that the under-resolved N_{x_1}-16 case under-predicted the von Karman constant, κ, and the additive constant, β, when plotting the logarithmic 'law-of-the-wall' velocity profiles, where the logarithmic region is given by the relation

$$U^+ = \frac{1}{\kappa} \ln y^+ + \beta\,.$$

Similarly, all three turbulence statistics for the four cases studied were over-predicted, especially around the peak value, with low resolution. The same trends are apparent in Reynolds stress $-\overline{u'v'}$ with respect to streamwise resolution. The results from all cases studied in [118] are shown in Fig. 9.15. Also shown in this figure are points from a simulation at $\mathrm{Re}_\tau = 180$ reported in [95].

Lastly, the one-dimensional energy spectra for the different cases reported in [118] are plotted. The energy per streamwise Fourier mode on the two-dimensional spectral element plane is calculated, that is, a sum of the square of the modal coefficients of the velocities is integrated over the x_2–x_3 domain 'slice'. These spectra demonstrate how much of the energy is resolved in the flow (that is, decades in a log–log plot). The slope of the spectrum is a measure of how turbulent energy is being transferred from lower to higher modes; any 'curl-ups' in the slope are an indication of under-resolution (*aliasing*), as this would mean

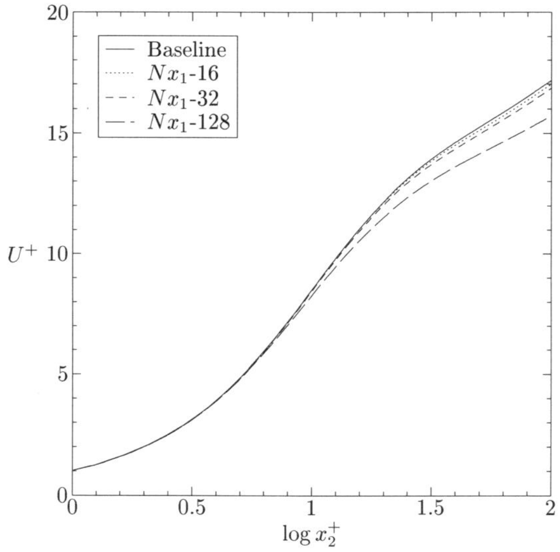

FIG. 9.14. Mean streamwise velocity plotted in wall coordinates for all cases in the streamwise resolution study. ([118].)

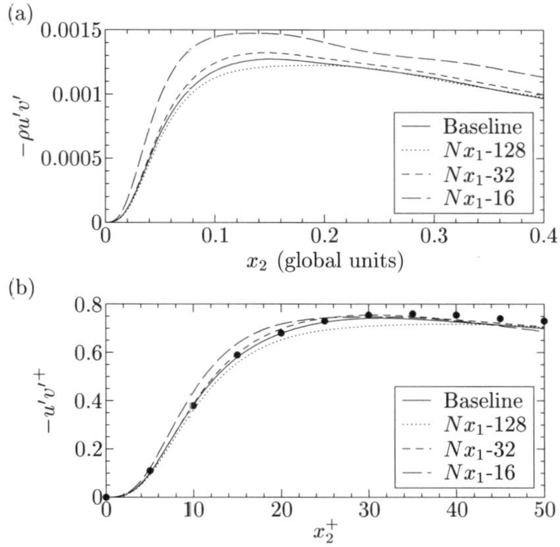

FIG. 9.15. Reynolds stress, plotted in (a) global, and (b) wall coordinates for all cases in the streamwise resolution study. Points plotted are taken from the DNS data of [95] ($\mathrm{Re}_\tau = 180$). ([118].)

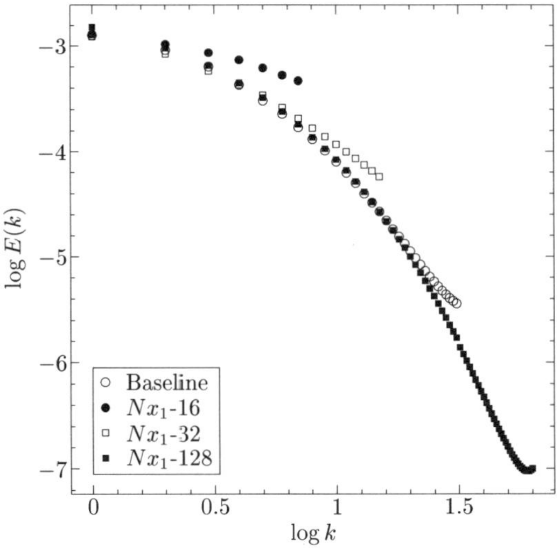

FIG. 9.16. One-dimensional energy spectra for all cases in the streamwise resolution study. ([118].)

that energy was being transferred from higher modes to lower modes. In Fig. 9.16, we plot the energy per mode for all modes except mode 0 (mean energy).

We see that the N_{x_1}-16 only captures energy in one half-decade in the spectrum compared to four decades of N_{x_1}-128. We also see that the most energetic modes, i.e., modes one to eight, appear to have the same energy and decay rate in cases N_{x_1}-32, N_{x_1}-64, and N_{x_1}-128. The baseline case, N_{x_1}-64, covers 2.5 decades in the energy spectrum. There is a saturation that occurs with the N_{x_1}-128 case at mode ≈ 60 where the energy level flattens out at $\mathcal{O}(10^{-7})$. Finally, we note that the energy in each of the Fourier modes is also dependent on the energetic scales resolved in the *two-dimensional* plane. We now turn our attention to these resolution studies.

A separate study was performed in [118] to determine the differences between two meshes for the turbulent flow in a channel using the same geometry, boundary conditions, and simulation parameters as explained in the previous section. The first mesh had a relatively low number of elements ($N_{\text{el}} = 120$) and a relatively high polynomial expansion order ($P = 10$) per element, and the second mesh had a much larger number of elements ($N_{\text{el}} = 480$) and a lower expansion order ($P = 7$). In both cases the meshes were designed to keep the approximate number of degrees of freedom on the two-dimensional plane the same. The first mesh is shown in Fig. 9.17(a); it is the same mesh as used for the streamwise resolution tests. The second mesh, Fig. 9.17(b), was constructed by splitting the rows and columns of the first mesh in half, thus doubling the total elements in each direction. A comparison of the grid spacing of these meshes is shown in Table 9.3. For these simulations the streamwise resolution was kept constant at $N_{x_1} = 64$ (thirty-two Fourier modes).

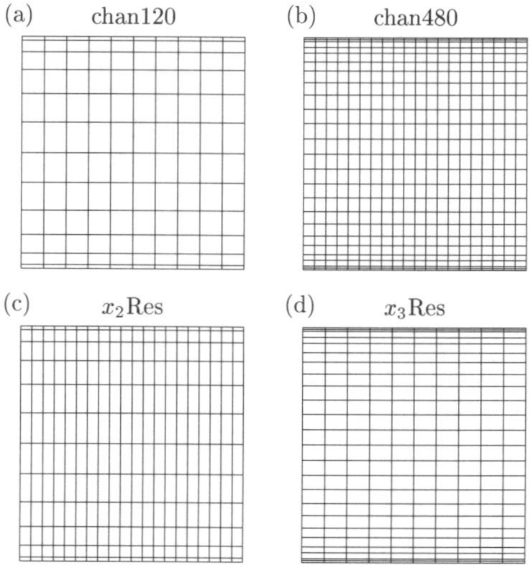

FIG. 9.17. Spectral element meshes of the two-dimensional solution domain, Ω_m, used for x_2-resolution and x_3-resolution studies in [118] for turbulent flow in a channel; flow is perpendicular to Ω_m.

TABLE 9.3. Numerical simulation parameters from simulation tests of a spectral element resolution study reported in [118].

Case	N_{x_2}	N_{x_3}	$\Delta x_2 \times 10^{-2}$	$\Delta x_3 \times 10^{-2}$	Re_τ
chan120	120	100	$0.11 \leqslant \Delta x_2 \leqslant 3.83$	$0.66 \leqslant \Delta x_3 \leqslant 2.96$	207
chan480	144	120	$0.14 \leqslant \Delta x_2 \leqslant 3.03$	$0.85 \leqslant \Delta x_3 \leqslant 2.34$	206.4
x_2Res	72	120	$0.29 \leqslant \Delta x_2 \leqslant 6.17$	$0.85 \leqslant \Delta x_3 \leqslant 2.34$	211.3
x_3Res	144	60	$0.14 \leqslant \Delta x_2 \leqslant 3.03$	$1.73 \leqslant \Delta x_3 \leqslant 4.71$	203

Mean quantities between the two meshes vary by less than 0.5 per cent. Overall, the differences in the two cases are quite small. The velocities were adequately approximated using both tenth-order ($P = 10$) and sixth-order ($P = 6$) polynomials. The previous plots show that the sizes of Δx_2 and Δx_3 are important factors in turbulence statistics prediction for this simulation. A closer comparison of the two cases shows that there are approximately 5000 more local degrees of freedom in the x_2–x_3 plane ($N_{x_2} \times N_{x_3}$) for *chan480* than *chan120.* Since the turbulence statistics for the two cases are almost identical, this indicates that the *chan120* case is sufficient to adequately resolve the turbulence statistics at Reynolds numbers of $\mathcal{O}(5000)$. Increasing the total degrees of freedom on the x_2–x_3 plane will not significantly change the statistical results.

In order to determine the behaviour of the simulation data with larger differences in the grid spacing and degrees of freedom, two more cases were studied and are listed in Table 9.3 as x_2Res and x_3Res. These meshes are shown in Figs 9.17(c) and 9.17(d), respectively. In these simulations *chan480* is taken as a baseline case, and then the x_2 and x_3 resolution is tested independently by halving the number of elements in each direction, one direction halved per mesh. In this way, the total number of grid points is halved whilst the grid point spacing is doubled, as shown in Table 9.3.

Measurements of Re_τ have a direct relationship to mean flow calculations. From Re_τ in Table 9.3, we see that under-resolution in the x_2 direction over-predicts the mean flow data, while under-resolution in the x_3 direction under-predicts the mean flow data. Plots of the logarithmic law-of-the-wall mean velocity profiles also showed discrepancies. The value of the U^+ intercept in the law-of-the-wall velocity plot for the logarithmic region of the flow, β, was over-predicted in the x_2Res case, while it was under-predicted for the x_3Res case; the von Karman constant κ was the same for all three cases but the additive constant β was not. These results are shown in Fig. 9.18.

From the turbulence statistics results [118], it is evident that under-resolution in the normal direction results in over-prediction of w_{rms} and v_{rms}, but under-prediction of u_{rms}. The spanwise resolution had a smaller effect but now under-resolution in the span caused a slight under-prediction in w_{rms} and v_{rms}, and a slight over-prediction in u_{rms}. This trend in resolution requirements, however, is not the same for the Reynolds stress. This is shown in Fig. 9.19. In these plots we see that under-resolution in the x_2 direction results in over-prediction

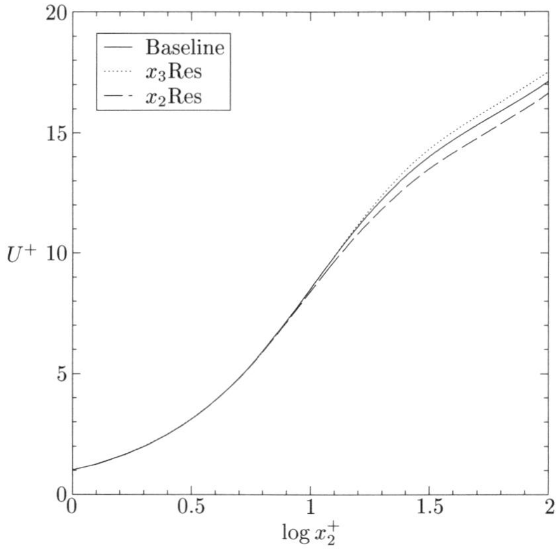

FIG. 9.18. Mean streamwise velocity plotted in wall coordinates for all cases in the two-dimensional plane resolution study. ([118].)

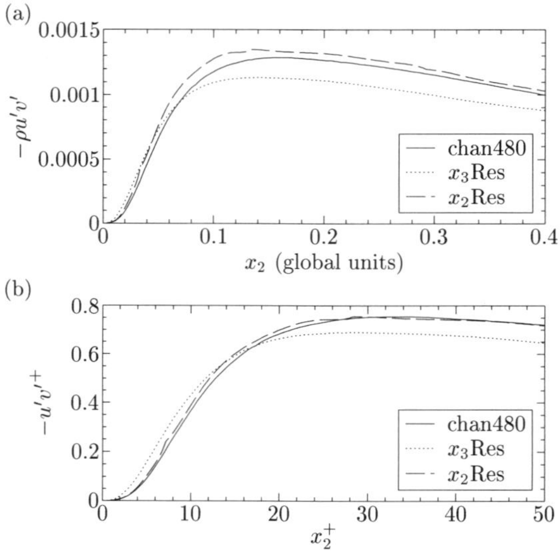

FIG. 9.19. Reynolds stress plotted in (a) global, and (b) wall coordinates for all cases of the two-dimensional plane resolution study. ([118].)

of the Reynolds stress, and under-resolution in the x_3 direction results in under-prediction. It is also apparent from these plots that under-resolution in x_3 has the larger deviation from the baseline case. At first, this may seem unexpected, as it was the x_2 resolution that had the greatest effect on the $u_{\rm rms}$ and $v_{\rm rms}$ results. One must, however, recall that Reynolds stress is a measure of the correlation between u' and v', and cannot be deduced from the intensities.

The first DNS of turbulence at $\text{Re}_\tau > 1000$ was performed by Iwamoto and Kasagi (2003, University of Tokyo; private communications). This was done on the Earth Simulator, the fastest supercomputer in the world at that time, where such DNS were performed with a nominal speed of over 40 TFLOPs [152]. Specifically, they performed DNS of fully-developed turbulent channel flow at $\text{Re}_\tau = 1160$. The numerical method used is similar to the one used in [268], i.e., Chebyshev discretisation along the inhomogeneous direction (x_2 direction) with Fourier expansions along the homogeneous directions. A fourth-order Runge–Kutta scheme and a second-order Crank–Nicolson scheme were used for time discretisation of the nonlinear terms and the viscous terms, respectively. The flow was driven by maintaining a constant pressure gradient. The size of the computational domain is $6\pi\delta \times 2\delta \times 2\pi\delta$ (δ is the half-channel width), with corresponding discretisation $1152 \times 513 \times 1024$ in x_1, x_2, and x_3 directions, respectively. The $3/2$ rule was applied to avoid aliasing errors. The number of total grid points was about 2 billion, and the effective computational speed reached about 1.4 TFLOPs, with 512 CPUs and 600 GB main memory on the Earth Simulator. For very high order in the Chebyshev expansion, the maximum error in the computed first derivative grows as $\mathcal{O}(N^2)$ due to loss of orthogonality. Iwamoto and Kasagi (private communications) developed a new scheme for reducing this error to $\mathcal{O}(\log N)$. If a spectral element discretisation is employed in the x_2 direction then such problems are avoided.

Figure 9.20 shows the mean streamwise velocity profiles at $\text{Re}_\tau = 1160$, 590, and 180. The profile of $\text{Re}_\tau = 1160$ has a viscous sublayer at $0 < y^+ < 5$, a logarithmic layer at $30 < y^+ < 250$, and a wake layer at $y^+ > 250$. The profiles of $\text{Re}_\tau = 1160$ and 590 agree with each other at $y^+ < 200$. However, the low Reynolds number profile $\text{Re}_\tau = 180$ deviates from the profile at higher Reynolds numbers beyond $y^+ = 10$ due to a low Reynolds number effect. At $\text{Re}_\tau = 1160$, the results of Iwamoto and Kasagi show that, in the region from $y^+ \approx 70$ to $y^+ \approx 600$, there are deviations from the logarithmic law, which is in agreement with experimental results at high Reynolds number.

9.3.2 *Stabilisation at high Reynolds number*

In this section we first demonstrate the effect of over-integration or super-collocation in dealing with the nonlinear terms (see also Section 6.5.3). For a quadratic nonlinearity we observe a $3/2$ rule, similar to the rule typically applied for de-aliasing Fourier collocation methods [350]. We then use implicit filtering in the form of spectral vanishing viscosity (SVV) to obtain robust DNS, but at low resolution.

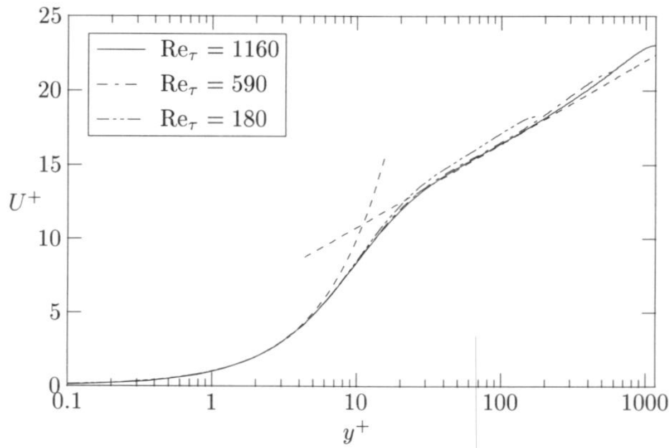

FIG. 9.20. Mean streamwise velocity profiles at $\mathrm{Re}_\tau = 1160$, 590, and 180. Dashed lines represent $\overline{u}^+ = y^+$ and $\overline{u}^+ = 1/0.41 \cdot \ln y^+ + 5.17$. (Courtesy of K. Iwamoto and N. Kasagi.)

9.3.2.1 *Polynomial de-aliasing*

We start by demonstrating the effect of under-integration and associated aliasing errors. As an example, we simulate transition to turbulence of incompressible flow in a duct, with its cross-section being an equilateral triangle. The laminar fully-developed solution is known analytically, see Section 9.1. We first introduce some random disturbances in the flow and then integrate in time until these disturbances start decaying or growing in time. All simulations were performed in the domain shown in Fig. 9.21, with the cross-section discretised using one triangular element only and sixteen Fourier modes (thirty-two collocation points) in the streamwise (homogeneous) direction. The Reynolds number is defined as $\mathrm{Re} = UD_e/\nu$, where U is the average velocity and D_e is the equivalent (hydraulic) diameter. For $\mathrm{Re} \leqslant 500$ all disturbances decay, but for $\mathrm{Re} = 1250$ the flow goes through transition and a turbulent state is sustained.

When using a tensorial triangular expansion we typically use the collapsed coordinate system defined in Section 3.2.1.1. If the inner product of the nonlinear terms are evaluated using under-integration on this coordinate system then the result is not rotationally symmetric due to the asymmetry of the collocation projection. Consistent integration, however, removes the non-rotationally symmetric error associated with the collocation projection. The exact Galerkin integral is then recovered, and so the symmetry of the expansion basis ensures the rotational symmetry of the inner product with respect to the basis orientation.

To demonstrate this point we compare two simulations at $\mathrm{Re} = 1250$ corresponding to three different combinations of polynomial and quadrature order. In the first simulation, shown in Fig. 9.22(a), we consider the case where $Q = P+1$, and where $P = 16$. The forces on the three walls of the duct are plotted as a function of time. From symmetry considerations, we expect that the statistical

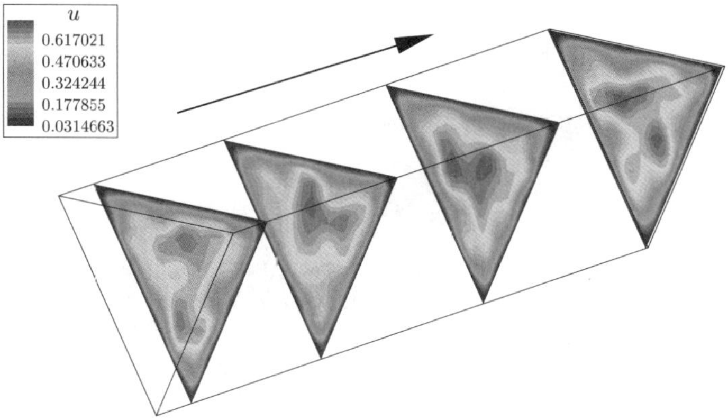

FIG. 9.21. Duct flow domain. The cross-section is an equilateral triangle and the streamwise length is three times the triangle edge. Shown is a snapshot of streamwise velocity contours at Re = 1250.

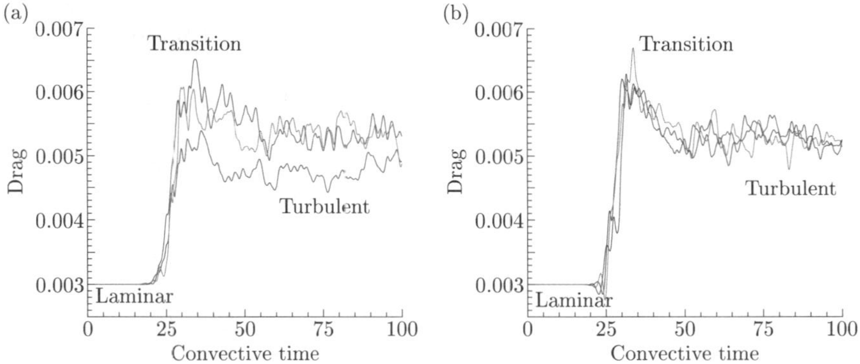

FIG. 9.22. Wall shear forces as a function of time for (a) $Q = P+1$, and (b) $Q = 3P/2$.

averages of the three forces are identical, but clearly the symmetry is not preserved. In Fig. 9.22(b) we plot the forces for the case with $Q = 3P/2$; similar results are obtained in the case with $Q = 2P$.

As a further indication of the role of aliasing, we consider the case of a double shear layer from the work of Warburton *et al.* [486] based on the cases considered by Brown and Minion [80]. This problem consists of an initial value problem in a periodic box of length two. Two thin shear layers are perturbed and subsequently roll up into two vortices with trailing arms, where the initial conditions are defined as

$$
\begin{aligned}
u &= \begin{cases} \tanh(\epsilon(x_2 + 0.5)) \,, & x_2 \leqslant 0 \,, \\ \tanh(\epsilon(0.5 - x_2)) \,, & x_2 > 0 \,, \end{cases} \\
v &= \delta \cos(\pi x_1) \,, \\
\epsilon &= 40.0 \,, \\
\delta &= 0.05 \,.
\end{aligned}
$$

In this example non-tensorial nodal triangular spectral expansions, based upon the Fekete points (see Section 3.3), were compared with a similar implementation, as in the tensorial expansion of Section 3.2.3 [486]. In both cases an integration order consistent with the exact integration of the linear terms was adopted, and so aliasing is present. We see in Fig. 9.23 comparisons, at time 1.87, of the vorticity field for the simulation run with $P = 9$, 11, and 14 on a mesh of 12×12 square regions each subdivided into two triangular elements. The plots on the left show the nodal results, and the plots on the right show the modal results.

In Fig. 9.23(a) we observe that four spurious vortices are created due to lack of resolution. In contrast, the modal version in Fig. 9.23(b) has created several weaker spurious vortices, and the roll-up of the arms has been distorted. Both these cases confirm that lack of resolution can cause non-physical phenomena. It is also interesting to note that the different triangle types have notably different properties in this regime.

Comparing Figs 9.23(c) and 9.23(d), we see that for $P = 11$ the modal version still exhibits a spurious vortex on the trailing arm of the lower vortex. The nodal version in Fig. 9.23(c) does not show any strong spurious vortices, but there are still some residual oscillations on the trailing arms. The results from further increasing the resolution are shown in Figs 9.23(e,f), and we see that the anomalous vortices are removed and the correct symmetries in the solutions are recovered.

Based on the above observations and our experience with such simulations, we can state the following semi-empirical rule.

De-aliasing rule. In general, for quadratic nonlinearities, employing consistent ♦39 integration/super-collocation with $3/2\,P$ grid (quadrature) points per direction, where P is the polynomial order per direction, followed by a Galerkin projection, leads to a de-aliased simulation on non-uniform spectral grids.

9.3.2.2 *Spectral vanishing viscosity*

Turbulence simulations using *monotonicity-preserving* schemes have addressed problems in homogeneous turbulence as well as wall-bounded flows using low-order methods, see [174, 175, 381, 475]. Representative algorithms, such as the piecewise parabolic method (PPM) and flux corrected transport (FCT), were originally developed for aerodynamic compressible flows; see Chapter 10. They

[39] De-aliasing rule for quadratic nonlinearities.

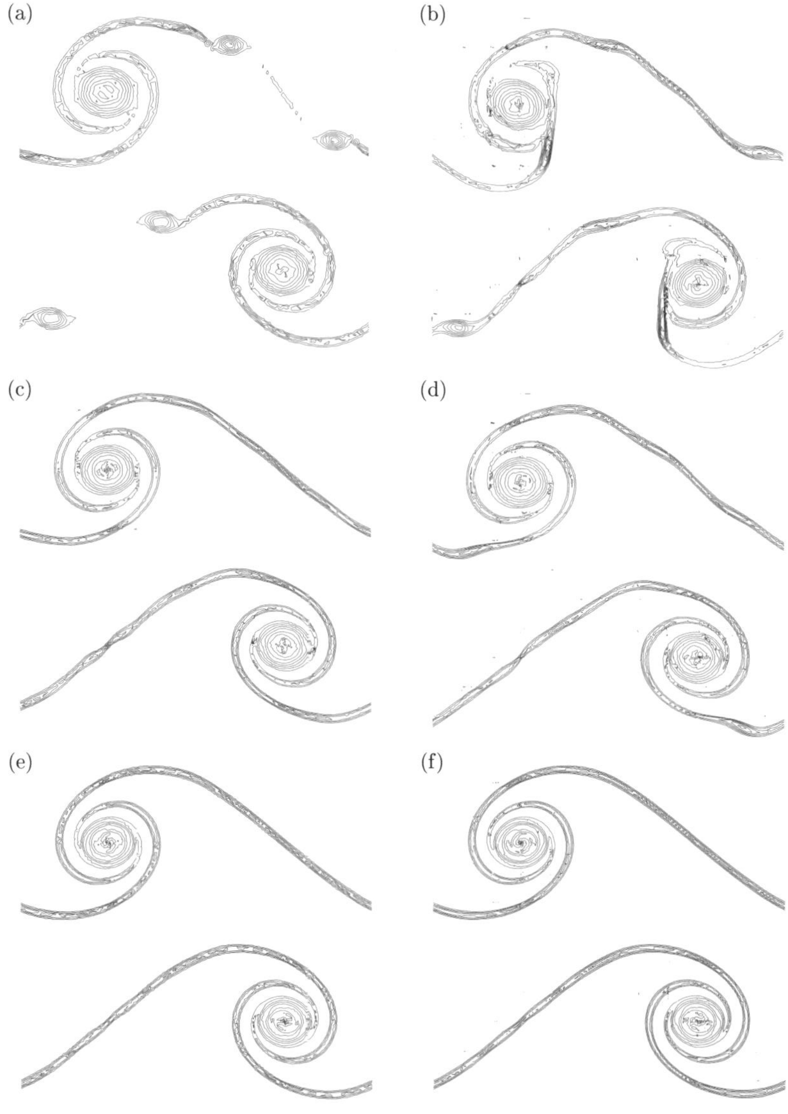

FIG. 9.23. Vorticity contours for the double shear layer flow (Re = 10 000, $t = 1.87$). Fekete nodes: (a) $P = 9$, (c) $P = 11$, and (e) $P = 14$. Modified tensorial basis $\phi_{pq} = \psi_p^a \psi_{pq}^b$: (b) $P = 9$, (d) $P = 11$, and (f) $P = 14$. (Courtesy of T. C. Warburton.)

employ nonlinear limiters and maintain monotonicity locally, while preserving at least second-order accuracy both in phase and amplitude. The intriguing feature of the monotonically-integrated LES (large-eddy simulations) (or MILES) approach [174] is the activation of the limiter on the convective fluxes and its role in generating implicitly a tensorial form of eddy viscosity that acts to stabilise the flow and suppress oscillations. From empirical evidence, it seems that if the resolution is fine enough to ensure that the cut-off wavenumber lies in the inertial range, then the simulation results seem to be independent of the generated viscosity.

The spectral vanishing viscosity (SVV) method which we presented in Section 6.5.2 introduces artificial dissipation selectively, i.e., for the correct parameters it is sufficiently large to suppress oscillations, yet small enough not to affect the solution accuracy. In the context of spectral discretisations, SVV can be viewed as a compromise between the classical total variation bounded (TVB) viscosity approximation and the exponentially accurate, yet unstable, spectral approximation. In [259] the SVV approach was first used in the context of simulating incompressible turbulent flows. The equations used are the *unfiltered* Navier–Stokes equations, which are enhanced on the right-hand side with a spectrally-vanishing viscous operator.

It is instructive to compare the spectral vanishing viscosity to the spectral eddy viscosity introduced by Chollet [96], Kraichnan [284], and Lesieur and Metais [296]. We recall from Section 6.5.2 that the spectral vanishing viscosity has the form

$$\hat{F}_k(k, M, P) = \exp\left[-\frac{(k-P)^2}{(k-M)^2}\right], \quad k > M\,,$$

where M is the wavenumber above which the viscosity is activated. The spectral eddy viscosity has the non-dimensional form [96]

$$\nu\left(\frac{k}{P}\right) = K_0^{-3/2}\left[0.441 + 15.2\exp\left(-3.03\frac{P}{k}\right)\right],$$

where $K_0 = 2.1$.

Comparing the Fourier analogue of this eddy viscosity employed in large-eddy simulations LES [296] to the viscosity kernel $F_k(k, M, P)$, Fig. 9.24 shows both viscosity kernels normalised by their maximum value at $k = P$. For SVV two different values of the cut-off wavenumber are considered:

$$M = C\sqrt{P} \quad \text{for } C = 0, 5\,. \tag{9.3.1}$$

This range has been used in most of the numerical experiments and is consistent with the theoretical results in [449]. In Fig. 9.24 it is shown that, in general, the two forms of viscosity have similar distributions, but the SVV form does not affect the first one-third or one-half of the spectrum (viscosity-free portion); also, it increases faster than the Kraichnan/Chollet–Lesieur eddy viscosity in the higher wavenumbers range, e.g., in the second-half of the spectrum.

In the following we present simulations of a turbulent channel flow to appreciate the effect of SVV in improving the quality of under-resolved DNS. Specifically, we want to demonstrate the effect of the SVV parameters, namely the viscosity

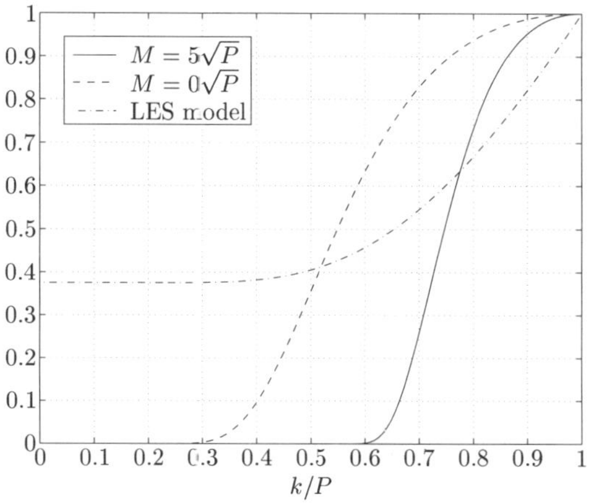

FIG. 9.24. Normalised viscosity kernels for the spectral vanishing viscosity (dashed line $C = 0$ and solid line $C = 5$) and the Kraichnan/Chollet–Lesieur viscosity (dashed–dotted line).

amplitude ϵ and the wavenumber cut-off M. The convolution is applied to the orthogonal basis which results from a 'rotation' of the semi-orthogonal basis, the latter being used in the C^0 formulation of incompressible flows, see [271].

Specifically, channel flow at $\mathrm{Re}_\tau = 180$ is simulated, with periodic boundary conditions in the streamwise and spanwise directions, following the benchmark solutions of Kim *et al.* [268]. The size of the computational domain is $L_x = 5$, $L_y = 2$, and $L_z = 2$. The mesh has 5×5 elements in the cross-flow plane, with polynomial order $P = 8$, while in the streamwise (Fourier) direction we have sixty-four points. This resolution is chosen in order to test carefully the effect of SVV, which acts on the cross-flow planes but not in the streamwise direction; the latter employs Fourier modes. We note that the resolution in the y direction in the current runs involves only thirty-five points, compared to 129 in the benchmark simulation of [268].

In Fig. 9.25 we plot the turbulence intensities versus the distance from the wall for three different cases. Results without SVV using the aforementioned low-resolution under-estimate the streamwise velocity component and over-estimate the cross-flow, as shown in Fig. 9.25 (solid line). The best results shown in Fig. 9.25 (dashed–dotted line) correspond to

$$M = 5 \quad \text{and} \quad \epsilon = \frac{3}{8}.$$

The other curves in the figure correspond to $M = 2$, $\epsilon = 1/8$ (dashed line), and they also show improvement compared to the untreated case (solid line). Similar simulations with $M = 2$, $\epsilon = 2/8$ produced results similar to the dashed line. In addition, another simulation with $M = 5$, $\epsilon = 1/8$ gave results similar to

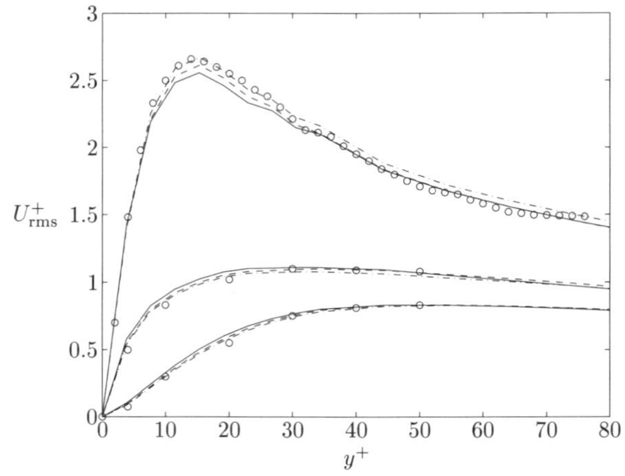

FIG. 9.25. Turbulence intensities for the turbulent channel flow. The solid line corresponds to the under-resolved DNS, the dashed line to $M = 2$, $\epsilon = 1/8$, and the dashed–dotted line to $M = 5$, $\epsilon = 3/8$. ([271].)

the untreated case. Although case-specific, these results confirm the theoretical results that only the upper one-third of the spectrum should be treated with SVV, and that there is an optimum (but unknown) viscosity amplitude level. This amplitude level should be computed dynamically, and in the next section we will present the main ideas for a dynamic SVV turbulence simulation.

9.4 Large-eddy simulations—LES

Large-eddy simulations (LES) of turbulence are based on appropriately filtered Navier–Stokes equations [296]. The first LES was performed by Deardorff in 1970 for a turbulent channel flow using the Smagorinsky model and a grid of 6720 points, without resolving the near-wall region [126]. LES methods have been developed as a computationally more efficient alternative to DNS. The discretisations employed are similar to DNS methods, but at lower resolution. The equations of motion govern the resolved scales only, but contain subgrid scale (SGS) stresses that need to be modelled analytically.

In this section we concentrate on incompressible flows and discuss several issues related to LES of complex-geometry flows. The main issues in this field relate to the filter, especially on non-uniform grids, and the type of subgrid model, especially close to wall boundaries. In the following section, we first present the basic equations for LES and discuss filters, and, subsequently, we will review representative subgrid models. Alternative LES approaches rely on built-in limiters that preserve monotonicity without explicit subgrid modelling. In these so-called MILES (monotonicity-integrated LES) methods, the nonlinear limiters generate implicitly a certain amount of anisotropic dissipation that depends strongly on the grid that stabilises the simulation. Several review articles on alternative LES approaches can be found in [204].

9.4.1 *Governing equations and filters*

In LES we make use of the notion of hierarchy of scales in turbulent flow structures, i.e., their decomposition into large and small scales. In a similar manner to Reynolds averaging (ensemble averaging), we decompose the velocity field into its filtered value (large scale) and its fluctuation (subgrid component). For a scalar quantity in one dimension the following decomposition is performed:

$$f(\xi) = \overline{f}(\xi) + f'(\xi)\,,$$

where we use a filter G of constant width Δ to perform the filtering

$$\overline{f}(\xi) = \frac{1}{\Delta}\int_{-\infty}^{\infty} G\left(\frac{\xi-\eta}{\Delta}\right) f(\eta)\,\mathrm{d}\eta\,.$$

Typical filter functions G are the top hat, Gaussian, and spectral cut-off, [7], all of which depend on the filter width Δ, which is of the order of the grid spacing. Some spectral filters are also presented in Section 6.5.1. We first consider a filter with a constant width Δ and perform filtering of the incompressible Navier–Stokes equations to obtain

$$\frac{\partial \overline{v}}{\partial t} + \frac{\partial(\overline{v}_i\overline{v}_j)}{\partial x_j} = -\frac{\partial \overline{p}}{\partial x_i} + \nu\frac{\partial^2 \overline{v}_i}{\partial x_j^2} - \frac{\partial \tau_{ij}}{\partial x_j}\,, \tag{9.4.1a}$$

$$\frac{\partial \overline{v}_i}{\partial x_i} = 0\,, \tag{9.4.1b}$$

where the subgrid scale (SGS) stress τ_{ij} is defined as

$$\tau_{ij} = \overline{v_i v_j} - \overline{v}_i\overline{v}_j\,.$$

It is the analytic modelling of τ_{ij} which is of interest in LES. The choice of filter influences the SGS stress, as both its type and width are important. If a top hat, a Gaussian, or any other *positive filter* is used, then the SGS tensor satisfies realisability conditions similar to the Reynolds stress tensor in Reynolds average equations [480]. This, in its turn, leads to the physically correct result that the subgrid kinetic energy is positive. The spectral cut-off filter does not satisfy such conditions as it achieves negative values.

The effect of different filtering approaches in spectral/*hp* element discretisations has been studied by Blackburn and Schmidt [57]. In particular, low-pass filtering can be implemented efficiently using the discrete polynomial transforms of Section 4.1.5. For the modal basis the implementation is straightforward and can be done using the Boyd–Vandeven filter, see Section 6.5.1 and also [57,299]. For the nodal basis a transformation to modal space employing the Legendre polynomial basis is required first, and then the Boyd–Vandeven filter can be applied on the new set of coefficients. An alternative approach for the nodal spectral elements is to use the projection method described in Section 6.5.1, i.e., transfer the solution to a coarser grid (Gauss–Lobatto–Legendre (GLL) points) and then interpolate back to the finer grid. In Section 6.5.1 we argued that we only affect the $(P-2)$th mode by transferring from $P+1$ GLL points to $P-1$ GLL points. However, for LES we typically project to the closest integer given by $(P+1)/2$. From the three filtering approaches, only two preserve the end-values, and thus the inter-elemental C^0 continuity is not affected. However, transformation and filtering in the Legendre basis affects the boundary values.

In Fig. 9.26 we present a comparison of two-dimensional filtering applied to the streamwise velocity field from a spectral element LES reported by Blackburn and Schmidt [57]. In the low-passing filter (cases (b) and (c)) the zero-lag Boyd–Vandeven filter of order $P = 9$ was employed. Also, sixty elements of order $P = 8$ were used in the discretisation, so in the projection filtering (case (d)) a coarse grid corresponding to $P = 4$ was employed. We see that the smoothing is similar in all three cases, but in case (c) the inter-element continuity is destroyed.

We next consider the extra errors produced by employing *non-uniform filters*. If the filter width is based on the elemental dimension then it will necessarily be variable if different-sized elements are applied. As discussed in [258], one approach to avoid this issue is to use a filter width based on a global scale. For this approach, even if the mesh is non-uniform, the filter width is constant; however, this also requires that every element is capable of resolving the global scale. The variable size of energetically-significant eddies due to flow anisotropy may necessitate the use of a variable filter width $\delta(\boldsymbol{x})$. Filtering the Navier–Stokes equations with a filter of variable width leads to extra terms in the equations, which involve higher-order spatial derivatives. This is because a filtering operation with

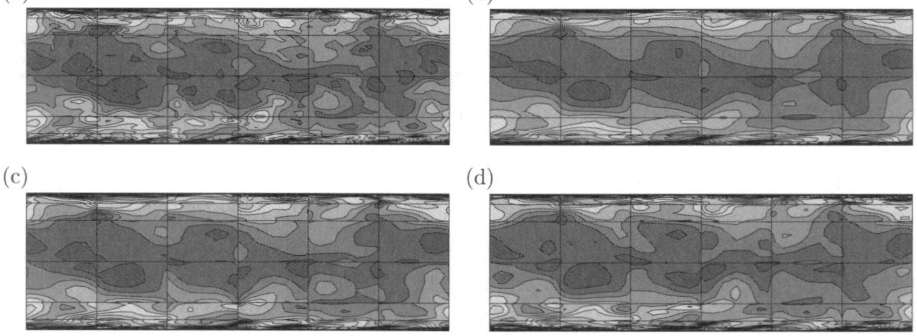

FIG. 9.26. Effect of filtering on spectral/hp element LES of turbulent channel flow. Shown are contours of streamwise velocity on planes normal to the spanwise direction (a) unfiltered data, (b) modal Jacobi basis, (c) nodal to Legendre basis, and (d) projection. (Courtesy of H. M. Blackburn.)

a non-uniform filter may not commute with the differentiation operation [184], i.e.,

$$\overline{\frac{df}{dx}} \neq \frac{d\overline{f}}{dx} .$$

To derive the correct governing equations, we follow the procedure outlined in [184]. We introduce a mapping function

$$\xi = g(x)$$

so that any non-uniform grid maps to a uniform one, i.e., $x \in [a, b] \mapsto \xi \in [-\infty, +\infty]$, with filter width $\delta(x)$ (variable) and Δ (constant), where

$$\delta(x) = \frac{\Delta}{g'(x)} .$$

The filtering operation is first defined by making a change of variables from the non-uniform grid (x) to the uniform grid (ξ) to obtain $f(g^{-1}(\xi))$. This function is then filtered using the standard procedure with the uniform filter, and finally we transform back to obtain

$$\overline{f}(x) = \frac{1}{\Delta} \int_a^b G\left(\frac{g(x) - g(y)}{\Delta}\right) f(y) g'(y)\, dy ,$$

which defines the operation with the variable filter width.

Applying, for example, this filtering operation to the linear advection equation $v_t + v_x = 0$ results in

$$\overline{v}_t + \overline{v}_x = \widehat{\nu}\, \overline{v}_{xx} ,$$

where $\widehat{\nu} = \alpha\delta^2(\delta'/\delta)$, and $\alpha = \int_{-\infty}^{\infty} \zeta^2 G(\zeta)\, d\zeta$. The order of the equation has increased due to the diffusive term on the right-hand side, and therefore extra boundary conditions are needed to solve the filtered equation. In essence, filtering of the linear equation produces a field $\overline{v}(x, t)$, which is smeared by an amount proportional to the filter width at increasing rates as the wave moves to the right. The diffusion coefficient $\widehat{\nu}$ is small, and thus only the high wavenumbers are affected.

Typically, the commutation error is proportional to Δ^2, but high-order approximations can be constructed [184]. A second-order approximation $\mathcal{O}(\Delta^2)$ is appropriate for second-order spatial discretisation since the filter width and the grid spacing are comparable. In spectral/*hp* element discretisations, however, high-order terms need to be included, which will produce several terms with high-order derivatives.

Similarly, in dealing with the Navier–Stokes equations, the extra terms due to the non-commuting of the filtering and differentiation operator have to be taken into account. The governing equations are then derived based on a new operator so that

$$\overline{\partial f} = \mathcal{D}\overline{f}\,,$$

where

$$\mathcal{D} \equiv \partial - \Delta^2 \boldsymbol{\Gamma} \partial^2 - \Delta^4 \mathbf{H} \partial^4 + \dots .$$

Here $\boldsymbol{\Gamma}$ and H are tensors that contain the information associated with the nonlinear mapping in three dimensions between the variable filter width $\delta(\boldsymbol{x})$ and the constant filter width Δ. The filtered (variable width) incompressible Navier–Stokes equations are then

$$\partial_t \overline{v}_i + \mathcal{D}_j(\overline{v}_i \overline{v}_j) = -\mathcal{D}_i \overline{p} - \mathcal{D}_j \tau_{ij} + \nu \mathcal{D}_k \mathcal{D}_k \overline{v}_i\,, \tag{9.4.2a}$$

$$\mathcal{D}_i \overline{v}_i = 0\,. \tag{9.4.2b}$$

To avoid complications with the extra boundary conditions required due to the introduction of high-order derivatives, we can expand the velocity field in terms of the small parameter Δ^2 as follows:

$$\overline{v}_i(\boldsymbol{x},t) = \overline{v}_i^0(\boldsymbol{x},t) + \Delta^2 \overline{v}_i^1(\boldsymbol{x},t) + \Delta^4 \overline{v}_i^2(\boldsymbol{x},t) + \dots .$$

The lowest-order equation for $\overline{v}^0(\boldsymbol{x},t)$ satisfies the (uniform filter) LES equations (see eqn (9.4.1a)), while the higher-order equations are derived by matching equal powers of Δ^2. For example, the first-order equations are

$$\frac{\partial \overline{v}_i^1}{\partial t} + \frac{\partial(\overline{v}_i^0 \overline{v}_j^1)}{\partial x_j} + \frac{\partial(\overline{v}_i^1 \overline{v}_j^0)}{\partial x_j} = -\frac{\partial \overline{p}^1}{\partial x_i} + \nu \frac{\partial^2 \overline{v}_i^1}{\partial x_j^2} - \frac{\partial \tau_{ij}^1}{\partial x_j} + F_i^1\,, \tag{9.4.3a}$$

$$\frac{\partial \overline{v}_i^1}{\partial x_i} = 0\,, \tag{9.4.3b}$$

with homogeneous boundary conditions for $\overline{v}^1(\boldsymbol{x},t)$. Here F_i^1 is a forcing term which is known from the solution at the previous level. For fourth-order accuracy, the computational work is increased by a factor of three compared to the standard LES equation, where a constant filter width Δ is used. The exact computational complexity depends on the tensors $\boldsymbol{\Gamma}$ and H which are associated with the nonlinear mapping.

The fix to increase the order of accuracy, proposed in [184], is equivalent to adding an extra term to the governing equations so that the leading-order commutation error is cancelled. For example, to make the commutation error of fourth-order in the filter width Δ, we have to add a term proportional to

$$\frac{g_{xx}}{g_x^3} \Delta^2 \frac{\mathrm{d}^2 \overline{v}}{\mathrm{d}x^2}\,,$$

which is a diffusion-like term, with a coefficient that depends on the mapping, i.e., the stretching of the grid. This term can then be positive or negative and potentially lead to numerical instabilities, as reported in [59]. Alternative approaches to controlling the commutation error in non-uniform grids have been suggested in [59, 476], but they apply to either infinite domains or periodic boundary conditions. It is difficult to formulate a satisfactory approach for general discretisations in order to control or to compute the commutation error in terms of filtered quantities.

9.4.2 *Subgrid models*

We now consider the subgrid scale stress (SGS) τ_{ij}, which requires similar modelling as the Reynolds stress $-\overline{u'v'}$ in the ensemble averaging approach. The subgrid models commonly used in LES studies can be broadly divided into two groups:

1. eddy-viscosity models, and
2. scale-similarity models.

The eddy-viscosity models assume a direct correlation between the subgrid scale stress and the large-scale (resolved) strain rate tensors. The scale-similarity model involves double filtering and it is based on the idea that the important interactions between resolved and unresolved scales are caused by the smallest eddies of the resolved state and the largest eddies of the unresolved state.

In the following, we review Smagorinsky-based eddy-viscosity models, as well as a mixed model that combines the properties of eddy-viscosity and scale-similarity approaches. Other models based on the idea of spectral eddy viscosity and its equivalent representation in physical space, that is, the structure-function eddy-viscosity model, are reviewed in [296, 329].

9.4.2.1 *Smagorinsky model and dynamic procedure*

The simple model first proposed by Smagorinsky is the analogue of the mixing length model of Prandtl. It assumes that the SGS stresses are proportional to an eddy viscosity times the strain rate of the resolved field, that is,

$$\tau_{ij} - \frac{1}{3}\delta_{ij}\tau_{kk} = -2\nu_e \overline{S}_{ij}\,, \tag{9.4.4a}$$

$$\overline{S}_{ij} = \frac{1}{2}\left(\frac{\partial \overline{v}_i}{\partial x_j} + \frac{\partial \overline{v}_j}{\partial x_i}\right), \tag{9.4.4b}$$

$$\nu_e = C_S \Delta^2 |\overline{S}|\,, \tag{9.4.4c}$$

where $|\overline{S}| = (2\overline{S}_{mn}\overline{S}_{mn})^{1/2}$, and the constant C_S can be computed from the Kolmogorov constant C_K assuming an inertial range $k^{-5/3}$, i.e.,

$$C_S = \frac{1}{\pi^2}\left(\frac{3C_K}{2}\right)^{-3/2},$$

which for $C_K \approx 1.4$ gives a value $C_S \approx 0.033$. This produces an overly dissipative model, so, in practice, the value of $C_S \approx 0.01$ is used [296].

The fundamental assumption of this model is that the principal axes of the SGS stress and the resolved strain rate tensors are aligned. From the practical point of view, this model generates eddy viscosity, even for laminar shear flows. It also assumes that the proportionality parameter C_S is constant everywhere in the flow, an assumption which cannot be true for shear flows. Another implicit assumption of this model is that it accommodates a one-way interaction between the resolved (large) scales and the subgrid scales; this interaction is expressed through the dissipation term $2\tau_{ij}\overline{S}_{ij}$ in the transport equation for the kinetic energy of the resolved flow, which is equal to $-2\nu_e|\overline{S}|^2$. With C_S always positive, the eddy viscosity ν_e is always positive and thus the dissipation term is always negative, which implies that energy flows only from the large scales to the small scales (forward scatter). However, from analysis with DNS data (see, for example, [377]) it has been found that inverse flow of energy is also possible (backscatter). Its magnitude depends strongly on the filter size and type, and in some cases backscatter may account for up to 40% of the total subgrid energy transfer.

A method for computing a variable parameter $C_S(\boldsymbol{x}, t)$ from information contained in the resolved field was proposed in [180]. This method requires the use of two filters: the grid filter denoted by $\overline{G}$ and the test filter denoted by $\widehat{G}$; the corresponding widths are chosen such that $\widehat{\Delta} > \Delta$. The equations for the test-filtered velocity field contain the test-level SGS stress term, i.e.,

$$T_{ij} = \widehat{\overline{v_i v_j}} - \widehat{\overline{v}}_i \widehat{\overline{v}}_j .$$

The standard $\widehat{\tau}_{ij}$ SGS terms are related to T_{ij} through the identity

$$L_{ij} = T_{ij} - \widehat{\tau}_{ij} ,$$

where the Leonard stresses $L_{ij} = \widehat{\overline{v}_i \overline{v}_j} - \widehat{\overline{v}}_i \widehat{\overline{v}}_j$ can also be computed by '$\widehat{G}$-filtering' the large-eddy velocity field. The extra assumption is that the test-level SGS stresses are also described by a Smagorinsky-type model with the *same parameter* C_S, that is,

$$T_{ij} - \frac{1}{3}\delta_{ij} T_{kk} = -2C_S \widehat{\Delta}^2 |\widehat{\overline{S}}| \widehat{\overline{S}}_{ij} .$$

Combining the above equations, we can derive equations for determining C_S:

$$L_{ij} - \frac{1}{3}\delta_{ij} L_{kk} = \alpha_{ij} C_S - \widehat{\beta_{ij} C_S} , \tag{9.4.5}$$

where

$$\alpha_{ij} = -2\widehat{\Delta}^2 |\widehat{\overline{S}}| \widehat{\overline{S}}_{ij} , \quad \beta_{ij} = -2\Delta^2 |\overline{S}| \overline{S}_{ij} .$$

In this expression, C_S appears inside the filter and thus further assumptions are needed to extract a unique value and construct a computationally efficient method for it. For simple flows with homogeneous flow directions, the assumption was made that C_S is constant, and thus it can be taken out of the filtering operation [180]. This procedure reduces the system of five independent integral equations in (9.4.5) to an algebraic equation for C_S, from which we obtain

$$C_S(\boldsymbol{x}, t) = -\frac{L_{ij}\overline{S}_{ij}}{M_{mn}\overline{S}_{mn}},$$

where $M_{ij} = \alpha_{ij} - \widehat{\beta}_{ij}$. An alternative way to deal with the redundancy of integral equations was developed in [302], where the 'best' C_S is chosen so as to minimise the difference between the actual and the modelled resolved turbulent stresses in the least-squares context. This approach gives

$$C_S(\boldsymbol{x}, t) = -\frac{L_{ij}M_{ij}}{M_{mn}M_{mn}}.$$

The values for C_S obtained from either of these expressions can be either positive or negative, which, as explained above, will represent forward or backward scatter of subgrid level energy, consistent with the flow physics. However, experience has shown that these values of C_S so obtained oscillate very rapidly between positive and negative values. This oscillatory distribution and, in particular, the negative values (negative dissipation) may lead to numerical instabilities.

An *ad hoc* way of avoiding this is to average over homogeneous directions or form local averages over grid cells and arbitrarily exclude the negative values, thus excluding backscatter. A justification of the smoothing approach was given in [183] for flows with at least one homogeneous direction by considering the constrained variational problem, extending the work in [302]. To account for the backscattering, a separate equation involving the subgrid kinetic energy was proposed. A transport equation for this kinetic energy, similar to the one adopted in ensemble averaging, is used to close the system of equations. To this end, the use of a positive filter, as mentioned earlier, is advantageous as it honours realisability conditions; this point is proved in [183]. The numerical tests which have to date been performed are inconclusive regarding the effect of backscattering in large-eddy simulations. Uncontrolled backscattering may lead to increases in the total subgrid energy and thus to explosive instabilities.

In summary, the Smagorinsky model with the parameter C_S computed from the resolved field is self-calibrating, given the aforementioned assumptions. It is an improvement to Smagorinsky's original proposal, as it produces no eddy viscosity in laminar flows, and it shows the right 'y^3' near-wall behaviour for the eddy viscosity. An adjustable parameter in the model is the ratio of the two filter widths $\widehat{\Delta}/\Delta$, which is typically taken as 2. This, in conjunction with the fundamental assumption that the same model can be used for subgrid stresses both at the grid as well as at the test filter level, makes the model an empirical rather than a theoretical one. Its use in complex-geometry flows with no homogeneous direction is an open issue, but time-averaging can instead be applied. However, to comply with Galilean invariance, the time-averaging should be formulated in a Lagrangian frame of reference [329].

We now consider an example of a spectral element LES using the dynamic Smagorinsky procedure. The results are adopted from [57] using Fourier expansions (thirty-two modes) along the spanwise direction, ten (two-dimensional) elements in the normal direction, and six elements in the streamwise direction, with $P = 8$. A nodal Legendre basis was used, with the first element closer to the wall within the viscous sublayer ($y^+ \leqslant 10$) and the second element covering the buffer layer, i.e., $10 \leqslant y^+ \leqslant 35$. The Reynolds number based on the wall shear velocity is $\mathrm{Re}_\tau = 650$. The filter width was taken to be $\Delta = (\Delta_1\Delta_2\Delta_3)^{1/3}$, and a combination of averaging along the homogeneous direction in conjunction with low-pass filtering in time was employed. In particular, the mixing length $l_S^2 = C_S\Delta^2$ was computed from

$$l_S^{2(n+1)} = \epsilon l_S^{2(0)} + (1-\epsilon) l_S^{2(n)} \,,$$

where $l_S^{2(0)}$ is an initial estimate computed by homogeneous averaging, and $l_S^{2(n)}$ is the value from the previous time-step. The degree of filtering is dictated by the constant $\delta = -\ln(1-\epsilon)/\Delta t$, and in the simulation of [57] this value was set to $\delta^+ = \delta(L_{x_2}/2)/u_\tau = 446$.

Figure 9.27 shows simulation results for the mean velocity (a) and streamwise and wall-normal velocity fluctuations (b), computed using the three aforementioned filtering techniques; the DNS data are taken from [338]. For the mean velocity, filtering in the Legendre basis seems to give the best agreement with the DNS data. For the velocity fluctuations the other two filtering techniques (i.e., Jacobi modal basis and projection) give better agreement. One of the interesting results in the simulations of [57] is the variation of the mixing length l_S along the wall normal; close to the element boundaries there is a substantial decrease in its value which reflects the clustering of the grid points there.

The cost of the dynamic procedure is about *fifty per cent* higher than that of the equivalent static Smagorinsky model, which, in turn, is *twice* as expensive as the simulation without an explicit SGS model. In the simulations of [57] there are twenty-four filtering operations per time-step.

9.4.2.2 *Scale-similarity and mixed models*

The assumption of scale invariance is used in a strong sense in the scale-similarity model, where the subgrid stress tensor is evaluated using a double filtering, i.e.,

$$\tau_{ij} = \overline{\overline{v}_i\overline{v}_j} - \overline{\overline{v}}_i\overline{\overline{v}}_j \,.$$

The second averaging could be done with a test filter greater than Δ, but most of the simulation results suggest only slight or no improvements for greater filter widths. What is important, however, is to avoid using cut-off filters since they introduce oscillations and have a non-local impact in physical space. This, in turn, may induce scrambling of the spatial relationship between the subgrid stress tensor and the simulated one, see [329]. It appears that, while the eddy-viscosity term models distant interactions, i.e., at scales below Δ and much larger

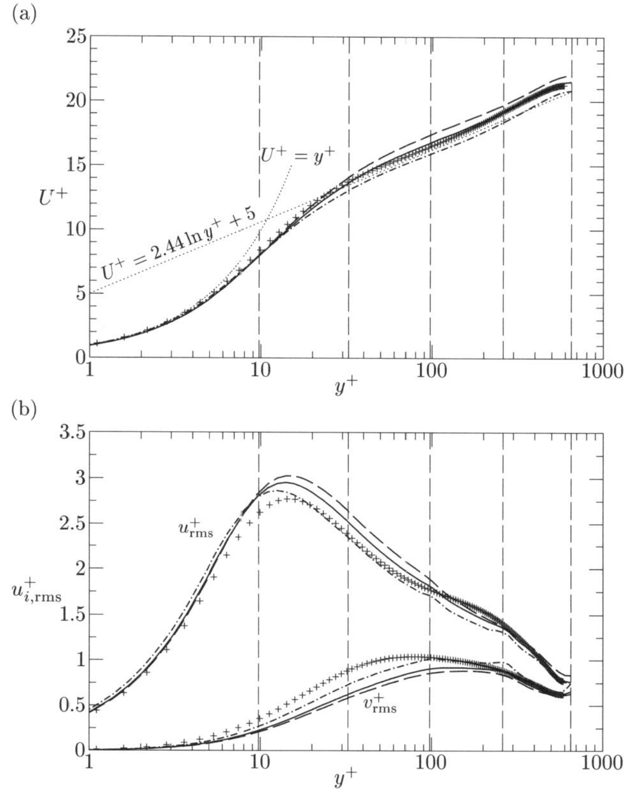

FIG. 9.27. LES at $\mathrm{Re}_\tau = 650$. (a) Mean velocity, and (b) streamwise and wall-normal velocity fluctuations. The circles denote DNS data, the solid line denotes the projection filtering technique, the small-dashed line denotes the Legendre basis technique, and the long-dashed line denotes the modal Jacobi basis technique. (Courtesy of H. M. Blackburn.)

than Δ, the scale-similarity term models local interactions, i.e., scales below Δ and marginally larger than Δ, see [329] for details.

In order to review the main ideas, we first use the classical decomposition of the SGS stress tensor:

$$\tau_{ij} = L_{ij} + C_{ij} + R_{ij} \,,$$

where L_{ij} is the Leonard stress (defined earlier), $C_{ij} = \overline{\overline{v}_i v'_j + v'_i \overline{v}_j}$ is the cross term, and $R_{ij} = \overline{v'_i v'_j}$ is the SGS Reynolds stress. The scale-similarity model proposed by Bardina [34] provides a closure for the SGS Reynolds stress as

$$R_{ij} = C_B(\overline{v}_i - \overline{\overline{v}}_i)(\overline{v}_j - \overline{\overline{v}}_j) \,,$$

where C_B is a constant between 2.0 and 9.0 obtained using different DNS data sets. This model was shown to exhibit excessive backscatter and, to this end, a filtered model was proposed by Horiuti [243]:

$$R_{ij} = C_B \overline{(\overline{v}_i - \overline{\overline{v}}_i)(\overline{v}_j - \overline{\overline{v}}_j)}\,.$$

While this model seems to represent backscattering of energy more accurately, it hardly dissipates any energy, and it is therefore necessary to combine it with an eddy-viscosity model. This is the motivation behind *mixed models*, see [410, 509]. Horiuti [243] has constructed such a model by introducing two coefficients (C_B, C_S), which he computes using the dynamic procedure described above. In particular, he proposed a decomposition of the SGS stress tensor as follows:

$$\tau_{ij} = L_{ij} + C_B L_{ij}^R - 2C_S \Delta^2 |\overline{S}| \overline{S}_{ij}\,,$$

where the new term L_{ij}^R combines both the cross terms and the SGS Reynolds stress contributions, i.e.,

$$L_{ij}^R = \overline{\overline{v'_i v'_j}} - \overline{\overline{v'_i}}\,\overline{\overline{v'_j}} = \overline{(\overline{v}_i - \overline{\overline{v}}_i)(\overline{v}_j - \overline{\overline{v}}_j)} - \overline{(\overline{v}_i - \overline{\overline{v}}_i)}\,\overline{(\overline{v}_j - \overline{\overline{v}}_j)}\,.$$

The first term on the right-hand side of the above equation is identical to the filtered Bardina model, and thus it can be modelled accurately using concepts of the scale-similarity approach.

LES results presented by Horiuti [243] and comparisons with other eddy-viscosity and mixed models showed that the two-parameter mixed model with a consistent dynamic procedure to compute (C_B, C_S) obtains better agreement with DNS for channel turbulent flow and mixing layers.

The many filtering operations involved in the scale-similarity model require substantial computational cost. To this end, an alternative model can be constructed by expanding the filtered field in a Taylor series and performing the filtering analytically. This results in a *tensor-viscosity model* of the form

$$\tau_{ij} = \overline{\overline{v}_i \overline{v}_j} - \overline{\overline{v}}_i \overline{\overline{v}}_j \approx C_{nl} \Delta^2 \frac{\partial \overline{v}_i}{\partial x_k} \frac{\partial \overline{v}_j}{\partial x_k}\,, \tag{9.4.6}$$

where the constant C_{nl} depends on the filter type. The tensor-diffusivity model includes many terms, but here we only consider the first term that is an approximation to the scale-similarity model [494]. The above equation is applicable only to filters with finite second moments; therefore, it is not applicable to the cut-off filter. Detailed tests based on a priori analysis of Gaussian-filtered DNS of isotropic turbulence were performed in [68]. For a Gaussian filter, given the filtered velocity field without truncation to a discrete grid, the full velocity field can be obtained by defiltering, see [292]. This provides justification of the non-linear model, as the defiltered field gives a subgrid stress equal to that in eqn (9.4.6) but with $C_{nl} = 1/12$.

9.4.2.3 *Dynamic SVV model*

We now return to the spectral vanishing viscosity (SVV), which seems to be a natural candidate for enforcing monotonicity but also for constructing subgrid scale models in spectral/*hp* discretisations. In the dynamic approach the viscosity amplitude in the SVV kernel varies as a function of space and time. We apply this idea of dynamic SVV to the Navier–Stokes equations using a variable viscosity amplitude given by

$$c(\boldsymbol{x}, t) = \frac{|S|}{||\,|S|\,||_\infty},$$

where

$$|S| = \sqrt{\mathrm{Tr}(S_{ij}S_{ij})}, \quad S_{ij} = \frac{1}{2}\left(\frac{\partial u_i}{\partial x_j} + \frac{\partial u_j}{\partial x_i}\right).$$

In order to smoothly affect the flow at the wall, we incorporate the Panton function [360] given by

$$g(y^+) = \frac{2}{\pi}\tan^{-1}\left(\frac{2ky^+}{\pi}\right)\left[1 - \exp\left(-\frac{y^+}{C^+}\right)\right]^2,$$

where all quantities are expressed in viscous wall units denoted by +. This function is multiplied pointwise by the coefficient $c(\mathbf{x}, t)$.

In Fig. 9.28 we plot a segment of the mesh around an airfoil and contours of the SVV amplitude computed using the aforementioned procedure. The SVV–LES is for a two-dimensional flow past an airfoil at 10 degrees angle of attack. The mesh involved 3888 triangular elements with fourth-order ($P = 5$) polynomial interpolation. It is clear that the SVV is nonzero in regions of high vorticity,

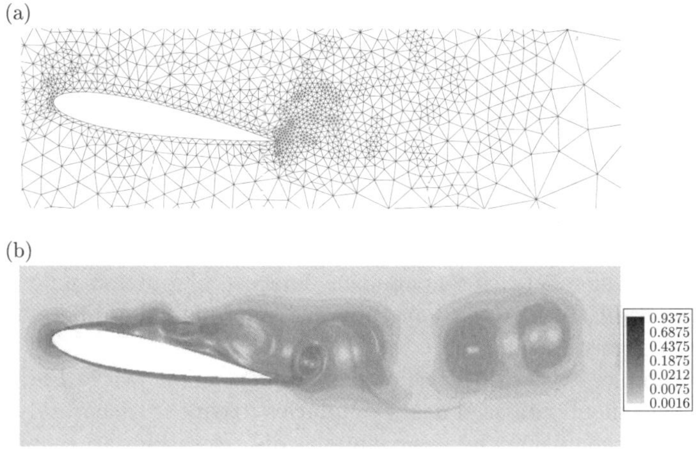

FIG. 9.28. (a) Segment of the unstructured mesh for flow past an airfoil at 10 degrees angle of attack and Re = 10 000. (b) SVV amplitude at one time instant.

which are the most probable candidates for under-resolution; more details can be found in [270].

9.5 Dynamic (dDNS) versus static DNS

The resolution requirements for DNS of turbulent flows are usually estimated based on the Kolmogorov theory of small scales. At large Reynolds number the eddies are significantly excited at scales ranging in size from L, at which energy input takes place, down to $\eta = L/\mathrm{Re}^{3/4}$, at which viscous dissipation becomes appreciable. Assuming that at least one mode is required to describe each scale of turbulence, the number of required modes in three dimensions is proportional to $\mathrm{Re}^{9/4}$ for a single time-step. In standard (semi-implicit) DNS algorithms the size of the time-step is dictated by the CFL number, and thus it should be less than η/v_{rms} (where v_{rms} represents velocity fluctuations). The Navier–Stokes equations are to be integrated for a time which is equivalent to at least one large-eddy turning time, i.e., proportional to L/v_{rms}, and therefore the number of time-steps required is proportional to $L/\eta \propto \mathrm{Re}^{3/4}$. Therefore, the total computational work should scale as $\mathrm{Re}^{9/4} \times \mathrm{Re}^{3/4} = \mathrm{Re}^3$ for a complete simulation cycle.

This type of rapid increase in resolution and corresponding increase in computational work requirements is a challenge for DNS at high Reynolds number. However, the above arguments apply if the same accuracy is required everywhere in the domain at all times. In practice, flow structures are very localised and coherent, and they often evolve in time in an almost deterministic manner. This suggests that a computation can be performed following some of these structures in time and space, in a Lagrangian-type description, without retaining uniformly fine resolution everywhere in the domain. In other words, a *resolution steering* procedure can be set up that will allow finite resolution resources to be distributed intelligently in the domain.

This type of steering or dynamic computing does not require the aforementioned rapid increase in resolution requirements, at least to the extent that *static* DNS of turbulence does. Its implementation, however, depends strongly on the ability to perform adaptive discretisations efficiently. The spectral/*hp* method provides such a possibility, as their dual path of convergence allows for a limited refinement or de-refinement process (h-convergence) of the physical domain and a selective spectral refinement or de-refinement for a fixed skeleton of the mesh, i.e., fixed elements (p-refinement). The use of polymorphic elements is particularly useful as they can naturally accommodate the different resolution requirements at any specific time.

As an example, we consider a flow past an aerofoil shown in Fig. 9.29 where quadrilateral elements are used near boundaries to capture accurately the vorticity generation mechanism, whereas triangular discretisations are employed downstream to follow detached vortices. Based on this mixed discretisation scheme, a hybrid computation approach can be developed where a DNS approach is followed around the body and in the near-wake, whereas coarser discretisation in

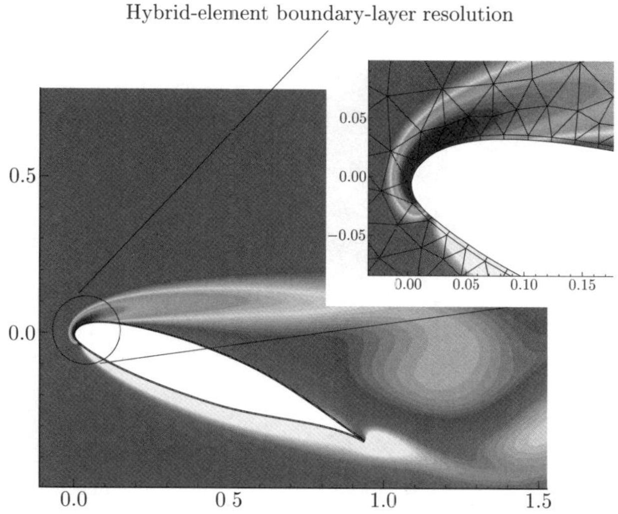

FIG. 9.29. Hybrid mesh discretisation for flow around an aerofoil.

conjunction with subgrid modelling (LES approach) can be followed in the far-wake. Inside the quadrilateral elements, very high spectral order may be used to capture the boundary layer, even at very high Reynolds number. This hybrid DNS–LES approach using polymorphic spectral discretisations may be the key to high Reynolds number turbulence simulations in complex-geometry domains.

9.5.1 *p-refinement and p-threads*

Another important consideration from the computational efficiency standpoint is the effective use of *multi-level* parallel algorithms to counteract the load imbalance in dynamic DNS. To this end, the use of posix threads (or p-threads) for local p-refinement has been shown in [139] to be particularly effective. For example, in a turbulent wake simulation, deploying a number of p-threads in proportion to the increase in polynomial order P resulted in a constant wall-clock computation.

The parallel paradigm for spectral/hp element methods is similar to the finite element methods, with the extra advantage of the increased *locality* in computational work. That is, a domain decomposition approach is followed where a group of elements is assigned to processors, while the implementation is achieved typically via MPI (message passing interface). Multiple levels of MPI can be implemented, i.e., across the entire domain and within the subgroup of elements, to achieve better parallel efficiency, especially for a large number of processors.

To counterbalance the increase in computational cost due to p-refinement, which scales as $\mathcal{O}(P^{d+1})$, we can employ *multiple threads* (or p-threads) within the MPI processes, giving rise to a hybrid MPI/threads dual-level parallelism. Figure 9.30 provides a schematic of the hybrid approach with domain decompo-

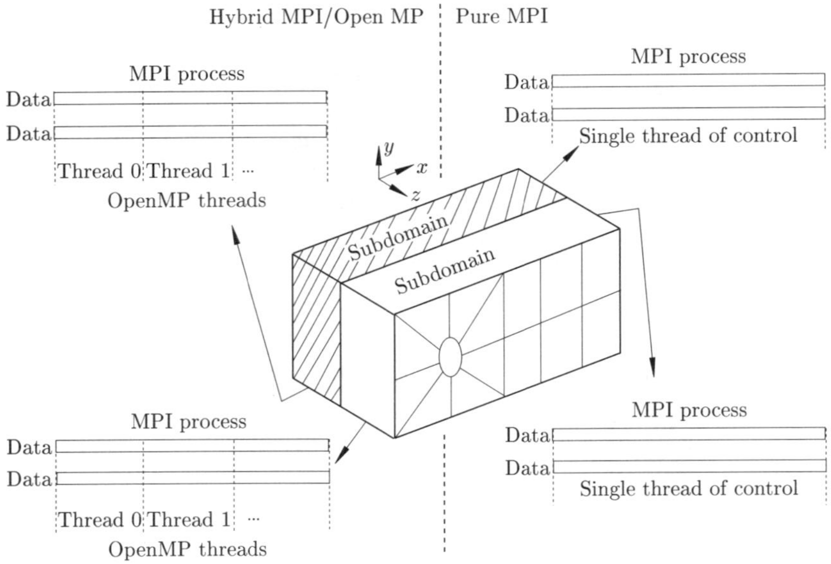

FIG. 9.30. Schematic diagram showing the pure MPI and hybrid MPI/OpenMP paradigms. In both approaches the MPI processes are responsible for different subdomains. However, only a single thread computes in pure MPI, while multiple threads split the work in each subdomain in hybrid MPI/OpenMP.

sition and multi-threading, as well as the pure MPI approach. At the outer level, multiple MPI processes are responsible for different subdomains of the flow problem. At the inner level, multiple threads within each MPI process conduct the computations of that subdomain in parallel. Data exchange across domains is implemented with MPI library calls, while conflicts within each subdomain caused by accessing shared objects by multiple threads are resolved with the synchronisation primitives provided by the thread library. In contrast, in the pure MPI approach a single thread in the MPI process performs the computations in the subdomain.

Specifically, it is highly effective to employ the hybrid MPI/OpenMP dual-level parallelism for a special class of three-dimensional unsteady flow problems in which the flow domain contains at least one homogeneous direction, while the non-homogeneous two-dimensional 'slice' domain is of arbitrary geometric complexity. Examples of this type of application include the flow between parallel plates or past long circular cylinders, as well as axisymmetric flows. The flow itself is fully three-dimensional due to intrinsic three-dimensional instabilities, but the geometry is homogeneous along that one direction.

A combined spectral/*hp* element–Fourier discretisation can be used to accommodate the requirements of high order, as well as the efficient handling of a multiply-connected computational domain in the x–y plane. Using multiple

FFTs, the three-dimensional problem is decomposed into two-dimensional problems of the Fourier modes. The flow domain is decomposed in the homogeneous direction. Each MPI process computes a certain number of Fourier modes. Typically, one process is assigned to one Fourier mode corresponding to two (real and imaginary parts) spectral/*hp* element planes. Multiple threads share the computations in the two-dimensional spectral element slices within the subdomain.

It is preferable to take a coarse-grain approach to the shared memory parallelism with OpenMP threads in MPI processes. A single parallel region encloses the main time-integration loops at the topmost level. Thus, multiple threads are created after the pre-processing stage. This avoids the overhead associated with frequent thread creations and destructions inherent in fine-grain programs. The number of OpenMP threads within each MPI process is set and can be varied dynamically by invoking the OpenMP library functions. This can be dictated by the expansion polynomial order or by the requirement of offsetting the load imbalance in different subdomains. Thus, multi-threading can be readily used for *dynamic* p-refinement, with the order P of different elements changing as the time integration proceeds to adapt to the flow characteristics.

Multiple threads share the computations within the subdomain by working on disjoint sections of the vectors, disjoint groups of elements, or disjoint computations within the element. Thus, the vector length, the element number, and the number of entries in linked lists are split based on the number of threads on the team. For runs with fixed polynomial order these computations can be done only once at the pre-processing stage, and the results are stored in a shared table. Different threads reference the results using their IDs. For dynamic p-refinement the results need to be updated when the polynomial order or the number of threads on the team is changed.

Increasing the number of Fourier modes in the homogeneous direction can be accomplished without increasing the amount of memory required per MPI process if the number of processes is increased accordingly. However, the p-refinement in the non-homogeneous two-dimensional spectral element planes (slices) induces an absolute increase in computational cost, and cannot be balanced by increasing the number of MPI processes if each process is already assigned to only one Fourier mode. The p-refinement directly causes an increase in the number of modes and quadrature points, the size of the vectors and various matrices (mass, differential, stiffness), and hence the cost. This cost increase can be compensated for by *dynamically* deploying a number of OpenMP threads to share the computations within the MPI processes.

Next, we demonstrate this hybrid MPI/OpenMP approach for a DNS of turbulent flow past a long circular cylinder at Reynolds number $\mathrm{Re} = 3900$ based on the inflow velocity and the cylinder diameter. The physics of this flow has been studied systematically in [316], where the parallel computation was involved using an MPI-only approach. The results presented here are taken from [139]. The computational domain in the x–y plane consists of 902 triangular prisms (elements). In the homogeneous z direction thirty-two Fourier modes (i.e., sixty-four planes) are employed. The flow domain extends from $-15D$ (where D is the cylinder diameter) at the inflow to $25D$ at the outflow, and from $-9D$ to $9D$ in the cross-flow direction. The spanwise length is fixed at $L_z/D = \pi$. A constant *uniform* flow is prescribed at the inflow. Neumann boundary conditions (zero flux) are applied at the outflow and on the sides of the domain. Periodic boundary conditions are used in the homogeneous direction.

The number of MPI processes is thirty-two, so that each process is responsible for only one Fourier mode (two spectral element planes). The p-refinement is performed systematically in the x–y plane, with the polynomial order increasing from $P = 8$ to $P = 16$ on all elements. This is a typical process in verifying new results in a large-scale flow simulation, i.e., after some successful initial runs p-refinement is undertaken to establish convergence and resolution-independent physical results. To demonstrate the effect of multi-threading, the histories of the wall-clock time per step for all the polynomial orders are collected. We start with the lowest order $P = 8$, and increase the polynomial order by two each time until the highest order $P = 16$ is reached. For each polynomial order a history of 450 time-steps is recorded. The timing data with a single thread per MPI process are first collected. Then, as the polynomial order is increased the number of OpenMP threads per process is increased in proportion to the cost increase in the single-thread case. Corresponding to the polynomial orders $P = 8$, 10, 12, 14, and 16, we deploy 1, 2, 4, 6, and 8 OpenMP threads per process, respectively.

In reality, the physical 'SMP' (symmetric multiprocessor) nodes in supercomputers or workstation clusters usually come with relatively few processors. So we also collected the timing data with two OpenMP threads per process, this being the most likely situation in practice. We plot the timing history as a function of the time-step for single-, two-, and multiple-thread cases in Fig. 9.31. For the case of a single thread per process, the wall-clock time per step increases from about four seconds to about thirty seconds. When two OpenMP threads are used per process, the wall-clock time per step is reduced essentially by half. As the number of threads per process increases in proportion, the wall-clock time per step decreases dramatically, and essentially remains *constant* for all the polynomial orders.

Returning now to a dynamic mode of computation, it is clear that effective adaptive strategies have to be devised in order to pursue *surgical* refinement/de-refinement. To this end, the residual-type methods developed by Oden and his associates in the context of *hp*-finite element methods can be employed. Alternatively, for a quick implementation, local residuals of the conservation laws based on 'derived' quantities (e.g., the vorticity equation) over the spectral elements can be used as rough error measures. However, more rigorous procedures are needed for robust and *self-steering* computing. In the following, we briefly review two representative approaches for *hp* and spectral element methods.

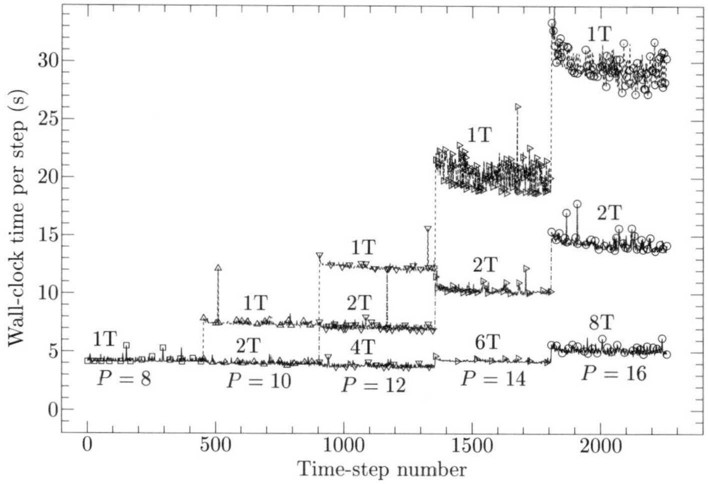

FIG. 9.31. Wall-clock time for p-refinement with one thread per process (dashed lines), two threads per process (dashed–dotted lines), and multiple threads per process (solid lines). The labels indicate the number of OpenMP threads per process, e.g., '2T' means two threads per process. (Timings obtained on the IBM SP3.) ([139].)

9.5.2 *The three-step Texas algorithm*

An *hp* strategy that has been implemented successfully for elliptic problems was first presented in [390]. A fundamental result was produced, which states that *for an h-refinement in an elliptic problem the error reaches a minimum if it is distributed uniformly among all elements.* For the more complicated *hp* adaptive refinement, the main idea is the use of both a priori and a posteriori error estimates. For example, for a scalar elliptic problem on a two-dimensional domain, that is,

$$\begin{aligned} -\nabla^2 u &= f \quad \text{in } \Omega\,, \\ u &= 0 \quad \text{on } \partial\Omega\,, \end{aligned}$$

the a priori error estimate (see Chapter 2) can be written in terms of local estimates, that is,

$$||\varepsilon||_1^2 = ||u - u^\delta||_1^2 \leqslant \sum_{e=1}^{N_{\text{el}}} \frac{(h^e)^{2s}}{(P^e)^{2\sigma}} \Lambda^e\,,$$

where $\Lambda^e = C||f||_\rho$, with $f \in H^\rho(\Omega)$ and $u \in H^k$. Here h^e and P^e represent the size and order of element e, respectively, and k measures the solution regularity. To estimate the convergence rate we assume a smooth domain, so we have

$$s = \min(P^e, k-1)\,, \quad \sigma = k - 1\,,$$

which may vary from one element to another; here, for simplicity, we assume them to be constant. The adaptive scheme is based on the assumption that the above asymptotic estimate is valid as an equality rather than an inequality. This, in turn, assumes that a relatively fine mesh should be used initially to obtain a first guess.

The a posteriori error estimate is evaluated based on residuals computed on individual elements from a variational problem that depends on the partial differential equation considered. For the Navier–Stokes equations, the a posteriori estimate is computed based on the solution pair (ϕ^e, ψ^e) of the following elemental residual problem (see [348]):

$$\begin{aligned} A^e(\phi^e, \mathbf{w}^e) = f^e(\mathbf{w}^e) - a^e(\mathbf{v}^e, \mathbf{w}^e) - b^e(p^e, \mathbf{w}^e) \\ - c^e(\mathbf{v}^e, \mathbf{v}^e, \mathbf{w}^e) + \oint_{\partial\Omega} \boldsymbol{n}^e \cdot \sigma(\mathbf{v}, p) \cdot \mathbf{w}^e \,\mathrm{d}\boldsymbol{x}\,, \end{aligned} \tag{9.5.1a}$$

$$B(\psi^e, q^e) = -b^e(q^e, \mathbf{v}^e)\,, \tag{9.5.1b}$$

where $\mathbf{w}^e$ and q^e are appropriate test functions in the velocity and pressure spaces, respectively (see Chapter 8). Here we define

$$A(\mathbf{v}, \mathbf{w}) = a(\mathbf{v}, \mathbf{w}) = \sum_{e=1}^{N_{\mathrm{el}}} A^e(\mathbf{v}, \mathbf{w})\,, \quad \mathbf{v}, \mathbf{w} \in \mathcal{U}\,,$$

$$B(q, s) = (q, s) = \sum_{e=1}^{N_{\mathrm{el}}} (q, s)\,, \quad q, s \in \mathcal{M}\,.$$

The boundary term of eqn (9.5.1a) (last term with the Cauchy stress σ) is computed using a flux balancing technique so that the elemental residuals and the boundary residuals are balanced [3]. The rest of the terms are defined in Chapter 8. Having obtained the elemental residual solution (ϕ^e, ψ^e), then the velocity error, ε_v, and the pressure error, ε_p, satisfy the inequality

$$|||(\varepsilon_v, \varepsilon_p)|||^2 \leqslant \sum_{e=1}^{N_{\mathrm{el}}} |||(\phi^e, \nabla \cdot \mathbf{v}^e)|||^2\,,$$

where

$$|||(\phi^e, \nabla \cdot \mathbf{v}^e)|||^2 = A^e(\phi^e, \phi^e) + \int_{\partial\Omega} (\nabla \cdot \mathbf{v}^e)^2 \,\mathrm{d}\boldsymbol{x}\,.$$

Following [344], we consider a scalar elliptic to present more clearly a three-step adaptivity algorithm. We assume that the elemental residuals (ϕ^e, ψ^e) can be obtained by solving equations similar to eqns (9.5.1a) and (9.5.1b). We first introduce some notation following [390]. We denote by $\epsilon = ||\varepsilon||_1/||u||_1$, the relative error, and by ϵ_T and ϵ_I the corresponding target and intermediate error levels, respectively, with a pre-specified ratio. For example, the target error may be five to ten times smaller than the intermediate error. We also denote by θ^e the local error indicator computed from the elemental residuals, for example, $(\theta^e)^2 = A^e(\phi^e, \phi^e)$, and the global error indicator $\theta^2 = \sum_{e=1}^{N_{\rm el}} (\theta^e)^2$. The three main steps of this algorithm are as follows.

- *Step 1*
  Generate an initial (relatively fine) mesh of ${N_{\rm el}}^0$ elements, which we denote by $({N_{\rm el}}^0, P_0, h_0)$. On this mesh we obtain a solution u_0 and compute an a posteriori error indicator

$$||\varepsilon||_1 \approx \theta_0^2 = \sum_{e=1}^{{N_{\rm el}}^0} (\theta_0^e)^2 \, .$$

  We can also obtain an estimate of the H^1 norm of the solution using the orthogonality property of the error, that is,

$$||u||_1^2 \approx ||u_0||_1^2 + \theta_0^2 \, ,$$

  and thus we have an initial error indicator $\epsilon_0 = ||\theta_0||/(||u_0||^2 + \theta_0^2)^{1/2}$. Assuming that the asymptotic convergence estimate holds for the initial mesh, we can now compute $\Lambda^2 = \sum_{e=1}^{N_{\rm el}} (\Lambda^e)^2$ from the convergence rate:

$$\theta_0^2 \approx ||\varepsilon||_1^2 = \frac{(h_0^e)^{2s}}{P_0^{2\sigma}} \sum_{e=1}^{N_{\rm el}} \Lambda_e^2 = \frac{h^{2s}}{P_0^{2\sigma}} \Lambda^2 \, ,$$

  and thus $\Lambda = \theta_0 P_0^\sigma / h^s$. Using now the definitions for the global Λ and θ_0, we can compute $\Lambda^e = \theta_0^e \Lambda / \theta_0$.

- *Step 2*
  Generate an intermediate mesh $({N_{\rm el}}^I, h_I^e, P_0)$ by using only h-refinement, keeping $P = P_0$ fixed. We first compute the intermediate error indicator $\theta_I = \varepsilon_I ||u||_1$ and assume an optimum h-refinement, which implies that the error is equidistributed. Based on this assumption, we can compute the new (extra) number of elements locally:

$$n_I^e = \left[\frac{(\Lambda^e)^2 {N_{\rm el}}^I h_0^{2s}}{P_0^{2\sigma} \theta_I^2} \right]^{1/(s+1)} \, , \qquad (9.5.2)$$

  where ${N_{\rm el}}^I$ is the number of elements in the intermediate mesh and n_I^e is the number of new elements covering the region previously occupied by element e. Therefore,

$${N_{\rm el}}^I = \sum_{e=1}^{{N_{\rm el}}^0} n_I^e \, . \qquad (9.5.3)$$

  To obtain the pair $({N_{\rm el}}^I, n_I^e)$, a few Newton–Raphson iterations are needed to solve the two coupled eqns (9.5.2) and (9.5.3). If $n_I^e > 1$ then each element

e is refined appropriately, otherwise if $n_I^e < 1$ then de-refinement is needed. In the new mesh $({N_{\mathrm{el}}}^I, h_I^e, P_0)$ we solve the problem again and compute an a posteriori error estimate, that is, a new θ_I at the global level and a new θ_I^e at the element level.

- *Step 3*
  Generate a third mesh by computing a new expansion order P_T^e for each element; this third mesh will produce the target error ϵ_T. We first compute the actual error distribution based on the intermediate mesh ${N_{\mathrm{el}}}^I$:

$$(\Lambda^e)^2 = \frac{P_0^{2s}(\theta_I)^2\Lambda^2}{(h_I^e)^{2s}(\theta_I^e)^2}.$$

The target error indicator obeys the relationship

$$\theta_T^2 = \epsilon_T ||u||_1^2 = \sum_{e=1}^{{N_{\mathrm{el}}}^I} \frac{(h_I^e)^{2s}(\Lambda^e)^2}{(P_T^e)^{2\sigma}},$$

and thus we compute the variable order for each element:

$$(P_T^e)^{2\sigma} = \frac{(h_I^e)^{2s}(\Lambda^e)^2 {N_{\mathrm{el}}}^I}{\theta_T^2}.$$

On this third mesh we can now compute the error and compare it with the target error; if it is larger then we repeat steps 2 and 3 with a higher intermediate error indicator ϵ_I.

The various levels and corresponding intermediate meshes required for an *hp* adaptive technique make this approach ineffective in practice. In [128] a two-grid solver is developed for elliptic problems. Given a mesh, the next optimally-refined mesh is determined by maximising the rate of decrease of the *hp* interpolation error for a reference solution. In [365] a similar but computationally more friendly approach was developed using nonlinear programming techniques. This effectively reduces the above three-step procedure to a two-step method. Constraints related to the specific problem may be incorporated relatively easily in the refinement procedure as part of a mixed integer nonlinear program (MINLP).

Several examples of *hp* refinement for elliptic problems, both with very smooth solutions and solutions of finite regularity, are given in [347], and for steady Navier–Stokes problems examples are given in [344]. Bernardi has reported new algorithms for *hp* adaptive strategies based on rigorous theoretical estimates for both structured and unstructured spectral elements [43]. The book by Ainsworth and Oden [4] provides a comprehensive treatment of *a posterior error estimation*, with some results applied to fluid flows in the last chapter.

9.5.3 *Non-conforming spectral element refinement*

In high Reynolds number flows the rigorous a priori estimates for elliptic problems are not as useful, so heuristic algorithms are typically used in practice. While the refinement criteria are usually application-specific, the use of the spectrum of the spectral expansion can provide some guidance in the refinement process. In this section we review a procedure developed by Mavriplis [325] and also by Henderson [226] for viscous incompressible flows. The discretisation is performed in the context of the mortar spectral element method presented in Section 7.4.

Let us consider a two-dimensional x–y domain and expand the velocity field as

$$u(x,y) = \sum_{n=0}^{\infty}\sum_{m=0}^{\infty} a_{n,m} L_n(x) L_m(y)\,,$$

where $u(x,y)$ is the x velocity component. The expansion coefficients are given by

$$a_{n,m} = \frac{1}{C_n C_m}\int_{1}^{1}\int_{-1}^{1} u(x,y) L_n L_m |J| \,\mathrm{d}x\,\mathrm{d}y\,,$$

where $C_k = (2k+1)/2$. The magnitude of the Jacobian $|J|$ is included in order to incorporate the effects of curvilinear boundaries and the element size.

We can then obtain an estimate of the approximation by estimating the approximation error $||u - u^\delta||$ in terms of the tail of the spectrum if we truncate the above expansion to order P. To this end, we first average over polynomials in the x and y directions in order to produce a representative one-dimensional spectrum of the form

$$\overline{a}_P = |a_{P,P}| + \sum_{i=0}^{P-1}(|a_{i,P}| + |a_{P,i}|)\,. \tag{9.5.4}$$

Next, assuming that we are in the asymptotic regime (typically $P \geqslant 5$), we approximate the discrete spectrum $\overline{a}_P$ with a decaying exponential

$$\tilde{a}(n) = C\mathrm{e}^{-\alpha n}\,,$$

where the above function $\tilde{a}(n)$ is a least-squares fit to the last few points in the spectrum $\overline{a}_P$. The refinement criterion proposed by Mavriplis [325] is then

$$\tilde{a}(P)^2 + \int_P^{\infty} \tilde{a}(n)^2\,\mathrm{d}n \leqslant \epsilon^2 ||u^\delta||^2\,. \tag{9.5.5}$$

This integral converges as long as $\alpha > 0$, which is true for non-singular solutions. If the above criterion is not met in a element-wise search, then the element is flagged for immediate refinement. Clearly, this approach needs some empirical 'calibration', but it can be used in a straightforward manner in *hp* refinement studies.

Henderson [226] modified the above procedure slightly as follows:

$$|a_{P,P}| + \sum_{i=0}^{P-1}(|a_{i,P}| + |a_{P,i}|) \leqslant \epsilon ||u^\delta||\,. \tag{9.5.6}$$

This is based on the observation that the main contribution to the left-hand side of eqn (9.5.5) comes from the coefficients of order P. There are several practical issues discussed in [226] that need to be considered, including geometric effects (i.e., corners and curvilinear boundaries) as well as effective solvers for non-conforming discretisations employed in this approach. In particular, an important issue is the distinction between singular solutions and steep gradients. When the solution is smooth, refinement based on gradient information or the tail of the spectrum leads to approximately similar refined meshes. However, gradient information alone is inadequate in detecting singularities, see also [417].

We conclude with an example given in [226] for the lid-driven cavity flow that illustrates the use of the *spectrum tail* as an effective refinement criterion. With our emphasis on high Reynolds number flow, we are interested in investigating how the solution and corresponding meshes vary as the Reynolds number increases. The driving lid velocity is $V = 1$ and the tolerance imposed is $\epsilon = 10^{-6}$, while the polynomial order is maintained at $P = 7$. Therefore, only h-refinement is applied in this example where singularities arise due to discontinuous boundary conditions between the side wall and the driven lid. The following refinement procedure was implemented in [226] using mortar elements. The solution is obtained after some initial integration in time and, subsequently, the refinement criterion is applied to both components of velocity. The previous solution is then projected onto the new mesh and the next iteration step begins.

Figure 9.32 (left) shows a series of refined meshes produced by Henderson [226] for the lid-driven cavity flow as a function of the Reynolds number. At low Reynolds number diffusion dominates and only a few elements are employed for refining the corners affected by the discontinuities in the velocity boundary conditions. However, as the Reynolds number increases steep vorticity gradients are developed along the walls (see the right-hand plots of Fig. 9.32), and the refined meshes reflect that change. This problem has also been considered by Mavriplis [325], who employed directional splitting for anisotropic refinement.

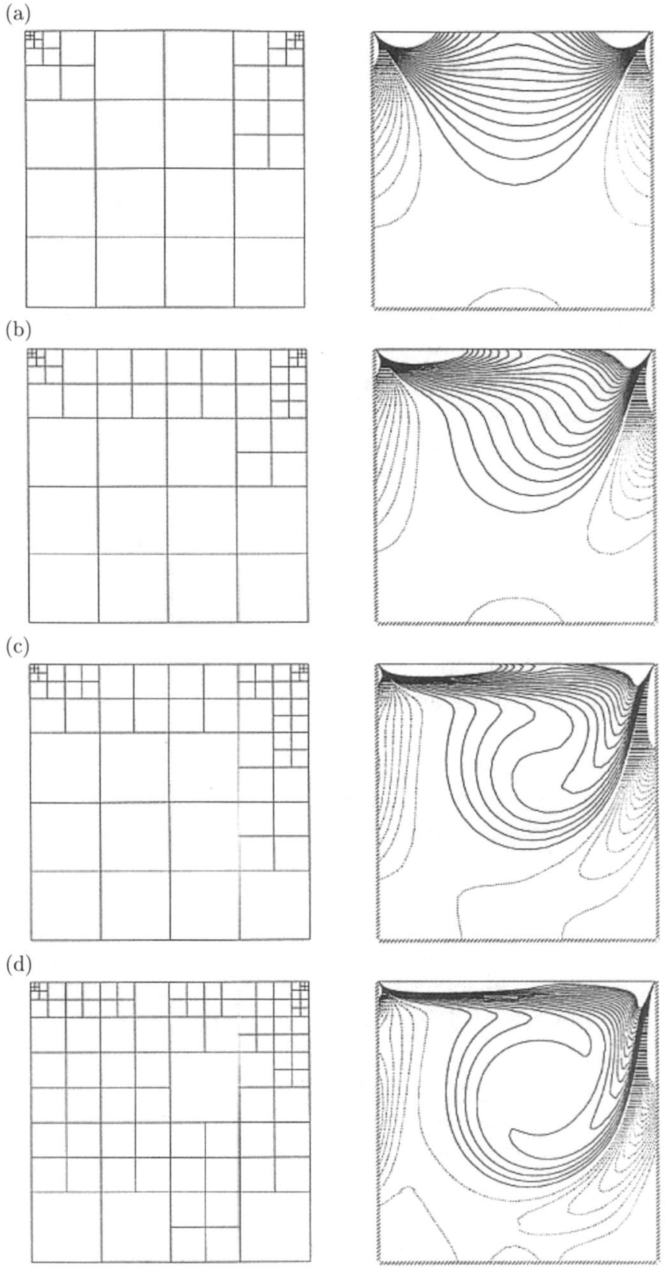

FIG. 9.32. Refined meshes (left) and corresponding vorticity fields (right) for lid-driven cavity flow. (a) Re = 10, (b) Re = 100, (c) Re = 250, and (d) Re = 500. (Courtesy of R. D. Henderson.)

10

HYPERBOLIC CONSERVATION LAWS

Spectral and multi-domain spectral methods are particularly effective in modelling propagation phenomena governed by hyperbolic conservations laws. For example, in atmospheric and ocean modelling as well as in seismic modelling, resolution of high frequencies and high wavenumbers is crucial, especially for long-term forecasting. Spectral methods have inherently small numerical dispersion and superb resolution properties, requiring typically four to six points (or modes) per wavelength. In addition, spectral elements provide several computational advantages, including nearest-neighbour communications and cache-blocked computations. This has been recognised by application scientists, and new scalable codes for large-scale simulations of atmospheric general circulation models and global seismology have been designed based on spectral element methods [273, 308].

However, these methods, and in general high-order methods, have not been used extensively in some other applications governed by hyperbolic conservation laws, e.g., for simulations of compressible flows or magneto-hydrodynamics, especially in the transonic and supersonic regimes. The main reason for this is the degradation of convergence of spectral-based methods for non-smooth solutions due to the Gibbs phenomenon, e.g., in the presence of shocks, contact discontinuities, etc. We have already discussed this issue in Section 6.5, where we presented several procedures in dealing with monotonicity and restoring the accuracy of the spectral solution. Given such enhancements and with emphasis shifted towards viscous solutions rather than simply inviscid solutions of compressible fluid dynamics, the standard advantages of high-order methods apply here as well. As an example, Pruett and Street [388] compared the results of a Chebyshev discretisation of the compressible boundary-layer equation to that of a second-order finite difference discretisation. They found that to compute the wall temperature for an adiabatic wall 77 points were required for the spectral method, compared to 400 points in the finite difference method to obtain the same five-digit accuracy.

In this chapter we first revisit the important issues of conservativity and monotonicity, thus extending the material presented earlier in Section 6.5. Our emphasis in this chapter is primarily on *systems* of hyperbolic conservations laws, which we solve in conservative form employing characteristic decomposition. The issue of boundary conditions is crucial in wave propagation phenomena and is often misunderstood for compressible flows and magneto-hydrodynamics. To this end, we discuss this issue in many different formulations that we present, including a flux-corrected transport (FCT) approach, a discontinuous Galerkin method, and the penalty formulation.

On the application side, we first consider the Euler equations in Section 10.3. Following this, in Section 10.4 we present the formulation of the spectral/*hp* element algorithm for shallow-water equations with applications to atmospheric circulation models. In Section 10.5 we consider the Navier–Stokes equations, and formulate appropriate solution algorithms. We then discuss shock-fitting techniques appropriate for high Mach number flows in Section 10.6. Finally, in Section 10.7 we consider the equations for magneto-hydrodynamics (MHD) and present ways of dealing with the divergence-free condition imposed on the magnetic flux.

10.1 Conservative formulation

Let us consider the scalar hyperbolic conservation law

$$\frac{\partial u}{\partial t} + \frac{\partial f(u)}{\partial x} = 0\,, \tag{10.1.1}$$

with proper initial and boundary conditions.

A conservative formulation is typically necessary for discontinuous solutions so that the shocks travel at the correct speed. In finite volume formulations, eqn (10.1.1) can be readily put into conservative form by considering the velocity at the middle of the finite volume cell and evaluating the fluxes (that is, terms involving spatial derivatives) at the faces of the volume cell. We have shown in Section 6.2.2.1 that the discontinuous Galerkin method, which can be thought of as an analogy of finite volumes to high order, is also conservative. However, in pseudo-spectral (collocation) discretisations we do not have these volumes available and thus we need to define appropriate average velocities and fluxes. To this end, we develop in the following appropriate cell averaging and reconstruction procedures for the average velocity and point values of fluxes, respectively. This formulation, which relies on a staggered-grid approach, is useful for *nodal* spectral element methods.

10.1.1 *Cell-averaging procedure*

Considering a non-uniform grid and adopting the terminology illustrated in Fig. 10.1, the cell-averaged scalar quantity $\bar{u}_i$ is given by

$$\bar{u}_i \equiv \bar{u}(x_i, t) = \frac{1}{x_{i^+} - x_{i^-}} \int_{x_{i^-}}^{x_{i^+}} u(x,t)\,\mathrm{d}x\,. \tag{10.1.2}$$

Given this definition, eqn (10.1.1) can be integrated along a cell extending from i^- to i^+ as follows:

$$\frac{\mathrm{d}\bar{u}_i}{\mathrm{d}t} + \frac{f(u_{i^+}) - f(u_{i^-})}{\Delta x_i} = 0\,, \tag{10.1.3a}$$

where we have also defined

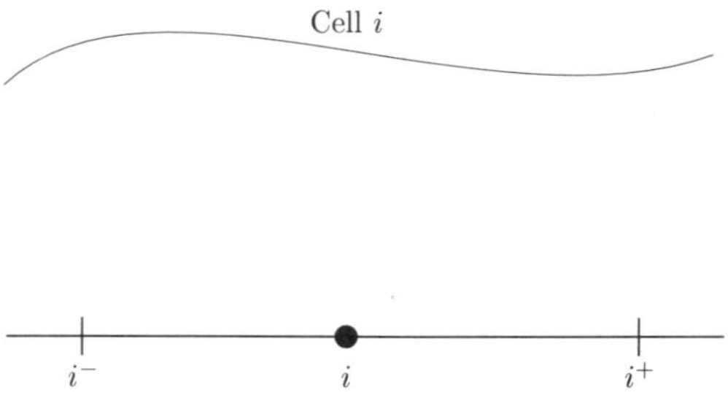

FIG. 10.1. Definition of a cell i. The cell $i+1$ has a left end-point $(i+1)^-$, which coincides with i^+.

$$\Delta x_i \equiv x_{i^+} - x_{i^-} \, . \tag{10.1.3b}$$

The above equation suggests that the fluxes $f(u)$ should be evaluated at the ends of the cell, $i^\pm$, using *de-averaged* (reconstructed) velocity values. This formulation leads to the conservative (or flux) form of the semi-discrete equation.

In the following, we define cell-averaged quantities for three particular discretisations. First, spectral (Fourier) discretisations; second, spectral (Chebyshev) discretisations; and third, spectral element (Chebyshev) discretisations. Proceeding with the first case, we refer to Fig. 10.1, where the set of points i at which cell-averaged quantities are defined are simply the midpoints of the cell. Using the definition of eqn (10.1.2), a spectral expansion of the form

$$u(x) = \sum_{-P}^{P} a_k \mathrm{e}^{\mathrm{i}kx} \tag{10.1.4a}$$

corresponds to the following cell-averaged quantity:

$$\bar{u}(x) = \sum_{k=-P}^{P} \bar{a}_k \mathrm{e}^{\mathrm{i}kx} \, , \tag{10.1.4b}$$

where $\bar{a}_k = \sigma_k a_k$ and the (Dirichlet) kernel σ_k is given by

$$\sigma_k = \frac{\sin(k\Delta x/2)}{k\Delta x/2} \, , \tag{10.1.4c}$$

with $\Delta x = \pi/P$.

A spectral-Chebyshev expansion corresponds to a non-uniform distribution of points with cells of variable size Δx_i. We assume, for simplicity, that we deal with a standard element so that $x \in [-1, 1]$ and, following the formulation of Cai *et al.* [82], we select the set of points i to be the Gauss–Chebyshev points defined by

$$x_i = \cos\left[\left(i - \frac{1}{2}\right)\Delta\theta\right], \quad \Delta\theta = \frac{\pi}{P}, \ 1 \leqslant i \leqslant P \, , \tag{10.1.5a}$$

while the end-points i^+ and i^- of each cell are the Gauss–Lobatto points defined as

$$x_{i^\pm} = \cos(i^\pm \Delta\theta)\,, \quad 0 \leqslant i^\pm \leqslant P\,. \tag{10.1.5b}$$

Using these two sets of points and the definition of eqn (10.1.2), a Chebyshev spectral expansion is of the form

$$u(x) = \sum_{k=0}^{P} a_k T_k(x)\,, \tag{10.1.6a}$$

which, after averaging, becomes

$$\bar{u}(x) = \sum_{k=0}^{P} a_k\ \bar{T}_k(x)\,.$$

In the above the cell-averaged Chebyshev polynomial is defined by

$$\begin{aligned}
\bar{T}_0 &= 1\,,\\
\bar{T}_1 &= \frac{1}{2}\sigma_1 U_1(x)\,,\\
\bar{T}_k &= \frac{1}{2}[\sigma_k U_k(x) - \sigma_{k-2}U_{k-2}(x)]\,, \quad \forall\ k \geqslant 2\,,
\end{aligned}$$

and

$$\sigma_k = \frac{\sin((k+1)\Delta\theta/2)}{(k+1)\sin(\Delta\theta/2)}\,. \tag{10.1.6b}$$

We have also introduced $U_k(x) = (1/(k+1))T'_{k+1}(x)$ to be the *second kind* of the Chebyshev polynomials.

In the spectral/*hp* element discretisation the domain is broken up into several macro-elements (similar to cells) within which the velocity is expanded in terms of a Lagrange polynomial (or another suitable basis):

$$u^e(x) = \sum_{n=0}^{P} u_n^e h_n(x)\,, \tag{10.1.7a}$$

defined on the Gauss–Lobatto–Chebyshev points. After the application of the averaging operator, $\bar{u}^e$ takes the form

$$\bar{u}^e(x) = \sum_{n=0}^{P} u_n^e \bar{h}_n(x)\,, \tag{10.1.7b}$$

where u_n^e are the nodal unknowns for element e; $h_n(x)$ and $\bar{h}_n(x)$ are the Gauss–Lobatto–Chebyshev Lagrangian interpolant and its corresponding cell-averaged function, respectively. They are obtained from

$$h_n(x) = \frac{2}{P}\sum_{p=0}^{P} \frac{1}{\bar{c}_n \bar{c}_p} T_p(x_n) T_p(x), \quad 0 \leqslant n \leqslant P,$$

$$\bar{h}_n(x) = \frac{2}{P}\sum_{p=0}^{P} \frac{1}{\bar{c}_n \bar{c}_p} T_p(x_n) \bar{T}_p(x), \quad 0 \leqslant n \leqslant P,$$

where $\bar{c}_n = 1$ if $n \neq 0, P$ and $\bar{c}_n = 2$ otherwise.

In matrix form the above cell-averaging procedure can be written as

$$\bar{\mathbf{u}}_i^e = \mathbf{A}_{in}^e \mathbf{u}_n^e, \quad 0 \leqslant n \leqslant P,\ 0 \leqslant i \leqslant P, \tag{10.1.8a}$$

where the cell-averaging matrix is defined as $A_{in}^e = \bar{h}_n(x_i)$; here x_i refers to the local coordinate. Based on the (nodal) cell-averaged values obtained from eqn (10.1.8a), the corresponding polynomial can be constructed using Lagrangian interpolation, that is,

$$\bar{u}(x) = \sum_{i=1}^{P} \bar{u}_i^e g_i(x), \tag{10.1.8b}$$

where the Gauss–Chebyshev–Lagrangian interpolant is given by

$$g_j(x) = \frac{T_P(x)}{T_P'(x_j)(x - x_j)}, \quad 1 \leqslant j \leqslant P. \tag{10.1.8c}$$

Having constructed a cell-averaging procedure for the spectral element discretisation, we first extend it to two dimensions and, subsequently, present the inverse operation of de-averaging and recovering point values for the evaluation of fluxes in eqn (10.1.3a).

Two-dimensional formulation

Let us now extend the cell-averaging formulation to two dimensions by considering the scalar hyperbolic equation

$$u_t + f(u)_x + g(u)_y = 0. \tag{10.1.9a}$$

To simplify the notation we redefine $x \equiv x_1$ and $y \equiv x_2$. In order to construct the cell averages in two dimensions we first integrate eqn (10.1.9a) in x, that is, $\int_{x_{i-}}^{x_{i+}}$, and subsequently in y, from y_{i-} to y_{i+}. The final *conservative* form of the equation on a *rectangular* two-dimensional mesh is

$$\frac{\partial}{\partial t}\bar{\bar{u}}_{i,j} + \frac{\bar{\mathbf{f}}_{i+,j} - \bar{\mathbf{f}}_{i-,j}}{\Delta x_i} + \frac{\bar{\mathbf{g}}_{i,j+} - \bar{\mathbf{g}}_{i,j-}}{\Delta y_j} = \mathbf{0}, \tag{10.1.9b}$$

where we have defined the following:

$$\bar{\mathbf{f}}_{i\pm,j} = \frac{1}{\Delta y_j}\int_{y_{j-}}^{y_{j+}} f\left(u(x_{i\pm}, y)\right) \mathrm{d}y\,,$$

$$\bar{\mathbf{g}}_{i,j\pm} = \frac{1}{\Delta x_i}\int_{x_{i-}}^{x_{i+}} g\left(u(x, y_{j\pm})\right) \mathrm{d}x\,,$$

$$\bar{\bar{u}}_{i,j} = \frac{1}{\Delta x_i}\frac{1}{\Delta y_j}\int_{y_{j-}}^{y_{j+}}\int_{x_{i-}}^{x_{i+}} u(x,y)\,\mathrm{d}x\,\mathrm{d}y\,.$$

Here, the average quantities are evaluated at the Gauss points (i, j), while the one-dimensional average quantities are evaluated at a *mixed set* of Gauss and Gauss–Lobatto points. De-averaging of the state $\bar{\bar{u}}$ is obtained by successive application of the one-dimensional reconstruction procedure described in the next section.

10.1.2 *Reconstruction procedure*

The reconstruction operation can also be put into matrix form, as we show in this section. We consider first the polynomial describing the cell-averaged values:

$$\bar{u}(x) = \sum_{j=1}^{P} \bar{u}_j^e g_j(x)\,. \tag{10.1.10a}$$

An alternative to expression (10.1.8c) for the Gauss–Chebyshev Lagrangian interpolant is

$$g_j(x) = \sum_{p=0}^{P-1} \frac{2}{P\bar{c}_p} T_p(x_j) T_p(x)\,. \tag{10.1.10b}$$

We can also express $g_j(x)$ in terms of the *second-kind* Chebyshev polynomials; to this end, we recall that

$$T_p(x) = \frac{1}{2}[U_p(x) - U_{p-2}(x)]\,, \quad \forall\, p \geqslant 2\,. \tag{10.1.11a}$$

Using the above equation we can rewrite $g_j(x)$ as

$$g_j(x) = \sum_{p=0}^{P-1} \lambda_p^j U_p(x)\,, \quad 1 \leqslant j \leqslant P\,, \tag{10.1.11b}$$

where

$$\lambda_p^j = \frac{1}{P} T_p(x_j)\,, \quad p = P-2,\, P-1\,,$$

$$\lambda_p^j = \frac{1}{P}[T_p(x_j) - T_{p+2}(x_j)]\,, \quad 0 \leqslant p \leqslant P-3\,.$$

The interpolating polynomial corresponding to point values (Gauss–Chebyshev–Lobatto points) can then be constructed using the *de-averaged* Lagrangian interpolants G_j as follows:

$$u(x) = \sum_{j=1}^{P} \bar{u}_j^e G_j(x) . \tag{10.1.12a}$$

The cell-averaged second-kind Chebyshev polynomial is obtained using the definition (10.1.2) (see details in [82]):

$$\bar{U}_p(x) = \sigma_p U_p(x) , \tag{10.1.12b}$$

with σ_p obtained from eqn (10.1.6b). To determine $G_j(x)$ we therefore consider eqns (10.1.12a) and (10.1.12b) and eqns (10.1.10a)–(10.1.11b) to obtain

$$G_j(x) = \sum_{p=0}^{P-1} \frac{\lambda_p^j}{\sigma_p} U_p(x) . \tag{10.1.12c}$$

To recover the point values u_i we simply set $x = x_i$ in the interpolating polynomial $u(x)$. In matrix form the reconstruction procedure (at an elemental level) can be written in the form

$$\mathbf{u}_{i^\pm} = \mathbf{g}^*_{i^\pm j} \bar{\mathbf{u}}_j , \tag{10.1.13a}$$

for

$$1 \leqslant j \leqslant P , \quad 1 \leqslant i^\pm \leqslant P ,$$

where

$$g^*_{i^\pm j} = G_j(x_{i^\pm}) . \tag{10.1.13b}$$

Based on these P point values, the interpolating polynomial $u(x)$ can then be constructed from eqn (10.1.12a). This local reconstruction procedure is then repeated for all elements. To form a global interpolating polynomial, however, we need to impose an appropriate continuity condition at elemental interfaces, as we discuss next.

10.1.3 *Interfacial constraint*

The interpolation polynomial $u(x)$ constructed using the Gauss–Lobatto–Chebyshev points is of degree P (requiring $P + 1$ values to be determined), while we only obtain P point values from the reconstruction procedure (eqn (10.1.13a)). The additional information needed to uniquely define $u(x)$ comes from requiring continuity of the solution at the interfacial nodal points. For the eth element, for example, we require that the rightmost nodal value be equal to some value u_γ. This can be accomplished by adding an extra term to the $(P-1)$th-order polynomial. Denoting the local coordinate by ξ, we get

$$u^{e+1}(\xi) = \sum_{j=1}^{P} \bar{u}_i^{e+1} G_j(\xi) + (1 - \xi) T'_P(\xi) \frac{\delta u^e}{2P^2} , \tag{10.1.14a}$$

where the jump δu^e is defined by

$$\delta u^e = u_\gamma - \sum_{j=1}^{P} \bar{u}_i^{e+1} G_j(-1)\,. \tag{10.1.14b}$$

Here $\xi \in [-1, 1]$ is the interval and $\xi = \pm 1$ are the end-points of the mapped (standard) element. Also, γ refers to the interface between adjacent elements.

Note that the expression $(1 - \xi)T'_P(\xi)$ attains zero values at all the Gauss–Lobatto–Chebyshev points of the $(e+1)$th element, except for the leftmost point. Therefore, implementing eqns (10.1.14a) and (10.1.14b) is equivalent to requiring only the leftmost point value to be equal to u_γ. The rest of the point values of the $(e + 1)$th element remain unchanged. The same will be true for the eth element and its rightmost point value.

The only undetermined quantity is u_γ. The value of u_γ should be set either to $\sum_{j=1}^{P} \bar{u}_j^{e+1} G_j(-1)$ or to $\sum_{j=1}^{P} \bar{u}_j^{e} G_j(1)$, depending on the direction of the advection velocity V. Therefore, $u_\gamma = u_0^{e+1}$ if $V < 0$, otherwise $u_\gamma = u_P^e$. In the case of a nonlinear advection equation, similar considerations are made. For example, in the case of the Burgers equation

$$\frac{\delta u}{\partial t} + \frac{\partial}{\partial x}\left(\frac{u^2}{2}\right) = 0\,,$$

we can linearise around a certain state and examine the sign of an appropriately defined advection velocity. In particular, in the Burgers equation, which requires no linearisation, the choice is the average state, and thus the advection velocity is $V = \frac{1}{2}(u_P^e + u_0^{e+1})$. For hyperbolic systems we need to consider the Jacobian of the system, as we will discuss later in this chapter.

10.1.4 *Non-oscillatory approximation*

The main difficulty in applying spectral methods to approximating discontinuous solutions is the Gibbs phenomenon. If a discontinuous function is approximated by a spectral expansion (Chebyshev, Fourier, and others), then the approximation is only $\mathcal{O}(1/P)$ accurate in smooth regions and contains $\mathcal{O}(1)$ oscillations near the discontinuity. When spectral methods are applied to *nonlinear* partial differential equations with discontinuous solutions, the Gibbs phenomenon may also lead to numerical instability.

An interesting approach in constructing a non-oscillatory spectral approximation to a discontinuous function was proposed in [83]. Let $u(x)$ be a piecewise C^∞ function, with a jump discontinuity at the point x_s and with a jump $[u]_{x_s}$. The key idea in [83] was to augment the Fourier (or Chebyshev) spectral space with step and saw-tooth functions. It was shown that the approximation using the augmented spectral space will be non-oscillatory if the saw-tooth function approximates the magnitude and the location of the discontinuity with *second-order* accuracy.

The discontinuity parameters can be determined with specified accuracy based on the spectral expansion coefficients; for example, by considering the upper one-third of the spectrum [198]. In subsequent work, it was pointed out that a *first-order* accurate approximation of the discontinuity magnitude also leads to non-oscillatory behaviour. In [438] this approach was extended to spectral element methods.

Let us denote, for simplicity, the entire array of cell averages defined at the Gauss–Chebyshev points (see Fig. 10.1) (regardless of which element they belong to) by

$$\bar{u}_i\,, \quad 1 \leqslant i \leqslant KP\,,$$

where $K = N_{\rm el}$ is the number of spectral elements covering the domain and $P+1$ is the number of Gauss–Lobatto–Chebyshev points in each element. We also denote by $u_{i^\pm}$ the entire array of point values defined at the Gauss–Lobatto–Chebyshev points (as shown in Fig. 10.1), that is,

$$u_{i^\pm}\,, \quad 0 \leqslant i^\pm \leqslant KP\,.$$

Reconstruction of point values

We assume that the cell-averaged values of a discontinuous function $\bar{u}_i$ are given. Following [438], we present a non-oscillatory reconstruction algorithm based on a simple procedure for estimating the discontinuity parameters with specified accuracy and incorporating this information in the numerical process. The main steps of the algorithm are as follows.

Algorithm R

- *Step 1*
  Find a cell i_s such that

$$|\bar{u}_{i_s+1} - \bar{u}_{i_s-1}| = \max_{2 \leqslant i \leqslant KP-1} |\bar{u}_{i+1} - \bar{u}_{i-1}|\,.$$

- *Step 2*
  Determine the discontinuous component

$$\bar{u}_i^d = \begin{cases} \bar{u}_{i_{s-1}}\,, & 1 \leqslant i \leqslant i_s - 1\,, \\ \bar{u}_{i_s}\,, & i = i_s\,, \\ \bar{u}_{i_{s+1}}\,, & i_s + 1 \leqslant i \leqslant KP\,. \end{cases}$$

- *Step 3*
  Determine the continuous component

$$\bar{u}_i^c = \bar{u}_i - \bar{u}_i^d\,,$$

for

$$1 \leqslant i \leqslant KP\,.$$

- *Step 4*
  Find the indices i_s^- and i_s^+ such that i_s denotes the cell corresponding to the

interval $[x_{i_s^-}, x_{i_s^+}]$, and thus define the reconstructed values of discontinuous part of the solution:

$$u^d_{i^\pm} = \begin{cases} \bar{u}_{i_s-1}\,, & 0 \leqslant i^\pm \leqslant i_s^-\,, \\ u_{i^-_{s+1}}\,, & i_s^+ \leqslant i^\pm \leqslant KP\,. \end{cases}$$

- *Step 5*
  Obtain point values $u^c_{i^\pm}$ from $\bar{u}^c_i$ using the procedure presented in the previous section (eqn (10.1.13a)).
- *Step 6*
  Obtain
  $$u_{i^\pm} = u^c_{i^\pm} + u^d_{i^\pm}\,,$$
  for
  $$0 \leqslant i^\pm \leqslant KP\,.$$

It is obvious that the pair of cell averages $\bar{u}_{i_s-1}$ and $\bar{u}_{i_s+1}$ represents the discontinuity magnitude with first-order accuracy. It has been shown in [219], however, that three cell-averaged values $\bar{u}_{i_s-1}$, $\bar{u}_{i_s}$, and $\bar{u}_{i_s+1}$ contain information about the discontinuity location up to second-order accuracy. However, this algorithm does not require explicit information about the discontinuity location. This algorithm is based entirely on the cell-averaged values in the physical space and not on the coefficients of a spectral expansion. Therefore, it fits naturally into the context of the nodal version (that is, collocation version) of the spectral element method.

Cell averages

As in the smooth case, we assume that the point values of a function are known and that the function contains a single jump discontinuity at the point x_s. Again, we decompose a given discrete function into two parts: a discontinuous and a smooth part. The smooth part can be averaged using the procedure described in the previous section. The cell averages corresponding to the discontinuous part can be computed directly using the following algorithm.

Algorithm A

- *Step 1*
  Find point values indices i_s^+ and i_s^- such that
  $$x_{i_s^-} \leqslant x_s \leqslant x_{i_s^+}\,.$$
- *Step 2*
  Evaluate the discontinuous component
  $$u^d_{i^\pm} = \begin{cases} u_{i_s^-}\,, & 0 \leqslant i^\pm \leqslant i_s^-\,, \\ u_{i_s^+}\,, & i_s^+ \leqslant i^\pm \leqslant KP\,. \end{cases}$$

- *Step 3*
  Obtain the continuous component
  $$u^c_{i^\pm} = u_{i^\pm} - u^d_{i^\pm}\,,$$
  for
  $$0 \leqslant i^\pm \leqslant KP\,.$$
- *Step 4*
  For the cell corresponding to the interval $[x_{i_s^-}, x_{i_s^+}]$, defined by i_s, determine
  $$\bar{u}^d_i = \begin{cases} u_{i_s^-}\,, & 1 \leqslant i < i_s\,,\\ [u_{i_s^-}(x_s - x_{i_s^-}) + u_{i_s^+}(x_{i_s^+} - x_s)]/(x_{i_s^+} - x_{i_s^-})\,, & i = i_s\,,\\ u_{i_s^+}\,, & i_s < i \leqslant KP\,. \end{cases}$$
- *Step 5*
  Evaluate $\bar{u}^c_i$ for $1 \leqslant i \leqslant KP$, applying the averaging procedure to $u^c_{i^\pm}$ ($0 \leqslant i^\pm \leqslant KP$).
- *Step 6*
  Define
  $$\bar{u}_i = \bar{u}^c_i + \bar{u}^d_i\,,$$
  for
  $$1 \leqslant i \leqslant KP\,.$$

Next, we present an example from [438] of a non-oscillatory averaging and reconstruction, and compute the pointwise error as a function of the spectral order P. Given are the point values of the following function

$$u(x) = \frac{4}{1 + \cos^2 y} - 3\,, \quad \text{where } y = \begin{cases} \dfrac{\pi}{40}(x-7)\,, & 0 \leqslant x \leqslant 5\,,\\ \dfrac{\pi}{40}(x-9)\,, & 5 < x \leqslant 10\,. \end{cases} \qquad (10.1.15)$$

This function has a jump discontinuity at $x_s = 5$. First, we evaluate cell averages of the given function at discrete points using *Algorithm A*, and then we reconstruct the point values from the cell averages using *Algorithm R*. In Fig. 10.2 we plot on the logarithmic scale the errors corresponding to three different discretisations. In each case the discretisation consists of five spectral elements ($N_{\text{el}} = K = 5$), with each element containing $P = 20$, 40, and 80 points. The Vandeven filter of orders $p = 5$, 10, and 20 (see Section 6.5.1) was applied for each case, respectively, to the Chebyshev spectrum of the smooth component on each element, both in averaging and the reconstruction procedure. This filtering was essential in order to obtain the exponential accuracy shown in Fig. 10.2 away from the discontinuity.

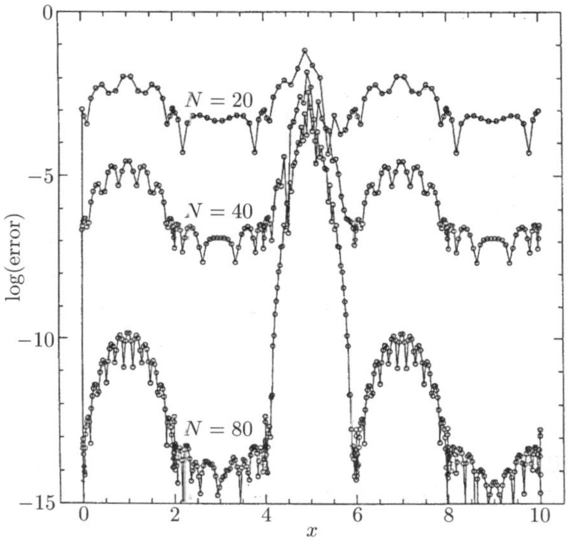

FIG. 10.2. Reconstruction of a discontinuous function (described by eqn (10.1.15)). Five elements were used in the discretisation ($N = P + 1$). ([438].)

10.2 Monotonicity

In Section 6.5 we discussed the Gibbs phenomenon associated with high-order schemes in the presence of discontinuities. This causes loss of monotonicity in the solution of hyperbolic conservation laws and thus appropriate remedies are needed. The straightforward incorporation of artificial dissipation to control oscillation is not typically appropriate for high-order methods. However, high-frequency oscillations of the numerical solution can be controlled by filters or flux-limiting procedures similar to second-order finite difference/volume schemes. These procedures, as well as other stabilisation techniques based on spectral vanishing viscosity (SVV), super-collocation, and consistent integration, have been described in Section 6.5. Here we focus on *nonlinear limiters*, which are typically employed in systems of hyperbolic conservation laws. Among the various monotone methods used in low-order discretisation methods, we describe here a linear hybridisation method and a total variational bounded (TVB) method.

10.2.1 *Flux-corrected transport (FCT)*

One approach to controlling oscillations in high-order methods is to use a nonlinear, monotone, positivity-preserving method that will eliminate the Gibbs phenomenon and produce physically acceptable solutions. This idea was first proposed by Boris and Book [66, 67], who used the advection equation (10.1.1) as a model problem.

The FCT algorithm of Boris and Book consists mainly of two stages. A transport–diffusive stage and an anti-diffusive or corrective stage. An equivalent but more descriptive interpretation of the FCT algorithm given later by Zalesak [506] suggested that the fluxes to be included in eqn (6.5.1) can be considered as nonlinear weighted averages of fluxes that can be computed by two distinct discretisation schemes corresponding to different properties: one of first-order that preserves monotonicity (according to Godunov's theorem [194]), and the other of high-order that corrects the solution and *dictates* the accuracy. This latter observation is what motivated Zalesak in [505] to incorporate a sixteenth-order finite difference formula and a pseudo-spectral method as high-order discretisation schemes in a finite volume framework. McDonald [328] has also studied the pseudo-spectral method as part of an FCT algorithm in solving scalar hyperbolic equations. In particular, he demonstrated through numerical experimentation superiority in accuracy of phase and group velocities, compared to finite difference schemes of all orders. Here, we review the main ideas of the method in the context of spectral/*hp* methods, as first presented in [186].

We consider the linearised advection equation (6.1.1a), and we adopt the approach of Zalesak [506] in formulating an FCT algorithm, where a low-order solution denoted by $\bar{u}_i^{\mathrm{td}}$ is computed first, followed by the application of a flux-limiting procedure which leads to the final field $\bar{u}_i^{n+1}(x,t)$; here n refers to the level of the time-step up to which the solution is known. The staggered grid described in the previous section is used to ensure conservation.

In particular, the main steps of the proposed spectral element–FCT method are as follows.

The initial condition $u(x,0)$ is cell-averaged to obtain the field $\bar{u}_i(0)$ on the Gauss–Chebyshev mesh. At time-step $n\Delta t$ perform the following operations.

- Compute the transportive flux $f_{i\pm}$ corresponding to a low-order scheme at each Gauss–Lobatto–Chebyshev nodal point.
- Advance (explicitly) the low-order transported and diffusive solution to obtain $\bar{u}_i^{\mathrm{td}}$.
- Compute the transportive flux F_i corresponding to the spectral element discretisation scheme; again, the Gauss–Lobatto–Chebyshev points are employed in the discretisation.
- Compute and limit the anti-diffusive flux $A_{i\pm}^c$:

$$A_{i\pm}^c = C_{i\pm}(F_{i\pm} - f_{i\pm}), \quad 0 \leqslant C_{i\pm} \leqslant 1. \tag{10.2.1}$$

- Through reconstruction obtain the de-averaged values $u_{i\pm}$.
- Update (explicitly) the final solution based on the limited anti-diffusive fluxes, that is,

$$\bar{u}_i^{n+1} = \bar{u}_i^{\mathrm{td}} - \frac{\Delta t}{\Delta x_i} \sum_{q=0}^{2} (A_{i+}^c - A_{i-}^c)^{n-q}. \tag{10.2.2}$$

The simplest low-order, positivity-preserving scheme is upwind differencing, which is used in almost all previous FCT methods. In our notation for the linear problem, we obtain

$$f_{i^+} = V_i \bar{u}_i^n\,, \quad V_i \geqslant 0\,, \tag{10.2.3a}$$

$$f_{i^-} = V_i \bar{u}_i^n\,, \quad V_i < 0\,. \tag{10.2.3b}$$

For the nonlinear problem, at elemental interfaces the appropriate sign is determined using the Roe speed [400], which is an average state between boundary values of adjacent elements. The advancement of the low-order solution therefore proceeds as follows:

$$\bar{u}_i^{\mathrm{td}} = \bar{u}_i^n - \Delta t \sum_{q=0}^{2} \left[\frac{f_{i^+}^n - f_{i^-}^n}{\Delta x_i} \right]^{n-q}\,. \tag{10.2.4}$$

The transportive–diffusive field $\bar{u}_i^{\mathrm{td}}$ is computed at the Gauss–Chebyshev points, i, which are the midpoints (in the transformed θ space) of the cells.

To be consistent with the integration scheme in eqns (10.2.2) and (10.2.4), here we use a third-order Adams–Bashforth time-stepping scheme, that is,

$$\bar{u}_i^{n+1} = \bar{u}_i^n - \frac{\Delta t}{\Delta x_i} \sum_{q=0}^{2} \beta_q [F_{i^+} - F_{i^-}]^{n-q}\,, \tag{10.2.5}$$

where β_q are appropriate weight coefficients (see Section 5.3). The high-order flux $F_{i^\pm} = V_{i^\pm} u_{i^\pm}$ is computed at the Gauss–Lobatto–Chebyshev points. A reconstruction operation is then involved to recover point values from the cell-averaged field $\bar{u}_i$.

A key component of a successful FCT algorithm is the flux limiter employed as it conveys an appropriate amount of dissipation from the low-order scheme to the high-order scheme, so that monotonicity is preserved and undesirable overshoots are avoided. Experimentation with the proposed limiters in [66, 328, 506] reveals some noticeable differences. Here, we present the original limiter of Boris and Book with the small modification introduced in [328].

- Determine the appropriate sign:

$$s_1 = \mathrm{sign}(A_{i^+})\,, \quad s_2 = \mathrm{sign}(\bar{u}_{i+1}^{\mathrm{td}} - \bar{u}_i^{\mathrm{td}})\,, \quad s_3 = \frac{1}{2}(s_1 + s_2)\,.$$

- Compute magnitudes:

$$a_1 = s_1(\bar{u}_{j+2}^{\mathrm{td}} - \bar{u}_{j+1}^{\mathrm{td}})\Delta x_j\,, \quad a_2 = s_1(\bar{u}_i^{\mathrm{td}} - \bar{u}_{i-1}^{\mathrm{td}})\Delta x\,.$$

- Compute limited flux:

$$a_3 = s_1 \max(0.\min(s_3 A_{i^+} \Delta t, a_1, a_2))\,, \quad A_{i^+}^c = \frac{a_3}{\Delta t}\,.$$

The key idea in the FCT method is to eliminate as much diffusion during the anti-diffusion stage as possible, without the introduction of new extrema or the accentuation of existing extrema. For example, the anti-diffusive flux before limiting tends to decrease $\bar{u}_i^{n+1}$ and to increase $\bar{u}_{i+1}^{n+1}$. The flux limiter then adjusts that flux so that it cannot push $\bar{u}_i^{n+1}$ below $\bar{u}_{i-1}^{n+1}$, which could produce a new minimum, or push $\bar{u}_{i+1}^{n+1}$ above $\bar{u}_{i+2}^{n+1}$, which could lead to a new maximum.

We present next an example of the application of an FCT limiter for the advection equation of a combined square–semicircle waveform. This problem has also been used as a benchmark result in comparing different monotone methods in [328, 505] and serves as a test for the flux limiter. In particular, here the effect of a non-uniform point distribution becomes especially important since no attempt was made to correct the limiter for such discretisations. Typically, a strong limiter creates a large number of terraces for smoothly-varying functions. Initially, the leftmost point of the combined waveform is located at $x = 0.5$; the solution is obtained in the periodic domain $x \in [0, 10]$. In Fig. 10.3 we plot the solution after 12 000 time-steps for (a) a Fourier scheme, and (b,c) a Chebyshev scheme. For (b) $N_{\text{el}} = 1$, and for (c) there are $N_{\text{el}} = 2$ spectral elements. These results should be compared with the solution obtained using a second-order finite difference–FCT scheme shown in (d).

The new feature in this solution is the pronounced terracing effect on the semicircle of the finite difference solution; this effect is less obvious in the spectral solution. The problem of terracing in smooth parts of the solution was first analysed by Grandjouan [202] in the context of the FCT limiting procedure, who attributed this phenomenon to dispersion errors. He subsequently corrected it by modifying the weights in the FCT scheme for uniform meshes. Further work was presented in [186,187], where dispersion errors were linked to the time integrator. A high-order integrator was employed which, in smooth regions, proved effective in eliminating the terracing effect.

10.2.2 *Local projection limiting*

We first present a nonlinear limiter developed in [100, 103, 104] and used in conjunction with the discontinuous Galerkin method already presented in Section 6.2.2. Let us consider the one-dimensional advection equation (6.1.1a) and assume that we expand the numerical solution in terms of Legendre polynomials in the standard element e:

$$u^e(x) = \sum_{i=0}^{P} \hat{u}_i L_i(x) , \quad x \in [-1, 1] . \tag{10.2.6a}$$

The basic idea is to use a flux limiter in order to preserve monotonicity of the solution averages $\hat{u}_0^e$, without changing the high-order coefficients $\hat{u}_i$, for $i \geqslant 2$. To this end, we need to adjust the interface values accordingly, so we limit the quantity $\delta u^e(z)$, where

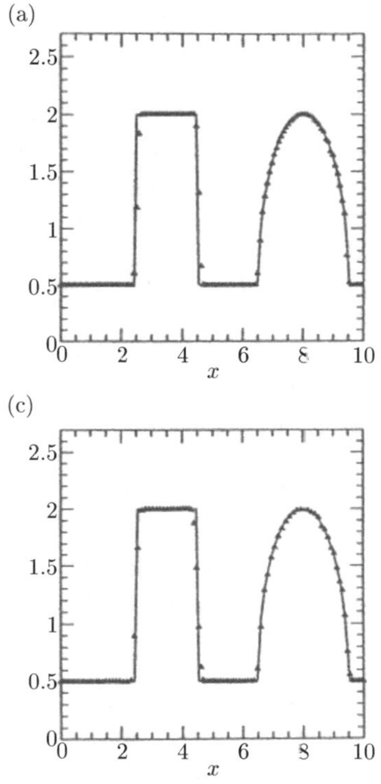

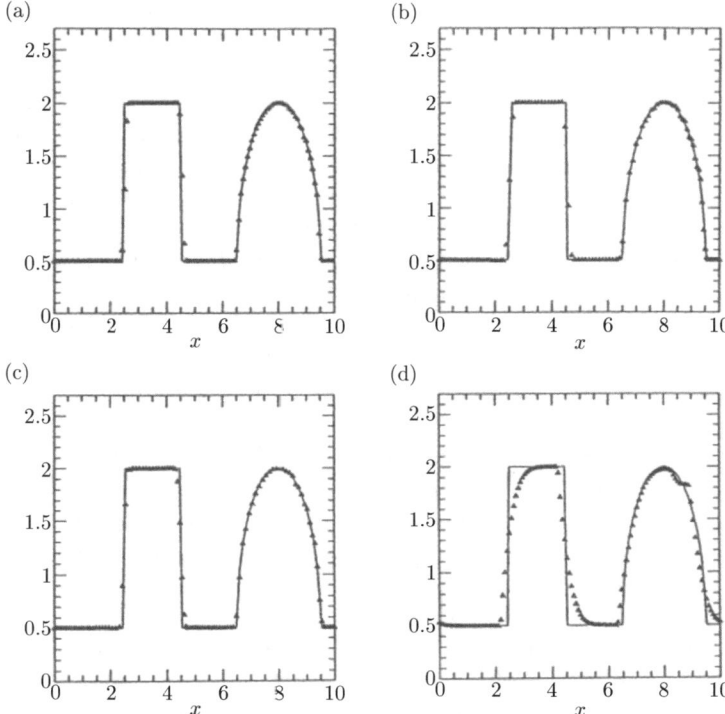

FIG. 10.3. Linear advection of a combined square–semicircle wave. The solution is obtained using various FCT schemes: (a) Fourier, (b) Chebyshev ($N_{el} = 1$), and (c) Chebyshev ($N_{el} = 2$). The total number of nodes is 128 for all simulations, while the solution is plotted after 12 000 time-steps ($\Delta t = 10^{-3}$). In (d) the corresponding solution using a second-order finite difference scheme with FCT is shown. ([186].)

$$\delta u^e(z) = \sum_{i=1}^{P} \hat{u}_i L_i(z), \quad z = -1, 1. \tag{10.2.6b}$$

The limiter is then constructed based on the average values of the numerical solution in the adjacent elements e and $e+1$, which requires a new value for $\delta u^e(z)$, that is,

$$\delta u^e(1) \to \text{minmod}\{\delta u^e(1),\ \hat{u}_0^e - \hat{u}_0^{e-1},\ \hat{u}_0^{e+1} - \hat{u}_0^e\}, \tag{10.2.6c}$$

$$-\delta u^e(-1) \to \text{minmod}\{-\delta u^e(-1),\ \hat{u}_0^e - \hat{u}_0^{e-1},\ \hat{u}_0^{e+1} - \hat{u}_0^e\}, \tag{10.2.6d}$$

where

$$\text{minmod}\{a, b, c\} = \begin{cases} \text{sgn}(a)\min(|a|, |b|, |c|), & \text{if } \text{sgn}(a) = \text{sgn}(b) = \text{sgn}(c), \\ 0, & \text{otherwise}. \end{cases} \tag{10.2.6e}$$

Having determined the new value for $\delta u^e(z)$, we can then compute uniquely the rest of the coefficients $\hat{u}_i^e$, for $i \leqslant 2$, from eqn (10.2.6b). However, high-order coefficients are undetermined and one possibility suggested by Cockburn and Shu [103] is to set them to zero. This, of course, implies that the approximation order will be lowered locally, and such an approach should therefore only be followed by activating the limiter only in regions of discontinuities. In one dimension, this limiter leads to a total variation bounded (TVB) algorithm.

In two dimensions, the same procedure can be followed but weaker conditions for a maximum norm boundedness hold. Here, the numerical flux along the boundary of an element 'e' is computed by a quadrature of a consistent order of accuracy. At each quadrature point, since there are two values of the numerical solution on u^e, one within the element $u^e_{\gamma^i}$ and the other outside $u^e_{\gamma^o}$, the numerical flux must be defined. Again, we can limit the values $u^e_{\gamma^i}$ (and $u^e_{\gamma^o}$), using the approximate gradient determined by the mean of u^e in this element and in adjacent elements. In practice, two ways of limiting can be used. The first is quick and easy, namely to require that $u_{\gamma^i} - \bar{u}^0$ be of the same sign and no bigger than $\bar{u}^1 - \bar{u}^0$ (see Fig. 10.4). If this is not true, then we limit the value of u_{γ^i} to make it true.

Even though this is an easy fix and has the advantage that it can be made to satisfy the maximum principle, it has the disadvantage that accuracy may be compromised, even in smooth, monotone regions of the solution. The main reason is that, in general, $u_{\gamma^i} - \bar{u}^0$ and $\bar{u}^1 - \bar{u}^0$ are directional derivatives in different directions. Another more sophisticated limiting is to use the approximate gradient determined by $\bar{u}^1 - \bar{u}^0$ and $\bar{u}^2 - \bar{u}^0$ to decide about an approximate directional derivative in the same direction as $u_{\gamma_i} - \bar{u}^0$, and then limit the latter if necessary. This has the advantage that accuracy can be kept in smooth, monotone regions (and, with care, also near extrema), but in general it is not maximum principle satisfying. What has been achieved in [100] is a limiter which has the advantage of both: maximum principle satisfying and high-accuracy preserving. It uses the

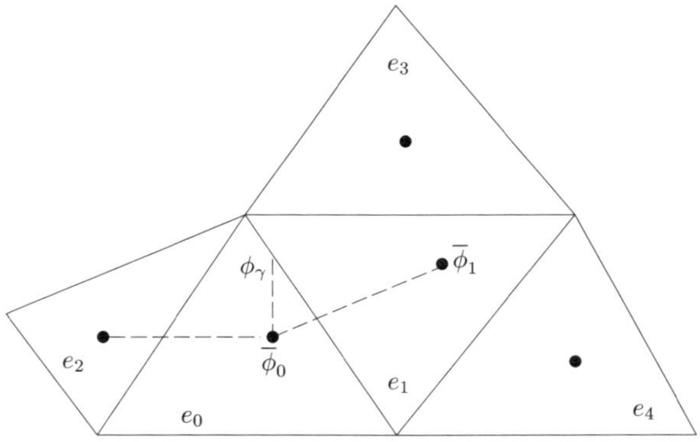

FIG. 10.4. Sketch for limiter construction on triangles.

information of the means in a wider stencil (also e_3 and e_4 in Fig. 10.4). Practical aspects of limiters are further discussed in [100, 104].

An alternative implementation, which is perhaps more appropriate for high-order elements, is to limit the *solution moments*, as suggested in [51]. The solution moments for Legendre expansions can be obtained by using eqn (10.2.6a) as

$$\int_{-1}^{1} u^e(x) L_i(x)\,\mathrm{d}x = \frac{2}{2i+1}\hat{u}_i^e\,. \tag{10.2.7a}$$

In order to keep the ith moment monotone, we must keep the coefficients $\hat{u}_i^e$ monotone on adjacent elements. Therefore, we need to update $\hat{u}_{i+1}^e$ as

$$(2i+1)\hat{u}_{i+1}^e = \operatorname{minmod}\{(2i+1)\hat{u}_{i+1}^e\,,\ \hat{u}_i^{e+1} - \hat{u}_i^e\,,\ \hat{u}_i^e - \hat{u}_i^{e-1}\}\,. \tag{10.2.7b}$$

First, the high-order coefficient is obtained and, subsequently, the limiter is applied to lower-order coefficients when the next higher coefficient on the interval has been changed by the limiting. The limiter is used adaptively where it is needed and it does not affect smooth regions. The same limiter can be used in two dimensions by splitting directions and with alternate directions starting again from the highest order. However, there is no formulation for limiting the cross-terms, but numerical experiments suggest that they may be of secondary importance.

An example of the performance of the average limiter versus the moment limiter is shown in Fig. 10.5, taken from [51]. The problem computed is Burger's equation in a periodic domain with initial condition $u(x,0) = \frac{1}{2}(1+\sin(\pi x))$. The discretisation corresponds to $N_{\mathrm{el}} = 32$ elements and an upwind numerical flux of the Lax–Friedrichs type is used (see Section 10.3). The two limiters perform identically for $P = 0, 1$, but for the quadratic elements ($P = 2$) we see a better performance with the moment limiter. As the number of elements increases, the expected convergence rate in the L^1 norm is recovered, that is, third order.

10.3 Euler equations

10.3.1 *One-dimensional equations*

The system of Euler equations for a polytropic ideal gas in one dimension is given by

$$\mathbf{u}_t + \mathbf{f}(\mathbf{u})_x = 0\,, \tag{10.3.1a}$$

with

$$\mathbf{u} = \begin{pmatrix} \rho \\ m \\ E \end{pmatrix}, \quad \mathbf{f} = \begin{pmatrix} \rho u \\ um + p \\ u(p+E) \end{pmatrix}, \tag{10.3.1b}$$

$$p = (\gamma - 1)\left(E - \frac{1}{2}\rho u^2\right), \tag{10.3.1c}$$

where ρ denotes the density, u is the velocity, p is the pressure, E is the total energy, $m = \rho u$ is the momentum, and γ is the ratio of the specific heats of a polytropic gas.

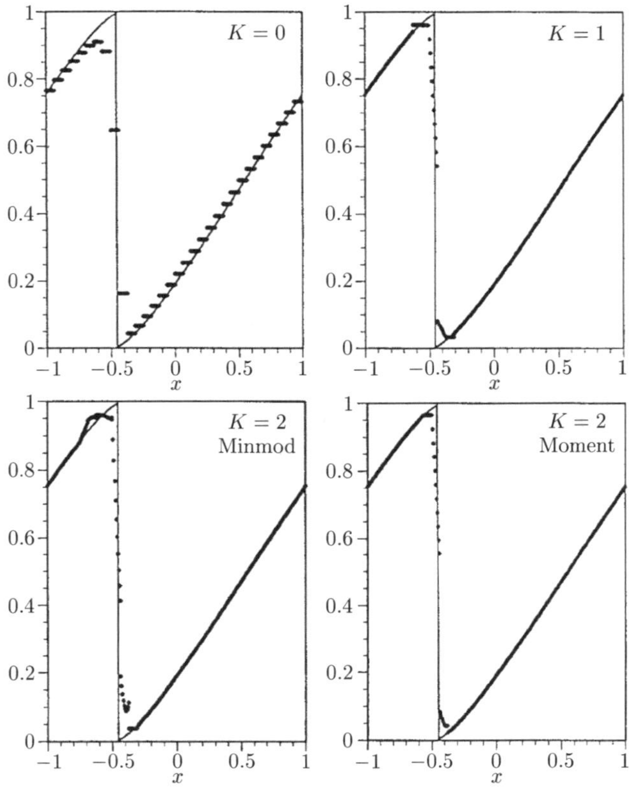

FIG. 10.5. Performance of minmod and moment limiters for the one-dimensional Burgers periodic solution. The solid line corresponds to the exact solution. (Courtesy of R. Biswas, K. Devine, and J. Flaherty.)

Interfacial and boundary conditions

A correct treatment of the interface and boundary conditions is very important for multi-domain spectral methods. The work of Gottlieb and his collaborators [136, 197] has demonstrated the sensitivity of global spectral methods to the boundary treatment. It has been conclusively established that, for stable spectral calculations, boundary conditions should be imposed on the *characteristic variables*. Multi-domain spectral methods, although not as sensitive as global spectral methods, should also base the interface as well as boundary conditions on characteristic treatment.

Consider the Jacobian matrix of the system given by $\mathbf{A}(\mathbf{u}) = \partial\mathbf{f}/\partial\mathbf{u}$. The *right* eigenvectors of $\mathbf{A}$ are

$$\mathbf{r}_1(\mathbf{u}) = \begin{pmatrix} 1 \\ u - c \\ H - uc \end{pmatrix}, \quad \mathbf{r}_2(\mathbf{u}) = \begin{pmatrix} 1 \\ u \\ \frac{1}{2}u^2 \end{pmatrix}, \quad \mathbf{r}_3(\mathbf{u}) = \begin{pmatrix} 1 \\ u + c \\ H + uc \end{pmatrix}, \tag{10.3.2a}$$

where $c = \sqrt{\gamma p/\rho}$ is the speed of sound, and H represents enthalpy and is defined by

$$H = \frac{E + p}{\rho} = \frac{c^2}{\gamma - 1} + \frac{1}{2}u^2 . \tag{10.3.2b}$$

The *left* eigenvectors of $\mathbf{A}$ are

$$\begin{aligned} \mathbf{l}_1(\mathbf{u}) &= \frac{1}{2c^2}\Big((2c + u(\gamma - 1))\frac{u}{2}, -c - u(\gamma - 1), \gamma - 1\Big), \\ \mathbf{l}_2(\mathbf{u}) &= \frac{1}{c^2}\left(c^2 - (\gamma - 1)\frac{u^2}{2}, (\gamma - 1)u, -(\gamma - 1)\right), \\ \mathbf{l}_3(\mathbf{u}) &= \frac{1}{2c^2}\Big(-(2c - u(\gamma - 1))\frac{u}{2}, c - u(\gamma - 1), \gamma - 1\Big). \end{aligned} \tag{10.3.2c}$$

Let us denote the matrix of right eigenvectors of the Jacobian $\mathbf{A} = \mathbf{A}(\tilde{\mathbf{u}})$ as

$$\mathbf{R} = (\mathbf{r}_1(\tilde{\mathbf{u}}), \mathbf{r}_2(\tilde{\mathbf{u}}), \mathbf{r}_3(\tilde{\mathbf{u}})), \tag{10.3.2d}$$

and the matrix of left eigenvectors as

$$\mathbf{L} = \begin{pmatrix} \mathbf{l}_1(\tilde{\mathbf{u}}) \\ \mathbf{l}_2(\tilde{\mathbf{u}}) \\ \mathbf{l}_3(\tilde{\mathbf{u}}) \end{pmatrix}, \tag{10.3.2e}$$

where $\tilde{\mathbf{u}}$ is the Roe-averaged state between the states $\mathbf{u}_P^e$ and $\mathbf{u}_0^{e+1}$. These states represent the last grid point (at the interface) of the element e, on the left, and the first grid point (at the interface) of the element $e + 1$, on the right. The Roe-averaged state is then computed from

$$\tilde{\mathbf{u}} = \frac{\sqrt{\rho_P^e}\,\mathbf{u}_P^e + \sqrt{\rho_0^{e+1}}\,\mathbf{u}_0^{e+1}}{\sqrt{\rho_P^e} + \sqrt{\rho_0^{e+1}}} .$$

More details on the Roe-averaged state for two-dimensional systems are presented in Section 10.3.3.

We then decompose the matrix

$$\mathbf{D} \equiv \mathbf{L} \cdot \mathbf{A} \cdot \mathbf{R} = \begin{pmatrix} \lambda_1 & 0 & 0 \\ 0 & \lambda_2 & 0 \\ 0 & 0 & \lambda_3 \end{pmatrix}, \tag{10.3.2f}$$

where the eigenvalues of $\mathbf{A}$ are given by

$$\lambda_1 = u - c, \quad \lambda_2 = u, \quad \lambda_3 = u + c. \tag{10.3.2g}$$

The characteristic variables $\mathbf{v} = \mathbf{L} \cdot \mathbf{u}$ can be introduced for both states as

$$\begin{aligned} \mathbf{v}_0^{e+1} &= \mathbf{L} \cdot \mathbf{u}_0^{e+1} , \\ \mathbf{v}_P^e &= \mathbf{L} \cdot \mathbf{u}_P^e . \end{aligned} \tag{10.3.2h}$$

Then the values to be imposed at the interface can be defined by

$$(\mathbf{v}_\gamma)_i = \begin{cases} (\mathbf{v}_0^{e+1})_i , & \lambda_i < 0 , \\ (\mathbf{v}_P^e)_i , & \lambda_i \geqslant 0 , \end{cases} \qquad i = 1, 2, 3 , \tag{10.3.3a}$$

and can be transformed back to the physical variables by

$$\mathbf{u}_\gamma = \mathbf{R} \cdot \mathbf{v}_\gamma . \tag{10.3.3b}$$

Alternatively, the numerical flux at the interface γ between elements e and $e+1$ can be computed directly, using, for example, Roe's flux [400]

$$\mathbf{f}_\gamma = \frac{\mathbf{f}(\mathbf{u}_0^{e+1}) + \mathbf{f}(\mathbf{u}_P^e)}{2} - \mathbf{R} \cdot |\mathbf{D}| \cdot \mathbf{L} \frac{\mathbf{u}_0^{e+1} - \mathbf{u}_P^e}{2} . \tag{10.3.4a}$$

Other possibilities for the numerical flux include the Osher flux [357] or the simpler Lax–Friedrichs flux

$$\mathbf{f}_\gamma = \frac{\mathbf{f}(\mathbf{u}_0^{e+1}) + \mathbf{f}(\mathbf{u}_P^e)}{2} - \alpha(\mathbf{u}_0^{e+1} - \mathbf{u}_P^e) , \tag{10.3.4b}$$

where α is a constant which can be different for different characteristic fields and is related to the largest eigenvalue of the Jacobian matrix $\mathbf{A}(\tilde{\mathbf{u}})$.

A stable penalty method

A novel way of imposing boundary and interface conditions in multi-domain spectral methods is through the penalty method [172,233,235,236]. The key idea is to collocate the differential equation, not only at the interior points but also at the boundary points. This idea exploits the fact that spectral derivatives can be obtained at all points in the domain, including the boundary points. Therefore, the appropriate boundary conditions can be introduced into the equation as penalty terms.

Let us consider first, for simplicity, the one-dimensional linear scalar advection equation (6.1.1a), i.e.,

$$\frac{\partial u}{\partial t} + V \frac{\partial u}{\partial x} = 0$$

in the interval $x \in [-1, 1]$, assuming that $V > 0$ and that the boundary condition at the upstream boundary is $u(-1, t) = g(t)$. The advection equation is then solved at a set of collocation points, in this case the Gauss–Lobatto–Legendre points, in the form

$$u_t + Vu_x + \tau_L Q_L(x)[u(-1,t) - g(t)] = 0\,, \tag{10.3.5}$$

where τ_L is the penalty parameter and

$$Q_L(x) = \frac{(1-x)L'_P(x)}{2L'_P(-1)}\,.$$

Here the function $Q_L(x)$ is chosen as a delta-function at the boundary, but other choices are also suitable. The choice of the penalty parameter is important as it determines the stability and the eigenspectrum of the advection operator, which, in turn, defines the allowable CFL number.

Funaro and Gottlieb [172] proved that for Legendre collocation schemes

$$\tau_L \geqslant \frac{P(P+1)V}{4}\,,$$

ensuring asymptotic stability of the semi-discrete approximation; for the Chebyshev collocation the stability condition is $\tau_L \geqslant P^2V/2$. Unfortunately, the penalty term adversely modifies the eigenspectrum, as shown by Hesthaven and Gottlieb [236]. In particular, it was shown that the maximum CFL number is reduced by a factor of two for Runge–Kutta integrators; for Chebyshev collocation there is a reduction factor of almost three. These estimates are rather conservative, and the numerical experiments of Hesthaven and Gottlieb [236] show that the penalty parameters can be reduced by almost a factor of four without affecting the stability of the approximation. This is true for advection–diffusion equations only, but for the advection equation the estimates for the penalty parameter are sharp.

For a *diagonal system* of hyperbolic equations we consider the form

$$\mathbf{u}_t^{(i)} = (-1)^i \boldsymbol{\mathcal{A}}^{(i)} \mathbf{u}_x^{(i)}\,,$$

where $i = 1, 2$, and $\boldsymbol{\mathcal{A}}^{(i)}$ are positive diagonal matrices. The coupling is given through the boundary conditions as

$$\mathbf{u}^{(1)}(-1,t) = \boldsymbol{\mathcal{L}}\mathbf{u}^{(2)}(-1,t) + \mathbf{g}_1(t)\,, \tag{10.3.6a}$$

$$\mathbf{u}^{(2)}(-1,t) = \boldsymbol{\mathcal{R}}\mathbf{u}^{(1)}(-1,t) + \mathbf{g}_2(t)\,, \tag{10.3.6b}$$

where $\boldsymbol{\mathcal{R}}$ and $\boldsymbol{\mathcal{L}}$ are matrices that represent the reflection of the ingoing characteristic variables in terms of the outgoing characteristics at the boundaries, with norms r and l, respectively, respectively so that $0 < rl < 1$ for solutions that do not grow in time. The penalty formulation for the diagonal system is similar to the scalar case, with $\tau^{(i)}$ being the corresponding penalty parameters.

It was shown in [172] that, for a stable collocation approximation, the penalty parameters should be chosen so that

$$\tau^{(i)} = \frac{\mathcal{A}^{(i)} P(P+1)}{2\sqrt{rl}} .$$

In the case of a general hyperbolic system, with the matrix $\mathcal{A}$ having both negative and positive eigenvalues, a characteristic decomposition can be applied to diagonalise the system and, subsequently, the theory applied on the diagonal system.

For multi-domain spectral collocation schemes, the interface conditions can be constructed similarly. For example, let us consider the inter-elemental boundary γ and assume that $V > 0$ again. The equations we need to solve in the left and right elements, respectively, are

$$(u_t^e)_P + V(u_x^e)_P = 0 , \qquad (10.3.7a)$$

$$(u_t^{e+1})_0 + V(u_x^{e+1})_0 + \tau_0^{e+1}[(u^{e+1})_0 - (u^e)_P] = 0 . \qquad (10.3.7b)$$

Thus, the patching enforces continuity of the function in a weak sense; however, as it is shown in [233], global spectral accuracy is maintained for the proper choice of τ_0, as before. The same ideas can be extended to the one-dimensional system by characteristic decomposition. We will return to this point in Section 10.5, where we will consider a dissipative model system to represent the compressible Navier–Stokes equations.

We note that the penalty method as outlined here bears great similarity to the discontinuous Galerkin method, see Section 6.2.2, where the penalty terms can be thought of as appropriate modifications of the numerical flux.

A spectral element–FCT algorithm

The spectral element–FCT method was formulated for scalar conservation laws in Section 10.2.1. Here, we extend it for the one-dimensional system (10.3.1a). The key idea is to use *characteristic decomposition* for inter-elemental as well as boundary conditions.

Following [187], the main steps of the proposed algorithm are as follows.

- *Step 1*
  Evaluate the field of cell averages corresponding to the initial condition.
- *Step 2*
  Compute the transportive fluxes corresponding to the low-order scheme. The low-order positive-type scheme used here is Roe's scheme based on the cell-averaged values. The low-order flux $\mathbf{f}_{i+}$ is defined as follows:

$$\mathbf{f}_{i+} = \frac{\mathbf{f}(\bar{\mathbf{u}}_{i+1}) + \mathbf{f}(\bar{\mathbf{u}}_i)}{2} - \mathbf{R} \cdot |\mathbf{D}| \cdot \mathbf{R}^{-1} \frac{(\bar{\mathbf{u}}_{i+1} - \bar{\mathbf{u}}_i)}{2} , \qquad (10.3.8)$$

  where $\mathbf{R}$ is the Jacobian matrix consisting of the right eigenvectors of the Euler system linearised around the Roe-averaged state between $\bar{\mathbf{u}}_{i+1}$ and $\bar{\mathbf{u}}_i$.

- *Step 3*
  Advance (explicitly) cell averages in time using low-order fluxes to obtain the low-order transported and diffusive solution $\bar{\mathbf{u}}_i^{\mathrm{td}}$. This is done using the third-order Adams–Bashforth scheme.
- *Step 4*
  Compute the transportive fluxes $\mathbf{F}_i$ corresponding to the spectral element discretisation.
- *Step 5*
  Compute the anti-diffusive fluxes $\mathbf{A}_i = \mathbf{F}_i - \mathbf{f}_i$ and limit them to obtain $\mathbf{A}_i^c$. It is crucial that the limiter be applied to the characteristic anti-diffusive fluxes and not the component-wise fluxes.
- *Step 6*
  Update (explicitly) the cell averages at the new time-level using the limited anti-diffusive fluxes $\bar{\mathbf{u}}_i^{n+1}$ (using the third-order Adams–Bashforth scheme).
- *Step 7*
  Reconstruct point values from the cell averages at the new time-level.
- *Step 8*
  If the target time is not achieved then go to *Step 2*.

Several numerical examples and comparisons with corresponding FCT–finite volume schemes are presented in [185], where it is demonstrated that the formal order of the scheme is important in resolving contact discontinuities and rarefaction fans for the standard Sod and Lax shock-tube benchmark problems used extensively in the literature. Here, we compare two particular simulations taken from [185] for the Sod problem, which is defined as

$$\begin{aligned} &\rho_L = 1\,, && u_L = 0\,, \quad p_L = 1\,, && -1 < x < 0\,, \\ &\rho_R = 0.125\,, && u_R = 0\,, \quad p_R = 0.1\,, && 0 < x < 1\,. \end{aligned} \tag{10.3.9}$$

The results shown in Fig. 10.6 are superior to results obtained with low-order schemes for the same number of points. However, as explained in [187], there are instances where, for highly-distorted meshes and strong shocks as in the case of the Lax problem, unphysical spikes (loss of monotonicity) may appear around elemental interfaces as the shock passes through. This effect seems to be due to the low-order scheme during the diffusive stage being used in a strongly non-uniform grid. It also appears in finite volume schemes if the non-uniform grid expands or contracts super-linearly.

10.3.2 *Two-dimensional equations*

The system of Euler equations for a polytropic ideal gas in two dimensions (where we have again used $x_1 = x$ and $x_2 = y$) is given by

$$\mathbf{u}_t + \mathbf{f}(\mathbf{u})_x + \mathbf{g}(\mathbf{u})_y = 0\,, \tag{10.3.10a}$$

with

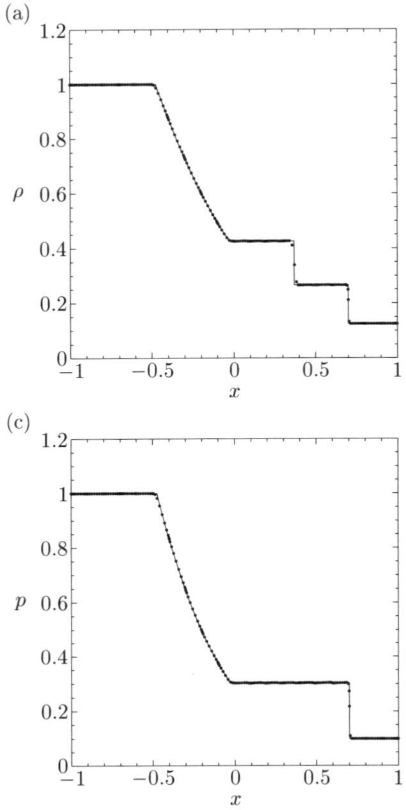

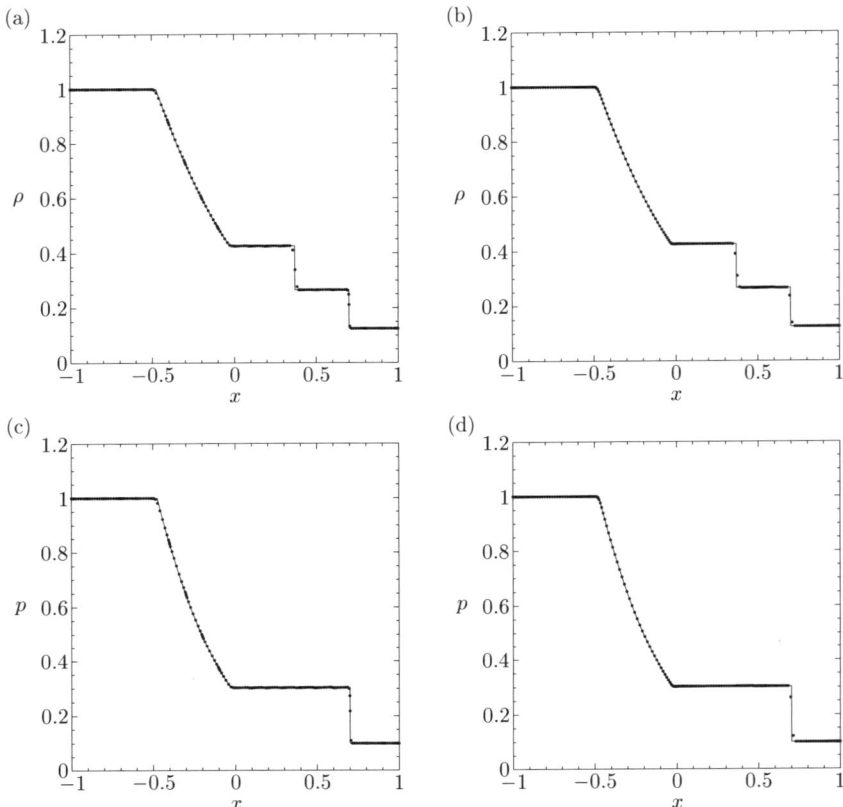

FIG. 10.6. Spectral element–FCT simulation for the 'Sod' shock-tube problem, with discretisation pairs $(N_{\mathrm{el}}, P) = (20, 11)$ (a,c), and $(N_{\mathrm{el}}, P) = (4, 51)$ (b,d). An Adams–Bashforth third-order time-stepping scheme was used with a CFL number of 0.2. Presented are (a,b) the density, and (c,d) the pressure at time $t = 0.4$. ([185].)

$$\mathbf{u} = \begin{pmatrix} \rho \\ m \\ n \\ E \end{pmatrix}, \quad \mathbf{f} = \begin{pmatrix} m \\ m^2/\rho + p \\ mn/\rho \\ (p+E)m/\rho \end{pmatrix}, \quad \mathbf{g} = \begin{pmatrix} n \\ mn/\rho \\ n^2/\rho + p \\ (p+E)n/\rho \end{pmatrix}, \tag{10.3.10b}$$

where $m = \rho u$, $n = \rho v$, and $p = (\gamma - 1)\{E - \frac{1}{2}(m^2 + n^2)/\rho\}$.

In two dimensions we introduce an approximation known as *dimensional splitting*, whereby each direction is treated independently of the other, in a sequential manner. In doing so, we can use the formalism developed in the previous sections. This means that we can still use the one-dimensional approximate Riemann solvers, as we have seen before. While there has been some research effort devoted to constructing multi-dimensional Riemann solvers ([300, 407]), numerical experiments have shown that they are less efficient and less robust, and do not clearly improve the solution quality.

In multi-domain spectral methods, in two and three dimensions we also need to consider how to treat the corners of elements. Such a treatment has been developed in [279] and in [48]. In two dimensions, for example, the problem is divided into two one-dimensional problems. The corresponding Riemann invariants are

$$R^+ = u + \frac{2}{\gamma - 1}c, \quad R^- = u - \frac{2}{\gamma - 1}c,$$
$$S^+ = v + \frac{2}{\gamma - 1}c, \quad S^- = v - \frac{2}{\gamma - 1}c.$$

We assume locally isentropic flow in the neighbourhood of the corner, and obtain the entropy value of the corner from the element that flow is leaving. We also define a domain of influence, and choose the calculated values of Riemann invariants from the corresponding elements, which lie in the domain of dependence. Figure 10.7 shows the domain of influence at the corners of four elements. Since the flow is leaving element 1, we get the entropy and Riemann invariants of R^+ and S^+ from element 1. The Riemann invariants R^- and S^- are obtained from elements 2 and 4, respectively. The flow variables are calculated as follows:

$$\begin{aligned} u &= \frac{1}{2}\left(R^+ + R^-\right), & v &= \frac{1}{2}\left(S^+ + S^-\right), \\ c &= \frac{\gamma - 1}{4}\left(R^+ - R^- + S^+ - S^-\right) - c_{\text{up}}, & s &= s_{\text{up}}, \end{aligned} \tag{10.3.11}$$

where the subscript 'up' denotes the upstream values (that is, values from the element that flow is leaving). Assuming an ideal gas and locally isentropic flow, the state of the gas is fixed by solving the local pressure and density using the

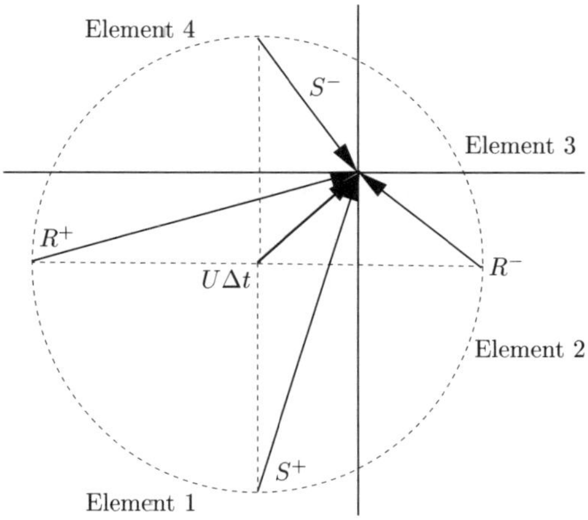

FIG. 10.7. Domain of influence at the corners of four elements, and corresponding Riemann invariants.

speed of sound and the entropy (p/ρ^γ = constant, $c^2 = \gamma p/\rho$), and thus the conservative variables and local temperature can be calculated.

An obvious drawback of dimensional splitting is that, although the original Euler equations are rotationally invariant, there is no guarantee that dimensional splitting preserves that condition. In addition, if one-dimensional limiters are employed, as in the spectral element–FCT formulation, then circular symmetry can be affected. Such an example was presented in [185], where a rotated form of the one-dimensional Sod problem presented in the previous section was computed. In particular, the following problem was solved using a Cartesian version of the Euler equations in the domain $(x, y) \in [0,1] \times [0,1]$, with initial conditions

$$\begin{aligned} \rho &= 1\,, & u &= 0\,, & p &= 1\,, & (x-0.5)^2 + (y-0.5)^2 &< 0.25^2\,, \\ \rho &= 0.125\,, & u &= 0\,, & p &= 0.1\,, & (x-0.5)^2 + (y-0.5)^2 &> 0.25^2\,. \end{aligned}$$

Three different discretisations were employed, corresponding to a single element (51×51), two elements, and four elements, as shown in Fig. 10.8. Density contours are presented in the figure at times $t = 0.5$ and $t = 1.0$. The effect of directional splitting is more evident at the later time in the case of multi-element discretisation. This effect is particularly visible in low-order schemes and has been investigated in some detail in [298]. In general, the spectral element–FCT schemes perform very well in two dimensions, despite the splitting; a number of classical benchmark results can be found in [185].

10.3.3 *Discontinuous Galerkin method*

By introducing the flux vector $\mathbf{F} = (\mathbf{f}, \mathbf{g})$, we can rewrite the two-dimensional Euler equations in compact form:

$$\mathbf{u}_t + \nabla \cdot \mathbf{F}(\mathbf{u}) = 0\,,$$

or in quasi-linear form:

$$\mathbf{u}_t + \mathbf{A}_x \frac{\partial \mathbf{u}}{\partial x} + \mathbf{A}_y \frac{\partial \mathbf{u}}{\partial y} = 0\,,$$

where $\mathbf{A}_x$ and $\mathbf{A}_y$ are the one-dimensional Jacobian matrices associated with the x and y directions, respectively, defined by

$$\mathbf{A}_x \equiv \frac{\partial \mathbf{f}}{\partial \mathbf{u}} \quad \text{and} \quad \mathbf{A}_y \equiv \frac{\partial \mathbf{g}}{\partial \mathbf{u}}\,.$$

The discontinuous Galerkin formulation is then obtained in a similar manner to the scalar advection equation, see Section 6.2.2, for each element e, i.e.,

$$(\mathbf{u}, \mathbf{v})_e + \int_{\partial\Omega^e} \mathbf{v}[\tilde{\mathbf{f}}(\mathbf{u}_i, \mathbf{u}_o - \mathbf{F}(\mathbf{u}_i)] \cdot \boldsymbol{n}\, \mathrm{d}s + (\nabla \cdot \boldsymbol{f}(\mathbf{u}), \mathbf{v})_e = 0\,, \qquad (10.3.12)$$

where i refers to the region interior to the element e and o to the region outside element e. The solution $\mathbf{u} \in (\mathcal{X}^\delta)^n$ satisfies this equation for all $\mathbf{v} \in (\mathcal{V}^\delta)^n$ (the

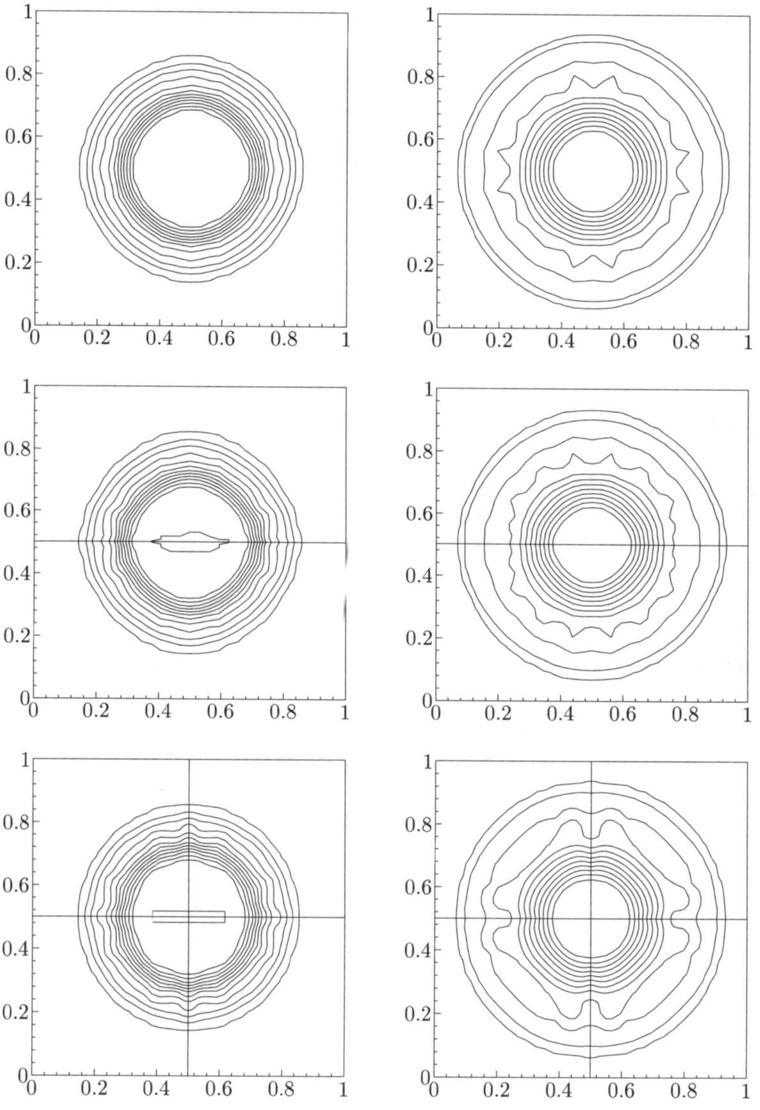

FIG. 10.8. The circular blast problem. A cylinder of high density and pressure is initially surrounded by the same fluid at a lower thermodynamic state. Density contours are shown at two different time instances for (top row) 1 element, (middle row) 2 elements, and (lower row) 4 elements. ([185].)

approximation space is the same as the test space), where n is the size of the vector $\mathbf{u}$, which is 4 in two dimensions and 5 in three dimensions.

The numerical surface flux $\tilde{\mathbf{f}}(\mathbf{u}_i, \mathbf{u}_o)$ is computed by an approximate Riemann solver, similar to the one-dimensional case, using dimensional splitting. This, in essence, will produce two intermediate states $\mathbf{u}_1$ and $\mathbf{u}_2$ such that

$$\mathbf{f}(\mathbf{u}_1) - \mathbf{f}(\mathbf{u}_2) = \mathbf{A}_x(\mathbf{u}_1 - \mathbf{u}_2)\,, \tag{10.3.13}$$

where the Jacobian matrix $\mathbf{A}_x$ for the conservative variables is

$$\mathbf{A}_x = \begin{pmatrix} 0 & 1 & 0 & 0 \\ \frac{1}{2}\left[(\gamma-3)u^2 + (\gamma-1)v^2\right] & (3-\gamma)u & -(\gamma-1)v & \gamma-1 \\ -uv & v & u & 0 \\ u[(\gamma-1)\mathbf{u}\cdot\mathbf{u} - \gamma e] & \gamma e - \frac{1}{2}(\gamma-1)(v^2+3u^2) & -(\gamma-1)uv & \gamma u \end{pmatrix}, \tag{10.3.14}$$

with $e = E/\rho$. The Jacobian matrix $\mathbf{A}_y$ is obtained by cyclic permutation of u and v. Each of the Jacobian matrices can then be diagonalised as in the one-dimensional case, e.g.,

$$\mathbf{A}_x(\bar{\mathbf{u}}) = \mathbf{RDL}\,,$$

where the matrices on the right-hand side are given in Appendix E. In order to apply this directional splitting, we attach at the interface of two adjacent elements a local coordinate system, as shown in Fig. 10.9, and align the edge with the y direction. Here $\bar{\mathbf{u}}$ is some average state around the point O shown in Fig. 10.9.

The question now becomes: what are the appropriate values for $\bar{\mathbf{u}}$ for computing the matrix $\mathbf{A}_x$? According to [240], the unique value of $\bar{\mathbf{u}}$ that satisfies eqn (10.3.13) is the *Roe-average* state defined by the following conditions on the average values of density, velocity, and enthalpy:

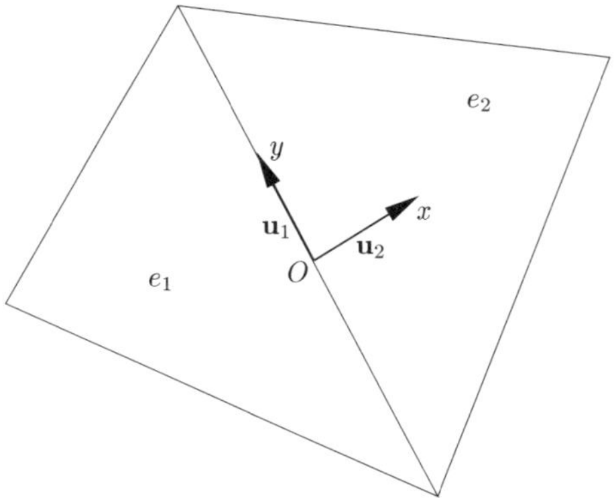

FIG. 10.9. Notation for the construction of an approximate interface flux for the discontinuous Galerkin method.

$$\bar{\rho} = \sqrt{\rho_2 \rho_1}\,,$$
$$\bar{u} = \frac{(u\sqrt{\rho})_2 + (u\sqrt{\rho})_1}{\sqrt{\rho_2} + \sqrt{\rho_1}}\,,$$
$$\bar{H} = \frac{(H\sqrt{\rho})_2 + (H\sqrt{\rho})_1}{\sqrt{\rho_2} + \sqrt{\rho_1}}\,.$$

These conditions are applied between the two adjacent elements e_1 and e_2.

Having computed the average state and having performed the characteristic decomposition per direction, we can then evaluate the numerical flux involved in the discontinuous Galerkin statement using any of the following approaches.

1. *Upwind flux*
$$\tilde{\mathbf{f}}(\mathbf{u}_1, \mathbf{u}_2) = \mathbf{F}(\mathbf{R}\mathbf{D}^+\mathbf{L}\mathbf{u}_1 + \mathbf{R}\mathbf{D}^-\mathbf{L}\mathbf{u}_2)\,,$$
where $\mathbf{D}^{\pm} = \frac{1}{2}(\text{sign}(\mathbf{D}) \pm 1)$.
2. *Roe splitting flux*
$$\tilde{\mathbf{f}}(\mathbf{u}_1, \mathbf{u}_2) = \frac{1}{2}[\mathbf{F}(\mathbf{u}_1) + \mathbf{F}(\mathbf{u}_2)] - \frac{1}{2}\mathbf{R}|\mathbf{D}|\mathbf{L}(\mathbf{u}_2 - \mathbf{u}_1)\,.$$
3. *Lax–Friedrichs flux*
$$\tilde{\mathbf{f}}(\mathbf{u}_1, \mathbf{u}_2) = \frac{\mathbf{F}(\mathbf{u}_1) + \mathbf{F}(\mathbf{u}_2) - \alpha_{\max}(\mathbf{u}_2 - \mathbf{u}_1)}{2}\,,$$
where $\alpha_{\max} = \max(|\mathbf{D}|_{i,i})$, i.e., it corresponds to a maximum absolute value of the eigenvalues.

Numerical experiments in [309] have shown that the upwind flux is more robust, especially in the presence of discontinuities.

A benchmark problem that tests the accuracy of the discontinuous Galerkin method described above for the Euler equations is inviscid flow over a semi-circular bump in a channel. For adiabatic and irrotational flow conditions the entropy should remain zero everywhere in the domain. In practice, this is difficult to achieve with low-order methods because of the numerical boundary-layer creation near the walls due to discretisation error. The convergence of the numerically-obtained entropy approaching zero should then determine the order of accuracy of the method. ♦[40]

Figure 10.10(a) shows the computational domain used in the current test, which consists of a parallel channel with a semicircular bump on the lower wall. The flow parameters at inflow are chosen to approximate a Mach 0.3 flow in air at standard temperature and pressure. The top and bottom walls are specified as reflecting boundary conditions. These were imposed by computing an adjacent element flux computed from a flow condition with zero normal velocity and pressure, identical to that of the element on the inner surface of the wall. Inflow

[40] Benchmark problem for the validation of Euler equations.

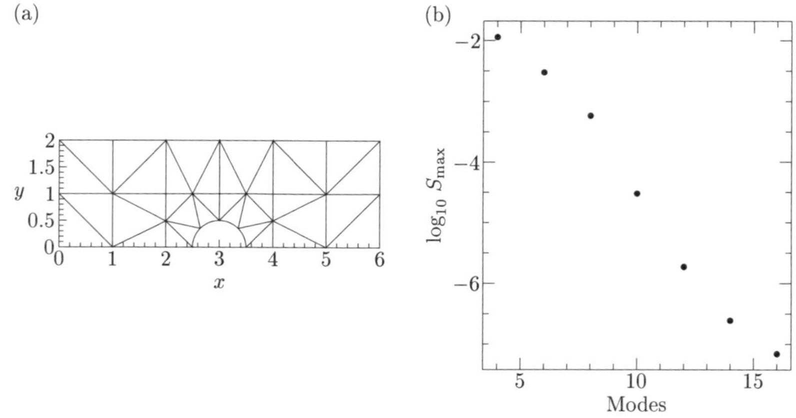

FIG. 10.10. (a) Computational domain for internal flow over a semicircular bump consisting of twenty-six triangular elements. The radius of the bump is 0.5, and the flow is from left to right.(b) Maximum entropy (logarithm) as a function of the expansion order per element. ([310].)

and outflow boundary conditions were imposed by specifying an adjacent flux computed using the reference flow conditions. The relative error in maximum entropy is plotted in Fig. 10.10(b), and it demonstrates that the numerical error approaches zero exponentially fast. We note here that no limiting or other filtering is applied and that the expansion order is the same in each element.

In *three dimensions*, a similar directional splitting is applied and the flux construction at the interface is based on a local coordinate system for each interface (quadrature) point, as in Fig. 10.9 for the two-dimensional case. The matrices involved in the characteristic decomposition are presented in Appendix E.

10.4 Shallow-water equations

In many hydrodynamic applications of interest to hydraulic, coastal, and ocean engineering, the characteristic length-scale of the problem is large compared to the vertical scale. In this case the flow can be regarded as uniform in depth and approximated by the shallow-water equations. The use of two-dimensional shallow-water systems provides an approximate simulation of the three-dimensional wave motion in computational domains which are much larger than those possible for models based on truly three-dimensional equations. Important from an implementation point of view is that the problem of a space–time-varying domain, caused by the moving free water surface, is avoided. Spectral/*hp* element methods have, however, also been applied to the fully-nonlinear free surface problems in moving domains [242, 397, 398, 484].

In oceanic and atmospheric sciences the shallow-water equations are typically used as a stepping stone towards constructing fully three-dimensional models based on primitive equations. Over the last decade these communities have had an increasing interest in spectral/hp element methods for solving the shallow-water equations, e.g., [189, 190, 252, 315, 450, 461]. The motivation behind this has been the fact that spectral/hp element methods—in addition to being able to handle complex geometries—generally give significant savings in computational time, compared to low-order methods, especially when long-time integration is necessary. More recently, applications of spectral/hp discontinuous Galerkin methods to the shallow-water equations have also been reported [147, 156, 190], which will also be discussed in Section 10.4.2.

10.4.1 *Governing equations*

The shallow-water equations are a two-dimensional system of nonlinear partial differential equations of hyperbolic type. The governing equations, expressed in terms of conservative variables, are written as

$$\mathbf{u}_t + \nabla \cdot \mathbf{F}(u) = \mathbf{s}(u)\,, \tag{10.4.1}$$

in which $\mathbf{F} = (\mathbf{f}, \mathbf{g})$ is the flux vector and

$$\mathbf{u} = \begin{pmatrix} H \\ uH \\ vH \end{pmatrix}, \quad \mathbf{f} = \begin{pmatrix} uH \\ u^2H + gH^2/2 \\ uvH \end{pmatrix}, \quad \mathbf{g} = \begin{pmatrix} vH \\ uvH \\ v^2H + gH^2/2 \end{pmatrix}, \tag{10.4.2}$$

where $H(x_1, x_2, t) = \zeta(x_1, x_2, t) + d(x_1, x_2)$ is the total water depth, $d(x_1, x_2)$ is the still-water depth, and $\zeta(x_1, x_2, t)$ is the free-surface elevation. In definition (10.4.2), $u(x_1, x_2, t)$ and $v(x_1, x_2, t)$ are the depth-averaged velocities in the x_1 and x_2 directions, respectively, and g is the acceleration due to gravity. The source term $\mathbf{s}$ usually accounts for forcing due to friction, bed slopes, Coriolis effects, etc. Note that, if the source term contains (to the lowest order) third-order mixed derivatives, then we have a Boussinesq-type equation—an interesting family of nonlinear and dispersive wave equations which we will consider in Section 10.4.4.

10.4.2 *Discontinuous Galerkin formulation*

Similar to the Euler formulation discussed in Section 10.3.3, the weak Galerkin formulation in each element e can be written

$$(\mathbf{v}, \mathbf{u}_t)_e - (\nabla \mathbf{v}, \mathbf{F}(u))_e + \langle \mathbf{F}(\mathbf{u}), \mathbf{v} \cdot \boldsymbol{n} \rangle_e = (\mathbf{v}, \mathbf{s})_e\,, \tag{10.4.3}$$

where the parentheses denote the standard inner product over Ω^e and the angle brackets denote boundary integrals along $\partial\Omega^e$. Following the discontinuous Galerkin formulation (see Section 6.2.2), coupling between the elemental regions is achieved by replacing $\mathbf{F}(\mathbf{u})$ with an *upwinded flux* denoted by $\tilde{\mathbf{F}}(\mathbf{u}_i, \mathbf{u}_o)$, where the subscripts i and o denote inside and outside the element e, respectively. If we integrate eqn (10.4.3) by parts once more then we obtain the divergence form:

$$(\mathbf{v}, \mathbf{u}_t)_e + (\mathbf{v}, \nabla \cdot \mathbf{F}(u))_e + \langle \mathbf{v}, [\tilde{\mathbf{F}}(\mathbf{u}_i, \mathbf{u}_o) - \mathbf{F}(\mathbf{u}_i)] \cdot \boldsymbol{n} \rangle_e = (\mathbf{v}, \mathbf{s})_e \,. \tag{10.4.4}$$

From a computational convenience point of view, the 'divergence' form is attractive since it involves inner products of the same forms as classical continuous Galerkin schemes. Equation (10.4.3) or (10.4.4) is discretised by replacing the continuous test and trial spaces $\boldsymbol{v}$ and $\boldsymbol{u}$ with finite spaces $\boldsymbol{v}^\delta$ and $\boldsymbol{u}^\delta$, as discussed in Section 6.2.2. We also recall that in evaluating the numerical fluxes along the element edges it is advantageous to use the Gauss, rather than the Gauss–Lobatto, quadrature points in order to avoid having to evaluate a two-dimensional flux at the vertices.

Evaluation of the upwind normal flux, $\tilde{\mathbf{F}} \cdot \boldsymbol{n}$

The evaluation of the upwind flux $\tilde{\mathbf{F}}$ can be reduced to a one-dimensional Riemann problem in the normal direction to a given edge. We therefore start by introducing the rotation matrix and its inverse:

$$\mathbf{T} = \begin{pmatrix} 1 & 0 & 0 \\ 0 & n_{x1} & n_{x2} \\ 0 & -n_{x2} & n_{x1} \end{pmatrix}, \quad \mathbf{T}^{-1} = \begin{pmatrix} 1 & 0 & 0 \\ 0 & n_{x1} & -n_{x2} \\ 0 & n_{x2} & n_{x1} \end{pmatrix}, \tag{10.4.5}$$

and define $\mathbf{U} = (H, H\bar{u}, H\bar{v})^\top = \mathbf{T}\mathbf{u}$, where $\bar{u}$ and $\bar{v}$ are the velocities normal and tangential to the edge, respectively.

The upwind flux normal to the edge, $\tilde{\mathbf{F}} \cdot \boldsymbol{n}$, can then be written in terms of a one-dimensional flux $\tilde{\mathbf{f}}$ as

$$\tilde{\mathbf{F}}(\mathbf{u}_i, \mathbf{u}_o) \cdot \boldsymbol{n} = \mathbf{T}^{-1} \tilde{\mathbf{f}}(\mathbf{U}_i, \mathbf{U}_o) \,. \tag{10.4.6}$$

Now, to define $\tilde{\mathbf{f}}$ using a Riemann-based approach, we need to determine the characteristic form of the system and then apply an approximate Riemann solver based on the characteristic information. We start by constructing the Jacobian of the flux function, $\mathbf{A} = \partial \mathbf{f} / \partial \mathbf{U}$, which can be written

$$\mathbf{A} = \begin{bmatrix} 0 & 1 & 0 \\ c^2 - \bar{u}^2 & 2\bar{u} & 0 \\ -\bar{u}\bar{v} & \bar{v} & \bar{u} \end{bmatrix}, \tag{10.4.7}$$

where $c = \sqrt{gH}$ is the wave speed. The eigenvalues of $\mathbf{A}$ are

$$\lambda_1 = \bar{u} - c\,, \quad \lambda_2 = \bar{u}\,, \quad \lambda_3 = \bar{u} + c\,. \tag{10.4.8}$$

Using the definition of $\mathbf{A}$ and constructing the Roe-average variables,

$$\bar{u}^* = \frac{\bar{u}_i\sqrt{H_i} + \bar{u}_o\sqrt{H_o}}{\sqrt{H_i} + \sqrt{H_o}}, \quad \bar{v}^* = \frac{\bar{v}_i\sqrt{H_i} + \bar{v}_o\sqrt{H_o}}{\sqrt{H_i} + \sqrt{H_o}},$$
$$H^* = \sqrt{H_i H_o}\,, \qquad c^* = \sqrt{0.5(c_i^2 + c_o^2)}\,,$$

we can compute the Roe matrix $\mathbf{A}^* = \mathbf{A}(\mathbf{U}^*)$ and the corresponding left, $\mathbf{L}^*$, and right, $\mathbf{R}^*$, eigenmatrices. We also define the diagonal matrix, $\mathbf{D}^*$, as containing the eigenvalues λ_i^* such that

$$\mathbf{D}^* = \mathbf{L}^* \cdot \mathbf{A}^* \cdot \mathbf{R}^* .$$

Similar to the Euler equations discussed in Section 10.3.3, we can determine $\tilde{\mathbf{F}}$ using eqn (10.4.6) and determine the one-dimensional upwind flux, $\tilde{\mathbf{f}}$, from one of the following approaches.

1. *The Lax–Friedrichs flux*

$$\tilde{\mathbf{f}}(\mathbf{U}_i, \mathbf{U}_o) = \frac{1}{2}\left(\mathbf{f}(\mathbf{U}_i) + \mathbf{f}(\mathbf{U}_o)\right) - \frac{\max_j |\lambda_j^*|}{2}\left(\mathbf{U}_o - \mathbf{U}_i\right). \tag{10.4.9}$$

2. *The Roe flux*

$$\tilde{\mathbf{f}}(\mathbf{U}_i, \mathbf{U}_o) = \frac{1}{2}\left(\mathbf{f}(\mathbf{U}_i) + \mathbf{f}(\mathbf{U}_o)\right) - \frac{1}{2}\mathbf{R}^*|\mathbf{D}^*|\mathbf{L}^*\left(\mathbf{U}_o - \mathbf{U}_i\right). \tag{10.4.10}$$

3. *The HLLC flux*
This approximate Riemann solver is a modification of the Harten, Lax, and van Leer (HLL) solver [220] which avoids excessive smearing of shear waves. The HLLC solver is based on three wave speed estimates given by [468]

$$\begin{aligned} S_i &= \bar{u}_i - c_i\, s_i\,, \\ S_o &= \bar{u}_o + c_o\, s_o\,, \\ S_m &= \frac{S_i H_o\left(\bar{u}_o - S_o\right) - S_o H_i\left(\bar{u}_i - S_i\right)}{H_o\left(\bar{u}_o - S_o\right) - H_i\left(\bar{u}_i - S_i\right)}\,, \end{aligned}$$

where

$$s_{(i,o)} = \begin{cases} \sqrt{\dfrac{H_m^2 + H_m H_{(i,o)}}{2H_{(i,c)}^2}}\,, & H_m > H_{(i,o)}\,, \\ 1\,, & H_m \leqslant H_{(i,o)}\,, \end{cases} \tag{10.4.11}$$

and in which H_m, for example, can be given by the two-rarefaction Riemann solver [468]

$$H_m = \frac{1}{g}\left(\frac{1}{2}\left(\sqrt{gH_i} + \sqrt{gH_o}\right) + \frac{1}{4}\left(\bar{u}_i - \bar{u}_o\right)\right)^2 . \tag{10.4.12}$$

After the wave speeds have been computed, the HLLC flux is given by

$$\tilde{\mathbf{f}}(\mathbf{U}_i, \mathbf{U}_o) = \begin{cases} \mathbf{f}(\mathbf{U}_i)\,, & S_i \geqslant 0\,, \\ \mathbf{f}(\mathbf{U}_i) + S_i\left(\mathbf{U}_{mi} - \mathbf{U}_i\right), & S_i \leqslant 0 \leqslant S_m\,, \\ \mathbf{f}(\mathbf{U}_o) + S_o\left(\mathbf{U}_{mo} - \mathbf{U}_o\right), & S_m \leqslant 0 \leqslant S_o\,, \\ \mathbf{f}(\mathbf{U}_o)\,, & S_{\mathrm{R}} \leqslant 0\,, \end{cases} \tag{10.4.13}$$

where $\mathbf{U}_{mi}$ and $\mathbf{U}_{io}$ are obtained from

$$\mathbf{U}_{m(i,o)} = H_{(i,o)} \left(\frac{S_{(i,o)} - \bar{u}_{(i,o)}}{S_{(i,o)} - S_m} \right) \begin{bmatrix} 1 \\ S_m \\ \bar{v}_{(i,o)} \end{bmatrix} . \tag{10.4.14}$$

Boundary conditions

Finally, we also need to assign boundary conditions to all element edges aligned to a domain boundary. These boundary conditions are enforced in an analogous manner to the inter-element boundaries using a Riemann solver to upwind information through the flux evaluation. From an implementation point of view, it is common practice to introduce a dummy edge along all element edges aligned with the domain boundary.

At a wall boundary we can impose the no-permeability boundary condition: $(u, v) \cdot \boldsymbol{n} = 0$. This can be enforced by setting

$$H_o = H_i \,, \quad \bar{u}_o = -\bar{u}_i \,, \quad \bar{v}_o = \bar{v}_i \,, \tag{10.4.15}$$

where we recall that the subscripts i and o refer to the inside and outside of an element, respectively, and so the o conditions are set in the dummy edge along the wall.

At inflow boundaries, the desired inflow conditions can be assigned to the dummy edges, which means that the boundary condition is only enforced in a weak sense through the characteristic variables. When reflected outgoing waves appear at the inflow boundary, a non-reflecting flux function [411] can also be applied, which is found to work satisfactorily for higher polynomial order P if the outgoing waves are reasonably normal to the boundary.

Finally, for outflow boundaries, when the flow is reasonably close to the normal direction, the values at the dummy edge can be set to be equal to the undisturbed initial state.

10.4.3 *Examples*

In this section we show two examples of simulations carried out with a spectral/hp discontinuous Galerkin method [156] using the orthogonal modal basis, $\phi_{pq} = \widetilde{\psi}_p^a \widetilde{\psi}_{pq}^b$, and an explicit Runge–Kutta scheme for the time-stepping. The numerical flux was evaluated using the HLLC solver.

We start by considering the model problem of a linear standing wave in a rectangular frictionless basin with no Coriolis effect, which has an analytical solution of the form

$$\begin{aligned} H(x, y, t) &= d + a\cos(kx)\cos(\omega t) \,, \\ u(x, y, t) &= a\frac{\omega}{kd}\sin(kx)\sin(\omega t) \,, \\ v(x, y, t) &= 0 \,, \end{aligned}$$

where a is the amplitude, k is the wavenumber, and ω is the frequency such that $\omega^2 = gdk^2$. We consider a basin of dimension $200\,\text{m} \times 100\,\text{m}$ and set the still-water depth to $d = 10\,\mathrm{m}$. We have then computed one wave period for a standing wave of wavelength $L = 400\,\mathrm{m}$ with $a = 0.2\,\mathrm{m}$. The time-step was chosen to be sufficiently small so that temporal error was negligible compared to spatial error. From the solution of this problem, we can demonstrate the exponential convergence of the spectral/*hp* element discretisation, as presented in Fig. 10.11.

A more realistic problem is the simulation of wave disturbance in the Port of Visby, located on the island of Gotland in the Baltic Sea (see Fig. 10.12). We have simplified the problem by assuming the break-waters to be fully reflective. We start with an initially motionless fluid and apply incoming sinusoidal waves with a period of 10.1 s and heights of 0.5 m. Figure 10.12 shows a snapshot of the surface elevation after running the model for 500 s using elements of polynomial order 5. At berth no. 5 the maximum wave height during the simulation is roughly 0.35 m. If the outer break-water is elongated by 200 m—as intended in the original layout, but abandoned for economic reasons—then the wave height is decreased to approximately 0.25 m.

10.4.4 *Boussinesq equations*

If we consider the homogeneous shallow-water equations and add third-order mixed derivatives to the source term, i.e.,

$$\mathbf{s} = \frac{d^2}{3}\begin{pmatrix} 0 \\ (\nabla \cdot (H\mathbf{u})_t)_{x_1} \\ (\nabla \cdot (H\mathbf{u})_t)_{x_2} \end{pmatrix}, \tag{10.4.16}$$

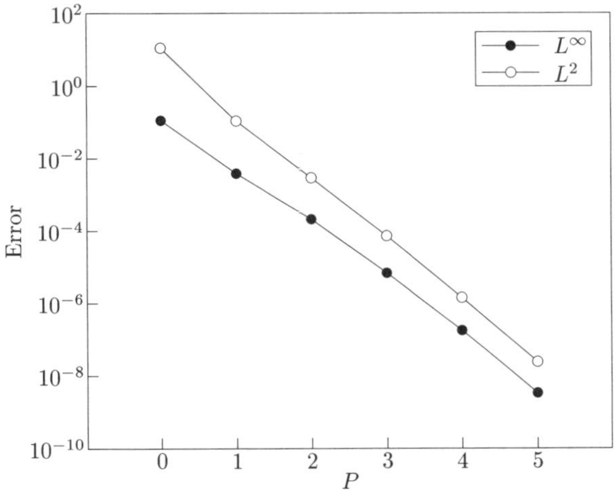

FIG. 10.11. Error of the H component versus expansion order for the case of a linear standing wave. ([156].)

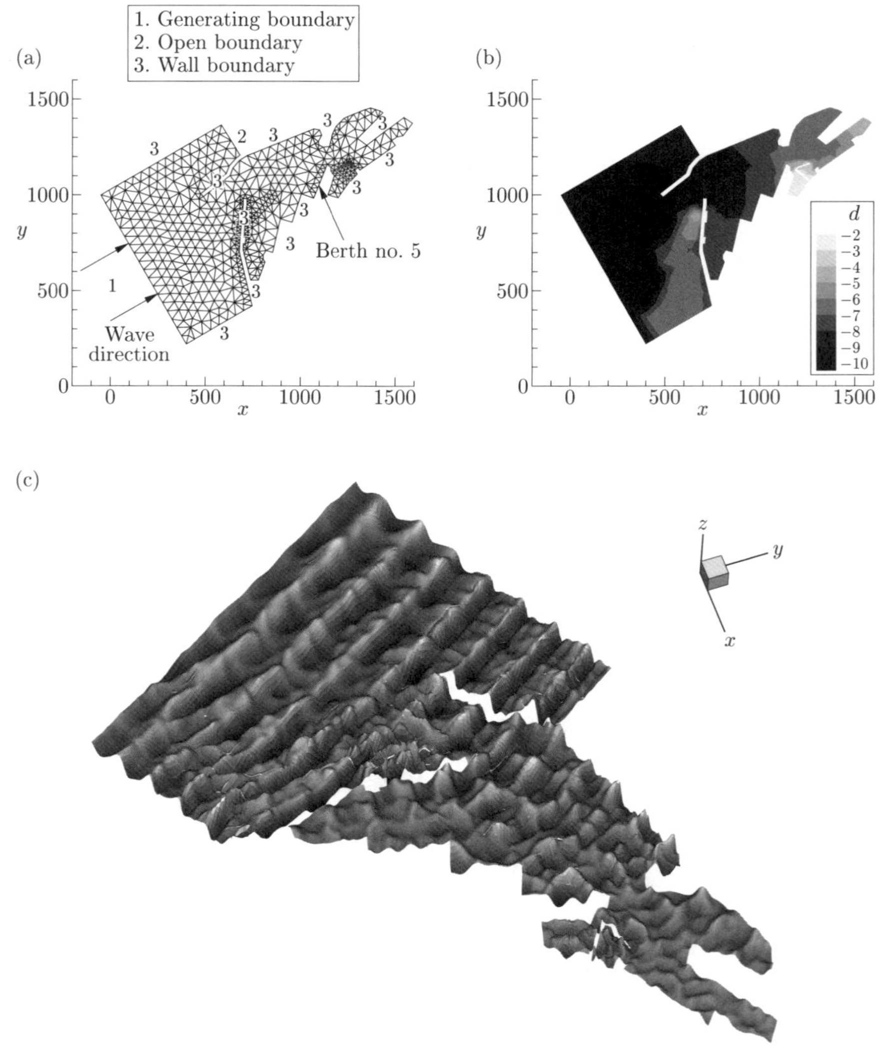

FIG. 10.12. Harbour problem: (a) the mesh and boundary conditions, (b) the depth of the harbour, and (c) the instantaneous surface elevation after 500 s. ([156].)

then we have a version of the classical Boussinesq equations of Peregrine [372], valid for weakly nonlinear and dispersive gravity waves in relatively shallow water. Usually Boussinesq-type equations are derived in primitive variables and, in this case, we lose in performance when going to conservative variables.

Recalling that there are no dispersive terms in the continuity equation, we can write the momentum equations as

$$\partial_t(H\mathbf{u}) - \frac{d^2}{3}\nabla(\nabla \cdot \partial_t(H\mathbf{u})) = \mathbf{a}. \quad (10.4.17)$$

Here $\mathbf{a}$ denotes the advective part, analogous to the shallow-water equations, associated with the momentum equations. Following [155], we introduce a new scalar variable of the time-rate-of-change of the momentum divergence, i.e., $z \equiv \nabla \cdot \partial_t(H\mathbf{u})$. An equivalent statement to problem (10.4.17) is

$$\partial_t(H\mathbf{u}) = \frac{d^2}{3}\nabla z + \mathbf{a}, \quad (10.4.18)$$

$$z - \nabla \cdot \partial_t(H\mathbf{u}) = 0. \quad (10.4.19)$$

Finally, substituting eqn (10.4.18) into eqn (10.4.19) and rearranging, we arrive at the Helmholtz equation

$$\nabla^2 z + \lambda z = \lambda(\nabla \cdot \mathbf{a}), \quad (10.4.20)$$

where $\lambda = -3/d^2$. Equation (10.4.20) can be interpreted as the divergence of eqn (10.4.17) expressed in terms of the time-rate-of-change of the momentum divergence, z. The solution of the Helmholtz problem using discontinuous Galerkin methods was previously discussed in Section 7.5; below we just briefly recall this formulation. We start by introducing the auxiliary flux variable $\mathbf{w} = \nabla z$, and so eqn (10.4.20) can be written as

$$\nabla \cdot \mathbf{w} + \lambda z = \lambda(\nabla \cdot \mathbf{a}), \quad (10.4.21)$$

$$\mathbf{w} - \nabla z = 0. \quad (10.4.22)$$

The elemental weak Galerkin formulation, defining s and $\mathbf{q}$ as test functions, then becomes

$$(s, \nabla \cdot \mathbf{w})_e + \langle s, [\tilde{\mathbf{w}} - \mathbf{w}] \cdot \boldsymbol{n}\rangle_e + \lambda(s, z)_e = \lambda(s, \nabla \cdot \mathbf{a})_e + \lambda\langle s, [\tilde{\mathbf{a}} - \mathbf{a}] \cdot \boldsymbol{n}\rangle_e,$$

$$(\mathbf{q}, \mathbf{w})_e - (\mathbf{q}, \nabla z)_e - \langle \mathbf{q} \cdot \boldsymbol{n}, [\tilde{z} - z]\rangle_e = 0,$$

where $\tilde{\mathbf{w}}$, $\tilde{\mathbf{a}}$, and $\tilde{z}$ contain the inter-element coupling through an appropriate flux balancing/upwinding between elements.

Having computed z, we return to the conservative variables $H\mathbf{u}$ using eqn (10.4.18), for which the discontinuous Galerkin formulation reads

$$(\boldsymbol{r}, (H\mathbf{u})_t)_e = (\boldsymbol{r}, \nabla z)_e + \langle \boldsymbol{r} \cdot \boldsymbol{n}, [\tilde{z} - z]\rangle_e + (\boldsymbol{r}, \mathbf{a})_e. \quad (10.4.23)$$

To summarise, if we use an explicit time-stepping scheme then for every time-step (or substep) do the following.

1. Compute the advective contribution using a suitable Riemann solver.
2. Solve the Helmholtz problem for the auxiliary variable z.
3. Return to the conservative variables using eqn (10.4.23).

Using the above algorithm with the HLLC solver for the advective fluxes and simple averaging for the dispersive fluxes, we obtain exponential convergence for the case of linear dispersive waves [155]. Another example of dispersive wave propagation is solitary waves. Consider a solitary wave with height 0.1 m propagating in a frictionless channel with an undisturbed depth of 1.0 m. The 100 m × 50 m domain, with periodic boundaries, is divided into sixty-four elements of order $P = 8$. The solution is then integrated for 20 s using 1000 time-steps. The solitary wave is initially centred at $x_1 = 20$ m and the initial condition is given by the sech-profile solitary wave solution with $\alpha = -1/3$,

$$\zeta = a_1 \operatorname{sech}^2(bx - ct) + a_2 \operatorname{sech}^4(bx - ct)\,,$$
$$u = a \operatorname{sech}^2(bx - ct)\,,$$

where

$$a = \frac{c^2 - gd}{c}\,, \qquad b = \sqrt{\frac{c^2 - gh}{4([\alpha + 1/3]gd^3 - \alpha d^2 c^2)}}\,,$$
$$a_1 = d\frac{c^2 - gd}{3([\alpha + 1/3]gd - \alpha d^2 c2)}\,, \quad a_2 = -\frac{2(c^2 - gd)}{2gdc^2} d \frac{[\alpha + 1/3]gd + 2\alpha c^2}{[\alpha + 1/3]gd - \alpha c^2}\,,$$

and the solitary wave speed c is given by the solution of a cubic equation in c^2:

$$2\alpha[c^2]^3 - \left(3\alpha + \frac{1}{3} + 2\alpha\frac{a}{d}\right)[c^2]^2 + 2\frac{a}{d}\left(\alpha + \frac{1}{3}\right)[c^2] + \alpha + \frac{1}{3} = 0\,.$$

These conditions were developed by Wei and Kirby [490] as a solution to the equations of Nwogu [340], but by considering the case $\alpha = -1/3$ they are valid as an approximation to the Peregrine equations. Figure 10.13 shows the water depths in a slice through $x_2 = 25$ m at different times. In the figure the

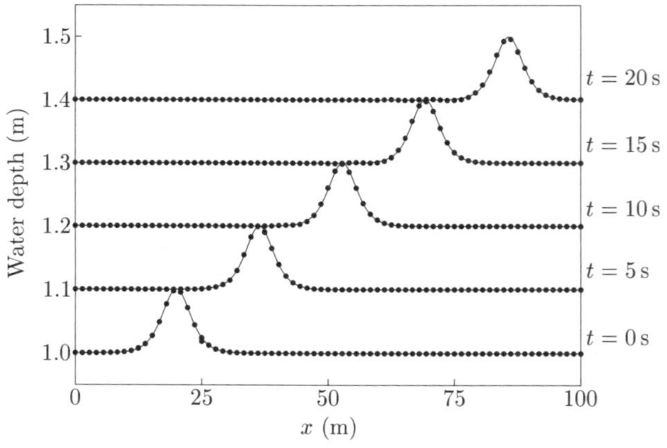

FIG. 10.13. Analytical (solid) and computed (dots) solitary wave propagation ([155].)

approximate analytical solutions are also plotted and agreement is seen to be reasonable.

10.5 Navier–Stokes equations

In this section we consider the Navier–Stokes equations for Newtonian gases, with the Stokes hypothesis valid for the second viscosity. In two dimensions they can be written in conservation form as

$$\frac{\partial}{\partial t}\begin{pmatrix}\rho\\ \rho u\\ \rho v\\ E\end{pmatrix}+\frac{\partial}{\partial x}\begin{pmatrix}\rho u\\ \rho u^2+p\\ \rho uv\\ (E+p)u\end{pmatrix}+\frac{\partial}{\partial y}\begin{pmatrix}\rho v\\ \rho vu\\ \rho v^2+p\\ (E+p)v\end{pmatrix}$$

$$=\frac{\partial}{\partial x}\begin{pmatrix}0\\ \frac{2}{3}\mu\left(2\frac{\partial u}{\partial x}-\frac{\partial v}{\partial y}\right)\\ \mu\left(\frac{\partial u}{\partial y}+\frac{\partial v}{\partial x}\right)\\ \frac{2}{3}\mu\left(2\frac{\partial u}{\partial x}-\frac{\partial v}{\partial y}\right)u+\mu\left(\frac{\partial u}{\partial y}+\frac{\partial v}{\partial x}\right)v+\kappa\frac{\partial T}{\partial x}\end{pmatrix}$$

$$+\frac{\partial}{\partial y}\begin{pmatrix}0\\ \mu\left(\frac{\partial u}{\partial y}+\frac{\partial v}{\partial x}\right)\\ \frac{2}{3}\mu\left(2\frac{\partial v}{\partial y}-\frac{\partial u}{\partial x}\right)\\ \frac{2}{3}\mu\left(2\frac{\partial v}{\partial y}-\frac{\partial u}{\partial x}\right)v+\mu\left(\frac{\partial u}{\partial y}+\frac{\partial v}{\partial x}\right)u+\kappa\frac{\partial T}{\partial y}\end{pmatrix}. \tag{10.5.1}$$

Here μ and κ are the dynamic viscosity and thermal conductivity, respectively. We can rewrite the equations by introducing the following reference quantities (the subscript '∞' denotes reference values): V_∞, ρ_∞, μ_∞, and κ_∞. The reference Prandtl number and Reynolds number are

$$\mathrm{Pr}_\infty=\frac{\mu_\infty c_p}{\kappa_\infty},\quad \mathrm{Re}_\infty=\frac{\rho_\infty V_\infty L}{\mu_\infty}. \tag{10.5.2}$$

In its simplest form, the dependence of the transport properties on the temperature is described by Sutherland's law:

$$\frac{\kappa}{\kappa_\infty}=\frac{\mu}{\mu_\infty}=\left(\frac{T}{T_\infty}\right)^{1/2}\frac{1+\theta}{1+\theta(T_\infty/T)}, \tag{10.5.3}$$

where $\theta=0.505$.

The Navier–Stokes equations for compressible flows now read as follows:

$$
\frac{\partial}{\partial t}\begin{pmatrix}\rho\\ \rho u\\ \rho v\\ E\end{pmatrix}+\frac{\partial}{\partial x}\begin{pmatrix}\rho u\\ \rho u^2+p\\ \rho uv\\ (E+p)u\end{pmatrix}+\frac{\partial}{\partial y}\begin{pmatrix}\rho v\\ \rho vu\\ \rho v^2+p\\ (E+p)v\end{pmatrix}
$$
$$
=\frac{1}{\mathrm{Re}_\infty}\left\{\frac{\partial}{\partial x}\begin{pmatrix}0\\ \frac{2}{3}\mu\left(2\frac{\partial u}{\partial x}-\frac{\partial v}{\partial y}\right)\\ \mu\left(\frac{\partial u}{\partial y}+\frac{\partial v}{\partial x}\right)\\ \frac{2}{3}\mu\left(2\frac{\partial u}{\partial x}-\frac{\partial v}{\partial y}\right)u+\mu\left(\frac{\partial u}{\partial y}+\frac{\partial v}{\partial x}\right)v+\kappa\frac{\gamma}{\mathrm{Pr}_\infty}\frac{\partial T}{\partial x}\end{pmatrix}\right.
$$
$$
\left.+\frac{\partial}{\partial y}\begin{pmatrix}0\\ \mu\left(\frac{\partial u}{\partial y}+\frac{\partial v}{\partial x}\right)\\ \frac{2}{3}\mu\left(2\frac{\partial v}{\partial y}-\frac{\partial u}{\partial x}\right)\\ \frac{2}{3}\mu\left(2\frac{\partial v}{\partial y}-\frac{\partial u}{\partial x}\right)v+\mu\left(\frac{\partial u}{\partial y}+\frac{\partial v}{\partial x}\right)u+\kappa\frac{\gamma}{\mathrm{Pr}_\infty}\frac{\partial T}{\partial y}\end{pmatrix}\right\}. \tag{10.5.4}
$$

The left-hand side coincides with the Euler equations of the previous section, and the right-hand side includes the effects of dissipation through the viscosity and thermal conductivity. We can rewrite in compact form the compressible Navier–Stokes equations as

$$
\mathbf{u}_t+\nabla\cdot\mathbf{F}=\mathrm{Re}_\infty^{-1}\nabla\cdot\mathbf{F}^\nu\,, \tag{10.5.5}
$$

where $\mathbf{F}$ and $\mathbf{F}^\nu$ correspond to inviscid and viscous fluxes, respectively. Splitting the Navier–Stokes operator in this form allows for a separate treatment of the inviscid and viscous contributions, which, in general, exhibit different mathematical properties. Alternatively, the viscous contributions could be thought of as corrections to the hyperbolic system of Euler equations, and thus a different numerical approach can be followed. In the following, we present the main ideas of both approaches, first using the discontinuous Galerkin and, subsequently, the penalty method frameworks.

10.5.1 *Mixed and discontinuous Galerkin formulations*

We will assume here that the Euler term $\nabla\cdot\mathbf{F}$ has been discretised first using a discontinuous Galerkin method, which implies that the solution is formally discontinuous, that is,

$$
\mathbf{u}\in L^2(\Omega)\,.
$$

With this in mind, we can subsequently discretise the viscous term $\nabla \cdot \mathbf{F}^{\nu}$ using the mixed and discontinuous Galerkin methods (DGM), which were first compared in Section 7.5.10. The *mixed* method uses two sets of functions: discontinuous and continuous. This method, although explicit, involves the inversion of the global mass matrix and, as a result, it is much less efficient than the one based on the *discontinuous Galerkin* formulation.

To illustrate the two approaches we consider the parabolic model problem for the scalar field $u(\boldsymbol{x}, t)$:

$$u_t = \nabla \cdot (\nu \nabla u) + f \quad \text{in } \Omega\,, \tag{10.5.6a}$$

$$u = g(\boldsymbol{x}, t) \quad \text{on } \partial\Omega\,. \tag{10.5.6b}$$

We introduce the flux variable

$$\mathbf{q} = \nu \nabla u\,,$$

as in Section 7.5.10, where $\mathbf{q} \in \mathbf{H}(\text{div}; \Omega)$ is defined as

$$\mathbf{H}(\text{div}; \Omega) = \{\mathbf{w} \in L^2(\Omega),\ \nabla \cdot w \in L^2(\Omega)\}\,.$$

We then discretise first in time using (for simplicity) a single-step integrator:

$$\frac{u^{n+1} - u^n}{\Delta t} = \nabla \cdot \mathbf{q}^m + f^m \quad \text{in } \Omega\,, \tag{10.5.7a}$$

$$\mathbf{q}^m = \nu \nabla u^m\,, \tag{10.5.7b}$$

$$u^{n+1} = g(\boldsymbol{x}, t^m) \quad \text{on } \partial\Omega\,, \tag{10.5.7c}$$

where $m = n$ or $n+1$ corresponds to explicit or implicit time integration, respectively. The variational form can now be derived similarly, as in the static case of Section 7.5.10.

10.5.1.1 *Mixed formulation*

For the *mixed* Galerkin formulation, with continuous test functions, we obtain the following.

Find $(\mathbf{q}, u) \in \mathbf{H}(\text{div}; \Omega) \times L^2(\Omega)$ such that

$$(u^{n+1}, v) = (u^n, v) + \Delta t[(\nabla \cdot \mathbf{q}^m, v) + (f^m, v)]\,, \quad \forall\, v \in L^2(\Omega)\,, \tag{10.5.8a}$$

$$\left(\frac{1}{\nu}\mathbf{q}^m, \mathbf{w}\right) = -(\nabla \cdot \mathbf{w}, u^m) + (g^m, \mathbf{w} \cdot \boldsymbol{n})_{\partial\Omega}\,, \quad \forall\, \mathbf{w} \in \mathbf{H}(\text{div}; \Omega)\,, \tag{10.5.8b}$$

where $\boldsymbol{n}$ is the outwards unit normal, and parentheses denote standard inner products.

For the implicit scheme a coupled system for $\mathbf{q}$ and u has to be solved, whereas for the explicit scheme the unknown u^{n+1} is obtained from the first equation by a simple projection. If a projection is not performed and the equation is solved in its strong form, then numerical evidence suggests that an instability may develop in time.

10.5.1.2 *Discontinuous Galerkin formulation*

In the *discontinuous Galerkin* method for the diffusion fluxes, the weak formulation of the problem is stated for each element e as follows.

Find $(\mathbf{q}, u) \in \mathbf{L}^2(\Omega) \times L^2(\Omega)$ such that

$$(u^{n+1}, v)_e = (u^n, v)_e + \Delta t \left[-(\mathbf{q}^m, \nabla v)_e + \langle v, \tilde{\mathbf{q}}_\gamma \cdot \boldsymbol{n} \rangle_e + (f^m, v)_e \right], \quad \forall\, v \in L^2(\Omega), \tag{10.5.9a}$$

$$-\frac{1}{\nu}(\mathbf{q}^m, \mathbf{w})_e = (u^m, \nabla \cdot \mathbf{w})_e - \langle \tilde{u}_\gamma, \mathbf{w} \cdot \boldsymbol{n} \rangle_e, \quad \forall\, \mathbf{w} \in \mathbf{L}^2(\Omega), \tag{10.5.9b}$$

and

$$u^{n+1} = g(\boldsymbol{x}, t^{n+1}) \quad \text{on } \partial\Omega,$$

where the parentheses denote the standard inner product in an element e and the angled brackets denote boundary integral terms on each element, with $\boldsymbol{n}$ denoting the outwards unit normal.

The selection of the interface fluxes was discussed extensively in Section 7.5.4.

By integrating by parts once more, we obtain an analytically equivalent formulation which, typically, is easier to implement. The modified weak formulation is

$$(u^{n+1}, v)_e = (u^n, v)_e + \Delta t \left[(\nabla \cdot \mathbf{q}^m, v)_e + \langle v, (\tilde{\mathbf{q}}_\gamma - \mathbf{q}_i) \cdot \boldsymbol{n} \rangle_e + (f^m, v)_e \right], \quad \forall\, v \in L^2(\Omega),$$

$$-\frac{1}{\nu}(\mathbf{q}^m, \mathbf{w})_e = (-\nabla u^m, \mathbf{w})_e - \langle \tilde{u}_\gamma - u_i, \mathbf{w} \cdot \boldsymbol{n} \rangle_e, \quad \forall\, \mathbf{w} \in \mathbf{L}^2(\Omega),$$

and

$$u^{n+1} = g(\boldsymbol{x}, t^{n+1}) \quad \text{on } \partial\Omega.$$

Here the subscript i denotes contributions evaluated at the interior side of the boundary.

10.5.1.3 *Implementation issues in DGM*

In this section we discuss details of the implementation of the explicit discontinuous Galerkin method in the context of spectral/*hp* element methods. This material is therefore directed towards writing and debugging a DGM formulation. This algorithm describes a first-order time-accurate method as first-order explicit integration is employed. However, second-order and third-order schemes can easily be constructed using multi-step time-integration, see [309]. We start by summarising the top level algorithm as follows.

Top-level DGM algorithm

- Read the initial conditions for the state vector $\mathbf{u}^0$, and project all fields to polynomial space.

- Begin the following time loop.
  - ⋆ Compute the Navier–Stokes operator $\mathbf{U}_f^n$.
  - ⋆ Advance in time: $\mathbf{u}^{n+1} = \mathbf{u}^n + \Delta t\, \mathbf{U}_f^n$. Update time: $t^{n+1} = (n+1)\Delta t$.
  - ⋆ Overwrite the wall boundary values: $\mathbf{u}_w^{n+1} = \mathbf{u}_w$.
  - ⋆ If $(n+1)\Delta t$ is less than the specified time then continue, else exit.
- Print output and store the results in polynomial space.

Next we present the algorithm to compute the Navier–Stokes operator $\mathbf{U}_f^n$.

Algorithm for the Navier–Stokes operator

- Compute the convection contribution: $\mathbf{U}_{f,c}$.
  - ⋆ Compute the characteristic boundary values: $\mathbf{U}_{f,\mathrm{cb}}^n$.
  - ⋆ Compute the divergence of the interior Euler flux: $\mathbf{U}_{f,c}^i$.
  - ⋆ Set $\mathbf{U}_{f,c} = \mathbf{U}_{f,\mathrm{cb}}^n + \mathbf{U}_{f,c}^i$.
- Save the values at the outer boundaries: $\mathbf{U}_{f,\mathrm{cb}}^n$.
- Compute the viscous contributions: $\mathbf{U}_{f,v}$.
- Set $\mathbf{U}_f = \mathbf{U}_{f,c} + \mathbf{U}_{f,v}$.
- Overwrite the outer boundary values: $\mathbf{U}_{f,b}^n = \mathbf{U}_{f,\mathrm{cb}}^n$.

All operations in this procedure are performed on the quadrature points, i.e., in physical space. However, since the initial conditions are in polynomial space, the above operations have not taken the arrays $\mathbf{U}$ out of the polynomial space. The only operation that can take this function out of the polynomial space is overwriting the boundary values (this is done by setting the functional values at the quadrature points on the boundary). In this case we should perform the projection of the field back to polynomial space. This is done locally, only in the elements which are adjacent to the boundary.

What quadrature to use

Let us consider polynomials of degree up to P employed in an element 'e', see Fig. 10.14. By employing exact Gaussian quadrature for a polynomial order of P, we choose $Q = P+2$ Gauss–Lobatto points in the η_1 direction and $Q = P+1$ Gauss–Radau points in the η_2 direction. In this case the quadrature rule is exact for polynomials of degree up to $2P$ in the interior of the elements (in non-curvilinear geometries).

All the boundary terms, such as the boundary integrals and boundary fluxes are computed at the $Q = P+1$ Gauss points on each edge, which requires interpolation from the Gauss–Lobatto or Gauss–Radau points. By choosing a similar number of Gauss points along an edge, it is possible to match the boundary flux computations between adjacent elements. If the orders in the elements are different, then the *higher degree* of edge quadrature points should be employed for flux matching. This choice is dictated by stability and accuracy considerations.

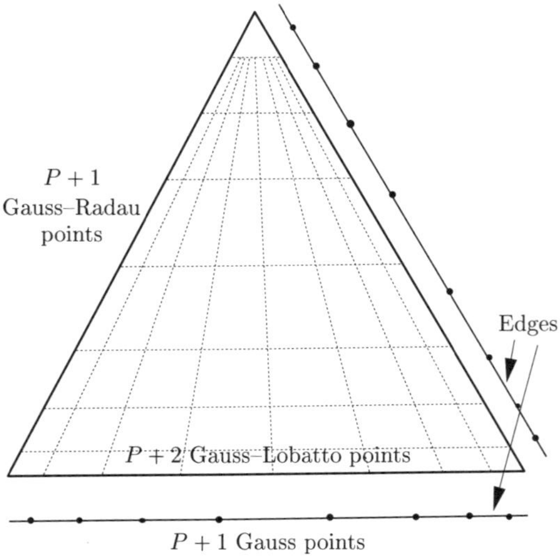

FIG. 10.14. Quadrature points used in a triangular element, $P = 7$.

However, to preserve the conservation property the edge flux can be projected onto the smaller of the polynomial spaces. Alternatively, as discussed in [425], the upwinded flux can be projected onto a Legendre expansion, which ensures that all high polynomial components are orthogonal to the mean component along an edge. We note that on the edges the quadrature is exact for polynomials of degree $2P + 1$. These conditions are necessary in order to prove the maximum possible accuracy of $\mathcal{O}(h^{P+1})$, see [100]. We will return to the issue of stability in Section 10.5.5, when we discuss consistent integration.

Computation of derivatives

All the derivatives are computed in a weak discontinuous Galerkin sense as follows. Let $u \in L^2(\Omega)$, then $q \in L^2(\Omega)$ is the x derivative of u, $q = u_x$, if

$$(q, w)_e = (u_x, w)_e + \langle w, (\tilde{u}_\gamma - u_i)n_x \rangle_e \,, \quad \forall\, w \in L^2(\Omega)\,,$$

where, for example, $\tilde{u}_\gamma = \frac{1}{2}(u_i + u_x)$, i.e., the arithmetic mean if a second derivative is computed. The derivatives are computed in a element-wise sense and $(\cdot, \cdot)_e$ denotes the standard inner product in an element e. The angled brackets denote the inner product over the elemental boundary, with n_x being the x component of the outward unit normal. All the operations are performed in physical space, that is, u_x is computed on the quadrature points and all the integrals are computed using the aforementioned quadrature rules. In these computations the same spaces for the functions and their derivatives are used. These spaces consist of functions which are polynomials of up to degree P inside the elements.

10.5.2 *Convergence and simulations*

In this section we present two analytical solutions and some validation simulations which can be used as benchmark cases in testing and debugging DGM or other spectral-based compressible Navier–Stokes solvers. First, we consider two dimensions on a square domain defined by the lower-left and upper-right corners $([-1,-1],[1,1])$ and discretised in eight triangles. On the left- and right-hand sides periodic boundary conditions are assumed, and on the top and bottom Dirichlet boundary conditions are prescribed. The analytical solution has the form

$$\rho = A + B\sin(\omega x)\,, \quad u = C + D\cos(\omega x)\sin(\omega y)\,, \quad T = E + Fy\,,$$

where we choose

$$\omega = \pi\,, \quad A = 1\,, \quad B = 0.1\,, \quad C = 1\,, \quad D = 0.04\,, \quad E = 84\,, \quad F = 28\,.$$

The Navier–Stokes equations are then integrated using a forcing term consistent with the above solution. Figure 10.15(a) shows the comparison of the L^∞ error for the total energy. The viscous component was computed using mixed and discontinuous Galerkin formulations. The two sets of results are in close agreement and show exponential convergence.

The exponential convergence of the method in a three-dimensional domain is shown in Fig. 10.15(b). The analytical solution for this case has the form

$$\rho = A + B\sin(\omega x)\,, \quad u = C + D\cos(\omega x)\sin(\omega y)\cos(\omega z)\,, \quad T = E + Fy + Gz^2\,,$$

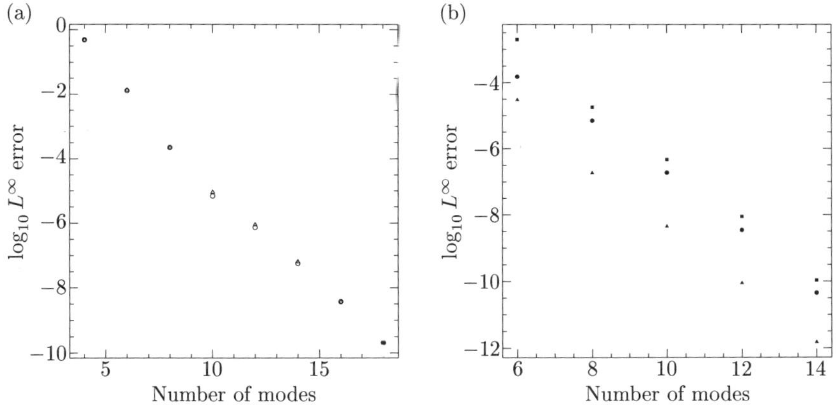

FIG. 10.15. (a) L^∞ error versus the expansion order for the energy of an analytic solution of the steady two-dimensional Navier–Stokes equations obtained by discontinuous (open triangles), and mixed Galerkin methods (open circles). (b) L^∞ error versus the expansion order for the energy (solid squares), momentum (solid circles), and density (solid triangles) of an analytic solution of the steady three-dimensional Navier–Stokes equations.

where here we choose

$$\omega = \frac{\pi}{2}, \quad A = 1, \quad B = 0.1, \quad C = 1, \quad D = 0.04,$$
$$E = 84, \quad F = 28, \quad G = 10.$$

Our next test considers a refinement study for a transonic flow past an aerofoil NACA0012 at an angle of attack $\alpha = 10°$, with free-stream Mach number Ma $= 0.8$, and Reynolds number (based on the free-stream velocity and the aerofoil chord) Re $= 73$. The wall temperature is equal to the free-stream total temperature. This problem was considered by Bassi and Rebay [38], and is one of the benchmark problems suggested in the GAMM (1986) Workshop [178]. Part of the mesh is shown in Fig. 10.16(a); it extends four chords downstream and

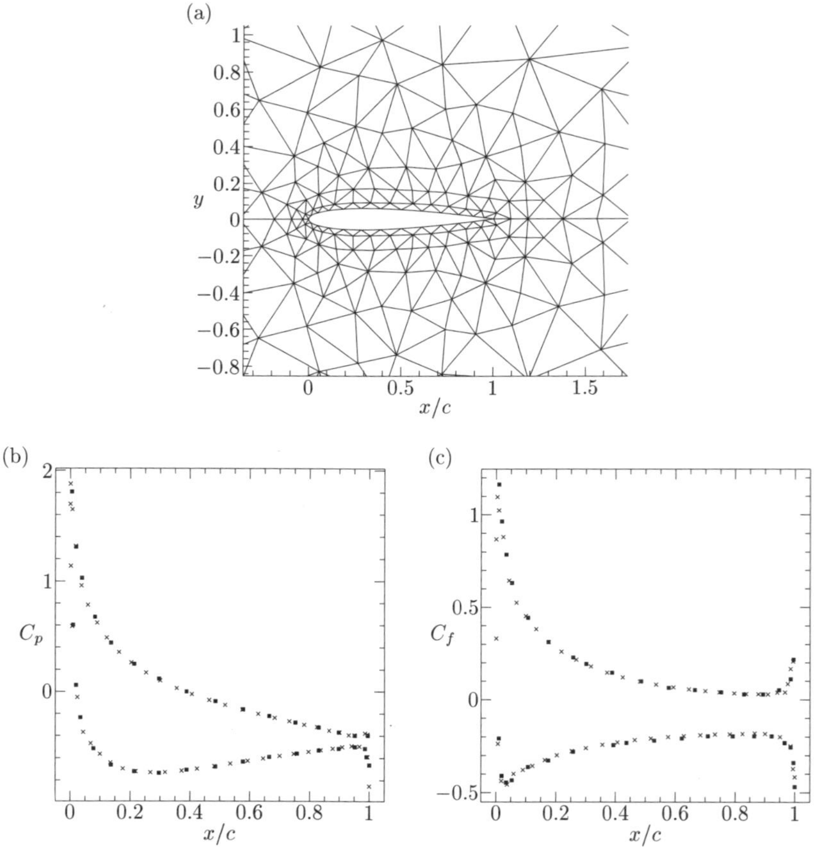

FIG. 10.16. (a) Discretisation around an NACA0012 aerofoil using 592 elements, (b) pressure, and (c) drag coefficients. Solid squares are data from [38] and crosses are from the spectral/*hp* DGM simulation for $P = 6$.

consists of 592 elements, which is about one-quarter of the number of elements used in [38]. Three different discretisations with P-refinement were used, corresponding to polynomial orders of 2, 4, and 6. A quantitative comparison is shown in Table 10.1 for the drag and lift coefficients computed on the three different polynomial orders. In Fig. 10.16(b) and (c) the distributions of the pressure and friction coefficients around the aerofoil are also plotted. Also, included in this figure are the results of Bassi and Rebay [38], which show very good agreement with the spectral/*hp* element method.

An example with a strong shock present is shown in Fig. 10.17(b). It describes flow around an NACA4420 aerofoil at an angle of attack, first presented in [310]. The corresponding mesh for this simulation is shown in Fig. 10.17(a). The triangulation was chosen to be rather irregular in order to demonstrate the discretisation flexibility and robustness of this method.

A final example is flow past a circular cylinder, which at subsonic state develops an unsteady vortex street, but at Mach number Ma > 1 a symmetric steady wake is developed. At the transonic state Ma $= 1$ the vortex street disappears, as shown in Fig. 10.18, but a large recirculation zone is present. For larger values of Mach number the flow is also steady, and the recirculation zone becomes very small at Ma $= 2$, see [310].

TABLE 10.1. For an NACA0012 aerofoil, drag and lift coefficients corresponding to different P-refinements.

Force coefficient	$P = 2$	$P = 4$	$P = 6$
C_d	0.6828	0.6786	0.6758
C_l	0.4762	0.5302	0.5317

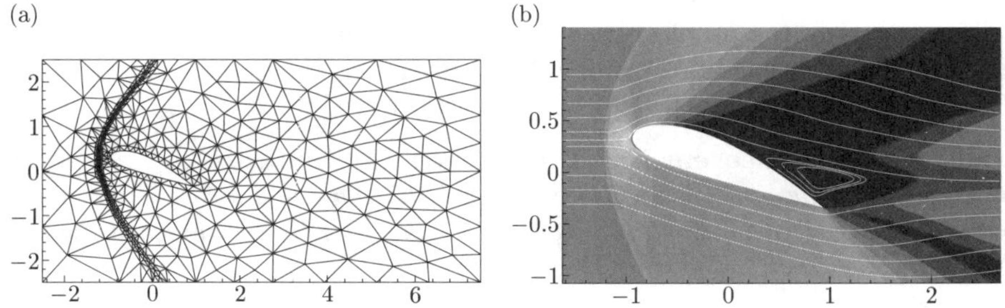

FIG. 10.17. (a) Unstructured mesh for the NACA4420 aerofoil supersonic flow, with 1128 elements which can each support a different order P. (b) Supersonic flow past an NACA4420 aerofoil at 20° angle of attack and Mach number Ma $= 2$. Density contours and streamlines are plotted. ([310].)

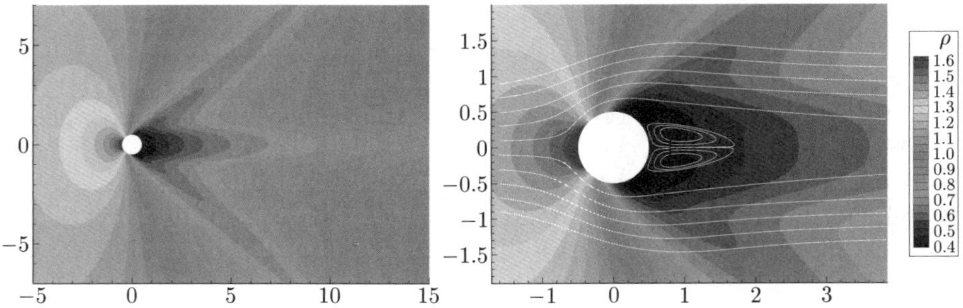

FIG. 10.18. Density contours and streamlines (close-up) for flow past a circular cylinder at Mach number 1 and Reynolds number 100 ([310]).

10.5.3 *A penalty formulation*

An alternative formulation for treating the Navier–Stokes equations was developed by Hesthaven [233]. The domain decomposition is based on imposing interface conditions using the penalty method, as already discussed in the previous section for hyperbolic conservation laws. For simplicity, we consider the viscous Burgers equation for the scalar $u(x,t)$:

$$u_t + uu_x = \nu u_{xx}\,, \quad x \in [x_L, x_R]\,, \tag{10.5.10a}$$

subject to the general boundary conditions

$$\alpha u(x_L,t) - \beta\nu u_x|_{x_L} = g_L(t)\,, \tag{10.5.10b}$$
$$\gamma u(x_R,t) + \delta\nu u_x|_{x_R} = g_R(t)\,, \tag{10.5.10c}$$

where α, β, γ, and δ are non-negative constants. In [286] it was shown that the well-posedness of the problem described above is equivalent to a problem with identical boundary conditions, but with the governing equation linearised around a constant state V. The following theorem was proved in [236].

Well-posedness theorem 10.5.1. (Hesthaven and Gottlieb [236]) *The linearised version of the problem described in eqn (10.5.10a), and subject to boundary conditions described in eqns (10.5.10b) and (10.5.10c), is well posed if the constants are such that*

$$\begin{aligned} &\alpha = V\,, \quad \beta = 1\,, \quad \gamma = 0\,, \quad \delta = 1\,, \quad V \geqslant 0\,,\\ &\alpha = 0\,, \quad \beta = 1\,, \quad \gamma = |V|\,, \quad \delta = 1\,, \quad V < 0\,. \end{aligned}$$

Based on this theorem, we can now construct the elemental equations for a multi-domain spectral discretisation, where the interface boundary conditions of element e with elements $e-1$ and $e+1$ play the role of boundary conditions. To this end, we use the penalty formulation as suggested in [233] by introducing the penalty parameters τ_L^e and τ_R^e for the left and right interface boundaries, respectively. Assuming, for example, that $V \geqslant 0$, we obtain for the left (donor) element

$$(u_t)_P^e + u_P^e(u_x)_P^e = \nu(u_{xx})_P^e - \tau_R^e \nu[(u_x)_P^e - (u_x)_0^{e+1}] . \tag{10.5.11a}$$

For the right (receiver) element we have

$$(u_t)_0^{e+1} + u_0^{e+1}(u_x)_0^e = \nu(u_{xx})_0^{e+1} - \tau_L^{e+1}[V_0^e(u_0^{e+1} - u_P^e) - \nu((u_x)_0^{e+1} - (u_x)_P^e)] . \tag{10.5.11b}$$

Continuity of the function as well as its derivatives is imposed in a weak sense. For stability, the penalty parameters τ_L^e and τ_R^e have to be chosen appropriately. By considering the asymptotic stability of the scheme, the following equations were obtained in [233]:

$$\tau_L^e = \frac{2}{\nu L^e \rho^e \beta}\left[\nu + 2\kappa_1 - 2\sqrt{\kappa_1^2 + \nu\kappa_1 - \frac{1}{2\nu\rho^e|V_0^e|}}\,\right],$$

$$\tau_R^e = \frac{2}{\nu L^e \rho^e \delta}\left[\nu + 2\kappa_2 - 2\sqrt{\kappa_2^2 + \nu\kappa_2 - \frac{1}{2\nu\rho^e|V_0^{e+1}|}}\,\right],$$

where L^e denotes the element length, and we have defined $\rho^e = (P^e)^{-2}$, $\kappa_1 = \alpha/\beta\rho^e$, and $\kappa_2 = \gamma/\delta\rho^e$. These expressions are valid for Chebyshev discretisations only. This choice of the penalty parameters leads to a spectrally accurate and stable approximation.

Extending these ideas to the Navier–Stokes equations, we consider eqn (10.5.5) and introduce the transformation derivatives for both the inviscid and viscous fluxes, that is,

$$A_i = \frac{\partial \mathbf{F}_i}{\partial \mathbf{u}}, \quad B_{ij} = \frac{1}{2}\left(\frac{\partial \mathbf{F}_i^\nu}{\partial \mathbf{u}_{,j}} + \frac{\partial \mathbf{F}_j^\nu}{\partial \mathbf{u}_{,i}}\right),$$

where subscripts refer to different directions and $\mathbf{u}_{,i}$ is the partial derivative of the state vector in the ith direction. We also define

$$A = (A_1, A_2)\cdot \boldsymbol{n}, \quad B_i = (B_{i1}, B_{i2})\cdot \boldsymbol{n}$$

for an arbitrary direction $\boldsymbol{n}$. Assuming a known state $\mathbf{u}_0$, a diagonalising similarity transformation can be applied to $A(\mathbf{u}_0)$ in the direction $\boldsymbol{n}$ to obtain the diagonal matrix

$$\mathbf{A}^{\boldsymbol{n}} = (\mathbf{S}^{\boldsymbol{n}})^{-1}\mathbf{A}\mathbf{S}^{\boldsymbol{n}},$$

where the elements of $\mathbf{S}^{\boldsymbol{n}}$ are given in [488]. The diagonal elements of $\mathbf{A}^{\boldsymbol{n}}$ are

$$\lambda_1 = \mathbf{u}_0\cdot\boldsymbol{n} + c_0, \quad \lambda_2 = \lambda_3 = \mathbf{u}_0\cdot\boldsymbol{n}, \quad \lambda_4 = \mathbf{u}_0\cdot\boldsymbol{n} - c_0,$$

where c_0 is the sound speed. These eigenvalues are the advection velocities of the characteristic functions $\mathbf{R}^{\boldsymbol{n}} = (\mathbf{S}^{\boldsymbol{n}})^{-1}\mathbf{u}$ along the $\boldsymbol{n}$ direction.

Similarly, we obtain the transformed matrices for the viscous terms $\mathbf{B}_i^{\boldsymbol{n}} = (\mathbf{S}^{\boldsymbol{n}})^{-1}\mathbf{B}_i\mathbf{S}^{\boldsymbol{n}}$, and construct the viscous correction vector [235]

$$\mathbf{G}^{\boldsymbol{n}} = \sum_{i=1}^{2} \mathbf{B}_i^{\boldsymbol{n}} \frac{\partial \mathbf{R}^{\boldsymbol{n}}}{\partial x_i}.$$

We can now write the compressible Navier–Stokes equations using a multi-domain approach by considering the elemental equations in element e and assuming that $\mathbf{u}_0 \cdot \boldsymbol{n} > 0$:

$$\begin{aligned}\mathbf{u}_t^e + \nabla \cdot \mathbf{F}^e = \mathrm{Re}_\infty^{-1} \nabla \cdot \mathbf{F}^{\nu,e} \\ - \tau_L^e Q_-^e(\xi_i^e) S^{\mathbf{n}} \left[\mathcal{R}_-^{e-1}(\mathbf{R}_0^e - \mathbf{R}_P^{e-1}) - \mathrm{Re}_\infty^{-1} \mathcal{G}_-(\mathbf{G}_0^e - \mathbf{G}_P^{e-1}) \right] \\ - \tau_R^e Q_+^e(\xi_i^e) S^{\mathbf{n}} \left[\mathcal{R}_+^e(\mathbf{R}_P^e - \mathbf{R}_0^{e+1}) + \mathrm{Re}_\infty^{-1} \mathcal{G}_+(\mathbf{G}_P^e - \mathbf{G}_0^{e+1}) \right],\end{aligned} \tag{10.5.12}$$

where ξ_i^e is the point along the direction $\boldsymbol{n}$, which coincides with either the x or the y direction. More general equations for the multi-dimensional case can be found in [235]. In addition, the boundary matrices $\mathcal{G}_\pm$ and $\mathcal{R}_\pm$ are diagonal, with $\mathcal{R}_-^e$ having the elements $[\lambda_1, \lambda_2, \lambda_3, \alpha\lambda_4]$, where $\alpha = 0$ for subsonic and $\alpha = 1$ for supersonic conditions. Also, $\mathcal{R}_+^e$ has all zeros but the last diagonal element, which is $(1-\alpha)\lambda_4$. In addition, $\mathcal{G}_-$ has all ones in the diagonal and $\mathcal{G}_+$ is similar, but the first entry is zero, reflecting the singular nature of the dissipative component of Navier–Stokes equations.

By analogy with the analysis of the Burgers equation, the penalty parameters have to be chosen so that the scheme is asymptotically stable. The required analysis was developed in [235] for general boundary conditions. Here, we summarise the results for the scheme described in eqn (10.5.12):

- subsonic and supersonic inflow:

$$\tau_L^e \geqslant \frac{2}{\rho^e \kappa L^e}\left(1 + \kappa - \sqrt{1+\kappa}\right);$$

- subsonic outflow:

$$\tau_R^e \geqslant \frac{2}{\rho^e \kappa L^e}\left(1 + \kappa - \sqrt{1+\kappa}\right);$$

- supersonic outflow:

$$\tau_R^e \geqslant \frac{2}{\rho^e L^e}\left(1 - \sqrt{\frac{1}{\kappa}}\right),$$

where ρ^e has been defined before, and

$$\kappa = \frac{k_0}{2\rho^e \mathrm{Re}_\infty \mathrm{Pr} \rho_0 |\mathbf{u}_0 \cdot \boldsymbol{n}|}.$$

In the time-stepping, the value of the state vector at the previous time-step can be used for linearisation. The above formulation is still valid in the limit of vanishing viscosity for the Euler equations. Examples of applications of this scheme, including subsonic and supersonic, steady and unsteady flows, can be found in [232, 233, 235]. In Fig. 10.19 we present an example taken from [232] showing compressible flow around a bluff body inside a narrow channel at Re = 250, Ma = 0.4, and a stagnation temperature of $T_0 = 300\,\mathrm{K}$.

10.5.4 *Moving domains*

We now turn our attention to moving domains encountered in flow-structure interaction problems, e.g., aeroelasticity. In the following, we will solve the Navier–Stokes equations in a time-dependent domain $\Omega(t)$ by discretising on a grid whose points may be moving with velocity $\mathbf{U}^g$, which is, in general, *different* to the local fluid velocity. We will follow the so-called arbitrary Lagrangian Eulerian (or ALE) formulation which reduces to the familiar Eulerian and Lagrangian forms when $\mathbf{U}^g = 0$ and $\mathbf{U}^g = \mathbf{u}$, respectively [138, 241, 242, 249, 339, 479].

Using the Reynolds transport theorem, we can write the Euler equations in the ALE framework:

$$\mathbf{u}_t + G_{i,i} = -U^g_{i\ ,i}\mathbf{u}\,, \tag{10.5.13}$$

where the ALE flux term is defined as

$$G_i = (u_i - U^g_i)\mathbf{u} + p[0, \delta_{1i}, \delta_{2i}, \delta_{3i}, u_i]\,, \quad i = 1, 2, 3\,.$$

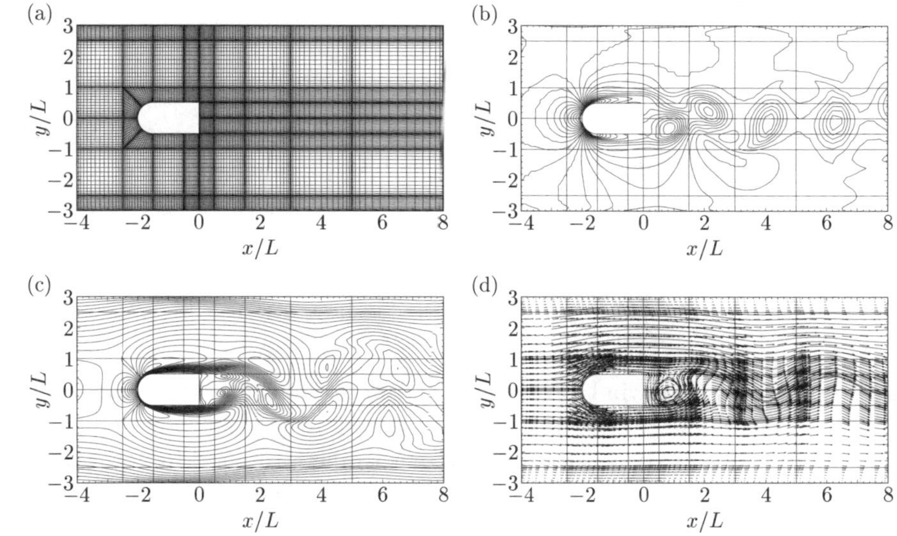

FIG. 10.19. Compressible flow around a bluff body inside a narrow channel. (a) A segment of the grid, employing 104 elements, is shown, as is (b) the instantaneous density, ρ/ρ_0, (c) the Mach number, Ma, and (d) the velocity field, $\mathbf{u}$. (Courtesy of J. Hesthaven.)

We can recover the *Euler flux* $\mathbf{F}$ by simply setting $\mathbf{U}^g = 0$, and, in general, we have that $G_i = F_i - U_i^g \mathbf{u}$. Now, if we write the ALE Euler equations in terms of the *Euler flux* then the source term on the right-hand side of eqn (10.5.13) is eliminated and we obtain

$$\mathbf{u}_t + F_{i,i} - U_i^g \mathbf{u}_{,i} = 0\,, \tag{10.5.14}$$

which can then be recast in the standard quasi-linear form

$$\mathbf{U}_t + [\mathbf{A}_i - U_i^g \mathbf{I}]\mathbf{u}_{,i} = 0\,,$$

where $\mathbf{A}_i = \partial \mathbf{F}_i / \partial \mathbf{u}$ $(i = 1, 2, 3)$ is the flux Jacobian and $\mathbf{I}$ is the unit matrix. In this form it is straightforward to obtain the corresponding characteristic variables since the ALE Jacobian matrix can be written as

$$\mathbf{A}_i^{\mathrm{ALE}} \equiv [\mathbf{A}_i - U_i^g \mathbf{I}] = \mathbf{R}_i \cdot [\mathbf{D}_i - U_i^g \mathbf{I}] \cdot \mathbf{L}_i\,,$$

where brackets denote matrices. Here the diagonal matrix $\mathbf{D}$ contains the eigenvalues of the original Euler Jacobian matrix $\mathbf{A}$, and $\mathbf{R}$ and $\mathbf{L}$ are the right-eigenvector and left-eigenvector matrices, respectively, containing the corresponding eigenvectors of $\mathbf{A}$. Notice that the *shifted eigenvalues* of the ALE Jacobian matrix do not change the corresponding eigenvectors in this characteristic decomposition.

To explain the discontinuous Galerkin ALE formulation we consider the two-dimensional equation for advection of a conserved scalar ϕ in a region $\Omega(t)$:

$$\frac{\partial \phi}{\partial t} + \nabla \cdot \mathbf{F}(\phi) - \mathbf{U}^g \cdot \nabla \phi = 0\,.$$

In the discontinuous Galerkin framework, we test the equation above with discontinuous test functions v separately on each element e to obtain

$$(v, \partial_t \phi)_e + (v, \nabla \cdot \mathbf{F}(\phi))_e - (v, \mathbf{U}^g \cdot \nabla \phi)_e + \int_{\partial \Omega^e} v[\tilde{\boldsymbol{f}}(\phi_i, \phi_o) - \mathbf{F}(\phi) - (\phi_{\mathrm{up}} - \phi_i) \cdot \mathbf{U}^g] \cdot \boldsymbol{n} \,\mathrm{d}s = 0\,, \tag{10.5.15}$$

where ϕ_{up} denotes an upwind flux. Here $(\cdot, \cdot)$ denotes the inner product evaluated over each element, and $\tilde{\boldsymbol{f}}$ is a numerical boundary flux; the notation is explained in Fig. 10.20. Notice that this form is different to the form used in, for example, the work of [242, 275], where the time derivative is applied to the inner product, i.e.,

$$\partial_t (v, \phi)_e + (v, \nabla \cdot \mathbf{F}(\phi))_e - (v, \mathbf{U}^g \cdot \nabla \phi)_e - (v, \phi \nabla \cdot \mathbf{U}^g)_e + \int_{\partial \Omega^e} v[\tilde{\boldsymbol{f}}(\phi_i, \phi_o) - \mathbf{F}(\phi) - (\phi_{\mathrm{up}} - \phi_i) \cdot \mathbf{U}^g] \cdot \boldsymbol{n} \,\mathrm{d}s = 0\,. \tag{10.5.16}$$

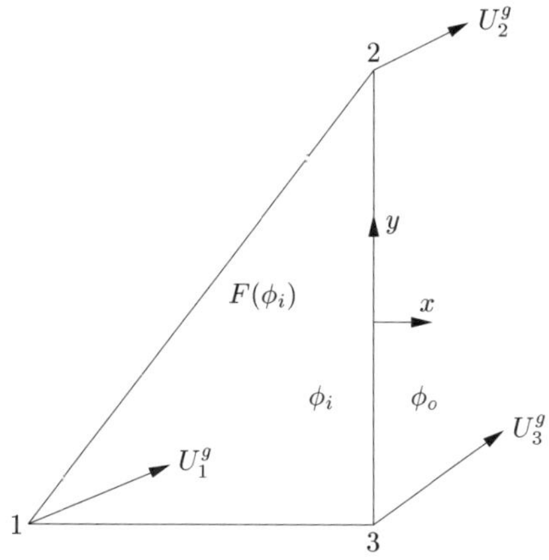

FIG. 10.20. Notation for a triangular element 'e'. The subscripts i and e denote interior and exterior quantities, respectively, and $\mathbf{U}^g$ is the grid velocity.

From the Reynolds transport theorem, we have that

$$\partial_t \int_{\Omega(t)} \phi \, d\Omega = \int_{\Omega(\bar{t})} (\phi_t + \phi \nabla \cdot \mathbf{U}^g) \, d\Omega \,,$$

where the partial time derivative on the right-hand side is with respect to the moving ALE grid. The difference between the forms in eqns (10.5.15) and (10.5.16) is that when the time derivative is taken outside the inner product a new term is introduced in eqn (10.5.16) that involves the divergence of the grid velocity. While the two forms are equivalent in the continuous case, they are not necessarily equivalent in the discrete case; see [212] for an interpretation in terms of the *geometric conservation law.*

To compute the boundary terms, we follow an upwind treatment based on *characteristics*, similar to the formulation in stationary domains, including the term representing the grid motion. To this end, we need to linearise the ALE Jacobian *normal to the surface*, i.e.,

$$[\mathbf{A} - U_n^g \mathbf{I}] = \mathbf{R} \, [\mathbf{D} - U_n^g \mathbf{I}] \, \mathbf{L} \,,$$

where U_n^g is the velocity of the grid in the direction $\boldsymbol{n}$. The term $\phi_{\text{up}} - \phi_i$ expresses a jump in the variable at the inflow edges of the element resulting from an upwind treatment. In the case of a system of conservation laws the numerical flux $\tilde{\boldsymbol{f}}$ is computed from an approximate Riemann solver, as in the stationary domains.

The grid velocity $\mathbf{U}^g$ can be computed from the solution of Laplace equations with boundary conditions of the velocity at the moving body boundaries. However, an alternative faster algorithm is based on graph theory and details can be found in [270].

10.5.5 *Stability and over-integration*

In Section 6.5.3 we studied the issue of the accuracy of quadratic and cubic nonlinearities and the removal of polynomial aliasing via super-collocation based on consistent integration. In this section, we present some examples of how this instability manifests itself in simulations of high Reynolds number flows.

It has been observed in numerical experiments using a discontinuous Galerkin solver that iso-parametric representation of geometry can lead to a weak instability [38]. Experiments with the spectral/*hp* discontinuous Galerkin method did not demonstrate this problem, even for very low polynomial order discretisations. However, it was observed that *consistent* integration in computing inner products in the weak formulation is important in obtaining stable results for flows with steep gradients, such as high Reynolds number flows.

Let us re-examine the quadrature rules that we employ in the discrete variational DGM problem. The Gauss–Jacobi quadrature is exact for standard inner products in non-curvilinear geometries if the quadrature order $Q_{2\mathrm{D}}$, as defined in Fig. 10.14, is based on the order of the spectral basis P. In fact, the quadrature orders shown in Fig. 10.14 only satisfy the linear theory [100]. However, this is not sufficient in nonlinear problems, especially at marginal resolutions, as demonstrated in [270, 309] for high Reynolds number simulations.

Let us consider the flow past an NACA0012 aerofoil studied in [309]. Simulations at Reynolds number (based on the chord length) Re $=$ 1000 led to asymptotically stable results at $Q_{2\mathrm{D}} = P = 3$. However, increasing the Reynolds number to Re $=$ 10 000 led to instabilities. This is shown in Fig. 10.21 where we plot the case $Q_{2\mathrm{D}} = P = 3$ in (a) and the case $Q_{2\mathrm{D}} = 4$, $P = 3$ in (b). We see that the latter is stable, while the former develops very steep gradients close to the leading edge that eventually render the computation unstable. This

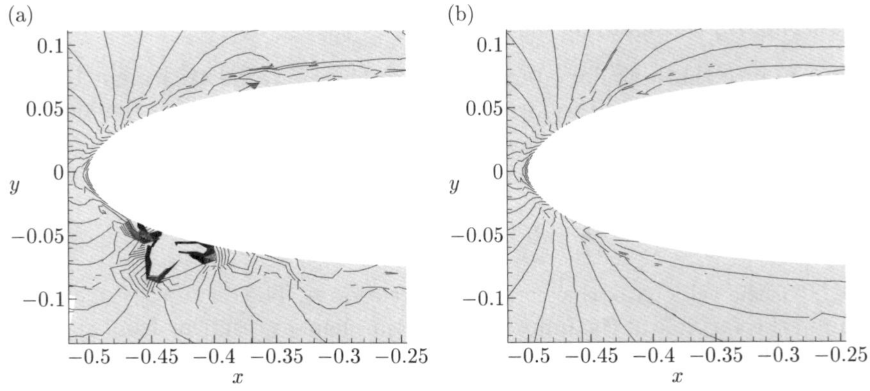

FIG. 10.21. Flow past an NACA0012 aerofoil [309]. Density contours for Re $=$ 10 000 and Ma $=$ 0.2. The simulation in (a) was performed with $Q_{2\mathrm{D}} = P = 3$, and in (b) with $Q_{2\mathrm{D}} = 4$ and $P = 3$.

is documented more clearly in Fig. 10.22 where we plot the time histories of the modal advection contributions from the boundary and the interior of an element computed at a point close to the leading edge of the aerofoil.

If we simply increase both the interpolation order and the quadrature order so that $Q_{2D} = P = 4$ the method still diverges, which reinforces further the finding on consistent integration. It is also of interest to determine if the source

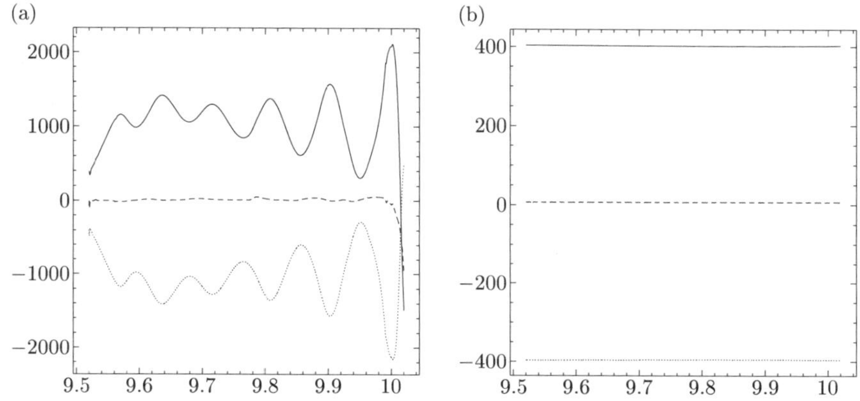

FIG. 10.22. Time history of advection boundary (dotted lines) and interior (solid lines) contributions of an element close to the leading edge for Re = 10 000 and Ma = 0.2 (the dashed line denotes the sum of the two contributions). The simulation in (a) was performed with $Q_{2D} = P = 3$, and in (b) with $Q_{2D} = 4$ and $P = 3$.

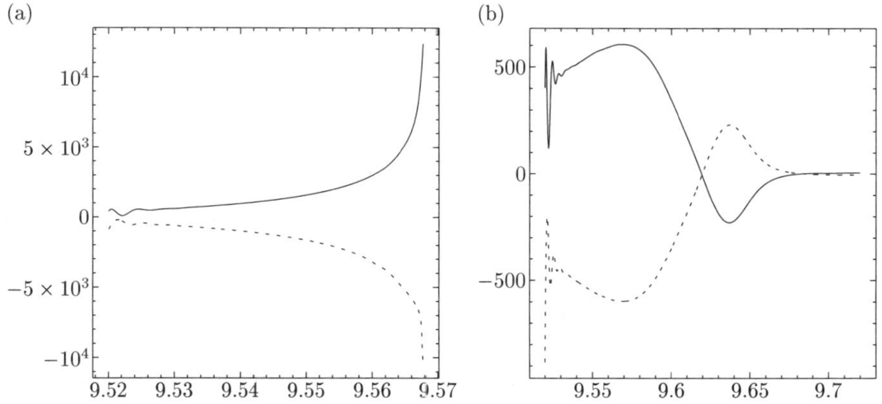

FIG. 10.23. Time history of advection boundary (dotted lines) and interior (solid lines) contributions of an element close to the leading edge for Euler simulations at Ma = 0.2. The simulation in (a) was performed with $Q_{2D} = P = 3$, and in (b) with $Q_{2D} = 4$ and $P = 3$.

of the instability comes from the treatment of the advection terms or the diffusion terms. To this end, two Euler simulations were performed in [309] for the same problem: one with $Q_{2D} = P = 3$ and the second case with $Q_{2D} = 4$ and $P = 3$. As initial conditions, the Navier–Stokes solution at Re = 10 000 was used in both cases, as it had more complex initial structure than a uniform state and thus instabilities were triggered earlier. An unstable computation was obtained in the former case but a stable one in the latter case, as shown in Fig. 10.23 that plots the histories of the same corresponding boundary and interior contributions, as before.

Stable results were obtained in the previous example by simply increasing the quadrature order by one. However, numerical experiments have shown that for more complex cases this quadrature is inadequate as well. For example, numerical simulations performed in [270] with a rapidly-pitching NACA0015 aerofoil at Re = 45 000 revealed that, for stability, over-integration with quadrature order $Q_{2D} = 2P$ is required to resolve the cubic nonlinearities. This was justified in Section 6.5.3 as effective polynomial de-aliasing, where the following semi-empirical rule was proposed for simulating compressible flows (cubic nonlinearities).

- Super-collocation with $2P$ grid (quadrature) points per direction followed by a Galerkin projection leads to a de-aliased simulation of compressible flows at high Reynolds number.

10.6 Shock-fitting techniques

For supersonic flows at high Mach numbers, shocks are usually very sharp and can be modelled as discontinuities. The range for which such an assumption is valid has been determined by Moretti and Salas [336], who constructed appropriate plots as a function of Mach and Reynolds number. For a sharp shock, a shock-fitting technique can be applied where the shock is treated as a free boundary and differentiation across the shock is avoided. Spectral methods work very well and provide uniformly high accuracy in these cases. Salas *et al.* [409] were the first to apply spectral methods to a problem involving a shock interacting with a vortex, and Hussaini *et al.* [250] considered supersonic flow around a cylinder.

In these early spectral simulations, as well as the spectral simulations for flow around a sphere reported in [481], filtering was employed and a low Courant number was used to stabilise the method. In addition, the solutions were not converged fully to steady states, even after a large number of time-steps. The use of explicit filtering in spectral simulations of smooth solutions or smoothing through artificial dissipation is usually suggestive of improper enforcement of boundary conditions. This was indeed the case for the early work, where a simple expression for the shock speed, described in [86], was employed. Although appropriate for finite differences, this formula is not appropriate for spectral-based discretisations. We therefore follow the analysis in [282], where the correct relation for the shock speed was derived using the theory of characteristics.

Let us consider a supersonic flow past a blunt body and denote upstream (known) conditions by the subscript '1' and downstream conditions by the sub-

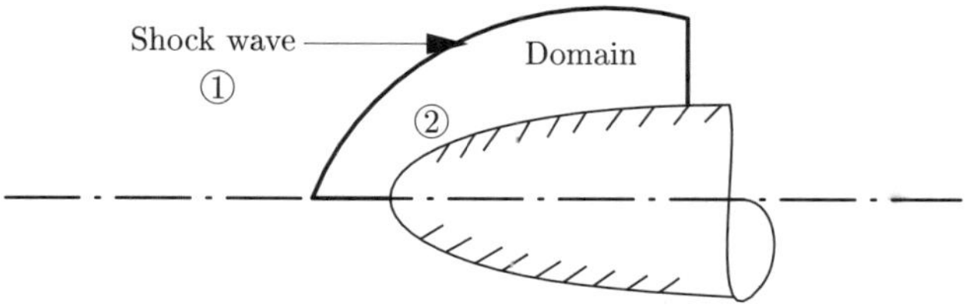

FIG. 10.24. Schematic showing the computational domain for a blunt body geometry.

script '2'. The computational domain is then between the shock (free boundary) and the blunt body (see Fig. 10.24). The shock speed is computed by integrating the shock acceleration in time, and the shock position is updated by integrating the shock speed in time. Assuming that all variables are normalised with corresponding values at upstream infinity, we apply the Rankine–Hugoniot relations that describe the pressure and normal velocity changes across the shock:

$$P_2 = P_1 + \ln\left(V_1^2 - \frac{\gamma - 1}{2}\right) - \ln\left(\frac{\gamma + 1}{2}\right), \tag{10.6.1a}$$

$$V_2 = \frac{\gamma - 1}{\gamma + 1} V_1 + \frac{2\gamma}{\gamma + 1} V_1^{-1}, \tag{10.6.1b}$$

where $P = \ln p$, $V_i = \mathbf{u}_i \cdot \boldsymbol{n} - w n_1$ is the normal gas velocity relative to the shock with w being the shock speed, and the vector $\boldsymbol{n} = (n_1, n_2)$ is normal to the body. (In a polar coordinate system $n_1 = \mathbf{r} \cdot \boldsymbol{n}$.) Taking derivatives in time, we obtain

$$\dot{p}_2 = G\dot{V}_1, \quad \dot{V}_2 = F\dot{V}_1,$$

where we have defined

$$F = \frac{\gamma - 1}{\gamma + 1} - \frac{2\gamma}{(\gamma + 1)V_1^2}, \quad G = \frac{2V_1}{V_1^2 - (\gamma - 1)/2},$$

with the additional assumption that the inflow conditions are time independent. Substituting these expressions into the pressure equation derived with a characteristic condition (see [282]), we obtain the shock acceleration

$$\dot{w} = \frac{\mathbf{u}_1 \cdot \dot{\boldsymbol{n}}(c_0 G - \gamma F) - w\dot{n}_1(c_0 G - \gamma F + \gamma) + \gamma \mathbf{u}_2 \cdot \dot{\boldsymbol{n}} + c_0 C}{n_1(c_0 G - \gamma F + \gamma)}, \tag{10.6.2}$$

where

$$C = (\mathbf{u}_1 - c_0 \boldsymbol{n}) \cdot \nabla P + R_p - \frac{\gamma}{c_0}(n_1 R_u + n_2 R_v),$$

with R_p, R_u, and R_v being the characteristics for pressure and the two velocity components. All quantities on the right-hand side can be computed from spectral derivatives of flow variables and the shock speed. The time derivatives of the shock normal are also related to the shock speed. The approximation of shock normals and their derivatives are also computed spectrally.

If, instead of the characteristic compatibility condition being used to calculate the waves intersecting the shock from downstream, the original differential equation is used, then a simpler relation for the shock acceleration is obtained:

$$\dot{w} = \frac{G(\mathbf{u}_1 \cdot \dot{\boldsymbol{n}} - w\dot{n}_1) - \dot{P}_2}{Gn_1}. \tag{10.6.3}$$

It was demonstrated in [282] for a blunt-nose corner problem that the use of eqn (10.6.3) in spectral methods may lead to instabilities. Corresponding results are reproduced in Fig. 10.25, where we see that after about 300 iterations the solution obtained with the shock acceleration of eqn (10.6.3) diverges.

Multi-domain spectral methods have been developed by Kopriva [278, 279] and Kopriva and Kolias [281] in both single and staggered grids. The interface conditions are based on characteristic decomposition and approximate Riemann solvers, as discussed in Section 10.3. A new type of interface points occurs where the shock is divided at an elemental interface. The difficulty at those points is that there are two values of the shock normal vector corresponding to two adjacent elements. The correct choice is to use the normal vector and shock acceleration computed on the upstream side of the interface to determine the flow. For Navier–Stokes simulations the diffusion terms can be treated with a penalty method as before (see Section 10.5), with the penalty parameter being proportional to the square of the number of collocation points. This will impose a flux continuity weakly. For sharp shocks the diffusion terms can be neglected close to the shock.

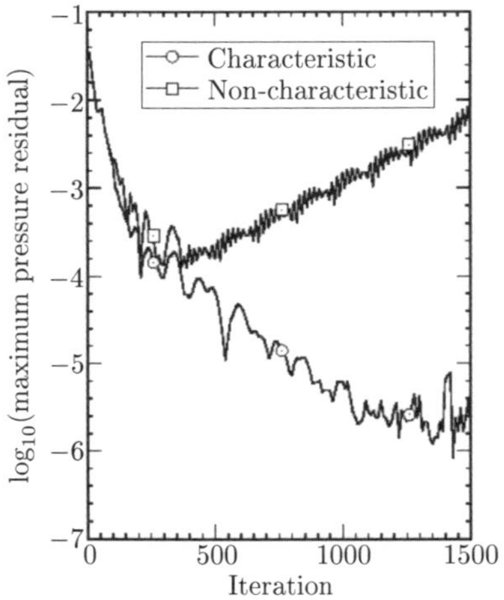

FIG. 10.25. Comparison of the pressure residual for the two shock acceleration formulae for a blunt-nose corner problem. ([282].)

The shock-fitting multi-domain spectral method is particularly suitable for solving high Mach number flow problems. We present here the flow past a spherically-blunted 15° half-cone, simulated in [280]. The Mach number is 10.6 and the Reynolds number (based on the nose radius) is Re = 83 300. While the wake is turbulent at this Reynolds number, the boundary layer is still laminar. The spectral element mesh is shown in Fig. 10.26; it consists of fourteen elements and it was designed so that it resolves both the boundary layer and the entropy layer. Figure 10.27(a) shows a comparison of the pressure coefficient along the surface of the cone with an inviscid simulation and experimental results. The arrows show the position of elemental interfaces. Figure 10.27(b) shows the heat flux normalised to the value at the nose for a constant temperature prescribed along the surface of the cone. The computed solution is in very good agreement with experimental results; usually the heat flux is the most sensitive quantity to discretisation.

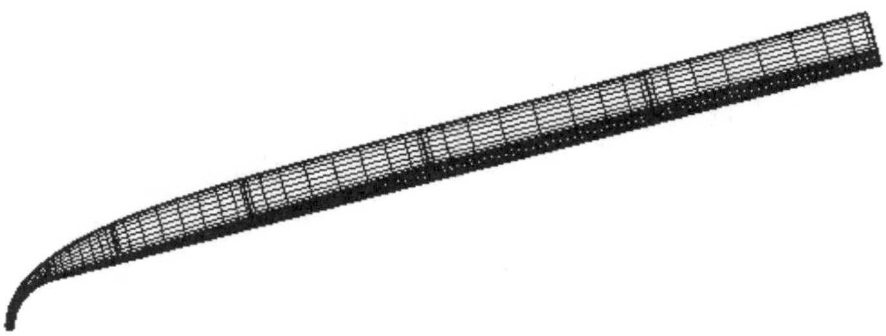

FIG. 10.26. Spectral element mesh for the half-cone simulation. (Courtesy of D. A. Kopriva.)

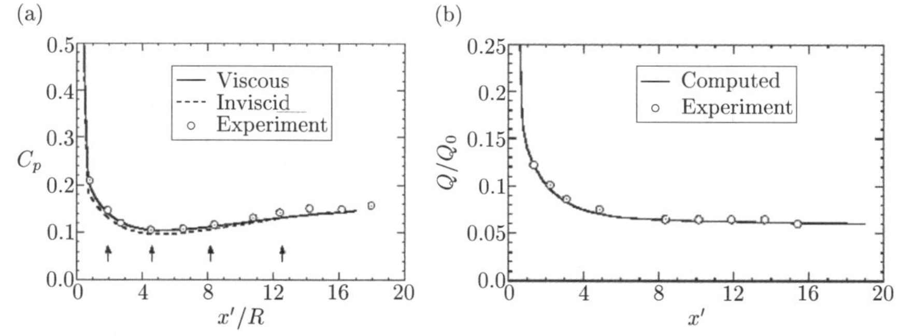

FIG. 10.27. (a) Pressure coefficient along the cone surface for Ma = 10.6, Re = 83 300, Pr = 0.78, and T_∞ = 85.2°R. (b) Normalised heat flux along the cone surface. (Courtesy of D. A. Kopriva.)

10.7 Magneto-hydrodynamics (MHD)

Plasmas can be modelled accurately using kinetic theory, especially partially ionised plasmas. However, this involves solutions of the seven-dimensional Boltzmann equation coupled with Maxwell's equations, which is prohibitively expensive. Particle-based methods, such as direct simulation Monte Carlo (DSMC) [50], are possible alternatives but, for efficiency, they need to be coupled with continuum fluid equations. Such hybrid kinetic–continuum methods are not very mature, and open issues with the DSMC remain the treatment of electrons as well as the modelling of charged particle collisions.

Continuum-based, i.e., purely fluid, approaches have been successful in describing the macroscopic features of high-density plasmas in many diverse applications [106, 121, 374, 382, 504]. They are derived from the Boltzmann equation by taking appropriate moments for each species. The standard mathematical description is that of single-fluid MHD with magnetic and gas dynamic viscous effects. However, a single-fluid MHD description has its limitations as it cannot account for local thermodynamic non-equilibrium effects and cannot consider non-neutral regions and sheath interactions. To this end, two-fluid plasma models and corresponding solvers have been under development more recently [437]. They can overcome certain limitations of the single-fluid MHD model such as the Hall effect and diamagnetic terms, which model contributions to the ion current and the finite Larmor radius of the plasma constituents. However, they still assume local thermodynamic equilibrium within each fluid. From the computational standpoint, the two-fluid model is a much more complex system to solve, especially for large values of the Hall parameter [437].

In the following, we present a discontinuous Galerkin method for single-fluid plasmas based on work of [483] and [303]. In addition to this approach, other efforts to develop effective high-order methods for plasma flows have been reported in [176, 177, 192].

Discontinuous Galerkin methods address two of the main difficulties in employing a high-order discretisation for the solution of hyperbolic conservation laws:

1. maintaining monotonicity for non-smooth solutions; and
2. preserving conservativity.

In the MHD framework, such difficulties are compounded by the imposition of the *divergence-free* condition for the magnetic field, which results in a loss of the hyperbolicity of the ideal MHD equations. This condition has been dealt with by employing staggered grids; however, such an approach cannot be easily incorporated in high-order discretisations. Alternative approaches include the locally divergence-free discontinuous Galerkin method developed by Li and Shu [301] and the development of an extended Riemann solver by Powell [382]; the latter is easily extended to multiple dimensions and also to high-order discretisations. In some approaches, the divergence-free condition is not imposed directly during time-stepping, but the initial conditions are projected in the divergence-free space. The assumption that the *flux* of the divergence of the magnetic field satisfies a homogeneous discrete parabolic equation with homogeneous boundary conditions will lead to zero discrete divergence at all times. However, in practice, discretisation errors or other inconsistencies may trigger large divergence errors for such cases.

10.7.1 *Governing equations*

The governing equations for single-fluid compressible magneto-hydrodynamics (MHD) can be expressed in conservative form as follows.

Mass conservation:

$$\frac{\partial \rho}{\partial t} = -\nabla \cdot (\rho \mathbf{u}) . \tag{10.7.1a}$$

Momentum conservation:

$$\frac{\partial (\rho \mathbf{u})}{\partial t} = -\nabla \cdot \left[\rho \mathbf{u}\mathbf{u}^{\top} - \mathbf{B}\mathbf{B}^{\top} - \left(p + \frac{1}{2}|\mathbf{B}|^2 \right) \mathbf{I} - \frac{1}{S_v} \boldsymbol{\tau} \right] . \tag{10.7.1b}$$

Magnetic field:

$$\frac{\partial \mathbf{B}}{\partial t} = -\nabla \times \left[\mathbf{B} \times \mathbf{u} + \frac{1}{S_r} \nabla \times \mathbf{B} \right] . \tag{10.7.1c}$$

Total energy conservation:

$$\begin{aligned} \frac{\partial E}{\partial t} = -\nabla \cdot & \left[(E+p)\mathbf{u} + \left(\frac{1}{2}|\mathbf{B}|^2 \mathbf{I} - \mathbf{B}\mathbf{B}^{\top} \right) \cdot \mathbf{u} - \frac{1}{S_v} \mathbf{u} \cdot \boldsymbol{\tau} \right] \\ & + \nabla \cdot \left[\frac{1}{S_r} \left(\mathbf{B} \cdot \nabla \mathbf{B} - \nabla \left(\frac{1}{2} |\mathbf{B}|^2 \right) \right) - \frac{\nabla T}{S_v \mathrm{Pr}} \right] . \end{aligned} \tag{10.7.1d}$$

Magnetic flux constraint:

$$\nabla \cdot \mathbf{B} = 0 . \tag{10.7.1e}$$

Simplified Ohm's law:

$$E = \eta \mathbf{J} - \mathbf{u} \times \mathbf{B} . \tag{10.7.1f}$$

Stress tensor:

$$\boldsymbol{\tau} = (\partial_j u_i + \partial_i u_j) - \frac{2}{3} \nabla \cdot \mathbf{u} \delta_{ij} . \tag{10.7.1g}$$

In eqns (10.7.1a)–(10.7.1g) all the parameters are as defined in Table 10.2.

Alternatively, in flux form, they can be expressed compactly as follows:

$$\begin{aligned} \frac{\partial \mathbf{U}}{\partial t} &= -\frac{\partial \mathbf{F}_x}{\partial x} - \frac{\partial \mathbf{F}_y}{\partial y} - \frac{\partial \mathbf{F}_z}{\partial z} + \frac{\partial \mathbf{F}_x^{\nu}}{\partial x} + \frac{\partial \mathbf{F}_y^{\nu}}{\partial y} + \frac{\partial \mathbf{F}_z^{\nu}}{\partial z} + S_{\mathrm{MHD}} , \\ \nabla \cdot \mathbf{B} &= 0 , \\ \mathbf{U} &= (\rho, \rho u, \rho v, \rho w, B_x, B_y, B_z, E) , \end{aligned}$$

where detailed expressions of all flux and source terms can be found in [485].

TABLE 10.2. Variables and parameters used in the equations of single-fluid compressible MHD.

Variable	Description
$\rho(\boldsymbol{x},t)$	density
$\mathbf{u}(\boldsymbol{x},t) = (u,v,w)(\boldsymbol{x},t)$	velocity
$\mathbf{B}(\boldsymbol{x},t) = (B_x,B_y,B_z)(\boldsymbol{x},t)$	magnetic field
$E = p/(\gamma-1) + \frac{1}{2}(\rho\mathbf{u}\cdot\mathbf{u} + \mathbf{B}\cdot\mathbf{B})$	total energy
p	pressure
$T = p/\rho$	temperature
$\mathrm{Pr} = c_p\mu/\kappa$	Prandtl number
η	magnetic resistivity
μ	viscosity
$S_v = \rho_0 V_A L_0/\mu$	viscous Lundquist number
$S_r = V_A L_0/\eta$	resistive Lundquist number
c_p	specific heat at constant pressure
$V_A^2 = \mathbf{B}\cdot\mathbf{B}/\rho$	Alfven wave speed
$A = \sqrt{V_A^2/V_0^2}$	Alfven number
e	electron charge
n_e	density of electrons
c	velocity of light

10.7.2 $\nabla \cdot \mathbf{B} = 0$ *constraint*

The presence of the $\nabla \cdot \mathbf{B} = 0$ constraint implies that the equations do not have a strictly hyperbolic character. It has been shown in [73] that even a small divergence in the magnetic fields can dramatically change the character of results from numerical simulations. An effective approach was developed by Powell in [382]. The idea is to reformulate the Jacobian matrix to include an 'extra wave', i.e., the divergent mode that corresponds to the velocity u. This way, the degeneracy associated with the divergence-free condition is avoided while the rest of the eigenvalues of the Jacobian remain the same.

The primitive Jacobian matrix $\mathbf{A}_p$ for *single-fluid/one-temperature equations* has the following form in three dimensions:

$$\mathbf{A}_p = \begin{bmatrix} u & \rho & 0 & 0 & 0 & 0 & 0 & 0 \\ 0 & u & 0 & 0 & -B_x/\rho & B_y/\rho & B_z/\rho & 1/\rho \\ 0 & 0 & u & 0 & -B_y/\rho & -B_x/\rho & 0 & 1/\rho \\ 0 & 0 & 0 & u & -B_z/\rho & 0 & -B_x/\rho & 0 \\ 0 & 0 & 0 & 0 & 0 & 0 & 0 & 0 \\ 0 & B_y & -B_x & 0 & -v & u & 0 & 0 \\ 0 & B_z & 0 & -B_x & -w & 0 & u & 0 \\ 0 & \gamma p & 0 & 0 & -(\gamma-1)\mathbf{u}\cdot\mathbf{B} & 0 & 0 & u \end{bmatrix}.$$

To modify the governing equations so as to make $\mathbf{A}_p$ non-singular, using Powell's criteria presented in [382], $\mathbf{A}_p$ is modified to be $\mathbf{A}'_p$:

$$\mathbf{A}'_p = \begin{bmatrix} u & \rho & 0 & 0 & 0 & 0 & 0 & 0 \\ 0 & u & 0 & 0 & 0 & B_y/\rho & B_z/\rho & 1/\rho \\ 0 & 0 & u & 0 & 0 & B_x/\rho & 0 & 1/\rho \\ 0 & 0 & 0 & u & 0 & 0 & B_x/\rho & 0 \\ 0 & 0 & 0 & 0 & u & 0 & 0 & 0 \\ 0 & B_y & -B_x & 0 & 0 & u & 0 & 0 \\ 0 & B_z & 0 & -B_x & 0 & 0 & u & 0 \\ 0 & \gamma p & 0 & 0 & 0 & 0 & 0 & u \end{bmatrix}.$$

This modification effectively corresponds to adding a source term proportional to $\nabla \cdot \mathbf{B}$,

$$\mathbf{S}_{\text{Powell}} = -(\nabla \cdot \mathbf{B})(0, B_x, B_y, B_z, u, v, w, \mathbf{u} \cdot \mathbf{B})^\top,$$

to the right-hand side of all evolution equations.

10.7.3 *A discontinuous Galerkin MHD solver*

In the following, we outline the main implementation steps for the compressible MHD system of equations using the discontinuous Galerkin method.

Implementation of the inviscid terms

The inviscid fluxes and their derivatives are evaluated in the interior of the elements and correction terms (jumps) for the discontinuities are added in the flux between any two adjacent elements. In order to evaluate the Euler flux at an element interface, we use a one-dimensional Riemann solver to supply an upwinded flux there. At a domain boundary, we provide far-field conditions and treat the exterior boundary as the boundary of a 'ghost' element, as also applied in the Euler equations. This way, we can use the same Riemann solver at all element boundaries.

We linearise the one-dimensional flux $\mathbf{F}_x$ in the normal direction to a shared element boundary using the average of the state vector at either side of the element boundary. That is, since $\mathbf{F}_x$ is a nonlinear function of the state vector, we use the average state to form an approximation to the conservative Jacobian of the flux vector $\mathbf{A}_c$, see eqn (10.7.2). The Jacobian matrix for the flux vector of the evolution equations expressed in primitive variables is simpler than in the conserved form. Thus, it is better to perform the linearisation about the primitive form and transform to the conserved form.

The left and right eigenvectors of the primitive Jacobian matrix $\mathbf{A}_p$ are as follows.

Entropy wave:

$$\begin{aligned}\lambda_e &= u\,,\\ l_e &= \left(1,0,0,0,0,0,0,-\frac{1}{a^2}\right),\\ r_e &= (1,0,0,0,0,0,0,0)^\top\,.\end{aligned}$$

Alfven waves:

$$\begin{aligned}\lambda_a &= u \pm \frac{B_x}{\sqrt{\rho}}\,,\\ l_a &= \frac{1}{\sqrt{2}}\left(0,0,-\beta_z,\beta_y,0,\pm\frac{\beta_z}{\sqrt{\rho}},\mp\frac{\beta_y}{\sqrt{\rho}},0\right),\\ r_a &= \frac{1}{\sqrt{2}}(0,0,-\beta_z,\beta_y,0,\pm\beta_z\sqrt{\rho},\mp\beta_y\sqrt{\rho},0)^\top\,.\end{aligned}$$

Fast waves:

$$\begin{aligned}\lambda_f &= u \pm c_f\,,\\ l_f &= \frac{1}{2a^2}\left(0,\pm\alpha_f c_f,\mp\alpha_s c_s\beta_x\beta_y,\mp\alpha_s c_s\beta_x\beta_z,0,\frac{\alpha_s\beta_y a}{\sqrt{\rho}},\frac{\alpha_s\beta_z a}{\sqrt{\rho}},\frac{\alpha_f}{\rho}\right),\\ r_f &= (\rho\alpha_f,\pm\alpha_f c_f,\mp\alpha_s c_s\beta_x\beta_y,\mp\alpha_s c_s\beta_x\beta_z,0,\alpha_s\beta_y a\sqrt{\rho},\alpha_s\beta_z a\sqrt{\rho},\alpha_f\gamma p)^\top\,.\end{aligned}$$

Slow waves:

$$\begin{aligned}\lambda_s &= u \pm c_s\,,\\ l_s &= \frac{1}{2a^2}\left(0,\pm\alpha_s c_s,\pm\alpha_f c_f\beta_x\beta_y,\pm\alpha_f c_f\beta_x\beta_z,0,-\frac{\alpha_f\beta_y a}{\sqrt{\rho}},-\frac{\alpha_f\beta_z}{\sqrt{\rho}},\frac{\alpha_s}{\rho}\right),\\ r_s &= (\rho\alpha_s,\pm\alpha_s c_s,\pm\alpha_f c_f\beta_x\beta_y,\pm\alpha_f c_f\beta_x\beta_z,0,-\alpha_f\beta_y a\sqrt{\rho},-\alpha_f\beta_z a\sqrt{\rho},\alpha_s\gamma p)^\top\,.\end{aligned}$$

Here,

$$(a^*)^2 = \frac{\gamma p + \mathbf{B}\cdot\mathbf{B}}{\rho}\,,\quad c_f^2 = \frac{1}{2}\left((a^*)^2 + \sqrt{(a^*)^4 - 4\frac{\gamma p B_x^2}{\rho^2}}\right),$$

$$c_s^2 = \frac{1}{2}\left((a^*)^2 - \sqrt{(a^*)^4 - 4\frac{\gamma p B_x^2}{\rho^2}}\right),$$

$$\alpha_f^2 = \frac{a^2 - c_s^2}{c_f^2 - c_s^2}\,,\quad \alpha_s^2 = \frac{c_f^2 - a^2}{c_f^2 - c_s^2}\,,$$

$$\beta_x = \operatorname{sgn}(B_x)\,,\quad \beta_y = \frac{B_y}{\sqrt{B_y^2 + B_z^2}}\,,\quad \beta_z = \frac{B_z}{\sqrt{B_y^2 + B_z^2}}\,.$$

We can transform between the primitive variables **W** and conserved variables **U** with the following transform:

$$\mathbf{A}_c = \frac{\partial \mathbf{U}}{\partial \mathbf{W}} \mathbf{A}_p \frac{\partial \mathbf{W}}{\partial \mathbf{U}}, \tag{10.7.2}$$

where

$$\mathbf{U} = (\rho, \rho u, \rho v, \rho w, B_x, B_y, B_z, E)$$

are the *conserved variables*, and

$$\mathbf{W} = (\rho, u, v, w, B_x, B_y, B_z, p)$$

are the *primitive variables*. This leads to

$$\frac{\partial \mathbf{U}}{\partial \mathbf{W}} = \begin{bmatrix} 1 & 0 & 0 & 0 & 0 & 0 & 0 & 0 \\ u & \rho & 0 & 0 & 0 & 0 & 0 & 0 \\ v & 0 & \rho & 0 & 0 & 0 & 0 & 0 \\ w & 0 & 0 & \rho & 0 & 0 & 0 & 0 \\ 0 & 0 & 0 & 0 & 1 & 0 & 0 & 0 \\ 0 & 0 & 0 & 0 & 0 & 1 & 0 & 0 \\ 0 & 0 & 0 & 0 & 0 & 0 & 1 & 0 \\ \frac{1}{2}\mathbf{u}\cdot\mathbf{u} & \rho u & \rho v & \rho w & B_x & B_y & B_z & 1/(\gamma-1) \end{bmatrix}$$

and

$$\frac{\partial \mathbf{W}}{\partial \mathbf{U}} = \begin{bmatrix} 1 & 0 & 0 & 0 & 0 & 0 & 0 & 0 \\ -u/\rho & 1/\rho & 0 & 0 & 0 & 0 & 0 & 0 \\ -v/\rho & 0 & 1/\rho & 0 & 0 & 0 & 0 & 0 \\ -w/\rho & 0 & 0 & 1/\rho & 0 & 0 & 0 & 0 \\ 0 & 0 & 0 & 0 & 1 & 0 & 0 & 0 \\ 0 & 0 & 0 & 0 & 0 & 1 & 0 & 0 \\ 0 & 0 & 0 & 0 & 0 & 0 & 1 & 0 \\ \frac{1}{2}\bar{\gamma}\mathbf{u}\cdot\mathbf{u} & -\bar{\gamma}u & -\bar{\gamma}v & -\bar{\gamma}w & -\bar{\gamma}B_x & -\bar{\gamma}B_y & -\bar{\gamma}B_z & \bar{\gamma} \end{bmatrix},$$

where $\bar{\gamma} = \gamma - 1$.

We are now in a position to evaluate the upwinded flux at the element boundaries, similar to the flux in compressible flows, i.e.,

$$\tilde{\mathbf{f}}(\mathbf{U}_i, \mathbf{U}_o) = \frac{1}{2}\left(\mathbf{F}(\mathbf{U}_i) + \mathbf{F}(\mathbf{U}_o) - \frac{\partial \mathbf{U}}{\partial \mathbf{W}} \sum_{k=1}^{k=9} \alpha_k |\lambda_k| \mathbf{r}_k\right),$$

$$\alpha_k = \mathbf{l}_k \cdot \frac{\partial \mathbf{W}}{\partial \mathbf{U}} (\mathbf{U}_o - \mathbf{U}_i),$$

where 'i' now denotes interior element and 'o' denotes exterior element. Here the $\mathbf{l}_k$ and $\mathbf{r}_k$ are the ordered *left* and *right* eigenvectors, respectively, of the *primitive Jacobian matrix*. We have to apply the $\partial\mathbf{U}/\partial\mathbf{W}$ operator to the right eigenvectors to calculate the conserved flux. The λ_ks are the wave speeds associated with the eigenvectors. In some cases, replacing the calculated fluxes for the magnetic fields with Lax–Friedrichs fluxes leads to a more stable and robust scheme, especially in unstructured meshes.

Implementation of the viscous terms

The viscous terms are evaluated in two steps. First, we obtain the spatial derivatives of the primitive variables using the discontinuous Galerkin approach. Then we repeat the process for each of the viscous fluxes using these derivatives. If we employ Dirichlet boundary conditions for the momentum and energy variables, then we set these terms explicitly after the fluxes have been evaluated and then project the result using the orthogonal basis. The Bassi–Rebay fluxes, i.e., the average of the variables and fluxes at the interface, lead to a stable and efficient scheme, although this approach leads to sub-optimal performance at low polynomial order P, see Section 7.5.

10.7.4 *Convergence and simulations*

In the following we first test the accuracy of the discontinuous Galerkin method using a three-dimensional analytical solution and verify its exponential convergence. We then present simulations of more complex flows that demonstrate how accurate solutions can be obtained and verified, without the need for remeshing, but by simply increasing the elemental polynomial order P.

Magneto-hydrostatic test case

A simple test for debugging MHD codes is to consider a steady irrotational magnetic field and zero velocity. The following exact solution was first presented in [383]:

$$\rho = 1\,, \quad u = 0\,, \quad v = 0\,,$$
$$E = 19.84 + \frac{e^{-2\pi y}}{2}\,, \quad B_x = -\cos(\pi x)e^{-\pi y}\,, \quad B_y = \sin(\pi x)e^{-\pi y}\,.$$

The irrotational magnetic field implies that the Lorentz force is zero, so the momentum equations are trivially satisfied. The magneto-viscous term is zero and the $\mathbf{u} \times \mathbf{B}$ term is also zero. Thus, the compressible MHD equations are satisfied.

The two-dimensional test case was extended by Warburton [483] to three dimensions. The test is performed as an initial value problem and the following exact solution:

$$\begin{aligned}
\rho = 1\,, \quad u = 0\,, \quad v &= 0\,, \quad w = 0\,,\\
B_x &= [\cos(\pi(y+1)) - \cos(\pi z)]\, e^{-\pi(x+1)}\,,\\
B_y &= \cos(\pi z)e^{-\pi(y+1)} + \sin(\pi(y+1))e^{-\pi(x+1)}\,,\\
B_z &= \sin(\pi z)(e^{-\pi(y+1)} - e^{-\pi(x+1)})\,,\\
E &= 5 + 0.5(B_x^2 + B_y^2 + B_z^2)
\end{aligned}$$

was used as the boundary conditions as well as the initial condition. The domain and discretisation, which consisted of a mix of prisms and hexahedra, is depicted in Fig. 10.28(a). Although such a hybrid discretisation consisting of heterogeneous elements is not needed here for this simple computational domain, this example demonstrates the flexibility of the unified spectral/*hp* basis in discretising complex-geometry domains using different types of elements. We also plot in Fig. 10.28(b) the numerical (maximum pointwise) error in the magnetic field, showing that it decreases exponentially fast to zero with increasing expansion order (*p*-order) while keeping the number of elements fixed.

p-refinement of the Orszag–Tang vortex

A series of detailed simulations was performed in [483] in order to investigate the small-scale structure exhibited in MHD turbulence. This problem was first studied by Orszag and Tang [356] in the incompressible case, and later extended by Dahlburg and Picone [120] to the compressible case. The initial conditions are non-random, periodic fields with the velocity field being solenoidal. The total initial pressure consists of the superposition of an appropriate incompressible pressure distribution upon a flat pressure field corresponding to an initial average Mach number below unity. It was found in [356] and [120] that the coupling of the two-dimensional flow with the magnetic field causes the formation of *singularities*, i.e., excited small-scale structure, which, although not as strong as the singularities in three-dimensional turbulence, they are certainly much stronger than two-dimensional hydrodynamic turbulence. Moreover, it was found in [120] that compressibility causes the formation of additional small-scale structure such as massive jets and bifurcation of eddies. It is of interest therefore to investigate if the spectral/*hp* element method can capture these fine features both on structured and unstructured meshes, as shown in Fig. 10.29.

The initial conditions used were as follows:

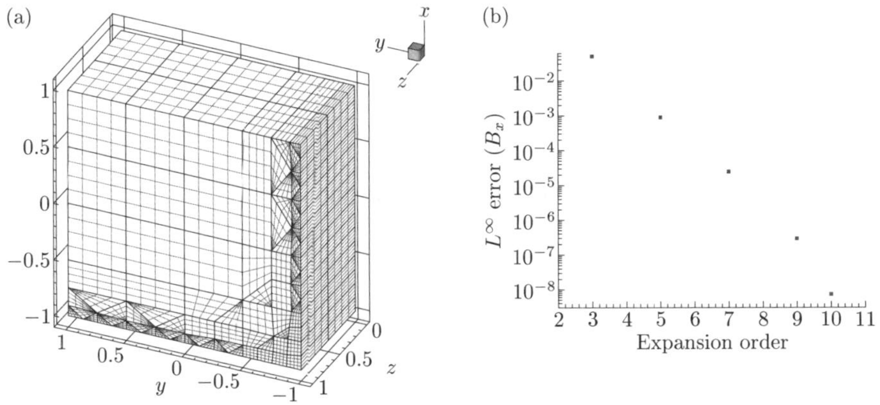

FIG. 10.28. Three-dimensional magneto-hydrostatic test case. (a) Mesh of prisms and hexahedra used. (b) Convergence plot showing the exponential decrease in maximum pointwise error with increasing polynomial order. ([485].)

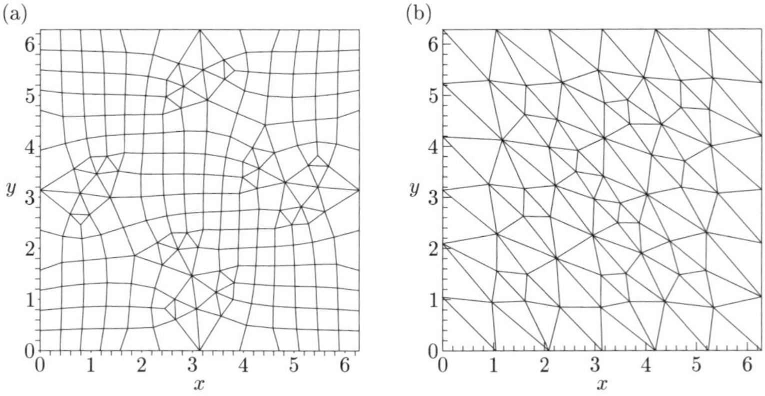

FIG. 10.29. (a) Hybrid mesh, and (b) unstructured mesh used for the Orszag–Tang vortex simulations.

$$\rho = 1\,, \quad u = -\sin\left(\frac{2\pi y}{L}\right), \quad v = \sin\left(\frac{2\pi x}{L}\right),$$

$$B_x = -\sin\left(\frac{2\pi y}{L}\right), \quad B_y = \sin\left(\frac{4\pi x}{L}\right),$$

$$p = C + \frac{1}{4}\cos\left(\frac{8\pi x}{L}\right) + \frac{4}{5}\cos\left(\frac{4\pi x}{L}\right)\cos\left(\frac{2\pi y}{L}\right) - \cos\left(\frac{2\pi x}{L}\right)\cos\left(\frac{2\pi y}{L}\right) + \frac{1}{4}\cos\left(\frac{4\pi y}{L}\right),$$

where C fixes the initial average Mach number and p is the instantaneous pressure for the equivalent incompressible flow.

We first present simulations on a hybrid grid consisting of quadrilaterals and triangles, as shown in Fig. 10.29. The parameters of this simulation are listed in Table 10.3. In Fig. 10.30 we plot streamlines of the compressible flow

TABLE 10.3. Simulation parameters for the compressible Orszag–Tang vortex problem (hybrid mesh).

Parameter	Value
Dimensions	2
S_v	100
S_r	100
A (Alfven number)	1
Mach number	0.4
P	12
Number of quadrilaterals	176
Number of triangles	64

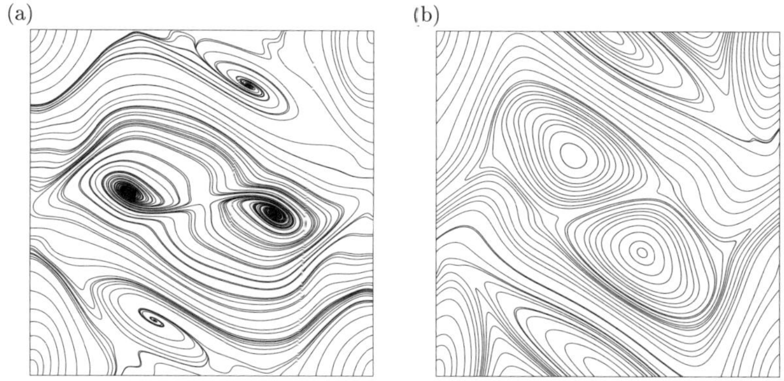

FIG. 10.30. Compressible Orszag–Tang vortex with instantaneous fields at non-dimensional time $t = 2$ and Mach number 0.4. (a) Flow streamlines, and (b) magnetic streamlines. ([485].)

at Mach number 0.4 and non-dimensional time $t = 2$. The compressible flow exhibits structures of finer features compared to the incompressible flow, but the differences in the magnetic field are less obvious [483].

We now consider the effect of p-refinement on accuracy using the unstructured mesh of Fig. 10.29(b). Vorticity is a good indicator of 'noise' in the solution of unsteady simulations as it reflects errors in derivatives. More specifically, we examine the *curl of momentum* as low-resolution simulations result in non-smoothness in this vorticity-like quantity. In Fig. 10.31 we compare the vorticity at time $t = 1$ for the unstructured mesh shown in Fig. 10.29(b) run with $P = 4$ and $P = 16$. Fig. 10.31(a,b) shows how the vorticity profile varies from the lower-left to the top-right diagonal. The vorticity in this direction should be symmetric about the midpoint. We see that at $P = 4$ the profile is noisy, the peaks are not well resolved, and the symmetry is not very well represented. The results are improved very quickly as we increase the polynomial order P. In Fig. 10.31(b) we present the final results at $P = 16$, and we see that symmetry is restored and the profile is very smooth. Results for the intermediate values of the polynomial order P can be found in [483].

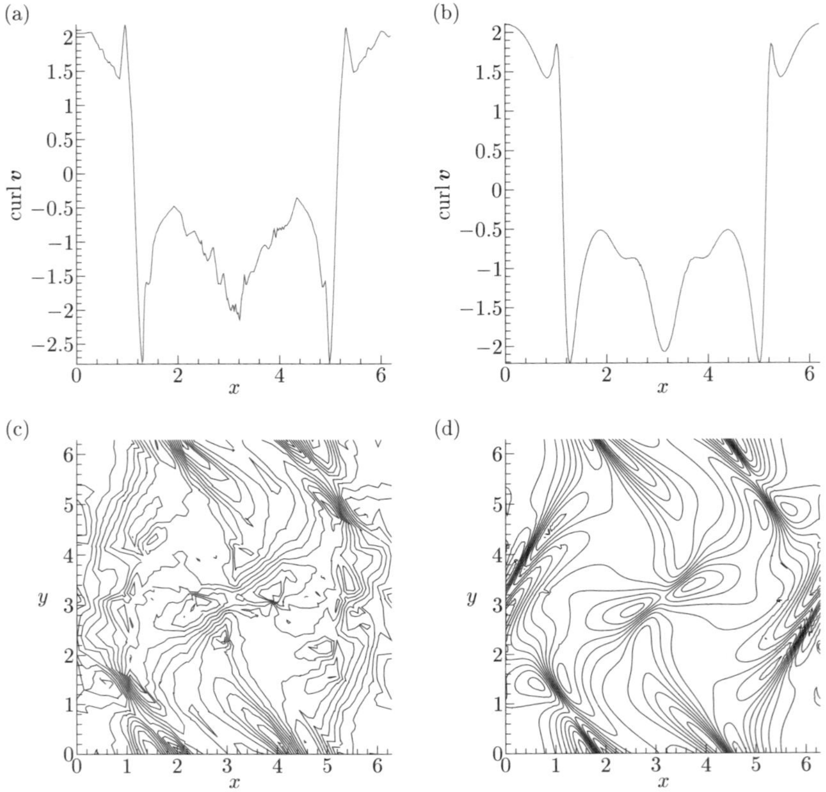

FIG. 10.31. Simulation of the compressible Orszag–Tang vortex on the unstructured mesh; instantaneous fields at $t = 1$, Mach number $= 0.2$, and $K = 132$. (a) Curl of the momentum along the diagonal ($P = 4$). (b) Curl of the momentum along the diagonal ($P = 16$). (c) Isocontours of the curl of the momentum ($P = 4$). (d) Isocontours of the curl of the momentum ($P = 16$). ([485].)

APPENDIX A

JACOBI POLYNOMIALS

A.1 Useful formulae for Jacobi polynomials

Jacobi polynomials $P_n^{\alpha,\beta}(x)$ are a family of polynomial solutions to the singular Sturm–Liouville problem. A significant feature of these polynomials is that they are orthogonal in the interval $[-1,1]$ with respect to the function $(1-x)^\alpha(1+x)^\beta$ $(\alpha,\beta > -1)$. Some useful formulae for Jacobi polynomials are listed below; further properties can be found in Abramowitz and Stegun [1, chapter 22] and Ghizzetti and Ossicini [182, section 3.4].

Rodriguez formula

$$P_n^{\alpha,\beta}(x) = \frac{(-1)^n}{2^n n!}(1-x)^{-\alpha}(1+x)^{-\beta}\frac{\mathrm{d}^n}{\mathrm{d}x^n}\left[(1-x)^{\alpha+n}(1+x)^{\beta+n}\right], \quad \alpha,\beta > -1\,.$$

Differential equation

$$(1-x)(1+x)\frac{\mathrm{d}^2 y(x)}{\mathrm{d}x^2} + [\beta - \alpha - (\alpha+\beta+2)x]\frac{\mathrm{d}y(x)}{\mathrm{d}x} = -\lambda_n y(x) \qquad \text{(A.1.1)}$$

or

$$\frac{\mathrm{d}}{\mathrm{d}x}\left[(1-x)^{1+\alpha}(1+x)^{1+\beta}\frac{\mathrm{d}y(x)}{\mathrm{d}x}\right] = -\lambda_n(1-x)^\alpha(1+x)^\beta y(x)\,, \qquad \text{(A.1.2)}$$

$$\lambda_n = n(n+\alpha+\beta+1)\,,$$

$$y(x) = P_n^{\alpha,\beta}(x)\,.$$

Special cases

Legendre polynomial $\quad (\alpha=\beta=0) \rightarrow L_n(x) = P_n^{0,0}(x)\,,$

Chebyshev polynomial $\quad \left(\alpha=\beta=-\frac{1}{2}\right) \rightarrow T_n(x) = \frac{2^{2n}(n!)^2}{(2n)!}P_n^{-1/2,-1/2}(x)\,.$

Recursion relations

$$\begin{aligned} P_0^{\alpha,\beta}(x) &= 1\,, \\ P_1^{\alpha,\beta}(x) &= \frac{1}{2}[\alpha - \beta + (\alpha+\beta+2)x]\,, \\ a_n^1 P_{n+1}^{\alpha,\beta}(x) &= (a_n^2 + a_n^3 x) P_n^{\alpha,\beta}(x) - a_n^4 P_{n-1}^{\alpha,\beta}(x)\,, \\ a_n^1 &= 2(n+1)(n+\alpha+\beta+1)(2n+\alpha+\beta)\,, \\ a_n^2 &= (2n+\alpha+\beta+1)(\alpha^2-\beta^2)\,, \\ a_n^3 &= (2n+\alpha+\beta)(2n+\alpha+\beta+1)(2n+\alpha+\beta+2)\,, \\ a_n^4 &= 2(n+\alpha)(n+\beta)(2n+\alpha+\beta+2)\,, \end{aligned} \tag{A.1.3}$$

$$\begin{aligned} b_n^1(x)\frac{\mathrm{d}}{\mathrm{d}x}P_n^{\alpha,\beta}(x) &= b_n^2(x)P_n^{\alpha,\beta}(x) + b_n^3(x)P_{n-1}^{\alpha,\beta}(x)\,, \\ b_n^1(x) &= (2n+\alpha+\beta)(1-x^2)\,, \\ b_n^2(x) &= n[\alpha-\beta-(2n+\alpha+\beta)x]\,, \\ b_n^3(x) &= 2(n+\alpha)(n+\beta)\,. \end{aligned} \tag{A.1.4}$$

Special values

$$P_n^{\alpha,\beta}(1) = \binom{n+\alpha}{n} = \frac{(n+\alpha)!}{\alpha!n!}\,, \tag{A.1.5}$$

$$P_n^{\alpha,\beta}(-x) = (-1)^n P_n^{\beta,\alpha}(x)\,. \tag{A.1.6}$$

Orthogonality relations

$$\int_{-1}^{1}(1-x)^\alpha(1+x)^\beta P_n^{\alpha,\beta}(x)P_m^{\alpha,\beta}(x)\,\mathrm{d}x = 0\,, \quad n \neq m\,, \tag{A.1.7}$$

$$\int_{-1}^{1}(1-x)^\alpha(1+x)^\beta P_n^{\alpha,\beta}(x)P_n^{\alpha,\beta}(x)\,\mathrm{d}x = \frac{2^{\alpha+\beta+1}}{2n+\alpha+\beta+1}\,\frac{\Gamma(n+\alpha+1)\Gamma(n+\beta+1)}{n!\Gamma(n+\alpha+\beta+1)}\,.$$

Miscellaneous relations

$$\frac{\mathrm{d}}{\mathrm{d}x}P_n^{\alpha,\beta}(x) = \frac{1}{2}(\alpha+\beta+n+1)P_{n-1}^{\alpha+1,\beta+1}(x)\,, \tag{A.1.8}$$

$$2n\int_{-1}^{x}(1-y)^\alpha(1+y)^\beta P_n^{\alpha,\beta}(y)\,\mathrm{d}y = -(1-x)^{\alpha+1}(1+x)^{\beta+1}P_{n-1}^{\alpha+1,\beta+1}(x)\,, \tag{A.1.9}$$

$$(2n+\alpha+\beta)P_n^{\alpha,\beta-1}(x) = (n+\alpha+\beta)P_n^{\alpha,\beta}(x) + (n+\alpha)P_{n-1}^{\alpha,\beta}(x)\,. \tag{A.1.10}$$

Lagrangian interpolants

Gauss–Jacobi Lagrange interpolation

We consider first the case where the collocation points x_i are the roots of the polynomial $P_Q^{\alpha,\beta}(x)$; they are determined as explained in Appendix B. Following the definition of Lagrangian interpolants of Section 2.3.4, we obtain

$$h_j(x) = \begin{cases} \dfrac{P_Q^{\alpha,\beta}(x)}{[P_Q^{\alpha,\beta}(x_j)]'(x - x_j)}, & x \neq x_j\,, \\ 1\,, & x = x_j\,. \end{cases} \tag{A.1.11}$$

As an example, the Chebyshev Lagrangian interpolant through Q points is

$$h_j(x) = \frac{T_Q(x)}{T_Q(x_j)'(x - x_j)}, \quad x \neq x_j\,,$$

which can also be written as

$$h_j(x) = \frac{2}{Q} \sum_{k=0}^{Q-1} \frac{1}{\bar{c}_k} T_k(x_j) T_k(x)\,,$$

where

$$\bar{c}_k = \begin{cases} 2\,, & k = 0 \text{ or } Q\,, \\ 1\,, & k \neq 0 \text{ or } Q\,. \end{cases}$$

Gauss–Lobatto interpolation

In this case, the collocation points are the zeros of $[P_{Q-2}^{\alpha,\beta}(x)]'$, plus the two end-points at $+1, -1$, as explained in Appendix B. The expression for the Lagrangian interpolant through Q points is then given by

$$h_j(x) = \begin{cases} \dfrac{(-1)^{Q-1}(Q-2)!\Gamma(\beta+2)}{(Q+\alpha+\beta)\Gamma(Q+\beta)}(x-1)[P_{Q-1}^{\alpha,\beta}(x)]'\,, & j = 0\,, \\ \dfrac{(x^2-1)[P_{Q-1}^{\alpha,\beta}(x)]'}{(Q-1)(Q+\alpha+\beta)P_{Q-1}^{\alpha,\beta}(x_j)(x-x_j)}\,, & 1 \leqslant j \leqslant Q-2\,, \\ \dfrac{(Q-2)!\Gamma(\alpha+2)}{(Q+\alpha+\beta)\Gamma(Q+\alpha)}(x+1)[P_{Q-1}^{\alpha,\beta}(x)]'\,, & j = Q-1\,. \end{cases}$$

For example, the Chebyshev Lagrangian interpolant through the $Q-1$ $(1 \leqslant j \leqslant Q-2)$ points is

$$h_j^C(x) = \frac{(-1)^{j+Q-1}(x^2-1)T'_{Q-1}(x)}{\bar{c}_j(Q-1)^2(x-x_j)}\,,$$

and the Legendre interpolant through the $Q-1$ $(1 \leqslant j \leqslant Q-2)$ points is

$$h_j^L(x) = \frac{(x^2-1)L'_{Q-1}(x)}{Q(Q-1)L_{Q-1}(x_j)(x-x_j)}\,.$$

A.2 The Askey scheme of hypergeometric orthogonal polynomials

We present here the Askey family of polynomials that includes the Jacobi polynomials, as well as many other continuous and discrete orthogonal polynomials. They can all be expressed in terms of a generalised hypergeometric series [19].

We first introduce the *Pochhammer symbol* $(a)_n$, defined by

$$(a)_n = \begin{cases} 1\,, & n = 0\,, \\ a(a+1)\cdots(a+n-1)\,, & n = 1, 2, 3, \ldots\,. \end{cases} \tag{A.2.1}$$

In terms of the gamma function, we have

$$(a)_n = \frac{\Gamma(a+n)}{\Gamma(a)}\,, \qquad n > 0\,.$$

The *generalised hypergeometric series* ${}_rF_s$ is defined by

$${}_rF_s(a_1, \ldots, a_r; b_1, \ldots, b_s; z) = \sum_{k=0}^{\infty} \frac{(a_1)_k \cdots (a_r)_k}{(b_1)_k \cdots (b_s)_k} \frac{z^k}{k!}\,, \tag{A.2.2}$$

where $b_i \neq 0, -1, -2, \ldots$ for $i = \{1, \ldots, s\}$ to ensure that the denominator factors in the terms of the series are never zero. Clearly, the ordering of the numerator parameters and of the denominator parameters is immaterial. The radius of convergence ρ of the hypergeometric series is

$$\rho = \begin{cases} \infty\,, & r < s+1\,, \\ 1\,, & r = s+1\,, \\ 0\,, & r > s+1\,. \end{cases} \tag{A.2.3}$$

Some elementary cases of the hypergeometric series are:

- the exponential series ${}_0F_0$;
- the binomial series ${}_1F_0$; and
- the Gauss hypergeometric series ${}_2F_1$.

If one of the numerator parameters a_i, $i = 1, \ldots, r$, is a negative integer, say $a_1 = -n$, then the hypergeometric series (A.2.2) terminates at the nth term and becomes a polynomial in z:

$${}_rF_s(-n, \ldots, a_r; b_1, \ldots, b_s; z) = \sum_{k=0}^{n} \frac{(-n)_k \cdots (a_r)_k}{(b_1)_k \cdots (b_s)_k} \frac{z^k}{k!}\,. \tag{A.2.4}$$

A system of polynomials $\{Q_n(x), n \in \mathcal{N}\}$, where $Q_n(x)$ is a polynomial of exact degree n and $\mathcal{N} = \{0, 1, 2, \ldots\}$ or $\mathcal{N} = \{0, 1, \ldots, N\}$ for a finite non-negative integer N, is an orthogonal system of polynomials with respect to some real positive measure ϕ if the following orthogonality relations are satisfied:

$$\int_S Q_n(x)Q_m(x)\,\mathrm{d}\phi(x) = h_n^2\delta_{nm}\,, \quad n,m \in \mathcal{N}\,, \tag{A.2.5}$$

where S is the support of the measure ϕ and the h_n are nonzero constants. The system is called orthonormal if $h_n = 1$.

The measure ϕ often has a density $w(x)$ or weights $w(i)$ at points x_i in the discrete case. The relations (A.2.5) then become

$$\int_S Q_n(x)Q_m(x)w(x)\,\mathrm{d}x = h_n^2\delta_{nm}\,, \quad n,m \in \mathcal{N}\,, \tag{A.2.6}$$

in the continuous case, or

$$\sum_{i=0}^{M} Q_n(x_i)Q_m(x_i)w(x_i) = h_n^2\delta_{nm}\,, \quad n,m \in \mathcal{N}\,, \tag{A.2.7}$$

in the discrete case where it is possible that $M = \infty$.

All orthogonal polynomials $\{Q_n(x)\}$ satisfy a *three-term recurrence relation*:

$$-xQ_n(x) = A_nQ_{n+1}(x) - (A_n + C_n)Q_n(x) + C_nQ_{n-1}(x)\,, \quad n \geqslant 1\,, \tag{A.2.8}$$

where $A_n, C_n \neq 0$ and $C_n/A_{n-1} > 0$. Together with $Q_{-1}(x) = 0$ and $Q_0(x) = 1$, all $Q_n(x)$ can be determined by the recurrence relation.

It is well known that continuous orthogonal polynomials satisfy the second-order differential equation

$$s(x)y'' + \tau(x)y' + \lambda y = 0\,, \tag{A.2.9}$$

where $s(x)$ and $\tau(x)$ are polynomials of at most second and first degree, respectively, and

$$\lambda = \lambda_n = -n\tau' - \frac{1}{2}n(n-1)s'' \tag{A.2.10}$$

are the eigenvalues of the differential equation; the orthogonal polynomials $y(x) = y_n(x)$ are the eigenfunctions.

In the discrete case, we introduce the forward and backward difference operators, respectively,

$$\Delta f(x) = f(x+1) - f(x) \quad \text{and} \quad \nabla f(x) = f(x) - f(x-1)\,. \tag{A.2.11}$$

The *difference equation* corresponding to the differential equation (A.2.9) is

$$s(x)\Delta\nabla y(x) + \tau(x)\Delta y(x) + \lambda y(x) = 0\,. \tag{A.2.12}$$

Again, $s(x)$ and $\tau(x)$ are polynomials of at most second and first degree, respectively; $\lambda = \lambda_n$ are eigenvalues of the difference equation, and the orthogonal polynomials $y(x) = y_n(x)$ are the eigenfunctions.

All orthogonal polynomials can be obtained by repeatedly applying the differential operator as follows:

$$Q_n(x) = \frac{1}{w(x)} \frac{\mathrm{d}^n}{\mathrm{d}x^n} \left[w(x) s^n(x) \right]. \tag{A.2.13}$$

In the discrete case, the differential operator $\mathrm{d}/\mathrm{d}x$ is replaced by the backward difference operator ∇. A constant factor can be introduced for normalisation. Equation (A.2.13) is referred to as the *generalised Rodriguez formula*, named after J. Rodriguez who first discovered the specific formula for Legendre polynomials.

The Askey scheme, which can be represented as a tree structure shown in Fig. A.1, classifies the hypergeometric orthogonal polynomials and indicates the limit relations between them. The 'tree' starts with the Wilson polynomials and the Racah polynomials on the top. They both belong to the class ${}_4F_3$ of the hypergeometric orthogonal polynomials (A.2.4). The Wilson polynomials are continuous polynomials and the Racah polynomials are discrete. The lines connecting different polynomials denote the limit transition relationships between them, which imply that polynomials at the lower end of the lines can be obtained by taking the limit of one parameter from their counterparts on the upper end. For example, the limit relation between Jacobi polynomials $P_n^{(\alpha,\beta)}(x)$ and Hermite polynomials $H_n(x)$ is

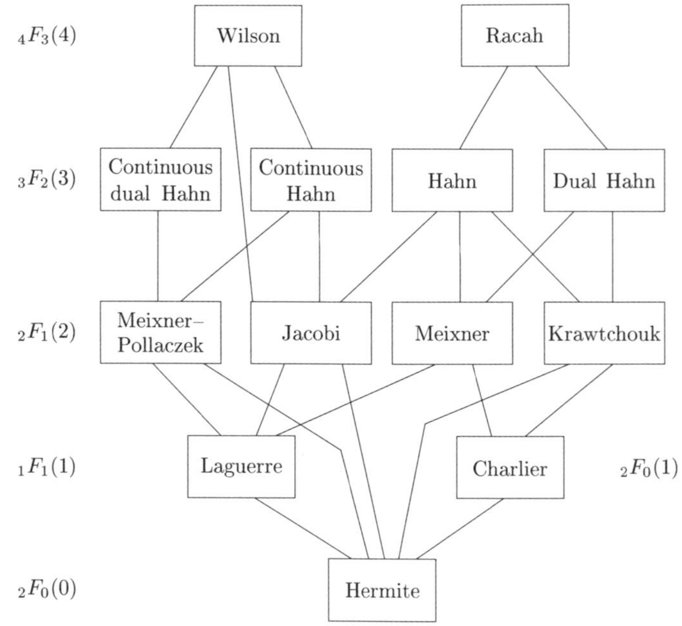

FIG. A.1. The Askey scheme of orthogonal polynomials.

$$\lim_{\alpha\to\infty} \alpha^{-n/2} P_n^{(\alpha,\alpha)}\left(\frac{x}{\sqrt{\alpha}}\right) = \frac{H_n(x)}{2^n n!}\,,$$

and between Meixner polynomials $M_n(x;\beta,c)$ and Charlier polynomials $C_n(x;a)$ it is

$$\lim_{\beta\to\infty} M_n\left(x;\beta,\frac{a}{a+\beta}\right) = C_n(x;a)\,.$$

A.2.1 *Examples*

Jacobi polynomial $P_n^{(\alpha,\beta)}(x)$

$$P_n^{(\alpha,\beta)}(x) = \frac{(\alpha+1)_n}{n!}\,{}_2F_1\left(-n, n+\alpha+\beta+1; \alpha+1; \frac{1-x}{2}\right).$$

Hermite polynomial $H_n(x)$

$$H_n(x) = (2x)^n\,{}_2F_0\left(-\frac{n}{2}, -\frac{n-1}{2}; \;; -\frac{1}{x^2}\right).$$

Orthogonality

$$\frac{1}{\sqrt{\pi}}\int_{-\infty}^{\infty} \mathrm{e}^{-x^2} H_m(x) H_n(x)\,\mathrm{d}x = 2^n n!\delta_{mn}\,.$$

Recurrence relation

$$H_{n+1}(x) - 2xH_n(x) + 2nH_{n-1}(x) = 0\,.$$

Rodriguez formula

$$\mathrm{e}^{-x^2} H_n(x) = (-1)^n \frac{\mathrm{d}^n}{\mathrm{d}x^n}\left(\mathrm{e}^{-x^2}\right).$$

Laguerre polynomial $L_n^{(\alpha)}(x)$

$$L_n^{(\alpha)}(x) = \frac{(\alpha+1)_n}{n!}\,{}_1F_1(-n; \alpha+1; x)\,.$$

Orthogonality

$$\int_0^{\infty} \mathrm{e}^{-x} x^{\alpha} L_m^{(\alpha)}(x) L_n^{(\alpha)}(x)\,\mathrm{d}x = \frac{\Gamma(n+\alpha+1)}{n!}\delta_{mn}\,, \quad \alpha > -1\,.$$

Recurrence relation

$$(n+1)L_{n+1}^{(\alpha)}(x) - (2n+\alpha+1-x)L_n^{(\alpha)}(x) + (n+\alpha)L_{n-1}^{(\alpha)}(x) = 0\,.$$

Rodriguez formula

$$\mathrm{e}^{-x} x^{\alpha} L_n^{(\alpha)}(x) = \frac{1}{n!}\frac{\mathrm{d}^n}{\mathrm{d}x^n}\left(\mathrm{e}^{-x} x^{n+\alpha}\right).$$

Charlier polynomial $C_n(x;a)$

$$C_n(x;a) = {}_2F_0\left(-n, -x; \ ; -\frac{1}{a}\right).$$

Orthogonality

$$\sum_{x=0}^{\infty} \frac{a^x}{x!} C_m(x;a) C_n(x;a) = a^{-n} \mathrm{e}^a n! \delta_{mn}, \quad a > 0.$$

Recurrence relation

$$-xC_n(x;a) = aC_{n+1}(x;a) - (n+a)C_n(x;a) + nC_{n-1}(x;a).$$

Rodriguez formula

$$\frac{a^x}{x!} C_n(x;a) = \nabla^n \left(\frac{a^x}{x!}\right),$$

where ∇ is the backward difference operator (A.2.11).

Krawtchouk polynomial $K_n(x;p,N)$

$$K_n(x;p,N) = {}_2F_1\left(-n, -x; -N; \frac{1}{p}\right), \quad n = 0, 1, \ldots, N.$$

Orthogonality

$$\sum_{x=0}^{N} \binom{N}{x} p^x (1-p)^{N-x} K_m(x;p,N) K_n(x;p,N) = \frac{(-1)^n n!}{(-N)_n} \left(\frac{1-p}{p}\right)^n \delta_{mn},$$

$$0 < p < 1.$$

Recurrence relation

$$\begin{aligned} -xK(x;p,N) &= p(N-n)K_{n+1}(x;p,N) \\ &\quad - [p(N-n) + n(1-p)]K_n(x;p,N) + n(1-p)K_{n-1}(x;p,N). \end{aligned}$$

Rodriguez formula

$$\binom{N}{x} \left(\frac{p}{1-p}\right)^x K_n(x;p,N) = \nabla^n \left[\binom{N-n}{x} \left(\frac{p}{1-p}\right)^x\right].$$

Meixner polynomial $M_n(x;\beta,c)$

$$M_n(x;\beta,c) = {}_2F_1\left(-n, -x; \beta; 1 - \frac{1}{c}\right).$$

Orthogonality

$$\sum_{x=0}^{\infty} \frac{(\beta)_x}{x!} c^x M_m(x;\beta,c) M_n(x;\beta,c) = \frac{c^{-n} n!}{(\beta)_n (1-c)^\beta} \delta_{mn}, \quad \beta > 0,\ 0 < c < 1.$$

Recurrence relation

$$(c-1)x M_n(x;\beta,c) = c(n+\beta) M_{n+1}(x;\beta,c) - [n + (n+\beta)c] M_n(x;\beta,c) + n M_{n-1}(x;\beta,c).$$

Rodriguez formula

$$\frac{(\beta)_x c^x}{x!} M_n(x;\beta,c) = \nabla^n \left[\frac{(\beta+n)_x c^x}{x!} \right].$$

Hahn polynomial $Q_n(x;\alpha,\beta,N)$

$$Q_n(x;\alpha,\beta,N) = {}_3F_2(-n, n+\alpha+\beta+1, -x; \alpha+1, -N; 1), \quad n = 0, 1, \ldots, N.$$

Orthogonality

For $\alpha > -1$ and $\beta > -1$ or for $\alpha < -N$ and $\beta < -N$,

$$\sum_{x=0}^{N} \binom{\alpha+x}{x} \binom{\beta+N-x}{N-x} Q_m(x;\alpha,\beta,N) Q_n(x;\alpha,\beta,N) = h_n^2 \delta_{mn},$$

where

$$h_n^2 = \frac{(-1)^n (n+\alpha+\beta+1)_{N+1} (\beta+1)_n n!}{(2n+\alpha+\beta+1)(\alpha+1)_n (-N)_n N!}.$$

Recurrence relation

$$-x Q_n(x) = A_n Q_{n+1}(x) - (A_n + C_n) Q_n(x) + C_n Q_{n-1}(x),$$

where

$$Q_n(x) := Q_n(x;\alpha,\beta,N)$$

and

$$A_n = \frac{(n+\alpha+\beta+1)(n+\alpha+1)(N-n)}{(2n+\alpha+\beta+1)(2n+\alpha+\beta+2)},$$

$$C_n = \frac{n(n+\alpha+\beta+N+1)(n+\beta)}{(2n+\alpha+\beta)(2n+\alpha+\beta+1)}.$$

Rodriguez formula

$$w(x;\alpha,\beta,N) Q_n(x;\alpha,\beta,N) = \frac{(-1)^n (\beta+1)_n}{(-N)_n} \nabla^n [w(x;\alpha+n, \beta+n, N-n)],$$

where

$$w(x;\alpha,\beta,N) = \binom{\alpha+x}{x} \binom{\beta+N-x}{N-x}.$$

APPENDIX B

GAUSS-TYPE INTEGRATION

Integration of $u(x)$ in the interval $[-1, 1]$ with respect to the function

$$(1-x)^{\alpha}(1+x)^{\beta}\,, \quad \alpha > -1,\ \beta > -1$$

can be numerically evaluated using Gauss-type integration in a discrete summation of the form

$$\int_{-1}^{1} (1-x)^{\alpha}(1+x)^{\beta} u(x)\,\mathrm{d}x = \sum_{i=0}^{Q-1} w_i^{\alpha,\beta} u(x_i) + R(u)\,, \tag{B.0.1}$$

where $R(u) = 0$ if $u(x) \in \mathcal{P}_{2Q-k}([-1, 1])$. The value of k is determined by the type of quadrature used, which can be either classical Gauss ($k = 1$), Gauss–Radau ($k = 2$), or Gauss–Lobatto ($k = 3$).

To derive eqn (B.0.1) we consider the case where $u(x)$ is a polynomial of order $2Q - k$ (that is, $u(x) \in \mathcal{P}_{2Q-k}([-1, 1])$) and decompose $u(x)$ into

$$u(x) = \sum_{i=0}^{Q-1} u(x_i)h_i(x) + s(x)r(x)\,, \tag{B.0.2}$$

where

$$\begin{aligned} h_i(x_j) &= \delta_{ij}\,, \quad & h(x) &\in \mathcal{P}_{Q-1}\,, \\ s(x_i) &= 0\,, & s(x) &\in \mathcal{P}_{Q}\,, \\ & & r(x) &\in \mathcal{P}_{Q-k}\,, \end{aligned}$$

and δ_{ij} denotes the Kronecker delta. The function $u(x)$ has been decomposed into $h_i(x_j)$ which is Lagrangian polynomial through the Q nodal points x_i, $s(x)$ which is a polynomial of order Q with roots at the nodal points x_i, and the remainder polynomial $r(x)$ of order $Q - k$. Integrating eqn (B.0.2) with respect to $(1-x)^{\alpha}(1+x)^{\beta}$ in the interval $[-1, 1]$ and comparing with eqn (B.0.1), we deduce

$$w_i^{\alpha,\beta} = \int_{-1}^{1} (1-x)^{\alpha}(1+x)^{\beta} h_i(x)\,\mathrm{d}x\,, \tag{B.0.3}$$

$$R(u) = \int_{-1}^{1} (1-x)^{\alpha}(1+x)^{\beta} s(x)r(x)\,\mathrm{d}x\,. \tag{B.0.4}$$

If $s(x)$ is specified then the nodal points x_i will be determined as the roots of $s(x)$ such that $s(x_i) = 0$. Specifying $s(x)$ is preferable to specifying the nodal points x_i because an appropriate choice of $s(x)$ will make $R(u) = 0$. For example, if $s(x) = P_Q^{\alpha,\beta}(x)$, where $P_Q^{\alpha,\beta}(x)$ is a Jacobi polynomial (see Appendix A), then

$$R(u) = \int_{-1}^{1} (1-x)^\alpha (1+x)^\beta P_Q^{\alpha,\beta}(x)\, r(x)\, \mathrm{d}x\,.$$

This integral is the orthogonality relationship for $P_Q^{\alpha,\beta}(x)$ (see eqn (A.1.7)), so, if $k = 1$, then $r(x)$ is a polynomial of order $Q - 1$ and therefore $R(u) = 0$. This choice of $s(x)$ determines the classical Gauss quadrature and therefore the nodal points are determined by $x_i = x_{i,P}^{\alpha,\beta}$, where $x_{i,P}^{\alpha,\beta}$ denote the m zeros of the Jacobi polynomial $P_P^{\alpha,\beta}$ such that

$$P_P^{\alpha,\beta}(x_{i,P}^{\alpha,\beta}) = 0\,, \quad i = 0, 1, \ldots, P-1\,, \tag{B.0.5}$$

where

$$x_{0,P}^{\alpha,\beta} < x_{1,P}^{\alpha,\beta} < \cdots < x_{P-1,P}^{\alpha,\beta}\,.$$

Gauss–Radau and Gauss–Lobatto integration both require that the nodal points include one or both of the end-points $x = \pm 1$. This restriction modifies the form of $s(x)$ and has the effect of altering the weight function $(1-x)^\alpha(1+x)^\beta$. Therefore the nodal points are determined by a Jacobi polynomial with different values of α and β.

B.1 Jacobi formulae

In this section we state the nodal values x_i and weights $w_i^{\alpha,\beta}$ for Gaussian integration with respect to the function $(1-x)^\alpha(1+x)^\beta$, where $\alpha, \beta > -1$. The Legendre case, when $\alpha = \beta = 0$, was introduced in Section 2.4.1 and can be considered as a special case of the following formulae. The nodal points x_i are determined in terms of the roots of Jacobi polynomials $x_{i,m}^{\alpha,\beta}$ defined by eqn (B.0.5). A complete derivation of the weight formulae can be found in Ghizzetti and Ossicini [182].

Gauss–Jacobi

For this type of quadrature no restrictions are placed on the nodal points and the nodal points are the roots of $P_Q^{\alpha,\beta}(x)$. The Gauss–Jacobi formulae are as follows:

$$\begin{aligned}
x_i &= x_{i,Q}^{\alpha,\beta}\,, &\quad i &= 0, \ldots, Q-1\,,\\
w_i^{\alpha,\beta} &= H_{i,Q}^{\alpha,\beta}\,, &\quad i &= 0, \ldots, Q-1\,,\\
R(u) &= 0\,, &\quad u(x) &\in \mathcal{P}_{2Q-1}([-1,1])\,,
\end{aligned}$$

$$H_{i,Q}^{\alpha,\beta} = \frac{2^{\alpha+\beta+1}\Gamma(\alpha+Q+1)\Gamma(\beta+Q+1)}{Q!\Gamma(\alpha+\beta+Q+1)[1-(x_i)^2]}\left[\frac{\mathrm{d}}{\mathrm{d}x}\left(P_Q^{\alpha,\beta}(x)\right)\Big|_{x=x_i}\right]^{-2},$$

$$R(u) \leqslant \frac{2^{\alpha+\beta+2Q+1}Q!\Gamma(\alpha+Q+1)\Gamma(\beta+Q+1)\Gamma(\alpha+\beta+Q+1)}{(2Q)!(\alpha+\beta+2Q+1)[\Gamma(\alpha+\beta+2Q+1)]^2} M_{2Q}$$

$(M_{2Q} = \sup_{-1\leqslant x\leqslant 1} |u^{2Q}(x)|)$.

Gauss–Radau–Jacobi—Bouzitat formulae of the first kind

Gauss–Radau-type integration requires that the nodal points include one of the end-points of the integration interval $x = \pm 1$. If we choose to include $x = -1$ then the value of $s(x)$ in eqn (B.0.2) has the form

$$s(x) = (1+x)s_1(x)\,,$$

where

$$s_1(x) \in \mathcal{P}_{Q-1}([-1,1])\,.$$

Therefore the relation for $R(u)$ given by eqn (B.0.4) becomes

$$R(u) = \int_{-1}^{1} (1-x)^{\alpha}(1+x)^{\beta+1} s_1(x) r(x)\,\mathrm{d}x\,.$$

If we let $s_1(x) = P_{Q-1}^{\alpha,\beta+1}(x)$ then $R(u) = 0$ if $k = 2$. The Gauss–Radau–Jacobi formulae are as follows:

$$x_i = \begin{cases} -1\,, & i = 0\,, \\ x_{i-1,Q-1}^{\alpha,\beta+1}\,, & i = 1,\ldots,Q-1\,, \end{cases}$$

$$w_i^{\alpha,\beta} = \begin{cases} (\beta+1)B_{0,Q-1}^{\alpha,\beta}\,, & i = 0\,, \\ B_{i,Q-1}^{\alpha,\beta}\,, & i = 1,\ldots,Q-1\,, \end{cases}$$

$$R(u) = 0\,, \quad u(x) \in \mathcal{P}_{2Q-2}([-1,1])\,,$$

$$B_{i,Q-1}^{\alpha,\beta} = \frac{2^{\alpha+\beta}\Gamma(\alpha+Q)\Gamma(\beta+Q)(1-x_i)}{(Q-1)!(\beta+Q)\Gamma(\alpha+\beta+Q+1)[P_{(Q-1)}^{\alpha,\beta}(x_i)]^2}\,,$$

$$R(u) \leqslant \frac{2^{\alpha+\beta+2Q}(Q-1)!\Gamma(\alpha+Q)\Gamma(\beta+Q+1)\Gamma(\alpha+\beta+Q+1)}{(2Q-1)!(\alpha+\beta+2Q)[\Gamma(\alpha+\beta+2Q)]^2} M_{2Q-1}$$

$(M_{2Q-1} = \sup_{-1\leqslant x\leqslant 1} |u^{2Q-1}(x)|)$.

Gauss–Lobatto–Jacobi—Bouzitat formulae of the second kind

Gauss–Lobatto-type integration requires that the nodal points include both of the end-points of the integration interval $x = \pm 1$. For this case the value of $s(x)$ in eqn (B.0.2) has the form

$$s(x) = (1-x)(1+x)s_2(x)\,,$$

where

$$s_2(x) \in \mathcal{P}_{Q-2}([-1,1])\,,$$

and the relation for $R(u)$ given by eqn (B.0.4) becomes

$$R(u) = \int_{-1}^{1} (1-x)^{\alpha+1}(1+x)^{\beta+1} s_2(x) r(x)\,\mathrm{d}x\,.$$

Therefore, if we let $s_2(x) = P^{\alpha+1,\beta+1}_{Q-2}(x)$ then $R(u) = 0$ if $k = 3$. The Gauss–Lobatto–Jacobi formulae are as follows:

$$x_i = \begin{cases} -1\,, & i = 0\,, \\ x^{\alpha+1,\beta+1}_{i-1,Q-2}\,, & i = 1,\ldots,Q-2\,, \\ 1\,, & i = Q-1\,, \end{cases}$$

$$w_i^{\alpha,\beta} = \begin{cases} (\beta+1)C^{\alpha,\beta}_{0,Q-2}\,, & i = 0\,, \\ C^{\alpha,\beta}_{i,Q-2}\,, & i = 1,\ldots,Q-2\,, \\ (\alpha+1)C^{\alpha,\beta}_{Q-1,Q-2}\,, & i = Q-1\,, \end{cases}$$

$$R(u) = 0\,, \quad u(x) \in \mathcal{P}_{2Q-3}([-1,1])\,,$$

$$C^{\alpha,\beta}_{i,Q-2} = \frac{2^{\alpha+\beta+1}\Gamma(\alpha+Q)\Gamma(\beta+Q)}{(Q-1)(Q-1)!\Gamma(\alpha+\beta+Q+1)[P^{\alpha,\beta}_{Q-1}(x_i)]^2}\,,$$

$$R(u) \leqslant \frac{2^{\alpha+\beta+2Q-1}(Q-2)!\Gamma(\alpha+Q)\Gamma(\beta+Q)\Gamma(\alpha+\beta+Q+1)}{(2Q-2)!(\alpha+\beta+2Q-1)[\Gamma(\alpha+\beta+2Q-1)]^2} M_{2Q-2}$$

$(M_{2Q-2} = \sup_{-1\leqslant x\leqslant 1} |u^{2Q-2}(x)|)$.

B.2 Evaluation of the zeros of Jacobi polynomials

The formulae for the *weights* in Section B.1 have a closed form in terms of the nodal points x_i. In general, however, there are no explicit formulae for the nodes. These are defined in terms of the roots of the Jacobi polynomial such that

$$x_i = x^{\alpha,\beta}_{i,m}\,,$$
$$P^{\alpha,\beta}_m(x^{\alpha,\beta}_{i,m}) = 0\,, \quad i = 0,1,\ldots,m-1\,.$$

The zeros $x_{i,m}^{\alpha,\beta}$ can be numerically evaluated using an iterative technique such as Newton–Raphson. However, we note that the zeros of the Chebyshev polynomial ($\alpha = \beta = -\frac{1}{2}$) do have an explicit form:

$$x_{i,m}^{-1/2,-1/2} = -\cos\left(\frac{2i+1}{2m}\pi\right), \quad i = 0, \ldots, m-1,$$

and so we can use $x_{i,m}^{-1/2,-1/2}$ as an initial guess to the iteration.

To ensure that we find a new root at each search we can apply *polynomial* deflation, where the known roots are factored out of the initial polynomial once they have been determined. This means that the root-finding algorithm is applied to the polynomial

$$f_{m-n}(x) = \frac{P_m^{\alpha,\beta}(x)}{\Pi_{i=0}^{n-1}(x - x_i)},$$

where x_i ($i = 0, \ldots, n-1$) are the known roots of $P_m^{\alpha,\beta}(x)$.

Noting that

$$\frac{f_{m-n}(x)}{f'_{m-n}(x)} = \frac{P_m^{\alpha,\beta}(x)}{[P_m^{\alpha,\beta}(x)]' - P_m^{\alpha,\beta}(x) \sum_{i=0}^{n-1} 1/(x - x_i)},$$

a root-finding algorithm to determine the m roots of $P_m^{\alpha,\beta}(x)$ using the Newton–Raphson iteration with polynomial deflation is as follows:

```
do k = 0, m − 1
   r = x_{k,m}^{−1/2,−1/2}
   if (k > 0) r = (r + x_{k−1})/2
   do j = 1, stop
      s = Σ_{i=0}^{k−1} 1/(r − x_i)
      δ = −P_m^{α,β}(r)/([P_m^{α,β}(r)]' − P_m^{α,β}(r)s)
      r = r + δ
      if (δ < ε) exit loop
   continue
   x_k = r
continue
```

Here ϵ is a specified tolerance. Numerically, we find that a better approximation for the initial guess is given by the average of $r = x_{k,m}^{-1/2,-1/2}$ and x_{k-1}. The values of $P_m^{\alpha,\beta}(x)$ and $[P_m^{\alpha,\beta}(x)]'$ can be generated using the recursion relationships (A.1.3) and (A.1.4).

APPENDIX C

COLLOCATION DIFFERENTIATION

The function $u(x) \in \mathcal{P}_P$, where $\mathcal{P}_P$ is the space of all polynomials of degree $\leqslant P$, can be written in terms of Lagrange polynomials $h_i(x)$ through a set of Q nodal points x_i $(0 \leqslant i \leqslant Q-1)$ as follows:

$$u(x) = \sum_{i=0}^{Q-1} u(x_i) h_i(x),$$

where $Q \geqslant P+1$. The derivative of $u(x)$ is therefore

$$\frac{\mathrm{d}u(x)}{\mathrm{d}x} = \sum_{i=0}^{Q-1} u(x_i) \frac{\mathrm{d}}{\mathrm{d}x} h_i(x).$$

Evaluating $\mathrm{d}u(x)/\mathrm{d}x$ at the nodal points x_i, we have

$$\left.\frac{\mathrm{d}u(x)}{\mathrm{d}x}\right|_{x=x_i} = \sum_{j=0}^{Q-1} \boldsymbol{D}_{ij}\, u(x_j),$$

where

$$\boldsymbol{D}_{ij} = \left.\frac{\mathrm{d}h_j(x)}{\mathrm{d}x}\right|_{x=x_i}.$$

The Lagrange polynomial is

$$h_i(x) = \frac{p_Q(x)}{p'_Q(x_i)(x-x_i)}, \quad p_Q(x) = \prod_{j=0}^{Q-1} (x - x_j);$$

the derivative of $h_i(x)$ is therefore

$$\frac{\mathrm{d}h_i(x)}{\mathrm{d}x} = \frac{p'_Q(x)(x-x_i) - p_Q(x)}{p'_Q(x_i)(x-x_i)^2}.$$

Noting that

$$\lim_{x \to x_i} \frac{\mathrm{d}h_i(x)}{\mathrm{d}x} = \lim_{x \to x_i} \frac{p''_Q(x)}{2p'_Q(x)} = \frac{p''_Q(x_i)}{2p'_Q(x_i)},$$

$\boldsymbol{D}_{ij}$ is defined as

$$\boldsymbol{D}_{ij} = \begin{cases} \dfrac{p'_Q(x_i)}{p'_Q(x_j)} \dfrac{1}{x_i - x_j}, & i \neq j, \\ \dfrac{p''_Q(x_i)}{2p'_Q(x_i)}, & i = j. \end{cases} \tag{C.0.1}$$

C.1 Jacobi formulae

In this section we shall find the specific forms of $p'_Q(x_i)$ and $p''_Q(x_i)$ corresponding to the nodal points of the general Gauss–Jacobi quadrature. To evaluate the derivative matrix $\boldsymbol{D}_{ij}$ these values should be substituted into eqn (C.0.1). The explicit form of this matrix for the case of the Gauss–Legendre quadrature points (that is, $\alpha = \beta = 0$) was given in Section 2.4.2.1. We recall that $x_{i,m}^{\alpha,\beta}$ is defined as

$$P_m^{\alpha,\beta}(x_{i,m}^{\alpha,\beta}) = 0, \quad i = 0, 1, \ldots, m-1,$$

where

$$x_{0,m}^{\alpha,\beta} < x_{0,m}^{\alpha,\beta} < \cdots < x_{m-1,m}^{\alpha,\beta},$$

and $P_m^{\alpha,\beta}$ is the Jacobi polynomial (see Appendix A).

Gauss–Jacobi

For this case the nodes are given by

$$x_i = x_{i,Q}^{\alpha,\beta},$$

and so

$$\begin{aligned} p_Q(x) &= P_Q^{\alpha,\beta}(x), \\ p'_Q(x) &= [P_Q^{\alpha,\beta}(x_i)]', \\ p''_Q(x) &= [P_Q^{\alpha,\beta}(x_i)]''. \end{aligned}$$

Evaluating eqn (A.1.1) at $x_i = x_{i,Q}^{\alpha,\beta}$ (that is, $P_Q^{\alpha,\beta}(x_i) = 0$), we can express $[P_Q^{\alpha,\beta}(x_i)]''$ in terms of $[P_Q^{\alpha,\beta}(x_i)]'$ $(0 \leqslant i \leqslant Q-1)$ to find the relation

$$[P_Q^{\alpha,\beta}(x_i)]'' = \frac{\alpha - \beta + (\alpha + \beta + 2)x_i}{1 - x_i^2}[P_Q^{\alpha,\beta}(x_i)]'. \tag{C.1.1}$$

The value of $[P_Q^{\alpha,\beta}(x_i)]'$ can be computed by applying eqn (A.1.8) and using the recursion relation (A.1.3).

Gauss–Radau–Jacobi

For this case the nodes are given by

$$x_i = \begin{cases} -1, & i = 0, \\ x_{i-1,Q-1}^{\alpha,\beta+1}, & i = 1, \ldots, Q-1, \end{cases}$$

and so

$$p_Q(x) = (1+x)P_{Q-1}^{\alpha,\beta+1}(x)\,,$$
$$p'_Q(x) = (1+x)[P_{Q-1}^{\alpha,\beta+1}(x)]' + P_{Q-1}^{\alpha,\beta+1}(x)\,,$$
$$p''_Q(x) = (1+x)[P_{Q-1}^{\alpha,\beta+1}(x)]'' + 2P_{Q-1}^{\alpha,\beta+1}(x)\,.$$

Using a similar argument to that used for eqn (C.1.1), the second derivative term $(1 \leqslant i \leqslant Q-1)$ can be expressed as

$$(1+x_i)[P_{Q-1}^{\alpha,\beta+1}(x_i)]'' = \frac{\alpha-\beta-1+(\alpha+\beta+3)x_i}{1-x_i}[P_{Q-1}^{\alpha,\beta+1}(x_i)]'\,.$$

Therefore $p'_Q(x_i)$ and $p''_Q(x_i)$ become

$$p'_Q(x_i) = \begin{cases} \dfrac{(-1)^{Q-1}\Gamma(Q+\beta+1)}{\Gamma(Q)\Gamma(\beta+2)}\,, & i=0\,,\\ (1+x_i)[P_{Q-1}^{\alpha,\beta+1}(x_i)]'\,, & i=1,\ldots,Q-1\,, \end{cases}$$

$$p''_Q(x_i) = \begin{cases} \dfrac{(-1)^{Q}(Q+\alpha+\beta+1)\Gamma(Q+\beta+1)}{\Gamma(Q-1)\Gamma(\beta+3)}\,, & i=0\,,\\ \dfrac{\alpha-\beta+1+(\alpha+\beta+1)x_i}{1-x_i}[p_{Q-1}^{\alpha,\beta+1}(x_i)]'\,, & i=1,\ldots,Q-1\,. \end{cases}$$

Here, we have used eqns (A.1.5), (A.1.6), and (A.1.8) to determine the values of $p_Q(-1)$ and $p'_Q(-1)$. To evaluate $[P_{Q-1}^{\alpha,\beta+1}(x_i)]'$ we could use the recursion relation (A.1.4) or apply eqn (A.1.8) and use the recursion relation (A.1.3). Alternatively, we note that $[P_{Q-1}^{\alpha,\beta+1}(x_i)]'$ can be expressed in terms of $P_{Q-1}^{\alpha,\beta}$ using the relation

$$[P_{Q-1}^{\alpha,\beta+1}(x_i)]' = \frac{2(Q+\beta)}{1-x_i^2}P_{Q-1}^{\alpha,\beta}(x_i)\,, \quad i=1,\ldots,Q-1\,,$$

which can be derived by from eqns (A.1.4) and (A.1.10) by substituting $\beta=\beta+1$, $n=Q-1$ and evaluating at $x_i = x_{i,Q-1}^{\alpha,\beta+1}$ (see Ghizzetti and Ossicini [182, p. 102]).

Gauss–Lobatto–Jacobi

For this case the nodes are given by

$$x_i = \begin{cases} -1\,, & i=0\,,\\ x_{i-1,Q-2}^{\alpha+1,\beta+1}\,, & i=1,\ldots,Q-2\,,\\ 1\,, & i=Q-1\,, \end{cases}$$

and so

$$
\begin{aligned}
p_Q(x) &= (1-x)(1+x)P_{Q-2}^{\alpha+1,\beta+1}(x)\,,\\
p_Q'(x) &= (1-x)(1+x)[p_{Q-2}^{\alpha+1,\beta+1}(x)]' - 2P_{Q-2}^{\alpha+1,\beta+1}(x)\,,\\
p_Q''(x) &= (1-x)(1+x)[p_{Q-2}^{\alpha+1,\beta+1}(x)]'' - 4x[p_{Q-2}^{\alpha+1,\beta+1}(x)]' - 2P_{Q-1}^{\alpha,\beta+1}(x)\,.
\end{aligned}
$$

Once again, following a similar argument to that used for eqn (C.1.1), the second derivative term ($1 \leqslant i \leqslant Q-2$) can be expressed as

$$
(1-x_i)(1+x_i)[P_{Q-2}^{\alpha+1,\beta+1}(x_i)]'' = [\alpha-\beta+(\alpha+\beta+4)x_i]\,[P^{\alpha+1,\beta+1}(x_i)]'\,.
$$

Therefore $p_Q'(x_i)$ and $p_Q''(x_i)$ become

$$
p_Q'(x_i) = \begin{cases}
\dfrac{(-1)^Q 2\Gamma(Q+\beta)}{\Gamma(Q-1)\Gamma(\beta+2)}\,, & i=0\,,\\
(1-x_i)(1+x_i)[P_{Q-2}^{\alpha+1,\beta+1}(x_i)]'\,, & i=1,\dots,Q-2\,,\\
\dfrac{-2\Gamma(Q+\alpha)}{\Gamma(Q-1)\Gamma(\alpha+2)}\,, & i=Q-1\,,
\end{cases}
$$

$$
p_Q''(x_i) = \begin{cases}
\dfrac{(-1)^Q 2[\alpha-(Q-1)(Q+\alpha+\beta)]}{\beta+2}\dfrac{\Gamma(Q+\beta)}{\Gamma(Q-1)\Gamma(\beta+2)}\,, & i=0\,,\\
(\alpha-\beta+(\alpha+\beta)x_i)[p_{Q-2}^{\alpha+1,\beta+1}(x_i)]'\,, & i=1,\dots,Q-2\,,\\
\dfrac{2(\beta-(Q-1)(Q+\alpha+\beta))}{\alpha+2}\dfrac{\Gamma(Q+\alpha)}{\Gamma(Q-1)\Gamma(\alpha+2)}\,, & i=Q-1\,.
\end{cases}
$$

The values of $p_Q(-1)$ and $p_Q'(-1)$ can be derived from eqns (A.1.5), (A.1.6), and (A.1.8). To evaluate $[P_{Q-2}^{\alpha+1,\beta+1}(x_i)]'$ we could use the recursion relation (A.1.3) or apply eqn (A.1.8) and then use the recursion relation (eqn (A.1.4)).

It is also possible to express $[P_{Q-2}^{\alpha+1,\beta+1}(x_i)]'$ in terms of $P_{Q-1}^{\alpha,\beta}$ via the relationship

$$
(1-x_i)(1+x_i)[P_{Q-2}^{\alpha+1,\beta+1}(x_i)]' = -2(Q-1)P_{Q-1}^{\alpha,\beta}(x_i)\,, \quad i=1,\dots,Q-2\,.
$$

This can be derived by substituting $n=Q-1$ into eqn (A.1.1) and applying eqn (A.1.8) to the derivative terms. This is followed by the evaluation of this expression at $x_i = x_{i,Q-2}^{\alpha+1,\beta+1}$, so $P_{Q-2}^{\alpha+1,\beta+1}(x_i) = 0$ and this gives the above relation (Ghizzetti and Ossicini [182, p. 107]).

APPENDIX D

C^0 CONTINUOUS EXPANSION BASES

In the following appendix we document the complete polynomial expansions for both the modal and nodal bases discussed in Chapter 3. This appendix supplements Chapter 3, where the bases are defined in a compact tensor form. The bases are defined in terms of three orders P_1, P_2, and P_3, corresponding to the local coordinate system. For the modal expansions we have indicated the indices over which the bases should be assembled in order to achieve the maximum polynomial space, although it is possible to reduce the interior, face, and edge expansion orders to produce a variable-order expansion, as discussed in Section 3.1.2. The nodal basis must be assembled over all the indicated indices for the expansion to span a complete polynomial space.

D.1 Modal basis

In the following section $P_p^{\alpha,\beta}(\xi)$ refers to the pth-order Jacobi polynomial defined in Section 2.3.3.1 and Appendix A.

D.1.1 *Two-dimensional expansions*

D.1.1.1 *Quadrilateral domain*

Using the labelling given in Fig. D.1, the quadrilateral basis is

$$\text{vertex } A: \quad \frac{1-\xi_1}{2}\frac{1-\xi_2}{2}, \qquad \text{vertex } B: \quad \frac{1+\xi_1}{2}\frac{1-\xi_2}{2},$$
$$\text{vertex } C: \quad \frac{1-\xi_1}{2}\frac{1+\xi_2}{2}, \qquad \text{vertex } D: \quad \frac{1+\xi_1}{2}\frac{1+\xi_2}{2},$$

$$\text{edge } AB: \quad \frac{1-\xi_1}{2}\frac{1+\xi_1}{2}P_{p-1}^{1,1}(\xi_1)\frac{1-\xi_2}{2}, \quad 0<p<P_1,$$
$$\text{edge } CD: \quad \frac{1-\xi_1}{2}\frac{1+\xi_1}{2}P_{p-1}^{1,1}(\xi_1)\frac{1+\xi_2}{2}, \quad 0<p<P_1,$$
$$\text{edge } AC: \quad \frac{1-\xi_1}{2}\frac{1-\xi_2}{2}\frac{1+\xi_2}{2}P_{q-1}^{1,1}(\xi_2), \quad 0<q<P_2,$$
$$\text{edge } BD: \quad \frac{1+\xi_1}{2}\frac{1-\xi_2}{2}\frac{1+\xi_2}{2}P_{q-1}^{1,1}(\xi_2), \quad 0<q<P_2,$$

$$\text{interior:} \quad \frac{1-\xi_1}{2}\frac{1+\xi_1}{2}P_{p-1}^{1,1}(\xi_1)\frac{1-\xi_2}{2}\frac{1+\xi_2}{2}P_{q-1}^{1,1}(\xi_2),$$
$$0<p,q,\ p<P_1,\ q<P_2.$$

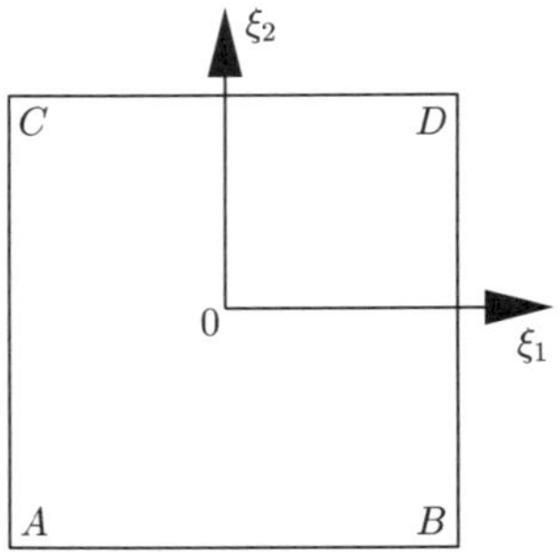

FIG. D.1. Vertex labelling for the standard quadrilateral region.

D.1.1.2 *Triangular domain*

Using the labelling given in Fig. D.2, the triangular basis is

$$\eta_1 = 2\frac{1+\xi_1}{1-\xi_2} - 1, \quad \eta_2 = \xi_2,$$

$$\text{vertex } A\text{: } \frac{1-\eta_1}{2}\frac{1-\eta_2}{2}, \quad \text{vertex } B\text{: } \frac{1+\eta_1}{2}\frac{1-\eta_2}{2}, \quad \text{vertex } C\text{: } \frac{1+\eta_2}{2},$$

$$\text{edge } AB\text{: } \frac{1-\eta_1}{2}\frac{1+\eta_1}{2} P_{p-1}^{1,1}(\eta_1)\left(\frac{1-\eta_2}{2}\right)^{p+1}, \quad 0 < p < P_1,$$

$$\text{edge } AC\text{: } \frac{1-\eta_1}{2}\frac{1-\eta_2}{2}\frac{1+\eta_2}{2} P_{q-1}^{1,1}(\eta_2), \quad 0 < q < P_2,$$

$$\text{edge } BC\text{: } \frac{1+\eta_1}{2}\frac{1-\eta_2}{2}\frac{1+\eta_2}{2} P_{q-1}^{1,1}(\eta_2), \quad 0 < q < P_2,$$

$$\text{interior: } \frac{1-\eta_1}{2}\frac{1+\eta_1}{2} P_{p-1}^{1,1}(\eta_1)\left(\frac{1-\eta_2}{2}\right)^{p+1} \frac{1+\eta_2}{2} P_{q-1}^{2p+1,1}(\eta_2),$$

$$0 < p, q, \ p < P_1, \ p+q < P_2, \ P_1 \leqslant P_2 .$$

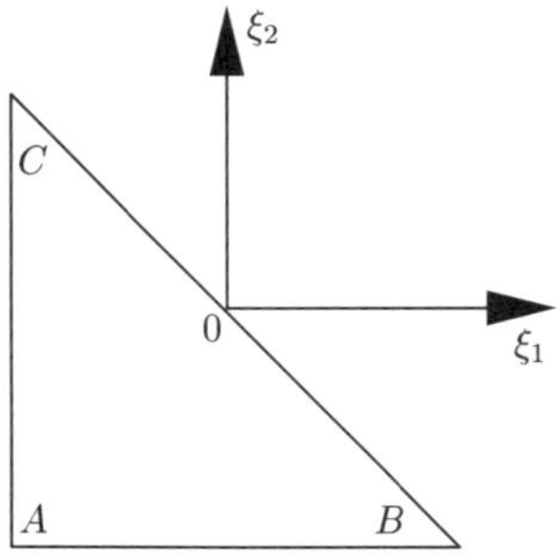

FIG. D.2. Vertex labelling for the standard triangular region.

D.1.2 *Three-dimensional expansions*

In Fig. D.3 we define the vertices of the four standard three-dimensional regions: hexahedrons, prisms, pyramids, and tetrahedrons. Using these vertex definitions it is then possible to identify each edge and face to define the standard expansions as follows.

D.1.2.1 *Hexahedral domain*

Using the labelling given in Fig. D.3(a), the hexahedral basis is

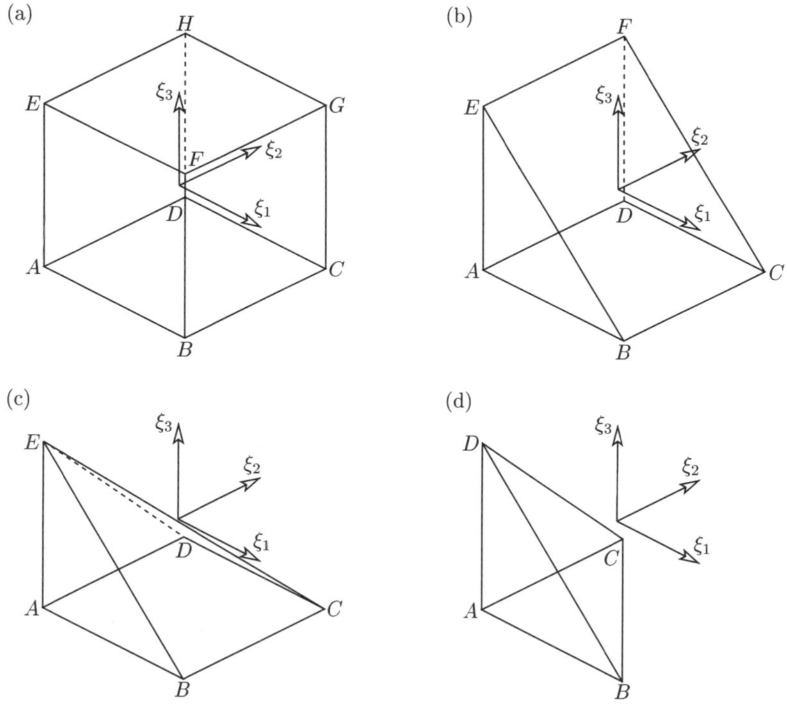

FIG. D.3. Vertex labelling for the three-dimensional standard regions: (a) a hexahedral domain, (b) a prismatic domain, (c) a pyramidic domain, and (d) a tetrahedral domain.

$$\text{vertex } A: \quad \frac{1-\xi_1}{2}\frac{1-\xi_2}{2}\frac{1-\xi_3}{2}, \quad \text{vertex } B: \quad \frac{1+\xi_1}{2}\frac{1-\xi_2}{2}\frac{1-\xi_3}{2},$$
$$\text{vertex } C: \quad \frac{1-\xi_1}{2}\frac{1+\xi_2}{2}\frac{1-\xi_3}{2}, \quad \text{vertex } D: \quad \frac{1+\xi_1}{2}\frac{1+\xi_2}{2}\frac{1-\xi_3}{2},$$
$$\text{vertex } E: \quad \frac{1-\xi_1}{2}\frac{1-\xi_2}{2}\frac{1+\xi_3}{2}, \quad \text{vertex } F: \quad \frac{1+\xi_1}{2}\frac{1-\xi_2}{2}\frac{1+\xi_3}{2},$$
$$\text{vertex } G: \quad \frac{1-\xi_1}{2}\frac{1+\xi_2}{2}\frac{1+\xi_3}{2}, \quad \text{vertex } H: \quad \frac{1+\xi_1}{2}\frac{1+\xi_2}{2}\frac{1+\xi_3}{2},$$

$$\text{edge } AB: \quad \frac{1-\xi_1}{2}\frac{1+\xi_1}{2}P^{1,1}_{p-1}(\xi_1)\frac{1-\xi_2}{2}\frac{1-\xi_3}{2}, \quad 0 < p < P_1,$$
$$\text{edge } CD: \quad \frac{1-\xi_1}{2}\frac{1+\xi_1}{2}P^{1,1}_{p-1}(\xi_1)\frac{1+\xi_2}{2}\frac{1-\xi_3}{2}, \quad 0 < p < P_1,$$
$$\text{edge } EF: \quad \frac{1-\xi_1}{2}\frac{1+\xi_1}{2}P^{1,1}_{p-1}(\xi_1)\frac{1-\xi_2}{2}\frac{1+\xi_3}{2}, \quad 0 < p < P_1,$$
$$\text{edge } GH: \quad \frac{1-\xi_1}{2}\frac{1+\xi_1}{2}P^{1,1}_{p-1}(\xi_1)\frac{1+\xi_2}{2}\frac{1+\xi_3}{2}, \quad 0 < p < P_1,$$
$$\text{edge } AC: \quad \frac{1-\xi_1}{2}\frac{1-\xi_2}{2}\frac{1+\xi_2}{2}P^{1,1}_{q-1}(\xi_2)\frac{1-\xi_3}{2}, \quad 0 < q < P_2,$$
$$\text{edge } BD: \quad \frac{1+\xi_1}{2}\frac{1-\xi_2}{2}\frac{1+\xi_2}{2}P^{1,1}_{q-1}(\xi_2)\frac{1-\xi_3}{2}, \quad 0 < q < P_2,$$
$$\text{edge } EG: \quad \frac{1-\xi_1}{2}\frac{1-\xi_2}{2}\frac{1+\xi_2}{2}P^{1,1}_{q-1}(\xi_2)\frac{1+\xi_3}{2}, \quad 0 < q < P_2,$$
$$\text{edge } FH: \quad \frac{1+\xi_1}{2}\frac{1-\xi_2}{2}\frac{1+\xi_2}{2}P^{1,1}_{q-1}(\xi_2)\frac{1+\xi_3}{2}, \quad 0 < q < P_2,$$
$$\text{edge } AE: \quad \frac{1-\xi_1}{2}\frac{1-\xi_2}{2}\frac{1-\xi_3}{2}\frac{1+\xi_3}{2}P^{1,1}_{r-1}(\xi_3), \quad 0 < r < P_3,$$
$$\text{edge } BF: \quad \frac{1+\xi_1}{2}\frac{1-\xi_2}{2}\frac{1-\xi_3}{2}\frac{1+\xi_3}{2}P^{1,1}_{r-1}(\xi_3), \quad 0 < r < P_3,$$
$$\text{edge } CG: \quad \frac{1+\xi_1}{2}\frac{1+\xi_2}{2}\frac{1-\xi_3}{2}\frac{1+\xi_3}{2}P^{1,1}_{r-1}(\xi_3), \quad 0 < r < P_3,$$
$$\text{edge } DH: \quad \frac{1+\xi_1}{2}\frac{1+\xi_2}{2}\frac{1-\xi_3}{2}\frac{1+\xi_3}{2}P^{1,1}_{r-1}(\xi_3), \quad 0 < r < P_3,$$

$$\text{face } ABCD: \quad \frac{1-\xi_1}{2}\frac{1+\xi_1}{2}P^{1,1}_{p-1}(\xi_1)\frac{1-\xi_2}{2}\frac{1+\xi_2}{2}P^{1,1}_{q-1}(\xi_2)\frac{1-\xi_3}{2},$$
$$\text{face } EFGH: \quad \frac{1-\xi_1}{2}\frac{1+\xi_1}{2}P^{1,1}_{p-1}(\xi_1)\frac{1-\xi_2}{2}\frac{1+\xi_2}{2}P^{1,1}_{q-1}(\xi_2)\frac{1+\xi_3}{2},$$
$$0 < p,q,\ p < P_1,\ q < P_2,$$

$$\text{face } ABEF: \quad \frac{1-\xi_1}{2}\frac{1+\xi_1}{2}P^{1,1}_{p-1}(\xi_1)\frac{1-\xi_2}{2}\frac{1-\xi_3}{2}\frac{1+\xi_3}{2}P^{1,1}_{r-1}(\xi_3),$$
$$\text{face } CDGH: \quad \frac{1-\xi_1}{2}\frac{1+\xi_1}{2}P^{1,1}_{p-1}(\xi_1)\frac{1+\xi_2}{2}\frac{1-\xi_3}{2}\frac{1+\xi_3}{2}P^{1,1}_{r-1}(\xi_3),$$
$$0 < p,r,\ p < P_1,\ r < P_3,$$

$$\text{face } ACEG\text{:}\quad \frac{1-\xi_1}{2}\frac{1-\xi_2}{2}\frac{1+\xi_2}{2}P^{1,1}_{q-1}(\xi_2)\,\frac{1-\xi_3}{2}\frac{1+\xi_3}{2}P^{1,1}_{r-1}(\xi_3)\,,$$

$$\text{face } BDFH\text{:}\quad \frac{1+\xi_1}{2}\frac{1-\xi_2}{2}\frac{1+\xi_2}{2}P^{1,1}_{q-1}(\xi_2)\,\frac{1-\xi_3}{2}\frac{1+\xi_3}{2}P^{1,1}_{r-1}(\xi_3)\,,$$

$$0<q,r,\ q<P_2,\ r<P_3\,,$$

$$\text{interior:}\quad \frac{1-\xi_1}{2}\frac{1+\xi_1}{2}P^{1,1}_{p-1}(\xi_1)\,\frac{1-\xi_2}{2}\frac{1+\xi_2}{2}P^{1,1}_{q-1}(\xi_2)\,\frac{1-\xi_3}{2}\frac{1+\xi_3}{2}P^{1,1}_{r-1}(\xi_3)\,,$$

$$0<p,q,r,\ p<P_1,\ q<P_2,\ r<P_3\,.$$

D.1.2.2 *Prismatic domain*

Using the labelling given in Fig. D.3(b) and noting that

$$\overline{\eta_1} = 2\frac{1+\xi_1}{1-\xi_3} - 1\,,$$

the prismatic basis is:

$$\text{vertex } A\text{:}\quad \frac{1-\overline{\eta_1}}{2}\frac{1-\xi_2}{2}\frac{1-\xi_3}{2}\,,\qquad \text{vertex } B\text{:}\quad \frac{1+\overline{\eta_1}}{2}\frac{1-\xi_2}{2}\frac{1-\xi_3}{2}\,,$$

$$\text{vertex } C\text{:}\quad \frac{1-\overline{\eta_1}}{2}\frac{1+\xi_2}{2}\frac{1-\xi_3}{2}\,,\qquad \text{vertex } D\text{:}\quad \frac{1+\overline{\eta_1}}{2}\frac{1+\xi_2}{2}\frac{1-\xi_3}{2}\,,$$

$$\text{vertex } E\text{:}\quad \frac{1-\xi_2}{2}\frac{1+\xi_3}{2}\,,\qquad \text{vertex } F\text{:}\quad \frac{1+\xi_2}{2}\frac{1+\xi_3}{2}\,,$$

$$\text{edge } AB\text{:}\quad \frac{1-\overline{\eta_1}}{2}\frac{1+\overline{\eta_1}}{2}P^{1,1}_{p-1}(\overline{\eta_1})\,\frac{1-\xi_2}{2}\left(\frac{1-\xi_3}{2}\right)^{p+1},\qquad 0<p<P_1\,,$$

$$\text{edge } CD\text{:}\quad \frac{1-\overline{\eta_1}}{2}\frac{1+\overline{\eta_1}}{2}P^{1,1}_{p-1}(\overline{\eta_1})\,\frac{1+\xi_2}{2}\left(\frac{1-\xi_3}{2}\right)^{p+1},\qquad 0<p<P_1\,,$$

$$\text{edge } AC\text{:}\quad \frac{1-\overline{\eta_1}}{2}\frac{1-\xi_2}{2}\frac{1+\xi_2}{2}P^{1,1}_{q-1}(\xi_2)\,\frac{1-\xi_3}{2}\,,\qquad 0<q<P_2\,,$$

$$\text{edge } BD\text{:}\quad \frac{1+\overline{\eta_1}}{2}\frac{1-\xi_2}{2}\frac{1+\xi_2}{2}P^{1,1}_{q-1}(\xi_2)\,\frac{1-\xi_3}{2}\,,\qquad 0<q<P_2\,,$$

$$\text{edge } EF\text{:}\quad \frac{1-\overline{\eta_1}}{2}\frac{1-\xi_2}{2}\frac{1+\xi_2}{2}P^{1,1}_{q-1}(\xi_2)\,\frac{1+\xi_3}{2}\,,\qquad 0<q<P_2\,,$$

$$\text{edge } AE\text{:}\quad \frac{1-\overline{\eta_1}}{2}\frac{1-\xi_2}{2}\frac{1-\xi_3}{2}\frac{1+\xi_3}{2}P^{1,1}_{r-1}(\xi_3)\,,\qquad 0<r<P_3\,,$$

$$\text{edge } BE\text{:}\quad \frac{1+\overline{\eta_1}}{2}\frac{1-\xi_2}{2}\frac{1-\xi_3}{2}\frac{1+\xi_3}{2}P^{1,1}_{r-1}(\xi_3)\,,\qquad 0<r<P_3\,,$$

$$\text{edge } CF\text{:}\quad \frac{1+\overline{\eta_1}}{2}\frac{1+\xi_2}{2}\frac{1-\xi_3}{2}\frac{1+\xi_3}{2}P^{1,1}_{r-1}(\xi_3)\,,\qquad 0<r<P_3\,,$$

$$\text{edge } DF\text{:}\quad \frac{1-\overline{\eta_1}}{2}\frac{1+\xi_2}{2}\frac{1-\xi_3}{2}\frac{1+\xi_3}{2}P^{1,1}_{r-1}(\xi_3)\,,\qquad 0<r<P_3\,,$$

$$\text{face } ABCD\text{:}\quad \frac{1-\overline{\eta_1}}{2}\frac{1+\overline{\eta_1}}{2}P^{1,1}_{p-1}(\overline{\eta_1})\,\frac{1-\xi_2}{2}\frac{1+\xi_2}{2}P^{1,1}_{q-1}(\xi_2)\left(\frac{1-\xi_3}{2}\right)^{p+1},$$
$$0<p,q,\ p<P_1,\ q<P_2\,,$$

$$\text{face } ADEF\text{:}\quad \frac{1-\overline{\eta_1}}{2}\frac{1-\xi_2}{2}\frac{1+\xi_2}{2}P^{1,1}_{q-1}(\xi_2)\,\frac{1-\xi_3}{2}\frac{1+\xi_3}{2}P^{1,1}_{r-1}(\xi_3)\,,$$
$$\text{face } BCEF\text{:}\quad \frac{1+\overline{\eta_1}}{2}\frac{1-\xi_2}{2}\frac{1+\xi_2}{2}P^{1,1}_{q-1}(\xi_2)\,\frac{1-\xi_3}{2}\frac{1+\xi_3}{2}P^{1,1}_{r-1}(\xi_3)\,,$$
$$0<q,r,\ q<P_2,\ r<P_3\,,$$

$$\text{face } ABE\text{:}\quad \frac{1-\overline{\eta_1}}{2}\frac{1+\overline{\eta_1}}{2}P^{1,1}_{p-1}(\overline{\eta_1})\,\frac{1-\xi_2}{2}\left(\frac{1-\xi_3}{2}\right)^{p+1}\frac{1+\xi_3}{2}P^{2q+1,1}_{r-1}(\xi_3)\,,$$
$$\text{face } CDF\text{:}\quad \frac{1-\overline{\eta_1}}{2}\frac{1+\overline{\eta_1}}{2}P^{1,1}_{p-1}(\overline{\eta_1})\,\frac{1+\xi_2}{2}\left(\frac{1-\xi_3}{2}\right)^{p+1}\frac{1+\xi_3}{2}P^{2q+1,1}_{r-1}(\xi_3)\,,$$
$$0<p,r,\ p<P_1,\ p+r<P_3,\ P_2\leqslant P_3\,,$$

interior:

$$\frac{1-\overline{\eta_1}}{2}\frac{1+\overline{\eta_1}}{2}P^{1,1}_{p-1}(\overline{\eta_1})\,\frac{1-\xi_2}{2}\frac{1+\xi_2}{2}P^{1,1}_{q-1}(\xi_2)\left(\frac{1-\xi_3}{2}\right)^{p+1}\frac{1+\xi_3}{2}P^{2q+1,1}_{r-1}(\xi_3)\,,$$
$$0<p,q,r,\ p<P_1,\ q<P_2,\ q+r<P_3,\ P_2\leqslant P_3\,.$$

D.1.2.3 *Pyramidic domain*

Using the labelling given in Fig. D.3(c), the pyramidic basis is

$$\overline{\eta_1}=2\frac{1+\xi_1}{1-\xi_3}-1\,,\quad \eta_2=2\frac{1+\xi_2}{1-\xi_3}-1\,,\quad \eta_3=\xi_3\,,$$

$$\text{vertex } A\text{:}\quad \frac{1-\overline{\eta_1}}{2}\frac{1-\eta_2}{2}\frac{1-\eta_3}{2},\qquad \text{vertex } B\text{:}\quad \frac{1+\overline{\eta_1}}{2}\frac{1-\eta_2}{2}\frac{1-\eta_3}{2},$$
$$\text{vertex } C\text{:}\quad \frac{1-\overline{\eta_1}}{2}\frac{1+\eta_2}{2}\frac{1-\eta_3}{2},\qquad \text{vertex } D\text{:}\quad \frac{1+\overline{\eta_1}}{2}\frac{1+\eta_2}{2}\frac{1-\eta_3}{2},$$
$$\text{vertex } E\text{:}\quad \frac{1+\eta_3}{2},$$

$$\text{edge } AB\text{:}\quad \frac{1-\overline{\eta_1}}{2}\frac{1+\overline{\eta_1}}{2}P^{1,1}_{p-1}(\overline{\eta_1})\,\frac{1-\eta_2}{2}\left(\frac{1-\eta_3}{2}\right)^{p+1},\quad 0<p<P_1\,,$$
$$\text{edge } CD\text{:}\quad \frac{1-\overline{\eta_1}}{2}\frac{1+\overline{\eta_1}}{2}P^{1,1}_{p-1}(\overline{\eta_1})\,\frac{1+\eta_2}{2}\left(\frac{1-\eta_3}{2}\right)^{p+1},\quad 0<p<P_1\,,$$
$$\text{edge } AC\text{:}\quad \frac{1-\overline{\eta_1}}{2}\frac{1-\eta_2}{2}\frac{1+\eta_2}{2}P^{1,1}_{q-1}(\eta_2)\left(\frac{1-\eta_3}{2}\right)^{q+1},\quad 0<q<P_2\,,$$
$$\text{edge } BD\text{:}\quad \frac{1+\overline{\eta_1}}{2}\frac{1-\eta_2}{2}\frac{1+\eta_2}{2}P^{1,1}_{q-1}(\eta_2)\left(\frac{1-\eta_3}{2}\right)^{q+1},\quad 0<q<P_2\,,$$

$$\begin{aligned}
&\text{edge } AE\text{:} && \frac{1-\overline{\eta_1}}{2}\frac{1-\eta_2}{2}\frac{1-\eta_3}{2}\frac{1+\eta_3}{2}\,P_{r-1}^{1,1}(\eta_3)\,, \quad 0 < r < P_3\,,\\
&\text{edge } BE\text{:} && \frac{1+\overline{\eta_1}}{2}\frac{1-\eta_2}{2}\frac{1-\eta_3}{2}\frac{1+\eta_3}{2}\,P_{r-1}^{1,1}(\eta_3)\,, \quad 0 < r < P_3\,,\\
&\text{edge } CE\text{:} && \frac{1+\overline{\eta_1}}{2}\frac{1+\eta_2}{2}\frac{1-\eta_3}{2}\frac{1+\eta_3}{2}\,P_{r-1}^{1,1}(\eta_3)\,, \quad 0 < r < P_3\,,\\
&\text{edge } DE\text{:} && \frac{1+\overline{\eta_1}}{2}\frac{1+\eta_2}{2}\frac{1-\eta_3}{2}\frac{1+\eta_3}{2}\,P_{r-1}^{1,1}(\eta_3)\,, \quad 0 < r < P_3\,,
\end{aligned}$$

$$\text{face } ABCD\text{:} \quad \frac{1-\overline{\eta_1}}{2}\frac{1+\overline{\eta_1}}{2}\,P_{p-1}^{1,1}(\overline{\eta_1})\,\frac{1-\eta_2}{2}\frac{1+\eta_2}{2}\,P_{q-1}^{1,1}(\eta_2)\left(\frac{1-\eta_3}{2}\right)^{p+q+1},$$
$$0 < p,q,\ p < P_1,\ q < P_2\,,$$

$$\begin{aligned}
&\text{face } ABE\text{:} && \frac{1-\overline{\eta_1}}{2}\frac{1+\overline{\eta_1}}{2}\,P_{p-1}^{1,1}(\overline{\eta_1})\,\frac{1-\eta_2}{2}\left(\frac{1-\eta_3}{2}\right)^{p+1}\frac{1+\eta_3}{2}\,P_{r-1}^{2p+1,1}(\eta_3)\,,\\
&\text{face } CDE\text{:} && \frac{1-\overline{\eta_1}}{2}\frac{1+\overline{\eta_1}}{2}\,P_{p-1}^{1,1}(\overline{\eta_1})\,\frac{1+\eta_2}{2}\left(\frac{1-\eta_3}{2}\right)^{p+1}\frac{1+\eta_3}{2}\,P_{r-1}^{2p+1,1}(\eta_3)\,,
\end{aligned}$$
$$0 < p,r,\ p < P_1,\ p+r < P_3,\ P_1 \leqslant P_3\,,$$

$$\begin{aligned}
&\text{face } ACE\text{:} && \frac{1-\overline{\eta_1}}{2}\frac{1-\eta_2}{2}\frac{1+\eta_2}{2}\,P_{q-1}^{1,1}(\eta_2)\left(\frac{1-\eta_3}{2}\right)^{q+1}\frac{1+\eta_3}{2}\,P_{r-1}^{2q+1,1}(\eta_3)\,,\\
&\text{face } BDE\text{:} && \frac{1+\overline{\eta_1}}{2}\frac{1-\eta_2}{2}\frac{1+\eta_2}{2}\,P_{q-1}^{1,1}(\eta_2)\left(\frac{1-\eta_3}{2}\right)^{q+1}\frac{1+\eta_3}{2}\,P_{r-1}^{2q+1,1}(\eta_3)\,,
\end{aligned}$$
$$0 < q,r,\ q < P_2,\ q+r < P_3,\ P_2 \leqslant P_3\,,$$

$$\text{interior:} \quad \frac{1-\overline{\eta_1}}{2}\frac{1+\overline{\eta_1}}{2}\,P_{p-1}^{1,1}(\overline{\eta_1})\,\frac{1-\eta_2}{2}\frac{1+\eta_2}{2}\,P_{q-1}^{1,1}(\eta_2)$$
$$\times\left(\frac{1-\eta_3}{2}\right)^{p+q+1}\frac{1+\eta_3}{2}\,P_{r-1}^{2p+2q+1,1}(\eta_3)\,,$$
$$0 < p,q,r,\ p < P_1,\ q < P_2,\ p+q+r < P_3,\ P_1,P_2 \leqslant P_3\,.$$

D.1.2.4 *Tetrahedral domain*

Using the labelling given in Fig. D.3(d), the tetrahedral basis is

$$\eta_1 = 2\frac{1+\xi_1}{-\xi_2-\xi_3} - 1\,, \quad \eta_2 = 2\frac{1+\xi_2}{1-\xi_3} - 1\,, \quad \eta_3 = \xi_3\,.$$

$$\begin{aligned}
&\text{vertex } A\text{:} \quad \frac{1-\eta_1}{2}\frac{1-\eta_2}{2}\frac{1-\eta_3}{2}\,, && \text{vertex } B\text{:} \quad \frac{1+\eta_1}{2}\frac{1-\eta_2}{2}\frac{1-\eta_3}{2}\,,\\
&\text{vertex } C\text{:} \quad \frac{1-\eta_1}{2}\frac{1+\eta_2}{2}\frac{1-\eta_3}{2}\,, && \text{vertex } D\text{:} \quad \frac{1+\eta_3}{2}\,,
\end{aligned}$$

edge AB: $\frac{1-\eta_1}{2}\frac{1+\eta_1}{2}\,P^{1,1}_{p-1}(\eta_1)\left(\frac{1-\eta_2}{2}\right)^{p+1}\left(\frac{1-\eta_3}{2}\right)^{p+1}, \quad 0<p<P_1,$

edge AC: $\frac{1-\eta_1}{2}\frac{1-\eta_2}{2}\frac{1+\eta_2}{2}\,P^{1,1}_{q-1}(\eta_2)\left(\frac{1-\eta_3}{2}\right)^{q+1}, \quad 0<q<P_2,$

edge BC: $\frac{1+\eta_1}{2}\frac{1-\eta_2}{2}\frac{1+\eta_2}{2}\,P^{1,1}_{q-1}(\eta_2)\left(\frac{1-\eta_3}{2}\right)^{q+1}, \quad 0<q<P_2,$

edge AD: $\frac{1-\eta_1}{2}\frac{1-\eta_2}{2}\frac{1-\eta_3}{2}\frac{1+\eta_3}{2}\,P^{1,1}_{r-1}(\eta_3), \quad 0<r<P_3,$

edge BD: $\frac{1+\eta_1}{2}\frac{1-\eta_2}{2}\frac{1-\eta_3}{2}\frac{1+\eta_3}{2}\,P^{1,1}_{r-1}(\eta_3), \quad 0<r<P_3,$

edge CD: $\frac{1+\eta_2}{2}\frac{1-\eta_3}{2}\frac{1+\eta_3}{2}\,P^{1,1}_{r-1}(\eta_3), \quad 0<r<P_3,$

face ABC:
$$\frac{1-\eta_1}{2}\frac{1+\eta_1}{2}\,P^{1,1}_{p-1}(\eta_1)\left(\frac{1-\eta_2}{2}\right)^{p+1}\frac{1+\eta_2}{2}$$
$$\times\,P^{2p+1,1}_{q-1}(\eta_2)\left(\frac{1-\eta_3}{2}\right)^{p+q+1},$$
$$0<p,q,\ p<P_1,\ p+q<P_2,\ P_1\leqslant P_2,$$

face ABE:
$$\frac{1-\eta_1}{2}\frac{1+\eta_1}{2}\,P^{1,1}_{p-1}(\eta_1)\left(\frac{1-\eta_2}{2}\right)^{p+1}\left(\frac{1-\eta_3}{2}\right)^{p+1}$$
$$\times\,\frac{1+\eta_3}{2}\,P^{2p+1,1}_{r-1}(\eta_3),$$
$$0<p,r,\ p<P_1,\ p+r<P_3,\ P_1\leqslant P_3,$$

face ACE: $\frac{1-\eta_1}{2}\frac{1-\eta_2}{2}\frac{1+\eta_2}{2}\,P^{1,1}_{q-1}(\eta_2)\left(\frac{1-\eta_3}{2}\right)^{q+1}\frac{1+\eta_3}{2}\,P^{2q+1,1}_{r-1}(\eta_3),$

face BCE: $\frac{1+\eta_1}{2}\frac{1-\eta_2}{2}\frac{1+\eta_2}{2}\,P^{1,1}_{q-1}(\eta_2)\left(\frac{1-\eta_3}{2}\right)^{q+1}\frac{1+\eta_3}{2}\,P^{2q+1,1}_{r-1}(\eta_3),$

$$0<q,r,\ q<P_2,\ q+r<P_3,\ P_2\leqslant P_3,$$

interior:
$$\frac{1-\eta_1}{2}\frac{1+\eta_1}{2}\,P^{1,1}_{p-1}(\eta_1)\left(\frac{1-\eta_2}{2}\right)^{p+1}\frac{1+\eta_2}{2}\,P^{2p+1,1}_{q-1}(\eta_2)$$
$$\times\left(\frac{1-\eta_3}{2}\right)^{p+q+1}\frac{1+\eta_3}{2}\,P^{2p+2q+1,1}_{r-1}(\eta_3),$$
$$0<p,q,r,\ p<P_1,\ p+q<P_2,\ p+q+r<P_3,\ P_1\leqslant P_2\leqslant P_3.$$

D.2 Nodal basis

D.2.1 *Tensorial expansions*

In the following section we define the notation $h_p^P(\xi)$ to be the Lagrange polynomial (see Section 2.3.4.1 and Appendix A) of order P through the $P+1$ zeros of the Gauss–Lobatto–Legendre integration scheme (see Section 2.4.1 and Appendix B).

D.2.1.1 *Quadrilateral domain*

Using the labelling given in Fig. D.4(a), the quadrilateral basis is

$$\text{vertex } A\text{:}\quad h_0^{P_1}(\xi_1)h_0^{P_2}(\xi_2)\,,\qquad \text{vertex } B\text{:}\quad h_{P_1}^{P_1}(\xi_1)h_0^{P_2}(\xi_2)\,,$$
$$\text{vertex } C\text{:}\quad h_0^{P_1}(\xi_1)h_{P_2}^{P_2}(\xi_2)\,,\qquad \text{vertex } D\text{:}\quad h_{P_1}^{P_1}(\xi_1)h_{P_2}^{P_2}(\xi_2)\,,$$

$$\text{edge } AB\text{:}\quad h_p^{P_1}(\xi_1)h_0^{P_2}(\xi_2)\,,\quad 0 < p < P_1\,,$$
$$\text{edge } CD\text{:}\quad h_p^{P_1}(\xi_1)h_{P_2}^{P_2}(\xi_2)\,,\quad 0 < p < P_1\,,$$
$$\text{edge } AC\text{:}\quad h_0^{P_1}(\xi_1)h_q^{P_2}(\xi_2)\,,\quad 0 < q < P_2\,,$$
$$\text{edge } BD\text{:}\quad h_{P_1}^{P_1}(\xi_1)h_q^{P_2}(\xi_2)\,,\quad 0 < q < P_2\,,$$

$$\text{interior:}\quad h_p^{P_1}(\xi_1)h_q^{P_2}(\xi_2)\,,\quad 0 < p, q,\ p < P_1,\ q < P_2\,.$$

D.2.1.2 *Hexahedral domain*

Using the labelling given in Fig. D.4(b), the hexahedral basis is

$$\text{vertex } A\text{:}\quad h_0^{P_1}(\xi_1)h_0^{P_2}(\xi_2)h_0^{P_3}(\xi_3)\,,\qquad \text{vertex } B\text{:}\quad h_{P_1}^{P_1}(\xi_1)h_0^{P_2}(\xi_2)h_0^{P_3}(\xi_3)\,,$$
$$\text{vertex } C\text{:}\quad h_0^{P_1}(\xi_1)h_{P_2}^{P_2}(\xi_2)h_0^{P_3}(\xi_3)\,,\qquad \text{vertex } D\text{:}\quad h_{P_1}^{P_1}(\xi_1)h_{P_2}^{P_2}(\xi_2)h_0^{P_3}(\xi_3)\,,$$
$$\text{vertex } E\text{:}\quad h_0^{P_1}(\xi_1)h_0^{P_2}(\xi_2)h_{P_3}^{P_3}(\xi_3)\,,\qquad \text{vertex } F\text{:}\quad h_{P_1}^{P_1}(\xi_1)h_0^{P_2}(\xi_2)h_{P_3}^{P_3}(\xi_3)\,,$$
$$\text{vertex } G\text{:}\quad h_0^{P_1}(\xi_1)h_{P_2}^{P_2}(\xi_2)h_{P_3}^{P_3}(\xi_3)\,,\qquad \text{vertex } H\text{:}\quad h_{P_1}^{P_1}(\xi_1)h_{P_2}^{P_2}(\xi_2)h_{P_3}^{P_3}(\xi_3)\,,$$

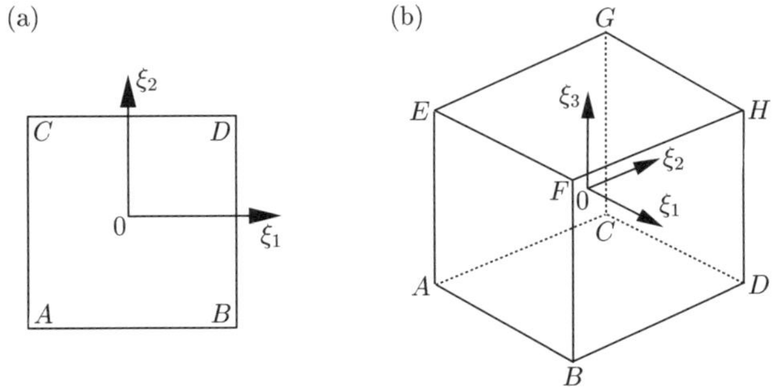

FIG. D.4. Vertex labelling for (a) the standard quadrilateral, and (b) the standard hexahedral region.

$$
\begin{array}{lll}
\text{edge } AB: & h_p^{P_1}(\xi_1)h_0^{P_2}(\xi_2)h_0^{P_3}(\xi_3)\,, & 0 < p < P_1\,, \\
\text{edge } CD: & h_p^{P_1}(\xi_1)h_{P_2}^{P_2}(\xi_2)h_0^{P_3}(\xi_3)\,, & 0 < p < P_1\,, \\
\text{edge } EF: & h_p^{P_1}(\xi_1)h_0^{P_2}(\xi_2)h_{P_3}^{P_3}(\xi_3)\,, & 0 < p < P_1\,, \\
\text{edge } GH: & h_p^{P_1}(\xi_1)h_{P_2}^{P_2}(\xi_2)h_{P_3}^{P_3}(\xi_3)\,, & 0 < p < P_1\,, \\
\text{edge } AC: & h_0^{P_1}(\xi_1)h_q^{P_2}(\xi_2)h_0^{P_3}(\xi_3)\,, & 0 < q < P_2\,, \\
\text{edge } BD: & h_0^{P_1}(\xi_1)h_q^{P_2}(\xi_2)h_0^{P_3}(\xi_3)\,, & 0 < q < P_2\,, \\
\text{edge } EG: & h_0^{P_1}(\xi_1)h_q^{P_2}(\xi_2)h_{P_3}^{P_3}(\xi_3)\,, & 0 < q < P_2\,, \\
\text{edge } FH: & h_0^{P_1}(\xi_1)h_q^{P_2}(\xi_2)h_{P_3}^{P_3}(\xi_3)\,, & 0 < q < P_2\,, \\
\text{edge } AE: & h_0^{P_1}(\xi_1)h_0^{P_2}(\xi_2)h_r^{P_3}(\xi_3)\,, & 0 < r < P_3\,, \\
\text{edge } BF: & h_{P_1}^{P_1}(\xi_1)h_0^{P_2}(\xi_2)h_r^{P_3}(\xi_3)\,, & 0 < r < P_3\,, \\
\text{edge } CG: & h_0^{P_1}(\xi_1)h_{P_2}^{P_2}(\xi_2)h_r^{P_3}(\xi_3)\,, & 0 < r < P_3\,, \\
\text{edge } DH: & h_{P_1}^{P_1}(\xi_1)h_{P_2}^{P_2}(\xi_2)h_r^{P_3}(\xi_3)\,, & 0 < r < P_3\,,
\end{array}
$$

$$
\begin{array}{ll}
\text{face } ABCD: & h_p^{P_1}(\xi_1)h_q^{P_2}(\xi_2)h_0^{P_3}(\xi_3)\,, \\
\text{face } EFGH: & h_p^{P_1}(\xi_1)h_q^{P_2}(\xi_2)h_{P_3}^{P_3}(\xi_3)\,, \\
& \qquad 0 < p, q,\ p < P_1,\ q < P_2\,,
\end{array}
$$

$$
\begin{array}{ll}
\text{face } ABEF: & h_p^{P_1}(\xi_1)h_0^{P_2}(\xi_2)h_r^{P_3}(\xi_3)\,, \\
\text{face } CDGH: & h_p^{P_1}(\xi_1)h_{P_2}^{P_2}(\xi_2)h_r^{P_3}(\xi_3)\,, \\
& \qquad 0 < p, r,\ p < P_1,\ r < P_3\,,
\end{array}
$$

$$
\begin{array}{ll}
\text{face } ACEG: & h_0^{P_1}(\xi_1)h_q^{P_2}(\xi_2)h_r^{P_3}(\xi_3)\,, \\
\text{face } CDGH: & h_{P_1}^{P_1}(\xi_1)h_q^{P_2}(\xi_2)h_r^{P_3}(\xi_3)\,, \\
& \qquad 0 < q, r,\ q < P_1,\ r < P_3\,,
\end{array}
$$

$$
\begin{array}{ll}
\text{interior:} & h_p^{P_1}(\xi_1)h_q^{P_2}(\xi_2)h_q^{P_3}(\xi_3)\,, \\
& \qquad 0 < p, q, r,\ p < P_1,\ q < P_2,\ r < P_3\,.
\end{array}
$$

D.2.2 *Non-tensorial expansions*

D.2.2.1 *Triangular domain*

We recall from Section 3.2.1.3 that the barycentric coordinates (l_1, l_2, l_3) have the property that

$$l_1 + l_2 + l_3 = 1\,,$$

and for the standard right-handed triangular region, as defined in Fig. 3.10, they can be expressed in terms of Cartesian coordinates ξ_1, ξ_2 as

$$l_1 = \frac{1}{2}(1 - \xi_1) - \frac{1}{2}(1 + \xi_2)\,,$$

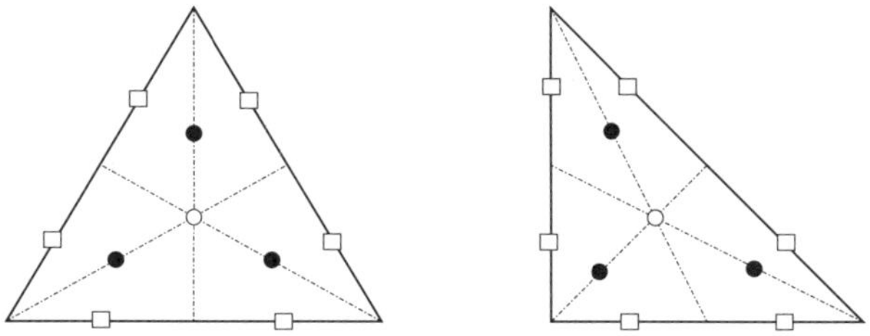

FIG. D.5. Symmetries of triangular nodal points in an equilateral and right-handed triangle. There is one centre-point with a single symmetry denoted by the open circle. Along the three dashed lines there is a three-fold symmetry, as indicated by the solid circles. For any other points there is a six-fold symmetry, as indicated by the open squares.

$$l_2 = \frac{1}{2}(1+\xi_1)\,,$$
$$l_3 = \frac{1}{2}(1+\xi_2)\,,$$

or, equivalently, we observe that

$$\xi_1 = 2l_2 - 1\,, \quad \xi_2 = 2l_3 - 1\,.$$

The two-dimensional collapsed coordinate system defined in Section 3.2.1.1 can also be written in terms of the area coordinates as

$$\eta_1 = \frac{2l_2}{1-l_3} - 1 = \frac{l_2 - l_1}{1-l_3}\,, \quad \eta_2 = 2l_3 - 1\,.$$

In the following we define the sets of barycentric coordinates that define the electrostatic and Fekete nodal points for use in non-tensorial nodal bases, as discussed in Section 3.3. As highlighted in Fig. D.5, the points have three different possible symmetries. A point in the centre of the triangle has a single symmetry, as indicated by the open circles in Fig. D.5. Points which lie on the lines from the vertices to the edge bisectors have a three-fold symmetry, as indicated by the solid circles in Fig. D.5. Finally, all other points have a six-fold symmetry, as indicated by the open squares in Fig. D.5. Using n_1, n_3, and n_6 to denote the single, three-fold, and six-fold symmetries, respectively, we can define the following points where points with more than one symmetry are only defined once and the other coordinates can be calculated through permutations.

Note that the dimension of expansion is $N_m = n_1 + 3n_3 + 6n_4 = (P+1)(P+2)/2$. We also observe that the barycentric coordinates l_2 and l_3 provide an indication of the ξ_1 and ξ_2 locations in a right-handed triangle.

Electrostatic points

The following points were provided by courtesy of Hesthaven [234].

P	n_1	n_3	n_6	l_1	l_2	l_3
1		1		1.0000000000	0.0000000000	0.0000000000
2		2		1.0000000000	0.0000000000	0.0000000000
				0.5000000000	0.5000000000	0.0000000000
3	1			0.3333333333	0.3333333333	0.3333333333
		1		1.0000000000	0.0000000000	0.0000000000
			1	0.7236067977	0.2763932023	0.0000000000
4		3		1.0000000000	0.0000000000	0.0000000000
				0.5000000000	0.5000000000	0.0000000000
				0.2371200168	0.2371200168	0.5257599664
			1	0.8273268354	0.1726731646	0.0000000000
5		3		1.0000000000	0.0000000000	0.0000000000
				0.4105151510	0.4105151510	0.1789696980
				0.1575181512	0.1575181512	0.6849636976
			2	0.8825276620	0.1174723380	0.0000000000
				0.6426157582	0.3573842418	0.0000000000
6	1			0.3333333333	0.3333333333	0.3333333333
		3		1.0000000000	0.0000000000	0.0000000000
				0.5000000000	0.5000000000	0.0000000000
				0.1061169285	0.1061169285	0.7877661430
			3	0.9151119481	0.0848880519	0.0000000000
				0.7344243967	0.2655756033	0.0000000000
				0.3097982151	0.5569099204	0.1332918645
7		4		1.0000000000	0.0000000000	0.0000000000
				0.4477725053	0.4477725053	0.1044549894
				0.2604038024	0.2604038024	0.4791923952
				0.0660520784	0.0660520784	0.8678958432
			4	0.9358700743	0.0641299257	0.0000000000
				0.7958500907	0.2041499093	0.0000000000
				0.6046496090	0.3953503910	0.0000000000
				0.2325524777	0.6759625951	0.0914849272
8		5		1.0000000000	0.0000000000	0.0000000000
				0.5000000000	0.5000000000	0.0000000000
				0.3905496216	0.3905496216	0.2189007568
				0.2033467796	0.2033467796	0.5933064408
				0.0469685351	0.0469685351	0.9060629298
			5	0.9498789977	0.0501210023	0.0000000000
				0.8385931398	0.1614068602	0.0000000000
				0.6815587319	0.3184412681	0.0000000000
				0.3617970895	0.5541643672	0.0840385433
				0.1801396087	0.7519065566	0.0679538347
9	1			0.3333333333	0.3333333333	0.3333333333
		4		1.0000000000	0.0000000000	0.0000000000
				0.4640303025	0.4640303025	0.0719393950
				0.1633923069	0.1633923069	0.6732153862
				0.0355775717	0.0355775717	0.9288448566
			7	0.9597669541	0.0402330459	0.0000000000
				0.8693869326	0.1306130674	0.0000000000
				0.7389624749	0.2610375251	0.0000000000
				0.5826394788	0.4173605212	0.0000000000
				0.3225890045	0.4968009397	0.1806100558
				0.2966333890	0.6349633653	0.0684032457
				0.1439089974	0.8031490682	0.0529419344

P	n_1	n_3	n_6	l_1	l_2	l_3
10		6		1.0000000000	0.0000000000	0.0000000000
				0.5000000000	0.5000000000	0.0000000000
				0.4232062312	0.4232062312	0.1535875376
				0.2833924371	0.2833924371	0.4332151258
				0.1330857076	0.1330857076	0.7338285848
				0.0265250690	0.0265250690	0.9469498620
			8	0.9670007152	0.0329992848	0.0000000000
				0.8922417368	0.1077582632	0.0000000000
				0.7826176635	0.2173823365	0.0000000000
				0.6478790678	0.3521209322	0.0000000000
				0.3934913008	0.5472380443	0.0592706549
				0.2707097521	0.5811217960	0.1481684519
				0.2462883939	0.6991456238	0.0545659823
				0.1163195333	0.8427538829	0.0409265838
11		6		1.0000000000	0.0000000000	0.0000000000
				0.4746683133	0.4746683133	0.0506633734
				0.3764778430	0.3764778430	0.2470443140
				0.2422268680	0.2422268680	0.5155462640
				0.1104863580	0.1104863580	0.7790272840
				0.0204278105	0.0204278105	0.9591443790
			10	0.9724496361	0.0275503639	0.0000000000
				0.9096396608	0.0903603392	0.0000000000
				0.8164380765	0.1835619235	0.0000000000
				0.6997654705	0.3002345295	0.0000000000
				0.5682764664	0.4317235336	0.0000000000
				0.3649298091	0.5060795425	0.1289906484
				0.3361965758	0.6146545276	0.0491488966
				0.2299163838	0.6466193522	0.1234642640
				0.2073828199	0.7483396985	0.0442774816
				0.0959885136	0.8715407373	0.0324707491
12	1			0.3333333333	0.3333333333	0.3333333333
		6		1.0000000000	0.0000000000	0.0000000000
				0.5000000000	0.5000000000	0.0000000000
				0.4452623516	0.4452623516	0.1094752968
				0.2091994115	0.2091994115	0.5816011770
				0.0156346170	0.0156346170	0.9687307660
				0.0926054449	0.0926054449	0.8147891102
			12	0.9766549233	0.0233450767	0.0000000000
				0.9231737823	0.0768262177	0.0000000000
				0.8430942345	0.1569057655	0.0000000000
				0.7414549105	0.2585450895	0.0000000000
				0.6246434651	0.3753565349	0.0000000000
				0.4145270405	0.5430117044	0.0424612551
				0.3292603428	0.4567841256	0.2139555316
				0.3177076680	0.5753716491	0.1069206829
				0.2890538328	0.6707932601	0.0401529071
				0.1969473706	0.6989024696	0.1041501598
				0.1765067641	0.7877603302	0.0357329057
				0.0799959921	0.8942989924	0.0257050155

P	n_1	n_3	n_6	l_1	l_2	l_3
13		7		1.0000000000	0.0000000000	0.0000000000
				0.0125158559	0.0125158559	0.9749682882
				0.4814041934	0.4814041934	0.0371916132
				0.0791069148	0.0791069148	0.8417861704
				0.1831015613	0.1831015613	0.6337968774
				0.4068245409	0.4068245409	0.1863509182
				0.2954002632	0.2954002632	0.4091994736
			14	0.9799675226	0.0200324774	0.0000000000
				0.9339005269	0.0660994731	0.0000000000
				0.8644342995	0.1355657005	0.0000000000
				0.7753197015	0.2246802985	0.0000000000
				0.6713620067	0.3286379933	0.0000000000
				0.5581659344	0.4418340656	0.0000000000
				0.3944786597	0.5151527485	0.0903685918
				0.3615540438	0.6022119116	0.0362340446
				0.2767319260	0.6304802457	0.0927878283
				0.2513189999	0.7149811088	0.0336998913
				0.1706952037	0.7399484931	0.0893563032
				0.1523482839	0.8179762698	0.0296754463
				0.2915934740	0.5215057084	0.1869008176
				0.0680753374	0.9108476566	0.0210770060
14		8		1.0000000000	0.0000000000	0.0000000000
				0.5000000000	0.5000000000	0.0000000000
				0.4587977340	0.4587977340	0.0824045320
				0.3671908093	0.3671908093	0.2656183814
				0.2595748224	0.2595748224	0.4808503552
				0.1603773315	0.1603773315	0.6792453370
				0.0685768901	0.0685768901	0.8628462198
				0.0103478215	0.0103478215	0.9793043570
			16	0.9826229633	0.0173770367	0.0000000000
				0.9425410221	0.0574589779	0.0000000000
				0.8817598450	0.1182401550	0.0000000000
				0.8031266027	0.1968733973	0.0000000000
				0.7103190274	0.2896809726	0.0000000000
				0.6076769777	0.3923230223	0.0000000000
				0.9234057833	0.0588664441	0.0177277726
				0.8416660290	0.1330619104	0.0252720607
				0.7721478696	0.1494014554	0.0784506750
				0.7521462903	0.2191316332	0.0287220764
				0.6781794781	0.2494595411	0.0723609808
				0.6425344401	0.3256595455	0.0318060144
				0.5834887797	0.2550816171	0.1614296032
				0.5652822600	0.3505856755	0.0841320645
				0.5379997781	0.4292596705	0.0327405513
				0.4692799339	0.3610693351	0.1696507310

P	n_1	n_3	n_6	l_1	l_2	l_3
15	1			0.3333333333	0.3333333333	0.3333333333
		7		1.0000000000	0.0000000000	0.0000000000
				0.4852528832	0.4852528832	0.0294942336
				0.4255990820	0.4255990820	0.1488018360
				0.2320836842	0.2320836842	0.5358326316
				0.1405179457	0.1405179457	0.7189641086
				0.0599963506	0.0599963506	0.8800072988
				0.0085857218	0.0085857218	0.9828285564
			19	0.9847840231	0.0152159769	0.0000000000
				0.9496002665	0.0503997335	0.0000000000
				0.8960041459	0.1039958541	0.0000000000
				0.8261943514	0.1738056486	0.0000000000
				0.7430297109	0.2569702891	0.0000000000
				0.6499152345	0.3500847655	0.0000000000
				0.5506631368	0.4493368632	0.0000000000
				0.9337196169	0.0512926813	0.0149877018
				0.8617284617	0.1169291070	0.0213424313
				0.8006726315	0.1325472687	0.0667800998
				0.7794706504	0.1957424109	0.0247869387
				0.7134213061	0.2207559648	0.0658227291
				0.6834546988	0.2890601702	0.0274851310
				0.6286428956	0.2270147509	0.1443423535
				0.6135939037	0.3123505510	0.0740555453
				0.5886608293	0.3824865582	0.0288526124
				0.5244220875	0.3265860871	0.1489918254
				0.5171616793	0.4145995177	0.0682388030
				0.4324830342	0.3299898010	0.2375271649
16		9		1.0000000000	0.0000000000	0.0000000000
				0.5000000000	0.5000000000	0.0000000000
				0.4692881367	0.4692881367	0.0614237267
				0.3932305456	0.3932305456	0.2135389089
				0.3025577746	0.3025577746	0.3948844509
				0.2078492229	0.2078492229	0.5843015542
				0.1284932701	0.1284932701	0.7430134598
				0.0529388222	0.0529388222	0.8941223556
				0.0072437614	0.0072437614	0.9855124772
			21	0.9865660883	0.0134339117	0.0000000000
				0.9554399980	0.0445600020	0.0000000000
				0.9078481256	0.0921518744	0.0000000000
				0.8455144903	0.1544855097	0.0000000000
				0.7706926997	0.2293073003	0.0000000000
				0.6860872168	0.3139127832	0.0000000000
				0.5947559868	0.4052440132	0.0000000000
				0.9419105656	0.0452492885	0.0128401458
				0.8772044870	0.1042200005	0.0185755125
				0.8214366546	0.1173927115	0.0611706338
				0.8060459745	0.1727546471	0.0211993784
				0.7464832409	0.1959218281	0.0575949309
				0.7198500717	0.2560058606	0.0241440677
				0.6657283975	0.1263493158	0.2079222867
				0.6578452111	0.2860606787	0.0560941102
				0.6256312013	0.3487602527	0.0256085459
				0.5755509622	0.2921362629	0.1323127749
				0.5620632423	0.3743524772	0.0635842805
				0.5343544917	0.4393591499	0.0262863584
				0.4883520007	0.2978352313	0.2138127680
				0.4806452274	0.3853613574	0.1339934152

D.2.2.2 *Fekete points*

The following points were provided by courtesy of Taylor *et al.* [451]. For $P = 12$, only the points with positive quadrature grids are provided.

P	n_1	n_3	n_6	l_1	l_2	l_3
1		1		1.0000000000	0.0000000000	0.0000000000
2		2		1.0000000000	0.0000000000	0.0000000000
				0.5000000000	0.5000000000	0.0000000000
3	1			0.3333333333	0.3333333333	0.3333333333
		1		1.0000000000	0.0000000000	0.0000000000
			1	0.7236067977	0.2763932023	0.0000000000
4		3		0.5669152707	0.2165423647	0.2165423647
				0.5000000000	0.5000000000	0.0000000000
				1.0000000000	0.0000000000	0.0000000000
			3	0.8273268354	0.1726731646	0.0000000000
5		3		1.0000000000	0.0000000000	0.0000000000
				0.7039610574	0.1480194713	0.1480194713
				0.4208255393	0.4208255393	0.1583489214
			2	0.8825276620	0.1174723380	0.0000000000
				0.6426157582	0.3573842418	0.0000000000
6	1			0.3333333333	0.3333333333	0.3333333333
		3		1.0000000000	0.0000000000	0.0000000000
				0.5000000000	0.5000000000	0.0000000000
				0.7873290632	0.1063354684	0.1063354684
			3	0.9151145777	0.0848854223	0.0000000000
				0.7344348598	0.2655651402	0.0000000000
				0.5665492870	0.3162697959	0.1171809171
7		4		1.0000000000	0.0000000000	0.0000000000
				0.8408229862	0.0795885069	0.0795885069
				0.4735817901	0.2632091050	0.2632091050
				0.4539853441	0.4539853441	0.0920293119
			4	0.9358700743	0.0641299257	0.0000000000
				0.7958500908	0.2041499092	0.0000000000
				0.6676923094	0.2433646431	0.0889430475
				0.6046496090	0.3953503910	0.0000000000
8		5		1.0000000000	0.0000000000	0.0000000000
				0.8444999609	0.0777500195	0.0777500195
				0.5000000000	0.5000000000	0.0000000000
				0.4683305115	0.4683305115	0.0633389771
				0.3853668203	0.3853668204	0.2292663593
			5	0.9498789977	0.0501210023	0.0000000000
				0.8385931398	0.1614068602	0.0000000000
				0.7172965409	0.2335581033	0.0491453558
				0.6815587319	0.3184412681	0.0000000000
				0.5853134902	0.2667701010	0.1479164088

P	n_1	n_3	n_6	l_1	l_2	l_3
9	1			0.3333333333	0.3333333333	0.3333333333
		4		1.0000000000	0.0000000000	0.0000000000
				0.9021308608	0.0489345696	0.0489345696
				0.6591363598	0.1704318201	0.1704318201
				0.4699587644	0.4699587644	0.0600824712
			7	0.9597669930	0.0402330070	0.0000000000
				0.8693870908	0.1306129092	0.0000000000
				0.7389628040	0.2610371960	0.0000000000
				0.7904339977	0.1543901944	0.0551758079
				0.6401193011	0.3010242110	0.0588564879
				0.5826397065	0.4173602935	0.0000000000
				0.4963227512	0.3252434900	0.1784337588
10		6		1.0000000000	0.0000000000	0.0000000000
				0.9145236987	0.0427381507	0.0427381507
				0.5331019411	0.2334490294	0.2334490294
				0.5000000000	0.5000000000	0.0000000000
				0.4814795342	0.4814795342	0.0370409316
				0.3800851251	0.3800851251	0.2398297498
			8	0.9670007152	0.0329992848	0.0000000000
				0.8922417368	0.1077582632	0.0000000000
				0.8150971991	0.1351329831	0.0497698178
				0.7826176635	0.2173823365	0.0000000000
				0.6778669104	0.2844305545	0.0377025351
				0.6759450113	0.2079572403	0.1160977485
				0.6478790678	0.3521209322	0.0000000000
				0.5222323306	0.3633472465	0.1144204229
11		6		1.0000000000	0.0000000000	0.0000000000
				0.9201760661	0.0399119670	0.0399119670
				0.8097416696	0.0951291652	0.0951291652
				0.4216558161	0.2891720920	0.2891720920
				0.4200100315	0.4200100315	0.1599799371
				0.4832770031	0.4832770031	0.0334459938
			10	0.9724496361	0.0275503639	0.0000000000
				0.9096396608	0.0903603392	0.0000000000
				0.8236881237	0.1452587341	0.0310531421
				0.8164380765	0.1835619235	0.0000000000
				0.7030268141	0.2021386640	0.0948345219
				0.6997654705	0.3002345295	0.0000000000
				0.6642752329	0.3066778199	0.0290469472
				0.5682764664	0.4317235336	0.0000000000
				0.5605605456	0.3510551601	0.0883842943
				0.5584153138	0.2661283688	0.1754563174

P	n_1	n_3	n_6	l_1	l_2	l_3
12	1			0.3333333333	0.3333333333	0.3333333333
		6		1.0000000000	0.0000000000	0.0000000000
				0.5000000000	0.5000000000	0.0000000000
				0.4005558261	0.4005558262	0.1988883477
				0.4763189598	0.2618405201	0.2618405201
				0.8385226450	0.0807386775	0.0807386775
				0.9326048528	0.0336975736	0.0336975736
			12	0.9766549233	0.0233450767	0.0000000000
				0.9231737823	0.0768262177	0.0000000000
				0.8430942345	0.1569057655	0.0000000000
				0.7414549105	0.2585450895	0.0000000000
				0.6246434651	0.3753565349	0.0000000000
				0.8530528428	0.1206826354	0.0262645218
				0.7260597695	0.2489279690	0.0250122615
				0.6376427136	0.2874821712	0.0748751152
				0.5625832895	0.4071849276	0.0302317829
				0.7415828366	0.1697134458	0.0887037176
				0.5954847541	0.2454317980	0.1590834479
				0.5072511952	0.3837518758	0.1089969290
13		7		1.0000000000	0.0000000000	0.0000000000
				0.8520024555	0.0739987723	0.0739987723
				0.7432264994	0.1283867503	0.1283867503
				0.5395738831	0.2302130584	0.2302130584
				0.4644144021	0.4644144021	0.0711711958
				0.4114075139	0.2942962431	0.2942962431
				0.3929739081	0.3929739081	0.2140521839
			14	0.9799675226	0.0200324774	0.0000000000
				0.9339005269	0.0660994731	0.0000000000
				0.9198937820	0.0580254800	0.0220807380
				0.8644342995	0.1355657005	0.0000000000
				0.8269495692	0.1490288338	0.0240215969
				0.7753197015	0.2246802985	0.0000000000
				0.7425432195	0.1884717952	0.0689849852
				0.7075668818	0.2676625888	0.0247705295
				0.6713620067	0.3286379933	0.0000000000
				0.6366434033	0.2298513909	0.1335052058
				0.5976134593	0.3320684992	0.0703180415
				0.5665760284	0.4108329044	0.0225910673
				0.5581659344	0.4418340656	0.0000000000
				0.5028853010	0.3560657231	0.1410489759

P	n_1	n_3	n_6	l_1	l_2	l_3
14		8		1.0000000000	0.0000000000	0.0000000000
				0.9545282960	0.0227358520	0.0227358520
				0.7279988378	0.1360005811	0.1360005811
				0.6146143335	0.1926928332	0.1926928332
				0.5000000000	0.5000000000	0.0000000000
				0.4704320969	0.4704320969	0.0591358062
				0.4062455674	0.2968772163	0.2968772163
				0.3997167456	0.3997167456	0.2005665089
			16	0.9826229633	0.0173770367	0.0000000000
				0.9425410221	0.0574589779	0.0000000000
				0.8994312852	0.0727092891	0.0278594257
				0.8817598450	0.1182401550	0.0000000000
				0.8235331194	0.1556596225	0.0208072581
				0.8159593721	0.1175577052	0.0664829227
				0.8031266027	0.1968733973	0.0000000000
				0.7277004755	0.2121104191	0.0601891053
				0.7103190274	0.2896809726	0.0000000000
				0.7017966273	0.2792485214	0.0189548513
				0.6286342673	0.2519081185	0.1194576142
				0.6152236423	0.3254238863	0.0593524714
				0.6076769777	0.3923230223	0.0000000000
				0.5599300851	0.4197058520	0.0203640629
				0.5124435882	0.2931807429	0.1943756690
				0.5038948451	0.3812965190	0.1148086359
15	1			0.3333333333	0.3333333333	0.3333333333
		7		1.0000000000	0.0000000000	0.0000000000
				0.9561008318	0.0219495841	0.0219495841
				0.8937652924	0.0531173538	0.0531173538
				0.7626684778	0.1186657611	0.1186657611
				0.5478917292	0.2260541354	0.2260541354
				0.4761452138	0.4761452137	0.0477095725
				0.4206960976	0.4206960976	0.1586078048
			19	0.9847840231	0.0152159769	0.0000000000
				0.9496002665	0.0503997335	0.0000000000
				0.8960041459	0.1039958541	0.0000000000
				0.8261943514	0.1738056486	0.0000000000
				0.7430297109	0.2569702891	0.0000000000
				0.6499152345	0.3500847655	0.0000000000
				0.5506631368	0.4493368632	0.0000000000
				0.9011419064	0.0816639421	0.0171941515
				0.8302595715	0.1162464503	0.0534939782
				0.8059790318	0.1759511193	0.0180698489
				0.7424917767	0.1978887556	0.0596194677
				0.6958779924	0.2844332752	0.0196887324
				0.6684861226	0.1975591066	0.1339547708
				0.6374939126	0.2749910734	0.0875150140
				0.6000365168	0.3524012205	0.0475622627
				0.5666808380	0.4176001732	0.0157189888
				0.5400834895	0.3013819154	0.1585345951
				0.5173966708	0.3853507643	0.0972525649
				0.4350225702	0.3270403780	0.2379370518

P	n_1	n_3	n_6	l_1	l_2	l_3
16		9		1.0000000000	0.0000000000	0.0000000000
				0.9082733123	0.0458633439	0.0458633439
				0.7814558654	0.1092720673	0.1092720673
				0.5061583246	0.2469208377	0.2469208377
				0.5000000000	0.5000000000	0.0000000000
				0.4752124798	0.4752124798	0.0495750405
				0.4264731058	0.4264731058	0.1470537884
				0.3970899287	0.3014550356	0.3014550356
				0.3866478326	0.3866478326	0.2267043349
			21	0.9865660883	0.0134339117	0.0000000000
				0.9554399980	0.0445600020	0.0000000000
				0.9485194031	0.0373109962	0.0141696007
				0.9078481256	0.0921518744	0.0000000000
				0.8743842850	0.1097814309	0.0158342841
				0.8512484463	0.0959614844	0.0527900693
				0.8455144903	0.1544855097	0.0000000000
				0.7826094027	0.1639276476	0.0534629497
				0.7790251588	0.2050142286	0.0159606126
				0.7706926997	0.2293073003	0.0000000000
				0.7045940167	0.2476638915	0.0477420917
				0.7008539071	0.1815134068	0.1176326861
				0.6860872168	0.3139127832	0.0000000000
				0.6644617368	0.3197919643	0.0157462988
				0.6229177644	0.2778907781	0.0991914575
				0.6010014909	0.3493750783	0.0496234308
				0.6009076040	0.2332142592	0.1658781368
				0.5947559868	0.4052440132	0.0000000000
				0.5481915845	0.4347608890	0.0170475265
				0.5160516828	0.3894384191	0.0945098981
				0.5043443688	0.3260983223	0.1695573088

D.2.2.3 *Tetrahedral domain*

As discussed in Section 3.2.1.3, the volume coordinates (l_1, l_2, l_3, l_4) have the property that

$$l_1 + l_2 + l_3 + l_4 = 1\,.$$

In the standard right-handed tetrahedral region, where the l_i are defined as having a unit value at the vertex i in Fig. 3.10, we can express them in terms of the Cartesian coordinates (ξ_1, ξ_2, ξ_3) as follows:

$$l_1 = \frac{-(1+\xi_1+\xi_2+\xi_3)}{2}, \quad l_2 = \frac{1+\xi_1}{2},$$
$$l_3 = \frac{1+\xi_2}{2}, \quad l_4 = \frac{1+\xi_3}{2},$$

or, equivalently,

$$\xi_1 = 2l_2 - 1\,, \quad \xi_2 = 2l_3 - 1\,, \quad \xi_3 = 2l_4 - 1\,.$$

The three-dimensional collapsed coordinate system for the tetrahedron can be defined in terms of the volume coordinates as

$$\eta_1 = \frac{2l_2}{1-l_3-l_4} = \frac{l_2 - l_1}{1-l_3-l_4}, \quad \eta_2 = \frac{2l_3}{1-l_4} - 1\,, \quad \eta_3 = 2l_4 - 1\,.$$

For the tetrahedral points the symmetry is such that a point may have a single, four-fold, six-fold, twelve-fold, or twenty-four-fold symmetry, and points are listed according to this symmetry. We also note that

$$N_m = n_1 + 4n_4 + 6n_6 + 12n_{12} + 24n_{24} = \frac{(P+1)(P+2)(P+3)}{6}.$$

Electrostatic points

The following points were provided by courtesy of Hesthaven and Teng [238].

P	n_1	n_4	n_6	n_{12}	n_{24}	l_1	l_2	l_3	l_4
1		1				1.0000000000	0.0000000000	0.0000000000	0.0000000000
2		1				1.0000000000	0.0000000000	0.0000000000	0.0000000000
			1			0.5000000000	0.5000000000	0.0000000000	0.0000000000
3		2				1.0000000000	0.0000000000	0.0000000000	0.0000000000
						0.3333333333	0.3333333333	0.3333333333	0.0000000000
				1		0.7236067977	0.2763932023	0.0000000000	0.0000000000
4	1					0.2500000000	0.2500000000	0.2500000000	0.2500000000
		1				1.0000000000	0.0000000000	0.0000000000	0.0000000000
			1			0.5000000000	0.5000000000	0.0000000000	0.0000000000
				2		0.8273268354	0.1726731646	0.0000000000	0.0000000000
						0.2371200168	0.2371200168	0.5257599664	0.0000000000
5		2				1.0000000000	0.0000000000	0.0000000000	0.0000000000
						0.1834903473	0.1834903473	0.1834903473	0.4495289581
				4		0.8825276620	0.1174723380	0.0000000000	0.0000000000
						0.6426157582	0.3573842418	0.0000000000	0.0000000000
						0.1575181512	0.1575181512	0.6849636976	0.0000000000
						0.4105151510	0.4105151510	0.1789696980	0.0000000000
6		3				1.0000000000	0.0000000000	0.0000000000	0.0000000000
						0.3333333333	0.3333333333	0.3333333333	0.0000000000
						0.1402705801	0.1402705801	0.1402705801	0.5791882597
			2			0.5000000000	0.5000000000	0.0000000000	0.0000000000
						0.3542052583	0.3542052583	0.1457947417	0.1457947417
				3		0.9151119481	0.0848880519	0.0000000000	0.0000000000
						0.7344243967	0.2655756033	0.0000000000	0.0000000000
						0.1061169285	0.1061169285	0.7877661430	0.0000000000
					1	0.3097982151	0.5569099204	0.1332918645	0.0000000000
7		3				1.0000000000	0.0000000000	0.0000000000	0.0000000000
						0.1144606542	0.1144606542	0.1144606542	0.6566180374
						0.2917002822	0.2917002822	0.2917002822	0.1248991534
				7		0.9358700743	0.0641299257	0.0000000000	0.0000000000
						0.7958500907	0.2041499093	0.0000000000	0.0000000000
						0.6046496090	0.3953503910	0.0000000000	0.0000000000
						0.0660520784	0.0660520784	0.8678958432	0.0000000000
						0.4477725053	0.4477725053	0.1044549894	0.0000000000
						0.2604038024	0.2604038024	0.4791923952	0.0000000000
						0.1208429970	0.1208429970	0.4770203357	0.2812936703
					1	0.2325524777	0.6759625951	0.0914849272	0.0000000000

P	n_1	n_4	n_6	n_{12}	n_{24}	l_1	l_2	l_3	b_4
8	1					0.2500000000	0.2500000000	0.2500000000	0.2500000000
		2				1.0000000000	0.0000000000	0.0000000000	0.0000000000
						0.0991203900	0.0991203900	0.0991203900	0.7026388300
			2			0.5000000000	0.5000000000	0.0000000000	0.0000000000
						0.3920531037	0.3920531037	0.1079468963	0.1079468963
				8		0.9498789977	0.0501210023	0.0000000000	0.0000000000
						0.8385931398	0.1614068602	0.0000000000	0.0000000000
						0.6815587319	0.3184412681	0.0000000000	0.0000000000
						0.0660520784	0.0660520784	0.8678958432	0.0000000000
						0.2033467796	0.2033467796	0.5933064408	0.0000000000
						0.3905496216	0.3905496216	0.2189007568	0.0000000000
						0.1047451941	0.1047451941	0.5581946462	0.2323149656
						0.2419418605	0.2419418605	0.4062097450	0.1099065340
					2	0.3617970895	0.5541643672	0.0840385433	0.0000000000
						0.1801396087	0.7519065566	0.0679538347	0.0000000000
9		4				1.0000000000	0.0000000000	0.0000000000	0.0000000000
						0.3333333333	0.3333333333	0.3333333333	0.0000000000
						0.0823287303	0.0823287303	0.0823287303	0.7530138091
						0.2123055477	0.2123055477	0.2123055477	0.3630833569
				11		0.9597669541	0.0402330459	0.0000000000	0.0000000000
						0.8693869326	0.1306130674	0.0000000000	0.0000000000
						0.7389624749	0.2610375251	0.0000000000	0.0000000000
						0.5826394788	0.4173605212	0.0000000000	0.0000000000
						0.0355775717	0.0355775717	0.9288448566	0.0000000000
						0.4640303025	0.4640303025	0.0719393950	0.0000000000
						0.1633923069	0.1633923069	0.6732153862	0.0000000000
						0.0873980781	0.0873980781	0.6297057875	0.1954980564
						0.0916714679	0.0916714679	0.4819523024	0.3347047619
						0.2040338880	0.2040338880	0.4996292993	0.0923029247
						0.3483881173	0.3483881173	0.2075502723	0.0956734931
					3	0.2966333890	0.6349633653	0.0684032457	0.0000000000
						0.1439089974	0.8031490682	0.0529419344	0.0000000000
						0.3225890045	0.4968009397	0.1806100558	0.0000000000
10		4				1.0000000000	0.0000000000	0.0000000000	0.0000000000
						0.0678316144	0.0678316144	0.0678316144	0.7965051568
						0.1805746957	0.1805746957	0.1805746957	0.4582759129
						0.3051527124	0.3051527124	0.3051527124	0.0845418628
			3			0.5000000000	0.5000000000	0.0000000000	0.0000000000
						0.3164336236	0.3164336236	0.1835663764	0.1835663764
						0.4219543801	0.4219543801	0.0780456199	0.0780456199
				11		0.9670007152	0.0329992848	0.0000000000	0.0000000000
						0.8922417368	0.1077582632	0.0000000000	0.0000000000
						0.7826176635	0.2173823365	0.0000000000	0.0000000000
						0.6478790678	0.3521209322	0.0000000000	0.0000000000
						0.0265250690	0.0265250690	0.9469498620	0.0000000000
						0.1330857076	0.1330857076	0.7338285848	0.0000000000
						0.4232062312	0.4232062312	0.1535875376	0.0000000000
						0.2833924371	0.2833924371	0.4332151258	0.0000000000
						0.1734555313	0.1734555313	0.5762731177	0.0768158196
						0.0724033935	0.0724033935	0.6893564961	0.1658367169
						0.0768451848	0.0768451848	0.5573732958	0.2889363346
					5	0.3934913008	0.5472380443	0.0592706549	0.0000000000
						0.2462883939	0.6991456238	0.0545659823	0.0000000000
						0.1163195334	0.8427538829	0.0409265838	0.0000000000
						0.2707097521	0.5811217960	0.1481684519	0.0000000000
						0.3019928872	0.4393774966	0.1776946096	0.0809350066

APPENDIX E

CHARACTERISTIC FLUX DECOMPOSITION

A hyperbolic conservation law, such as the Euler equations

$$\mathbf{u}_t + \mathbf{f}(\mathbf{u})_x = 0\,, \tag{E.0.1}$$

can be written in non-conservative form:

$$\mathbf{u}_t + \mathbf{A}\mathbf{u}_x = 0\,, \tag{E.0.2}$$

where $\mathbf{A}(u) = \partial\mathbf{f}/\partial\mathbf{u}$. This form is useful since we can decompose the system with Jacobian matrix $\mathbf{A}$ into its characteristic form to obtain a diagonal matrix of eigenvalues $\mathbf{D}$, that is,

$$\mathbf{L}\cdot\mathbf{A}\cdot\mathbf{R} = \mathbf{D} \quad \Rightarrow \quad \mathbf{A} = \mathbf{R}\cdot\mathbf{D}\cdot\mathbf{L}\,,$$

where $\mathbf{L}$ and $\mathbf{R}$ are the left and right eigenvectors of $\mathbf{A}$ and $\mathbf{RL} = \mathbf{I}$. The non-conservative equation (E.0.2) can therefore be written as

$$\mathbf{u}_t + \mathbf{RDL}\mathbf{u}_x = 0\,.$$

Finally, if we linearise this equation, typically about the Roe average, then we can treat the eigenvector matrices as constant and therefore obtain

$$\mathbf{R}^{-1}\mathbf{u}_t + \mathbf{DL}\mathbf{u}_x = \mathbf{L}\mathbf{u}_t + \mathbf{DL}\mathbf{u}_x = 0\,.$$

This is now a decoupled system in terms of the characteristic variables $\mathbf{v} = \mathbf{L}\mathbf{u}$ to which we are able to apply upwinding techniques or Riemann solvers.

E.1 One dimension

For the one-dimensional Euler system

$$\mathbf{u} = \begin{bmatrix} \rho \\ m \\ E \end{bmatrix}, \quad \mathbf{f} = \begin{bmatrix} \rho u \\ um + p \\ u(p+E) \end{bmatrix},$$

$$p = (\gamma - 1)\left(E - \frac{1}{2}\rho u^2\right),$$

where ρ denotes the density, u is the x component of the velocity, p is the pressure, E is the total energy, $m = \rho u$ is the x component of the momentum, and γ is the ratio of the specific heats of a polytropic gas. The right eigenvectors of $\mathbf{A}$ can be written as

$$\mathbf{R} = \begin{bmatrix} 1 & 1 & 1 \\ u-c & u & u+c \\ H-uc & \frac{1}{2}u^2 & H+uc \end{bmatrix},$$

where $c = \sqrt{\gamma p/\rho}$ is the speed of sound, and H represents enthalpy and is defined by

$$H = \frac{E+p}{\rho} = \frac{c^2}{\gamma-1} + \frac{1}{2}u^2 .$$

The left eigenvectors of $\mathbf{A}$ can be written as

$$\mathbf{L} = \begin{bmatrix} \dfrac{\frac{1}{2}u^2\beta c + u}{2c} & -\dfrac{1+u\beta c}{2c} & \dfrac{\beta}{2} \\ 1-\frac{1}{2}\beta u^2 & \beta u & -\beta \\ \dfrac{\frac{1}{2}u^2\beta c - u}{2c} & \dfrac{1-u\beta c}{2c} & \dfrac{\beta}{2} \end{bmatrix},$$

where $\beta = (\gamma-1)/c^2$. The diagonal matrix of eigenvalues is therefore

$$\mathbf{D} \equiv \mathbf{L}\cdot\mathbf{A}\cdot\mathbf{R} = \begin{bmatrix} u-c & & \\ & u & \\ & & u+c \end{bmatrix}.$$

E.2 Two dimensions

For the two-dimensional Euler equations the same diagonalisation may be applied to the Jacobian matrix $\mathbf{A}(\mathbf{u})$. To obtain a comparable system to eqn (E.0.1) we shall consider the vector $\mathbf{u}$ as being constant in one direction, and so the full Euler system only has x-component derivatives. Therefore we take

$$\mathbf{u} = \begin{bmatrix} \rho \\ m \\ n \\ E \end{bmatrix}, \quad \mathbf{f} = \begin{bmatrix} \rho u \\ um+p \\ un \\ u(p+E) \end{bmatrix},$$

$$p = (\gamma-1)\left(E - \frac{1}{2}\rho(\mathbf{v}\cdot\mathbf{v})\right),$$

where ρ, u, E , m, and γ are defined as before, $\mathbf{v} = [u,v]^\top$, and v and $n = \rho v$ are the y components of the velocity and momentum, respectively. The right eigenvectors of $\mathbf{A}$ can be written as

$$\mathbf{R} = \begin{bmatrix} 1 & 1 & 0 & 1 \\ u-c & u & 0 & u+c \\ v & v & 1 & v \\ H-uc & \frac{1}{2}(\mathbf{v}\cdot\mathbf{v}) & v & H+uc \end{bmatrix},$$

where $c = \sqrt{\gamma P/\rho}$ is the speed of sound, and H represents enthalpy and is defined by

$$H = \frac{E+p}{\rho} = \frac{c^2}{\gamma - 1} + \frac{1}{2}(\mathbf{v}\cdot\mathbf{v})\,.$$

The left eigenvectors of $\mathbf{A}$ can be written as

$$\mathbf{L} = \begin{bmatrix} \dfrac{\frac{1}{2}(\mathbf{v}\cdot\mathbf{v})\beta c + u}{2c} & -\dfrac{1+\beta uc}{2c} & -\dfrac{\beta v}{2} & \dfrac{\beta}{2} \\ 1 - \frac{1}{2}(\mathbf{v}\cdot\mathbf{v})\beta & \beta u & \beta v & -\beta \\ -v & 0 & 1 & 0 \\ \dfrac{\frac{1}{2}(\mathbf{v}\cdot\mathbf{v})\beta c - u}{2c} & \dfrac{1-\beta uc}{2c} & -\dfrac{\beta v}{2} & \dfrac{\beta}{2} \end{bmatrix},$$

where $\beta = (\gamma - 1)/c^2$. The diagonal matrix of eigenvalues is therefore

$$\mathbf{D} \equiv \mathbf{L}\cdot\mathbf{A}\cdot\mathbf{R} = \begin{bmatrix} u-c & & & \\ & u & & \\ & & u & \\ & & & u+c \end{bmatrix}.$$

E.3 Three dimensions

For the three-dimensional Euler equations the same diagonalisation may be applied to the Jacobian matrix $\mathbf{A}(\mathbf{u})$. To obtain a comparable system to eqn (E.0.1) we shall again consider the vector $\mathbf{u}$ as being constant in two directions, and so the full Euler system only has variation in the x direction. Therefore, we consider

$$\mathbf{u} = \begin{bmatrix} \rho \\ m \\ n \\ l \\ E \end{bmatrix}, \quad \mathbf{f} = \begin{bmatrix} \rho u \\ um + p \\ un \\ ul \\ u(p+E) \end{bmatrix},$$

$$p = (\gamma - 1)\left(E - \frac{1}{2}\rho(\mathbf{v}\cdot\mathbf{v})\right),$$

where ρ, u, v, E, m, n, and γ are defined as before, $\mathbf{v} = [u, v, w]^\top$, and w and $l = \rho w$ are the z components of the velocity and momentum, respectively. The right eigenvectors of $\mathbf{A}$ can be written as

$$\mathbf{R} = \begin{bmatrix} 1 & 1 & 0 & 0 & 1 \\ u-c & u & 0 & 0 & u+c \\ v & v & 1 & 0 & v \\ w & w & 0 & 1 & w \\ H - uc & \frac{1}{2}(\mathbf{v}\cdot\mathbf{v}) & v & w & H+uc \end{bmatrix},$$

where $c = \sqrt{\gamma p/\rho}$ is the speed of sound, and H represents enthalpy and is defined by

$$H = \frac{E+p}{\rho} = \frac{c^2}{\gamma - 1} + \frac{1}{2}(\mathbf{v}\cdot\mathbf{v})\,.$$

The left eigenvectors of $\mathbf{A}$ can be written as

$$\mathbf{L} = \begin{bmatrix} \dfrac{\frac{1}{2}(\mathbf{v}\cdot\mathbf{v})\beta c + u}{2c} & -\dfrac{1+\beta uc}{2c} & -\dfrac{\beta v}{2} & -\dfrac{\beta w}{2} & \dfrac{\beta}{2} \\ 1 - \frac{1}{2}(\mathbf{v}\cdot\mathbf{v})\beta & \beta u & \beta v & \beta w & -\beta \\ -v & 0 & 1 & 0 & 0 \\ -w & 0 & 0 & 1 & 0 \\ \dfrac{\frac{1}{2}(\mathbf{v}\cdot\mathbf{v})\beta c - u}{2c} & \dfrac{1-\beta uc}{2c} & -\dfrac{\beta v}{2} & -\dfrac{\beta w}{2} & \dfrac{\beta}{2} \end{bmatrix},$$

where $\beta = (\gamma - 1)/c^2$. The diagonal matrix of eigenvalues is therefore

$$\mathbf{D} \equiv \mathbf{L}\cdot\mathbf{A}\cdot\mathbf{R} = \begin{bmatrix} u-c & & & & \\ & u & & & \\ & & u & & \\ & & & u & \\ & & & & u+c \end{bmatrix}.$$

REFERENCES

[1] Abramowitz, M. and Stegun, I. A. (1972). *Handbook of mathematical functions.* Dover Publications, New York.

[2] Agmon, S., Douglis, A., and Nirenberg, L. (1964). Estimates near the boundary for solutions of elliptic partial differential equations satisfying general boundary conditions II. *Comm. Pure Appl. Math.*, **17**, 35.

[3] Ainsworth, M. and Oden, J. T. (1992). A procedure for a posteriori error estimation for *h-p* finite element methods. *Comp. Meth. Appl. Mech. Eng.*, **101**, 73.

[4] Ainsworth, M. and Oden, J. T. (2000). *A posteriori error estimation in finite element analysis.* Wiley, New York.

[5] Ainsworth, M. and Sherwin, S. J. (1999). Domain decomposition preconditioners for *p* and *hp* finite element approximation of Stokes equations. *Comp. Meth. Appl. Mech. Eng.*, **175**, 243.

[6] Akin, J. E. (1994). *Finite elements for analysis and design.* Academic Press, London.

[7] Aldama, A. A. (1990). *Filtering techniques for turbulent flow simulations.* Lecture Notes in Engineering 56. Springer-Verlag, Berlin.

[8] Alevsky, A. V. (1996). Spline-characteristic method for simulation of convective turbulence. *J. Comp. Phys.*, **123**, 466.

[9] AMTEC Engineering Inc. (1999). *Tecplot version 8.0 user's manual.* AMTEC Engineering Inc.

[10] Anagnostou, G. (1991). *Nonconforming sliding spectral element method for the unsteady incompressible Navier–Stokes equations.* Ph. D. thesis, Massachusetts Institute of Technology.

[11] Anagnostou, G., Maday, Y., and Patera, A. T. (1990). A sliding mesh method for partial differential equations in nonstationary geometries: application to the incompressible Navier–Stokes equations. Massachusetts Institute of Technology, Fluid mechanics laboratory report.

[12] Anderson, E., Bai, Z., Bischof, C., Demmel, J., Dongarra, J., Du Croz, J., Greenbaum, A., Hammarling, S., McKenney, A., Ostrouchov, S., and Sorensen, D. (1992). *LAPACK users' guide.* SIAM, Philadelphia, PA.

[13] Andersson, B., Falk, U., Babuška, I., and Von-Petersdorff, T. (1995). Reliable stress and fracture mechanics analysis of complex components using a *h-p* version of FEM. *Int. J. Numer. Meth. Eng.*, **38**, 2135.

[14] Argyris, J. H. (1960). *Energy theorems and structural analysis.* Butterworths, London. Reprinted from *Aircraft Eng.* **26–7**, 1954–1955.

[15] Arnold, D. N. (1982). An interior penalty finite element method with discontinuous elements. *SIAM J. Numer. Anal.*, **19**, 742.

[16] Arnold, D. N., Brezzi, F., Cockburn, B., and Marini, D. (2000). Discontinuous Galerkin methods for elliptic problems. In *Discontinuous Galerkin methods: theory, computation and applications* (ed. B. Cockburn, G. E. Karniadakis, and C.-W. Shu), p. 89. Springer-Verlag, Germany.

[17] Arnold, D. N., Brezzi, F., Cockburn, B., and Marini, L. D. (2002). Unified analysis of discontinuous Galerkin methods for elliptic problems. *SIAM J. Numer. Anal.*, **39**, 1749.

[18] Arnoldi, W. E. (1951). The principle of minimised iterations in the solution of the matrix eigenvalue problem. *Q. Appl. Math.*, **9**, 17.

[19] Askey, R. and Wilson, J. (1985). Some basic hypergeometric polynomials that generalize Jacobi polynomials. *Memoirs Amer. Math. Soc.*, **319**, 55. AMS, Providence, RI.

[20] Auteri, F., Guermond, J. L., and Parolini, N. (2001). Roles of the LBB condition in weak spectral projection methods. *J. Comp. Phys.*, **174**, 405.

[21] Axelsson, O. (1996). *Iterative solution methods.* Cambridge University Press.

[22] Babuška, I. (1973). The finite element method with Lagrangian multipliers. *Numer. Math.*, **20**, 179.

[23] Babuška, I., Baumann, C. E., and Oden, J. T. (1999). A discontinuous *hp* finite element method for diffusion problems: 1-D analysis. *Comp. Math. Appl.*, **37**, 103.

[24] Babuška, I., Craig, A. W., Mandel, J., and Pitkaranta, J. (1991). Efficient preconditioning for the *p*-version finite element method in two dimensions. *SIAM J. Numer. Anal.*, **28**, 624.

[25] Babuška, I., Guo, B., and Suri, M. (1989). Implementation of nonhomogeneous Dirichlet boundary conditions in the *p*-version of the finite element method. *Impact Comp. Sci. Eng.*, **1**, 36.

[26] Babuška, I. and Guo, B. Q. (1992). The *h*, *p* and *h-p* version of the finite element method—Basic theory and application. Technical note BN-1134, Institute for Physical Sciences and Technology, University of Maryland at College Park, May 1992.

[27] Babuška, I. and Guo, B. Q. (1996). Approximation properties of the for the *h-p* version of the finite element method. *Comp. Meth. Appl. Mech. Eng.*, **133**, 319.

[28] Babuška, I. and Miller, A. (1984). The post-processing approach in the finite element method. Part 2. The calculation of stress intensity factors. *Int. J. Numer. Meth. Eng.*, **20**, 1111.

[29] Babuška, I. and Oh, H.-S. (1990). The *p*-version of finite element method for domains with corners and for infinite domains. *Numer. Meth. PDEs*, **6**, 371.

[30] Babuška, I. and Suri, M. (1987). The *h-p* version of the finite element method with quasiuniform meshes. *Math. Mod. Numer. Anal.*, **21**, 199.

[31] Babuška, I. and Suri, M. (1994). The *p* and *h-p* versions of the finite element method, basic principles and properties. *SIAM Rev.*, **36**, 578.

[32] Babuška, I., Szabó, B. A., and Katz, I. N. (1981). The *p*-version of the finite element method. *SIAM J. Numer. Anal.*, **18**, 515.

[33] Babuška, I., Von-Petersdorff, T., and Andersson, B. (1994). Numerical treatment of vertex singularities and intensity factors for mixed boundary value problems for the Laplace equation in $\mathbb{R}^3$. *SIAM J. Numer. Anal.*, **31**, 1265.

[34] Bardina, J. (1983). *Improved turbulence models based on large eddy simulation of homogeneous, incompressible turbulent flows.* Ph. D. thesis, Stanford University.

[35] Barkley, D., Gomes, M. G. M., and Henderson, R. D. (2002). Three-dimensional instability in flow over a backward facing step. *J. Fluid Mech.*, **473**, 167.

[36] Barkley, D. and Henderson, R. D. (1996). Three-dimensional Floquet stability analysis of the wake of a circular cylinder. *J. Fluid Mech.*, **322**, 215.

[37] Basdevant, C., Deville, M., Haldenwang, P., Lacroix, J. M., Quazzani, J., Peyret, R., Orlandi, P., and Patera, A. T. (1986). Spectral and finite difference solutions of the Burgers equation. *Comp. Fluids*, **14**, 23.

[38] Bassi, F. and Rebay, S. (1997). A high-order accurate discontinuous finite element method for the numerical solution of the compressible Navier–Stokes equations. *J. Comp. Phys.*, **131**, 267.

[39] Batchelor, G. K. (1967). *An introduction to fluid dynamics.* Cambridge University Press.

[40] Baumann, C. E. (1997). *An hp-adaptive discontinuous finite element method for computational fluid dynamics.* Ph. D. thesis, The University of Texas at Austin.

[41] Baumann, C. E. and Oden, J. T. (1999). A discontinuous *hp* finite element method for

convection–diffusion problems. *Comp. Meth. Appl. Mech. Eng.*, **175**, 311.

[42] Bell, J. B., Colella, P., and Glaz, H. M. (1989). A second-order projection method for the incompressible Navier–Stokes equations. *J. Comp. Phys.*, **85**, 257.

[43] Bernardi, C. (1996). Indicateurs d'erreur en *h-n* version des elements spectraux. *Modelisation Math. Analyse Numer. (M2AN)*, **30**, 1.

[44] Bernardi, C., Debit, N., and Maday, Y. (1990). Coupling finite element and spectral methods: first results. *Math. Comput.*, **54**, 21.

[45] Bernardi, C. and Maday, Y. (1992). *Approximations spectrales de problemes aux limites elliptiques.* Springer, Paris.

[46] Bernardi, C., Maday, Y., Mavriplis, C., and Patera, A. T. (1989). The mortar element method applied to spectral discretizations. In *Finite element analysis in fluids, seventh international conference on finite element methods in flow problems* (ed. T. J. Chung and G. R. Karr). UAH Press, Huntsville, TX.

[47] Bernardi, C., Maday, Y., and Patera, A. T. (1992). A new nonconforming approach to domain decomposition: the mortar element method. In *Nonlinear partial differential equations and their applications* (ed. H. Brezis and J. L. Lions). Pitman and Wiley, Boston.

[48] Beskok, A. (1996). *Simulations and models for gas flows in microgeometries.* Ph.D. thesis, Princeton University, NJ.

[49] Bica, I. (1997). *Iterative substructuring algorithms for the p-version finite element method for elliptic problems.* Ph.D. thesis, New York University.

[50] Bird, G. (1994). *Molecular gas dynamics and the direct simulation of gas flows.* Oxford University Press, New York.

[51] Biswas, R., Devine, K., and Flaherty, J. (1994). Parallel, adaptive finite element methods for conservation laws. *Appl. Numer. Math.*, **14**, 255.

[52] Bjorstad, P. and Widlund, O. (1986). Iterative methods for the solution of elliptic problems on regions partitioned into substructures. *SIAM J. Numer. Anal.*, **23**, 1097.

[53] Blackburn, H. M. (2002). Three-dimensional instability and state selection in an oscillatory axisymmetric swirling flow. *Phys. Fluids*, **14**, 3983.

[54] Blackburn, H. M. and Lopez, J. M. (2003). On three-dimensional quasi-periodic Floquet instabilities of two-dimensional bluff body wakes. *Phys. Fluids*, **15**, 57.

[55] Blackburn, H. M. and Lopez, J. M. (2003). The onset of three-dimensional standing and modulated travelling waves in a periodically driven cavity flow. *J. Fluid Mech.*, **497**, 289.

[56] Blackburn, H. M., Marques, F., and Lopez, J. M. (2004). On three-dimensional instabilities of two-dimensional flows with a Z_2 spatio-temporal symmetry. *J. Fluid Mech.* Submitted.

[57] Blackburn, H. M. and Schmidt, S. (2003). Spectral element filtering techniques for large eddy simulation with dynamic estimation. *J. Comp. Phys.*, **186**, 610.

[58] Blackburn, H. M. and Sherwin, S. J. (2004). Formulation of a Galerkin spectral element–Fourier method for three-dimensional incompressible flow in cylindrical geometries. *J. Comp. Phys.* In press.

[59] Blaisdell, G. A. (1997). Commutation of discrete filters and differential operators for large eddy simulation. In *Proceedings of the first international conference on DNS/LES*, Louisiana Tech., Ruston, LA, 4–8 August, p. 333.

[60] Blum, H. and Dobrowolski, M. (1982). On finite element methods for elliptic equations on domains with corners. *Computing*, **28**, 53.

[61] Blyth, M. G. and Pozrikidis, C. (2005). A Lobatto interpolation grid over the triangle. *IMA J. Appl. Math.* In press.

[62] Bochev, P. B. (1997). Analysis of least-squares finite element methods for the Navier–Stokes equations. *SIAM J. Numer. Anal.*, **34**, 1817.

[63] Bochev, P. B. and Gunzburger, M. D. (1998). Finite element methods of least-squares type. *SIAM Rev.*, **40**, 789.

[64] Boffi, D. and Funaro, D. (1994). An alternative approach to the analysis and the approximation of the Navier–Stokes equations. *J. Sci. Comp.*, **9**, 1.

[65] Bonet, J. and Peraire, J. (1990). An alternating digital tree (ADT) algorithm for geometric searching and intersection problems. *Int. J. Numer. Meth. Eng.*, **31**, 11.

[66] Boris, J. P. and Book, D. L. (1973). Flux-corrected transport. I. SHASTA, a fluid transport algorithm that works. *J. Comp. Phys.*, **11**, 38.

[67] Boris, J. P. and Book, D. L. (1976). Flux-corrected transport. III. Minimal error FCT algorithms. *J. Comp. Phys.*, **20**, 397.

[68] Borue, V. and Orszag, S. A. (1998). Local energy flux and subgrid-scale statistics in three-dimensional turbulence. *J. Fluid Mech.*, **366**, 1.

[69] Bos, L. (1983). Bounding the Lebesgue function for Lagrange interpolation in a simplex. *J. Approx. Theory*, **38**, 43.

[70] Bos, L. P., Taylor, M. A., and Wingate, B. A. (2001). Tensor product Gauss–Lobatto points are Fekete points for the cube. *Math. Comput.*, **70**, 1543.

[71] Boyd, J. P. (1989). *Chebyshev and Fourier spectral methods.* Springer-Verlag, New York.

[72] Boyd, J. P. (1998). Two comments on filtering (artificial viscosity) for Chebyshev and Legendre spectral and spectral element methods: preserving the boundary conditions and interpretation of the filter as diffusion. *J. Comp. Phys.*, **143**, 283.

[73] Brackbill, J. U. and Barnes, D. C. (1980). The effect of nonzero $\nabla \cdot \boldsymbol{B}$ on the numerical solution of the magnetohydrodynamic equations. *J. Comp. Phys.*, **35**, 426.

[74] Braess, D. and Schwab, C. (2000). Approximation on simplices with respect to weighted Sobolev norms. *J. Approx. Theory*, **103**, 329.

[75] Brenner, S. C. and Scott, L. R. (1994). *The mathematical theory of finite element methods.* Springer, New York.

[76] Brezzi, F. (1974). On the existence, uniqueness and approximation of saddle-point problems arising from Lagrangian multipliers. *RAIRO*, **8**, 129.

[77] Brezzi, F., Manzini, G., Marini, D., Pietra, P., and Russo, A. (1999). Discontinuous finite elements for diffusion problems. In *Atti convegno in onore di F. Brioschi*, p. 197. Istituto Lombardo, Accademia di Scienze e Lettere, Milano, 1997.

[78] Brezzi, F. and Pitkaranta, J. (1984). On the stabilization of finite element approximations of the Stokes problem. In *Efficient solutions of elliptic systems, notes on numerical fluid mechanics* (ed. W. Hackbusch), Volume 10, p. 11. Vieweg, Wiesbaden.

[79] Brown, D. L., Cortez, R., and Minion, M. L. (2001). Accurate projection for the incompressible Navier–Stokes equations. *J. Comp. Phys.*, **168**, 464.

[80] Brown, D. L. and Minion, M. L. (1995). Performance of under-resolved two-dimensional incompressible flow simulations. *J. Comp. Phys.*, **122**, 165.

[81] Cahouet, J. and Chabard, J. P. (1986). Some fast 3D finite element solvers for the generalized Stokes problem. *Int. J. Numer. Meth. Fluids*, **8**, 869.

[82] Cai, W., Gottlieb, D., and Harten, A. (1990). Cell averaging Chebyshev methods for hyperbolic problems. Report no. 90-72, ICASE.

[83] Cai, W., Gottlieb, D., and Shu, C. W. (1989). Non-oscillatory spectral Fourier methods for shock wave calculations. *Math. Comput.*, **52**, 389.

[84] Canuto, C. (1994). Stabilization of spectral methods by finite element bubble functions. *Comp. Meth. Appl. Mech. Eng.*, **116**, 39.

[85] Canuto, C. and Funaro, D. (1988). The Schwarz algorithm for spectral methods. *SIAM*

J. Numer. Anal., **25**, 24.

[86] Canuto, C., Hussaini, M. Y., Quarteroni, A., and Zang, T. (1987). *Spectral methods in fluid dynamics.* Springer, New York.

[87] Canuto, C. and Quarteroni, A. (1982). Approximation results for orthogonal polynomials in Sobolov spaces. *Math. Comput.*, **38**, 55.

[88] Carey, G. F. and Oden, J. T. (1986). *Finite elements: fluid mechanics*, Volume 6. Prentice-Hall, New Jersey.

[89] Casarin, M. A. (1996). *Schwarz preconditioners for spectral and mortar finite element methods with applications to incompressible fluids.* Ph. D. thesis, New York University.

[90] Casarin, M. A. (1997). Quadi-optimal Schwarz methods for the conforming spectral element discretization. *SIAM J. Numer. Anal.*, **34**, 2482.

[91] Castillo, P. (2002). Performance of discontinuous Galerkin methods for elliptic PDEs. *SIAM J. Sci. Comp.*, **24**, 524.

[92] Chan, T. F. and Goovaerts, D. (1990). Schur complement domain decomposition algorithms for spectral methods. *Appl. Numer. Math.*, **6**, 53.

[93] Chen, Q. and Babuška, I. (1995). Approximate optimal points for polynomial interpolation of real functions in an interval and in a triangle. *Comp. Meth. Appl. Mech. Eng.*, **128**, 405.

[94] Chen, Q. and Babuška, I. (1996). The optimal symmetrical points for polynomial interpolation of real functions in a tetrahedron. *Comp. Meth. Appl. Mech. Eng.*, **137**, 89.

[95] Choi, H., Moin, P., and Kim, J. (1993). Direct numerical simulation of turbulent flow over riblets. *J. Fluid Mech.*, **255**, 503.

[96] Chollet, J. P. (1984). Two-point closures as a subgrid scale modelling for large eddy simulations. In *Turbulent shear flows IV* (ed. F. Durst and B. Launder), Lecture Notes in Physics. Springer-Verlag, Heidelberg.

[97] Chorin, A. J. (1967). The numerical solution of the Navier–Stokes equations for an incompressible fluid. *Bull. Amer. Math. Soc.*, **73**, 928.

[98] Chu, D. C. and Karniadakis, G. E. (1993). The direct numerical simulation of laminar and turbulent flow over riblets. *J. Fluid Mech.*, **250**, 1.

[99] Clercx, H. J. H. (1997). A spectral solver for the Navier–Stokes equations in the velocity–vorticity formulation for flows with two nonperiodic directions. *J. Comp. Phys.*, **137**, 186.

[100] Cockburn, B., Hou, S., and Shu, C.-W. (1990). The Runge–Kutta local projection discontinuous Galerkin finite element method for conservation laws IV: the multi-dimensional case. *J. Comp. Phys.*, **54**, 545.

[101] Cockburn, B., Kanschat, G., Perugia, I., and Schotzau, D. (2001). Superconvergence of the local discontinuous Galerkin method for elliptic problems on Cartesian grids. *SIAM J. Numer. Anal.*, **39**, 264.

[102] Cockburn, B., Karniadakis, G. E., and Shu, C.-W. (2000). The development of discontinuous Galerkin methods. In *Discontinuous Galerkin methods: theory, computation and applications* (ed. B. Cockburn, G. E. Karniadakis, and C.-W. Shu), p. 3. Springer, Germany.

[103] Cockburn, B. and Shu, C.-W. (1989). TVB Runge–Kutta local projection discontinuous Galerkin finite element method for conservation laws II: general framework. *Math. Comput.*, **52**, 411.

[104] Cockburn, B. and Shu, C.-W. (1991). P^1–RKDG method for two-dimensional Euler equations of gas dynamics. In *Proceedings of the fourth international symposium on computational fluid dynamics*, Davis, CA, 9–12 September.

[105] Cockburn, B. and Shu, C.-W. (1998). The local discontinuous Galerkin for convection–diffusion systems. *SIAM J. Numer. Anal.*, **35**, 2440.

[106] Colella, P., Dorr, M., and Wake, D. D. (1998). A conservative finite difference method for the numerical solution of plasma fluid equations. Technical report ucrl-jc-129912.

[107] Constantin, P. and Foias, C. (1988). *Navier–Stokes equations.* University of Chicago Press.

[108] Constantin, P., Foias, C., Manley, O. P., and Temam, R. (1985). Determining modes and fractal dimension of turbulent flows. *J. Fluid Mech.*, **150**, 427.

[109] Coppola, G., Sherwin, S. J., and Peiró, J. (2001). Non-linear particle tracking for high-order elements. *J. Comp. Phys.*, **172**, 356.

[110] Costabel, M. and Dauge, M. (1993). General edge asymptotics of solution of second-order elliptic boundary value problems I and II. *Proc. R. Soc. Edin.*, **123A**, 109.

[111] Costabel, M. and Stephan, E. (1983). Curvature terms in asymptotic expansions for solutions of boundary integral equations on curved polygons. *J. Integral Eqns*, **5**, 353.

[112] Courant, R. (1943). Variational methods for the solution of problems of equilibrium and vibration. *Bull. Amer. Math. Soc.*, **49**, 1.

[113] Courant, R. and Hilbert, D. (1989). *Methods of mathematical physics*, Volume 1. Wiley, New York.

[114] Couzy, W. (1995). *Spectral element discretization of the unsteady Navier–Stokes equations and its iterative solution on parallel computers.* Ph. D. thesis, Ecole Polytechnique Federale de Lausanne.

[115] Couzy, W. and Deville, M. O. (1994). Spectral element preconditioners for the Uzawa pressure operator applied to incompressible flows. *J. Sci. Comp.*, **9**, 107.

[116] Crandall, M. G. and Lions, P. L. (1983). Viscosity solutions of Hamilton–Jacobi equations. *Trans. Amer. Math. Soc.*, **61**, 629.

[117] Crawford, C. H. (1994). The structure and statistics of turbulent flow over riblets. Master's thesis, Princeton University, NJ.

[118] Crawford, C. H. (1996). *Direct numerical simulation of near-wall turbulence: passive and active control.* Ph. D. thesis, Princeton University, NJ.

[119] Cuthill, E. H. and McKee, J. (1969). Reducing the bandwidth of sparse symmetric systems. In *Proceedings of the 1969 24th national conference of the association for computing machinery*, p. 157. ACM Press, New York.

[120] Dahlburg, R. B. and Picone, J. M. (1989). Evolution of the Orszag–Tang vortex system in a compressible medium. I. Initial average subsonic flow. *Phys. Fluids B*, **1**, 2153.

[121] Dai, W. and Woodward, P. R. (1995). A simple Riemann solver and high-order Godunov schemes for hyperbolic systems of conservation laws. *J. Comp. Phys.*, **121**, 51.

[122] Darmofal, D. L. and Haimes, R. (1996). An analysis of 3-D particle path integration algorithms. *J. Comp. Phys.*, **123**, 182.

[123] Daube, O. (1992). Resolution of the 2D Navier–Stokes equations in velocity–vorticity form by means of an influence matrix technique. *J. Comp. Phys.*, **103**, 402.

[124] Dauge, M. (1988). *Elliptic boundary value problems in corner domains—smoothness and asymptotics of solutions.* Lecture Notes in Mathematics 1341. Springer-Verlag, Heidelberg.

[125] Deang, J. M. and Gunzburger, M. D. (1998). Issues related to least-squares finite element methods for the Stokes equations. *SIAM J. Sci. Comp.*, **20**, 878.

[126] Deardorff, J. W. (1970). A numerical study of three-dimensional channel flow at large Reynolds number. *J. Fluid Mech.*, **41**, 453.

[127] Demkowicz, L., Oden, J. T., Rachowicz, W., and Hardy, O. (1989). Toward a universal *h-p* adaptive finite element strategy. Part I. Constrained approximations and data structure. *Comp. Meth. Appl. Mech. Eng.*, **77**, 79.

[128] Demkowicz, L., Rachowicz, W., and Devloo, Ph. (2002). A fully automatic *hp* -adaptivity. *J. Sci. Comp.*, **17**, 117.

[129] Deville, M. O., Fischer, P. F., and Mund, E. H. (2002). *High-order methods for incompressible fluid flow.* Cambridge University Press.

[130] Deville, M. O., Kleiser, L., and Montigny-Rannou, F. (1984). Pressure and time treatment for Chebyshev spectral solution of a spectral problem. *Int. J. Numer. Meth. Fluids*, **4**, 1149.

[131] Deville, M. O. and Mund, E. H. (1985). Chebyshev pseudospectral solution of second-order elliptic equations with finite element preconditioning. *J. Comp. Phys.*, **60**, 517.

[132] Deville, M. O., Mund, E. H., and Patera, A. T. (1987). Iterative solution of isoparametric spectral element equations by low-order finite element preconditioning. *J. Comp. Appl. Math.*, **20**, 189.

[133] Dey, S. (1997). *Geometry-based three-dimensional hp finite element modelling and computations.* Ph. D. thesis, Rensselaer Polytechnic Institute, Troy, NY.

[134] Dey, S., O'Bara, R. M., and Shephard, M. S. (1999). Curvilinear mesh generation in 3D. In *Proceedings of the 8th international meshing roundtable*, South Lake Tahoe, CA, p. 407. Elsevier, New York.

[135] Dey, S., Shephard, M. S., and Flaherty, J. E. (1997). Geometry representation issues associated with *p*-version finite element computations. *Comp. Meth. Appl. Mech. Eng.*, **150**, 39.

[136] Don, W.-S. and Gottlieb, D. (1990). Spectral simulation of an unsteady compressible flow past a circular cylinder. *Comp. Meth. Appl. Mech. Eng.*, **80**, 39.

[137] Don, W.-S. and Gottlieb, D. (1998). Spectral simulation of supersonic reactive flow. *SIAM J. Numer. Anal.*, **35**, 2370.

[138] Donea, J., Giuliani, S., and Halleux, J. P. (1982). An arbitrary Lagrangian–Eulerian finite element method for transient dynamic fluid–structure interactions. *Comp. Meth. Appl. Mech. Eng.*, **33**, 689.

[139] Dong, S. and Karniadakis, G. E. (2003). *p*-refinement and *p*-threads. *Comp. Meth. Appl. Mech. Eng.*, **192**, 2191.

[140] Douglas, J., Jr and Dupont, T. (1976). *Interior penalty procedures for elliptic and parabolic Galerkin methods.* Lecture Notes in Physics 58. Springer-Verlag, Berlin.

[141] Drazin, P. G. and Reid, W. H. (1981). *Hydrodynamic stability.* Cambridge University Press.

[142] Driscoll, T. A. (1997). Eigenmodes of isospectral drums. *SIAM Rev.*, **39**, 1.

[143] Dubiner, M. (1991). Spectral methods on triangles and other domains. *J. Sci. Comp.*, **6**, 345.

[144] Duffy, M. G. (1982). Quadrature over a pyramid or cube of integrands with a singularity at a vertex. *SIAM J. Numer. Anal.*, **19**, 1260.

[145] Dukowicz, J. K. and Dvinsky, A. S. (1992). Approximate factorization as a high order splitting for the incompressible flow equations. *J. Comp. Phys.*, **102**, 336.

[146] Dunavant, D. A. (1985). High degree efficient symmetrical Gaussian quadrature rules for the triangle. *Int. J. Numer. Meth. Eng.*, **21**, 1129.

[147] Dupont, F. (2001). *Comparison of numerical methods for modelling ocean circulation in basins with irregular coasts.* Ph. D. thesis, McGill University, Montreal, Quebec, Canada.

[148] E, Weinan and Liu, J.-G. (1995). Projection method I: convergence and numerical boundary layers. *SIAM J. Numer. Anal.*, **32**, 1017.

[149] E, Weinan and Liu, J.-G. (1996). Projection method II: Godunov–Ryabenki analysis. *SIAM J. Numer. Anal.*, **33**, 1597.

[150] E, Weinan and Liu, J.-G. (2000). Gauge finite element method for incompressible flows. *Int. J. Numer. Meth. Fluids*, **34**, 701.

[151] E, Weinan and Liu, J.-G. (2003). Gauge method for viscous incompressible flows. *Comm. Math. Sci.*, **1**, 317.

[152] Earth Simulator (2003). `http://www.es.jamstec.go.jp/esc/eng/index.html`.

[153] Ehrenstein, U. (1996). On the linear stability of channel flows over riblets. *Phys. Fluids*, **8**, 3194.

[154] Emmons, H. W. (1970). Critique of numerical modeling of fluid-mechanics phenomena. *Ann. Rev. Fluid Mech.*, **2**, 15.

[155] Eskilsson, C. and Sherwin, S. J. (2004). Discontinuous Galerkin spectral/*hp* element modelling of dispersive shallow water systems. *J. Sci. Comp.* Submitted.

[156] Eskilsson, C. and Sherwin, S. J. (2004). A triangular spectral/*hp* discontinuous Galerkin method for modelling 2D shallow water equations. *Int. J. Numer. Meth. Fluids*, **45**, 605.

[157] Falcone, M. and Ferretti, R. (1998). Convergence analysis for a class of high-order semi-Lagrangian advection schemes. *SIAM J. Numer. Anal.*, **35**, 909.

[158] Fejér, L. (1932). Lagrangesche interpolation und die zugehörigen konjugierten punkte. *Math. Ann.*, **106**, 1.

[159] Fietier, N. and Deville, M. O. (2003). Time-dependent algorithms for the simulation of viscoelastic flows with spectral element methods: applications and stability. *J. Comp. Phys.*, **186**, 93.

[160] Finlayson, B. A. (1972). *The method of weighted residuals and variational principles.* Academic Press, New York.

[161] Fischer, P. F. (1997). An overlapping Schwarz method for spectral element solution of the incompressible Navier–Stokes equations. *J. Comp. Phys.*, **133**, 84.

[162] Fischer, P. F. and Mullen, J. (2001). Filter-based stabilization of spectral element methods. *Comptes Rendus Acad. Sci. Paris*, **332**, 265.

[163] Fletcher, C. A. J. (1984). *Computational Galerkin methods.* Springer, Berlin.

[164] Fornberg, B. (1996). *A practical guide to pseudospectral methods.* Cambridge University Press.

[165] Fortin, M., Peyret, R., and Temam, R. (1971). Resolution numerique des equations de Navier–Stokes pour un fluide incompressible. *J. Mech.*, **10**, 357.

[166] Franca, L. P., Hughes, T. J. R., and Stenberg, R. (1993). Stabilized finite element methods. In *Incompressible computational fluid dynamics* (ed. M. D. Gunzburger and R. A. Nicolaides), p. 87. Cambridge University Press.

[167] Frey, P. and George, P.-L. (1999). *Maillages.* Editions Hermes, France.

[168] Funaro, D. (1992). *Polynomial approximations of differential equations.* Springer-Verlag, Berlin.

[169] Funaro, D. (1997). Some remarks about the collocation method on a modified Legendre grid. *Comp. Math. Appl.*, **33**, 95.

[170] Funaro, D. (1997). *Spectral elements for transport-dominated equations.* Springer-Verlag, Berlin.

[171] Funaro, D. (2002). Superconsistent discretizations. *J. Sci. Comp.*, **17**, 67.

[172] Funaro, D. and Gottlieb, D. (1991). Convergence results for pseudospectral approximations of hyperbolic systems by a penalty-type boundary treatment. *Math. Comput.*, **57**, 585.

[173] Funaro, D., Quarteroni, A., and Zanolli, P. (1988). An iterative procedure with interface relaxation for domain decomposition methods. *SIAM J. Numer. Anal.*, **25**, 1213.

[174] Fureby, C. and Grinstein, F. F. (1999). Monotonically integrated large eddy simulation of free shear flows. *AIAA J.*, **37**, 544.

[175] Fureby, F. and Grinstein, F. F. (2000). Large eddy simulation of high Reynolds-number

free and wall-bounded flows. AIAA paper 2000-2307.

[176] Gaitonde, D. V. (2001). High-order solution procedure for three-dimensional nonideal magnetogasdynamics. *AIAA J.*, **39**, 2111.

[177] Gaitonde, D. V. and Poggie, J. (2002). Elements of a numerical procedure for 3-D MGD flow control analysis. In *AIAA-2002-0198, 40th AIAA aerospace sciences meeting and exhibit*, Reno, NV, 14–17 January. American Institute of Aeronautics and Astronautics, New York.

[178] GAMM workshop (1986). Numerical simulation of compressible Navier–Stokes equations—external 2D flows around a NACA0012 airfoil. Ed. INRIA, Centre de Rocquefort, de Rennes et de Sophia-Antipolis, Nice, France, 4–6 December 1985.

[179] Gear, C. W. (1971). *Numerical initial value problems in ordinary differential equations.* Prentice-Hall, Englewood Cliffs, NJ.

[180] Germano, M., Piomelli, U., Moin, P., and Cabot, W. H. (1991). A dynamic subgrid-scale eddy viscosity model. *Phys. Fluids A*, **3**, 1760.

[181] Gerritsma, M. I. and Phillips, T. N. (2000). Spectral element methods for axisymmetric Stokes problems. *J. Comp. Phys.*, **164**, 81.

[182] Ghizzetti, A. and Ossicini, A. (1970). *Quadrature formulae.* Academic Press, New York.

[183] Ghosal, S., Lund, T. S., Moin, P., and Akselvoll, K. (1995). A dynamic localization model for large-eddy simulation of turbulent flows. *J. Fluid Mech.*, **286**, 229.

[184] Ghosal, S. and Moin, P. (1995). The basic equations for the large-eddy simulation of turbulent flows in complex geometry. *J. Comp. Phys.*, **118**, 24.

[185] Giannakouros, I. G. (1994). *Spectral element/flux-corrected methods for unsteady compressible viscous flows.* Ph. D. thesis, Princeton University, NJ.

[186] Giannakouros, I. G. and Karniadakis, G. E. (1992). Spectral element–FCT method for scalar hyperbolic conservation laws. *Int. J. Numer. Meth. Fluids*, **14**, 707.

[187] Giannakouros, I. G. and Karniadakis, G. E. (1994). A spectral element–FCT method for the compressible Euler equations. *J. Comp. Phys.*, **115**, 65.

[188] Giraldo, F. X. (1998). The Lagrange–Galerkin spectral element method on unstructured quadrilateral grids. *J. Comp. Phys.*, **147**, 114.

[189] Giraldo, F. X. (2001). A spectral element shallow water model on spherical geodesic grids. *Int. J. Numer. Meth. Fluids*, **35**, 869.

[190] Giraldo, F. X., Hesthaven, J. S., and Warburton, T. C. (2002). Nodal high-order discontinuous Galerkin methods for the spherical shallow water equations. *J. Comp. Phys.*, **181**, 499.

[191] Girault, V. and Raviart, P. A. (1986). *Finite element approximation of the Navier–Stokes equations.* Springer-Verlag, Berlin.

[192] Glasser, A. H. and Tang, X. Z. (2003). Progress in the development of the SEL macroscopic modeling code. In *Sherwood fusion theory conference*, Corpus Christi, TX, 28–30 April.

[193] Goda, K. (1978). A multistep technique with implicit difference schemes for calculating two- or three-dimensional cavity flows. *J. Comp. Phys.*, **30**, 76.

[194] Godunov, S. K. (1959). A finite-difference method for the numerical computation of discontinuous solutions of the equations of fluid dynamics. *Matematicheskii Sbornik*, **47**, 271.

[195] Golub, G. and Van Loan, C. (1993). *Matrix computations* (2nd edn). Johns Hopkins University Press, Baltimore, MD.

[196] Gordon, W. J. and Hall, C. A. (1973). Transfinite element methods: blending function interpolation over arbitrary curved element domains. *Numer. Math.*, **21**, 109.

[197] Gottlieb, D., Gunzburger, M., and Turkel, E. (1982). On numerical boundary treatment for hyperbolic systems. *SIAM J. Numer. Anal.*, **19**, 671.

[198] Gottlieb, D., Lustman, L., and Orszag, S. A. (1981). Spectral calculations of one-dimensional inviscid compressible flows. *SIAM J. Sci. Stat. Comp.*, **2**, 296.

[199] Gottlieb, D. and Orszag, S. A. (1977). Numerical analysis of spectral methods: theory and applications. SIAM–CMBS.

[200] Gottlieb, D. and Shu, C.-W. (1997). On the Gibbs phenomenon and its resolution. *SIAM Rev.*, **30**, 644.

[201] Goublomme, A., Draily, B., and Crochet, M. J. (1992). Numerical prediction of extrudate swell of a high-density polyetilene. *J. Non-Newtonian Fluid Mech.*, **44**, 171.

[202] Grandjouan, N. (1990). The modified equation approach to flux-corrected transport. *J. Comp. Phys.*, **91**, 424.

[203] Gresho, P. M. and Chan, S. (1990). On the theory of semi-implicit projection methods for viscous incompressible flow and its implementation via a finite element method that also introduces a nearly consistent mass matrix. Parts I and II. *Int. J. Numer. Meth. Fluids*, **11**, 587.

[204] Grinstein, F. F. and Karniadakis, G. E. (2002). Special issue on alternative LES and hybrid RANS/LES for turbulent flows. *J. Fluids Eng.*, **124**, 821.

[205] Grisvard, P. (1985). *Elliptic problems in nonsmooth domains.* Pitman Advanced Publishing Program, Boston, MA.

[206] Grisvard, P. (1992). *Singularities in boundary value problems.* Masson, Paris.

[207] Guermond, J. L. and Shen, J. (2003). A new class of truly consistent splitting schemes for incompressible flows. *J. Comp. Phys.*, **192**, 262.

[208] Guermond, J. L. and Shen, J. (2003). On the error estimates for the rotational pressure-correction projection methods. *Math. Comput.* In press.

[209] Guermond, J. L. and Shen, J. (2003). Velocity-correction projection methods for incompressible flows. *SIAM J. Numer. Anal.*, **41**, 112.

[210] Guevermont, G., Habashi, W. G., and Hafez, M. M. (1990). Finite element solution of the Navier–Stokes equations by a velocity–vorticity method. *Int. J. Numer. Meth. Fluids*, **10**, 461.

[211] Gui, W. and Babuška, I. (1986). The h, p and h-p versions of the finite element methods in 1 dimension. Part 1. The error analysis of the p-version. *Numer. Math.*, **49**, 577.

[212] Guillarda, H. and Farhat, C. (2002). On the significance of the geometric conservation law for flow computations on moving meshes. *Comp. Meth. Appl. Mech. Eng.*, **190**, 1467.

[213] Guj, G. and Stella, F. (1993). A vorticity–velocity method for the numerical solution of 3D incompressible flows. *J. Comp. Phys.*, **106**, 286.

[214] Gunzburger, M., Mund, M., and Peterson, J. (1990). Experiences with finite element methods for the velocity–vorticity formulation of three-dimensional, viscous, incompressible flows. In *Computational methods in viscous aerodynamics* (ed. T. K. S. Murphy and C. A. Brebbia), Chapter 8, p. 231. Elsevier, UK.

[215] Gunzburger, M. D. and Nicolaides, R. A. (1993). *Incompressible computational fluid dynamics.* Cambridge University Press.

[216] Guo, B. (1996). The h-p version of finite elements in $\mathbb{R}^3$. Theory and algorithm. In *ICOSAHOM 95: Proceedings of the third international conference on spectral and high-order methods* (ed. A. V. Ilin and L. R. Scott), Houston, TX, June 5, 1995, p. 489. Houston Journal of Mathematics.

[217] Guo, B.-Y., Ma, H.-P., and Tadmor, E. (2001). Spectral vanishing viscosity method for nonlinear conservation laws. *SIAM J. Numer. Anal.*, **39**, 1254.

[218] Harten, A. (1974). The method of artificial compression. AEC research and development report C00-3077-50, New York University.

[219] Harten, A. (1989). ENO schemes with subcell resolution. *J. Comp. Phys.*, **83**, 148.

[220] Harten, A., Lax, P. D., and Van Leer, B. (1983). On upstream differencing and Godunov-type schemes for hyperbolic conservation laws. *SIAM Rev.*, **25**, 35.

[221] Hein, S. and Theofilis, V. (2004). On instability characteristics of isolated vortices and models of trailing-vortex systems. *Comp. Fluids*, **33**, 741.

[222] Heinrich, J. C. and Zienkiewicz, O. C. (1977). Quadratic finite element schemes for two-dimensional convective-transport problems. *Int. J. Numer. Meth. Eng.*, **11**, 1831.

[223] Heinrichs, W. (1998). Splitting techniques for the unsteady Stokes equations. *SIAM J. Numer. Anal.*, **35**, 1646.

[224] Henderson, R. D. (1994). *Unstructured spectral element methods: parallel algorithms and simulations.* Ph. D. thesis, Princeton University, NJ.

[225] Henderson, R. D. (1999). Adaptive spectral element methods. In *High-order methods for computational physics*, Lecture Notes in Computational Science and Engineering. Springer, Germany.

[226] Henderson, R. D. (1999). Dynamic refinement algorithms for spectral element methods. *Comp. Meth. Appl. Mech. Eng.*, **175**, 395.

[227] Henderson, R. D. and Karniadakis, G. E. (1991). Hybrid spectral element-low order methods for incompressible flows. *J. Sci. Comp.*, **6**, 79.

[228] Henderson, R. D. and Karniadakis, G. E. (1993). Unstructured spectral element methods for the incompressible Navier–Stokes equations. In *Finite elements in fluids: new trends and applications.* Pineridge Press, Swansea, UK.

[229] Henderson, R. D. and Karniadakis, G. E. (1995). Unstructured spectral element methods for simulation of turbulent flows. *J. Comp. Phys.*, **122**, 191.

[230] Henningson, D. S., Lundblach, A., and Johansson, A. V. (1993). A mechanism for bypass transition from localized disturbances in wall-bounded shear flows. *J. Fluid Mech.*, **250**, 169.

[231] Herrera, I., Keyes, D. E., Widlund, O. B., and Yates, R. (ed.) (2003). *Domain decomposition methods in science and engineering*, Proceedings of the fourteenth international conference on domain decomposition methods, Cocoyoc, Mexico. National Autonomous University of Mexico.

[232] Hesthaven, J. S. (1996). A stable spectral multi-domain method for the unsteady compressible Navier–Stokes equations. In *Proceedings of the ninth international conference on domain decomposition* (ed. P. E. Bjorstad, M. S. Espedal, and D. E. Keyes), Bergen, Norway, p. 121. Wiley, New York.

[233] Hesthaven, J. S. (1997). A stable penalty method for the compressible Navier–Stokes equations. II. One-dimensional domain decomposition schemes. *SIAM J. Sci. Comp.*, **18**, 658.

[234] Hesthaven, J. S. (1998). From electrostatics to almost optimal nodal sets for polynomial interpolation in a simplex. *SIAM J. Numer. Anal.*, **35**, 655.

[235] Hesthaven, J. S. (1999). A stable penalty method for the compressible Navier–Stokes equations. III. Multi-dimensional domain decomposition schemes. *SIAM J. Sci. Comp.*, **20**, 62.

[236] Hesthaven, J. S. and Gottlieb, D. (1996). A stable penalty method for the compressible Navier–Stokes equations. I. Open boundary conditions. *SIAM J. Sci. Comp.*, **17**, 579.

[237] Hesthaven, J. S. and Gottlieb, D. (1999). Stable spectral methods for conservation laws on triangles with unstructured grids. *Comp. Meth. Appl. Mech. Eng.*, **175**, 361.

[238] Hesthaven, J. S. and Teng, C. H. (2000). Stable spectral methods on tetrahedral elements. *SIAM J. Sci. Comp.*, **21**, 2352.

[239] Hesthaven, J. S. and Warburton, T. C. (2002). Nodal high-order methods on unstructured grids: 1. time-domain solution of Maxwell's equations. *J. Comp. Phys.*, **181**, 186.

[240] Hirsch, C. (1988). *Numerical computation of internal and external flows.* Wiley, New

York.

[241] Hirt, C. W., Amsden, A. A., and Cook, H. K. (1974). An arbitrary Lagrangian–Eulerian computing method for all flow speeds. *J. Comp. Phys.*, **14**, 27.

[242] Ho, L.-W. (1989). *A Legendre spectral element method for simulation of incompressible unsteady free-surface flows.* Ph. D. thesis, Massachusetts Institute of Technology.

[243] Horiuti, K. (1997). A new dynamic two-parameter mixed model for large-eddy simulation. *Phys. Fluids*, **9**, 3443.

[244] Houston, P., Schwab, C., and Suli, E. (2000). Discontinuous *hp*-finite element methods for advection–diffusion problems. Tech. report 00/15, Oxford University Computing Laboratory.

[245] Hu, F. Q. and Atkins, H. L. (2002). Eigensolution analysis of the discontinuous Galerkin method with nonuniform grids. I. One space dimension. *J. Comp. Phys.*, **182**, 516.

[246] Hu, F. Q., Hussaini, M. Y., and Rasetarinera, P. (1999). An analysis of the discontinuous Galerkin method for wave propagation problems. *J. Comp. Phys.*, **151**, 921.

[247] Hughes, T. J. R. (1987). *The finite element method.* Prentice-Hall, New Jersey.

[248] Hughes, T. J. R., Franca, L. P., and Balestra, M. (1986). A new finite element formulation for computational fluid dynamics: V. Circumventing the Babuška–Brezzi condition: a stable Petrov–Galerkin formulation of the Stokes problem accommodating equal-order interpolations. *Comp. Meth. Appl. Mech. Eng.*, **59**, 85.

[249] Hughes, T. J. R., Liu, W. K., and Zimmerman, T. K. (1981). Lagrangian–Eulerian finite element formulation for incompressible viscous flows. *Comp. Meth. Appl. Mech. Eng.*, **29**, 329.

[250] Hussaini, M. Y., Kopriva, D. A., Salas, M. D., and Zang, T. A. (1985). Spectral methods for the Euler equations. Part II. Chebyshev methods and shock-fitting. *AIAA J.*, **23**, 234.

[251] Isaacson, E. and Keller, H. B. (1966). *Analysis of numerical methods.* Wiley, New York.

[252] Iskandarani, M., Haidvogel, D. B., and Boyd, J. P. (1995). A staggered spectral element model with application to the oceanic shallow water equations. *Int. J. Numer. Meth. Fluids*, **20**, 393.

[253] Jiang, B.-N. (1998). *The least-squares finite element method: theory and applications in computational fluid dynamics and electromagnetics.* Springer, New York.

[254] Jiménez, J. and Moin, P. (1991). The minimal flow unit in near-wall turbulence. *J. Fluid Mech.*, **225**, 213.

[255] Johnson, C. (1994). *Numerical solution of partial differential equations by the finite element method.* Cambridge University Press.

[256] Kaiktsis, L., Karniadakis, G. E., and Orszag, S. A. (1991). Onset of three-dimensionality, equilibria, and early transition in flow over a backward-facing step. *J. Fluid Mech.*, **191**, 501.

[257] Kaiktsis, L., Karniadakis, G. E., and Orszag, S. A. (1996). Unsteadiness and convective instabilities in two-dimensional flow over a backward-facing step. *J. Fluid Mech.*, **321**, 157.

[258] Karamanos, G.-S. (1999). *Large eddy simulation using unstructured spectral/hp finite elements.* Ph. D. thesis, Imperial College London.

[259] Karamanos, G.-S. and Karniadakis, G. E. (2000). A spectral vanishing viscosity method for large-eddy simulations. *J. Comp. Phys.*, **162**, 22.

[260] Karlin, S. and McGregor, J. (1964). Some stochastic models in genetics. In *Stochastic models in medicine and biology.* University of Wisconsin Press, Madison, WI.

[261] Karniadakis, G. E. (1990). Spectral element–Fourier methods for incompressible turbulent flows. *Comp. Meth. Appl. Mech. Eng.*, **80**, 367.

[262] Karniadakis, G. E. and Amon, C. (1987). Stability calculations of wall bounded flows

in complex geometries. In *Proceedings of the 6th IMACS symposium on PDEs* (ed. R. Vichnevetsy and R. S. Stepleman), p. 525. IMACS, New Brunswick, NJ.

[263] Karniadakis, G. E., Bullister, E. T., and Patera, A. T. (1985). A spectral element method for solution of two- and three-dimensional time-dependent incompressible Navier–Stokes equations. In *Proceedings of the Europe–US symposium on finite element methods for nonlinear problems* (ed. K. J. Bergan, P. G. Bathe, and W. Wunderlich), Norway, August 12–16, 1985, p. 803. Springer-Verlag, Berlin.

[264] Karniadakis, G. E., Israeli, M., and Orszag, S. A. (1991). High-order splitting methods for incompressible Navier–Stokes equations. *J. Comp. Phys.*, **97**, 414.

[265] Karniadakis, G. E. and Orszag, S. A. (1993). Nodes, modes and flow codes. *Phys. Today*, **46**, 34.

[266] Karypis, G. and Kumar, V. (1995). Multilevel k-way partitioning scheme for irregular graphs. Technical report, Department of Computer Science, University of Minnesota.

[267] Kim, J. and Moin, P. (1985). Application of a fractional-step method to incompressible Navier–Stokes equations. *J. Comp. Phys.*, **59**, 308.

[268] Kim, J., Moin, P., and Moser, R. (1987). Turbulence statistics in fully developed channel flow at low Reynolds number. *J. Fluid Mech.*, **177**, 133.

[269] Kincaid, D. and Cheney, W. (1996). *Numerical analysis* (2nd edn). Brooks/Cole Publishing, Pacific Grove, CA.

[270] Kirby, R. M. (2003). *Toward dynamic spectral/hp refinement: algorithms and applications to flow–structure interactions.* Ph. D. thesis, Division of Applied Mathematics, Brown University, Providence, RI.

[271] Kirby, R. M. and Karniadakis, G. E. (2002). Coarse resolution turbulence simulations with SVV–LES. *ASME J. Fluids Eng.*, **124**, 886.

[272] Komatitsch, D., Martin, R., Tromp, J., Taylor, M. A., and Wingate, B. A. (2001). Wave propagation in 2-D elastic media using spectral element method with triangles and quadrangles. *J. Comp. Acoustics*, **9**, 703.

[273] Komatitsch, D., Ritsema, J., and Tromp, J. (2002). The spectral-element method, beowulf computing, and global seismology. *Science*, **298**, 1737.

[274] Kondratiev, V. A. (1967). Boundary value problems for elliptic equations in domains with conical or angular points. *Trans. Moscow Math. Soc.*, **16**, 227.

[275] Koobus, B. and Farhat, C. (1999). Second-order time-accurate and geometrically conservative implicit schemes for flow computations on unstructured dynamic meshes. *Comp. Meth. Appl. Mech. Eng.*, **170**, 103.

[276] Koornwinder, T. (1975). Two-variable analogues of the classical orthogonal polynomials. In *Theory and applications of special functions* (ed. R. Askey). Academic Press, San Diego.

[277] Kopriva, D. A. (1987). A practical assessment of spectral accuracy for hyperbolic problems with discontinuities. *J. Sci. Comp.*, **2**, 249.

[278] Kopriva, D. A. (1991). Multidomain spectral solution of the Euler gas-dynamics equations. *J. Comp. Phys.*, **96**, 428.

[279] Kopriva, D. A. (1992). Spectral solutions of inviscid supersonic flows over wedges and axisymmetric cones. *Comp. Fluids*, **21**, 247.

[280] Kopriva, D. A. (1996). Spectral solution of the viscous blunt body problem. II. multidomain approximation. *AIAA J.*, **34**, 560.

[281] Kopriva, D. A. and Kolias, J. H. (1996). A conservative staggered-grid Chebyshev multidomain method for compressible flows. *J. Comp. Phys.*, **125**, 244.

[282] Kopriva, D. A., Zang, T. A., and Hussaini, M. Y. (1991). Spectral methods for the Euler equations: the blunt body problem revisited. *AIAA J.*, **29**, 1458.

[283] Kovasznay, L. I. G. (1948). Laminar flow behind a two-dimensional grid. *Proc. Camb.*

Phil. Soc., **44**, 58.

[284] Kraichnan, R. H. (1976). Eddy viscosity in two and three dimensions. *J. Atmos. Sci.*, **33**, 1521.

[285] Krall, J. and Sheffer, I. (1967). Orthogonal polynomials in two variables. *Ann. Mat. Pura. Appl.*, **76**, 325.

[286] Kreiss, H. O. and Lorentz, J. (1989). *Initial-boundary value problem and the Navier–Stokes equations.* Series in Applied and Pure Mathematics. Academic Press, New York.

[287] Kreiss, H. O. and Oliger, J. (1973). *Methods for the approximate solution of time dependent problems.* World Meteorological Organization, Case Postale No. 1, CH-1211 Geneva 20, Switzerland.

[288] Lambert, J. D. (1993). *Numerical methods for ordinary differential systems: the initial value problem.* Wiley, New York.

[289] Lax, P. D. (1978). Accuracy and resolution in the computation of solutions of linear and nonlinear equations. In *Recent advances in numerical analysis: proceedings of a symposium conducted by the Mathematics Research Center* (ed. C. D. Boor and G. H. Golub), University of Wisconsin–Madison, p. 107. Academic Press, New York.

[290] Lee, C. K. (2001). On curvature element-size control in metric surface generation. *Int. J. Numer. Meth. Eng.*, **50**, 787.

[291] Lee, S.-J., Oh, H.-S., and Yun, J.-H. (2001). Extension of the method of auxiliary mapping for three-dimensional elliptic boundary value problems. *Int. J. Numer. Meth. Eng.*, **50**, 1103.

[292] Leonard, A. (1997). Large-eddy simulation of chaotic convection and beyond. AIAA paper 97-0204.

[293] Leray, J. (1933). Etude de diverses equtions integrales non lineaires et de quelques problemes que pose l'hydrodynamique. *J. Math. Pures Appl.*, **12**, 1.

[294] Leriche, E. and Labrosse, G. (2000). High-order direct Stokes solvers with or without temporal splitting: numerical investigations of their comparative properties. *SIAM J. Sci. Comp.*, **22**, 1386.

[295] Leriche, E., Perchat, E., Labrosse, G., and Deville, M. O. (2004). Numerical evaluation of the accuracy and stability properties of high order direct Stokes solver with or without temporal splitting. *J. Sci. Comp.* In press.

[296] Lesieur, M. and Metais, O. (1996). New trends in large-eddy simulations of turbulence. *Ann. Rev. Fluid Mech.*, **28**, 45.

[297] LeTallec, P. and Patra, A. (1997). Non-overlapping domain decomposition methods for adaptive *hp* approximations of the Stokes problem with discontinuous pressure fields. *Comp. Meth. Appl. Mech. Eng.*, **145**, 361.

[298] LeVeque, R. J. and Walder, R. (1991). Grid-refinement effects and rotated methods for computing complex flows in astrophysics. In *Proceedings of the 9th GAMM conference on numerical methods in fluid mechanics* (ed. J. B. Vos, A. Rizzi, and I. L. Ryhming), p. 376. Vieweg, Braunschweig.

[299] Levin, J. G., Iskandarani, M., and Haidvogel, D. B. (1997). A spectral filtering procedure for eddy-resolving simulations with a spectral element ocean model. *J. Comp. Phys.*, **137**, 130.

[300] Levy, D. W., Powell, K. G., and Van Leer, B. (1993). Use of a rotational Riemann solver for the two-dimensional Euler equations. *J. Comp. Phys.*, **106**, 201.

[301] Li, F. and Shu, C.-W. (2004). Locally divergence-free discontinuous Galerkin methods for MHD equations. *J. Sci. Comp.* In press.

[302] Lilly, D. K. (1992). A proposed modification to the Germano subgrid-scale closure method. *Phys. Fluids A*, **4**, 633.

[303] Lin, G. (2005). *Deterministic and stochastic simulations of plasma flows.* Ph. D. thesis,

Brown University, Providence, RI.

[304] Lin, R.-S. and Malik, M. R. (1996). On the stability of attachment-line boundary layers. Part 1. The incompressible swept Hiemenz flow. *J. Fluid Mech.*, **311**, 239.

[305] Lions, P. L. (1989). On the Schwarz alternating method III: a variant for non-overlapping subdomains. Houston, TX, SIAM, p. 202.

[306] Lions, P. L. (1996). *Mathematical topics in fluid mechanics.* Oxford Science Publications, Oxford.

[307] l'Isle, E. B. D. and George, P. L. (1993). Optimization of tetrahedral meshes. *Modeling, Mesh Gen. Adaptive Numer. Meth. PDEs*, **75**, 97.

[308] Loft, R. D., Thomas, S. J., and Dennis, J. M. (2001). Terascale spectral element dynamical code for atmospheric general circulation models. In *Proceedings of supercomputing 2001*, Denver, CO. IEEE/ACM. On CD-ROM.

[309] Lomtev, I. (1998). *A discontinuous Galerkin spectral/hp element method for compressible Navier–Stokes equations.* Ph. D. thesis, Brown University, Providence, RI.

[310] Lomtev, I., Quillen, C. B., and Karniadakis, G. E. (1998). Spectral/*hp* methods for viscous compressible flows on unstructured 2D meshes. *J. Comp. Phys.*, **144**, 325.

[311] Loth, F., Fischer, P., Arslan, N., Bertram, C. D., Lee, S. E., Royston, T. J., Song, R. H., Shaalan, W. E., and Bassiouny, H. S. (2003). Transitional flow at the venous anastomosis of an arteriovenous graft: potential relationship with activation of the ERK1/2 mechanotransduction pathway. *J. Biomech. Eng.*, **125**, 49.

[312] Lottes, J. W. and Fischer, P. F. (2004). Hybrid multigrid/Schwarz algorithms for the spectral element method. *J. Sci. Comp.* Submitted.

[313] Luo, X. J., Shephard, M. S., Remacle, J. F., O'Bara, R. M., Beall, M. W., Szabó, B., and Actis, R. (2002). *p*-version mesh generation issues. In *Proceedings of the 11th international meshing roundtable*, Ithaca, NY. Internal report no. 2002-3, `http://www.scorec.rpi.edu`.

[314] Lyons, S. L., Hanratty, T. J., and McLaughlin, J. B. (1991). Large-scale computer simulation of fully developed turbulent channel flow with heat transfer. *Int. J. Numer. Meth. Fluids*, **13**, 999.

[315] Ma, H. (1993). A spectral element basin model for the shallow water equations. *J. Comp. Phys.*, **109**, 133.

[316] Ma, X., Karamanos, G.-S., and Karniadakis, G. E. (2000). Dynamics and low-dimensionality in the turbulent near-wake. *J. Fluid Mech.*, **410**, 29.

[317] Maday, Y., Meiron, D., Patera, A. T., and Rønquist, E. M. (1993). Analysis of iterative methods for the steady and unsteady Stokes problem: application to spectral element discretizations. *SIAM J. Sci. Stat. Comp.*, **14**, 310.

[318] Maday, Y., Ould Kaber, S. M., and Tadmor, E. (1993). Legendre pseudospectral viscosity method for nonlinear conservation laws. *SIAM J. Numer. Anal.*, **30**, 321.

[319] Maday, Y. and Patera, A. T. (1989). Spectral element methods for the Navier–Stokes equations. In *State-of-the-art surveys in computational mechanics* (ed. A. K. Noor and J. T. Oden), Chapter 3, p. 71. ASME, New York.

[320] Maday, Y., Patera, A. T., and Rønquist, E. M. (1990). An operator–integration-factor splitting method for time-dependent problems: application to incompressible fluid flow. *J. Sci. Comp.*, **4**, 263.

[321] Maday, Y. and Tadmor, E. (1989). Analysis of the spectral vanishing method for periodic conservation laws. *SIAM J. Numer. Anal.*, **26**, 854.

[322] Mandel, J. (1991). Two-level decomposition preconditioning for the *p*-version finite element method in three dimensions. *Int. J. Numer. Meth. Eng.*, **29**, 1095.

[323] Marini, L. D. and Quarteroni, A. (1989). A relaxation procedure for domain decomposition methods using finite element methods. *Numer. Math.*, **55**, 575.

[324] Mavriplis, C. (1989). *Nonconforming discretizations and a posteriori error estimators for adaptive spectral element techniques.* Ph. D. thesis, Massachusetts Institute of Technology.

[325] Mavriplis, C. (1994). Adaptive mesh strategies for the spectral element method. *Comp. Meth. Appl. Mech. Eng.*, **116**, 77.

[326] Mavriplis, C. and Rosendale, J. V. (1993). Triangular spectral elements for incompressible fluid flow. AIAA-93-3346-CP, p. 540.

[327] McDonald, A. (1984). Accuracy of multi-upstream, semi-Lagrangian advective schemes. *Monthly Weather Rev.*, **112**, 1267.

[328] McDonald, B. E. (1989). Flux-corrected pseudospectral method for scalar hyperbolic conservation laws. *J. Comp. Phys.*, **82**, 413.

[329] Meneveau, C. and Katz, J. (2000). Scale-invariance and turbulence models for large-eddy simulation. *Ann. Rev. Fluid Mech.*, **32**, 1.

[330] Miliou, A. (2004). *Three-dimensional wake dynamics of curved cylinders.* Ph. D. thesis, Imperial College London.

[331] Miliou, A., Sherwin, S. J., and Graham, J. M. R. (2003). Wake topology of curved cylinders at low Reynolds numbers. *J. Flow Turb. Comb.*, **71**, 147.

[332] Minev, P. D. (2001). A stabilized incremental projection scheme for the incompressible Navier–Stokes equations. *Int. J. Numer. Meth. Fluids*, **36**, 441.

[333] Minev, P. D. and Gresho, P. (1998). A remark on pressure correction schemes for transient viscous incompressible flow. *Comm. Numer. Meth. Eng.*, **14**, 335.

[334] Moffatt, H. K. (1964). Viscous and resistive eddies near a sharp corner. *J. Fluid Mech.*, **18**, 1.

[335] Momeni-Masuleh, S. H. (2001). *Spectral methods for the three field formulation of incompressible fluid flow.* Ph. D. thesis, University of Aberystwyth.

[336] Moretti, G. and Salas, M. D. (1969). The blunt body problem for a viscous rarefied gas flow. AIAA paper 69-139.

[337] Morzynski, M. and Thiele, F. (1991). Numerical stability analysis of flow about a cylinder. *Z. Angew. Math. Mech. (ZAMM)*, **71**, T424.

[338] Moser, R. D., Kim, J., and Mansour, N. N. (1999). Direct numerical simulation of turbulent channel flow up to $\mathrm{Re}_\tau = 590$. *Phys. Fluids*, **11**, 943.

[339] Nomura, T. and Hughes, T. J. R. (1992). An arbitrary Lagrangian–Eulerian finite element method for interaction of fluid and a rigid body. *Comp. Meth. Appl. Mech. Eng.*, **95**, 115.

[340] Nwogu, O. (1993). Alternative form of Boussinesq equations for nearshore wave propagation. *J. Waterway, Port, Coastal Ocean Eng.*, **119**, 618.

[341] Oden, J. T. (1962). *Plate beam structures.* Ph. D. thesis, Oklahoma State University.

[342] Oden, J. T. (1991). Finite elements: an introduction. In *Handbook of numerical analysis* (ed. P. G. Ciarlet and J. L. Lions), Volume 2, p. 3. Elsevier, Amsterdam.

[343] Oden, J. T. (1992). Optimal *hp*-finite element methods. TICOM Report 92-09, University of Texas at Austin.

[344] Oden, J. T. (1993). Error estimation and control in computational fluid dynamics. The Zienkiewicz lecture.

[345] Oden, J. T., Babuška, I., and Baumann, C. E. (1998). A discontinuous *hp* finite element method for diffusion problems. *J. Comp. Phys.*, **146**, 491.

[346] Oden, J. T., Demkowicz, L., Rachowicz, W., and Westermann, T. (1989). Toward a universal *h-p* adaptive finite element strategy. Part II. A posteriori error estimation. *Comp. Meth. Appl. Mech. Eng.*, **77**, 113.

[347] Oden, J. T., Patra, A., and Feng, Y. (1992). An *hp* adaptive strategy. *Adaptive, Multilevel, Hierarchical Comp. Strat.*, **157**, 23. ASME.

[348] Oden, J. T., Wu, W., and Ainsworth, M. (1994). An a posteriori error estimate for finite element approximations of the Navier–Stokes equations. *Comp. Meth. Appl. Mech. Eng.*, **111**, 185.

[349] Oran, E. S. and Boris, J. P. (1987). *Numerical simulation of reactive flow.* Elsevier, New York.

[350] Orszag, S. A. (1971). On the elimination of aliasing in finite-difference schemes by filtering high wavenumber components. *J. Atmos. Sci.*, **28**, 1074.

[351] Orszag, S. A. (1974). Fourier series on spheres. *Monthly Weather Rev.*, **102**, 56.

[352] Orszag, S. A. (1980). Spectral methods for problems in complex geometries. *J. Comp. Phys.*, **37**, 70.

[353] Orszag, S. A., Israeli, M., and Deville, M. O. (1986). Boundary conditions for incompressible flows. *J. Sci. Comp.*, **1**, 75.

[354] Orszag, S. A. and Patera, A. T. (1983). Secondary instability of wall-bounded shear flows. *J. Fluid Mech.*, **128**, 347.

[355] Orszag, S. A. and Patterson, G. S. (1972). Numerical simulation of the three-dimensional homogeneous isotropic turbulence. *Phys. Rev. Lett.*, **28**, 76.

[356] Orszag, S. A. and Tang, C. (1979). Small-scale structure of two-dimensional magnetohydrodynamic turbulence. *J. Fluid Mech.*, **90**, 129.

[357] Osher, S. and Solomon, F. (1982). Upwind difference schemes for hyperbolic systems of conservation laws. *Math. Comput.*, **38**, 339.

[358] Owens, R. G. (1998). Spectral approximations on the triangle. *Proc. R. Soc. Lond. A*, **454**, 857.

[359] Pahl, S. S. (University of Witwatersrand, Johannesburg). *Schwarz type domain decomposition methods for spectral element discretizations.* Ph. D. thesis, 1993.

[360] Panton, R. L. (1997). A Reynolds stress function for wall layers. *J. Fluids Eng.*, **119**, 325.

[361] Papaharilaou, Y., Doorly, D. J., and Sherwin, S. J. (2002). The influence of out-of-plane geometry on pulsatile flow within a distal end-to-side anastomosis. *J. Biomech.*, **35**, 1225.

[362] Pasquetti, R. and Xu, C. J. (2002). Comments on 'Filter-based stabilization of spectral element methods'. *J. Comp. Phys.*, **182**, 646.

[363] Patera, A. T. (1984). A spectral method for fluid dynamics: laminar flow in a channel expansion. *J. Comp. Phys.*, **54**, 468.

[364] Pathria, D. and Karniadakis, G. E. (1995). Spectral element methods for elliptic problems in nonsmooth domains. *J. Comp. Phys.*, **122**, 83.

[365] Patra, A. and Gupta, A. (2001). A systematic strategy for simultaneous adaptive *hp* finite element mesh modification using nonlinear programming. *Comp. Meth. Appl. Mech. Eng.*, **190**, 3797.

[366] Patterson, G. S. and Orszag, S. A. (1971). Spectral calculations of isotropic turbulence: efficient removal of aliasing interaction. *Phys. Fluids*, **14**, 2358.

[367] Pavarino, L. and Widlund, O. (1996). A polylogarithmic bound for an iterative substructuring method for spectral elements in three dimensions. *SIAM J. Numer. Anal.*, **33**, 1303.

[368] Pavarino, L. F. and Warburton, T. C. (2000). Overlapping Schwarz methods for unstructured elements. *J. Comp. Phys.*, **160**, 298.

[369] Peano, A. G. (1976). Hierarchies of conforming finite elements for plate elasticity and plate bending. *Comp. Math. Appl.*, **2**, 211.

[370] Peiró, J., Giordana, S., Griffith, C., and Sherwin, S. J. (2002). High-order algorithms for vascular flow modelling. *Int. J. Numer. Meth. Fluids*, **40**, 137.

[371] Peiró, J., Peraire, J., and Morgan, K. (1994). FELISA system reference manual. Part I. Basic theory. Tech. report CR/821/94, Civil Engineering Department, University of Wales.

[372] Peregrine, D. H. (1967). Long waves on a beach. *J. Fluid Mech.*, **27**, 815.

[373] Perot, B. J. (1993). An analysis of the fractional-step method. *J. Comp. Phys.*, **108**, 51.

[374] Peterkin, R. E., Frese, M. H., and Sovinec, C. R. (1998). Transport magnetic flux in an arbitrary coordinate ALE code. *J. Comp. Phys.*, **140**, 148.

[375] Petersson, N. A. (2001). Stability of pressure boundary conditions for Stokes and Navier–Stokes equations. *J. Comp. Phys.*, **172**, 40.

[376] Phillips, T. N. and Roberts, G. W. (1993). The treatment of spurious pressure modes in spectral incompressible flow calculations. *J. Comp. Phys.*, **105**, 150.

[377] Piomelli, U., Cabot, W. H., Moin, P., and Lee, S. (1991). Subgrid-scale backscatter in turbulent and transitional flows. *Phys. Fluids A*, **3**, 1766.

[378] Pironneau, O. (1982). On the transport–diffusion algorithm and its application to Navier–Stokes equations. *Numer. Math.*, **38**, 309.

[379] Pironneau, O. (1989). *Finite element methods for fluids.* Wiley, Chichester.

[380] Pitt, R. E., Sherwin, S. J., and Theofilis, V. (2005). BiGlobal stability analysis of steady flow in constricted channel geometries. *Int. J. Numer. Meth. Fluids.* In press.

[381] Porter, D. H., Pouquet, A., and Woodward, P. R. (1994). Kolmogorov-like spectra in decaying three-dimensional supersonic flows. *Phys. Fluids*, **6**, 2133.

[382] Powell, K. G. (1994). An approximate Riemann solver for magnetohydrodynamics (that works in more than one dimension).

[383] Priest, E. (1982). *Solar magneto-hydrodynamics.* D. Reidel Publishing, Hingham, MA.

[384] Proot, M. (2003). *The least-squares spectral element method.* Ph. D. thesis, Delft University of Technology, The Netherlands.

[385] Proot, M. and Gerritsma, M. I. (2002). Least-squares spectral elements applied to the Stokes problem. *J. Comp. Phys.*, **181**, 454.

[386] Proriol, J. (1957). Sur une famille de polynomes á deux variables orthogonaux dans un triangle. *Comptes Rendus Acad. Sci. Paris*, **257**(2459).

[387] Prudhomme, S., Pascal, F., Oden, J. T., and Romkes, A. (2001). A priori error estimate for the Baumann–Oden version of the discontinuous Galerkin method. *Comptes Rendus Acad. Sci. Paris*, **332**, 851.

[388] Pruett, C. D. and Street, C. L. (1991). A spectral collocation method for compressible, non-similar boundary layers. *Int. J. Numer. Meth. Fluids*, **13**, 713.

[389] Quartapelle, L. (1993). *Numerical solution of the incompressible Navier–Stokes equations.* Birkhauser, Basel.

[390] Rachowicz, W., Oden, J. T., and Demkowicz, L. (1989). Toward a universal *h-p* adaptive finite element strategy. Part III. Design of *h-p* meshes. *Comp. Meth. Appl. Mech. Eng.*, **77**, 181.

[391] Rai, M. M. and Moin, P. (1991). Direct simulations of turbulent flow using finite-difference schemes. *J. Comp. Phys.*, **96**, 15.

[392] Reynolds, O. (1883). An experimental investigation of the circumstances which determine whether the motion of water shall be direct or sinuous, and of the law of resistance in parallel channels. *Phil. Trans. Roy. Soc.*, **174**, 935.

[393] Riviere, B., Wheeler, M. F., and Girault, V. (1999). Improved energy estimates for interior penalty, constrained and discontinuous Galerkin methods for elliptic problems. Part I. *Comp. Geosci.*, **3**, 337.

[394] Riviere, B., Wheeler, M. F., and Girault, V. (2001). A priori error estimates for finite element methods based on discontinuous approximation spaces for elliptic problems. *SIAM J. Numer. Anal.*, **39**, 902.

[395] Roache, P. J. (1982). *Computational fluid dynamics.* Hermosa Publishers, Albuquerque, NM.

[396] Robert, A. (1981). A stable numerical integration scheme for the primitive meteorological equations. *Atmos. Ocean*, **19**, 35.

[397] Robertson, I. (2000). *Free surface flow simulations using high order algorithms.* Ph. D. thesis, Imperial College London.

[398] Robertson, I. and Sherwin, S. J. (1999). Free-surface flow simulations using *hp*/spectral elements. *J. Comp. Phys.*, **155**, 26.

[399] Robichaux, J., Balachandar, S., and Vanka, S. (1999). Three-dimensional Floquet instability of the wake of a square cylinder. *Phys. Fluids*, **11**, 560.

[400] Roe, P. L. (1981). Approximate Riemann solvers, parameter vectors, and difference schemes. *J. Comp. Phys.*, **43**, 357.

[401] Romkes, A., Prudhomme, S., and Oden, J. T. (2003). A priori error analyses of a stabilized discontinuous Galerkin method. *Comp. Math. Appl.*, **46**, 1289.

[402] Rønquist, E. M. (1988). *Optimal spectral element methods for the unsteady three-dimensional incompressible Navier–Stokes equations.* Ph. D. thesis, Massachusetts Institute of Technology.

[403] Rønquist, E. M. (1992). A domain decomposition method for elliptic boundary value problems: application to unsteady incompressible fluid flow. In *SIAM, Proceedings of the fifth conference on domain decomposition methods for partial differential equations* (ed. D. Keyes, T. Chan, G. Meurant, J. Scroggs, and R. Voigt), Norfolk, VA, p. 545. SIAM, Philadelphia, PA.

[404] Rønquist, E. M. (1996). Convection treatment using spectral elements of different order. *Int. J. Numer. Meth. Fluids*, **22**, 241.

[405] Rønquist, E. M. and Patera, A. T. (1987). Spectral element multigrid. I. Formulation and numerical results. *J. Sci. Comp.*, **2**, 389.

[406] Ruas, V. (1991). Variational approaches to the two-dimensional Stokes system in terms of the vorticity. *Mech. Res. Comm.*, **18**, 359.

[407] Rumsey, C. L., Van Leer, B., and Roe, P. L. (1993). A multidimensional flux function with applications to the Euler and Navier–Stokes equations. *J. Comp. Phys.*, **105**, 306.

[408] Saad, Y. (1980). Variations on Arnoldi's method for computing eigenelements of large unsymmetric matrices. *Lin. Algebra Appl.*, **34**, 269.

[409] Salas, M. D., Zang, T. A., and Hussaini, M. Y. (1982). Shock-fitted Euler solutions to shock–vortex interactions. In *Proceedings of the 8th international conference on numerical methods in fluid dynamics* (ed. E. Krause), p. 461. Springer, Berlin.

[410] Salvetti, M. V. and Banerjee, S. (1995). A priori tests of a new dynamic subgrid-scale model for finite-difference large-eddy simulations. *Phys. Fluids*, **7**, 2831.

[411] Sanders, B. F. (2002). Non-reflecting boundary flux function for finite volume shallow-water models. *Adv. Water Res.*, **25**, 195.

[412] Savvides, A. (1997). *Application of two-dimensional spectral/finite-difference and spectral/hp finite-element methods to cylinder flows.* Ph. D. thesis, Imperial College London.

[413] Schotzau, D. and Schwab, C. (2001). Exponential convergence in a Galerkin least-squares *hp*-FEM for Stokes flow. *IMA J. Numer. Anal.*, **21**, 53.

[414] Schotzau, D., Schwab, C., and Toselli, A (2003). Mixed *hp*-DGFEM for incompressible flows. *SIAM J. Numer. Anal.*, **40**, 2171.

[415] Schubauer, G. B. and Skramstad, H. K. (1947). Laminar boundary layer oscillations and

stability of laminar flow. *J. Aero. Sci.*, **14**, 69.

[416] Schultz, W. W., Lee, N. Y., and Boyd, J. P. (1989). Chebyshev pseudospectral method of viscous flows with corner singularities. *J. Sci. Comp.*, **4**, 1.

[417] Schwab, C. (1998). *p- and hp-finite element methods.* Oxford University Press.

[418] Schwab, C. and Suri, M. (1999). Mixed *hp* finite element methods for Stokes and non-Newtonian flow. *Comp. Meth. Appl. Mech. Eng.*, **175**, 217.

[419] Sheard, C. J., Thompson, M. C., and Hourigan, K. (2003). From spheres to circular cylinders: the stability and flow structures of bluff ring wakes. *J. Fluid Mech.*, **492**, 147.

[420] Shen, J. (1992). On error estimates of projection methods for Navier–Stokes equations: first-order schemes. *SIAM J. Numer. Anal.*, **29**, 57.

[421] Shen, J. (1993). A remark on the projection-3 method. *Int. J. Numer. Meth. Fluids*, **16**, 249.

[422] Shen, J. (1996). On error estimates of projection methods for Navier–Stokes equations: second-order schemes. *SIAM J. Numer. Anal.*, **65**, 1039.

[423] Sherwin, S. J. (1995). *Triangular and tetrahedral spectral/hp element methods for fluid dynamics.* Ph. D. thesis, Princeton University, NJ.

[424] Sherwin, S. J. (1997). Hierarchical *hp* finite elements in hybrid domains. *Finite Elements Anal. Design*, **27**, 109.

[425] Sherwin, S. J. (1998). A high order Fourier/unstructured discontinuous Galerkin method for hyperbolic conservation laws. In *Seventh international conference on hyperbolic problems, theory numerics, applications* (ed. M. Fey and R. Jeltsch), Zurich, p. 875. Birkhäuser, Basel.

[426] Sherwin, S. J. (2000). Dispersion analysis of the continuous and discontinuous Galerkin formulations. In *Discontinuous Galerkin methods* (ed. B. Cockburn, G. E. Karniadakis, and C. W. Shu), p. 425. Springer, Berlin.

[427] Sherwin, S. J. (2003). A substepping Navier–Stokes splitting scheme for spectral/*hp* element discretisations. In *Parallel computational fluid dynamics: new frontiers and multi-disciplinary applications* (ed. T. Matuson, A. Ecer, J. Periaux, N. Satufka, and P. Fox), p. 43. Elsevier Science BV.

[428] Sherwin, S. J. and Ainsworth, M. (2000). Unsteady Navier–Stokes solvers using hybrid spectral/*hp* element methods. *Appl. Numer. Math.*, **33**, 357.

[429] Sherwin, S. J. and Casarin, M. (2001). Low-energy preconditioning for elliptic substructured solvers based on unstructured spectral/*hp* element discretisation. *J. Comp. Phys.*, **171**, 394.

[430] Sherwin, S. J. and Doorly, D. J. (2003). Flow dynamics within model distal arterial bypass grafts. In *Vascular grafts: experiments and modelling* (ed. A. Tura), Advances in Fluid Mechanics 34, Chapter 9, p. 327. WIT Press, Southampton.

[431] Sherwin, S. J. and Karniadakis, G. E. (1995). A new triangular and tetrahedral basis for high-order finite element methods. *Int. J. Numer. Meth. Eng.*, **38**, 3775.

[432] Sherwin, S. J. and Karniadakis, G. E. (1995). A triangular spectral element method; applications to the incompressible Navier–Stokes equations. *Comp. Meth. Appl. Mech. Eng.*, **123**, 189.

[433] Sherwin, S. J. and Karniadakis, G. E. (1996). Tetrahedral *hp* finite elements: algorithms and flow simulations. *J. Comp. Phys.*, **124**, 14.

[434] Sherwin, S. J. and Peiró, J. (2002). Mesh generation in curvilinear domains using high-order elements. *Int. J. Numer. Meth. Eng.*, **53**, 207.

[435] Sherwin, S. J., Shah, O., Doorly, D. J., Peiro, J., Papaharilaou, Y., Watkins, N., Caro, C. G., and Dumoulin, C. L. (2000). The influence of out-of-plane geometry on the flow within a distal end-to-side anastomosis. *ASME J. Biomech. Eng.*, **122**, 86.

[436] Shu, C.-W. (2001). Different formulations of the discontinuous Galerkin method for the

viscous terms. In *Advances in scientific computing* (ed. Z.-C. Shi, M. Mu, W. Xue, and J. Zou), p. 144. Science Press, Beijing.

[437] Shumlak, U. and Loverich, J. (2003). Approximate Riemann solver for the two-fluid plasma model. *J. Comp. Phys.*, **187**, 620.

[438] Sidilkover, D. and Karniadakis, G. E. (1993). Non-oscillatory spectral element Chebyshev method for shock wave calculations. *J. Comp. Phys.*, **107**, 1.

[439] Smith, B., Bjorstad, P., and Gropp, W. (1996). *Domain decomposition. Parallel multilevel methods for elliptic differential equations.* Cambridge University Press.

[440] Smolarkiewicz, P. K. and Pudykiewicz, J. (1992). A class of semi-Lagrangian approximations for fluids. *J. Atmos. Sci.*, **49**, 2082.

[441] Stenberg, R. and Suri, M. (1996). Mixed *hp* finite element methods for problems in elasticity and Stokes flows. *Numer. Math.*, **72**, 367.

[442] Stieltjes, T. J. (1885). Sur les polynômes de Jacobi. *Comptes Rendus Acad. Sci.*, **100**, 620.

[443] Strang, G. and Fix, G. J. (1973). *An analysis of the finite element method.* Prentice-Hall, Englewood Cliffs, NJ.

[444] Sun, J. and Tanner, R. J. (1994). Computation of steady flow past a sphere in a tube using a PTT integral model. *J. Non-Newtonian Fluid Mech.*, **54**, 179.

[445] Szabó, B. and Babuška, I. (1991). *Finite element analysis.* Wiley, New York.

[446] Szabó, B. and Yosibash, Z. (1996). Superconvergent extraction of flux intensity factors and first derivatives from finite element solutions. *Comp. Meth. Appl. Mech. Eng.*, **129**, 349.

[447] Szego, G. (1975). *Orthogonal polynomials.* American Mathematical Society, Providence, RI.

[448] Tadmor, E. Semi-discrete approximations to nonlinear systems of conservation laws; consistency and L^∞-stability imply convergence. Unpublished.

[449] Tadmor, E. (1989). Convergence of spectral methods for nonlinear conservation laws. *SIAM J. Numer. Anal.*, **26**, 30.

[450] Taylor, M., Tribbia, J., and Iskandarani, M. (1997). The spectral element method for the shallow water equations on a sphere. *J. Comp. Phys.*, **130**, 92.

[451] Taylor, M., Wingate, B. A., and Vincent, R. E. (2000). An algorithm for computing Fekete points in the triangle. *SIAM J. Numer. Anal.*, **38**, 1707.

[452] Temam, R. (1968). Une methode d'approximation de la solution des equations de Navier–Stokes. *Bull. Soc. Math., France*, **96**, 115.

[453] Temam, R. (1969). Sur l'approximation de la solution des equations de Navier–Stokes par la methode des fractionnarires II. *Arch. Rational Mech. Anal.*, **33**, 377.

[454] Temam, R. (1979). *Navier–Stokes equations: Theory and numerical analysis.* North-Holland, Amsterdam.

[455] Temam, R. (1991). Remark on the pressure boundary condition for the projection method. *Theor. Comp. Fluid Dyn.*, **3**, 181.

[456] Theofilis, V. (2000). Globally unstable basic flows in open cavities. AIAA paper 00-1965.

[457] Theofilis, V. (2003). Advances in global instability of nonparallel and three-dimensional flows. *Progress Aero. Sci.*, **39**, 249.

[458] Theofilis, V., Barkley, D., and Sherwin, S. J. (2002). Spectral/*hp* element technology for global instability and control. *Roy. Aero. J.*, **106**, 619.

[459] Theofilis, V., Duck, P., and Owen, J. (2004). Viscous linear stability analysis of rectangular duct and cavity flows. *J. Fluid Mech.*, **505**, 249.

[460] Theofilis, V., Fedorov, A., Obrist, D., and Dallmann, U. Ch. (2003). The extended Görtler–Hämmerlin model for linear instability of three-dimensional incompressible

swept attachment-line boundary layer flow. *J. Fluid Mech.*, **487**, 271.

[461] Thomas, S. J. and Loft, R. D. (2002). A semi-implicit spectral element atmospheric model. *J. Sci. Comp.*, **17**, 339.

[462] Thompson, J. F., Weatherill, N. P., and Soni, B. K. (ed.) (1999). *Handbook of grid generation.* CRC Press, Boca Raton, FL.

[463] Timmermans, L. J. P. (1994). *Analysis of spectral element methods with applications to incompressible flow.* Ph. D. thesis, Eindhoven University of Technology.

[464] Timmermans, L. J. P., Minev, P. D., and Van De Vosse, F. N. (1996). An approximation projection scheme for incompressible flow using spectral elements. *Int. J. Numer. Meth. Fluids*, **22**, 673.

[465] Tollmien, W. (1929). Über die Entstehung der Turbulenz. *Nach. Ges. Wiss. Göttingen*, 21.

[466] Tomboulides, A. G., Israeli, M., and Karniadakis, G. E. (1989). Efficient removal of boundary-divergence errors in time-splitting methods. *J. Sci. Comp.*, **4**, 291.

[467] Tomboulides, A. G., Orszag, S. A., and Karniadakis, G. E. (1993). Direct and large eddy simulations of axisymmetric wakes. In *AIAA-93-0546, 31st aerospace meeting*, Reno, NV.

[468] Toro, E. F. (2001). *Shock-capturing methods for free surface flows.* Wiley, Chichester.

[469] Toselli, A. and Schwab, C. (2003). Mixed *hp*-finite element approximations on geometric edge and boundary layer meshes in three dimensions. *Numer. Math.*, **94**, 771.

[470] Trefethen, A. E., Trefethen, L. N., and Schmid, P. J. (1999). Spectra and pseudospectra for pipe Poiseuille flow. *Comp. Meth. Appl. Mech. Eng.*, **175**, 413.

[471] Trefethen, L. N. and Trummer, M. R. (1987). An instability phenomenon in spectral methods. *SIAM J. Numer. Anal.*, **24**, 1008.

[472] Trujillo, J. (1998). *Velocity–vorticity formulations for high-order methods.* Ph. D. thesis, Princeton University, NJ.

[473] Trujillo, J. and Karniadakis, G. E. (1999). A penalty method for the vorticity–velocity formulation. *J. Comp. Phys.*, **149**, 32.

[474] Tuckerman, L. S. and Barkley, D. (2000). Bifurcation analysis for timesteppers. In *Numerical methods for bifurcation problems and large-scale dynamical systems* (ed. E. Doedel and L. S. Tuckerman), p. 453. Springer, Berlin.

[475] Urbin, G. and Knight, D. (1999). Large eddy simulation of the interaction of a turbulent boundary layer with a shock wave using unstructured grids. In *Second AFSOR international conference on DNS and LES*, Rutgers, NJ.

[476] Van der Ven, H. (1995). A family of large eddy simulation (LES) filters with non-uniform filter widths. *Phys. Fluids*, **7**, 1171.

[477] Van Kan, J. (1986). A second-order accurate pressure-correction scheme for viscous incompressible flow. *SIAM J. Sci. Stat. Comp.*, **7**, 870.

[478] Vandeven, H. (1991). Family of spectral filters for discontinuous problems. *J. Sci. Comp.*, **6**, 159.

[479] Venkatasubban, C. S. (1995). A new finite element formulation for ALE arbitrary Lagrangian Eulerian compressible fluid mechanics. *Int. J. Eng. Sci.*, **33**, 1743.

[480] Vreman, B., Geurts, B., and Kuerten, H. (1994). Realizability conditions for the turbulent stress tensor in large-eddy simulation. *J. Fluid Mech.*, **278**, 351.

[481] Wang, J.-P., Nakamura, Y., and Yashuhara, M. (1993). Global coefficient adjustment method for Neumann condition in explicit Chebyshev collocation method and its application to compressible Navier–Stokes equations. *J. Comp. Phys.*, **107**, 160.

[482] Wannier, G. W. (1950). A contribution to the hydrodynamics of lubrication. *Q. Appl. Math.*, **8**, 1.

[483] Warburton, T. C. (1998). *Spectral/hp methods on polymorphic multi-domains: algorithms and applications.* Ph. D. thesis, Brown University, Providence, RI.

[484] Warburton, T. C. and Karniadakis, G. E. (1997). Spectral simulations of flow past a cylinder close to a free surface. ASME FEDSM97-3389.

[485] Warburton, T. C. and Karniadakis, G. E. (1999). A discontinuous Galerkin method for the viscous MHD equations. *J. Comp. Phys.*, **152**, 608.

[486] Warburton, T. C., Pavarino, L., and Hesthaven, J. S. (2000). A pseudo-spectral scheme for the incompressible Navier–Stokes equations using unstructured nodal elements. *J. Comp. Phys.*, **164**, 1.

[487] Warburton, T. C., Sherwin, S. J., and Karniadakis, G. E. (1999). Spectral basis functions for 2D hybrid *hp* elements. *SIAM J. Sci. Comp.*, **20**, 1671.

[488] Warming, R. F., Beam, R. M., and Hyett, B. J. (1975). Diagonalization and simultaneous symmetrization of the gas-dynamics matrices. *Math. Comput.*, **29**, 1037.

[489] Webb, J. P. and Abouchacra, R. (1995). Hierarchical triangular elements using orthogonal polynomials. *Int. J. Numer. Meth. Eng.*, **38**, 245.

[490] Wei, G. and Kirby, J. T. (1995). Time-dependent numerical code for extended Bousinesq equations. *J. Waterway, Port, Coastal Ocean Eng.*, **121**, 251.

[491] Wheeler, M. F. (1978). An elliptic collocation–finite element method with interior penalties. *SIAM J. Numer. Anal.*, **15**, 152.

[492] Wilhem, D. and Kleiser, L. (2000). Stable and unstable formulations of the convection operator in spectral element simulations. *Appl. Numer. Math.*, **33**, 275.

[493] Williamson, C. H. K. (1988). The existence of two stages in the transition to three-dimensionality of a circular cylinder wake. *Phys. Fluids*, **31**, 3165.

[494] Winckelmans, G. S., Wray, A. A., Vasilyev, O. V., and Jeanmart, H. (2002). Explicit-filtering large-eddy simulation using the tensor-diffusivity model supplemented by a dynamic Smagorinsky term. *Phys. Fluids*, **13**, 1385.

[495] Wingate, B. A. and Boyd, J. P. (1996). Triangular spectral element methods for geophysical fluid dynamics applications. In *ICOSAHOM 95: Proceedings of the third international conference on spectral and high-order methods* (ed. A. V. Ilin and L. R. Scott), Houston, TX, June 5, 1995, p. 305. Houston Journal of Mathematics.

[496] Wingate, B. A. and Taylor, M. A. (1998). The natural function space for triangular and tetrahedral spectral elements. Tech. report LA-UR-98-1711, Los Alamos National Laboratory.

[497] Wintergerste, T. and Kleiser, L. (2000). Secondary stability analysis of nonlinear crossflow vortices. In *Proceedings of the IUTAM laminar–turbulent symposium V* (ed. W. Saric and H. Fasel), Sedona, AZ, p. 583. Springer-Verlag, Telos.

[498] Xiu, D. and Karniadakis, G. E. (2001). A semi-Lagrangian high-order method for Navier–Stokes equations. *J. Comp. Phys.*, **172**, 658.

[499] Xiu, D., Sherwin, S. J., Dong, S., and Karniadakis, G. E. (2004). Strong and auxiliary forms of the semi-Lagrangian method for incompressible flows. *J. Sci. Comp.* In press.

[500] Yosibash, Z. (1997). Numerical analysis of edge singularities in three-dimensional elasticity. *Int. J. Numer. Meth. Eng.*, **40**, 4611.

[501] Yosibash, Z. (2000). Computing singular solutions of elliptic boundary value problems in polyhedral domains using the *p*-FEM. *Appl. Numer. Math.*, **33**, 71.

[502] Yosibash, Z. and Szabó, B. (1995). Numerical analysis of singularities in two dimensions. Part 1. Computation of eigenpairs. *Int. J. Numer. Meth. Eng.*, **38**, 2055.

[503] Yu, J. (1983). Symmetric Gaussian quadrature formulae for tetrahedral regions. *Comp. Meth. Appl. Mech. Eng.*, **43**, 349.

[504] Zachary, A. L., Malagoli, A., and Colella, P. (1994). A higher-order Godunov method for multidimensional ideal magnetohydrodynamics. *SIAM J. Sci. Stat. Comp.*, **15**, 15.

[505] Zalesak, S.T. (1981). Very-high order and pseudospectral flux-corrected transport (FCT) algorithms for conservation laws. In *Advances in computer methods for partial differential equations IV* (ed. R. Vichnevetsky and R. S. Stepleman), Proceedings of the fourth IMACS international symposium on computer methods for partial differential equations, Lehigh University, Bethlehem, PA, 30 June–2 July, p. 126.

[506] Zalesak, S. T. (1979). Fully multidimensional flux-corrected transport algorithms for fluids. *J. Comp. Phys.*, **31**, 335.

[507] Zang, T. A. and Hussaini, M. Y. (1986). On spectral multigrid methods for the time-dependent Navier–Stokes equations. *Appl. Math. Comp.*, **19**, 359.

[508] Zang, T. A., Wong, Y. S., and Hussaini, M. Y. (1982). Spectral multigrid methods for elliptic equations. *J. Comp. Phys.*, **48**, 489.

[509] Zang, Y., Street, R. L., and Koseff, J. (1993). A dynamic mixed subgrid-scale model and its application to turbulent recirculating flows. *Phys. Fluids*, **5**, 3186.

[510] Zhang, M. and Shu, C.-W. (2003). An analysis of three different formulations of the discontinuous Galerkin method for diffusion equations. *Math. Models Meth. Appl. Sci.*, **13**, 395.

[511] Zienkiewicz, O. C. and Taylor, R. L. (1989). *The finite element method* (4th edn). McGraw-Hill, New York.

[512] Zumbusch, G. W. (1996). Symmetric hierarchical polynomials and the adaptive *hp* version. In *ICOSAHOM 95: Proceedings of the third international conference on spectral and high-order methods* (ed. A. V. Ilin and L. R. Scott), Houston, TX, June 5, 1995, p. 529. Houston Journal of Mathematics.

INDEX